EFFECTIVE COMPUTATIONAL METHODS FOR WAVE PROPAGATION

T0187681

Numerical Insights

Series Editor
A. Sydow, GMD-FIRST, Berlin, Germany

The Numerical Insights series aims to show how numerical simulations provide valuable insights into the mechanisms and processes involved in a wide range of disciplines. Such simulations provide a way of assessing theories by comparing simulations with observations. These models are also powerful tools which serve to indicate where both theory and experiment can be improved.

In most cases the books will be accompanied by software on disk demonstrating working examples of the simulations described in the text.

The editors will welcome proposals using modelling, simulation and systems analysis techniques in the following disciplines: physical sciences; engineering; environment; ecology; biosciences; economics.

EFFECTIVE COMPUTATIONAL METHODS FOR WAVE PROPAGATION

Edited by

Nikolaos A. Kampanis
Vassilios A. Dougalis
John A. Ekaterinaris

CRC Press
Taylor & Francis Group
Boca Raton London New York

CRC Press is an imprint of the
Taylor & Francis Group, an **informa** business

A CHAPMAN & HALL BOOK

CRC Press
Taylor & Francis Group
6000 Broken Sound Parkway NW, Suite 300
Boca Raton, FL 33487-2742

First issued in paperback 2019

© 2008 by Taylor & Francis Group, LLC
Chapman & Hall/CRC is an imprint of Taylor & Francis Group, an Informa business

No claim to original U.S. Government works

ISBN-13: 978-1-58488-568-9 (hbk)
ISBN-13: 978-0-367-38772-3 (pbk)

Library of Congress Cataloging-in-Publication Data

Kampanis, Nikolaos A.
  Effective computational methods for wave propagation / Nikolaos A. Kampanis, John A. Ekaterinaris, Vassilios Dougalis.
    p. cm. -- (Numerical insights)
  Includes bibliographical references and index.
  ISBN 978-1-58488-568-9 (alk. paper)
  1. Wave-motion, Theory of--Data processing. 2. Electromagnetic waves--Data processing. 3. Numerical analysis. I. Ekaterinaris, John A. II. Dougalis, Vassilios. III. Title. IV. Series.

QA927.K255 2008
530.12'4--dc22   2007030361

Visit the Taylor & Francis Web site at
http://www.taylorandfrancis.com

and the CRC Press Web site at
http://www.crcpress.com

Contents

11 Finite Element Methods with Discontinuous Displacement,
P. Joly and C. Tsogka   **331**

12 Fictitious Domains Methods for Wave Diffraction, P. Joly
and C. Tsogka   **359**

13 Space Time Mesh Refinement Methods, G. Derveaux, P.
Joly and J. Rodríguez   **385**

Preface

Waves are ubiquitous in nature. We are all familiar with water waves, sound waves, electromagnetic and seismic waves. It is fair to say that every area of science and technology has sources of problems involving some type of wave motion. As a result, a large arsenal of analytical techniques has been developed to describe and analyze linear and nonlinear wave phenomena. And, of course, the numerical simulation of wave motion, mainly the numerical solution of the partial differential equations of wave theory, has been at the center of attention of computational scientists since the advent of the modern computer era.

The aim of the volume at hand is to present some modern, effective computational methods used to describe wave propagation phenomena in selected areas of current interest in physics and technology. One cannot hope to be comprehensive in such an effort. It was the editors' choice to concentrate in four areas, which are, inevitably, close to their interests and expertise:

I. Nonlinear dispersive waves.

II. The Helmholtz equation and its paraxial approximations in underwater acoustics.

III. Numerical methods in elastic wave propagation.

IV. Waves in compressible flows.

Each part consists of chapters (or articles) on specific topics, written by internationally known experts. Part I begins with two articles on the simulation of nonlinear dispersive waves from nonlinear optics. In Chapter 1, X.-P.Wang reviews the dynamic rescaling technique and the iterative grid redistribution method for approximating singular (blow-up) and near-singular solutions of the Nonlinear Schrödinger equation. In Chapter 2, G.Fibich and S.Tsynkov consider the Nonlinear Helmholtz equation describing time-harmonic electromagnetic waves in Kerr media. They pose boundary-value problems for this type of equation with nonlocal artificial boundary conditions at the near- and far-boundaries of the half-space in the direction of which propagation of the wave mainly takes place, and radiation conditions on the transverse boundaries, and they solve them numerically by high-order finite difference methods. In Chapter 3, V. A. Dougalis and D. E. Mitsotakis, after reviewing the derivation and the well-posedness theory of a class of Boussinesq type systems that approximate the Euler equations of water wave theory and describe two-way propagation of long waves of small amplitude, they investigate by analytical and numerical means (using fully discrete Galerkin-finite element methods) solitary waves of these systems and their role in the evolution of general solutions.

Part II consists of four chapters on computational techniques for mathematical models of underwater sound propagation and scattering. In Chapter 4, D. A. Mitsoudis, N. A. Kampanis and V. A. Dougalis present a finite element method for the (linear) Helmholtz equation in a general fluid waveguide with range-dependent layer topography and concentrate on the implementation and the coupling of the finite element method with DtN type nonlocal boundary conditions at the inflow and outflow boundaries of the waveguide. The next three chapters concern various issues related to paraxial ('parabolic') approximations to the Helmholtz equation, that have been successfully used to model long-range propagation of sound in the sea. First, D. J. Thomson and G. H. Brooke in Chapter 5 present an overview of Parabolic Equation (PE) techniques in underwater acoustics, including an introduction to PE-based matched field processing techniques for source localization and for decomposing the acoustic field into its modal components. In the following Chapter 6, V. A. Dougalis, N. A. Kampanis, F. Sturm and G. E. Zouraris address modelling and numerical issues for the PE and its higher-order wide-angle analogs in waveguides with range-dependent interfaces and bottoms in axisymmetric and fully 3D environments, when range-dependent changes of variable are used to make the layers horizontal. Finally, in Chapter 7, G. H. Brooke, D. J. Thomson and the late T. W. Dawson, after reviewing various nonlocal boundary conditions that are applied at interfaces between the computational domain and an external half space (transverse to the direction of propagation) in the case of the PE, they study a particular half space with linear squared index of refraction and show how to construct nonlocal conditions for higher-order PE's as well.

Part III is a comprehensive introduction to modern numerical methods for time-dependent elastic wave propagation written by P. Joly and his collaborators. It consists of seven chapters. In the first one (Chapter 8) P. Joly provides a general introduction and an orientation to this part of the book. In Chapter 9 he continues with a presentation of the mathematical model for elastic wave propagation, i.e. the equations of linear elastodynamics. Chapter 10, also by P. Joly, is a detailed exposition of full discretizations of the elastodynamics equations using standard finite element subspaces in space. Chapter 11, by P. Joly and C. Tsogka, concerns mixed finite element techniques, while Chapter 12, by the same authors, covers fictitious domain (finite element) methods. In Chapter 13, G. Derveaux, P. Joly and J. Rodríguez analyze space-time mesh refinement techniques based on the principle of domain decomposition, while, in Chapter 14, P. Joly and C. Tsogka review the state of the art of two numerical methods for treating elastic waves in unbounded media, namely the discretization of local absorbing boundary conditions and the perfectly matched layer technique.

Part IV, by J. Ekaterinaris, is an overview of high-order, low numerical diffusion numerical methods for complex, compressible flows of aerodynamics. It consists of five chapters. Chapter 15 is introductory and Chapter 16 contains the governing equations of such flows. Chapter 17 concerns high-order finite difference schemes, Chapter 18 ENO and WENO schemes, while Chapter 19 provides an introduction of Discontinuous Galerkin methods for hyperbolic systems.

The editors would like to express their sincere thanks to the authors of the various chapters for contributing their work to this volume. They also wish to express their sincere thanks to their students G. Arabatzis, I. Toulopoulos, and S. Volanis for their help in the preparation of this volume.

Heraklion, June 2007

N. A. Kampanis
V. A. Dougalis
J. A. Ekaterinaris

Part I

Nonlinear Dispersive Waves

Chapter 1

Numerical Simulations of Singular Solutions of the Nonlinear Schrödinger Equations

Xiao-Ping Wang, The Hong Kong University of Science and Technology, mawang@ust.hk

---

1.1 Introduction

The nonlinear Schrödinger equation (NLS) with cubic nonlinearity

$$i\psi_t + \Delta\psi + |\psi|^2\psi = 0 \quad t > 0, \tag{1.1}$$

$$\psi(0, \mathbf{x}) = \psi_0(\mathbf{x}), \quad \mathbf{x} \in R^d$$

arises in various physical contest as an amplitude equation for weakly non-linear waves [23]. For a certain class of initial conditions, namely those for which the invariant $\mathbf{H} = \int_0^t (|\boldsymbol{\nabla}\psi|^2 - \frac{1}{2}|\psi|^2)d\mathbf{x}$ is negative, NLS has solutions that become singular in a finite time when the dimension of the space d is larger than or equal to two [32], [15]. In two space dimensions $d = 2$, NLS is also the model equation for the propagation of cw (continuous wave) laser beams in Kerr media, where ψ is the electric field envelope, t is axial distance in the direction of beam propagation, and $\Delta\psi$ is the diffraction term. It is also well known that when the power, or L^2 norm, of the input beam is sufficiently high, solutions of eq. (1.1) can self-focus and become singular in finite t [8], [29].

Because of the infinitely-large gradients that exist at the singularity, standard numerical methods break down after the solution undergoes moderate focusing. In this chapter, we review two methods which are effective for handling such singular solutions, namely, the *dynamic rescaling* method and the *iterative grid redistribution* method (IGR). The *dynamical rescaling* method was developed by McLaughlin, Papanicolaou, Sulem and Sulem [20] as part of the research effort during the eighties to find the blowup rate of the NLS, and was extended to NLS with non-isotropic initial conditions in [17], [25]. This method exploits the known self-similar structure of the collapsing part of the solution near the singularity, which relates the shrinking transverse width of the solution to the increase in its norm. Therefore, the solution is computed on a fixed computational grid, which in physical space corresponds to a grid that shrinks uniformly toward the singularity. The focusing rate of the grid points is determined dynamically from some norm of the solution ($\int |\nabla \psi|^2 \, dxdy$, $\max_{x,y} |\psi|$, etc). Because the focusing rate of the grid points can be chosen to be the same as the physical focusing rate, in the transformed variables the solution remains smooth and can thus be solved using 'standard' methods. The method of dynamic rescaling works extremely well for solutions of the NLS with radially-symmetric initial conditions, in which case focusing by 10^{10} or more can be easily realized.

Although the method of dynamic rescaling has been extended to NLS with non-isotropic initial conditions [17] and to perturbed NLS (e.g., NLS with normal time dispersion [10]), in such cases dynamic rescaling is considerably less efficient, because the solution does not focus uniformly and/or it is not clear how to extract the physical focusing rate from the solution. The iterative grid redistribution (IGR) method, developed by Ren and Wang [28], overcomes these difficulties by allowing the grid points to move independently (rather then uniformly) according to a general variational minimization principle and a grid iteration procedure which controls the grid density near the region of large solution variations. This method has been shown to be highly effective for solving solutions of PDEs with rather general singular structures, such as the NLS and the Keller-Segal equations with multiple blowup points.

1.2   Dynamic Rescaling Method

1.2.1   Radially Symmetric Case

The dynamic rescaling method is motivated by the scaling invariant of the solution of (1.1), i.e.,

$$\psi(x,t) \to \lambda^{-1}\psi(x/\lambda, t/\lambda^2) \tag{1.2}$$

where λ is a constant. To introduce the method, we consider radially symmetric solutions of the NLS. By analogy to λ in (1.2), let $L(t)$ be a positive function that will be defined later and

$$\xi = r/L(t), \quad \tau = \int_0^t \frac{1}{L^2(s)} ds, \quad u(\xi, \tau) = L(t)\psi(x,t) \tag{1.3}$$

One finds easily that $u(\xi, \tau)$ satisfies the equation

$$iu_\tau + u_{\xi\xi} + \frac{d-1}{\xi} u_\xi + |u|^2 u + ia(\xi u)_\xi = 0 \tag{1.4}$$

where $a = L\frac{dL}{dt} = \frac{1}{L}\frac{dL}{d\tau}$. In order to solve (1.4) numerically, we shall choose $L(t)$ such that the solution u of (1.4) is well behaved. There are various possibility for choosing $L(t)$. One proper choice of $L(t)$ is to ensure that $||\nabla u(\cdot, \tau)||_2^2$ constant in time. That is, $||\nabla u(\cdot, \tau)||_2^2 = ||\nabla u_0||_2^2$ which implies that

$$L(t) = [\frac{||\nabla u_0||_2^2}{||\nabla \psi(\cdot, t)||_2^2}]^{\frac{1}{p}}$$

where $p = 4 - d$. This leads to

$$a = \frac{p \int (1 - \xi^2)|u|^{2(p-1)} Im(u\Delta u^*) d\boldsymbol{\xi}}{\int |u|^{2p} d\boldsymbol{\xi}}$$

With the rescaling (1.3), the NLS has been transformed into equation (1.4) where $a(\tau)$ is given by an integral expression. The solution $u(\xi, \tau)$ behaves nicely for all τ. As $\tau \to \infty$, $t \to t^*$ and $L(t) \to 0$. The behavior of $a(t)$ as $\tau \to \infty$ determines the behavior of $L(t)$ as $t \to t^*$. For example, if $a(\tau) \to -A$ ($A > 0$), then $L(t) \approx (2A(t^* - t))^{1/2}$.

In the supercritical case $d = 3$, the singular solutions with a single blowup point can be easily computed numerically by the dynamic rescaling method [18]. The numerical solution $u(\xi, \tau)$ is shown to converges to a steady state solution Q and a converges to a negative constant which gives a blowup rate $(t^* - t)^{-1/2}$ and a self-similar character of the form;

$$\psi(\mathbf{x}, t) = \frac{1}{\sqrt{2K(t^* - t)}} Q(\frac{|\mathbf{x}|}{\sqrt{2K(t^* - t)}}) e^{\frac{i}{2K} \log \frac{t^*}{t^* - t}} \tag{1.5}$$

where Q satisfies the ordinary differential equation

$$Q_{\xi\xi} + \frac{d-1}{\xi} Q_\xi - Q + iK(\xi Q)_\xi + |Q|^2 Q = 0 \tag{1.6}$$

$$Q_\xi(0) = 0, \quad Q(0) \text{ real}, \quad Q(\infty) = 0$$

One also obtains a precise estimates for the constant $K \approx 0.9173$.

1.2.2 Anisotropic Dynamic Rescaling

We introduce a general change of dependent and independent variables in the nonlinear Schrödinger equation (1.1) in dimension d. Let $D(t)$ be a $d \times d$ matrix function of time, $\mathbf{x}_0(t)$ a vector function of time and $L(t)$ a nonnegative scalar function. We consider a change of variables

$$\boldsymbol{\xi} = D^{-1}(t)(\mathbf{x} - \mathbf{x}_0), \quad \tau = \int_0^t \frac{1}{L^2(s)} ds, \quad u(\boldsymbol{\xi}, \tau) = L(t)\psi(\mathbf{x}, \tau) \tag{1.7}$$

where the matrix $D(t)$ has the form

$$D(t) = O^T(t)\Lambda(t) \tag{1.8}$$

with $O(t)$ an orthogonal matrix and $\Lambda(t)$ a diagonal matrix whose diagonal elements are $\lambda_i (i = 1, ..., d)$. We will choose $D(t)$, $L(t)$ and $\mathbf{x}_0(t)$ so that the transformed solution u has desirable properties, such as boundedness.

Substituting (1.7) into (1.1) and noting that $D^T D = \Lambda^2$, the NLS becomes

$$i[u_\tau - L^{-1}L_\tau u + \mathbf{f} \cdot \nabla u] + L^2(\Lambda^{-2} : \nabla\nabla)u + |u|^2 u = 0 \qquad (1.9)$$

where

$$\mathbf{f} = -D^{-1}\frac{dD}{d\tau}\boldsymbol{\xi} - D^{-1}\frac{d\mathbf{x}_0}{d\tau}$$

and

$$\nabla\nabla = (\frac{\partial^2}{\partial\xi_i\partial\xi_j}) \quad i, j = 1, ..., d$$

The matrix product : is defined by

$$A : B = \sum_{i,j=1}^{d} a_{ij}b_{ij} \text{ where } A = (a_{ij}), B = (b_{ij})$$

Note that we may consider L, D and $\mathbf{x}_0$ as functions of t or as functions of τ, because of (1.7). These quantities can be chosen in several ways. One choice is so that the transformed solution is as close as possible to isotropy. Let p be a positive integer. We take $\mathbf{x}_0$ to be the centroid of $2p$ power of $|\psi|$, which, for large τ, is very likely to be the blowup point

$$x_0^i = \frac{\int x_i|\psi|^{2p}d\mathbf{x}}{\int |\psi|^{2p}d\mathbf{x}} \qquad (1.10)$$

We will use $p \geq 3$ in order to ensure accuracy in the numerical computation of the integrals. To make u as isotropic as possible we choose $D(t)$ so that the second moment of $|u|^{2p}$ is the identity matrix, that is

$$\frac{\int \xi_i\xi_j|u|^{2p}d\boldsymbol{\xi}}{\int |u|^{2p}d\boldsymbol{\xi}} = \delta_{ij} \qquad (1.11)$$

or, using (1.7),

$$D^{-1}S(D^{-1})^T = I \qquad (1.12)$$

where $S = (s_{ij})$ and

$$s_{ij} = \frac{\int (x^i - x_0^i)(x^j - x_0^j)|\psi|^{2p}d\mathbf{x}}{\int |\psi|^{2p}d\mathbf{x}} \qquad (1.13)$$

We also set

$$L(t) = \sqrt{\frac{d}{\sum 1/\lambda_i^2}} \qquad (1.14)$$

which makes the coefficients of the second order derivative terms in (1.9) bounded by 1.

Given ψ, we compute S from (1.13) and (1.10). Since S is symmetric and positive definite we have the decomposition $S = O^T \Lambda^2 O$. If S has distinct eigenvalues (i.e. ψ is anisotropic) then O is unique.

To see how $\mathbf{x}_0, D, L$ vary with time, we take derivatives of (1.10), (11.2), (1.13) and (11.3) with respect to the scaled time τ and using (1.1) and (1.7) we get

$$\frac{d\lambda_i}{d\tau} = -\lambda_i a_{ii}, \quad (i = 1, ..., d) \tag{1.15}$$

and

$$\frac{1}{L}\frac{dL}{d\tau} = -\frac{\sum a_{ii}/\lambda_i^2}{\sum 1/\lambda_i^2} \tag{1.16}$$

with

$$a_{ij} = \frac{p \int (\delta_{ij} - \xi_i\xi_j)|u|^{2(p-1)} Im(L^2\Lambda^{-2} : u\boldsymbol{\nabla}\boldsymbol{\nabla})u^* d\boldsymbol{\xi}}{\int |u|^{2p} d\boldsymbol{\xi}}. \tag{1.17}$$

We also deduce that

$$\mathbf{f} = B\boldsymbol{\xi} - 2\boldsymbol{\beta}$$

where $B = (b_{ij})$ is defined by

$$\begin{cases} b_{ii} = a_{ii} \\ b_{ij} = \frac{2\lambda_j^2}{\lambda_j^2 - \lambda_i^2} a_{ij} & (i \neq j) \end{cases} \tag{1.18}$$

and $\boldsymbol{\beta}$ is given by

$$\beta_j = \frac{p \int \xi_j |u|^{2(p-1)} Im(L^2\Lambda^{-2} : u\boldsymbol{\nabla}\boldsymbol{\nabla})u^* d\boldsymbol{\xi}}{\int |u|^{2p} d\boldsymbol{\xi}}. \tag{1.19}$$

The rotation O and the centriod $\mathbf{x}_0$ can be obtained by solving

$$\frac{d\mathbf{x}_0}{d\tau} = 2O^T\Lambda\boldsymbol{\beta} \tag{1.20}$$

$$\frac{dO}{d\tau} = -GO \tag{1.21}$$

where

$$g_{ii} = 0; \quad g_{ij} = \frac{2\lambda_i\lambda_j}{\lambda_i^2 - \lambda_j^2} a_{ij} \quad (\lambda_i \neq \lambda_j)$$

We observe that (1.9) and (1.15) are a closed system so that u and λ_i are determined by these equations alone without having to compute the rotation O or the centroid $\mathbf{x}_0$. This shows that the translation and rotation are not fundamental in the singularity formation. It is the local scaling that determines the collapse of the solution ψ. The secondary quantities $\mathbf{x}_0$ and O can be computed from (1.20) and (1.21) once u and λ_i are obtained by solving (1.9) and (1.15).

The above approach was used to solve the nonlinear Schrödinger equation with the initial conditions corresponding to a wavepacket without rotational symmetry [17]. We found that after an early transient, during which the anisotropy is significantly amplified, the solution becomes isotropic which shows that the isotropic singular solutions are dynamically stable with respect to a broad class of anisotropic initial perturbations. The method was also applied to solve the singular solutions for the Zakharov equations and the Dave Stewartson equations [24], [26].

1.3   Adaptive Method Based on the Iterative Grid Redistribution (IGR)

The dynamic rescaling method described in the previous section works well for solution with single blowup point and the method relies on the known self-similar structures of the solution. However, it is in general not applicable for solutions with more complicated structures such as solutions with multiple blowup points, or solutions to perturbed NLS (e.g., NLS with normal time dispersion [10]). More general solution adaptive methods are needed for computing the singular solutions for NLS. In fact such a method would work for other type of equations with singular behavior. Various solution-adaptive methods which eliminates the need to have a priori qualitative estimate of the solution, have been developed towards the above end, such as local adaptive mesh refinement [2], adaptive finite elements mesh refinement [21], [22], [1], adaptive node movement (see, e.g., [3], [16], [6], [19], and references therein), or methods based on attraction and repulsion pseudoforces between nodes [27].

In order to handle singular problems (in particular blowup solutions) more effectively, Ren and Wang [28] introduced the iterative grid redistribution (IGR) method to give an effective control of grid density near the region of large solution variation. The method has been successfully applied to many problems with singular behavior [28], [13], [9]. The method consists of the following three parts:

(1) A grid generation rule that determines the mesh mapping $\mathbf{x} = T(\xi)$.

(2) An iterative procedure that controls the grid distribution near the singular points.

(3) A procedure for solving PDEs.

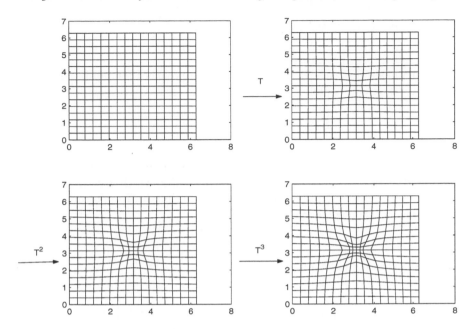

FIGURE 1.1: The effect of map iteration.

The step (2) is the key improvement introduced in [28] so that the method can be effective for the problem with singular behavior. It is a procedure that improves the grid distribution near singular region if the mapping T in step (1) cannot achieve enough resolution in the singular region. To understand this, let us assume that the mesh mapping T in (1) has the tendency to move the grids toward a point $\mathbf{x}_0$ in the domain as in Fig. 1.1 where $\mathbf{x}_0$ is the center of the domain. Then the grid points will continue to move toward $\mathbf{x}_0$ as we iterate the mapping T. This gives certain control of the density of the grid points near the point $\mathbf{x}_0$ and therefore improves over step (1). We now explain the three steps in details.

1.3.1 Grid Distribution Based on the Variational Principle

In two (or higher) spatial dimensions, mesh distribution is usually obtained using a variational approach, specifically by minimizing a functional of the coordinate mapping between the physical domain and the computational domain. The functional is chosen so that the minimum is suitably influenced by the desired properties of the solution $u(\mathbf{x})$ of the PDE itself.

Let $\mathbf{x}$ and ξ denote the physical and computational coordinates, respectively, on the computational domain $\Omega \in \mathbf{R}^d$. A one-to-one coordinate transformation on Ω is denoted by

$$\mathbf{x} = \mathbf{x}(\xi), \qquad \xi \in \Omega. \tag{1.22}$$

The functionals used in existing variational approaches for mesh generation and adaptation can usually be expressed in the form

$$E(\xi) = \int_\Omega \sum_{i,j,\alpha,\beta} g^{i,j} \frac{\partial \xi^\alpha}{\partial x^i} \frac{\partial \xi^\beta}{\partial x^j} d\mathbf{x}, \tag{1.23}$$

where $G = (g_{i,j})$, $G^{-1} = (g^{i,j})$ are symmetric positive definite matrices that are monitor functions in a matrix form. Normally, we choose $g_{i,j} = \delta_{i,j} + u_{x_i} u_{x_j}$. The coordinate transformation and the mesh are determined from the Euler-Lagrange equation

$$\nabla \cdot (G^{-1} \nabla \xi) = 0. \tag{1.24}$$

In one dimension, this is reduced to

$$x_\xi G = C. \tag{1.25}$$

where G is the monitor function and C is a constant. This is the familiar case of the equal-distribution principal introduced in [8]. We note that more terms can be added to the functional (1.23) to control other properties of the mesh, such as orthogonality of the mesh and the alignment of the mesh lines with a prescribed vector field [4].

Understanding how the monitor function influences the resulting mesh properties is crucial for the success of a mesh adaption method. In one dimension, such a relation is given by the equidistribution rule, but only in an average sense. In multiple dimensions, it is even more difficult to predict the overall resulting mesh behavior from the monitor function itself. In other words, it is difficult to have a precise control of the resulting grid distribution from (1.24), (1.25). Numerical experiments in [28] have shown that adaptivity is achieved only when the solution is moderate. Grid points stop moving towards (or even turn away) from the singular region when singularity is approached, i.e., adaptivity is lost when it is most needed (Fig. 1.2). Part of the reason is because the solution is so concentrated in a small region that it has little effect on the grids far away from the singularity. In a particular application, such a problem may be fixed partially by choosing a carefully designed monitor function when the structure of the singularity is available. But a method that works in general is desirable. Such a method must be based on a better grid redistribution procedure than those given above.

The phenomena observed in Fig. 1.2 may be explained by a asymptotic analysis for a one dimensional example [28]. Let's assume that we have a monitor function that is close to a self-similar singularity characterized by a parameter ε

$$G_\varepsilon(x) = \sqrt{1 + \frac{1}{\varepsilon^k} g(\frac{x}{\varepsilon})}$$

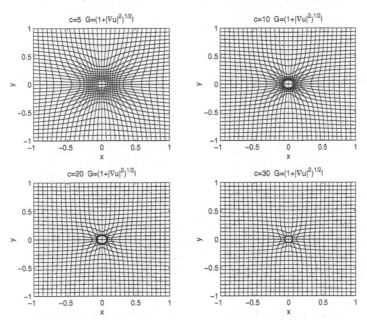

FIGURE 1.2: Grid behavior for $u = ce^{-c^2(x^2+y^2)}$ in 2D as c increases.

where $g(y)$ is positive, has maximum at $x = 0$ and decays rapidly (say, fast than $\frac{1}{y^n}$ for a large enough n) as $y \to \infty$. We want to study the behavior of the grid as $\varepsilon \to 0$. ¿From (1.25), we have

$$x_\xi = \frac{C(\varepsilon)}{G} = \frac{C(\varepsilon)}{\sqrt{1 + \frac{1}{\varepsilon^k} g(\frac{x}{\varepsilon})}} \tag{1.26}$$

where

$$C(\varepsilon) = \int_0^1 \sqrt{1 + \frac{1}{\varepsilon^k} g(\frac{x}{\varepsilon})} dx$$

Integration by parts, we have

$$C(\varepsilon) = x\sqrt{1 + \frac{1}{\varepsilon^k} g(\frac{x}{\varepsilon})} \Big|_0^1 - \int_0^1 \frac{x \frac{1}{\varepsilon^k} g'(\frac{x}{\varepsilon}) \frac{1}{\varepsilon}}{2\sqrt{1 + \frac{1}{\varepsilon^k} g(\frac{x}{\varepsilon})}} dx$$

$$= \sqrt{1 + \frac{1}{\varepsilon^k} g(\frac{1}{\varepsilon})} - \varepsilon^{1-\frac{k}{2}} \int_0^{\frac{1}{\varepsilon}} \frac{yg'(y)}{2\sqrt{\varepsilon^k + g(y)}} dy$$

Since $g(y)$ decays rapidly as $y \to \infty$, we have, as $\varepsilon \to 0$,

$$C(\varepsilon) \sim 1 + A\varepsilon^{1-\frac{k}{2}} \tag{1.27}$$

where
$$A = -\int_0^\infty \frac{yg'(y)}{2\sqrt{g(y)}}dy = \int_0^\infty \sqrt{g(y)}dy.$$

We thus have from (17.2) that for $x \neq 0$ (i.e., away from the singularity) $x_\xi \approx C(\varepsilon)$ as $\varepsilon \to 0$, and

$$x_\xi \to 1 \quad \text{when} \quad k < 2$$
$$x_\xi \to 1 + A \quad \text{when} \quad k = 2$$
$$x_\xi \to \infty \quad \text{when} \quad k > 2 \tag{1.28}$$

This implies that, in the case $k < 2$, we have from Taylor expansion, $\Delta X \approx \Delta \xi$ away from the singularity. Therefore, as $\varepsilon \to 0$, only grid points very close to 0 are allowed to move, i.e., the grid adaption is very limited near the place that function g is large. This also shows that one has to know the solution behavior in some detail in order to design the monitor function which can generate the desired grid distribution near the singularity.

1.3.2   An Iterative Grid Redistribution Procedure

Examples and analysis in the previous section have shown that grid generation based on (1.24) cannot give a satisfactory grid distribution adaptive to the solution in the singular case unless detailed singular behavior of the function is known a prior. The procedure designed in [28] gives a much better control of the grid adaptive to the solution.

Let us define the *grid mapping:*

$$\mathbf{T}: \quad (\mathbf{x}, u(\mathbf{x})) \to (\xi, v(\xi)) = (\xi, u(\mathbf{x}(\xi))).$$

Here $\mathbf{x} = \mathbf{x}(\xi)$ is determined from (1.24) with a monitor matrix involving $u(\mathbf{x})$. If the monitor matrix G is properly chosen (typically involve gradient of the function $u(\mathbf{x})$), the resulting mesh should concentrate more grid points in the regions with large variations. This also means that $v(\xi)$ should be better behaved than the original function $u(\mathbf{x})$ in the sense that the variation of the monitor function in the new variables is reduced. However, as shown in the previous section, this improvement might be very limited. A natural idea to improve further is to repeat the same procedure for $v(\xi)$. In fact, this process can be repeated until a satisfactory $v(\xi)$ is achieved as shown in Fig. 1.1. Based on this intuition, an iterative grid redistribution procedure is introduced by applying the grid mapping $\mathbf{T}$ iteratively:

- Let $u^k(\mathbf{x})$ be the function after k iterations.

- Determine the mapping $\mathbf{x}^{k+1}(\xi)$ from $u^k(\mathbf{x})$ according to (1.24) where monitor matrix w^k is defined using $u^k(x)$.

- Define $u^{k+1}(\xi) := u^k(\mathbf{x}(\xi))$.

For example, after two iterations, we have

$$(\mathbf{x}, u(\mathbf{x})) \to^T (\xi_1, v_1(\xi_1)) \to^{\mathbf{T}} (\xi_2, v_2(\xi_2))$$

In the first iteration, we determine a grid mapping $\mathbf{x}(\xi_1)$ and $v_1(\xi_1) = u(\mathbf{x}(\xi_1))$. In the second iteration, we have $\xi_1(\xi_2)$ and $v_2(\xi_2) = v_1(\xi_1(\xi_2))$.

The results of the iteration is to flatten out the monitor function gradually. In fact, if $u^k(\mathbf{x})$ and $\mathbf{x}^k(\xi)$ converge, then we must have $\mathbf{x}^k \to \mathbf{x}^*(\xi) = \xi$ and $u^k \to u^*(\mathbf{x})$.

How to obtain the grid mapping (1.22) from (1.24) efficiently is also an important issue. In [28], the mapping (1.22) is obtained by interchanging the dependent and independent variables in (1.24) and rewriting the equation (1.24) into an equivalent equations for $\mathbf{x}(\xi)$ which is then solved by heat flow. Although affordable in two dimensions, it is too expensive for three dimensional applications. The computational cost for time integration of the underline PDE in 3D is further increased by the fact the number of terms generated from the chain rule (when calculating derivatives in the computational variables) increases exponentially with the order of the derivatives. Therefore, in order to solve a 3d singular problem within a reasonable CPU time, we introduce a fast algorithm for grid generation in [30]. The algorithm is based on an intuition that the curvilinear coordinates or equivalently the grid points are just the intersection points of the iso-surfaces (contour lines in 2d) of the map $\xi(\mathbf{x})$ from the physical domain to the computational domain, which can be obtained easily from a linear decoupled elliptic system. Based on this intuition, we develop a fast algorithm consisting of solving the linear system for $\xi(\mathbf{x})$, constructing the iso-surfaces and finding the intersections. This direct grid generation method not only is very efficient but also eliminates the convergence and stability issues for the existing methods for solving the grid system. Furthermore, it can be parallelized effectively.

1.3.3 Adaptive Procedure for Solving Nonlinear Schrödinger Equations

We now incorporate the iterative remeshing into a static adaptive method for solving nonstationary NLS equations whose solution is $\psi = \psi(r, t, z)$. Recall that these are initial value problems in z, in which t plays the role of a third spatial variable.

The procedure is as follows:

(1) Given an initial condition $\psi(r, t, z = 0)$, the initial grid transforms $r(\rho, \tau)$, $t(\rho, \tau)$ are determined from the iterative remeshing, which in turn gives an initial condition in the computational domain $\psi(r(\rho, \tau), t(\rho, \tau), z = 0)$. solution $\psi(\rho, \tau, z^*)$ cannot meet a certain smoothness criterion.

(2) Generate a new mesh by the iterative remeshing, starting with $\psi(\rho, \tau, z^*)$. The remeshing iteration stops if the criterion in (1) is sat-

isfied. Interpolation is used to generate the solution at the new grid points.

(3) Go to (1) to continue the integration.

Remark 1. The stopping criterion in step (1) depends on the specific problem. In our numerical examples below, the criterion is set so that the maximum amplitude of the gradient is smaller than a given value TOL. The choice of TOL is flexible. However, it has to be larger than the integral average of the gradient of the solution over the domain. It is easy to see that a smaller TOL requires more remeshing iterations. In actual applications, one should include quantities of interests near the singularity in both the monitor function and in the stopping criterion. For example, in the fluid problem, vorticity might be the desired quantity to be included in the monitor function.

Remark 2. In most of the cases, only one remeshing iteration is needed when we start the iteration with $\vec{u}(\xi, \eta, t^*)$ in Step 2. Since we always start the iteration from the most recent solution in the computational domain, when the cycle (1)-(3) is repeated k times, effectively we have at least k remeshing iterations at t^* from the solution in the original physical variables.

Remark 3. Our grid movement method is static, i.e., the grids are held stationary during the evolution of PDEs until the stopping criterion is violated and are shifted to their new positions by our iterative procedure. The solution values are moved from old grid to the new grid by interpolation. The interpolation is carried out on the uniform mesh and using cubic polynomial interpolations. As pointed out in Remark 2, our iterative procedure was carried out gradually as the solution evolves toward the singularity and solution behavior in the computational domain is always controlled by the stopping criterion (e.g., $\max |\nabla \mathbf{u}| \leq TOL$). Therefore interpolation errors are also controlled.

1.4   Applications

1.4.1   Singular Solutions with Multiple Blowup Points

We first solve the nonlinear Schrödinger equation(NLS)

$$\begin{cases} i\psi_t + \Delta\psi + |\psi|^2\psi = 0, & (x, y) \in \Omega_p, t > 0, \\ \psi(x, y, t)|_{\partial\Omega_p} = 0 \end{cases} \tag{1.29}$$

in two dimensions. Let $(x(\xi, \eta), y(\xi, \eta))$ be the spatial coordinate transformations. Both Ω_p (the physical domain) and Ω_c (the computational domain) are chosen to be $[-1, 1] \times [-1, 1]$. In the iterative grid redistribution, the monitor function is taken to be $w(\xi, \eta) = (1 + \alpha |\phi(\xi, \eta)|^2 + \beta |\nabla \phi(\xi, \eta)|^2)$. The criterion is set so that the maximum amplitude of the gradient is smaller than a given value of TOL.

As the minimum mesh size is decreased, the time step is also decreased according to the CFL condition. We may also rescale the time, as in the dynamic rescaling, from t to τ as

$$\frac{d\tau}{dt} = \frac{1}{\lambda^2(t)} \tag{1.30}$$

Let

$$\psi = \frac{1}{L(t)} \phi,$$

where $L(t)$ is a scaling factor chosen to be

$$L(t) = \frac{1}{\max_{(x,y) \in \Omega_p} |\psi(x, y, t)|}. \tag{1.31}$$

To balance the coefficients in the transformed equation, we choose $\lambda = L$. In the coordinate system (ξ, η, τ), the NLS becomes

$$\phi_\tau - \frac{L_\tau}{L} \phi - i(\lambda^2 \Delta_B \phi + |\phi|^2 \phi) = 0 \tag{1.32}$$

together with

$$L_\tau = L^3 \mathrm{Im}(\phi^* \Delta_B \phi)|_{(\xi_0, \eta_0)} \tag{1.33}$$

where $(\xi_0(t), \eta_0(t))$ is the maximum point of $|\phi(\xi, \eta, t)|$, and

$$\Delta_B \phi = \frac{1}{J} \{ \frac{\partial}{\partial \xi} (\frac{b_{22} \phi_\xi - b_{12} \phi_\eta}{J}) + \frac{\partial}{\partial \eta} (\frac{b_{11} \phi_\eta - b_{12} \phi_\xi}{J}) \},$$

$$b_{11} = x_\xi^2 + y_\xi^2, \quad b_{12} = x_\xi x_\eta + y_\xi y_\eta, \quad b_{22} = x_\eta^2 + y_\eta^2.$$

J is the Jacobian of the coordinate transformation.

In the first example, we solve the NLS in the critical case with the initial condition

(i) $\qquad \psi(x, y, 0) = 10 e^{-4x^2 - 9y^2}$.

(1.32) and (1.33) are solved simultaneously on Ω_c with 100×100 grid points and $\alpha = 2$, $\beta = 1$, $TOL = 5$. About 20 remeshing steps are conducted before the computation reaches $\tau = 95.82$ (or $t = 0.024317104$) when the maximum value of the solution is 2.9514×10^5. Fig. 1.3 shows solutions in the computational and the physical variables respectively at $\tau = 95.82$. The mesh distribution is shown in Fig. 1.4 where one can see that the minimum mesh size is about 10^{-6}. The computation is well resolved with more than 40 grid points within the interval of size 10^{-5}.

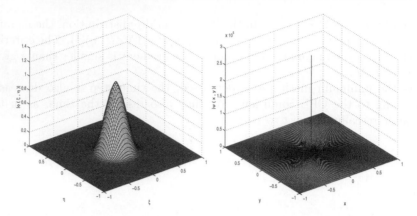

FIGURE 1.3: The solution ϕ (with initial value (i)) in the computational domain (left) and the physical domain (right) at $t = 0.024317104$ with the maximum amplitude at $2.9514e + 05$.

To verify the stable property of the isotropic singularity, a result first shown numerically by the dynamic rescaling method [17], we define two scaling factors in the x and y directions:

$$L_1 = \sqrt{\frac{\int_{\Omega_p} |\psi_x|^2 dxdy}{\int_{\Omega_p} |\psi|^2 dxdy}}$$

$$L_2 = \sqrt{\frac{\int_{\Omega_p} |\psi_y|^2 dxdy}{\int_{\Omega_p} |\psi|^2 dxdy}}$$

Fig. 1.5 shows the ratio of $\frac{L_1}{L_2}$ as a function of τ and the limit converges to 1 which shows that the solution converges to an isotropic singularity. It also shows $\frac{\lambda_\tau}{\lambda}(= \sigma\frac{L_\tau}{L})$ as function of τ, which approaches 0 slowly that leads to the blowup rate. All the results above are consistent with the results obtained from the dynamic rescaling method.

In the second example, we use the initial condition

(ii) $\psi(x, y, 0) = 20(e^{-20((x+0.5)^2+y^2)} + e^{-20((x-0.5)^2+y^2)})$

with two maximum points. Our results show that the solution blows up at two points. Such a calculation cannot be done by the dynamic rescaling method because one can only rescale around one point. We solve the equation with 160×160 grid points in the computational domain and $TOL = 7$. Fig. 1.6 show solutions in computational and physical variables respectively at $t = 0.005651$ when the maximum of the solution reaches 1.488×10^5. The grid distribution at the same time is shown in Fig. 1.7. The blowup structure of each singularity is the same as that of the solution with single blowup point.

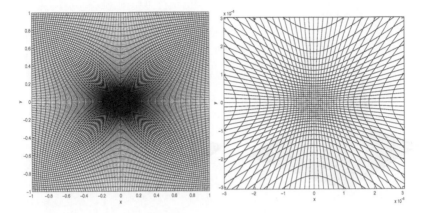

FIGURE 1.4: Mesh in Ω_p at t=0.024317104 ($\tau = 95.82$) corresponding to Fig. 1.3 (left) and an enlarged picture of the above mesh around the center (right).

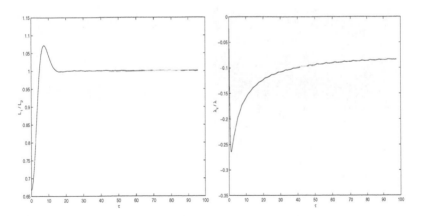

FIGURE 1.5: Ratio $\frac{L_1}{L_2}$ (left) and $a(\tau) = \frac{\lambda_\tau}{\lambda}$ (right) as a function of τ.

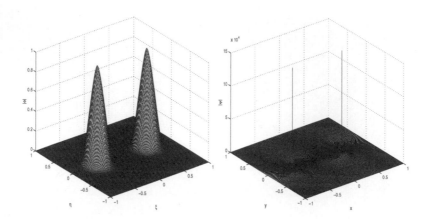

FIGURE 1.6: $\phi(\xi, \eta, t)$ in Ω_c (left) and in Ω_p (right) at $t = 0.005651$, $1/\lambda = 1.4880e + 05$.

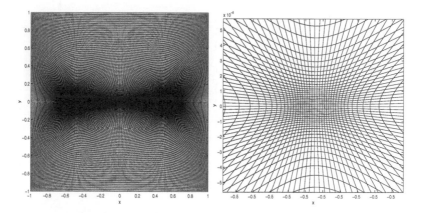

FIGURE 1.7: Mesh in Ω_p (left) at t=0.005651 corresponding to Fig. 1.6 and an enlarged view (right) of the above mesh in Ω_p around one blowup point.

1.4.2 Numerical Simulations of Self–Focusing of Ultrafast Laser Pulses

We now present some examples with more complicated singular structures. The NLS (1.29) does not include temporal effects. These effects become important in the case of ultrashort laser pulses, whose propagation can be modeled by the dimensionless nonstationary NLS [11]

$$i\psi_z(z, x, y, t) + \Delta_\perp \psi + |\psi|^2 \psi + -\epsilon_3 \psi_{tt} = 0, \tag{1.34}$$

where the propagation distance is denoted by z and the real time variable is denoted by t.

In Figures 1.8–1.10 we present simulations of the time-dispersive NLS (1.34) performed using the IGR method, with $\epsilon_3 = 1/32$ and the initial conditions

$$\psi_0 = A\sqrt{2}\exp\left(-\frac{x^2 + y^2}{2} - t^2\right). \tag{1.35}$$

We use the values of $A = 1.75$, $A = 2$, and $A = 3$, whose peak power at $t = 0$ is $1.65P_c$, $2.15P_c$ and $4.83P_c$, respectively, where P_c is the threshold power for collapse [8]. In these simulations we integrate the equation on a domain $[r, t] = [0, \quad 4] \times [0, \quad 12/\sqrt{32}]$ with an initial uniform mesh of 100×300. With the IGR we reached smallest mesh size of $\delta r = 2.1 \times 10^{-3}$, $\delta t = 1.4863 \times 10^{-4}$ for the simulation of Fig 1.8, $\delta r = 1.0197 \times 10^{-4}$, $\delta t = 6.4876 \times 10^{-6}$ for Fig 1.9 and $\delta r = 1.911 \times 10^{-6}$, $\delta t = 4.048 \times 10^{-7}$ for Fig 1.10. We also plot the solution in the computational temporal variable (Fig. 1.8–1.10, right columns) to show that there is enough resolution in the oscillatory and shock-like regions. Indeed, even after focusing by 10^5 the solution appears to be sufficiently smooth in the computational temporal and radial variables (Fig. 1.10).

The observed dynamics agrees with the simulation results of Germaschewski et al. [14] and of Coleman and Sulem [7], namely, that after the pulse splitting the two peaks continue to focus, but later decay into small temporal oscillations. The overall arrest of collapse by small normal time-dispersion can be clearly seen in Figs. 1.8–1.10. Since the simulations in [14] and in [7] were done using a different numerical method, this agreement provides strong support to the validity of the results obtained with these two methods.

1.4.3 Ring Blowup Solutions of the NLS

In [9], a new type of singular solutions of the critical nonlinear Schrödinger equation (NLS), that collapse with a quasi self-similar ring profile at a square root blowup rate, are discovered by both the dynamic rescaling method and the IGR method. This type of singular solutions happen when the initial input power is sufficiently high (the so-called super-Gaussian beam). We observe

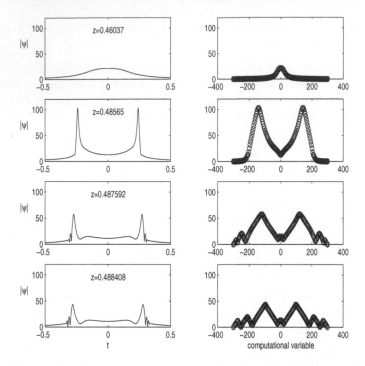

FIGURE 1.8: On-axis amplitude $|\psi(z, r = 0, t)|$ for the solution of eq. (1.34) with the initial condition (1.35) and $A = 1.75$. The right column shows the corresponding solution in the computational variable of t (only one-forth of the grid-points are shown).

that the self-similar ring profile is an attractor for a large class of radially-symmetric initial conditions, but is unstable under symmetry-breaking perturbations. Collapsing ring solutions are also observed in the supercritical NLS.

We first solve the radially-symmetric NLS

$$i\psi_t(t, r) + \psi_{rr} + \frac{1}{r}\psi_r + |\psi|^2\psi = 0, \quad \psi(0, r) = \psi_0(r), \qquad (1.36)$$

with the high-powered super-gaussian initial condition $\psi_0 = 15e^{-r^4}$ ($N(0) \simeq 38N_c$). As can be seen in Fig. 1.11, the solution collapses with a ring profile that becomes taller in amplitude and smaller in radius.

In addition, the results suggest that the ring solution is self-similar, i.e., of the form

$$|\psi(t, r)| \sim \frac{1}{L(t)}Q(\xi), \qquad \xi = \frac{r}{L(t)}, \qquad (1.37)$$

for some profile $Q(\xi)$. In order to check for self-similarity, we rescale the

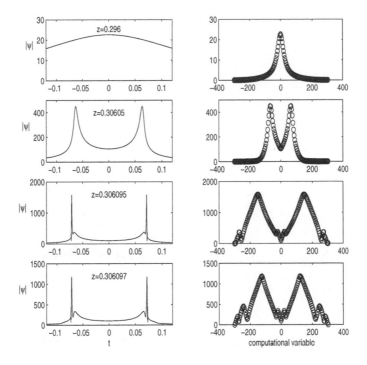

FIGURE 1.9: Same as Fig. 1.8 with $A = 2$.

numerical solution according to

$$\psi_{normalized}(t, r) = L(t)\psi\left(\frac{r}{L(t)}\right), \quad L(t) = \frac{1}{\max_r |\psi|}. \tag{1.38}$$

Fig. 1.12 shows the results of Fig. 1.11D-F, rescaled according to (1.38). All three normalized plots are indistinguishable, indicating that the collapsing ring solution is indeed self-similar while focusing over more than 10 orders of magnitude.

Fig. 1.13 shows that $\lim_{t \to T_c} LL_t \cong 0.085$ for the super-gaussian initial condition $\psi_0 = 15e^{-r^4}$, suggesting a square root blowup rate of $L(t) \sim \alpha\sqrt{T_c - t}$ with $\alpha \cong 0.41$. We can then construct an explicit self-similar ring blowup solution as follows. Let

$$\psi_Q^{(ex)} = \frac{1}{\alpha\sqrt{T_c - t}}Q\left(\frac{r}{\alpha\sqrt{T_c - t}}\right)e^{-i\frac{log(T_c - t)}{\alpha^2} - i\frac{r^2}{8(T_c - t)}}, \tag{1.39}$$

where $Q(\xi; \alpha)$ is a solution of

$$Q''(\xi) + \frac{Q'}{\xi} + \left[\frac{\alpha^4}{16}\xi^2 - 1\right]Q + Q^3 = 0, \quad 0 \neq Q(0) \in \mathbf{R}, \quad Q'(0) = 0. \tag{1.40}$$

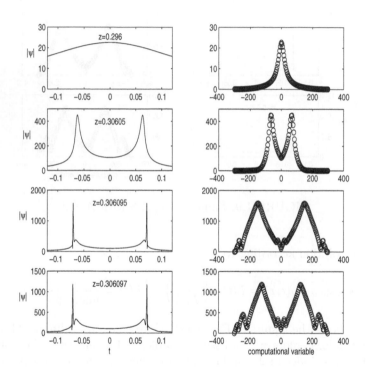

FIGURE 1.10: Same as Fig. 1.8 with $A = 3$.

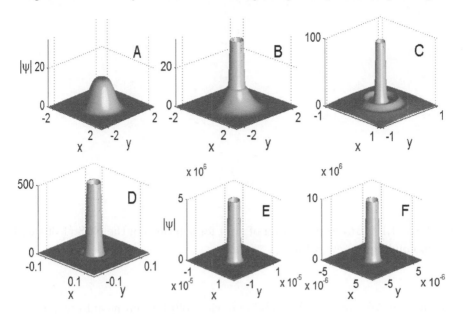

FIGURE 1.11: Solution of the NLS (1.36) with $\psi_0 = 15e^{-r^4}$. A: $t = 0$, $A(t) = 1$; B: $t = 0.020$, $A(t) = 1$; C: $t = 0.027$, $A(t) = 1$; D: $t - 0.0286$, $A(t) = 1$; E: $A(t) = 3.32 \cdot 10^5$; F: $A(t) = 6.64 \cdot 10^5$.

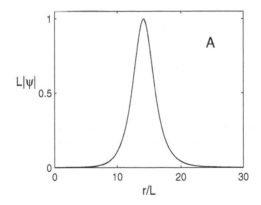

FIGURE 1.12: Results of Figure 1.11D-F normalized and superimposed; all three lines are indistinguishable.

Then, ψ_Q is an explicit blowup solution of the NLS whose blowup rate is $L(t) = \alpha\sqrt{T_c - t}$.

We now present numerical simulations to study the effect of radial-symmetry breaking on collapsing ring solutions. Our simulations above show that the collapsing self-similar single ring profile is stable in the radially sym-

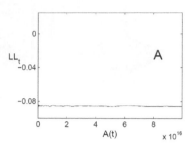

FIGURE 1.13: LL_t as a function of $A(t)$ for solutions of the NLS (1.36) with initial conditions: $\psi_0 = 15e^{-r^4}$.

metric case. We now test the stability of ring solutions in the anisotropic case, in which symmetry breaking is due to the introduction of small ellipticity in the initial condition. We first solve the radially-symmetric NLS (1.36) with the ring initial condition ψ_Q^0 of the form (1.39) up to some time t_0 such that the maximum of the solution $= 1000$. Then, we add small perturbation to the solution at t_0 and use it as an initial condition for the simulation of the NLS

$$i\psi_t(t, x, y) + \Delta\psi + |\psi|^2\psi = 0, \quad \psi(0, x, y) = \psi_0(x, y), \qquad (1.41)$$

In other words, we solve the NLS (1.41) with the initial condition

$$\psi_0(x, y) = \psi_Q(t_0, r = \sqrt{x^2 + 1.01y^2}). \qquad (1.42)$$

As Fig. 1.14 shows, the ring breaks into eight filaments located along a circle $r = r_{fil}$. Therefore, the collapsing ring profile is unstable as a solution of the NLS (1.41), i.e., with respect to perturbations that breakup the radial symmetry.

1.4.4   Keller-Seigel Equation: Complex Singularity

Our next example is the three dimensional Keller-Segel (KS) model for bacterial pattern formation:

$$\rho_t = \epsilon\Delta\rho - \nabla \cdot (\rho\nabla C), (x, y, z) \in \Omega_p, t > 0.$$
$$C_t = \Delta C + \rho.$$

Here ρ is the bacterial density and C is the attractant field. In some cases, the concentration of the attractant C draws the bacteria together and they achieve

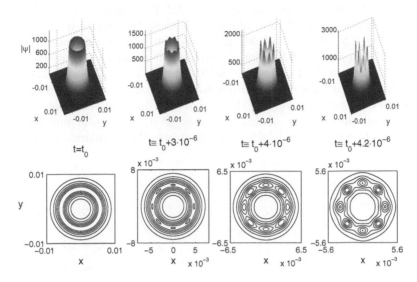

FIGURE 1.14: Solution of the NLS (1.41) with the slightly elliptic single ring profile (1.42). Top: surface plot. Bottom: level sets of $|\psi|$.

an infinite density with complicated geometric structures. It is observed in [5] that the high density regions initially collapse into cylindrical structures (line singularities) which subsequently destabilize and break up into spherical aggregates (point singularities). Here, we are interested in simulating the phenomena with the Keller-Segal(KS) model. We consider the case in which the initial attractant concentration on a straight line. Subsequent evolution shows a transition from line singularity to point singularities. The initial condition is an uniform density distribution and an attractant field with a small perturbation in the z direction on the domain $[0,1]^3$:

$$\rho(x, y, z, 0) = 1.0,$$
$$C(x, y, z, 0) = e^{-10((x-0.5)^2+(y-0.5)^2)}(1 + 0.01|\sin 2\pi z|).$$

The ϵ is taken to be 0.01.

The above equation is solved on a uniform mesh in Ω_c with $40 \times 40 \times 200$ grid points. The monitor function and the parameters α and β are chosen as that in the NLS point singularity problem, and the TOL is chosen to be 8.0. The computation is continued until $t = 0.7534794$ when the maximum density reaches about 4.6×10^4.

The density ρ, at first, increases gradually, showing a line singularity along the central line $(0.5, 0.5, z), z = 0, 1$ until t reaches about 0.70. Then, this line singularity changes to multiple points singularity on two points which are initially perturbed to the maximal: $(0.5, 0.5, 0.25)$ and $(0.5, 0.5, 0.75)$. The density increases quickly and blows up at these two points in the end (Fig. 1.15). The color Fig. 1.16 shows the transition from line singularity to point singularity in contour filling forms corresponding to times: $t_1 = 0.0023$; $t_2 = 0.736591$; $t_3 = 0.7511$; and $t_4 = 0.75347497$. Fig. 1.17 shows the cutting view of the local grids corresponding to line singularity and points singularity and the local sectional view of the grid around one singular point. Also, Fig. 1.15 shows the log plot of density values along the cental line $(0.5, 0.5, z)$ $(0 < z < 1)$ versus time, which again illustrates the singularity transition.

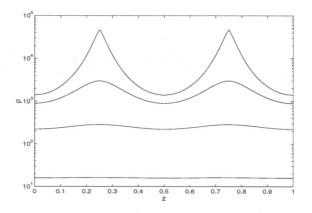

FIGURE 1.15: Transition from line singularity to point singularities for KS: ρ (log scale) along the center line $(0.5, 0.5, z)$.

1.5   Conclusion

In this chapter, we have reviewed two numerical methods for computing singular (blowup) and near singular solutions of the nonlinear Schrödinger equations, namely the dynamic rescaling method and the iterative grid redistribution method. The dynamic rescaling method is based on the scaling invariants of the solutions of the PDE and is effective for the self-similar single point singularity and it is easy to implement. The iterative grid redistribution method is more general and can handle singularities with more complicated structures. Our numerical examples have shown the effectiveness of our methods. We note that error analysis of the iterative grid redistribution method is difficult and requires further studies.

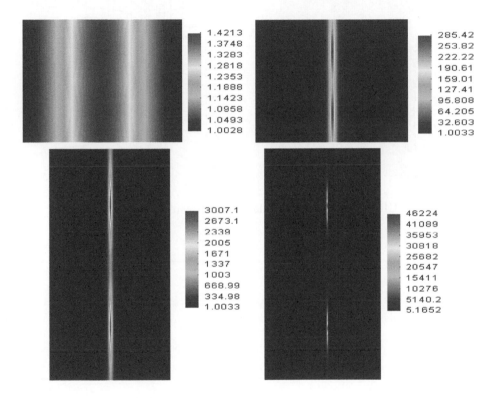

FIGURE 1.16: Cross sectional view of the color density plot for the same solution as in the previous figure.

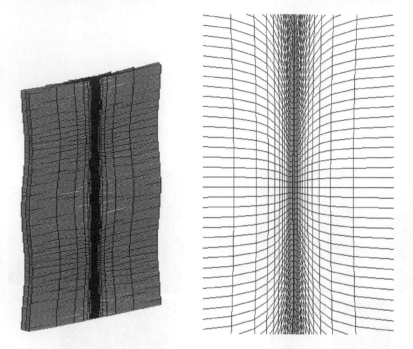

FIGURE 1.17: Local grid distribution near the line singularity.

Acknowledgments

I would like to thank G. Fibich for some helpful discussions.

References

[1] G. D. Akrivis, V. A. Dougalis, O. A. Karakashian, and W. R. McKinney. Numerical approximation of blow-up of radially symmetric solutions of the nonlinear Schrödinger equation. *SIAM J. Sci. Comput.*, 25(1):186–212, 2003.

[2] M. Berger and P. Colella. Local adaptive mesh refinement for shock hydrodynamics. *J. Comput. Physics*, 82:64–84, 1989.

[3] J. Brackbill. An adaptive grid with directional control. *J. Comput.*

Physics, 108:38–50, 1993.

[4] J. U. Brackbill and J. S. Saltzman. Adaptive zoning for singular problems in two dimensions. *J. Comput. Physics*, 46:342–368, 1982.

[5] M. P. Brenner, L. S. Levitov, and E. O. Budrene. Physical mechanisms for chemotactic pattern formation by bacteria. *Biophysical Journal*, 74:1677–1693, 1998.

[6] H. Ceniceros and T. Y. Hou. An efficient dynamically adaptive mesh for potentially singular solutions. *J. Comput. Physics*, 172:1–31, 2001.

[7] J. Coleman and C. Sulem. Numerical simulations of blow–up solutions of the vector nonlinear Schrödinger equation. *Phys. Rev. E*, 66:036701, 2002.

[8] C. De Boor. Good approximation by splines with variable knots II. *in Springer Lecture Notes Series.*, Vol. 363, Springer–Verlag, Berlin, 1973.

[9] G. Fibich, Nir Gavish, and X. P. Wang. New singular solutions of the nonlinear Schrodinger equation. accepted for publication in *Physica D*, 211:193–220, 2005.

[10] G. Fibich, V. M. Malkin, and G. C. Papanicolaou. Beam self–focusing in the presence of small normal time dispersion. *Phys. Rev. A*, 52(5):4218–4228, 1995.

[11] G. Fibich and G. C. Papanicolaou. Self-focusing in the presence of small time dispersion and nonparaxiality. *Opt. Lett.*, 22:1379-1-381, 1997.

[12] G. Fibich and G. C. Papanicolaou. Self-focusing in the perturbed and unperturbed nonlinear Schrödinger equation in critical dimension. *SIAM J. Applied Math.*, 60:183–240, 1999.

[13] G. Fibich, Weiqing Ren, and X. P. Wang. Numerical simulations of self focusing of ultrafast laser pulses. *Physical Review E*, 67:056603, 2003.

[14] K. Germaschewski, R. Grauer, L. Berge, V. K. Mezentsev, and J. J. Rasmussen. Splittings, coalescence, bunch and snake patterns in the 3D nonlinear Schrödinger equation with anisotropic dispersion. *Physica D*, 151:175–198, 2001.

[15] R. T. Glassey. On the blowing up of solutions to the Cauchy problem for nonlinear Schrödinger equations. *J. Math. Phys.* 18:1794–1797, 1977.

[16] W. Huang and R. D. Russell. Moving mesh strategy based upon gradient flow equation for two dimensional problems. *SIAM J. Sci. Comput.*, (to appear).

[17] M. Landman, G. C. Papanicolaou, C. Sulem, P. L. Sulem, and X. P. Wang. Stability of isotropic singularities for the nonlinear Schrödinger equation. *Physica D*, 47:393–415, 1991.

[18] B. LeMesurier, G. Papanicolaou, C. Sulem, and P. L. Sulem. *Physica*, 31D:78, 1988.

[19] R. Li, T. Tang and P.-W. Zhang. A moving mesh finite element algorithm for singular problems in two and three space dimensions. *J. Comput. Physics*, 177:365–393, 2002.

[20] D. W. McLaughlin, G. C. Papanicolaou, C. Sulem, and P. L. Sulem. Focusing singularity of the cubic Schrödinger equation. *Phys. Rev. A*, 34(2):1200–1210, 1986.

[21] K. Miller and R. N. Miller. Moving finite elements I. *SIAM J. Numer. Anal.*, 18:1019–1032, 1981.

[22] P. K. Moore and J. E. Flaherty. Adaptive local overlapping grid methods for parabolic systems in two space dimensions. *J. Comput. Physics*, 98:54–63, 1992.

[23] A. Newell. Solitons in Mathematics and Physics. *SIAM CBMS Appl. Math. Series*, 48, 1985.

[24] G. C. Papanicolaou, P. L. Sulem, C. Sulem, and X. P. Wang. Singular solutions of the Zakharov equations for Langmuir turbulence. *Physics Fluids B*, 3:969–980, 1991.

[25] G. Papanicolaou, C. Sulem, P. L. Sulem, and X. P. Wang. Dynamic rescaling for tracking point singularities: application to Nonlinear Schrödinger Equation and related problems. *Singularities in Fluids, Plasmas and Optics, NATO ASI Series*, Vol. 404, pages 265–279, 1992.

[26] G. Papanicolaou, C. Sulem, P. L. Sulem, and X. P. Wang. Focusing singularities of Davey-Stewartson equations for gravity capillary waves. *Physica D*, 72:61–86, 1994.

[27] M. M. Rai and D. A. Anderson. Grid evolution in time asymptotic problems. *J. Comput. Physics*, 43:327–344, 1981.

[28] W. Ren and X. P. Wang. An iterative grid redistribution method for singular problems in multiple dimensions. *J. Comput. Physics*, 159:246–273, 2000.

[29] C. Sulem and P. L. Sulem. *The nonlinear Schrödinger equation*. Springer, New-York, 1999.

[30] Desheng Wang and Xiao-Ping Wang. Three dimensional adaptive method based on iterative grid redistribution procedure. *J. Comput. Physics*, 199:423–436, 2004.

[31] A. Winslow. Numerical solution of the quasi-linear Poisson equation. *J. Comput. Physics*, 1:149–172, 1966.

[32] V. E. Zakharov, Zh. Eksp. Theor. Fiz, 18:1745, 1972, *Sov. Physics JETP*, 35:908, 1972.

Chapter 2

Numerical Solution of the Nonlinear Helmholtz Equation

G. Fibich, Department of Applied Mathematics, School of Mathematical Sciences, Tel Aviv University, Ramat Aviv, Tel Aviv 69978, Israel, fibich@math.tau.ac.il

S. Tsynkov, Department of Mathematics and Center for Research in Scientific Computation, North Carolina State University, Box 8205, Raleigh, NC 27695, USA, tsynkov@math.ncsu.edu

---

2.1 Introduction

2.1.1 Background

The objective of this chapter is to describe a new numerical algorithm for studying nonlinear self-focusing of time-harmonic electromagnetic waves. The physical mechanism that leads to self-focusing is known as the Kerr effect. At the microscopic level, the Kerr effect may originate from electrostriction, nonresonant electrons, or from molecular orientation. At the macroscopic level the Kerr effect is manifested through an increase in the index of refraction, which is proportional to the intensity of the electric field $|E|^2$. Since light rays bend toward regions with higher index of refraction, an impinging laser beam would become narrower as it propagates, a phenomenon known as self-focusing. For more information on self-focusing, see, e.g., [3, 8, 11].

Since nonlinear self focusing leads to nonuniformities in the refraction index, a part of the incoming beam gets reflected backwards, a phenomena known as nonlinear backscattering. At present, very little is known about it, except for the general belief that it is "small." Since, however, small-magnitude mechanisms can have a large effect in nonlinear self-focusing in bulk medium [8], one needs to be able to accurately quantify the magnitude of backscattering and study how it may affect the beam propagation. From the standpoint of applications in modern science and engineering, the capability to quantitatively analyze and predict the phenomena of nonlinear self-focusing and backscattering is extremely important for a large number of problems. Those range from remote atmosphere sensing (when an earth-based powerful laser sends pulses to the sky [14], and the backscattered radiation accounts for the detected signal), to laser surgery (propagation of laser beams in tissues), to transmitting information along optical fibers. There are other possible applications, e.g., all-optical switching in electrical/electronic circuits, that involve interactions ("collisions") between co-propagating or counter-propagating beams. It is well known that within the framework of a simpler model based on the nonlinear Schrödinger equation, these beams have the form of spatial solitary waves (solitons), and that collisions between solitons are elastic (i.e., involve no power losses). However, whether such collisions remain elastic in more comprehensive models, such as the nonlinear Helmholtz question, is currently an open problem.

2.1.2   The Nonlinear Helmholtz Equation

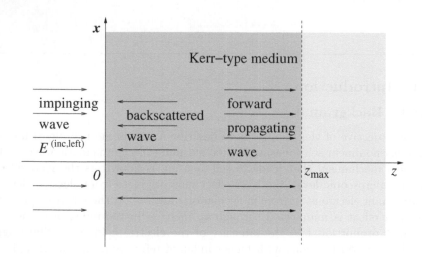

FIGURE 2.1: Schematic for the propagation of waves in Kerr media.

The simplest theoretical (and experimental) setup for the propagation of electromagnetic waves in Kerr media is shown in Figure 2.1. An incoming laser beam with known characteristics impinges normally on the planar interface $z = 0$ between the linear and nonlinear medium. The time-harmonic electric field $E = E(\boldsymbol{x}, z)$ is assumed to be linearly polarized and is governed by the scalar nonlinear Helmholtz equation (NLH):

$$(\partial_{zz} + \Delta_\perp)E + k^2 E = 0,$$
$$k^2 = k_0^2(1 + \epsilon|E|^{2\sigma}), \quad \sigma > 0, \quad \boldsymbol{x} \in \mathbb{R}^{D-1}, \quad z \geq 0, \tag{2.1}$$

where k_0 is the linear wavenumber, $\epsilon = 2n_2/n_0$, n_0 is the linear index of refraction, n_2 is the Kerr coefficient, and $\Delta_\perp = \partial_{x_1 x_1} + \cdots + \partial_{x_{D-1} x_{D-1}}$ is the $D-1$ dimensional transverse Laplacian (the diffraction term). The value of σ is equal to one for the physical Kerr effect. However, for reasons that will become apparent is Section 2.1.5, other values of σ are also considered in this study. Note also that in all the physical problems mentioned earlier, the nonlinearity in (2.1) is typically weak, i.e., $\epsilon |E^{(\text{inc, left})}(\boldsymbol{x})|^{2\sigma} \ll 1$, where $E^{(\text{inc, left})}$ is the impinging wave, see Figure 2.1. The reason for this is that the physical value of the Kerr index n_2 is so small, that even for high-power lasers the nonlinear change in the index of refraction is usually small compared with the linear index of refraction, i.e., $n_2 |E|^2 \ll n_0$.

The nonlinear medium occupies the half-space $z \geq 0$. Consequently, the NLH (2.1) should be supplemented with boundary conditions at $z = 0$ and at $z = +\infty$. Although global existence of its solutions is currently an open problem, there are various indications that, at least in some cases, the NLH is solvable (see Section 2.1.5). In fact, one of our primary long term goals in this study is to address the solvability issue with the help of our numerical methodology. Assuming that the solution does exist globally, for large propagation distances (i.e., as $z \longrightarrow +\infty$) it can either diffract, in which case the propagation becomes linear, or it can maintain its shape through a balance of the focusing nonlinearity and diffraction (i.e., converge to a soliton). In either case, as $z \longrightarrow +\infty$, E will have no left-propagating components, i.e., E will only be composed of the right-traveling waves. Since the actual numerical simulation is carried out on a truncated domain $0 \leq z \leq z_{\max}$, the desired behavior of the solution as $z \longrightarrow +\infty$ has to be captured by a far-field artificial boundary condition (ABC) at the artificial boundary $z = z_{\max}$. This boundary condition should guarantee the reflectionless propagation of all the waves traveling toward $z = +\infty$ through the interface $z = z_{\max}$. Often, boundary conditions designed to ensure the transparency of the outer boundary to the outgoing waves are referred to as radiation boundary conditions [12].

The situation at the interface $z = 0$ is more complicated, as the total field there, $E(\boldsymbol{x}, 0)$, is composed of a given incoming component $E^{(\text{inc, left})}(\boldsymbol{x})$ and the unknown backscattered (i.e., outgoing) component $E^{(\text{scat})}(\boldsymbol{x}, 0)$, i.e.,

$$E(\boldsymbol{x}, 0) = E^{(\text{inc, left})}(\boldsymbol{x}) + E^{(\text{scat})}(\boldsymbol{x}, 0) . \tag{2.2}$$

As such, the boundary condition at $z = 0$ has to provide for the reflectionless propagation of any outgoing (left-traveling) wave through this interface, and at the same time has to be able to correctly prescribe the incoming (right-traveling) field. A two-way ABC that possesses the required capabilities has been first implemented in [9].

In addition to the foregoing simplest setup, we can also use our algorithm to solve more elaborate problems. For example, instead of assuming that the nonlinear Kerr medium occupies the entire semi-space $z \geq 0$ and as such, that it can only be excited by waves coming from the left, we can consider the Kerr medium to be confined to the "slab" $0 \leq z \leq z_{\max}$, as is frequently the case in experiments. Then, the excitation can be provided by waves impinging on both interfaces $z = 0$ and $z = z_{max}$. As such, from the standpoint of setting the two-way ABCs these interfaces become equal — either has to be transparent for its corresponding outgoing waves while simultaneously being able to fully transmit the corresponding incoming field. The latter setup will be instrumental for simulations of coherent counter-propagating beams.

2.1.3   Transverse Boundary Conditions

Any of the foregoing formulations requires setting boundary conditions in the transverse direction(s) x_j, $j = 1, \ldots, D - 1$. Although in some cases the domain filled by the Kerr medium extends all the way to infinity in the transverse direction(s), for the purpose of practical computing, we truncate this domain at an external artificial boundary $\partial\Omega \times \mathbb{R}^+$ and set the boundary conditions for all $(\boldsymbol{x}, z)$ such that $\boldsymbol{x} \in \partial\Omega \subset \mathbb{R}^{D-1}$ and $z \geq 0$. In our work [6,7,9], we assumed that the electric field vanished at this lateral artificial boundary: $E(\boldsymbol{x}, z)\big|_{\boldsymbol{x} \in \partial\Omega, z \geq 0} = 0$. While the approach in [6, 7, 9] enabled us, apparently for the first time ever, to solve the NLH as a genuine boundary value problem, there have still been indications that it could be improved. Indeed, the Dirichlet boundary conditions, while being easy to implement, possess the non-physical property of reflecting the waves back from the lateral boundaries into the computational domain. Even though the solution with the Dirichlet transverse boundary conditions can be expected to approximate well the original infinite-domain solution in the domain of physical interest, namely in the central region of the computational domain, we still had to position the lateral boundaries sufficiently far away from the center so that to alleviate the undesirable reflections, see [6,7,9]. Therefore, in [10] we replaced the Dirichlet boundary condition with a local Sommerfeld-type radiation boundary condition at the transverse boundaries of the computational domain. This change facilitated major gains in performance, allowing us to considerably reduce the width of the computational domain, and also enabled the computation of solutions that could not be computed previously.

2.1.4   Paraxial Approximation and the Nonlinear Schrö-dinger Equation

Let r_0 be the initial radius of the impinging laser beam. We first introduce the dimensionless quantities $\tilde{\boldsymbol{x}}$, $\tilde{z}$, and A:

$$\tilde{\boldsymbol{x}} = \frac{\boldsymbol{x}}{r_0} , \quad \tilde{z} = \frac{z}{2L_{DF}} , \quad E = e^{ik_0 z}(\epsilon r_0^2 k_0^2)^{-1/2\sigma} A(\boldsymbol{x}, z) , \qquad (2.3)$$

where $L_{DF} = k_0 r_0^2$ is the diffraction length. Then, by dropping the tildes we obtain from (2.1):

$$iA_z + \Delta_\perp A + |A|^{2\sigma} A = -4f^2 A_{zz} , \qquad (2.4)$$

where $f = 1/r_0 k_0 \ll 1$ is the small nonparaxiality parameter.

The standard derivation of the nonlinear Schrödinger equation (NLS) is based on the assumption that the envelope A varies slowly. Then, one can neglect the right-hand side of (2.4) [i.e., apply the paraxial approximation] and obtain the NLS:

$$iA_z + \Delta_\perp A + |A|^{2\sigma} A = 0 . \qquad (2.5)$$

The NLS (2.5) is an evolution equation in which the variable z plays the role of "time." Hence, it only needs to be supplemented by the initial condition at $z = 0$:

$$A(\boldsymbol{x}, 0) = (\epsilon r_0^2 k_0^2)^{1/2\sigma} E^{(\text{inc, left})}(\boldsymbol{x}).$$

Subsequently, the NLS (2.5) is to be integrated by a "time"-marching algorithm for $z > 0$. We note that almost all self-focusing studies available in the Literature have used NLS-based models, rather than the NLH, the main reason being that the NLH is much a "tougher" object for analysis and for simulations. However, the NLH provides a more comprehensive physical model for self-focusing. In particular, it takes into account the effect of nonlinear backscattering, which the NLS model disregards. Indeed, once (2.5) has been integrated, the overall solution, according to (2.3), is the slowly varying amplitude A times the forward propagating oscillatory component $e^{ik_0 z}$.

The capability to account for the effects of nonparaxiality and backscattering makes the NLH model (2.1) more in-depth compared to the simpler NLS model (2.5). It is precisely the role of these two important effects in nonlinear wave propagation that the numerical methodology developed in this paper would enable us to focus on. In the overall perspective, however, the scalar NLH (2.1) itself does not represent the most general setup either. A yet more comprehensive model would be the vector NLH that also accounts for the vectorial nature of the electric field while still taking care of the phenomena of nonparaxiality and backscattering. Note that the vectorial effects, nonparaxiality, and backscattering are all of the same order of magnitude; see [5] for more detail.

2.1.5   Solitons and Collapse

It is well known that solutions of the NLS (2.5) exist globally when $\sigma(D - 1) < 2$, the subcritical NLS, but can become singular (collapse) at finite propagation distances, when either $\sigma(D - 1) > 2$, the supercritical NLS, or when $\sigma(D - 1) = 2$, the critical NLS [11]. As shown by Weinstein [13], a necessary condition for singularity formation in the critical NLS is that the input beam power (i.e., $\|A(\cdot, 0)\|_2^2$) exceeds the critical power N_c. The value of N_c is equal to the power of the ground-state solitary wave solution of the NLS; this value can be calculated explicitly for $D = 2$ and numerically for $D > 2$.

In our subsequent numerical simulations in Section 2.4 we will focus on two cases. The first one is $D = 2$ and $\sigma = 2$ (quintic nonlinearity), for which the NLS (2.5) is critical. Although the actual physical setting for the critical NLS is $D = 3$ and $\sigma = 1$, which corresponds to propagation of waves in a bulk Kerr medium, one can expect the role of nonparaxiality and backscattering to be quite similar for both cases. This expectation should, in particular, pertain to the question, which has been open for many years, of whether the more comprehensive NLH model for nonlinear self-focusing eliminates the singular behavior that characterizes collapsing solutions of the critical NLS.

The second case that we will consider is $D = 2$ and $\sigma = 1$ (cubic nonlinearity), which physically corresponds to nonlinear propagation in a planar waveguide (i.e., there is only one transverse dimension). In this subcritical case solutions to the NLS do not collapse. Instead, the laser beam can propagate in the Kerr medium over very long distances without changing its profile — the type of behavior often referred to as spatial soliton. In the past, solitons have primarily been studied as solutions to the NLS. Although it is generally expected that in the subcritical case the NLH will have similarly looking solutions, until now it was not actually possible to study the effect of nonparaxiality and backscattering on solitons. Even more so, the NLH appears particularly well suited for simulating interactions between counter-propagating solitons. Indeed, in the NLH framework the counter-propagating case can be naturally formulated as a boundary-value problem on the slab $0 \le z \le z_{\max}$, see Figure 2.1, whereas in the NLS framework the two counter-propagating solitons will imply two opposite directions of marching.

2.2   Algorithm — Continuous Formulation

For clarity of presentation, we first outline the algorithm using a continuous formulation. We consider from now on the case $D = 2$ that corresponds to propagation in a planar waveguide. Therefore, equation (2.1) becomes

$$(\partial_{zz} + \partial_{xx})E(x, z) + k^2 E = 0,$$
$$k^2 = k_0^2(1 + \epsilon |E|^{2\sigma}), \quad \sigma > 0, \quad x \in \mathbb{R}, \quad z \ge 0. \tag{2.6}$$

Compared with the case $D = 3$, the two-dimensional setup offers a considerable reduction of the computational effort while preserving all the essential physical and numerical effects. For simplicity and with no substantial loss of generality, for all our numerical experiments we will choose the rectangular computational domain $\{-x_{\max} \le x \le x_{\max}, \, 0 \le z \le z_{\max}\}$. At the "upstream" and "downstream" boundaries $z = 0$ and $z = z_{\max}$ we will set the nonlocal two-way ABCs. At the transverse boundaries $x = \pm x_{\max}$ we will set local radiation boundary conditions of the Sommerfeld type.

2.2.1   Iteration Scheme

We solve the NLH (2.6) iteratively by freezing the nonlinearity and reducing the NLH to a linear variable-coefficient equation on every iteration

$$\Delta E^{(j)} + k_0^2(1 + \epsilon|E^{(j-1)}|^{2\sigma})E^{(j)} = 0, \qquad j = 1, 2, \ldots, \qquad (2.7)$$

where $\Delta = \partial_{zz} + \partial_{xx}$. The initial guess is typically chosen as $E^{(0)} \equiv 0$. The sequence (2.7) will be referred to as the *outer iteration loop* or the *nonlinear iteration loop*. Then, for every j the corresponding linear equation (2.7) is also solved iteratively as

$$\Delta E^{(j,k)} + k_0^2 E^{(j,k)} = -k_0^2\epsilon \left|E^{(j-1,K)}\right|^{2\sigma} E^{(j,k-1)}, \qquad k = 1, 2, \ldots, K, \quad (2.8)$$

where $E^{(j,0)} = E^{(j-1,K)}$. We will refer to sequence (2.8) as to the *inner iteration loop*. It is clear that finding $E^{(j,k)}$ for every k in (2.8) amounts to solving the standard constant-coefficient inhomogeneous Helmholtz equation:

$$\Delta E + k_0^2 E = \Phi(x, z), \qquad (2.9)$$

with $E = E^{(j,k)}(x, z)$ and $\Phi(x, z) \equiv -k_0^2\epsilon \left|E^{(j-1,K)}(x, z)\right|^{2\sigma} E^{(j,k-1)}(x, z)$.

Equation (2.9) is solved repeatedly [for updated $\Phi(x, z)$] on the rectangular computational domain via the separation of variables. In so doing, the boundary conditions in either the longitudinal direction z or the transverse direction x are specified for the linear constant-coefficient equation (2.9), rather than for the original nonlinear equation (2.6), see Section 2.2.2. The reason is that the key role of the boundary conditions is to distinguish between the waves propagating in different directions, see Section 2.1.2, and this is done most naturally in the linear framework. Once the iterations converge, the resulting solution is assumed to inherit the desired wave radiation properties that are built into the methodology for solving equation (2.9).

2.2.2   Separation of Variables and Boundary Conditions

The Sommerfeld radiation boundary conditions in the transverse direction x are set based on factorization of the one-dimensional second-order Helmholtz operator into the product of two first-order factors:

$$\partial_{xx} + k_0^2 I = (\partial_x + ik_0 I)(\partial_x - ik_0 I), \qquad (2.10)$$

where $i = \sqrt{-1}$, I is the operator identity, and $k_0 = \sqrt{k_0^2} > 0$. While the equation $(\partial_{xx} + k_0^2 I)E = 0$ admits two linearly independent solutions, e^{ik_0x} and e^{-ik_0x}, each of the first-order factors on the right-hand side of (2.10) selects only one solution from the foregoing pair. Therefore, the boundary conditions

$$E_x - ik_0 E\big|_{x=x_{\max}} = 0 \qquad \text{and} \qquad E_x + ik_0 E\big|_{x=-x_{\max}} = 0 \qquad (2.11)$$

correspond to propagation of waves only upward or only downward, respectively, see Figure 2.1. In other words, they guarantee that the upper artificial boundary $x = x_{\max}$ be completely transparent for the plane waves $e^{ik_0 x}$ traveling with normal incidence in the positive x direction, and the lower artificial boundary $x = -x_{\max}$ be completely transparent for the plane waves $e^{-ik_0 x}$ traveling with normal incidence in the negative x direction.

Let us now introduce the following eigenvalue problem on $[-x_{\max}, x_{\max}]$ for the transverse Laplacian $\Delta_\perp \equiv \partial_{xx}$, subject to the radiation boundary conditions (2.11):

$$\psi_{xx} = \lambda\psi, \quad \psi_x - ik_0\psi\big|_{x=x_{\max}} = 0, \quad \psi_x + ik_0\psi\big|_{x=-x_{\max}} = 0. \tag{2.12}$$

It is easy to show that the eigenfunctions $\psi = \psi^{(l)}(x)$ of (2.12) are given by $\psi = \cosh(\sqrt{\lambda}x)$ and $\psi = \sinh(\sqrt{\lambda}x)$, and that the eigenvalues $\lambda = \lambda^{(l)}$ can be obtained by solving the transcendental equation:

$$e^{2\sqrt{\lambda}x_{\max}} = \pm\frac{\sqrt{\lambda} + ik_0}{\sqrt{\lambda} - ik_0}, \tag{2.13}$$

where the plus and minus signs correspond to $\psi = \cosh(\sqrt{\lambda}x)$ and to $\psi = \sinh(\sqrt{\lambda}x)$, respectively.

Equation (2.12) is not a classical Sturm-Liouville problem, because the operator ∂_{xx} subject to boundary conditions (2.11) is not self-adjoint. As such, one should not expect its eigenvalues to be real. Indeed,

Proposition 1 ([10]) *Let λ be an eigenvalue of (2.12). Then, $\mathrm{Im}(\lambda) > 0$.*

Furthermore, since the boundary-value problem (2.12) is not self-adjoint, its eigenfunctions $\{\psi^{(l)}(x)\}$ are, generally speaking, nonorthogonal. Therefore, one cannot separate the variables in equation (2.9) using standard Fourier expansion of its solution in terms of the eigenfunctions of (2.12). It, however, turns out that one can still build the expansion $E(x, z) = \sum_l \hat{E}_l(z)\psi^{(l)}(x)$ with the help of the following "real orthogonality" property of the eigenfunctions $\{\psi^{(l)}(x)\}$ (see, e.g., [1] for further detail) summarized in Proposition 2:

Proposition 2 ([10]) *Let $\psi^{(m)}$ and $\psi^{(n)}$ be the eigenvectors of (2.12) with corresponding eigenvalues $\lambda_m \neq \lambda_n$. Then,*

$$\int_{-x_{\max}}^{x_{\max}} \psi^{(m)}(x)\psi^{(n)}(x)\,dx = 0, \qquad \int_{-x_{\max}}^{x_{\max}} \psi_m^2(x)\,dx \neq 0. \tag{2.14}$$

Note that as the eigenfunctions $\psi^{(m)}$ and $\psi^{(n)}$ are complex, expression (2.14) does not yield a genuine inner product. However, Proposition 2 still indicates that the eigenfunctions of (2.12) can be rescaled so that

$$\int_{-x_{\max}}^{x_{\max}} \psi^{(m)}(x)\psi^{(n)}(x)\,dx = \delta_{mn}.$$

Therefore, for all $z \in [0, z_{max}]$, solution to the linear Helmholtz equation (2.9) can, in principle, be obtained by expanding E and Φ on $[-x_{max}, x_{max}]$ in terms of the rescaled eigenfunctions $\psi^{(l)}(x)$ of (2.12), i.e.,

$$E(x, z) = \sum_l \hat{E}_l(z)\psi^{(l)}(x), \qquad \Phi(x, z) = \sum_l \hat{\Phi}_l(z)\psi^{(l)}(x), \qquad (2.15)$$

where

$$\hat{E}_l(z) = \int_{-x_{max}}^{x_{max}} E(x, z)\psi^{(l)}(x)\, dx, \qquad \hat{\Phi}_l(z) = \int_{-x_{max}}^{x_{max}} \Phi(x, z)\psi^{(l)}(x)\, dx. \qquad (2.16)$$

Substituting representation (2.15) into equation (2.9) and taking the "real inner product" (2.14) with $\psi^{(l)}$ we obtain the following set of ordinary differential equations with respect to the unknown quantities $\hat{E}_l(z)$:

$$\frac{d^2 \hat{E}_l}{dz^2} + (k_0^2 + \lambda^{(l)})\hat{E}_l = \hat{\Phi}_l. \qquad (2.17)$$

Since Proposition 1 shows that $\text{Im}(\lambda^{(l)}) > 0$, we conclude from equation (2.17) that introduction of the Sommerfeld radiation boundary conditions at the lateral boundaries results in the addition of a positive linear "damping" in the z direction. This conclusion has an intuitive explanation, since as the beam propagates in the z direction, the transverse radiation at $\pm x_{max}$ will obviously cause a power drain away from the interval $[-x_{max}, x_{max}]$. As shown in [7], introduction of linear damping has a regularizing effect on the NLH (which, e.g., has allowed us to solve the linearly damped NLH for initial conditions that led to singularity formation in the corresponding linearly damped NLS). However, unlike in [7], where linear damping was essentially motivated by the physical process of absorption of waves by the medium through which they propagate, the "damping" in (2.17) has nothing to do with actual physical absorption or with power losses. This "damping" in the z direction is rather a manifestation of the power radiation at $\pm x_{max}$. Another important difference is that unlike the case of physical damping, the magnitude of "damping" in (2.17) is determined by the transverse eigenvalue $\lambda^{(l)}$, i.e., it changes with the mode number l.

To complete the current illustrative section, we yet have to describe a key component of the algorithm — nonlocal ABCs at the boundaries $z = 0$ and $z = z_{max}$. Each equation (2.17) is to be considered on the interval $[0, z_{max}]$ and is to be supplemented by the boundary conditions at the endpoints $z = 0$ and $z = z_{max}$. Assuming that supp $\hat{\Phi}_l(z) \subseteq [0, z_{max}]$, i.e., that if extended beyond the interval $[0, z_{max}]$ the solution $\hat{E}_l(z)$ would be governed by the homogeneous counterpart of equation (2.17), we employ the same considerations as those that led to the Sommerfeld conditions (2.11) and obtain:

$$\frac{d\hat{E}_l}{dz} + i\sqrt{k_0^2 + \lambda^{(l)}}\,\hat{E}_l\bigg|_{z=0} = 0 \quad \text{and} \quad \frac{d\hat{E}_l}{dz} - i\sqrt{k_0^2 + \lambda^{(l)}}\,\hat{E}_l\bigg|_{z=z_{max}} = 0. \qquad (2.18)$$

The boundary conditions (2.18) guarantee that all outgoing waves will leave the domain $[0, z_{max}]$ with no reflection. Indeed, the left-traveling waves $e^{-i\sqrt{k_0^2 + \lambda^{(l)}}z}$ will propagate freely through the endpoint $z = 0$, and the right-traveling waves $e^{i\sqrt{k_0^2 + \lambda^{(l)}}z}$ will propagate freely through the endpoint $z = z_{max}$.

Having obtained the ABCs (2.18) that allow for reflectionless propagation of all the outgoing waves, we now need to take into account the given incoming wave as well, see formula (2.2) and Figure 2.1. The incoming field $E^{(inc, \, left)}(x) = \sum_l \hat{E}_l^{(inc, \, left)} \psi^{(l)}(x)$ that impinges on the interface $z = 0$ from the left gives rise to a solution of the homogeneous constant-coefficient linear Helmholtz equation of the form

$$E(x, z) = \sum_l \hat{E}_l^{(inc, \, left)} e^{i\sqrt{k_0^2 + \lambda^{(l)}}z} \psi^{(l)}(x). \qquad (2.19)$$

After the separation of variables, each individual component $\hat{E}_l = \hat{E}_l^{(inc, \, left)} e^{i\sqrt{k_0^2 + \lambda^{(l)}}z}$ can be substituted into the first relation of (2.18). Since this component satisfies $\frac{d\hat{E}_l}{dz} = i\sqrt{k_0^2 + \lambda^{(l)}}\hat{E}_l$ for any z, the aforementioned substitution yields the two-way boundary condition at $z = 0$:

$$\frac{d\hat{E}_l}{dz} + i\sqrt{k_0^2 + \lambda^{(l)}}\hat{E}_l \Big|_{z=0} = 2i\sqrt{k_0^2 + \lambda^{(l)}}\hat{E}_l^{(inc, \, left)}. \qquad (2.20a)$$

Symmetrically, if there is also incoming radiation at the right interface $z = z_{max}$: $E^{(inc, \, right)}(x) = \sum_l \hat{E}_l^{(inc, \, right)} \psi^{(l)}(x)$, then a similar procedure leads to the two-way boundary condition at $z = z_{max}$ as well:

$$\frac{d\hat{E}_l}{dz} - i\sqrt{k_0^2 + \lambda^{(l)}}\hat{E}_l \Big|_{z=z_{max}} = -2i\sqrt{k_0^2 + \lambda^{(l)}}\hat{E}_l^{(inc, \, right)}. \qquad (2.20b)$$

Note that while being able to correctly prescribe the given incoming wave(s), boundary conditions (2.20) still retain the full radiation capability of boundary conditions (2.18). Indeed, any solution of type (2.2) identically satisfies boundary condition (2.20a). A similar property will obviously hold for boundary condition (2.20b) as well. The capability of properly handling the waves propagating through a given interface in both directions has prompted us in [9] to call boundary conditions (2.20) *the two-way ABCs*. In contradistinction to that, relations of type (2.18) that only guarantee the radiation of waves in one particular direction, are often referred to in the literature as *the one-way Helmholtz equations* [9, 12]. Let us also emphasize that as the two-way ABCs (2.20) are obtained in the transformed space individually for every mode l, they would become *nonlocal* if transformed back to the original space.

2.3   Algorithm — Finite-Difference Formulation

We now describe the finite-difference implementation of the algorithm. The NLH (2.6) is approximated on the uniform two-dimensional Cartesian grid with mesh sizes $h_x = x_{\max}/M$ and $h_z = z_{\max}/N$, so that:

$$
\begin{aligned}
x_m &= m \cdot h_x, \qquad m = -M, \ldots, 0, \ldots, M, \\
z_n &= n \cdot h_z, \qquad n = 0, \ldots, N.
\end{aligned}
\tag{2.21}
$$

2.3.1   Fourth Order Scheme

The discrete implementation of the algorithm is carried out with the fourth order of accuracy. As we shall see, the construction of the algorithm and its analysis are more complex for our fourth-order discretization than they would have been for a second order discretization. Nevertheless, this was a price worth paying, since higher-order approximations offer the possibility to take fewer points per wavelength. In addition, higher order offers the capability to better resolve the small-scale phenomenon of backscattering at the background of the larger forward-propagating wave.

For our algorithm, we have chosen the standard central-difference fourth-order scheme on the grid (2.21):

$$
\begin{aligned}
&\frac{-E_{m-2,n} + 16E_{m-1,n} - 30E_{m,n} + 16E_{m+1,n} - E_{m+2,n}}{12h_x^2} \\
&+ \frac{-E_{m,n-2} + 16E_{m,n-1} - 30E_{m,n} + 16E_{m,n+1} - E_{m,n+2}}{12h_z^2} \\
&+ k_0^2 \left(1 + \epsilon |E_{m,n}|^{2\sigma}\right) E_{m,n} = 0.
\end{aligned}
\tag{2.22}
$$

Scheme (2.22) is written on the stencil that extends five grid nodes wide in each coordinate direction. Therefore, scheme (2.22) can only be written for the interior nodes of the grid (2.21) that are at least two nodes away from the boundary. Alternatively, two ghost nodes can be added to grid (2.21) from each side in each direction. In either case, the finite-difference equations to be solved near the boundaries of the computational domain will differ from the interior equations (2.22), and these special near-boundary equations shall be interpreted as discrete boundary conditions for the scheme (2.22).

To keep all the notations straightforward, let us introduce the ghost nodes (x_m, z_n) for $m = \pm(M+1)$ and $\pm(M+2)$, and $n = -2, -1, N+1$, and $N+2$. From here on, we will assume that the special near-boundary treatment shall apply to these ghost nodes, while the finite-difference equations (2.22) can keep their form on the entire grid (2.21). It is important to emphasize that system (2.22) itself is a system of fourth order finite-difference equations, and therefore it requires more boundary conditions than the original second order differential equation does, even though the former approximates the latter. More precisely, we will see that scheme (2.22) will require two boundary conditions at each boundary, whereas the original differential equation requires only one, see Section 2.2.2.

As indicated in Section 2.2, the boundary conditions are to be set for the linear constant coefficient equation solved repeatedly on the inner loop of the nested iteration scheme. Therefore, for the nonlinear finite-difference system (2.22) we introduce an iterative solver fully analogous to the one outlined in Section 2.2.1 for the continuous case [cf. formula (2.8)]:

$$\frac{-E_{m-2,n}^{(j,k)} + 16E_{m-1,n}^{(j,k)} - 30E_{m,n}^{(j,k)} + 16E_{m+1,n}^{(j,k)} - E_{m+2,n}^{(j,k)}}{12h_x^2}$$

$$+\frac{-E_{m,n-2}^{(j,k)} + 16E_{m,n-1}^{(j,k)} - 30E_{m,n}^{(j,k)} + 16E_{m,n+1}^{(j,k)} - E_{m,n+2}^{(j,k)}}{12h_z^2} \qquad (2.23)$$

$$+k_0^2 E_{m,n}^{(j,k)} = -k_0^2 \epsilon \left| E_{m,n}^{(j-1,K)} \right|^{2\sigma} E_{m,n}^{(j,k-1)},$$

where $j = 1, 2, \ldots$; $k = 1, 2, \ldots, K$; $E_{m,n}^{(0,0)} = 0$; and $E_{m,n}^{(r,0)} = E_{m,n}^{(j-1,K)}$. Next, denoting $E_{m,n} = E_{m,n}^{(j,k)}$ and $\Phi_{m,n} = -k_0^2 \epsilon \left| E_{m,n}^{(j-1,K)} \right|^{2\sigma} E_{m,n}^{(j,k-1)}$, we arrive at the fourth-order central-difference approximation to the linear constant coefficient Helmholtz equation (2.9):

$$\frac{-E_{m-2,n} + 16E_{m-1,n} - 30E_{m,n} + 16E_{m+1,n} - E_{m+2,n}}{12h_x^2}$$

$$+\frac{-E_{m,n-2} + 16E_{m,n-1} - 30E_{m,n} + 16E_{m,n+1} - E_{m,n+2}}{12h_z^2} \qquad (2.24)$$

$$+k_0^2 E_{m,n} = \Phi_{m,n}.$$

System (2.24) is to be supplemented by the boundary conditions and solved repeatedly in the course of the iteration (2.23) for updated $\Phi_{m,n}$.

2.3.2   Transverse Boundary Conditions

To set the discrete Sommerfeld radiation boundary conditions at $m = \pm M$, we first need to identify the waves that propagate upward and downward in the corresponding discrete framework. To do that, let us consider the one-dimensional discrete homogeneous Helmholtz equation (for clarity, we suppress here the subscript n)

$$\frac{-E_{m-2} + 16E_{m-1} - 30E_m + 16E_{m+1} - E_{m+2}}{12h_x^2} + k_0^2 E_m = 0. \qquad (2.25)$$

Equation (2.25) has a four-dimensional fundamental set of solutions: $\{q_1^m, q_1^{-m}, q_2^m, q_2^{-m}\}$, where q_1, $1/q_1$, q_2, and $1/q_2$ are the four roots of the algebraic characteristic equation that corresponds to the discretization (2.25):

$$-1 + 16q + (12\alpha_x^2 - 30)q^2 + 16q^3 - q^4 = 0. \tag{2.26}$$

They are given explicitly by

$$q_1 = \frac{d_1 + \sqrt{d_1^2 - 4}}{2}, \qquad q_1^{-1} = \frac{d_1 - \sqrt{d_1^2 - 4}}{2},$$

$$q_2 = \frac{d_2 - \sqrt{d_2^2 - 4}}{2}, \qquad q_2^{-1} = \frac{d_2 + \sqrt{d_2^2 - 4}}{2}. \tag{2.27}$$

where

$$d_1 = 8 - 6\sqrt{1 + \alpha_x^2/3} \qquad \text{and} \qquad d_2 = 8 + 6\sqrt{1 + \alpha_x^2/3}.$$

The parameter

$$\alpha_x = h_x k_0 \tag{2.28}$$

is a measure of how well the waves are resolved by the transverse grid. In [9] we have shown that when $0 < \alpha_x \ll 1$ (i.e., when the wavenumber k_0 is well resolved on the transverse grid), the roots q_1 and q_1^{-1} have unit magnitudes, $|q_1| = |q_1^{-1}| = 1$, and the solutions q_1^m and q_1^{-m} approximate the genuine traveling waves $e^{ik_0 x}$ and $e^{-ik_0 x}$, respectively, with the fourth order of accuracy:

$$q_1^m = e^{ik_0 h_x m} + O\left(\alpha_x^4\right), \qquad q_1^{-m} = e^{-ik_0 h_x m} + O\left(\alpha_x^4\right).$$

Regarding the second pair of roots, q_2 and q_2^{-1}, it was shown in [9] that $|q_2| < 1$ and $|q_2^{-1}| > 1$. Accordingly, solutions q_2^m and q_2^{-m} are always evanescent. These solutions are numerical artifacts, as they do not exist in the continuous context, see Section 2.2.2, and only appear because the discretization is fourth order, whereas the original differential equation is second order. Still, the presence of a second evanescent pair of waves requires special treatment at the boundary and necessitates setting an additional pair of the boundary conditions.

Consistently with the idea of Section 2.2.2, to guarantee the reflectionless propagation of the discrete outgoing waves through the lateral artificial boundaries, we need to require that the boundary at $m = M$ be transparent for the waves q_1^m traveling upward, and the boundary at $m = -M$ be transparent for the waves q_1^{-m} traveling downward. This requirement constitutes the Sommerfeld radiation principle in the fourth order discrete framework, and it cannot be either altered or relaxed in any way. As, however, concerns the second, evanescent, pair of waves, q_2^m and q_2^{-m}, more flexibility can be exercised toward their near-boundary treatment. Indeed, these waves are nonphysical and as such, we only need to ensure stability of the resulting overall discretization once the corresponding second pair of the boundary conditions has been chosen.

The fourth-order discrete Sommerfeld radiation conditions are obtained in [10] in the form of one-way Helmholtz equations:

$$E_{M+1} = C_1 E_M + C_2 E_{M-1}, \qquad E_{-M-1} = C_1 E_{-M} + C_2 E_{-M+1},$$
$$E_{M+2} = C_3 E_M + C_4 E_{M-1}, \qquad E_{-M-2} = C_3 E_{-M} + C_4 E_{-M+1}, \tag{2.29}$$

where

$$C_1 = \frac{q_2 + q_2^2 - q_1^2(1 + q_2^3)}{q_2 + q_2^2 - q_1(1 + q_2^3)}, \qquad C_3 = \frac{(1 + q_2^3)(1 - q_1^3)}{q_2 + q_2^2 - q_1(1 + q_2^3)},$$
$$C_2 = -\frac{(q_2 + q_2^2)(q_1 - q_1^2)}{q_2 + q_2^2 - q_1(1 + q_2^3)}, \qquad C_4 = -\frac{q_1(1 + q_2^3) - q_1^3(q_2 + q_2^2)}{q_2 + q_2^2 - q_1(1 + q_2^3)}. \tag{2.30}$$

The fundamental property of boundary conditions and (2.29), (2.30) is that they render the reflectionless radiation of the physical waves q_1^m and q_1^{-m}. Moreover, they yield an overall stable discretization and possess additional symmetries which become useful when proving properties of the transverse eigenvectors in Section 2.3.4.

2.3.3   Discrete Eigenvalue Problem

Let us now formulate a discrete eigenvalue problem analogous to (2.12). For that purpose we first introduce a $(2M+1) \times (2M+1)$ transverse discretization matrix $\boldsymbol{A}$ that would contain the fourth order central difference approximation of ∂_{xx} at the interior nodes, see (2.24) or (2.25), and would also incorporate the boundary conditions (2.29). By eliminating the ghost variables E_{-M-2}, E_{-M-1}, E_{M+1}, and E_{M+2}, we obtain:

$$\boldsymbol{A} = \frac{1}{12h_x^2} \begin{bmatrix} -30 + C_5 & 16 + C_6 & -1 & 0 & 0 & 0 & \cdots \\ 16 - C_1 & -30 - C_2 & 16 & -1 & 0 & 0 & \cdots \\ -1 & 16 & -30 & 16 & -1 & 0 & \cdots \\ \vdots & \ddots & \ddots & \ddots & \ddots & & \vdots \\ \cdots & 0 & -1 & 16 & -30 & 16 & -1 \\ \cdots & 0 & 0 & -1 & 16 & -30 - C_2 & 16 - C_1 \\ \cdots & 0 & 0 & 0 & -1 & 16 + C_6 & -30 + C_5 \end{bmatrix},$$

$$\tag{2.31a}$$

where

$$C_5 = 16C_1 - C_3, \qquad C_6 = 16C_2 - C_4. \tag{2.31b}$$

The discrete counterpart of the eigenvalue problem (2.12) can now be formulated as follows:

$$\boldsymbol{A}\psi = \lambda\psi. \tag{2.32}$$

In order to be able to separate the variables in the finite-difference system (2.24) using the eigenvectors of $\boldsymbol{A}$, we need to make sure that these eigenvectors form a basis. To do that, we will use the results of the forthcoming

Proposition 3 to prove in Proposition 4 that the $(2M+1) \times (2M+1)$ matrix $\boldsymbol{A}$ of (2.31) has $(2M+1)$ distinct eigenvalues. This implies that $\boldsymbol{A}$ has $(2M+1)$ linearly-independent eigenvectors. Then we will prove in Proposition 7 that these eigenvectors can be rescaled so that they posses the real orthogonality property, that allows us to arrive at formula (2.37) for the inverse of the matrix of the eigenvectors; the latter is needed for the actual implementation of the separation of variables.

Let us recall that the roots of the characteristic equation (2.26), and consequently, the entries of the matrix $\boldsymbol{A}$ of (2.31), are functions of α_x. In particular, it is easy to see that in the fine grid/long waves limit $\alpha_x \longrightarrow 0+$, we have $q_1 \longrightarrow 1$. For $q_1 = 1$ formulae (2.30) imply that $C_1 = C_4 = 1$ and $C_2 = C_3 = 0$. As such, boundary conditions (2.29), (2.30) for $\alpha_x = 0$ reduce to

$$E_{M+1} = E_M, \qquad E_{-M-1} = E_{-M},$$
$$E_{M+2} = E_{M-1}, \qquad E_{-M-2} = E_{-M+1}. \tag{2.33}$$

Equalities (2.33) imply symmetry with respect to the points $\pm(M + 1/2)$, which can also be interpreted as a fourth order approximation of the homogeneous Neumann boundary conditions. Moreover, the matrix $\boldsymbol{A}$ of (2.31) acquires a particularly convenient form for $\alpha_x = 0$:

$$\boldsymbol{A}(0) = \frac{1}{12h_x^2}
\begin{bmatrix}
-14 & 15 & -1 & 0 & 0 & 0 & \cdots \\
15 & -30 & 16 & -1 & 0 & 0 & \cdots \\
-1 & 16 & -30 & 16 & -1 & 0 & \cdots \\
\vdots & \ddots & \ddots & \ddots & \ddots & \ddots & \vdots \\
\cdots & 0 & -1 & 16 & -30 & 16 & -1 \\
\cdots & 0 & 0 & -1 & 16 & -30 & 15 \\
\cdots & 0 & 0 & 0 & -1 & 15 & -14
\end{bmatrix}, \tag{2.34}$$

so that the corresponding eigenvalue problem (2.32) can be solved analytically.

Proposition 3 *The eigenvalues* $\lambda^{(l)}(0)$ *and eigenvectors* $\psi^{(l)}(0) = \left[\psi_{-M}^{(l)}(0), \ldots, \psi_0^{(l)}(0), \ldots, \psi_M^{(l)}(0) \right]^T$ *of the matrix* $\boldsymbol{A}(0)$ *of (2.34) are given by:*

$$\lambda^{(l)}(0) = -\frac{16}{3h_x^2} \sin^2 \left(\frac{1}{2} \frac{l\pi}{2M+1} \right) + \frac{1}{3h_x^2} \sin^2 \left(\frac{l\pi}{2M+1} \right), \tag{2.35a}$$

$$\psi_m^{(l)}(0) = b_l \cos \left[l \left(\frac{\pi m}{2M+1} + \frac{\pi}{2} \right) \right], \tag{2.35b}$$

where $l = 0, 1, \ldots, 2M$; $b_0 = \frac{1}{\sqrt{2M+1}}$, *and* $b_l = \sqrt{\frac{2}{2M+1}}$ *for* $l > 0$.

The eigenvectors $\psi^{(l)}(0)$ *of (2.35b) are orthonormal, and the eigenvalues* $\lambda^{(l)}(0)$ *of (2.35a) are distinct.*

Proposition 4 *Let* $\alpha_x \ll 1$. *Then, the* $(2M+1) \times (2M+1)$ *matrix* $\boldsymbol{A}$ *of (2.31), with* $\{C_j\}_{j=1}^4$ *given by (2.30), has* $2M+1$ *distinct eigenvalues* $\lambda^{(l)} = \lambda^{(l)}(\alpha_x)$.

Proposition 4 clearly implies that the matrix $\boldsymbol{A}$ of (2.31), (2.30) has $2M+1$ linearly independent eigenvectors. Our next goal is to show that this basis of eigenvectors has the real orthogonality property (Proposition 7).

Proposition 5 ([10]) *The matrix $\boldsymbol{A}$ of (2.31), where $\{C_j\}_{j=1}^4$ are given by (2.30), is real-symmetric, i.e., $\boldsymbol{A}^T = \boldsymbol{A}$.*

Proposition 6 ([10]) *Let ϕ and ψ be eigenvectors of the matrix $\boldsymbol{A}$ of (2.31) with the corresponding eigenvalues λ_ϕ and λ_ψ. If $\lambda_\phi \neq \lambda_\psi$, then, $\phi^T \psi = 0$.*

Remark. Proposition 6 is the discrete analogue of Proposition 2.

Since the eigenvectors of $\boldsymbol{A}$ are, generally speaking, complex, the operation $\phi^T \psi$ is not a genuine inner product of the two vectors. Hence, it is not obvious whether any nontrivial eigenvector ψ would satisfy $\psi^T \psi \neq 0$. However, for $\alpha_x \ll 1$ a continuation argument can be employed to prove the following:

Proposition 7 *Let $\alpha_x \ll 1$. Then, the eigenvectors $\psi^{(l)}$, $l = 0, \ldots, 2M$, of the matrix $\boldsymbol{A}$ of (2.31), where $\{C_j\}_{j=1}^4$ are given by (2.30), can be normalized so that $\left(\tilde{\psi}^{(l)}\right)^T \tilde{\psi}^{(m)} = \delta_{lm}$.*

With the Dirichlet boundary conditions set at the transverse boundaries, we have used the discrete Fourier transform to separate the variables in system (2.24). In the Sommerfeld case we rather need to use the transformation by means of the eigenvectors of $\boldsymbol{A}$. This transformation will be a discrete analogue of (2.15) and (2.16). Moreover, as formula (2.37) will show, its inverse is numerically straightforward and inexpensive.

Let us introduce a square matrix of order $2M + 1$ that would have the original non-normalized eigenvectors $\psi^{(l)}$ as its columns:

$$\boldsymbol{\Psi} = \left[\psi^{(0)}, \ldots, \psi^{(2M)}\right]. \tag{2.36}$$

From Proposition 7 it follows that $\tilde{\boldsymbol{\Psi}}^T \tilde{\boldsymbol{\Psi}} = \boldsymbol{I}$ and

$$\boldsymbol{\Psi}^{-1} = \left(\boldsymbol{\Psi}^T \boldsymbol{\Psi}\right)^{-1} \boldsymbol{\Psi}^T = \begin{bmatrix} 1/\mu^{(0)} & 0 & 0 & 0 \\ 0 & 1/\mu^{(1)} & 0 & 0 \\ & & \ddots & \\ 0 & \cdots & 0 & 1/\mu^{(2M)} \end{bmatrix} \boldsymbol{\Psi}^T. \tag{2.37}$$

The easy formula (2.37) is used in the computations of Section 2.4 to invert the matrix of eigenvectors $\boldsymbol{\Psi}$.

2.3.4   Separation of Variables

We now use the transformation matrices $\boldsymbol{\Psi}$ of (2.36) and $\boldsymbol{\Psi}^{-1}$ of (2.37) to separate the variables in the discrete system (2.24).

Let us define the $2M + 1$-dimensional vectors $\mathcal{E}_n$ and Φ_n for $n = 0, \ldots, N$:

$$\mathcal{E}_n = [E_{-M,n}, \ldots, E_{0,n}, \ldots, E_{M,n}]^T,$$

$$\Phi_n = [\Phi_{-M,n}, \ldots, \Phi_{0,n}, \ldots, \Phi_{M,n}]^T.$$

Then, system (2.24) subject to the transverse radiation boundary conditions (2.29), (2.30) can be recast as follows:

$$A\mathcal{E}_n + \frac{-\mathcal{E}_{n-2} + 16\mathcal{E}_{n-1} - 30\mathcal{E}_n + 16\mathcal{E}_{n+1} - \mathcal{E}_{n+2}}{12h_z^2} + k_0^2\mathcal{E}_n = \Phi_n. \qquad (2.38)$$

In analogy with (2.15) and (2.16), let $\mathcal{E}_n = \Psi\hat{\mathcal{E}}_n \iff \hat{\mathcal{E}}_n = \Psi^{-1}\mathcal{E}_n$ and $\Phi_n = \Psi\hat{\Phi}_n \iff \hat{\Phi}_n = \Psi^{-1}\Phi_n$. In so doing, system (2.38) transforms into:

$$\Lambda\hat{\mathcal{E}}_n + \frac{-\hat{\mathcal{E}}_{n-2} + 16\hat{\mathcal{E}}_{n-1} - 30\hat{\mathcal{E}}_n + 16\hat{\mathcal{E}}_{n+1} - \hat{\mathcal{E}}_{n+2}}{12h_z^2} + k_0^2\hat{\mathcal{E}}_n = \hat{\Phi}_n. \qquad (2.39)$$

The variables in (2.39) have separated, as the diagonal structure of the matrix Λ enables a natural decomposition of system (2.39) into a set of $2M + 1$ separate one-dimensional systems that would govern individual components of $\hat{\mathcal{E}}_n = \left[\hat{E}_{0,n}, \ldots, \hat{E}_{2M,n}\right]^T$:

$$\frac{-\hat{E}_{l,n-2} + 16\hat{E}_{l,n-1} - 30\hat{E}_{l,n} + 16\hat{E}_{l,n+1} - \hat{E}_{l,n+2}}{12h_z^2} + (k_0^2 + \lambda^{(l)})\hat{E}_{l,n} = \hat{\Phi}_{l,n},$$

$$l = 0, \ldots, 2M, \qquad (2.40)$$

here $\hat{\Phi}_{l,n}$, $l = 0, \ldots, 2M$, are components of the vector $\hat{\Phi}_n$.

Each system (2.40) needs to be solved independently of the others, after which the solution in the original space can be reconstructed as $\mathcal{E}_n = \Psi\hat{\mathcal{E}}_n$ for every $n = 0, \ldots, N$. For any $l = 0, \ldots, 2M$, the corresponding system (2.40) is still subdefinite unless supplemented by the boundary conditions at $n = 0$ and $n = N$. We describe these discrete two-way ABCs in Section 2.3.6.

2.3.5 Properties of Eigenvalues

In Section 2.2, we have shown that the introduction of the Sommerfeld transverse radiation boundary conditions can be interpreted as linear "damping" in the z direction. We have an equivalent result at the discrete level (cf. Proposition 1):

Proposition 8 ([10]) *Let $\lambda = \lambda(\alpha_x)$ be an eigenvalue of the matrix A of (2.31), where $\{C_j\}_{j=1}^4$ are given by (2.30). Then,*

$$\left.\frac{d\lambda}{d\alpha_x}\right|_{\alpha_x=0} = i\nu \quad and \quad \nu > 0.$$

2.3.6   Nonlocal ABCs

Having separated the variables in system (2.24), we obtain a set of one-dimensional finite-difference equations (2.40). Each of these equations will be supplemented by 1D boundary conditions at $n = 0$ and at $n = N$ in much the same way as done in Section 2.3.2 for equation (2.25). Thus, let q_1, $1/q_1$, q_2 and $1/q_2$ be the four roots of the algebraic characteristic equation that corresponds to the homogeneous counterpart of equation l of (2.40)[1] (cf. equation (2.26)):

$$-1 + 16q + (12\alpha_z^2 - 30)q^2 + 16q^3 - q^4 = 0, \tag{2.41}$$

where (cf. formula (2.28))

$$\alpha_z = h_z \sqrt{k_0^2 + \lambda^{(l)}}. \tag{2.42}$$

The desired finite-difference two-way ABCs at the left boundary are given by (see [10] for detail):

$$\begin{aligned}
\hat{E}_{l,-1} &- (q_1 + q_2)\hat{E}_{l,0} + q_1 q_2 \hat{E}_{l,1} \\
&= \hat{\Phi}_{l,0} \left[G^{-1} - (q_1 + q_2)G^0 + q_1 q_2 G^1 \right] \\
&+ \hat{E}_l^{(\text{inc, left})} \left[q_1^{-1} - (q_1 + q_2) + q_1^2 q_2 \right]
\end{aligned} \tag{2.43a}$$

and

$$\begin{aligned}
\hat{E}_{l,-2} &- [(q_1 + q_2)^2 - q_1 q_2]\hat{E}_{l,0} + (q_1 + q_2)q_1 q_2 \hat{E}_{l,1} \\
&= \hat{\Phi}_{l,0} \left[G^{-2} - [(q_1 + q_2)^2 - q_1 q_2]G^0 + (q_1 + q_2)q_1 q_2 G^1 \right] \\
&+ \hat{E}_l^{(\text{inc, left})} \left[q_1^{-2} - [(q_1 + q_2)^2 - q_1 q_2] + (q_1 + q_2)q_1^2 q_2 \right].
\end{aligned} \tag{2.43b}$$

Similarly, at the right boundary we have:

$$\begin{aligned}
\hat{E}_{l,N+1} &- (q_1 + q_2)\hat{E}_{l,N} + q_1 q_2 \hat{E}_{l,N-1} \\
&= \hat{\Phi}_{l,N} \left[G^1 - (q_1 + q_2)G^0 + q_1 q_2 G^{-1} \right] \\
&+ \hat{E}_l^{(\text{inc, right})} q_1^{-N} \left[q_1^{-1} - (q_1 + q_2) + q_1^2 q_2 \right]
\end{aligned} \tag{2.44a}$$

and

$$\begin{aligned}
\hat{E}_{l,N+2} &- [(q_1 + q_2)^2 - q_1 q_2]\hat{E}_{l,N} + (q_1 + q_2)q_1 q_2 \hat{E}_{l,N-1} \\
&= \hat{\Phi}_{l,N} \left[G^2 - [(q_1 + q_2)^2 - q_1 q_2]G^0 + (q_1 + q_2)q_1 q_2 G^{-1} \right] \\
&+ \hat{E}_l^{(\text{inc, right})} q_1^{-N} \left[q_1^{-2} - [(q_1 + q_2)^2 - q_1 q_2] + (q_1 + q_2)q_1^2 q_2 \right].
\end{aligned} \tag{2.44b}$$

---

[1] Note that unlike in Section 2.3.2, the parameter α_z of (2.42) and the roots of equation (2.41) all depend on the transverse mode number l.

Altogether, to obtain the solution on every iteration, we now need to solve equations (2.40) for $n = 0, \ldots, N$ and for all l, subject to the boundary conditions (2.43) that supplement the ghost nodes $n = -2$ and $n = -1$, and boundary conditions (2.44) that supplement the ghost nodes $n = N + 1$ and $n = N + 2$.

The ABCs (2.43) and (2.44) are fourth order discrete counterparts of the continuous boundary conditions (2.20). We emphasize that the right-hand sides in relations (2.43) and (2.44) are obtained by local formulae and as such, are numerically inexpensive, see [10]. We also emphasize that similarly to boundary conditions (2.20), the discrete two-way ABCs (2.43) and (2.44) are obtained in the transformed space independently for each one-dimensional system (2.40), and would therefore become nonlocal if transformed back to the original space by means of the matrix $\boldsymbol{\Psi}$. However, we never actually need to explicitly transform the ABCs to the original physical space, because system (2.24) is solved on each iteration (2.23) using the separation of variables (2.39) and consequently, the boundary conditions only need to be applied to variables $\hat{E}_{l,n}$ in the transformed space.

Each one-dimensional difference equation (2.40) subject to boundary conditions (2.43) and (2.44) is normally solved using LU decomposition. In our latest paper [2], we proposed an alternative approach based on an efficient summation rule that evaluates convolution with the discrete Green's function. The latter has the boundary conditions (2.43) and (2.44) automatically built in.

2.4   Results of Computations

In this section we present computational results obtained with the help of the algorithm of Section 2.3. In order to judge whether the iterations (2.23) converge, we monitor the following three quantities on the outer loop:

1. The difference between successive iterations: $\max_{m,n} |E_{m,n}^{(j)} - E_{m,n}^{(j-1)}|$.

2. The maximal residual: $\max_{m,n} |R_{m,n}^{(j)}|$, where

$$
R_{m,n}^{(j)} = \frac{-E_{m-2,n}^{(j)} + 16E_{m-1,n}^{(j)} - 30E_{m,n}^{(j)} + 16E_{m+1,n}^{(j)} - E_{m+2,n}^{(j)}}{12h_x^2}
$$
$$
+ \frac{-E_{m,n-2}^{(j)} + 16E_{m,n-1}^{(j)} - 30E_{m,n}^{(j)} + 16E_{m,n+1}^{(j)} - E_{m,n+2}^{(j)}}{12h_z^2}
$$
$$
+ k_0^2 \left[1 + \epsilon \left| E_{m,n}^{(j)} \right|^{2\sigma} \right] E_{m,n}^{(j)}.
$$

3. The mean residual: $\frac{1}{MN} \sum_m \sum_n |R_{m,n}^{(j)}|$.

2.4.1   Critical Case

We solve the 2D NLH (2.6) in the critical case $\sigma = 2$ with $k_0 = 8$ and with the incident Gaussian input beam given by:

$$E^{(\text{inc, left})}(x,0) = e^{-x^2}.$$

The fractional critical power of this input beam is equal to, see [9],

$$p = \sqrt{\frac{2\epsilon}{3\pi}} k_0.$$

Therefore, $\epsilon = 0.06$ corresponds to $p = 90\%$, i.e., when the power of the corresponding NLS solution is 90% of the threshold power for collapse N_c. In Figure 2.2, we show a solution to the critical NLH obtained for $k_0 = 8$ and $p = 90\%$ on the domain with $x_{\max} = 13$ and $z_{\max} = 40$. As in the NLS model, the solution undergoes mild self-focusing and then decays (diffracts) with propagation.

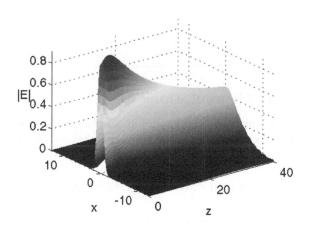

FIGURE 2.2: Solution of the critical NLH for $k_0 = 8$ and $p = 90\%$.

In Figure 2.3 we show the results that correspond to higher input powers and as such, allow one to observe stronger self-focusing. These results are obtained for $k_0 = 8$ on the domain $z_{\max} = 60$, $x_{\max} = 20$, $h_z = \lambda_0/20$, and $h_x = \lambda_0/8$ [Figure 2.3 (left)], and for $k_0 = 30$ on the domain $z_{\max} = 60$, $x_{\max} = 15$, $h_z = \lambda_0/5$, and $h_x = \lambda_0/4$ [Figure 2.3 (right)]. In the computation with

$p = 99.6\%$ the plot of the solution looks more stretched in the longitudinal direction, because we used $k_0 = 30$, which implies a larger diffraction length $L_{DF} = k_0 r_0^2$.

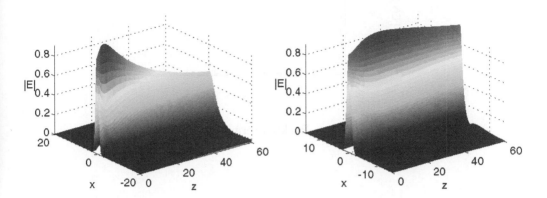

FIGURE 2.3: Solution of the critical NLH subject to the Sommerfeld transverse boundary conditions for $k_0 = 8$ and $p = 99\%$ (left), and $k_0 = 30$ and $p = 99.6\%$ (right).

One of the advantages of our method is its ability to calculate the magnitude of *backscattering*. This is most easily done at $z = 0$, where the backscattered way is given by $|E^{(\text{computed})}(x, 0) - E^{(\text{inc, left})}(x)|$. In Figure 2.4 we show the corresponding graph for $\epsilon = 0.04$, $z_{\max} = 20$, $x_{\max}/z_{\max} = 1$, $h_z = \lambda_0/10$, and $h_x = \lambda_0/4$. From Figure 2.4 we conclude that most backscattering occurs around the axis of symmetry $x = 0$, and that the magnitude of backscattering there is about 1.2% of the incoming power.

2.4.2 Subcritical Case

We recall that the subcritical NLS (2.5) for $\sigma = 1$ admits a well-known soliton solution

$$A = \frac{e^{iz/2}}{\cosh(x/\sqrt{2})}.$$

To see how this soliton would propagate in the framework of the nonparaxial Helmholtz equation, we solve the two-dimensional NLH (2.6) with $\sigma = 1$ and with the incoming beam specified as

$$E^{(\text{inc,left})}(x) = \frac{(\epsilon k_0^2)^{-1/2}}{\cosh(x/\sqrt{2})}.$$

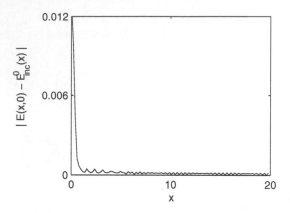

FIGURE 2.4: Backscattering for $\epsilon = 0.04$.

To maximize the effect of nonparaxiality and backscattering, we set $k_0 = 5$. As we can say that $r_0 \approx \sqrt{2}$, the diffraction length is $L_{DF} = k_0 r_0^2 \approx 10$. We also set $z_{\max} = 70$, which corresponds to propagation over 7 diffraction lengths.

In order to see how well our boundary conditions can treat the incoming waves that are not parallel to the z axis, we solve the subcritical NLH with the incoming beam

$$E^{(\text{inc,left})}(x,0) = \frac{\exp\left(-ik_0 \sin\theta(x-5)\right)\left(\epsilon k_0^2\right)^{-1/2}}{\cosh\left((x-5)\cos\theta/\sqrt{2}\right)}, \quad \theta = \arctan(1/4),$$

which corresponds to a solitonic initial condition impinging on the $z = 0$ interface at an angle $\theta \neq 0°$. The numerical parameters are as before, except that we set $x_{\max} = 15$. As can be seen in Figures 2.5, 2.6, and 2.7, the solitonic propagation is unaffected by the non-normal incidence angle. In particular, there is no indication of spurious numerical reflections from the artificial boundaries, showing that our algorithm is capable of handling well the waves not aligned with the grid/boundaries.

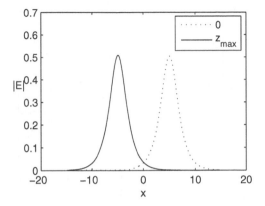

FIGURE 2.5: Soliton at an angle: Cross-section profiles at $z = 0$ and $z = z_{max}$.

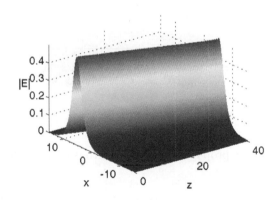

FIGURE 2.6: Soliton at an angle: 3D mesh plot.

Another significant advantage of our NLH solver is *its natural ability to solve,* apparently for the first time ever, *for counter-propagating beams without making any additional assumptions.* This is in contrast with the NLS model that admits propagation in one direction only and as such would require numerical marching in two opposite directions when analyzing the counter-propagation case. Note that although coupled NLS models have been used for counter-propagation, see [4], this approach involves some approximation of the Kerr nonlinearity induced by two coherent beams. In Figures 2.8 and 2.9

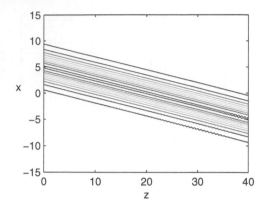

FIGURE 2.7: Soliton at an angle: Level curves.

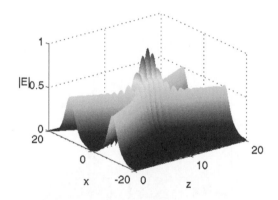

FIGURE 2.8: Counter-propagation of two solitons at an angle: 3D mesh plot.

we show our numerical solution of the subcritical NLH for the case of two counter-propagating incoming beams:

$$E^{(\text{inc,left})}(x, 0) = E^{(\text{inc,right})}(x, z_{\max}) = \frac{\exp\left(-ik_0 \sin\theta(x - 10)\right)\left(\epsilon k_0^2\right)^{-1/2}}{\cosh\left(\cos\theta(x - 10)/\sqrt{2}\right)},$$

where $\theta = \pi/4$. The computational domain is $x_{\max} = 20$, $z_{\max} = 20$, and the grid sizes are $h_z = \lambda_0/60$ and $h_x = \lambda_0/16$. The centers of the two input beams are located at $(z = 0, x = 10)$ and $(z = 20, x = 10)$, the beams propagate downward at an angle of 45°, interact nonlinearly, and then continue to propagate as solitons. The periodic oscillations in the interaction region result from the phase differences between the two counter-propagating beams, they disappear as the beams move further away from each other.

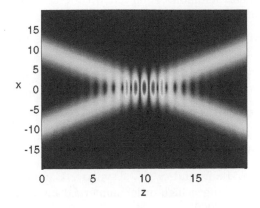

FIGURE 2.9: Counter-propagation of two solitons at an angle.

2.5 Summary

In this chapter we described an algorithm for solving the NLH as a true nonlinear boundary value problem. This methodology allows, apparently for the first time ever, to calculate the backscattered wave. Several extensions that we developed over the years for these method, which are nor presented in this chapter, are the addition of linear damping [7], an integral Green's function formulation for implementing the two-way boundary conditions [2], and an extension to the cylindrically symmetric three-dimensional case [2]. Most recently [10], we have also introduced a compact fourth order discretization and a Newton-type nonlinear solver for the one-dimensional NLH. The new algorithm has allowed us to compute the solutions for the media with material discontinuities and for very high levels of nonlinearity.

References

[1] M. S. Agranovich, B. Z. Katsenelenbaum, A. N. Sivov, and N. N. Voitovich. *Generalized Method of Eigenoscillations in Diffraction Theory.* WILEY-VCH Verlag Berlin GmbH, Berlin, 1999. (Translated from the Russian manuscript by Vladimir Nazaikinskii.)

[2] G. Baruch, G. Fibich, and S. V. Tsynkov. High-order numerical solution

of the nonlinear helmholtz equation with axial symmetry. *J. Comput. Applied Math.*, submitted for publication.

[3] R. W. Boyd. *Nonlinear Optics.* Academic Press, Boston, 2nd Edition, 2003.

[4] O. Cohen, R. Uzdin, T. Carmon, J. W. Fleischer, M. Segev, and S. Odoulov. Collisions between optical spatial solitons propagating in opposite directions. *Phys. Rev. Lett.*, 89:133901, 2002.

[5] G. Fibich and B. Ilan. Vectorial and random effects in self-focusing and in multiple filamentation. *Physica D*, 157:112–146, 2001.

[6] G. Fibich, B. Ilan, and S. V. Tsynkov. Computation of nonlinear backscattering using a high-order numerical method. *J. Sci. Comput.*, 17:351–364, Dec. 2002.

[7] G. Fibich, B. Ilan, and S. V. Tsynkov. Backscattering and nonparaxiality arrest collapse of damped linear waves. *SIAM J. Applied Math.*, 63(5):1718–1736, 2003.

[8] G. Fibich and G. C. Papanicolaou. Self-focusing in the perturbed and unperturbed nonlinear Schrödinger equation in critical dimension. *SIAM J. Appl. Math.*, 60:183–240, 1999.

[9] G. Fibich and S. V. Tsynkov. High-order two-way artificial boundary conditions for nonlinear wave propagation with backscattering. *J. Comput. Physics*, 171:632–677, 2001.

[10] G. Fibich and S. V. Tsynkov. Numerical solution of the nonlinear Helmholtz equation using nonorthogonal expansions. *J. Comput. Physics*, 210(1):183–224, 2005.

[11] C. Sulem and P. L. Sulem. *The nonlinear Schrödinger equation.* Springer, New-York, 1999.

[12] S. V. Tsynkov. Numerical solution of problems on unbounded domains. A review. *Appl. Numer. Math.*, 27:465–532, 1998.

[13] M. I. Weinstein. Nonlinear Schrödinger equations and sharp interpolation estimates. *Comm. Math. Physics*, 87:567–576, 1983.

[14] L. Wöste, C. Wedekind, H. Wille, P. Rairoux, B. Stein, S. Nikolov, C. Werner, S. Niedermeier, F. Ronnenberger, H. Schillinger, and R. Sauerbrey. Femtosecond atmospheric lamp. *Laser und Optoelektronik*, 29:51–53, 1997.

Chapter 3

Theory and Numerical Analysis of Boussinesq Systems: A Review

V. A. Dougalis, Mathematics Department, University of Athens, 15784 Zographou, Greece *and* Foundation for Research and Technology-Hellas, Institute of Applied and Computational Mathematics, 71110 Heraklion, Greece, doug@math.uoa.gr

D. E. Mitsotakis, Mathematics Department, University of Athens, 15784 Zographou, Greece *and* Foundation for Research and Technology-Hellas, Institute of Applied and Computational Mathematics, 71110 Heraklion, Greece, dmitsot@gmail.com

---

3.1 Introduction

In this chapter we shall be mainly concerned with the numerical solution of the Boussinesq systems of equations that occur in water wave theory. These systems, derived in their general form in [11], may be written in nondimensional, unscaled variables as

$$\eta_t + u_x + (\eta u)_x + a u_{xxx} - b\eta_{xxt} = 0,$$
$$u_t + \eta_x + u u_x + c\eta_{xxx} - d u_{xxt} = 0, \tag{3.1}$$

where $\eta = \eta(x,t)$, $u = u(x,t)$ are real functions defined for $x \in \mathbb{R}$ and $t \geq 0$. In addition

$$a = \tfrac{1}{2}(\theta^2 - \tfrac{1}{3})\nu, \quad b = \tfrac{1}{2}(\theta^2 - \tfrac{1}{3})(1 - \nu),$$
$$c = \tfrac{1}{2}(1 - \theta^2)\mu, \quad d = \tfrac{1}{2}(1 - \theta^2)(1 - \mu), \tag{3.2}$$

where ν and μ are constants and $0 \leq \theta \leq 1$. Our main goal in this paper is to review the existing mathematical theory for the systems (3.1) and solve them numerically for specific choices of a, b, c, d in order to study interesting features of their solutions.

The systems are named after J. V. Boussinesq, in whose pioneering work on water waves in the 1870's, [19], [20], systems similar to (3.1) (and approximations of the type that we will be considered later in this section) played a significant role; cf. e.g. [44] for references and historical comments.

The family of systems in (3.1) is an approximation of the two-dimensional Euler equations that describe the irrotational, free surface flow of an incompressible, inviscid fluid in a uniform horizontal channel, when cross-channel variations of the flow are negligible. As opposed to one-way models, such as the Korteweg-de Vries equation, the equations (3.1) describe two-way propagation of surface waves. The independent variables x and t are proportional to position along the channel and time, respectively, while the dependent variables $\eta(x, t)$, respectively $u(x, t)$, are proportional to the values at (x, t) of the height of the free surface of the fluid above an undisturbed level, and to the horizontal velocity of the fluid at some height above the bottom, respectively. In Section 3.2 of this chapter we review the derivation of (3.1) from the Euler equations and specify the range of parameters for which the approximation is valid.

Our aim in Section 3.3 is to review the existing well-posedness theory for the initial-value problem and for some initial-boundary-value problems for (3.1). Being approximations to the Euler equations, the systems (3.1) are expected to have *solitary wave* solutions. In Section 3.4 we discuss known results of existence, uniqueness and stability for these important special solutions. In Section 3.5 we construct and outline the properties of a fully discrete Galerkin-finite element numerical method that we will use to approximate the solutions of the periodic initial-value problem for (3.1) in a bounded interval. We shall use this scheme in Section 3.6 to simulate numerically phenomena such as two-way propagation of solutions and the generation, stability and interaction of solitary waves. Finally, in Section 3.7 we shall briefly outline the analogous Boussinesq models in two space dimensions and review aspects of the well-posedness theory of their initial-value problems, and of their numerical analysis.

3.2   Derivation and Examples of Boussinesq Systems

In this section we shall derive the Boussinesq equations (3.1) from the Euler equations in the appropriate parameter regime. This has been done in [11] using the potential flow form of the Euler equations. Here, following [48], [11], and [4], we derive (3.1) using the primitive variables.

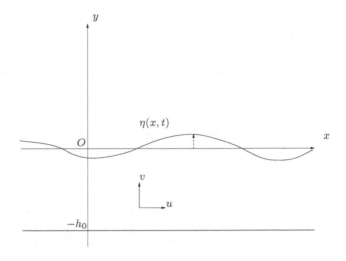

FIGURE 3.1: Euler equations variables.

The 2D Euler equations for inviscid, incompressible, irrotational flow with a free surface over a horizontal bottom at height $y = -h_0$ (the undisturbed level of fluid is at $y = 0$) may be written as follows in dimensional, unscaled variables: Let $\eta(x,t)$ be the deviation of the free surface of the fluid above its level of rest and consider the domain $\Omega_t = \{(x,y) : -\infty < x < \infty, -h_0 \leq y \leq \eta(x,t)\}$. Then, for $(x,y) \in \Omega_t$ and $t \geq 0$ we have

$$u_t + uu_x + vu_y + \frac{1}{\rho}p_x = 0, \tag{3.3}$$

$$v_t + uv_x + vv_y + \frac{1}{\rho}p_y = -g, \tag{3.4}$$

$$u_x + v_y = 0, \tag{3.5}$$

$$u_y = v_x, \tag{3.6}$$

where g is the acceleration of gravity, $u = u(x,y,t)$, respectively $v = v(x,y,t)$, denotes the horizontal, respectively vertical, velocity component, ρ is the (constant) density of the fluid, and $p = p(x,y,t)$ the pressure. (If the irrotationality condition (3.6) holds at $t = 0$, then it will hold for all $t > 0$, by Kelvin's theorem). The system (3.3)–(3.6) is supplemented by the free surface kinematic and dynamic boundary conditions

$$\eta_t + u\eta_x = v, \ \text{ at } \ y = \eta(x,t), \tag{3.7}$$

$$p = 0, \ \text{ at } \ y = \eta(x,t). \tag{3.8}$$

At the bottom $y = -h_0$ we assume that the normal component of the velocity vanishes, i.e., that

$$v = 0, \quad \text{at } y = -h_0. \tag{3.9}$$

We also assume that initial conditions for η and u have been specified and let

$$\eta(x, 0) = \phi(x), \quad x \in \mathbb{R}, \tag{3.10}$$

$$u(x, y, 0) = \psi(x, y), \quad (x, y) \in \Omega_0, \tag{3.11}$$

where η_0 and u_0 are given real functions. (We assume that u_0 satisfies the compatibility condition $\frac{\partial \psi}{\partial y}(x, y) = -\int_{h_0}^{y} \frac{\partial^2 \psi}{\partial x^2}(x, y')dy'$ in Ω_0, which follows by assuming that (3.5), (3.6) and (3.9) hold at $t = 0$.)

The first step in deriving (3.1) from the Euler equations is to nondimensionalize and scale the problem (3.3)–(3.11) in an appropriate way. We are interested in the regime of long surface waves of small amplitude, wherein

$$\varepsilon := \frac{A}{h_0} \ll 1, \quad \sigma := \frac{h_0}{\lambda} \ll 1, \tag{3.12}$$

where A is a typical amplitude and λ a typical wavelength of the waves. Accordingly, cf. [48], we make the change of variables

$$\begin{aligned} x^* &= \tfrac{\sigma}{h_0}x, \quad y^* = \tfrac{1}{h_0}y, \quad t^* = \tfrac{\sigma g}{c_0}t, \quad \eta^* = \tfrac{1}{\varepsilon h_0}\eta, \\ u^* &= \tfrac{1}{\varepsilon c_0}u, \quad v^* = \tfrac{1}{\varepsilon \sigma}\tfrac{v}{c_0}, \quad p^* = \tfrac{1}{\rho c_0^2}p, \end{aligned} \tag{3.13}$$

where $c_0 := \sqrt{gh_0}$. Then, the problem (3.3)–(3.11) is transformed into the equivalent problem that follows, in which the dependent variables and initial conditions are of order one, while powers of ε and σ signify the order of magnitude of terms that they multiply. We seek $u^* = u^*(x^*, y^*, t^*)$, $v^* = v^*(x^*, y^*, t^*)$, $p^* = p^*(x^*, y^*, t^*)$, $\eta^* = \eta^*(x^*, t^*)$, defined for $-\infty < x^* < \infty$, $-1 \le y^* \le \varepsilon\eta^*(x^*, t^*)$, $t^* \ge 0$, such that,

$$\varepsilon u_{t^*}^* + \varepsilon^2 u^* u_{x^*}^* + \varepsilon^2 v^* u_{y^*}^* + p_{x^*}^* = 0, \tag{3.14}$$

$$\varepsilon \sigma^2 v_{t^*}^* + \varepsilon^2 \sigma^2 u^* v_{x^*}^* + \varepsilon^2 \sigma^2 v^* v_{y^*}^* + p_{y^*}^* = -1, \tag{3.15}$$

$$u_{x^*}^* + v_{y^*}^* = 0, \tag{3.16}$$

$$u_{y^*}^* - \sigma^2 v_{x^*}^* = 0. \tag{3.17}$$

This system is supplemented with the free surface and bottom boundary conditions, which now take the form:

$$\eta_{t^*}^* + \varepsilon u^* \eta_{x^*}^* = v^*, \quad \text{for } y^* = \varepsilon\eta^*(x^*, t^*), \tag{3.18}$$

$$p^* = 0, \quad \text{for } y^* = \varepsilon\eta^*(x^*, t^*), \tag{3.19}$$

$$v^* = 0, \quad \text{for } y^* = -1, \tag{3.20}$$

and the initial conditions

$$\eta^*(x^*,0) = \phi^*(x^*), \quad u^*(x^*,y^*,0) = \psi^*(x^*,y^*), \tag{3.21}$$

where, in terms of the functions ϕ, ψ in (3.10)–(3.11) we have

$$\phi^*(x^*) := \frac{1}{h_0\varepsilon}\phi\left(\frac{x^*h_0}{\sigma}\right), \quad \psi^*(x^*,y^*) := \frac{1}{\varepsilon c_0}\psi\left(\frac{h_0}{\sigma}x^*, h_0y^*\right). \tag{3.22}$$

The incompressibility and irrotationality conditions (3.16), (3.17) constitute an elliptic system for u and v that may be solved, using the boundary condition (3.20), in terms, for example, of the function U^* defined by $U^*(x^*,t^*) := u^*(x^*,0,t^*)$. We easily obtain by an iterative process that

$$u^*(x^*,y^*,t^*) = U^* - \sigma^2(y^* + \frac{1}{2}(y^*)^2)U^*_{x^*x^*} + O(\sigma^4), \tag{3.23}$$

$$v^*(x^*,y^*,t^*) = -(y^*+1)U^*_{x^*} + \frac{\sigma^2}{6}(3(y^*)^2 + (y^*)^3 - 2)U^*_{x^*x^*x^*} + O(\sigma^4). \tag{3.24}$$

Using (3.23) and (3.24) in the vertical momentum equation (3.15), and also (3.19), we obtain

$$p^*(x^*,y^*,t^*) = -y^* + \varepsilon\eta^* + \varepsilon\sigma^2 y^*(1 + \frac{1}{2}y^*)U^*_{x^*t^*} + O(\varepsilon\sigma^4, \varepsilon^2\sigma^2). \tag{3.25}$$

Substituting now these expressions in the horizontal momentum equation (3.14) yields

$$U^*_{t^*} + \eta^*_{x^*} + \varepsilon U^* U^*_{x^*} = O(\sigma^4, \varepsilon\sigma^2). \tag{3.26}$$

Another equation coupling U^* and η^* is obtained by evaluating u^* and v^*, given by (3.23)–(3.24) at $y^* = \varepsilon\eta^*$ and substituting in (3.18). This gives

$$\eta^*_{t^*} + U^*_{x^*} + \varepsilon(\eta^* U^*)_{x^*} + \frac{\sigma^2}{3}U^*_{x^*x^*x^*} = O(\varepsilon\sigma^2, \sigma^4). \tag{3.27}$$

If we now assume that the dispersive and nonlinear terms in the system (3.26)–(3.27) are of equal importance, i.e. that the Stokes number $S = \frac{\varepsilon}{\sigma^2} = O(1)$, and put for definiteness $\varepsilon = \sigma^2$, we obtain the Boussinesq-type system

$$\begin{cases} \eta^*_{t^*} + U^*_{x^*} + \varepsilon(\eta^* U^*)_{x^*} + \frac{\varepsilon}{3}U^*_{x^*x^*x^*} = 0, \\ U^*_{t^*} + \eta^*_{x^*} + \varepsilon U^* U^*_{x^*} = 0, \end{cases} \tag{3.28}$$

in which we replaced the right-hand side (which by our assumption is of $O(\varepsilon^2)$) by zero.

One may obtain a plethora of Boussinesq-type systems by choosing other velocity variables. For example, defining the mean velocity with respect to depth as

$$\overline{u^*}(x^*,t^*) = \frac{1}{1+\varepsilon\eta^*}\int_{-1}^{\varepsilon\eta^*} u^*(x^*,y^*,t^*)dy^*,$$

Peregrine, [48], obtains, expressing $\overline{u^*}$ in terms of U^* in a series expansion of powers of ε and σ, a different system, formally equivalent to (3.28) up to $O(\varepsilon^2)$ terms, which is the so-called 'classical' Boussinesq system, see (3.35) below.

Alternatively, following [11], we may choose as velocity variable the horizontal velocity of the fluid $u_\theta^*(x^*, t^*)$ at height $y^* = -1 + \theta(1 + \varepsilon\eta^*)$, for some $\theta \in [0, 1]$, i.e. put $u_\theta^*(x^*, t^*) = u^*(x^*, -1 + \theta(1 + \varepsilon\eta^*), t^*)$. Taylor expansions and the use of (3.23) with $\sigma^2 = \varepsilon$ yield

$$u_\theta^* = U^* + \frac{1}{2}\varepsilon(1 - \theta^2)U_{x^*x^*}^* + O(\varepsilon^2),$$

which we may invert, using e.g., the Fourier transform, to obtain

$$U^* = u_\theta^* - \frac{1}{2}\varepsilon(1 - \theta^2)u_{\theta,x^*x^*}^* + O(\varepsilon^2).$$

Substituting this expression in the equations of the system (3.28) we obtain the Boussinesq system, valid for $-\infty < x^* < \infty$, $t^* \geq 0$,

$$\begin{cases} \eta_{t^*}^* + u_{\theta,x^*}^* + \varepsilon(\eta^* u_\theta^*)_{x^*} + \frac{\varepsilon}{2}(\theta^2 - \frac{1}{3})u_{\theta,x^*x^*x^*}^* = 0, \\ u_{\theta,t^*}^* + \eta_{x^*}^* + \varepsilon u_\theta^* u_{\theta,x^*}^* - \frac{\varepsilon}{2}(1 - \theta^2)u_{\theta,x^*x^*t^*}^* = 0, \end{cases} \tag{3.29}$$

where we have replaced the $O(\varepsilon^2)$ right-hand side by zero. This system is supplemented for $x^* \in \mathbb{R}$ by the initial conditions

$$\eta^*(x^*, 0) = \eta_0^*(x^*) := \phi^*(x^*) \tag{3.30}$$

and

$$u_\theta^*(x^*, 0) = u_{\theta,0}^*(x^*) := \psi^*(x^*, -1 + \theta(1 + \varepsilon\eta_0^*(x^*))), \tag{3.31}$$

where the functions ϕ^* and ψ^* are given in terms of the original Euler data in (3.22). Once the u_θ^* and η^* are determined by (3.29)–(3.31), we can then compute U^*, and u^*, v^* and p^* by the appropriate formulas given above.

Following now [11] (see also [8]) we observe that the equations (3.29) give

$$\eta_{t^*}^* + u_{\theta,x^*}^* = O(\varepsilon),$$

$$u_{\theta,t^*}^* + \eta_{x^*}^* = O(\varepsilon),$$

which imply that the third order derivatives of u_θ^* in (3.29) may be expressed in terms of third order derivatives of η^* with an error of $O(\varepsilon)$. For example, we may write

$$u_{\theta,x^*x^*x^*}^* = -\eta_{x^*x^*t^*}^* + O(\varepsilon),$$

and, more general, using a modelling parameter ν

$$u_{\theta,x^*x^*x^*}^* = \nu u_{\theta,x^*x^*x^*}^* - (1 - \nu)\eta_{x^*x^*t^*}^* + O(\varepsilon).$$

Using such expressions in (3.29) we obtain a family of Boussinesq systems in nondimensional, scaled variables given by

$$\begin{cases} \eta^*_{t^*} + u^*_{\theta,x^*} + \varepsilon(\eta^* u^*_\theta)_{x^*} + \varepsilon(a u^*_{\theta,x^*x^*x^*} - b\eta^*_{x^*x^*t^*}) = 0, \\ u^*_{\theta,t^*} + \eta^*_{x^*} + \varepsilon u^*_\theta u^*_{\theta,x^*} + \varepsilon(c\eta^*_{x^*x^*x^*} - d u^*_{\theta,x^*x^*t^*}) = 0, \end{cases} \qquad (3.32)$$

where the coefficients a, b, c, d are given for $0 \le \theta \le 1$ and $\nu, \mu \in \mathbb{R}$ by (3.2), and where, again, the $O(\varepsilon^2)$ right-hand side has been replaced by zero. The system is supplemented with the initial conditions (3.30)–(3.31).

In the sequel, we shall write the system (3.32) in the following equivalent nondimensional but unscaled form, whenever the parameter ε plays no essential role in the analytical or numerical investigation. Performing the change of variables $\tilde{u} = \varepsilon u^*_\theta$, $\tilde{\eta} = \varepsilon \eta^*$, $\tilde{x} = \varepsilon^{-1/2} x^*$, $\tilde{t} = \varepsilon^{-1/2} t^*$, we obtain the system

$$\begin{cases} \tilde{\eta}_{\tilde{t}} + \tilde{u}_{\tilde{x}} + (\tilde{\eta}\tilde{u})_{\tilde{x}} + a\tilde{u}_{\tilde{x}\tilde{x}\tilde{x}} - b\tilde{\eta}_{\tilde{x}\tilde{x}\tilde{t}} = 0, \\ \tilde{u}_{\tilde{t}} + \tilde{\eta}_{\tilde{x}} + \tilde{u}\tilde{u}_{\tilde{x}} + c\tilde{\eta}_{\tilde{x}\tilde{x}\tilde{x}} - d\tilde{u}_{\tilde{x}\tilde{x}\tilde{t}} = 0, \end{cases} \quad \tilde{x} \in \mathbb{R}, \ \tilde{t} \ge 0, \qquad (3.33)$$

where the initial conditions

$$\tilde{\eta}(x,0) = \tilde{\eta}_0(\tilde{x}), \quad \tilde{u}(x,0) = \tilde{u}_0(\tilde{x}) \qquad (3.34)$$

are obtained by (3.30), (3.31) using the last-mentioned change of scale. (In terms of the original Euler dimensional variables we have for example $\tilde{\eta}_0(\tilde{x}) = \frac{1}{h_0}\phi(\tilde{x}h_0)$, and an analogous formula for $\tilde{u}_0(\tilde{x})$. Since (3.33) is intended to model waves of small amplitude and large wavelength, we have that $\tilde{\eta}_0 = O(A/h_0) = O(\varepsilon)$. Similarly we can see that $\tilde{u}_0 = O(\varepsilon)$, i.e., that the system (3.33) should be integrated with small initial data. It should also be stressed that in the approximation procedures that lead to (3.33), one-way propagation assumptions were nowhere made. Hence, these systems may be used to study two-way propagation of long surface waves of small amplitude. If one-way assumptions are made, cf. [59], [48], one recovers well-known one-way model equations such as the Korteweg-de Vries (KdV) and Benjamin-Bona-Mahoney (BBM) equations. In the sequel, we shall drop, for simplicity of notation, the tilde from the variables in (3.33)–(3.34), reverting back to the notation of the system (3.1).

As it will be seen in the sequel, although all systems of the family (3.1) are, in a sense, formally equivalent (being $O(\varepsilon^2)$ formal approximations to the Euler equations), there are choices of the parameters θ, μ and ν leading to Boussinesq systems that:

(i) *Have favorable mathematical properties.* As will be seen in Section 3.3, some of these systems are linearly ill-posed and should be excluded from further study as useful model equations. Among the linearly well-posed ones one should consider systems whose Cauchy problem is at least locally (nonlinearly) well-posed with long enough temporal interval of existence of solutions. In addition, existence of solutions to *initial-boundary-value* problems is an important requirement. We shall briefly review the literature on the well-posedness of such problems in Section 3.3. Finally, the systems in their scaled form should be rigorously justified to be good approximations of the Euler equations as $\varepsilon \to 0$. (See the remarks at the end of this section.)

(ii) *Have solitary wave solutions.* Being approximations to the Euler equations, the Boussinesq systems are expected to have *solitary wave* solutions. Attention must be paid therefore to systems that have solitary waves whose uniqueness, stability and other properties should be also studied. A review of the theory of solitary waves for these systems will be made in Section 3.4.

(iii) *Can be easily solved numerically to high accuracy.* The scarcity of closed form analytical solutions and the need to study, e.g., the stability and the interactions of the solitary waves of some of these systems, and the characteristics of the long-time evolution of their solutions make it imperative that the systems should be solved numerically by fully discrete methods. A brief study of some of these systems from the point of view of their numerical analysis will be done in Section 3.5.

A fourth (and perhaps the most important) criterion for selecting a 'good' system is, of course, the favorable comparison of its solutions with experimental data of propagation of surface waves. We will not discuss this issue of modelling, referring instead to the relevant references cited, e.g., in [11] and [10].

We now list some examples of particular Boussinesq systems of the form (3.1) that we will have occasion to refer to in the sequel of this paper. The initial-value problem for all these systems has been shown to be at least non-linearly well-posed locally in time, cf. [12].

(i) The 'classical' Boussinesq system ($\mu = 0$, $\theta^2 = 1/3$, i.e., $a = b = c = 0$, $d = 1/3$), whose initial-value problem is globally well-posed, cf. [3], [50]. The system is given by

$$\eta_t + u_x + (\eta u)_x = 0,$$
$$u_t + \eta_x + u u_x - \tfrac{1}{3} u_{xxt} = 0. \tag{3.35}$$

(ii) The BBM-BBM system ($\nu = \mu = 0$, $\theta^2 = 2/3$, i.e., $a = c = 0$, $b = d = 1/6$), whose initial-value problem is locally well-posed, cf. [12], [10]

$$\eta_t + u_x + (\eta u)_x - \tfrac{1}{6}\eta_{xxt} = 0,$$
$$u_t + \eta_x + u u_x - \tfrac{1}{6} u_{xxt} = 0. \tag{3.36}$$

(iii) The Bona-Smith system ($\nu = 0$, $\mu = (4 - 6\theta^2)/3(1 - \theta^2)$, i.e., $a = 0$, $b = d = (3\theta^2 - 1)/6$, $c = (2 - 3\theta^2)/3$, $2/3 < \theta^2 < 1$), whose initial-value problem is globally well-posed, cf. [17]. The limiting form of this system as $\theta \to 1$, corresponding to $a = 0$, $b = d = 1/3$, $c = -1/3$, is the system actually studied by Bona and Smith, [17]. These systems are

$$\eta_t + u_x + (\eta u)_x - \frac{3\theta^2 - 1}{6}\eta_{xxt} = 0,$$
$$u_t + \eta_x + uu_x + \frac{2 - 3\theta^2}{3}\eta_{xxx} - \frac{3\theta^2 - 1}{6}u_{xxt} = 0. \qquad (3.37)$$

(iv) The KdV-KdV system ($\nu = \mu = 1$, $\theta^2 = 2/3$, i.e., $a = c = 1/6$, $b = d = 0$), whose initial-value problem is locally well-posed, cf. [11], [15]

$$\eta_t + u_x + (\eta u)_x + \frac{1}{6}u_{xxx} = 0,$$
$$u_t + \eta_x + uu_x + \frac{1}{6}\eta_{xxx} = 0. \qquad (3.38)$$

As we will see in Section 3.5, the systems (i)–(iii) possess solitary wave solutions, while system (iv) has mainly *generalized* solitary waves.

It is worthwhile to mention that in [14] a new family of Boussinesq systems was derived. These are the *symmetric* Boussinesq of general form

$$\eta_t + u_x + \frac{1}{2}(\eta u)_x + au_{xxx} - b\eta_{xxt} = 0,$$
$$u_t + \eta_x + \frac{1}{2}\eta\eta_x + \frac{3}{2}uu_x + c\eta_{xxx} - du_{xxt} = 0, \qquad (3.39)$$

where a, b, c, d are given by (3.2). These systems, like the usual Boussinesq systems (3.1) are formally $O(\varepsilon^2)$ approximations of the Euler equations, when written in nondimensional, scaled variables in a form similar to that of (3.32), wherein their nonlinear and dispersive terms are multiplied by ε. These systems have several favorable mathematical properties that are due to the fact that their hyperbolic part (the system which is obtained by omitting the dispersive, third-order derivative terms) is symmetric and L^2-conservative.

We close this section with some brief comments on the rigorous justification of the scaled form of system (3.1), i.e., (3.32), and the analogous form of the symmetric system (3.39), as approximation to the Euler equations as $\varepsilon \to 0$. On this topic, the literature is large and growing, cf. e.g., [27], [14], [2], and we cannot describe it in a satisfactory way here. Let us just mention that since the right-hand side of (3.32) is really of $O(\varepsilon^2)$, it is expected that the solutions of (3.32) with suitable initial data approximate for $t > 0$ appropriate smooth solutions of the Euler equations, in the same scaling, with an error of $O(\varepsilon^2 t)$. In [14] and [2] it was proved that the error is indeed of $O(\varepsilon^2 t)$ uniformly for $t \in [0, T_\varepsilon]$, where $T_\varepsilon = O(1/\varepsilon)$ for the Boussinesq systems of the form (3.1) or (3.39) that are locally well-posed. (The proof holds for one or two space dimensions and it includes, [2], a local existence proof of smooth solutions of the scaled Euler equations, written in Hamiltonian form, in temporal intervals with $T_\varepsilon = O(1/\varepsilon)$.)

3.3   Well-Posedness Theory

We consider the initial-value problem for the general Boussinesq system (3.1), i.e., we seek $\eta = \eta(x,t)$, $u = u(x,t)$ defined for $x \in \mathbb{R}$, $t \geq 0$, such that for $x \in \mathbb{R}$, $t > 0$

$$\begin{aligned}
\eta_t + u_x + (\eta u)_x + a u_{xxx} - b \eta_{xxt} = 0, \\
u_t + \eta_x + u u_x + c \eta_{xxx} - d u_{xxt} = 0,
\end{aligned} \tag{3.40}$$

with given initial conditions

$$u(x,0) = u_0(x), \quad \eta(x,0) = \eta_0(x), \quad x \in \mathbb{R}. \tag{3.41}$$

We always assume that the constants a, b, c, d are given by the relations (3.2).

A first step in the analysis of the initial-value problem (3.40)–(3.41) is the study, via the Fourier transform, of the analogous problem for the *linearized* equations,

$$\begin{aligned}
\eta_t + u_x + a u_{xxx} - b \eta_{xxt} = 0, \\
u_t + \eta_x + c \eta_{xxx} - d u_{xxt} = 0,
\end{aligned} \tag{3.42}$$

for $x \in \mathbb{R}$, $t > 0$ with the initial conditions

$$u(x,0) = u_0(x), \quad \eta(x,0) = \eta_0(x), \quad x \in \mathbb{R}. \tag{3.43}$$

The problem (3.42)–(3.43) has been thoroughly analyzed in [11]. We give here a synopsis of the theory of its well-posedness in L^2; for the L^p, $1 \leq p \leq \infty$, theory and related questions, cf. [11]. (By "well-posedness" we mean, as usual, existence and uniqueness of solutions and continuous dependence on the initial data in specific function spaces.)

In the sequel, we shall denote $L^p = L^p(\mathbb{R})$, $1 \leq p \leq \infty$, and let, for real s, $H^s = H^s(\mathbb{R})$ denote the L^2-based Sobolev spaces of functions on $\mathbb{R}$. Defining the Fourier transform of a function $f \in L^2$ for $k \in \mathbb{R}$ by

$$\widehat{f}(k) = \int_{-\infty}^{\infty} e^{-ikx} f(x) dx,$$

and taking the Fourier transform with respect to x of the system (3.42) we obtain

$$\frac{d}{dt} \begin{pmatrix} \widehat{\eta} \\ \widehat{u} \end{pmatrix} + ik A(k) \begin{pmatrix} \widehat{\eta} \\ \widehat{u} \end{pmatrix} = 0, \tag{3.44}$$

where

$$A(k) = \begin{pmatrix} 0 & w_1(k) \\ w_2(k) & 0 \end{pmatrix},$$

with $w_1(k) = \frac{1-ak^2}{1+bk^2}$, $w_2(k) = \frac{1-ck^2}{1+dk^2}$. If we let $\sigma = |w_1 w_2|^{1/2}$ and take into account (3.43), we conclude that the solution of (3.44) is

$$\begin{pmatrix} \widehat{\eta}(k,t) \\ \widehat{u}(k,t) \end{pmatrix} = e^{-ikA(k)t} \begin{pmatrix} \widehat{\eta}_0(k) \\ \widehat{u}_0(k) \end{pmatrix}, \tag{3.45}$$

where, provided that $w_1(k)w_2(k) \geq 0$,

$$e^{-ikA(k)t} = \begin{pmatrix} \cos(k\sigma(k)t) & -i\sin(k\sigma(k)t)\frac{w_1(k)}{\sigma(k)} \\ -i\sin(k\sigma(k)t)\frac{w_2(k)}{\sigma(k)} & \cos(k\sigma(k)t) \end{pmatrix}.$$

Using an elementary result of the theory of Fourier multipliers we may conclude that the linearized problem (3.42)–(3.43) is well-posed if the matrix $e^{-ikA(k)t}$ is bounded in finite intervals of t. It turns out that this happens if and only if one of the following conditions hold:

(C1) $b \geq 0$, $d \geq 0$, $a \leq 0$, $c \leq 0$,

(C2) $b \geq 0$, $d \geq 0$, $a = c > 0$,

(C3) $b = d < 0$, $a = c > 0$,

i.e. for those a, b, c, d for which the rational function $\frac{w_1(k)}{w_2(k)}$ has neither zeros nor poles on the real axis. Each of these conditions implies that $w_1 w_2 \geq 0$ for all k. We shall henceforth call the Boussinesq systems for which (C1) or (C2) or (C3) holds, "admissible." We conclude that there are indeed systems of the type (3.40) which are not even linearly well-posed (for example the Bona-Smith systems for $\theta^2 < 2/3$). Note that the systems (3.35)–(3.37) satisfy (C1), while the KdV-KdV system (3.38) satisfies (C2).

The representation (3.45) is also useful in determining the well-posedness of the initial-value problem (3.42)–(3.43) is specific pairs of Sobolev spaces. Indeed, from (3.45) there follows that

$$|\hat{\eta}|^2 + \left(\frac{w_1}{w_2}\right)|\hat{u}|^2 = |\hat{\eta}_0|^2 + \left(\frac{w_1}{w_2}\right)|\hat{u}_0|^2,$$

from which one may infer information about the well-posedness of the problem from the behavior of $\frac{w_1}{w_2}$ as $|k| \to \infty$. If the *order* of $\sqrt{\frac{w_1}{w_2}}$ is the integer ℓ such that $\left(\frac{w_1(k)}{w_2(k)}\right)^{1/2} \sim c|k|^\ell$, as $|k| \to \infty$, it is proved in [11] that if $m_1 = \max(0, -\ell)$, $m_2 = \max(0, \ell)$, then the initial-value problem (3.42)–(3.43) is well-posed for $(\eta, u) \in H^{s+m_1} \times H^{s+m_2}$ for any $s \geq 0$. For example, the BBM-BBM or KdV-KdV systems, for which $\ell = 0$, are well posed in $H^s \times H^s$, the Bona-Smith systems for $2/3 < \theta^2 \leq 1$ ($\ell = -1$) are well-posed in $H^{s+1} \times H^s$, while the 'classical' Boussinesq system ($\ell = 1$) is well-posed in $H^s \times H^{s+1}$.

We consider now the question of well-posedness of the *nonlinear* initial-value problem (3.40)–(3.41). In contrast to the case of the linearized problem (3.42)–(3.43), where the well-posedness results are of course global, i.e., existence and uniqueness of solutions hold for all $t > 0$, the nonlinear problem can be shown to be well-posed in general only locally, i.e. on finite intervals $[0, T]$, where $T < \infty$ depends on the coefficients a, b, c, d and the initial data η_0, u_0. Local well-posedness has been well established thus far for a subclass of the linearly well-posed Boussinesq systems.

A systematic investigation of local well-posedness has been carried out in [12]. It is shown that if the coefficients a, b, c, d, defined by (3.2), satisfy in addition one of the conditions (C1), (C2), then the corresponding nonlinear system (3.40)–(3.41) is locally well-posed in appropriate Sobolev spaces. For example, for the BBM-BBM system (3.36) (whose coefficients satisfy (C1)), it is shown in [12] that given $s \geq 0$ and any pair $(\eta_0, u_0) \in H^s(\mathbb{R})^2$, there exists a $T > 0$ and a unique solution (η, u) that satisfies $(\eta, u) \in C(0, T; H^s(\mathbb{R}))^2$. (Here, for a Banach space X, the space $C(0, T; X)$ denotes the functions $v : [0, T] \to X$ for which the mapping $t \mapsto v(t)$ is continuous from $[0, T]$ into X. For the KdV-KdV system (3.38) a similar result holds if $s > 3/4$. The proof of these results rests in general on fixed point arguments in appropriate function spaces, which are used in integral equation formulations of (3.40) effected by the Duhamel principle acting on the nonlinear terms considered as right-hand sides. In [14] it was proved that the subclass of the symmetric systems consisting of systems of the form (3.39) with $a = c$ and $b, d \geq 0$ (the *completely* symmetric systems) are locally well-posed in $H^s \times H^s$, $s > 3/2$.

A few systems have been shown to possess unique solutions for all $t > 0$ under mild restrictions on the initial data. Such global well-posedness results hold, for example, for the classical Boussinesq system (3.35), cf. [3], [50], [12]. In fact, if $(\eta_0, u_0) \in H^s \times H^{s+1}$, $s \geq 1$, and $\inf_{x \in \mathbb{R}} [1 + \eta_0(x)] > 0$, then the system (3.35) possesses a unique solution $(\eta, u) \in C(0, T; H^s) \times C(0, T; H^{s+1})$ for *any* $T > 0$.

It can be readily seen that all systems of the form (3.40) with $b = d$ are *Hamiltonian*. In fact, the 'energy' functional

$$E(t) := \int_{-\infty}^{\infty} (\eta^2 + (1 + \eta)u^2 - c\eta_x^2 - au_x^2)dx, \tag{3.46}$$

is a time-independent quantity of the system and may serve as a Hamiltonian. The invariance of E leads to a proof of global well-posedness of the associated system (if $b = d > 0$ and $a \leq 0$, $c < 0$, cf. [12]), under mild assumptions on the initial data. For example, in the case of the Bona-Smith systems, wherein $a = 0$, $c < 0$, it was shown in [17] (for the case $\theta^2 = 1$, but the proof may be readily extended to the whole family, i.e. for $2/3 < \theta^2 \leq 1$), that if the initial data in (3.41) are such that $(\eta_0, u_0) \in H^{s+1} \times H^s$, $s \geq 0$ and if $\inf_{x \in \mathbb{R}} [1 + \eta_0(x)] > 0$ and $E(0) = \int_{-\infty}^{\infty} (\eta_0^2 + |c|(\eta_0')^2 + (1 + \eta_0)u_0^2)dx < 2|c|^{1/2}$, then there is a unique solution (η, u) of the initial-value problem, which for any $T > 0$ belongs to $C(0, T; H^{s+1}) \times C(0, T; H^s)$. As explained in [17], the basic step in this proof is to establish an *a priori* $H^1 \times L^2$ estimate of (u, η). This estimate follows from the fact that the functional E is invariant for $t \geq 0$ and positive, since the restrictions $\eta_0 > -1$ and $E(0) < 2|c|^{1/2}$ ensure that $1 + \eta(x, t)$, and, consequently, $E(t)$ remain positive for $x \in \mathbb{R}$, $t \geq 0$. (Recall that in the unscaled, nondimensional variables $1 + \eta(x, t) > 0$ means that there is water in the channel at x at time t; so, the initial restriction $\inf_{x \in \mathbb{R}} (1 + \eta_0(x)) > 0$ required for global existence for the Bona-Smith and the classical Boussinesq systems has physical meaning.)

For the purposes of solving numerically the Boussinesq systems or comparing their solutions with experimental data, it is important to establish well-posedness, at least locally in time, for *initial-boundary-value* problems (ibvp's) for (3.40). In this direction, it has been proved in [10] that the ibvp for the BBM-BBM system (3.36) with Dirichlet boundary data at the endpoints of a finite interval $[A, B]$, i.e., when η and u are given functions of t at $x = A$ and B, is locally well-posed.

In the case of the Bona-Smith family of systems (3.37) with $2/3 < \theta^2 \leq 1$, it has been shown in [4] that the periodic initial-value problem on the spatial interval $[-L, L]$, $L > 0$, is well-posed. For example, if $(\eta_0, u_0) \in C_\pi^3 \times C_\pi^2$ (where $C_\pi^k = C_\pi^k(-L, L)$, for integer $k \geq 0$, is the space of k-times continuous differentiable periodic functions on $[-L, L]$), $\eta_0(x) > -1$ for $-L \leq x \leq L$, and $\int_{-L}^{L} (\eta_0^2 + |c|(\eta_0')^2 + (1 + \eta_0)u_0^2) dx < \frac{2L|c|^{1/2}}{L+|c|^{1/2}}$, then there is a unique, global classical solution (η, u) for the periodic initial-value problem for (3.37), which, for each $T > 0$, belongs to the space $C(0, T; C_\pi^3) \times C(0, T; C_\pi^2)$. (In this case it is easy to check that the energy functional, defined now as $E(t) = \int_{-L}^{L} [\eta^2 + |c|\eta_x^2 + (1 + \eta)u^2] dx$ is invariant, and the proof proceeds much as in the case of the Cauchy problem.)

Other types of boundary conditions, cf. [4], may also be shown to lead to well-posed initial-boundary-value problems in the case of the Bona-Smith system. Notable among them is the problem with *reflective* boundary conditions, i.e., $\eta_x = 0$, $u = 0$ at both endpoints of a finite interval, which is globally well-posed under similar, mild restrictions on the initial data. (The analogous associated energy functional is conserved in this case too.) In the case of homogeneous Dirichlet boundary conditions on η and u at the endpoints, however, the energy functional is no longer conserved and the problem is only locally well-posed.

We close this section with a remark on the temporal interval of local existence of solutions. Consider the initial-value problem for the system (3.1) written in nondimensional, scaled variables, i.e., in the form (3.32), that we rewrite for convenience using simpler notation as

$$\eta_t + u_x + \varepsilon \left[(\eta u)_x + au_{xxx} - b\eta_{xxt} \right] = 0,$$
$$u_t + \eta_x + \varepsilon \left(uu_x + c\eta_{xxx} - du_{xxt} \right) = 0. \tag{3.47}$$

If the system (3.47) is locally well-posed (recall that a sufficient condition for this is that (C1) or (C2) is satisfied, [12]), its solution exists on a finite interval $[0, T_\varepsilon]$, where T_ε is a function of ε. For modelling purposes, T_ε should be large for small ε. In [30] it was proved (in the case of Boussinesq systems in two space dimensions but the proof is valid for one dimension too) that if $b > 0$, $d > 0$ and $a \leq 0$, $c \leq 0$ or $a = c \geq 0$, then $T_\varepsilon = O(1/\varepsilon^\alpha)$ for any $\alpha < 1/2$. This result shows that $T_\varepsilon \to \infty$ as $\varepsilon \to 0$ but it is not very satisfactory as it is expected that T_ε should be at least of $O(1/\varepsilon)$. It is worthwhile to note that for the completely symmetric systems (in one or two space dimensions) it was proved in [14] that $T_\varepsilon = O(1/\varepsilon)$. The recent results of [2] imply that, in addition, $T_\varepsilon = O(1/\varepsilon)$ for the Boussinesq systems of the form (3.47) that are locally well-posed.

3.4   Solitary Waves

Being approximations of the Euler equations, the Boussinesq equations may be reasonably expected to possess *solitary wave* solutions. It is well-known of course, cf., e.g., [59], that there is a wealth of information on the solitary waves of one-way counterparts of the Boussinesq systems such as the KdV equation, which is integrable, and the BBM equation. The analogous theory for the Boussinesq systems is less well-developed; in this section we shall give a brief review of some aspects of this theory.

The solitary waves that we will study are travelling wave solutions of the system (3.1) of the form

$$\eta_s(x,t) = \eta_s(x + x_0 - c_s t), \quad u_s(x,t) = u_s(x + x_0 - c_s t), \tag{3.48}$$

where c_s is the (constant) speed of propagation of the travelling wave. The functions $\eta_s = \eta_s(\xi)$, $u_s = u_s(\xi)$, $\xi \in \mathbb{R}$, will be supposed to be smooth, positive and even, with a single maximum located at $\xi = 0$, and decaying monotonically to zero, along with all their derivatives, as $\xi \to \pm\infty$. We will call such solutions *classical solitary waves*. (Other kinds of solitary waves, such as generalized and multi-pulsed solitary waves will be mentioned briefly in the sequel.)

Substituting (3.48) in (3.1) one obtains, after integrating once and setting the integration constants equal to zero, the system of nonlinear ordinary differential equations (o.d.e.'s)

$$\begin{aligned}
-c_s\eta + u + \eta u + au'' - bc_s\eta'' &= 0, \\
-c_s u + \eta + \tfrac{1}{2}u^2 + c\eta'' - dc_s u'' &= 0,
\end{aligned} \tag{3.49}$$

where we denote $(u_s(\xi), \eta_s(\xi))$ simply by $(u(\xi), \eta(\xi))$ and $' = \frac{d}{d\xi}$. Note that if (u, η) is a solution of (3.49) corresponding to some $c_s > 0$, then $(-u, \eta)$ is also a solution propagating with speed $-c_s$, i.e., to the left as t increases. In the sequel we shall usually assume that $c_s > 0$.

A few solitary wave solutions may be found in closed form. Following e.g., [25], if we assume a solution of the special type

$$\begin{aligned}
\eta(\xi) &= \eta_0 \text{sech}^2(\lambda\xi), \\
u(\xi) &= B\eta(\xi),
\end{aligned} \tag{3.50}$$

where η_0, B, λ are real constants and $x_0 \in \mathbb{R}$ is arbitrary, and substitute into the system (3.49), we obtain a set of algebraic equations for η_0, B, λ and c_s

that has solutions under certain conditions, cf. [25]. For example, in the case of the Bona-Smith systems (3.37), it turns out that for each $\theta^2 \in (7/9, 1)$ there is precisely one solitary wave solution of the form (3.50) of the corresponding system; the parameters η_0, c_s, B, λ are given by the formulas

$$\eta_0 = \frac{9}{2} \cdot \frac{\theta^2 - 7/9}{1 - \theta^2}, \qquad c_s = \frac{4(\theta^2 - 2/3)}{\sqrt{2(1-\theta^2)(\theta^2 - 1/3)}},$$

$$\lambda = \frac{1}{2}\sqrt{\frac{3(\theta^2 - 7/9)}{(\theta^2 - 1/3)(\theta^2 - 2/3)}}, \qquad B = \sqrt{\frac{2(1-\theta^2)}{\theta^2 - 1/3}}. \tag{3.51}$$

Figure 3.2 shows the solitary wave profile (3.50)–(3.51) for the Bona-Smith system with $\theta^2 = 9/11$.

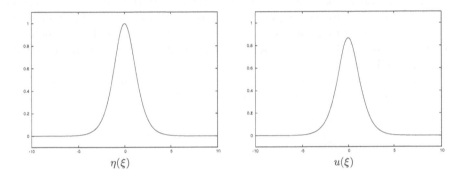

$$\eta(\xi) \qquad\qquad\qquad\qquad u(\xi)$$

FIGURE 3.2: Exact η- and u-solitary wave profiles of the form (3.50)–(3.51). Bona-Smith system, $\theta^2 = 9/11$.

It is worthwhile to note that there are no solitary waves of the form (3.50) for the Bona-Smith systems if $\theta^2 \notin (7/9, 1)$; in particular no exact solutions of this type may be found for the BBM-BBM system (3.36). Moreover, the classical Boussinesq, (3.35), and the KdV-KdV, (3.38), systems do not possess exact solitary wave solutions of this type. There exist however closed-form formulae for exact travelling wave solutions (which are not solitary waves) for many Boussinesq systems, cf. [25]. These solutions are useful for testing numerical methods for the system (3.1).

We now turn to the problem of existence of solutions of the general o.d.e. system (3.49). We first examine the case where the phase speed c_s is greater than but close to one, and then comment on known existence results for larger values of c_s. In the first case, following the exposition in [40], we write first (3.49) as a first-order 4×4 o.d.e. system of the form

$$U' = V(U; c_s) := L(c_s)U + R(U; c_s), \tag{3.52}$$

where $U = (\eta, \eta', u, u')^T$, $D := c_s^2 bd - ac$ (assumed to be nonzero), and

$$L(c_s) = \begin{pmatrix} 0 & 1 & 0 & 0 \\ \frac{c_s^2 d + a}{D} & 0 & -\frac{c_s(d+a)}{D} & 0 \\ 0 & 0 & 0 & 1 \\ -\frac{c_s(b+c)}{D} & 0 & \frac{c_s^2 b + c}{D} & 0 \end{pmatrix} \quad \text{and} \quad R(U; c_s) = \begin{pmatrix} 0 \\ \frac{a}{2D} u_3^2 - \frac{c_s d}{D} u_1 u_3 \\ 0 \\ -\frac{c_s b}{2D} u_3^2 + \frac{c}{D} u_1 u_3 \end{pmatrix}.$$

The analysis in [40] is based on *normal form theory* for nonlinear systems of o.d.e.'s. (See, e.g., [35] for a convenient reference). This is a local theory which may be applied, under certain hypotheses, in the neighborhood of an equilibrium point of the nonlinear system. It constructs, in a systematic way, a series of transformations that simplify the system reducing it to a 'normal form,' in which emphasis is placed on those nonlinear terms that contribute most to the structure of orbits near the equilibrium. The qualitative behavior of the transformed system is determined by the analogous behavior of the associated linearized system.

In our case, the linearization of (3.52) about the equilibrium point $U = 0$ is the linear system $U' = L(c_s)U$. The eigenvalues of the matrix $L(c_s)$ satisfy the characteristic equation $\lambda^4 - \beta\lambda^2 + \alpha = 0$, where $\beta = \frac{\Sigma}{D}$, $\alpha = \frac{c_s^2 - 1}{D}$, and $\Sigma := a + c + (b + d)c_s^2$. Hence, the set of eigenvalues is $\{\pm\lambda_-, \pm\lambda_+\}$, where $\lambda_\pm = \frac{1}{\sqrt{2}}\sqrt{\beta \pm \sqrt{\beta^2 - 4\alpha}}$.

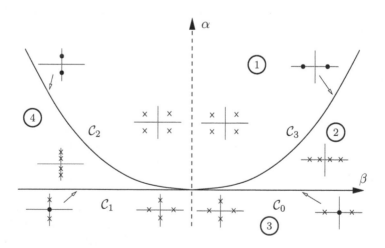

FIGURE 3.3: Spectrum of the matrix $L(c_s)$ of the linearized solitary wave equation of the Boussinesq systems. $\times$ = simple eigenvalue, $\bullet$ = double eigenvalue.

Characteristic equations of the type $\lambda^4 - \beta\lambda^2 + \alpha = 0$, with real α, β, appear often in the study of solitary waves of nonlinear dispersive wave equations that are derived e.g. from Euler equations of free surface flow or from other related systems, cf., e.g., [36], [21], [22], [23]. As in these references, one may consider Figure 3.3, in which we distinguish four regions of the (β, α) plane, labeled 1 to 4, depending on the type of eigenvalues of L that they yield. These four regions are separated by four curves: (i) Curve $\mathcal{C}_0 = \{(\beta, \alpha), \alpha = 0, \beta > 0\}$, on which L has a double eigenvalue equal to zero and two simple, real eigenvalues. (ii) Curve $\mathcal{C}_1 = \{(\beta, \alpha), \alpha = 0, \beta < 0\}$, on which we have two imaginary eigenvalues and one double eigenvalue equal to zero. (iii) Curve $\mathcal{C}_2 = \{(\beta, \alpha), \alpha > 0, \beta = -2\sqrt{\alpha}\}$, on which the spectrum consists of two double imaginary eigenvalues. (iv) Curve $\mathcal{C}_3 = \{(\beta, \alpha), \alpha > 0, \beta = 2\sqrt{\alpha}\}$, on which we have two double real eigenvalues. At the origin L has a quadruple eigenvalue equal to zero. The type of eigenvalues in the four regions between the curves is shown in Figure 3.3 in a schematic way.

It is now possible to study, using the properties of the nonlinear term $V(U; c_s)$ in the right-hand side of (3.52), following the general plan and the references found, e.g., in the review article [21], the question of existence of solitary waves at least for positive and small values of $c_s - 1$. This is done in detail in [40]. It turns out that we may distinguish two classes of Boussinesq systems:

The first class is typified by the Bona-Smith systems (3.37) for $2/3 \leq \theta^2 \leq 1$, which includes the BBM-BBM system as well. It is straightforward to check that for these systems $\beta > 0$ for all $c_s \geq 1$ and $\alpha > 0$ for $c_s > 1$. When $c_s = 1$, $\alpha = 0$, i.e., $(\beta, \alpha) \in \mathcal{C}_0$, while for $c_s > 1$ (β, α) belongs to the region 2, i.e. the spectrum of L consists of four distinct, real eigenvalues. The normal form theory yields then that, at least for small $c_s - 1$, these systems possess classical solitary wave solutions.

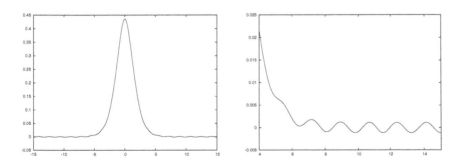

FIGURE 3.4: Generalized solitary wave for the KdV-KdV system (3.38).

A typical member of the second class is the KdV-KdV system (3.38), for which $\beta < 0$ for $c_s \geq 1$ and $\alpha > 0$ for $c_s > 1$. Hence, when $c_s = 1$, (β, α) belongs to $\mathcal{C}_1$, while for $c_s > 1$, with $c_s - 1$ small, the matrix L has four distinct imaginary eigenvalues ((β, α) in region 4). One may prove then that the system does not have classical solitary wave solutions. Using the theory of Lombardi, [37], it was verified in [15] that the system possesses "generalized" solitary wave solutions (see Figure 3.4). A generalized solitary wave has a main pulse which does not tend to zero as $|\xi|$ increases but to small-amplitude periodic solutions extending to $\pm\infty$. For more properties of the generalized solitary waves of the KdV-KdV systems we refer the reader to [15].

In Table 3.1, taken from [40], we list the type of solitary waves that the admissible Boussinesq systems possess. (Recall that the admissible systems satisfy one of the conditions (C1), (C2), or (C3) in Section 3.3. They are the systems for which the associated initial-value problem is linearly well-posed.) The first column of the table denotes the order of the function $\sigma(k) = |w_1(k)w_2(k)|^{1/2}$ introduced in Section 3.3. The second column describes the set of parameters of the admissible systems and the third specifies the type of solitary wave that each system possesses for small $c_s - 1 > 0$ and the number of the region in which (β, α) belongs for such c_s.

Table 3.1: Classification of the admissible Boussinesq systems relatively to the type of solitary waves that they possess for small and positive $c_s - 1$. (CSW=classical solitary wave, GSW=generalized solitary wave.)

Order of $\sigma(k)$	Admissible system	Type of solitary wave
2	$a = c > 0$, $b = d = 0$ (KdV-KdV)	GSW(3)
1	$a, c < 0, b = 0, d > 0$	GSW(3)
	$b = 0, a = c > 0, d > 0$	GSW(3)
	$d = 0, a = c > 0, b > 0$	GSW(3)
0	$a, c < 0,\ b, d > 0,\ bd - ac > 0$	CSW(2)
	$a, c < 0,\ b, d > 0,\ bd - ac < 0$	GSW(3)
	$a = c > 0, b, d > 0,\ bd - ac > 0$	CSW(2)
	$a = c > 0, b, d > 0,\ bd - ac < 0$	GSW(3)
	$a = c > 0, b = d < 0$	GSW(3)
-1	$a = 0, c < 0, b, d > 0$ (BS)	CSW(2)
	$c = 0, a < 0, b, d > 0$	CSW(2)
-2	$a = c = 0, b, d > 0$ (BBM-BBM)	CSW(2)

Missing from this table are the admissible but *degenerate* Boussinesq systems for which $D = 0$ for all c_s. (The classical Boussinesq system (3.35) belongs to this category, for example.) The problem of existence of solitary

waves for these systems is easier to study, and, typically, one may show existence of solitary waves for all values of $c_s > 1$. For the solitary waves of the
classical Boussinesq system cf., e.g., [44], [24]; for other such systems cf. [24].

We now turn to the problem of existence of solutions to the o.d.e. system
(3.49) for larger values of $c_s > 1$. Here the available theoretical results are
sparser. Toland, [53], established the existence of solitary wave solutions for
the Bona-Smith system (3.37), in the case $\theta^2 = 1$, for all $c_s > 1$. Existence for
some other Boussinesq systems follows from a more general result of Toland,
[54], which may be formulated as follows, using the notation of [24]. Consider
a system of o.d.e.'s of the form

$$S_1 \mathbf{u}'' + S_2 \mathbf{u} + \nabla g(u, \eta) = 0, \tag{3.53}$$

where $\mathbf{u} = (u, \eta)^T$. (Note that the system (3.49) is of this form, with

$$S_1 = -6 \begin{pmatrix} \frac{a}{c_s} & b \\ d & \frac{c}{c_s} \end{pmatrix}, \quad S_1 = 6 \begin{pmatrix} -\frac{1}{c_s} & 1 \\ 1 & -\frac{1}{c_s} \end{pmatrix},$$

and $g(u, \eta) = -\frac{3}{c_s} u^2 \eta$.)

THEOREM 3.1 (Toland) *Let S_1 and S_2 be symmetric and let $g \in C^2(\mathbb{R}^2)$
be such that g, ∇g, g_{uu}, $g_{u\eta}$, $g_{\eta\eta}$ are zero at $(0,0)$. Moreover, if $Q(\mathbf{u}) =
\mathbf{u}^T S_1 \mathbf{u}$ and $f(u, \eta) = \mathbf{u}^T S_2 \mathbf{u} + 2g$, assume that*

(I) $\det S_1 < 0$, *and there exist two linearly independent vectors $\mathbf{u}_1 =
(u_1, \eta_1)^T$ and $\mathbf{u} = (u_2, \eta_2)^T$ such that $Q(\mathbf{u}_1) = Q(\mathbf{u}_2) = 0$.*

(II) *There exists a closed plane curve $\mathcal{F}$ that passes through $(0,0)$ such that*

    (i) $f = 0$ *on $\mathcal{F}$, and $\mathcal{F} \setminus \{(0,0)\}$ belongs to the set $\{(u, \eta) : Q(u, \eta) <
0\}$.*

    (ii) $f(u, \eta) > 0$ *in the (nonempty) interior of $\mathcal{F}$.*

    (iii) $\mathcal{F} \setminus \{(0,0)\}$ *is strictly convex i.e. $f_{uu} f_\eta^2 - 2f_{u\eta} f_u f_\eta + f_{\eta\eta} f_u^2 <
0$ on $\mathcal{F} \setminus \{(0,0)\}$.*

    (iv) $\nabla f(u, \eta) = 0$ *on $\mathcal{F}$ if and only if $(u, \eta) = (0,0)$.*

*Then, there exists an orbit γ of (3.53) which is homoclinic to the origin.
Moreover,*

(a) $(u(0), \eta(0)) \in \Gamma$, *where Γ is the segment of $\mathcal{F}$ not including the origin
between P_1 and P_2, with P_1 satisfying $\nabla f(P_i) \cdot \mathbf{u}_i = 0$, $i = 1, 2$.*

(b) u, η *are even functions on $\mathbb{R}$.*

(c) $(u(\xi), \eta(\xi))$ *is in the interior of $\mathcal{F}$ for all $\xi \neq 0$.*

(d) γ *is monotone in the sense that $u(\xi) \leq u(s)$, $\eta(\xi) \leq \eta(s)$ if $\xi \geq s \geq 0$.*

Toland's theorem was applied by Chen, [24], to establish existence of solitary wave solutions for any $c_s > 1$ to several specific examples of Boussinesq systems, including the BBM-BBM system (3.36). It is also not hard to check, cf. [29], that Toland's theorem gives existence of solitary waves for any $c_s > 1$ for all Bona-Smith systems (3.37) for any $\theta^2 \in [2/3, 1]$. It also follows from (a) of Theorem 3.1 that for a solitary wave, the peak $(u(0), \eta(0))$ lies on the open segment $\Gamma = P_1 P_2$ (that does not include the origin) of the branch of the curve $f(u, \eta) = 0$ in the first quadrant of the u, η-plane (cf. Figure 3.5), where, for the Boussinesq system, $f(u, \eta) = \frac{6}{c_s}(-u^2(1 + \eta) - \eta^2 + 2c_s u \eta)$. It follows that the speed c_s of a solitary wave satisfies the equation $c_s = \frac{\mu^2(1+\eta(0))+1}{2\mu}$, $\mu := \frac{u(0)}{\eta(0)}$.

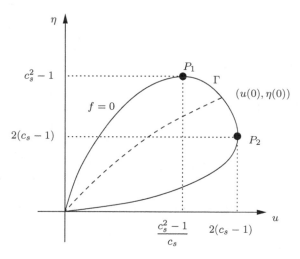

FIGURE 3.5: Locus of possible $(u(0), \eta(0))$ and orbits.

It also follows from Toland's theory that $\eta(0) > u(0)$ and, from the above, that $\frac{\partial c_s}{\partial \eta(0)} > 0$, $\frac{\partial c_s}{\partial u(0)} > 0$, i.e. that c_s increases with the height of the solitary wave. In Figure 3.5 half of the solitary wave profile (i.e. the orbit $(u(\xi), \eta(\xi))$, $0 \leq \xi < \infty$) is represented by the dashed line.

Toland also studied the uniqueness of these solitary waves in the specific case of the Bona-Smith system with $\theta^2 = 1$, [55]. He proved that (classical) solitary waves for this system are unique for any speed if $u(0) \leq 1$ and in general if $1 < c_s \leq 3/2$ or $c_s \gg 1$. In the case of the general Bona-Smith system it is possible to show uniqueness of solitary waves if $\theta^2 \in \left(\frac{2+\sqrt{0.2}}{3}, 1 \right]$ and $1 < c_s < c_{s\,\max}(\theta) := \min\left\{ \frac{12(3\theta^2 - 2)}{21\theta^2 - 13}, \frac{2(3\theta^2 - 2)}{\sqrt{3(1-\theta^2)(3\theta^2 - 1)}} \right\}$, cf. [29]. This result means that for the class of Bona-Smith systems corresponding to $\theta^2 \in \left(\frac{2+\sqrt{0.2}}{3}, 1\right]$ and given any value of the speed in the interval $(1, c_{s\,max}(\theta))$ there exists precisely one pair $(u(0), \eta(0))$ on Γ from which there issues a solitary wave of the corresponding system. In our numerical experiments we did not detect any lack of uniqueness for any $\theta^2 \in [2/3, 1]$ for any speed $c_s > 1$, that we tried.

Solitary waves apparently play an important role in the long-time asymptotics of the solution of the initial-value problem for the Boussinesq system. Numerical experiments (cf. Section 3.6) suggest that arbitrary initial data, that decay sufficiently fast at infinity, are in general resolved as time increases into two wavetrains each consisting of sequences of solitary waves followed by a decaying, small-amplitude, dispersive oscillatory tail. This property of *resolution* into solitary waves has been rigorously proved for the (one-way, integrable) KdV equation via the inverse scattering transform, cf. [51], and has been observed numerically in the case of many nonintegrable one-way or two-way models, cf., e.g., [14], [46], [47], [10], [5], [4], [6].

Related to the resolution property is the *stability* of solitary waves under small perturbations, famous since Scott Russell's first observation of a solitary wave in the 1830s. The *orbital* (or *shape*) *stability* property of the solitary wave (which implies that an initial profile, that is a small perturbation of a solitary wave $w_s(x)$, will evolve into a solution $w(x,t)$, which, for all t, will remain close to the family of translated profiles $w_s(x + \xi)$, $\xi \in \mathbb{R}$), was established by Benjamin, [7] (see also [9]), in the case of the KdV and BBM equations, by a variational theory in which the solitary wave is viewed as an extremal of an invariant of the equation under the constraint that another conserved functional is held fixed. This classical method (whose origins go back again to Boussinesq, [20]) was subsequently extended and applied to establish the orbital stability (or instability) of solitary waves of many examples of nonlinear dispersive wave equations, cf. [56], [57], [34], [18]. It turns out, cf. [28], that as in the case of other systems, cf., e.g., [16], [42], [52], the classical theory fails for the Boussinesq systems due to the indefiniteness in an infinite number of directions of the Hessian of a naturally defined functional of the solution of the system, of which the solitary waves are critical points. More detailed information about the long-time asymptotic behavior of the solution of the initial-value problem that emanates from a slightly perturbed initial solitary wave profile is provided by *asymptotic stability* theories of solitary waves. Such theories have been established, in the case of one-way models such as the KdV and BBM equations, cf., e.g., [43], [39], [58], [38], [33]. For such models it has been proved that a slight perturbation of an initial solitary wave profile w_s of speed c_s gives rise to a solution w which, for large t, satisfies $w(x,t) = w_{s_\infty}(x - c_\infty t + \xi_\infty) + z(x,t)$, where w_{s_∞} is a solitary wave with speed c_∞ close to c_s, ξ_∞ is a small phase shift, and $z(x,t)$ consists of smaller, slower solitary waves and small amplitude oscillatory dispersive tails, that appear to be convected to the left relative to the main solitary wave w_{s_∞}. Again, no such results of (nonlinear) asymptotic stability have been established in the case of Boussinesq systems. However, Pego and Weinstein, [44], have developed, in the case of the classical Boussinesq system (3.35) and for other equations, an approximate *linearized* theory of asymptotic stability of solitary waves, in which the residual z does not satisfy the exact relevant nonlinear evolution equation but its linearized version. A thorough numerical study of asymptotic stability of solitary wave solutions of the Bona-Smith system has been conducted in [28]; see Section 3.6 in the sequel. The conclusion is that the solitary waves of this system are apparently asymptotically stable.

We finally note that, in addition to the classical solitary waves, various Boussinesq systems also possess "multi-pulsed" (multi-hump) solitary waves. These solitary waves have been studied numerically, e.g., in [24], [40]. They appear to be unstable under small perturbations.

3.5 Numerical Methods

In this section we will consider fully discrete Galerkin/finite element numerical methods for the numerical solution of the Boussinesq system (3.1) on a finite interval. For simplicity, we shall mainly consider the periodic initial-value problem, which will be discretized in space by the standard Galerkin method with cubic splines and in time by Runge-Kutta methods of fourth-order of accuracy. We would like to mention that other methods have also been used in the literature for the numerical solution of Boussinesq systems. For example, in the mathematical literature, optimal-order error estimates were proved in [60] for a nonstandard semidiscrete Galerkin method for the initial-boundary-value problem for the Bona-Smith with $\theta^2 = 1$ with Dirichlet boundary conditions for η and u. In [10] a finite difference scheme of fourth order of accuracy in space and time was analyzed, in the case of the BBM-BBM system with Dirichlet boundary conditions for η and u, and used in numerical experiments with that system. In [45] a fully discrete spectral method was analyzed for the periodic initial-value problem of the Bona-Smith system ($\theta^2 = 1$); numerical results with this spectral method were presented for several Boussinesq systems in [46].

The numerical methods that we will consider in the sequel are fully discrete standard Galerkin methods for the discretization of the periodic initial-value problem for (3.1) on the spatial interval $[-L, L]$, given $2L$-periodic initial conditions η_0 and u_0. Let S_h be the space of smooth, periodic cubic splines on a uniform mesh on $[-L, L]$ with meshlength h. We approximate the solution of (η, u) of the system by a pair $(\eta_h, u_h) \in S_h \times S_h$, satisfying, for $t > 0$, the semidiscrete equations

$$
\begin{aligned}
\mathcal{A}_b(\eta_{ht}, \chi) + (u_{hx}, \chi) + ((\eta_h u_h)_x, \chi) - a(u_{hxx}, \chi') = 0, \; \forall \chi \in S_h, \\
\mathcal{A}_d(u_{ht}, \phi) + (\eta_{hx}, \phi) + (u_h u_{hx}, \phi) - c(\eta_{hxx}, \phi') = 0, \; \forall \phi \in S_h.
\end{aligned}
\tag{3.54}
$$

In this section we let $(\phi, \chi) := \int_{-L}^{L} \phi(x)\chi(x)dx$, and $\|\cdot\| := (\cdot, \cdot)^{1/2}$ be the L^2 inner product, respectively the L^2 norm on $[-L, L]$. $\mathcal{A}_b$ is the bilinear form defined by $\mathcal{A}_b(\phi, \chi) := (\phi, \chi) + b(\phi', \chi')$. We take as initial values for (3.54) the functions

$$
\eta_h(0) = \Pi_h \eta_0, \quad u_h(0) = \Pi_h u_0,
\tag{3.55}
$$

where $\Pi_h v$ is any of the usual optimal-order approximations of v in S_h (interpolant, L^2-projection etc.), for which we assume that $\|v - \Pi_h v\| = O(h^4)$ for any smooth enough $2L$-periodic function v. For several Boussinesq systems, including those of Bona-Smith and BBM-BBM type ($b > 0$, $d > 0$, $a = 0$, $c \leq 0$), and the classical Boussinesq system (3.35), it can be shown, cf. [4], [6], that, the o.d.e. initial-value problem has a unique solution on any temporal interval $[0, T]$ on which a smooth solution (η, u) of the underlying continuous periodic initial-value problem exists. Moreover, the semidiscrete approximations for these systems satisfy the optimal-order L^2 error estimate

$$
\max_{0 \leq t \leq T} (\|\eta_h(t) - \eta(t)\| + \|u(t) - u_h(t)\|) \leq Ch^4,
$$

where C will denote positive constants independent of the discretization parameters.

The o.d.e. initial-value problem (3.54)–(3.55) has various properties depending on the coefficients a, b, c, d. For example, for the Bona-Smith family of systems, it turns out that the system is not stiff; consequently it may by discretized in time by an explicit time-stepping method. One such method, whose accuracy matches the spatial order of accuracy that the cubic splines yield, is the classical, fourth-order, four-stage, explicit Runge-Kutta method, whose tableau of coefficients is

$$
\begin{array}{cccc|c}
0 & 0 & 0 & 0 & 0 \\
\frac{1}{2} & 0 & 0 & 0 & \frac{1}{2} \\
0 & \frac{1}{2} & 0 & 0 & \frac{1}{2} \\
0 & 0 & 1 & 0 & 1 \\
\hline
\frac{1}{6} & \frac{1}{3} & \frac{1}{3} & \frac{1}{6} &
\end{array}
\tag{3.56}
$$

For this method, if $k = T/J$ is the (constant) time step, where J is a positive integer, $t^n = nk$, $n = 0, 1, 2, \ldots$, and (η_h^n, u_h^n) denotes the fully discrete approximation of $(\eta(t^n), u(t^n))$, one may show, cf. [4], for the Bona-Smith systems for $\frac{2}{3} \leq \theta^2 \leq 1$ that

$$
\max_{0 \leq n \leq J} \|\eta_h^n - \eta(t^n)\| + \|u_h^n - u(t^n)\| \leq C(k^4 + h^4),
$$

without assuming any mesh condition between k and h. The classical Boussinesq system (for which $a = b = c = 0$) is slightly more stiff; it may be shown, [4], that a similar $O(k^4 + h^4)\, L^2$ error estimate holds if k/h is sufficiently small. The latter condition is not really restrictive.

However, for the KdV-KdV system (for which $b = d = 0$, $a = c > 0$) the o.d.e. system is highly stiff and its stable integration in time requires an implicit, nonlinearly stable o.d.e. solver. On the basis of experience gained in [14], the implicit, two-stage, fourth-order accurate, Gauss-Legendre Runge-Kutta method was used in [15] for this purpose. The tableau of this method is

$$
\begin{array}{cc|c}
\frac{1}{4} & \frac{1}{4} - \frac{1}{2\sqrt{3}} & \frac{1}{2} - \frac{1}{2\sqrt{3}} \\
\frac{1}{4} + \frac{1}{2\sqrt{3}} & \frac{1}{4} & \frac{1}{2} + \frac{1}{2\sqrt{3}} \\
\hline
\frac{1}{2} & \frac{1}{2} &
\end{array}
$$

The application of this implicit scheme to the nonlinear system (3.54)–(3.55) requires solving at each time step a nonlinear system of algebraic equations. This is effected via Newton's method; the attendant linear systems are solved by special iterative schemes, cf. [15]. The resulting fully discrete scheme is numerically shown to be stable and fourth-order accurate in space and time.

It has long been known, cf. [32], that for (nonperiodic) initial-boundary-value problems for p.d.e.'s that are of odd order in space, the standard Galerkin semidiscretization suffers from spatial order of convergence reduction. (This is precisely the reason why nonstandard Galerkin methods were used in [60]; however the nonstandard methods have other disadvantages). In [4], the initial-boundary-value problem for the Bona-Smith system with homogeneous Dirichlet or reflective boundary conditions at the endpoints $\pm L$ was discretized by the standard Galerkin method with cubic splines and the fourth-order explicit, Runge-Kutta time-stepping scheme (3.56); its error in L^2 was found to be of $O(k^4 + h^3)$. Although suboptimal in h, this level of error is acceptable for accurate computations for small enough k and h.

In the rest of this section we shall present the results of various preliminary numerical experiments that we have performed, integrating the periodic initial-value problem for the Bona-Smith systems using the standard Galerkin method with cubic splines and the fourth-order, explicit Runge-Kutta scheme (3.56). We are mainly interested in assessing the accuracy of this fully discrete numerical scheme when we approximate solitary wave solutions of the system. Note that if the spatial interval of integration is taken to be large enough relative to the effective support of the (exponentially decaying as $|\xi| \to \infty$) solitary waves, the solution of the periodic initial-value problem approximates well the solution of the initial-value problem up to some $T > 0$.

In a first series of numerical experiments we considered an exact solitary wave given by (3.50)–(3.51) for the Bona-Smith system with $\theta^2 = 9/11$; this has $\eta_0 = 1$, $B = \frac{\sqrt{3}}{2} \cong 0.86602540$, $c_s = \frac{5\sqrt{3}}{6} \cong 1.44337567$, $\lambda = \frac{1}{4}\sqrt{\frac{33}{5}} \cong 0.64226163$. Taking into account the width of the solitary wave, we took $x_0 = 100$ (i.e., initially the solitary wave was centered at $x = -100$) and integrated numerically the periodic initial-value problem for the system on the spatial interval $[-150, 150]$ with initial conditions given by (3.50) at $t = 0$. By $t = 100$ the solitary wave had completed almost one half of a revolution.

To verify the order of accuracy of the scheme we took five values of h, and $k = h/10$ in each case, and computed the normalized L^2 errors

$$E_\eta(t^n) = \frac{\|\eta_h^n - \eta_s(\cdot, t^n)\|}{\|\eta_s(\cdot, 0)\|}, \quad E_u(t^n) = \frac{\|u_h^n - u_s(\cdot, t^n)\|}{\|u_s(\cdot, 0)\|}$$

as functions of t^n. Table 3.2 shows the values of these errors and the corresponding experimental rates of convergence (defined as usual as $\frac{\ln(E(h_1)/E(h_2))}{\ln(h_1/h_2)}$, where h_1 is the listed value of h in the corresponding line and h_2 the previous value) at $t = 100$. (All computations were performed in double precision using the GNU Fortran compiler.)

Table 3.2: Normalized L^2 errors and convergence rates for solitary waves, $\theta^2 = 9/11$, $k = h/10$.

h	$E_\eta(100)$	rate for E_η	$E_u(100)$	rate for E_u
0.50	0.4762(-3)	–	0.4822(-3)	–
0.25	0.9687(-5)	5.619	0.9709(-5)	5.634
0.10	0.1809(-6)	4.344	0.1809(-6)	4.347
0.08	0.7263(-7)	4.089	0.7264(-7)	4.090
0.05	0.1083(-7)	4.050	0.1083(-7)	4.050

The rates in Table 3.2 approach 4 as h decreases and are therefore consistent with the theoretical error estimate. In the computations that led to this table the initial values η_h^0, u_h^0 of the fully discrete scheme were the elliptic projections of η_0, u_0, respectively, i.e. the projections of the functions η_0, u_0 onto S_h with respect to the inner product $A_b(\cdot, \cdot)$. The rates that correspond to taking the cubic spline interpolants of η_0, u_0 also approach 4 as h decreases.

As our primary interest is the study of solitary waves, it is appropriate to compute as functions of h and t a few more error indicators that are pertinent (cf. [14]) to the approximation of solitary waves, such as amplitude, speed, phase and shape errors.

The (normalized) *amplitude error* of, say, η_h^n is defined as the quantity

$$AE_\eta(t^n) = \left| \frac{\eta_0 - \eta_h^n(x^*)}{\eta_0} \right|,$$

where x^*, the point where η_h^n achieves its maximum, is found as follows: A first approximation x_0^* of x^* is furnished by that quadrature node at which η_h^n

is maximized relatively to its values at the quadrature nodes. (We use Gauss numerical quadrature with five nodes per mesh interval for computing the L^2 inner products in the finite element scheme.) Taking x_0^* as starting value, we compute x^* by Newton's method with a few iterations, as the nearby root of the equation $\frac{d}{dx}\eta_h^n(x) = 0$. The first two lines in Table 3.3 show the amplitude errors $AE_\eta(t^n)$, $AE_u(t^n)$ evaluated at $t^n = 0, 20, \ldots, 80$ for $h = 0.1$. (Recall from Table 3.2 that the normalized L^2 errors were of order of magnitude 10^{-7} for that value of h.) The amplitude errors fluctuate, but remain of the order of 10^{-7} in the temporal interval $[0, 100]$. The analogous errors for $h = 0.5$ and $h = 0.05$ (not shown in Table 3.3) are of $O(10^{-4})$ and $O(10^{-8})$, respectively, like the corresponding L^2 errors.

Table 3.3: Normalized amplitude errors (AE), speed errors (CE), shape errors (SE), and phase errors (PE) for η and u, and normalized error of the invariant E (IE) for solitary waves, $\theta^2 = 9/11$, $h = 0.1$, $k = 0.01$.

	$t^n = 0$	$t^n = 20$	$t^n = 40$	$t^n = 60$	$t^n = 80$
AE_η	0.3840(-6)	0.1741(-6)	0.2170(-6)	0.3732(-6)	0.1273(-6)
AE_u	0.3840(-6)	0.1755(-6)	0.2184(-6)	0.3719(-6)	0.1286(-6)
CE_η	–	0.2883(-5)	0.1082(-5)	0.2007(-5)	0.2845(-5)
CE_u	–	0.2883(-5)	0.1082(-5)	0.2007(-5)	0.2845(-5)
SE_η	0.1711(-6)	0.1712(-6)	0.1712(-6)	0.1712(-6)	0.1712(-6)
SE_u	0.1711(-6)	0.1712(-6)	0.1712(-6)	0.1712(-6)	0.1712(-6)
PE_η	–	-0.1418(-7)	-0.2670(-7)	-0.4026(-7)	-0.5489(-7)
PE_u	–	-0.1418(-7)	-0.2670(-7)	-0.4026(-7)	-0.5489(-7)
IE	0.5069(-10)	0.4062(-9)	0.7618(-9)	0.1117(-8)	0.1473(-8)

The normalized *speed error* in the approximation of, say, the η solitary wave is defined at $t = t^n$ as

$$CE_\eta(t^n) = \left| \frac{c_s - c_\eta(t^n)}{c_s} \right|,$$

where c_s is the speed of the exact solitary wave and $c_\eta(t^n)$ is computed as the quotient $\frac{x^*(t^n) - x^*(t^n - 10)}{10}$, where $x^*(t^n)$ is the position of the peak of η_h^n, found in the course of computing the amplitude error. The temporal interval $\Delta t = 10$ (an integer multiple of k in our computation) in this quotient proved sufficiently large for the purpose of smoothing away oscillations in the discrete approximation of the speed. In Table 3.3 we show these speed errors of the approximations of η and u as functions of t^n in the case $h = 0.1$; they fluctuate a bit but stay within $O(10^{-6})$ throughout the computation. The analogous orders of magnitude of the maximum speed errors for $h = 0.5$ and $h = 0.05$

were 10^{-4} and 10^{-7} respectively; accordingly, the discrete speeds retained about 4, 6 and 7 correct constant digits in the temporal interval $[0, 100]$ for $h = 0.5$, 0.1, and 0.05, respectively.

We turn now to the measurement of an L^2 based, normalized *shape error* of the discrete solitary waves. This is defined at each t^n for, say, the η solitary wave as

$$SE_\eta(t^n) = \inf_\tau \|\eta_h^n - \eta_s(\cdot, \tau)\| / \|\eta_s(\cdot, 0)\|.$$

To this end we compute τ^* as the point near t^n where $\frac{d}{d\tau}\xi^2(\tau^*) = 0$, with $\xi(\tau) := \|\eta_h^n - \eta_s(\cdot, \tau)\| / \|\eta_s(\cdot, 0)\|$, using Newton's method with a few iterations and $\tau^0 = t^n - k$ as initial guess. We then set $SE_\eta(t^n) = \xi(\tau^*)$; the associated *phase error* is defined as $PE_\eta(t^n) = \tau^* - t^n$. Both of these errors are shown in Table 3.3 for the computation with $h = 0.1$. The shape errors stay practically constant throughout the computation and are of $O(10^{-7})$. (For $h = 0.5$ and $h = 0.05$ the analogous errors, also practically constant in t, were of $O(10^{-4})$ and $O(10^{-8})$, respectively.) The phase errors were negative and of $O(10^{-8})$, with magnitudes increasing linearly with t. (Their corresponding orders of magnitude were 10^{-4} and 10^{-9} for $h = 0.5$ and $h = 0.05$, respectively.) The last line of Table 3.3 shows the normalized error IE in the Hamiltonian E defined by $E = \int_{-L}^{I_r} [\eta^2 + |c|\eta_x^2 + (1 + \eta)u^2]dx$. This is defined as $|E - E_h(t^n)|/|E|$, where $E_h(t^n)$ is the numerical value of the invariant for $\eta = \eta_h^n$, $u = u_h^n$. The table shows the values of IE for $h = 0.1$. The orders of magnitude of IE were 10^{-6} and 10^{-11} for $h = 0.5$ and $h = 0.05$, respectively.

3.6   Numerical Experiments

In this section, we describe some numerical experiments that we performed for periodic initial-value problems for several Boussinesq systems with the fully discrete Galerkin scheme previously described; we used cubic splines on a uniform grid for the spatial discretization and the explicit Runge-Kutta scheme (3.56) for time-stepping, except otherwise noted.

With such an evolution code in hand, one may *construct* solitary waves, when no analytical formula is available, or *isolate* solitary waves produced from the resolution of general initial profiles, in order to study their properties. This may be accomplished, for example, by performing a long-time simulation on a large spatial interval and by isolating a solitary wave, cutting it away after it has distanced itself sufficiently from the rest of the solution. Alternatively, one may 'clean' the approximate solitary wave by an iterative process, as done, e.g., in [10]. In [28] one may find a detailed description and quantitative comparisons of both procedures. Here, we will confine ourselves to describing the cleaning procedure in one typical case.

We consider the Bona-Smith system with $\theta^2 = 0.8$ posed with periodic boundary conditions on the interval $[-150, 150]$. As initial conditions we used the functions

$$\eta_0(x) = A\text{sech}^2(\sqrt{\frac{3A}{c_s}}(x + 100)), \quad u_0(x) = \eta_0(x) - \frac{1}{4}\eta_0^2(x), \qquad (3.57)$$

with $A = 1.2$, $c_s = 1 + \frac{1}{2}A$. (These are special initial conditions designed to produce an approximate solitary wave that propagates mainly in one direction [1]. The function η_0 is represented by the dotted line graph in Figure 3.6.) We integrate the system with the evolution code with $h = 0.1$ and $k = 0.01$. One main pulse is produced (of amplitude about 1.00121) and travels to the right with a speed equal to about 1.4424. It is followed by a small amplitude dispersive tail. Another dispersive tail, of larger amplitude, forms a wavetrain that travels to the left. At $t = 100$, cf. Figure 3.6 (which shows the η component of the solution), the main pulse has distanced itself from its trailing

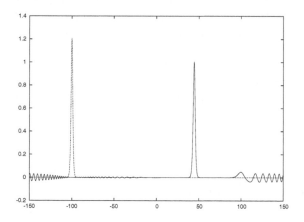

FIGURE 3.6: Evolution of $\eta(x, t)$ from initial conditions (3.57), $t = 100$; Bona-Smith system, $\theta^2 = 0.8$.

tail and has not yet interacted with the left-travelling dispersive tail that has wrapped itself around the boundary due to periodicity. At this time we cut the main pulse off the rest of the solution by setting η_h and u_h equal to zero in the intervals $[-150, 28.6]$ and $[64, 150]$, center it again at $x = -100$ by translation and, using it as new initial condition, integrate again numerically up to $t = 100$, cf. Figure 3.7(a). During this first iteration the amplitudes of the $\eta-$ and $u-$ components of the wave maintained 6 constant digits and were equal to $A_\eta = 1.00121$, $A_u = 0.86271$, respectively while the speed of the wave (computed as explained in Section 3.5) was equal to $c_s = 1.4424$

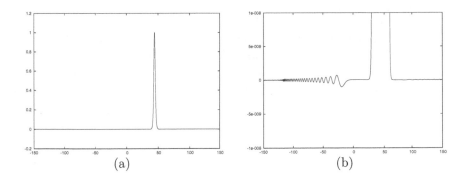

FIGURE 3.7: η-wave, first iteration, $t = 100$. Bona-Smith system $\theta^2 = 0.8$.

maintaining 5 digits. From Figure 3.7 (b) (a magnification of (a)) we can see that the dispersive tail that this pulse sheds has an amplitude of $O(10^{-9})$. If we repeat this procedure, cutting at $t = 100$ the new pulse off the rest of the solution and using it again as initial condition centered at $x = 100$, we obtain, at $t = 100$ (cf. Figure 3.8), a new pulse that maintains amplitudes $A_\eta = 1.00121$, $A_u = 0.86271$, and speed $c_s = 1.4424$. The oscillations behind this wave are of $O(10^{-11})$. Hence, this procedure produces a practically 'clean' solitary wave in two iterations.

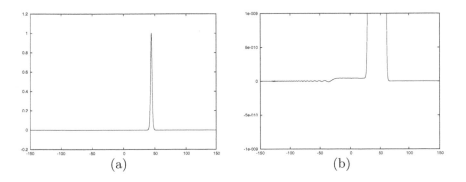

FIGURE 3.8: η-wave, second iteration, $t = 100$. Bona-Smith system $\theta^2 = 0.8$.

It should be pointed out that more iterations are needed to 'clean' smaller solitary waves, cf. [28]. Cleaning works reasonably well for the BBM-BBM system, cf. [10], and for the classical Boussinesq system, cf. [5]. However, for the KdV-KdV system (3.38), which, as we saw in Section 3.4, does not possess classical, but only generalized solitary waves, i.e., orbits homoclinic to small amplitude oscillatory solutions, 'cleaning' gives of course a picture like the one shown in Figure 3.9. This figure was produced, cf. [15], after seven iterations, starting from an initial profile with $\eta_0(x) = 0.3e^{-(x+100)^2/25}$ and $u_0(x) = 0$ on $[-150, 150]$, with the cubic spline-Gauss-Legendre implicit Runge-Kutta scheme.

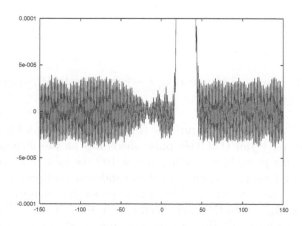

FIGURE 3.9: Evolution after 7 iterations, $t = 120$, KdV-KdV system.

As it was mentioned in Section 3.2, the Boussinesq systems model two-way surface wave propagation. This property, as well as the property of resolution into solitary waves, is illustrated in the next two Figures 3.10 and 3.11, which show the evolution that ensues from the initial condition

$$\eta_0(x) = e^{-x^2/15}, \quad u_0(x) = 0, \quad x \in [-200, 200]. \tag{3.58}$$

(The initial free surface elevation η_0 is shown with a dotted line in the graphs. All numerical computation were made with the spline-explicit Runge-Kutta scheme with $h = 0.1$, $k = 0.01$.) In Figure 3.10 we show the evolution of this initial heap of fluid under the Bona-Smith ($\theta^2 = 1$) system; Figure 3.10 (a), respectively (b), shows the profiles of $\eta(x,t)$, respectively $u(x,t)$, that have developed at $t = 140$. Two symmetric wave trains are produced, travelling in opposite directions. By $t = 140$ these wavetrains consist of apparently of two solitary waves (the larger of which has an approximate amplitude A_η and speed c_s equal, respectively, to about 0.60305 and 1.2864) followed by a dispersive tail. The same initial conditions with the same code and discretization parameters were used to integrate forward the BBM-BBM system (η profile at $t = 140$ in Figure 3.11(a)) and the classical Boussinesq

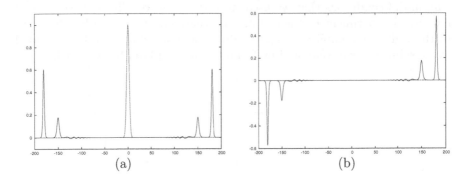

FIGURE 3.10: Evolution of initial condition (3.58), Bona-Smith, ($\theta^2 = 1$),
(a) $\eta(x, t)$ (b) $u(x, t)$, $t = 140$.

system (η profile at $t = 140$ in Figure 3.11(b)).

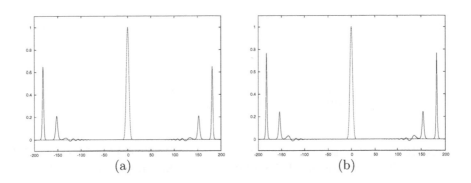

FIGURE 3.11: Elevation $\eta(x, t)$, produced by initial conditions (3.58) at $t = 140$. (a) BBM-BBM system, (b) classical Boussinesq system.

The picture of the evolution is similar. In the BBM-BBM case the leading solitary wave has an amplitude A_η of about 0.64639 and a speed c_s approximately equal to 1.29159. The analogous magnitudes for the classical Boussinesq system are $A_\eta = 0.76690$, $c_s = 1.2953$. Of course, the shapes of the solitary waves and the magnitude and form of the dispersive tails differ.

In the next group of Figures 3.12–3.14 we show the interaction of two colliding, unequal η- solitary waves of the Bona-Smith system with $\theta^2 = 0.8$.

The initial conditions that we used were the cleaned solitary wave from the numerical experiment of Figures 3.6–3.8 centered at $x = -100$ (this moves to the right with amplitude $A_{\eta_1} = 1.00121$ and speed $c_{s_1} = 1.4424$), and the exact solitary wave computed by formulas (3.50)–(3.51) for $\theta^2 = 0.8$.

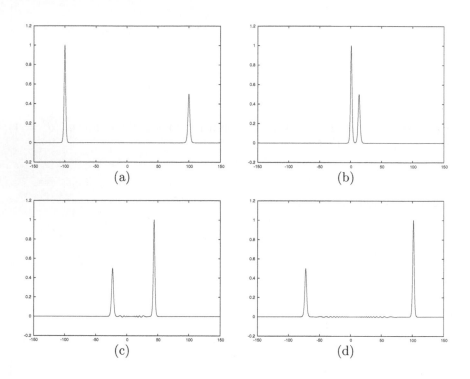

FIGURE 3.12: Collision of two solitary waves (a) $t = 0$, (b) $t = 70$, (c) $t = 100$, (d) $t = 140$, η-component of the solution, Bona-Smith system, $\theta^2 = 0.8$.

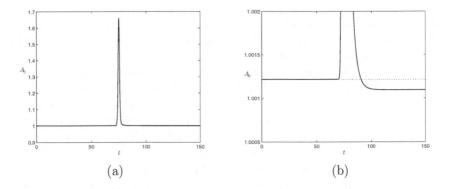

FIGURE 3.13: Total amplitude of the η-component of the solution as a function of t. (Interaction of Figure 3.12.) (b) is a magnification of (a).

This wave, initially centered at $x = 100$, was given a negative initial velocity, so it moves to the left with $A_{\eta_2} = 0.5$ and speed $c_{s_2} = 1.2345$. Figure 3.12(a)–(d) shows the η-component of the ensuing evolution. (Integration with $h = 0.1$, $k = 0.01$). The two waves start colliding at about $t = 69.54$, they interact and emerge basically unchanged.

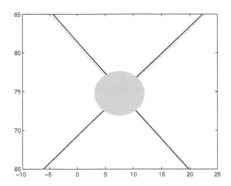

FIGURE 3.14: Paths of solitary waves in x, t-plane. (Interaction of Figure 3.12.)

The collision is inelastic and nonlinear and small-amplitude dispersive tails are produced after the interaction. The total amplitude of the η-component of the solution is shown in Figure 3.13(a). The maximum amplitude during the interaction is equal to 1.65931, thus exceeding the sum of the amplitudes of the two waves by 10.53%. Note (Fig. 3.13(b)) that the total amplitude just after the interaction dips below and returns, as t grows, back to its previous value. In Figure 3.14 we plot in the x, t-plane the positions of the centers of the two solitary waves as functions of t before and after the collision. The solid lines represent the true paths while the dotted lines represent the paths were there no interaction. Both solitary waves are slightly delayed after the interaction, suffering small phase shifts. Of course, interactions of this type are well-understood (for solitary waves travelling to one direction, with the larger (and faster) wave overtaking the smaller one) in the case of the integrable KdV equation, where the phenomenon may be studied by the inverse scattering transform. For nonintegrable equations and systems (such as the Bona-Smith model) this study can be performed only numerically for the time being.

Figures 3.15–3.16 depict evolutions that are produced by the Bona-Smith system with $\theta^2 = 0.8$ when initial solitary wave profiles are perturbed. They are meant to illustrate asymptotic stability properties of solitary waves. An

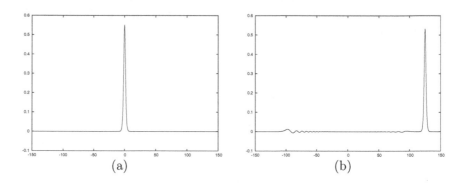

(a)   (b)

FIGURE 3.15: Evolution of a perturbed solitary wave, $r = 1.1$, Bona-Smith system, $\theta^2 = 0.8$.

exact solitary wave (η_s, u_s) is computed by (3.50)–(3.51) for $\theta^2 = 0.8$. This has an amplitude $A_\eta = 0.5$ and speed $c_s = 1.2345$. The initial condition that we take is

$$\eta_0(x) = r\eta_s(x, 0), \quad u_0(x) = u_s(x, 0). \tag{3.59}$$

In Figure 3.15 we show the evolution of the η-component of the system when the perturbation factor r is equal to 1.1. After some initial phase of adjustment, one solitary wave is produced that travels to the right with an amplitude of $A_\eta = 0.531200$ and speed $c_s = 1.24809$, followed by a dispersive tail. Another dispersive tail travels to the left. As the perturbation factor r increases (cf. Fig. 3.16 for the same problem with $r = 1.8$) more solitary waves appear. A detailed numerical study of the stability of solitary waves for the

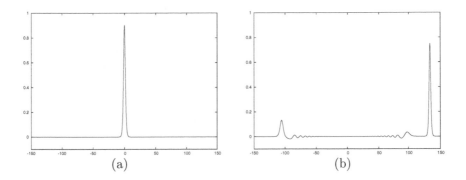

FIGURE 3.16: Evolution of a perturbed solitary wave, $r = 1.8$, Bona-Smith system, $\theta^2 = 0.8$.

Bona-Smith system is made in [28], where a variety of initial perturbations and phenomena of evolution are studied. The conclusion of [28] is that the solitary waves of the Bona-Smith systems appear to be asymptotically stable.

When, however, solitary waves are subjected to larger, special perturbations, the solution may become singular in finite time. Consider, for example, the evolution depicted in Figures 3.17 and 3.18. They show the η-profiles that evolve, under the Bona-Smith system with $\theta^2 = 9/11$, from initial conditions of the type

$$\eta_0(x) = \eta_0\text{sech}^2(\lambda(x + 100))(1 + \mu\tanh(0.5(x + 100))),$$
$$u_0(x) = B\eta_0\text{sech}^2(\lambda(x + 100)), \tag{3.60}$$

on $[-150, 150]$, where η_0, B, λ are given by the formulas (3.51). In Figure 3.17 the system was integrated with $h = 0.1$, $k = 0.01$ and $\mu = 6.2$. Note that η_0 dips below the bottom level of -1, i.e., the conditions for well-posedness of the Bona-Smith system are violated for this 'nonphysical' profile. However, despite the large-amplitude dispersive tail that is produced in the early stages of the evolution, the oscillations have an amplitude less than one and the solution of the problem settles down and apparently continues to evolve bounded and smooth. When the pertubation factor μ is slightly increased to 6.28, the oscillations exceed and stay below -1 in their negative excursion and the solution apparently blows up in finite time. (Of course one cannot study the details of blow-up with a code that uses constant h and k: For

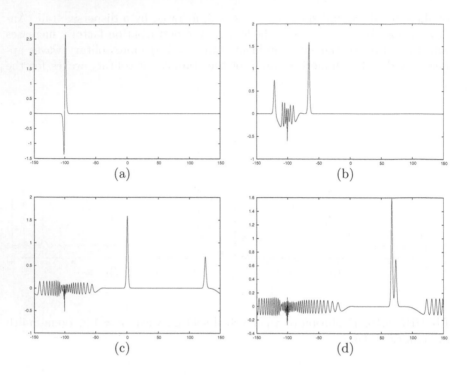

FIGURE 3.17: Evolution of the η profile, Bona-Smith, $\theta^2 = 9/11$. Initial conditions (3.60), $\mu = 6.2$. (a) $t = 0$, (b) $t = 20$, (c) $t = 60$, (d) $t = 100$.

this purpose adaptive refinement of the mesh parameters is needed, cf. [14]. However, when we repeated the experiment of Figure 3.18 with smaller h and k we obtained the same picture.)

In [28] we study numerically some details of this transition to blow-up. The initial value of η_0 need not have a negative excursion below -1 for the blow-up to occur; for example, one can design an experiment where a positive heap η_0 of water is given a large positive initial velocity u_0. This push empties the channel a little distance behind the wave, and blow-up ensues. (Of course if $\eta(x,t)$ remains greater that -1 in a given temporal interval, then there can be no blow-up since the boundedness of the invariant (Hamiltonian) E in this case implies the boundedness of the H^1 norm of η and, consequently, the boundedness of its L^∞ norm by Sobolev's theorem.) The key to understanding the blow-up mechanism therefore in the case of the Bona-Smith systems, seems to be the growth of the negative term $\int_{-\infty}^{\infty}(1+\eta)u^2dx$ relative to the quantity $\int_{-\infty}^{\infty}(\eta^2 + (\theta^2 - 2/3)\eta_x^2)dx$.

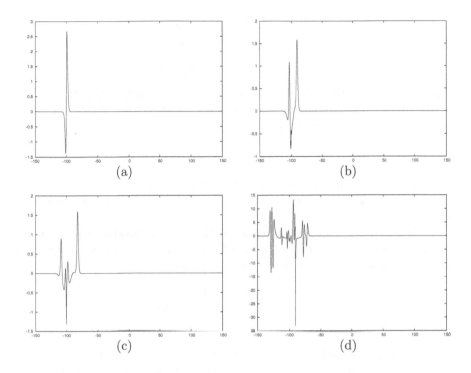

FIGURE 3.18: Evolution of the η profile, Bona-Smith, $\theta^2 = 9/11$. Initial conditions (3.60), $\mu = 6.28$. (a) $t = 0$, (b) $t = 5$, (c) $t = 10$, (d) $t = 17$.

3.7 Boussinesq Systems in Two Space Dimensions

In this section we give a brief sketch of some aspects of the theory and numerical analysis of the analogous to (3.1) Boussinesq systems in two space dimensions.

The 3-D Euler equations for inviscid, incompressible, irrotational flow with a free surface over a horizontal bottom at height $z = -h_0$, are given, in dimensional, unscaled variables in the following form: Let $\eta = \eta(x, y, t)$ be the deviation of the free surface of the fluid above its level of rest $z = 0$ and let $\Omega_t = \{(x, y, z) : (x, y) \in \mathbb{R}^2, -h_0 \leq z \leq \eta(x, y, t)\}$. Then, for $(x, y, z) \in \Omega_t$ and $t \geq 0$ we have

$$u_t + uu_x + vu_y + wu_z + \frac{1}{\rho}p_x = 0,$$

$$v_t + uv_x + vv_y + wv_z + \frac{1}{\rho}p_y = 0,$$

$$w_t + uw_x + vw_y + ww_z + \frac{1}{\rho}p_z = -g,$$

$$u_x + v_y + w_z = 0,$$

$$w_y = v_z, \quad w_x = u_z, \quad v_x = u_y,$$

where (u, v, w) is the velocity, ρ the (constant) density and p the pressure of the fluid. The system is supplemented by the (free surface) boundary conditions, valid for $z = \eta(x, y, t)$:

$$\eta_t + u\eta_x + v\eta_y = w, \quad p = 0,$$

the bottom boundary condition $w = 0$ at $z = -h_0$, and the initial conditions $\eta = \eta_0$, $u = u_0$, $v = v_0$ for $t = 0$, where η_0, u_0, v_0 are functions defined on $\mathbb{R}^2$.

With a process entirely analogous to the one-dimensional case, one may derive Boussinesq systems in the two-dimensional case. In nondimensional, unscaled variables, these systems are written as

$$\begin{aligned} \eta_t + \nabla \cdot \mathbf{v} + \nabla \cdot \eta\mathbf{v} + a\Delta\nabla \cdot \mathbf{v} - b\Delta\eta_t &= 0, \\ \mathbf{v}_t + \nabla\eta + \tfrac{1}{2}\nabla|\mathbf{v}|^2 + c\Delta\nabla\eta - d\Delta\mathbf{v}_t &= 0, \end{aligned} \tag{3.61}$$

where $\mathbf{x} = (x, y)$ represents the position, t is proportional to elapsed time, $\eta = \eta(\mathbf{x}, t)$ is proportional to the deviation of the free surface from its rest position $z = 0$, while $\mathbf{v}(\mathbf{x}, t) = (v_1(\mathbf{x}, t), v_2(\mathbf{x}, t))$ is proportional to the horizontal velocity of the fluid at some height expressed in terms of a parameter $\theta \in [0, 1]$ above the bottom $z = -1$. In (3.61) $\nabla = (\partial_x, \partial_y)$, $|\mathbf{v}|^2 = v_1^2 + v_2^2$, and the constants a, b, c, d multiplying the third-order dispersive terms are given again by (3.2). (As in the one-dimensional case, in the scaled variables case, when the Stokes number is equal to 1, the small parameter $\varepsilon = A/h_0$ multiplies the nonlinear and dispersive terms in (3.61), while the right-hand side, of $O(\varepsilon^2)$, is approximated by zero.) The system (3.61) is supplemented by initial conditions of the form

$$\eta(\mathbf{x}, 0) = \eta_0(\mathbf{x}), \quad \mathbf{v}(\mathbf{x}, 0) = \mathbf{v}_0(\mathbf{x}), \quad \mathbf{x} \in \mathbb{R}^2, \tag{3.62}$$

when the initial-value problem for (3.61) is being considered.

In [30] it was proved that some of the systems (3.61) give rise to initial-value problems, which are locally well-posed in appropriate Sobolev spaces. In particular, it was proved that for $s > 0$

(i) If $b > 0$, $d > 0$, $a < 0$, $c < 0$ or $b > 0$, $d > 0$, $a = c \geq 0$, (3.61)–(3.62) is locally well-posed in $H^s \times (H^s)^2$. (This means that given $(\eta_0, \mathbf{v}_0) \in H^s \times (H^s)^2$, there exists $T > 0$ and a unique solution $(\eta, \mathbf{v}) \in C(0, T; H^s) \times C(0, T; (H^s)^2)$ of (3.61) with initial conditions (3.62).)

(ii) If $b > 0$, $d > 0$, $a = 0$, $c < 0$, (3.61)–(3.62) is locally well-posed in $H^{s+1} \times (H^s)^2$.

(iii) If $b > 0$, $d > 0$, $a < 0$, $c = 0$, (3.61)–(3.62) is locally well-posed in $H^{s-1} \times (H^s)^2$.

(Here $H^s = H^s(\mathbb{R}^2)$ is the usual Sobolev space of (equivalence classes of) functions on $\mathbb{R}^2$.)

It is worthwhile to note that if $b = d$, the systems (3.61) are Hamiltonian. As invariant (Hamiltonian) we may now take the functional

$$H(\eta, \mathbf{v}) = \frac{1}{2} \int_{\mathbb{R}^2} [-c|\nabla\eta|^2 - a|\nabla\mathbf{v}|^2 + (1+\eta)|\mathbf{v}|^2 + \eta^2]d\mathbf{x},$$

where $|\nabla\mathbf{v}|^2 = v_{1,x}^2 + v_{1,y}^2 + v_{2,x}^2 + v_{2,y}^2$. However, contrary to the one-dimensional case, the conservation of H does not imply a uniform H^1 bound on $(\eta, \mathbf{v})$ for $a < 0$, $c < 0$.

Some systems of the form (3.61) posses *line solitary wave* solutions: Seek travelling wave solutions of (3.61) of the type $\eta(\mathbf{x}, t) = \eta(\xi)$, $\mathbf{v}(\mathbf{x}, t) = \mathbf{v}(\xi)$, where $\xi = \boldsymbol{\alpha} \cdot \mathbf{x} - c_s t - r_0$, $r_0 \in \mathbb{R}$, i.e. solutions that travel with constant speed c_s without change of form along the direction $\boldsymbol{\alpha} = (\alpha_1, \alpha_2)$, $|\boldsymbol{\alpha}| = 1$. If we assume that $\eta(\xi)$, $\mathbf{v}(\xi)$ and their derivatives tend to zero as $|\xi| \to \infty$ and the component of the velocity $\alpha_2 v_1 - \alpha_1 v_2$ perpendicular to $\boldsymbol{\alpha}$ vanishes, then substitution into (3.61) and integration with respect to ξ yield for the velocity component $u = u(\xi) = \boldsymbol{\alpha} \cdot \mathbf{v}(\xi)$, and $\eta = \eta(\xi)$ precisely the equations (3.49) of the one-dimensional solitary waves; we may therefore study existence of line solitary waves with the methods of Section 3.4.

In addition to the well-posedness of the initial-value problem (3.61)–(3.62), one may prove well-posedness for some initial-boundary value problems on finite domains for some systems. In this direction, a system of the form (3.61) with $a = c = 0$ and $b > 0$, $d > 0$, i.e., a BBM-BBM type system, was considered in [31]. The system was posed on a bounded open set $\Omega \in \mathbb{R}^2$ with smooth boundary $\partial\Omega$. At $t = 0$ initial conditions of η and $\mathbf{v}$ were given on Ω. Then, it was proved in [31] that the initial-boundary-value problem for the system with zero Dirichlet data for η and $\mathbf{v}$ on $\partial\Omega$ is locally well-posed in $(H_0^1)^3$. (Local well-posedness in $H^1 \times (H_0^1)^2$ also holds for the initial-boundary-value problem with boundary conditions $\frac{\partial\eta}{\partial n} = 0$ and $\mathbf{v} = 0$ on $\partial\Omega$, where $\frac{\partial}{\partial n}$ is the normal derivative at the boundary.)

All results on the dependence of T_ε on ε, where T_ε is the maximum time of local existence of solutions of (3.61)–(3.62) in its scaled form, that were quoted at the end of Section 3.3 hold as well for the two-dimensional analog of the system. In the case however of the initial-boundary-value problems considered in [31], one can only prove so far that T_ε may be taken independent of ε.

The fully discrete Galerkin-finite element methods considered in Section 3.5 in the one-dimensional case may be readily extended in 2-D for the numerical solution of initial-boundary-value problems for BBM-BBM type systems on bounded plane domains. For example, if Ω is a convex plane domain, and one triangulates it by a regular, quasiuniform partition with triangles of maximum size h, and considers, e.g., the space S_h of piecewise linear, continuous functions defined with respect to this triangulation and vanishing on the boundary, one may descretize the initial-boundary value problem for (3.61) when $a = c = 0$, $b, d > 0$ by the standard Galerkin method to obtain the semidiscrete system

$$\mathcal{A}_b(\eta_{ht}, \phi) + (u_{hx}, \phi) + (v_{hy}, \phi) + ((\eta_h u_h)_x, \phi) + ((\eta_h v_h)_y, \phi) = 0, \quad \forall \phi \in S_h,$$

$$\mathcal{A}_d(u_{ht}, \chi) + (\eta_{hx}, \chi) + (u_h u_{hx}, \chi) + (v_h v_{hx}, \chi) = 0, \quad \forall \chi \in S_h,$$

$$\mathcal{A}_d(v_{ht}, \psi) + (\eta_{hy}, \psi) + (u_h u_{hy}, \psi) + (v_h v_{hy}, \psi) = 0, \quad \forall \psi \in S_h,$$

where $\mathcal{A}_\lambda(\phi, \chi) = (\phi, \chi) + \lambda(\nabla\phi, \nabla\chi)$. (The approximation to the velocity $\mathbf{v}$ is denoted by (u_h, v_h)). The above system of o.d.e.'s is integrated in time, starting from initial conditions taken, e.g., as the elliptic projections (i.e., projections with respect to the bilinear form $\mathcal{A}_\lambda$) of η_0, $\mathbf{v}_0$ on S_h, by an explicit, second-order Runge-Kutta method with uniform time step k. It is not hard to prove, cf. [30], [40], that the resulting fully discrete scheme is unconditionally stable and has L^2 error bounds of $O(k^2 + h^2)$ for η_h, u_h and v_h. The large sparse linear systems that the finite element method gives at every time step are solved by the conjugate gradient method with various preconditioners.

In the case of systems of the Bona-Smith type (i.e., with $b > 0$, $d > 0$, $a = 0$, $c < 0$ in (3.61)) one either needs higher order elements of higher continuity (e.g. bicubic splines or tensor products of C^1 quadratic splines on rectangles) or to define an intermediate projection of $\Delta\eta$ in order to handle the $\Delta\nabla\eta$ term. Some error estimates are also available in this case; we refer the reader to [30] and [40].

We now give two examples of finite element solutions of the BBM-BBM system ($a = c = 0$, $b = d = 1/6$ in (3.61)) on bounded polygonal domains using continuous P_1 elements. First, we consider this system with radial initial conditions

$$\eta_0(x, y) = 3e^{-(x^2+y^2)/5}, \quad \mathbf{v}_0(x, y) = 0 \tag{3.63}$$

on the square $[-40, 40] \times [-40, 40]$, with boundary conditions $\partial\eta/\partial n = 0$ on $\partial\Omega$ and $u = v = 0$ for $x = -40$ and $x = 40$, and $\partial u/\partial n = \partial v/\partial n = 0$ on $y = -40$ and $y = 40$. We integrate the system in time using the second-order accurate explicit Runge-Kutta scheme known as 'improved Euler method'. Figure 3.19 shows the expanding η-wave that emerges and its reflection from the boundaries that are parallel to the y-axis. In this example we used a uniform triangulation with 84992 triangles, and $k = 0.1$.

In the next example we used as initial profile the line wave

$$\eta_0(x, y) = A\mathrm{sech}^2\left(\tfrac{1}{2}\sqrt{\tfrac{3A}{c_s}}(y + 10)\right),$$
$$u_0(x, y) = 0, \quad v_0(x, y) = \eta_0(x, y) - \tfrac{1}{4}\eta_0^2(x, y), \tag{3.64}$$

on the rectangle $[-15, 15] \times [-30, 50]$. We consider the complex domain shown in Figure 3.20 and assume reflection boundary conditions along the perimeter

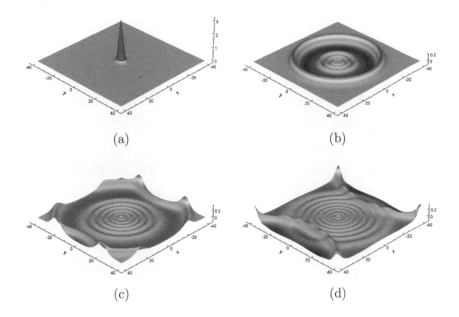

(a) (b)

(c) (d)

FIGURE 3.19: Expanding radial wave. BBM-BBM system, initial conditions (3.63). (a) $t = 0$, (b) $t = 25$, (c) $t = 40$, (d) $t = 50$.

of the white 'wave-breaker' structures for $15 \leq y \leq 20$ that form the entrance to the 'harbor' $[-15, 15] \times [20, 50]$. Outside the port Neumann boundary conditions hold for η and $\mathbf{v}$, while inside we assume reflective boundary conditions. The wave travels towards the positive y direction resembling a line solitary wave, impinges on the entrance of the harbor, is diffracted inside and partly reflected back. Figure 3.21 shows one dimensional η-graphs of the solution at the time instances of Figure 3.20 along the axis of symmetry $x = 0$, $-30 \leq y \leq 50$ of the domain. For this simulation we used a uniform triangulation with 143360 triangles and $k = 0.1$.

Due to space limitations we have not mentioned in this article Boussinesq systems for surface wave propagation above a variable bottom, a problem of central importance in coastal hydrodynamics. The engineering and mathematical literature on this topic is quite large and growing. We refer the reader e.g. to the papers [41], [49], [26] for modelling, computations and further references on this important topic.

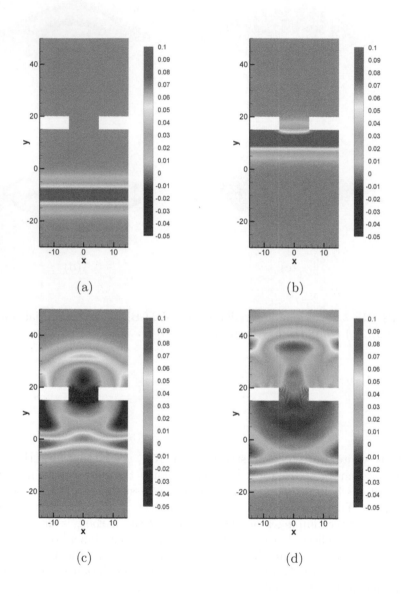

FIGURE 3.20: Line wave impinging on a 'harbor.' BBM-BBM system, initial conditions (3.64). (a) $t = 0$, (b) $t = 20$, (c) $t = 40$, (d) $t = 50$.

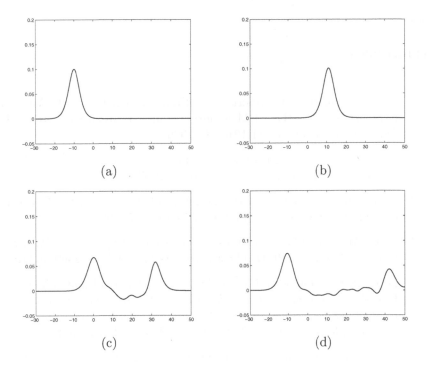

(a)

(b)

(c)

(d)

FIGURE 3.21: Cross-section along $x = 0$ of figure 3.20. BBM-BBM system, initial conditions (3.64). (a) $t = 0$, (b) $t = 20$, (c) $t = 40$, (d) $t = 50$.

Acknowledgment

This work was supported by a "Pythagoras" EPEAEK II grant to the Department of Mathematics, University of Athens, which was co-funded by the E.U. European Social Fund and national resources.

The authors would like to record their sincere thanks to Dr. D. Antonopoulos and Professors J. L. Bona, M. Chen, A. Duran, M.-A. Lopez Marcos, B. Pelloni and J.-C. Saut for many discussions and suggestions on the theory and numerical analysis of the Boussinesq systems over the years.

References

[1] A. A. Alazman, J. P. Albert, J. L. Bona, M. Chen, and J. Wu. Comparisons between the BBM equation and a Boussinesq system. *Adv. Differential Equations*, 11:121-166, 2006.

[2] B. Alvarez-Samaniengo and D. Lannes. Large time existence for 3D water-waves and asymptotics. (preprint 2007).

[3] C. J. Amick. Regularity and uniqueness of solutions to the Boussinesq system of equations. *J. Diff. Eqns.*, 54:231-247, 1984.

[4] D. C. Antonopoulos. *The Boussinesq system of equations: Theory and numerical analysis*. Ph.D. Thesis, *University of Athens*, 2000 (in Greek).

[5] D. C. Antonopoulos and V. A. Dougalis. Numerical approximation of Boussinesq systems. in *Proceedings of the 5th International Conference on Mathematical and Numerical Aspects of Wave Propagation*, ed. by A. Bermudez et al., p. 265-269. SIAM, Philadelphia 2000, pp. 265–269.

[6] D. C. Antonopoulos, V. A. Dougalis, and D. E. Mitsotakis. Theory and numerical analysis of the Bona-Smith type systems of Boussinesq equations (to appear).

[7] T. B. Benjamin. The stability of solitary waves. *Proc. R. Soc. London*, 328:153-183, 1972.

[8] T. B. Benjamin, J. L. Bona, and J. J. Mahony. Model equations for long waves in nonlinear dispersive systems. *Philos. Trans. Roy. Soc. London Ser. A*, 272:47-78, 1972.

[9] J. L. Bona. On the stability theory of solitary waves. *Proc. Roy. Soc. London*, 344:363-374, 1975.

[10] J. L. Bona and M. Chen. A Boussinesq system for two-way propagation of nonlinear dispersive waves. *Physica D*, 116:191-224, 1998.

[11] J. L. Bona, M. Chen, and J.-C. Saut. Boussinesq equations and other systems for small-amplitude long waves in nonlinear dispersive media: I. Derivation and Linear Theory. *J. Nonlinear Sci.*, 12:283-318, 2002.

[12] J. L. Bona, M. Chen, and J.-C. Saut. Boussinesq equations and other systems for small-amplitude long waves in nonlinear dispersive media: II. The nonlinear theory. *Nonlinearity*, 17:925-952, 2004.

[13] J. L. Bona, T. Colin, and D. Lannes. Long wave approximations for water waves. *Arch. Rational Mech. Anal.*, 178:373-410, 2005.

[14] J. L. Bona, V. A. Dougalis, O. A. Karakashian, and W. R. McKinney. Conservative, high-order numerical schemes for the generalized Korteweg-de Vries equation. *Philos. Trans. Royal Soc. London* A, 351:107-164, 1995.

[15] J. L. Bona, V. A. Dougalis, and D. E. Mitsotakis. Numerical solution of KdV-KdV systems of Boussinesq equations I: The numerical scheme and generalized solitary waves. *Math. Comput. Simulation*, 74:214-228, 2007.

[16] J. L. Bona and R. L. Sachs. The existence of internal solitary waves in a two-fluid system near the KdV limit. *Geophys. Astrophys. Fluid Dynamics*, 48:25-51, 1989.

[17] J. L. Bona and R. Smith. A model for the two-way propagation of water waves in a channel. *Math. Proc. Camb. Phil. Soc.*, 79:167-182, 1976.

[18] J. L. Bona, P. E. Souganidis, and W. A. Strauss. Stability and instability of solitary waves of KdV type. *Proc. Roy. Soc. London* A, 411:395-412, 1987.

[19] J. V. Boussinesq. Théorie des ondes et des remous qui se propagent le long d' un canal rectangulaire horizontal, en communiquant au liquide contenu dans ce canal des vitesses sensiblement pareilles de la surface au fond. *J. Math. Pures. Appl.*, 17:55-108, 1872.

[20] J. V. Boussinesq. Essai sur la théorie des eaux courants. *Mém. prés. div. sav. Acad. des Sci. Inst. Fr. (sér. 2)*, 23:1-680, 1877.

[21] A. R. Champneys. Homoclinic orbits in reversible systems and their applications in mechanics, fluids and optics. *Physica D*, 112:158-186, 1998.

[22] A. R. Champneys. Homoclinic orbits in reversible systems II: Multibumps and saddle-centres. *CWI Quarterly*, 12:185-212, 1999.

[23] A. R. Champneys and M. D. Groves. A global investigation of solitary wave solutions to a two-parameter model for water waves, *J. Fluid Mech.*, 342:199-229, 1997.

[24] M. Chen. Solitary-wave and multi pulsed traveling-wave solutions of Boussinesq systems. *Applic. Analysis*, 75:213-240, 2000.

[25] M. Chen. Exact traveling-wave solutions to bi-directional wave equations. *Int. J. Theor. Phys.*, 37:1547-1567, 1998.

[26] M. Chen. Equations for bi-directional waves over an uneven bottom. *Math. Comput. Simulation*, 62:3-9, 2003.

[27] W. Craig. An existence theory for water waves and the Boussinesq and Korteweg-de Vries scaling limits. *Commun. Partial Diff. Equations*, 10:787-1003, 1985.

[28] V. A. Dougalis, A. Duran, M. A. Lopez-Marcos, and D. E. Mitsotakis. A numerical study of the stability of solitary waves of the Bona-Smith system (to appear in *J. Nonlinear Sci.*).

[29] V. A. Dougalis and D. E. Mitsotakis. Solitary waves of the Bona-Smith system, in *Advances in scattering theory and biomedical engineering.* ed. by D. Fotiadis and C. Massalas, World Scientific, p. 286-294, New Jersey, 2004.

[30] V. A. Dougalis, D. E. Mitsotakis, and J.-C. Saut. On some Boussinesq systems in two space dimensions: Theory and numerical analysis (to appear in *Math. Model. Num. Anal.*).

[31] V. A. Dougalis, D. E. Mitsotakis, and J.-C. Saut. On initial-boundary value problems for a Boussinesq system of BBM-BBM type in a plane domain (preprint 2007).

[32] T. Dupont. Galerkin methods for first order hyperbolics: an example. *SIAM J. Numer. Anal.*, 10:890-899, 1973.

[33] K. El Dika. Asymptotic stability of solitary waves for the Benjamin-Bona-Mahony equation. *Discrete Contin. Dyn. Syst.*, 13:583-622, 2005.

[34] M. Grillakis, J. Shatah, and W. A. Strauss. Stability of solitary waves in the presence of symmetry I. *J. Funct. Anal.*, 74:170-197, 1987.

[35] G. Iooss and M. Adelmeyer. *Topics in Bifurcation Theory and Applications.* World Scientific, Singapore 1998.

[36] K. Kirchgässner. Nonlinearly resonant surface waves and homoclinic bifurcation. *Adv. Appl. Mech.*, 26:135-181, 1988.

[37] E. Lombardi. *Oscillatory Integrals and Phenomena Beyond all Algebraic Orders, with Applications to Homoclinic Orbits in Reversible Systems.* Lecture Notes in Mathematics, p. 1741, Springer-Verlag, Berlin, 2000.

[38] Y. Martel and F. Merle. Asymptotic stability of solitons for subcritical generalized KdV equations. *Arch. Rat. Mech. Anal.*, 157:219-254, 2001.

[39] J. R. Miller and M. I. Weinstein. Asymptotic stability of solitary waves for the Regularized Long-Wave equation, *Comm. Pure Appl. Math.*, 49:399-441, 1996.

[40] D. E. Mitsotakis. *Theory and numerical analysis of nonlinear dispersive wave equations: Boussinesq systems in one and two dimensions.* Ph.D. Thesis, *University of Athens*, 2007 (in Greek).

[41] O. Nwogu. Alternative form of Boussinesq equations for nearshore wave propagation. *J. Waterway, Port, Coastal, and Ocean Eng.*, 119:618-638, 1993.

[42] R. L. Pego, P. Smereka, and M. I. Weinstein. Oscillatory instability of solitary waves in a continuum model of lattice vibrations. *Nonlinearity*, 8:921-941, 1995.

[43] R. L. Pego and M. I. Weinstein. Asymptotic stability of solitary waves. *Commun. Math. Phys.*, 164:305-349, 1994.

[44] R. L. Pego and M. I. Weinstein. Convective linear stability of solitary waves for Boussinesq equations. *Studies in Appl. Math.*, 99:311-375, 1997.

[45] B. Pelloni. *Spectral methods for the numerical solution of nonlinear dispersive wave equations.* Ph.D. Thesis, Yale University, 1996.

[46] B. Pelloni and V. A. Dougalis. Numerical modelling of two-way propagation of non-linear dispersive waves. *Math. Comput. Simulation*, 55:595-606, 2001.

[47] B. Pelloni and V. A. Dougalis. Numerical solutions of some nonlocal, nonlinear, dispersive wave equations. *J. Nonlinear Sci.*, 10:1-22, 2000.

[48] D. H. Peregrine. Equations for water waves and the approximations behind them, in *Waves on Beaches and Resulting Sediment Transport.* ed. by R. E. Meyer, p. 95-121, Academic Press, New York, 1972.

[49] H. A. Schäffer and P. A. Madsen. Further enhancements of Boussinesq-type equations. *Coast. Eng.*, 26:1-14, 1995.

[50] M. E. Schonbek. Existence of solutions for the Boussinesq system of equations. *J. Diff. Eqns.*, 42:325-352, 1981.

[51] P. C. Schuur, *Asymptotic Analysis of Soliton Problems - an Inverse Scattering Approach.* Lecture Notes in Mathematics, v.1232, Springer-Verlag, Berlin, Heidelberg, 1986.

[52] P. Smereka. A remark on the solitary wave stability for a Boussinesq equation. in *Nonlinear Dispersive Wave Systems*, ed. by L. Debnath, p. 255-263, World Scientific, Singapore, 1992.

[53] J. F. Toland. Solitary wave solutions for a model of the two-way propagation of water waves in a channel. *Math. Proc. Camb. Phil. Soc.*, 90:343-360, 1981.

[54] J. F. Toland. Existence of symmetric homoclinic orbits for systems of Euler-Lagrange equations. *A.M.S. Proceedings of Symposia in Pure Mathematics*, volume 45, Pt. 2, p. 447-459, 1986.

[55] J. F. Toland. Uniqueness and a priori bounds for certain homoclinic orbits of a Boussinesq system modelling solitary water waves. *Commun. Math. Physics*, 94:239-254, 1984.

[56] M. I. Weinstein. Lyapounov stability of ground states of nonlinear dispersive evolution equations. *Commun. Pure Appl. Math.*, 39:51-68, 1986.

[57] M. I. Weinstein, Existence and dynamic stability of solitary-wave solutions of equations arising in long wave propagation. *Commun. Partial Differential Equations*, 12:1133-1173, 1987.

[58] M. I. Weinstein. Asymptotic stability of nonlinear bound states in conservative dispersive systems. *Contemp. Math.*, 200:223-235, 1996.

[59] G. B. Whitham. *Linear and Non-linear Waves*. Wiley, New York, 1974.

[60] R. Winther. A finite element method for a version of the Boussinesq equations. *SIAM J. Numer. Anal.*, 19:561-570, 1982.

Part II

The Helmholtz Equation and Its Paraxial Approximations in Underwater Acoustics

Chapter 4

Finite Element Discretization of the Helmholtz Equation in an Underwater Acoustic Waveguide

D. A. Mitsoudis, Department of Applied Mathematics, University of Crete, 71409 Heraklion, Greece *and* Foundation for Research and Technology-Hellas, Institute of Applied and Computational Mathematics, 71110 Heraklion, Greece, dmits@tem.uoc.gr

N. A. Kampanis, Foundation for Research and Technology-Hellas, Institute of Applied and Computational Mathematics, 71110 Heraklion, Greece, kampanis@iacm.forth.gr

V. A. Dougalis, Department of Mathematics, University of Athens, 15784 Zographou, Greece *and* Foundation for Research and Technology-Hellas, Institute of Applied and Computational Mathematics, 71110 Heraklion, Greece, doug@math.uoa.gr

---

4.1 Introduction

The numerical solution of time-harmonic problems governed by the Helmholtz equation remains a task of central importance and an active area of research for over thirty years. In this chapter we are mainly interested in applications in underwater acoustics, where the Helmholtz equation models the propagation of the sound field in the sea at any angle and taking into account backscattering effects. We consider specifically an axially symmetric waveguide and we use a cylindrical coordinate system, whose vertical axis passes through the source. The waveguide consists of two fluid layers of different acoustic parameters and is bounded in depth by a rigid bottom while it is unbounded in range. The precise boundary-value problem that we consider is stated in the next section.

We are interested in solving the problem numerically with the finite element method. This choice is reasonable when one has to deal with range-dependent features such as complex interface and bottom topographies. Of course it is impossible to discretize an unbounded domain, so the original problem has to be replaced by one posed in a bounded domain. A common approach in underwater acoustics is to assume that the inhomogeneities of the medium and the irregularities of the interface and the bottom are localized, in the sense that near the source and far from the source the speed of sound is independent of r, and the interface and the bottom are horizontal. Then, we may introduce two artificial boundaries at some appropriate values of r, near the source and far from the source, in order to truncate the semi-infinite waveguide. The associated p.d.e. problems, in the near-field and the far-field regions, may now be solved by separation of variables and series representations of the near-field and the far-field solutions may be obtained. Next, the Dirichlet-to-Neumann (DtN) map of the field in the exterior region may be posed as a nonlocal, nonreflecting condition on the outer artificial boundary, and similarly a nonhomogeneous DtN-type condition may be posed as a boundary condition on the inner artificial boundary in order to take into account the effects of the source and of the backscattering from the rest of the waveguide.

For the case of the far-field boundary, this approach was first adopted, in the context of underwater acoustics, by Fix and Marin in [13], where they considered a single-layered axisymmetric waveguide. They pointed out the inappropriateness of the classical Sommerfeld radiation condition for waveguide problems, and they suggested that the infinite sum appearing in the DtN operator may be truncated to include all the terms that correspond to the so-called propagating modes of the problem. Goldstein in [19] analyzed a finite element method coupled with a nonlocal boundary condition for a single-layered problem with constant coefficients in Cartesian coordinates and proved error estimates of optimal rate. Furthermore, he studied the influence that brings on the error the truncation of the series in the nonlocal operator. The term "Dirichlet-to-Neumann" map was introduced by Keller and Givoli in [25], where they constructed nonlocal nonreflecting boundary conditions for the Helmholtz equation for various exterior problems in two and three dimensions. For information on related work we refer the reader to the book of Givoli, [18]. Kampanis and Dougalis in [24] presented a code called FENL, which implemented a standard Galerkin discretization of the Helmholtz equation coupled to a DtN boundary condition on a far-field artificial boundary. For an extension of this method in order to handle a near-field DtN-type boundary condition we refer to [27].

We would like to mention that other traditional ways of representing the effect of exterior infinite domains include the use of absorbing layers and infinite elements; these techniques and earlier finite element methods are reviewed in the survey by Buckingham, [9]. We refer to [20] for a finite element PML approach for time-harmonic acoustic exterior problems. We would also like to note that a more realistic model of sound propagation in an underwater environment should include other features such as attenuation (absorption) in the medium and the replacement of the rigid bottom assumption by a more appropriate one. Hence, a complete formulation of the problem would entail some special treatment of the bottom, e.g., posing another suitable absorbing boundary condition at the bottom.

An important issue in the numerical solution of the Helmholtz equation is that the error in the finite element solution grows with the wavenumber k (which is proportional to the frequency). Therefore in order to achieve a given accuracy level for the desired computation the discretization parameter h and the wavenumber k should be adjusted. Many authors follow a "rule of the thumb" of the form $kh = $ const., taking a constant number of elements (usually 10) per wavelength λ (i.e., $\lambda/h = $ const.). However, it has been found that this rule does not suffice as k increases. For example, Bayliss et al., [5], [6], use piecewise linear finite elements and present numerical results for two-dimensional model problems in Cartesian coordinates, indicating that the error in the L^2 norm grows with k, when kh is kept constant, whereas it remains bounded for $k^3 h^2 = $ const. Aziz et al., [3], proved a rigorous convergence theorem for a one-dimensional Helmholtz problem, under the assumption that $k^2 h$ is sufficiently small. Analytical estimates, see e.g., Ihlenburg and Babuška, [21], show that for a one-dimensional model problem it is sufficient to keep $k^3 h^2$ constant (to control the so-called 'pollution error' term) in order to avoid a growth of the relative L^2-error. The reader may refer to the book of Ihlenburg, [22], for details. This kind of relationship between k and h has been numerically verified for more general problems and non-uniform meshes by Oberai and Pinsky in [28], where a numerical comparison between three finite element methods is performed. The difficulty of achieving accuracy with a reasonable amount of computer time and memory is identified by Zienkiewicz in [35] as a hard and challenging problem. Turkel in [146] mentions some of the difficulties that arise while solving numerically time harmonic equations. For a review and recent advances in finite element methods for general time-harmonic acoustic problems we refer the reader to the survey by Thompson, [32], and the references therein.

Another class of methods that also assume analytic representations of the field near and far from the source but adopt a different approach for the intermediate region, and have been used for a long time in underwater acoustics is normal mode methods. Over the last two decades the classical normal mode expansion of the acoustic field in a horizontally stratified medium, see, e.g., [8], [23], has been extended in order to provide approximations to the solution of the propagation and scattering problem in range-dependent acoustic waveguides. Evans in [10] constructed a coupled-mode model and an associated code called COUPLE, see [11], which subdivides the waveguide into a finite number of adjacent columns (using the staircase approximation to discretize variable interfaces and bottom) and represents the acoustic field as a normal-mode series in each column. Then, the coefficients are obtained by matching the expansions at the inter-element vertical interfaces; this gives a full coupling between the modes. However, in case of highly irregular interface or bottom topographies one has to assume a large number of horizontal staircase steps for accuracy purposes. For this purpose, other coupled-mode approaches with improved treatment of irregular bottom or interface topographies have been proposed, for example, by Rutherford and Hawker, [29], by Fawcett, [12], by Taroudakis et al., [31], and by Athanassoulis and Belibassakis, [1].

The contents of this chapter are as follows: In Section 4.2 we specify the geometry of the domain, and we formulate the original problem in the semi-infinite waveguide. Next we show that the introduction of two artificial boundaries and the construction of appropriate boundary conditions allow us to define a boundary-value problem, equivalent to the original one, posed in a bounded domain. In Section 4.3 we introduce the suitable function space setting and the finite element method that is used to discretize the boundary-value problem. Then, we briefly outline the code that implements the finite element discretization and the various software packages to which it is coupled. In Section 4.4 we present results of some numerical experiments that we performed with the code. Specifically, we present comparisons with an analytic solution in a simple one-layer model problem in order to check the accuracy of the method, and a numerical investigation of the dependence of the error on the wavenumber k under constraints of the type $k^\alpha h^\beta = \text{const.}$, comparisons with COUPLE in a more realistic environment simulating an underwater trench, and a computation in a domain with complex interface and bottom topographies. For full details on the formulation of the boundary-value problem and the implementation of the finite element method we refer the reader to [24], [2], [27].

4.2 Formulation of the Problem

In this section we present the boundary-value problem that we are going to study in the sequel. We consider a cylindrical waveguide which is axially symmetric and consists of a water layer and, for simplicity, of one fluid sediment layer, see Figure 4.1. The density is assumed to be constant in each layer and is equal to ρ_w in the water and ρ_s ($\rho_s > \rho_w$) in the sediment. A cylindrical coordinate system (r, θ, z) is introduced, with its origin located at the free surface. Following the usual convention, the depth coordinate, z, is considered positive downwards. The acoustic field is generated by a harmonic point source placed in the water column at $r = 0$ at a depth equal to z_s. Let $0 < r^N < r^F$. We assume that the two layers are separated by a piecewise smooth interface, denoted by Γ_{int}, which intersects the lines $r = r^N$ and $r = r^F$ at depths $z = h^N$ and $z = h^F$, respectively. The sediment layer is bounded below by a piecewise smooth bottom curve, denoted by Γ_{bot}, intersecting the lines $r = r^N$ and $r = r^F$ at depths $z = H^N$ and $z = H^F$, respectively. The sea surface at $z = 0$ is assumed to be horizontal. The parts of the interface and bottom lying inside the cylinder $r = r^N$ and outside the cylinder $r = r^F$ are

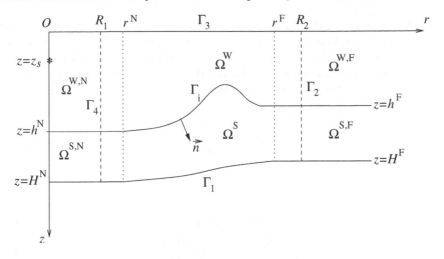

FIGURE 4.1: Geometric configuration and basic notation. The asterisk denotes the point source.

also assumed to be horizontal.

The waveguide Ω consists of three parts:

(i) The near-field bounded subdomain Ω^N, consisting of the layers $\Omega^{W,N} = \{(r, z) : 0 < r < r^N, \, 0 < z < h^N\}$ (water) and $\Omega^{S,N} = \{(r, z) : 0 < r < r^N, \, h^N < z < H^N\}$ (sediment).

(ii) The intermediate bounded subdomain ($r^N \le r \le r^F$) containing the range-dependent part of the bottom and the interface.

(iii) The far-field semi-infinite subdomain Ω^F, consisting of the layers $\Omega^{W,F} = \{(r, z) : r > r^F, \, 0 < z < h^F\}$ (water) and $\Omega^{S,F} = \{(r, z) : r > r^F, \, h^F < z < H^F\}$ (sediment).

We assume that the sound speed is range-independent in Ω^N and Ω^F, i.e. $c = c_N(z)$ and $c = c_F(z)$, respectively, and varies smoothly from its near-field to its far-field value within each layer in the intermediate region.

The formulation of the acoustic propagation and scattering problem in this environment is classical, see, e.g., [23], [8]. In each layer the acoustic field (acoustic pressure) satisfies the Helmholtz equation

$$\Delta p(r, z) + k^2(r, z)\, p(r, z) = -\frac{1}{2\pi} \frac{\delta(r)}{r} \delta(z - z_s), \qquad (4.1)$$

where $\Delta p = p_{rr} + \frac{1}{r} p_r + p_{zz}$ and $k(r, z) = \omega/c(r, z)$ is the wavenumber. In order to solve for p we have to specify appropriate boundary and interface conditions. We may consider the sea surface as a pressure-release boundary, since the density of the air is much smaller than that of the water. At the bottom we pose a Neumann condition simulating a hard bottom. The interface condition requires the pressure and the normal component of the particle velocity to be continuous across the interface. Therefore, the p.d.e. (4.1) is supplemented by the boundary conditions

$$p = 0, \quad \text{for } z = 0 \tag{4.2}$$

$$\frac{\partial p}{\partial n} = 0, \quad \text{on } \Gamma_{\text{bot}}, \tag{4.3}$$

the interface conditions

$$p \text{ continuous across } \Gamma_{\text{int}}, \tag{4.4}$$

$$\frac{1}{\rho_{\text{w}}} \left. \frac{\partial p}{\partial n} \right|_{\Gamma_{\text{int}-}} = \frac{1}{\rho_{\text{s}}} \left. \frac{\partial p}{\partial n} \right|_{\Gamma_{\text{int}+}}, \tag{4.5}$$

and the radiation condition

$$p(r, z) \text{ behaves like an outgoing cylindrical wave as } r \to \infty. \tag{4.6}$$

The normal derivative on Γ_{int}, appearing in (4.5), is the outward normal of the water layer, while the $\mp$ symbols denote that the functions are evaluated at Γ_{int} by their limits from Ω^{w} and Ω^{s}, respectively. We would like to note that the outgoing radiation condition (4.6) is the appropriate for a waveguide of this type, in contrast to the usual Sommerfeld radiation condition posed at infinity, [13].

4.2.1   Reformulation of the Problem in a Bounded Domain

In order to use a direct numerical method to solve the problem we have to truncate the original semi-infinite domain and reformulate the problem (4.1)–(4.6) in a bounded region. To this end, we introduce at distances $r = R_1 < r^{\text{N}}$ and $r = R_2 > r^{\text{F}}$ two artificial boundaries, denoted by Γ_4 and Γ_2, respectively (see Figure 4.1). We shall also denote by Γ_1, Γ_3 and Γ_i the parts of the bottom, the surface and the interface, respectively, lying between the lines $r = R_1$ and $r = R_2$. Then, our computational domain, denoted by Ω, consists of the water layer Ω^{w} and the sediment layer Ω^{s} separated by the interface Γ_i, and is confined by the irregular part of the bottom Γ_1, the artificial outflow boundary Γ_2, the part of the surface Γ_3, and the artificial inflow boundary Γ_4. On Γ_2 a transparent DtN boundary condition may be posed in order to avoid spurious reflections from the artificial boundary. At the inflow boundary Γ_4 we suppose that either the field is given, or that it satisfies an inflow boundary condition relating the solution in the domains Ω and Ω^{N}. Then the original problem may be written as follows:

We seek a complex-valued function $p(r, z)$, $(r, z) \in \overline{\Omega}$, satisfying

$$\Delta p + k^2(r, z)p \equiv \frac{\partial^2 p}{\partial r^2} + \frac{1}{r}\frac{\partial p}{\partial r} + \frac{\partial^2 p}{\partial z^2} + k^2(r, z)\,p = 0 \quad \text{in } \Omega^{\mathrm{W}} \cup \Omega^{\mathrm{s}}, \qquad (4.7)$$

$$p\,|_{\Gamma_i-} = p\,|_{\Gamma_i+}, \qquad \frac{1}{\rho_{\mathrm{w}}}\frac{\partial p}{\partial n}\bigg|_{\Gamma_i-} = \frac{1}{\rho_{\mathrm{s}}}\frac{\partial p}{\partial n}\bigg|_{\Gamma_i+}, \qquad (4.8)$$

$$p = g \text{ on } \Gamma_4, \qquad (4.9)$$

$$\text{or} \quad \frac{\partial p}{\partial r} = R(p) + S \text{ on } \Gamma_4. \qquad (4.9')$$

$$\frac{\partial p}{\partial n} = 0 \text{ on } \Gamma_1, \quad p = 0 \text{ on } \Gamma_3, \qquad (4.10)$$

$$\frac{\partial p}{\partial r} = T(p) \text{ on } \Gamma_2. \qquad (4.11)$$

In the above (4.9) and (4.9′) are the two alternatives for the boundary condition posed on the near field boundary Γ_4. In (4.9) g is a given complex function of z approximating at $r = R_1$ the acoustic field generated by the time-harmonic point source of frequency f placed at $r = 0$. In (4.9′), which is a nonhomogeneous DtN-type condition relating the incident field at Γ_4 to the backscattered field from the rest of the waveguide, R is a nonlocal integral operator and S is a function of z defined at $r = R_1$. In (4.11) T is also a nonlocal integral operator associated with the DtN map of the exterior wave field at Γ_2. In the next section we shall construct representations of these DtN maps.

4.2.2 Construction of the Nonlocal Conditions at the Artificial Boundaries

In order to construct the nonlocal integral operator T in the outflow boundary condition (4.11), we consider the following associated exterior interface problem. We seek $p(r, z)$ in Ω^{F}, satisfying

$$\Delta p + k_{\mathrm{F}}^2(z)p = 0 \quad \text{in } \Omega^{\mathrm{W,F}} \cup \Omega^{\mathrm{S,F}}, \qquad (4.12)$$

$$p\,|_{\Gamma_i^{\mathrm{F}}-} = p\,|_{\Gamma_i^{\mathrm{F}}+}, \qquad \frac{1}{\rho_{\mathrm{w}}}\frac{\partial p}{\partial z}\bigg|_{\Gamma_i^{\mathrm{F}}-} = \frac{1}{\rho_{\mathrm{s}}}\frac{\partial p}{\partial z}\bigg|_{\Gamma_i^{\mathrm{F}}+}, \qquad (4.13)$$

$$\frac{\partial p}{\partial z}\bigg|_{\Gamma_1^{\mathrm{F}}} = 0, \qquad p\,|_{\Gamma_3^{\mathrm{F}}} = 0, \qquad (4.14)$$

$$p \text{ is outgoing as } r \text{ increases}, \qquad (4.15)$$

where by Γ_1^{F}, Γ_3^{F} and Γ_i^{F} we denote the (horizontal) parts of the bottom, the surface and the interface, respectively, lying outside the cylinder $r = r^{\mathrm{F}}$. In (4.12) $k_{\mathrm{F}}(z) = \omega/c_{\mathrm{F}}(z)$, where $c_{\mathrm{F}}(z)$ denotes the speed of sound in Ω^{F}, in which a possible jump discontinuity is allowed at $z = h^{\mathrm{F}}$.

The boundary-value problem (4.12)–(4.15) is separable, so we may use the separation of variables technique to represent its solution in the form

$$p(r, z) = \sum_{n=1}^{\infty} \alpha_n H_0^{(1)}(\sqrt{\lambda_n^{\mathrm{F}}} r) Z_n^{\mathrm{F}}(z), \quad (r, z) \in \Omega^{\mathrm{F}},$$

where $H_0^{(1)}$ is the Hankel function of the first kind and zero order, and λ_n^{F}, $Z_n^{\mathrm{F}}(z)$, $n = 1, 2, \ldots$, are the (real) eigenvalues and eigenfunctions, respectively, of the two-point vertical eigenvalue problem

$$\frac{d^2 Z_n^{\mathrm{F}}}{dz^2} + (k_{\mathrm{F}}^2(z) - \lambda_n^{\mathrm{F}}) Z_n^{\mathrm{F}} = 0 \quad \text{in } [0, h^{\mathrm{F}}) \cup (h^{\mathrm{F}}, H^{\mathrm{F}}], \tag{4.16}$$

$$Z_n^{\mathrm{F}}(h^{\mathrm{F}}-) = Z_n^{\mathrm{F}}(h^{\mathrm{F}}+), \tag{4.17}$$

$$\frac{1}{\rho_{\mathrm{W}}} \frac{dZ_n^{\mathrm{F}}(h^{\mathrm{F}}-)}{dz} = \frac{1}{\rho_{\mathrm{S}}} \frac{dZ_n^{\mathrm{F}}(h^{\mathrm{F}}+)}{dz}, \tag{4.18}$$

$$Z_n^{\mathrm{F}}(0) = \frac{dZ_n^{\mathrm{F}}(H^{\mathrm{F}})}{dz} = 0. \tag{4.19}$$

We assume that the problem (4.16)–(4.19) is such that $\lambda_n^{\mathrm{F}} \neq 0$, for all n. In the specific case of the environment of Figure 4.1, where $\Omega^{\mathrm{W,F}}$, $\Omega^{\mathrm{S,F}}$ are homogeneous layers of thicknesses $h_1 = h^{\mathrm{F}}$ and $h_2 = H^{\mathrm{F}} - h^{\mathrm{F}}$, respectively, and the sound speeds are constant and equal to c_1 in the water layer, and c_2 in the sediment, we may compute the eigenvalues λ_n^{F} as roots of the equation (see e.g., [8])

$$\frac{h_2}{h_1} \frac{\rho_{\mathrm{S}}}{\rho_{\mathrm{W}}} \frac{\Lambda_1(\lambda)}{\Lambda_2(\lambda)} \cos \Lambda_1 \cos \Lambda_2 = \sin \Lambda_1 \sin \Lambda_2, \tag{4.20}$$

where

$$\Lambda_1(\lambda_n^{\mathrm{F}}) = h_1 \sqrt{(\omega/c_1)^2 - \lambda_n^{\mathrm{F}}}, \quad \text{and} \quad \Lambda_2(\lambda_n^{\mathrm{F}}) = h_2 \sqrt{(\omega/c_2)^2 - \lambda_n^{\mathrm{F}}}. \tag{4.21}$$

We also assume that the eigenfunctions Z_n^{F} are orthonormal with respect to the weighted L^2-inner product

$$(w, u)_{L_\rho^2(0, H^{\mathrm{F}})} := \int_0^{h^{\mathrm{F}}} w \overline{u} \, dz + \rho \int_{h^{\mathrm{F}}}^{H^{\mathrm{F}}} w \overline{u} \, dz,$$

where $\rho := \frac{\rho_{\mathrm{W}}}{\rho_{\mathrm{S}}}$.

Since Z_n^{F}, $n = 1, 2, \ldots$, form a complete orthonormal system in $L^2(0, H^{\mathrm{F}})$ equipped with the inner product $(\cdot, \cdot)_{L_\rho^2(0, H^{\mathrm{F}})}$, we deduce that for $r \geq r^{\mathrm{F}}$,

$$\alpha_n = \frac{1}{H_0^{(1)}(\sqrt{\lambda_n^{\mathrm{F}}} r)} \left(p(r, \cdot), Z_n^{\mathrm{F}} \right)_{L_\rho^2(0, H^{\mathrm{F}})}.$$

The eigenvalues λ_n^{F}, $n = 1, 2, \ldots$, form a decreasing sequence such that

$$\max_{z \in [0, H^{\mathrm{F}}]} k_{\mathrm{F}}^2(z) \geq \lambda_1^{\mathrm{F}} \geq \lambda_2^{\mathrm{F}} \geq \ldots \geq \lambda_n^{\mathrm{F}} \to -\infty.$$

Therefore, only a finite number of the λ_n^{F}'s is expected to be positive. Let us assume that N_p is a positive integer such that $\lambda_n^{\mathrm{F}} > 0$, for $n = 1, \ldots, N_p$, and $\lambda_n^{\mathrm{F}} < 0$, for $n = N_p + 1, N_p + 2, \ldots$. Then the eigenfunctions Z_n^{F}, $n = 1, 2, \ldots, N_p$, correspond to the *propagating* modes, while the rest of them to the *evanescent* modes. Here and in the sequel we follow the convention that the square root $\sqrt{\mu}$, for μ real, is taken to be equal to $\sqrt{\mu}$, if $\mu \geq 0$, and to $i\sqrt{-\mu}$, if $\mu < 0$.

The DtN map of the acoustic field in Ω^{F}, evaluated on Γ_2, is simply the matching condition

$$\frac{\partial p}{\partial r}(R_2, z) = T(p)(z) := \sum_{n=1}^{\infty} a_n(p)\, Z_n^{\mathrm{F}}(z), \qquad (4.22)$$

with

$$a_n(p) = \sqrt{\lambda_n^{\mathrm{F}}}\, \frac{\sigma_n'}{\sigma_n} \left(p(R_2, \cdot), Z_n^{\mathrm{F}} \right)_{L_\rho^2(0, H^{\mathrm{F}})},$$

and

$$\sigma_n' = \frac{dH_0^{(1)}}{dr}(\sqrt{\lambda_n^{\mathrm{F}}} R_2), \quad \sigma_n = H_0^{(1)}(\sqrt{\lambda_n^{\mathrm{F}}} R_2).$$

Therefore the nonlocal operator T in (4.11) is defined by (4.22).

The hypothesis that the acoustic pressure is given on Γ_4, i.e., that condition (4.9) holds on Γ_4, means that we have accounted not only for the incident field on Γ_4 due to the radiation of the source but also for the effect of the backscattering on Γ_4 from the rest of the waveguide. Hence, a more correct formulation of the boundary-value problem entails posing on Γ_4 another DtN-type inflow boundary condition relating the field in Ω to that in the strip Ω^{N}. Since in Ω^{N} the problem is also separable, a series representation of the field involving both incoming and outgoing wave terms may be derived, and a single set of coefficients may be evaluated by assuming that as $r \to 0^+$ the field agrees asymptotically with the outgoing field produced by the source in the range-independent environment of the strip $0 < r < r^{\mathrm{N}}$, see, e.g., [10] or [23]. Therefore we may obtain the following representation of the field for $(r, z) \in \Omega^{\mathrm{N}}$

$$p(r, z) = \sum_{n=1}^{\infty} \beta_n\, J_0(\sqrt{\lambda_n^{\mathrm{N}}} r)\, Z_n^{\mathrm{N}}(z) + \frac{i}{4\rho_{\mathrm{w}}} \sum_{n=1}^{\infty} H_0^{(1)}(\sqrt{\lambda_n^{\mathrm{N}}} r)\, Z_n^{\mathrm{N}}(z_s)\, Z_n^{\mathrm{N}}(z), \qquad (4.23)$$

where J_0 is the Bessel function of zero order, and λ_n^{N}, $Z_n^{\mathrm{N}}(z)$, $n = 1, 2, \ldots$, are the (real) eigenvalues and orthonormal eigenfunctions, respectively, of the analogous to (4.12)–(4.15) two-point vertical eigenvalue problem, where $h^{\mathrm{F}}, H^{\mathrm{F}}, \lambda_n^{\mathrm{F}}, Z_n^{\mathrm{F}}$ are now replaced by $h^{\mathrm{N}}, H^{\mathrm{N}}, \lambda_n^{\mathrm{N}}, Z_n^{\mathrm{N}}$, respectively. The first series in (4.23) represents the scattering term, i.e., a solution of the homogeneous counterpart of (4.1), while the second series the source term, i.e., the solution of (4.1) obtained under the assumption that the whole waveguide has the properties of the near-field region, and which satisfies the radiation condition (4.6).

If we multiply (4.23) by Z_m^{N}, integrate along Γ_4, and use the orthonormality of the eigenfunctions Z_n^{N}, $n = 1, 2, \ldots$, with respect to the weighted L^2–inner product

$$(w, u)_{L^2_\rho(0, H^{\mathrm{N}})} := \int_0^{h^{\mathrm{N}}} w \,\overline{u}\, dz + \rho \int_{h^{\mathrm{N}}}^{H^{\mathrm{N}}} w \,\overline{u}\, dz,$$

we conclude that the coefficients β_n are given by

$$\beta_n = \frac{1}{J_0(\sqrt{\lambda_n^{\mathrm{N}}} r)} \left[\left(p(r, \cdot), Z_n^{\mathrm{N}} \right)_{L^2_\rho(0, H^{\mathrm{N}})} - \frac{i}{4\rho_{\mathrm{w}}} H_0^{(1)}(\sqrt{\lambda_n^{\mathrm{N}}} r) \, Z_n^{\mathrm{N}}(z_s) \right].$$

Differentiating then (4.23) with respect to r, using properties of the special functions involved, and evaluating the resulting expression at $r = R_1$, we deduce that the nonlocal inflow condition on Γ_4 may be written in the form

$$\frac{\partial p}{\partial r}(R_1, z) = R(p)(z) + S(z), \tag{4.24}$$

where

$$R(p)(z) := \sum_{n=1}^{\infty} b_n(p) \, Z_n^{\mathrm{N}}(z),$$

$$S(z) := -\frac{1}{2\pi \rho_{\mathrm{w}} R_1} \sum_{n=1}^{\infty} \frac{1}{J_0(\sqrt{\lambda_n^{\mathrm{N}}} R_1)} \, Z_n^{\mathrm{N}}(z_s) \, Z_n^{\mathrm{N}}(z),$$

with

$$b_n(p) = \sqrt{\lambda_n^{\mathrm{N}}} \, \frac{J_0'(\sqrt{\lambda_n^{\mathrm{N}}} R_1)}{J_0(\sqrt{\lambda_n^{\mathrm{N}}} R_1)} \left(p(R_1, \cdot), Z_n^{\mathrm{N}} \right)_{L^2_\rho(0, H^{\mathrm{N}})}.$$

Thus the complete formulation of the problem in the bounded domain Ω is given by (4.7)–(4.11), wherein the nonlocal inflow boundary condition (4.9′) replaces the nonhomogeneous Dirichlet condition (4.9). We shall assume that this problem has a unique solution. For analysis and references on the theory of waveguide problems of this kind, cf. [19].

4.3   The Finite Element Method

Let $\overset{\mathrm{o}}{\mathcal{H}}(\Omega, S)$ be the energy space associated with our problem, i.e., the space of complex-valued functions u of r and z, defined on $\overline{\Omega}$, such that

$$\int_{\Omega^{\mathrm{W}}} |u|^2 \, r \, dr dz + \rho \int_{\Omega^{\mathrm{S}}} |u|^2 \, r \, dr dz < \infty$$

and

$$\int_{\Omega^{W}} (|u_r|^2 + |u_z|^2)\, r\, drdz + \rho \int_{\Omega^{S}} (|u_r|^2 + |u_z|^2)\, r\, drdz < \infty,$$

and vanishing on a subset S of $\partial\Omega$. The weak formulation of the boundary-value problem (4.7)–(4.11), where (4.9) is considered as a b.c. on Γ_4, is then the following:

Seek $p \in \overset{0}{\mathcal{H}}(\Omega, \Gamma_3)$ satisfying $p = g$ on Γ_4, and

$$-(\nabla p, \nabla v)_{L_\rho^2(\Omega)} + (k^2 p, v)_{L_\rho^2(\Omega)} + (T(p), v)_{L_\rho^2(\Gamma_2)} = 0, \qquad (4.25)$$

for each $v \in \overset{0}{\mathcal{H}}(\Omega, \Gamma_3 \cup \Gamma_4)$, where the operator T is defined by (4.22), $\nabla :=$ $(\frac{\partial}{\partial r}, \frac{\partial}{\partial z})$, and

$$(u, v)_{L_\rho^2(\Omega)} := \int_{\Omega^{W}} u\bar{v}\, r\, drdz + \rho \int_{\Omega^{S}} u\bar{v}\, r\, drdz,$$

$$(u, v)_{L_\rho^2(\Gamma_2)} := \int_{0}^{h^{F}} u\bar{v}\, R_2\, dz + \rho \int_{h^{F}}^{H^{F}} u\bar{v}\, R_2\, dz.$$

The analogous weak formulation of the problem (4.7) when we impose (4.9′) as a boundary condition on Γ_4 is the following:

Seek $p \in \overset{0}{\mathcal{H}}(\Omega, \Gamma_3)$ such that

$$-(\nabla p, \nabla v)_{L_\rho^2(\Omega)} + (k^2 p, v)_{L_\rho^2(\Omega)} + (T(p), v)_{L_\rho^2(\Gamma_2)}$$
$$- (R(p), v)_{L_\rho^2(\Gamma_4)} = (S, v)_{L_\rho^2(\Gamma_4)}, \quad (4.26)$$

for all $v \in \overset{0}{\mathcal{H}}(\Omega, \Gamma_3)$.

4.3.1 The Finite Element Discretization

The boundary-value problems given by the variational equations (4.25) and (4.26) are discretized by the standard Galerkin/finite element method with continuous in Ω, piecewise linear functions defined on a triangulation $\mathcal{T}_h$ of Ω with triangles of maximum sidelength h and nodes on the interface Γ_i. For simplicity, we confine ourselves here to the problem given by (4.9), (4.25). For a detailed study of the problem (4.9′), (4.26) we refer the reader to [27]. We also assume that the interface and the bottom consist of straight line segments; thus Ω^{W} and Ω^{S} are polygonal domains. We define the finite element spaces

$$S_h = \{\phi : \phi \in C(\overline{\Omega}),\ \phi = 0 \text{ on } \Gamma_3,\ \phi|_\tau \in \mathbb{P}_1\ \forall \tau \in \mathcal{T}_h\},$$
$$S_h^0 = \{\phi : \phi \in C(\overline{\Omega}),\ \phi = 0 \text{ on } \Gamma_3 \cup \Gamma_4,\ \phi|_\tau \in \mathbb{P}_1\ \forall \tau \in \mathcal{T}_h\}.$$

The spaces S_h and S_h^0 are finite dimensional subspaces of $\overset{0}{\mathcal{H}}(\Omega, \Gamma_3)$ and $\overset{0}{\mathcal{H}}(\Omega, \Gamma_3 \cup \Gamma_4)$, respectively.

With this notation in place we may write the discrete analog of (4.25) as follows: Seek $p_h \in S_h$, such that $p_h |_{\Gamma_4} = \Pi_h g |_{\Gamma_4}$, and for every $\phi \in S_h^0$,

$$-(\nabla p_h, \nabla \phi)_{L_\rho^2(\Omega)} + (k^2 p_h, \phi)_{L_\rho^2(\Omega)} + (T_h p_h, \phi)_{L_\rho^2(\Gamma_4)} = 0. \qquad (4.27)$$

Here $\Pi_h g$ is the piecewise linear interpolant of g on the grid induced by $\mathcal{T}_h$ on Γ_4, and T_h is a discrete approximation of T evaluated as follows. We truncate the series in (4.22) retaining only a finite number of terms, say M, containing all the propagating and the most significant evanescent modes. Then the discrete DtN map T_h is defined by the formula

$$T_h(p_h)(z) := \sum_{n=1}^{M} a_{h,n}(p_h) \, Z_{h,n}^{\mathrm{F}}(z), \quad 0 \le z \le H^{\mathrm{F}}, \qquad (4.28)$$

with

$$a_{h,n}(p_h) = A_{h,n} \, (p_h(R_2, \cdot), Z_{h,n}^{\mathrm{F}})_{L_\rho^2(0, H^{\mathrm{F}})}, \qquad (4.29)$$

where

$$A_{h,n} = \sqrt{\lambda_{h,n}^{\mathrm{F}}} \, \frac{\sigma_{h,n}'}{\sigma_{h,n}}, \qquad (4.30)$$

and

$$\sigma_{h,n}' = \frac{dH_0^{(1)}}{dr}(\sqrt{\lambda_{h,n}^{\mathrm{F}}} \, R_2), \quad \sigma_{h,n} = H_0^{(1)}(\sqrt{\lambda_{h,n}^{\mathrm{F}}} \, R_2).$$

Here $\{\lambda_{h,n}^{\mathrm{F}}, Z_{h,n}^{\mathrm{F}}\}_{n=1,2,\dots,M}$ are numerically obtained eigenpairs. In the case where k_{F}^2 depends on z they may be found by a finite element discretization of the eigenvalue problem (4.16)–(4.19) with continuous piecewise linear elements on the partition induced on the interval $[0, H^{\mathrm{F}}]$ by the triangulation $\mathcal{T}_h$, cf. [4].

The error estimate for the discretization (4.27), proved by Goldstein in [19] in the case of a homogeneous single-layer problem in Cartesian coordinates with a homogeneous Dirichlet boundary condition on Γ_1, was extended in [26] to the case of cylindrical coordinates for axisymmetric problems. If $T(p)$ is defined by (4.22), it may be shown that the L^2 norm of the error $p - p_h$ is of $O(h^2)$. If the series in the right-hand side of (4.22) is truncated, so that the sum extends over all the propagating and sufficiently many of the evanescent modes, it may be shown that the H^1 norm of $p - p_h$ is of $O(h)$ plus a term of $O(\exp[-\frac{1}{2}\sqrt{\lambda_J^{\mathrm{F}}}(R_2 - r^{\mathrm{F}})])$, where J is the order of the first evanescent term that is ignored. It is expected that a similar theory holds for the two-layer problem for solutions p that are smooth in each layer and satisfy the transmission conditions (4.4) and (4.5).

4.3.2   Implementation Issues

The finite element method outlined above has been implemented in the Fortran code FENL; see [24] for a detailed description. The main steps of FENL are:

1. In order to triangulate the domain Ω we use mesh generation techniques from the MODULEF library [7], especially the modules APNOPO and TRIGEO, cf. also [17]. The generated mesh for range-varying interface topography is non-uniform.

2. A subroutine is called which reads the NOPO data structure produced by MODULEF and provides the information required for the assembly of the finite element matrices.

3. The next step is the numerical solution of the eigenvalue problem (4.16)–(4.19) with the standard Galerkin/finite element method with continuous, piecewise linear functions on the partition induced on $[0, H^{\mathrm{F}}]$ by the triangulation $\mathcal{T}_h$. To solve it, we use routines from EISPACK, [30].

4. We continue with the assembly of the stifness matrix S, the mass matrix Q, and the associated with the nonlocal condition on Γ_2 matrix B, with elements, respectively,

$$S_{ij} = (\nabla \phi_j, \nabla \phi_i)_{L^2_\rho(\Omega)}, \quad Q_{ij} = (k^2 \phi_j, \phi_i)_{L^2_\rho(\Omega)}, \quad B_{ij} = (T_h \phi_j, \phi_i)_{L^2_\rho(\Gamma_2)},$$

where ϕ_i, $i = 1, \ldots, N_h$ are the basis functions of the finite element space. S and Q are real, symmetric, sparse matrices, while B is complex symmetric (not hermitian).

5. The resulting linear system is large, sparse, indefinite and complex symmetric, and is solved with methods from the QMRPACK software package, [16]. QMPRPACK contains implementations of various Quasi-Minimal Residual (QMR) iterative algorithms, cf. [14]. These are iterative schemes that belong to the general class of Krylov subspace methods, involve few–term recurrences, and are applicable to systems with nonsingular, complex, not necessarily Hermitian matrices, [33]. In our tests we have mainly used the double precision complex version of CPL (i.e., the nonsymmetric QMR algorithm based on the coupled two-term Lanczos with look-ahead, see [15]), and QMR (i.e., a QMR algorithm based on the coupled three-term Lanczos with look-ahead), combined with the built-in preconditioners two-sided SSOR or two-sided ILUT. We would like to note here that most of the computational effort of the finite element method is devoted to solving the linear system.

6. Finally, to produce one- or two-dimensional plots we used MATLAB's PDE Toolbox, [36], with the aid of a routine which exports the data (triangulation information and solution at the nodes) from FENL in the appropriate format needed by the graphics module.

4.4   Numerical Experiments

Example 1   In a first experiment our aim is to check the accuracy of the method. We consider a single-layer waveguide and we take Ω to be the rectangle $R_1 < r < R_2$, $0 < z < H$. We assume that $k = k_0 = \frac{2\pi f}{c_0}$, and let the exact solution to be given by the function

$$p(r,z) = \sum_{n=1}^{M} H_0^{(1)}(\sqrt{\lambda_n}r)\, Z_n(z),$$

where $\lambda_n = k^2 - \left(\frac{(2n-1)\pi}{2H}\right)^2$, $Z_n(z) = \sin\frac{(2n-1)\pi z}{2H}$, $n = 1,2,3,\ldots$. We take $g(z) = p(R_1,z)$, $0 < z < H$ (i.e., on the near-field boundary a condition of the form (4.9) holds), and use the code FENL to solve numerically the problem in a uniform triangulation of right isosceles triangles with perpendicular side of size h. We let $\xi := c_0/(fh)$, i.e., ξ denotes the number of meshlengths per wavelength λ ($\lambda = c_0/f$), and we denote by ε the value of the normalized ℓ_2-error

$$\varepsilon := \left(\frac{\sum |p_h(Q) - p(Q)|^2}{\sum |p(Q)|^2}\right)^{1/2},$$

where the sums are taken over all the nodes Q of the triangulation.

As a first try we let $R_1 = 20$ m, $R_2 = 120$ m, $H = 50$ m, $c_0 = 1500$ m/sec and $\rho_{\mathrm{w}} = 1.0$ g/cm^3. In this and the subsequent tests of Example 1, we used QMR as the linear system solver, combined with the two-sided ILUT preconditioner with parameters fill-in=1, tol=10^{-3}. (The calculated errors reported in the sequel include the errors of the iterative method.) Table 4.1 contains the results obtained for frequencies $f = 30$, 60 and 90 Hz. NE denotes the number of elements used in each case. From these results one can verify that the ℓ_2-error is of order 2, as expected. It is also immediate to note that the error in the finite element solution grows with k (i.e., as the frequency increases).

In Table 4.2 we use the data of Table 4.1 to compute the dependence of the error ε on the wavenumber k for the values of h=1.25, 1.00, 0.50 and 0.40. We assume that $\varepsilon(k) \approx c(h)\, k^\nu$ and compute the exponent ν as $\nu = (\log\frac{\varepsilon(k_1)}{\varepsilon(k_2)})/(\log\frac{k_1}{k_2})$. We observe that it is approximately equal to three. Even though the variation of k is limited in our experiment ($k = 0.126, 0.251, 0.377$) we note that the results of the two tables are compatible with a dependence of the form $\varepsilon(k,h) \approx ck^3h^2$, that has been reported by many authors, see, e.g., the book by Ihlenburg, [22].

We shall consider another related example in order to investigate the dependence of the error on the wavenumber k given constraints of the form $k^\alpha h^\beta = const$. Now our domain Ω is taken to be the square $(r,z) \in$

TABLE 4.1: Normalized ℓ_2-error ε as a function of h for frequencies f=30,60 and 90 Hz.

		$f = 30$ Hz $(k = 0.126)$		$f = 60$ Hz $(k = 0.251)$		$f = 90$ Hz $(k = 0.377)$	
h	NE	ξ	ε	ξ	ε	ξ	ε
1.2500	6,400	40	0.0086	20	0.0914	13	0.3124
1.0000	10,000	50	0.0055	25	0.0591	17	0.2074
0.7143	19,600	70	0.0028	85	0.0305	23	0.1095
0.5000	40,000	100	0.0014	50	0.0150	33	0.0545
0.4545	48,400	110	0.0011	55	0.0124	37	0.0452
0.4000	62,500	125	0.0009	63	0.0096	42	0.0351

TABLE 4.2: Dependence of the error ε on the wavenumber k, $\varepsilon \approx c(h)\,k^{\nu}$, for h=1.25, 1.00, 0.50 and 0.40.

	h=1.25			h=1.00			h=0.50			h=0.40	
k	ε	ν		ε	ν		ε	ν		ε	ν
0.126	0.008583			0.005506			0.001382			0.000885	
		3.41			3.42			3.44			3.44
0.251	0.091367			0.059135			0.015004			0.009621	
		3.03			3.09			3.18			3.19
0.377	0.312370			0.207368			0.054528			0.035070	

$[10, 60] \times [0, 50]$, and it is subdivided into $2N^2$ equal orthogonal isosceles triangles with perpendicular side of length $h := 50/N$. We shall examine three cases: a) $kh = \text{const.} = 100\pi/c_0$, b) $k^3h^2 = \text{const.} = \pi^3/c_0$, and c) $k^2h = \text{const.} = 16200\pi^2/c_0^2$. These constants have been chosen so that, when N varies in the interval $[3,180]$, the values of the frequency f in the three cases lie between 14 Hz to 112 Hz. Hence we take the corresponding frequencies a) $f = N$, b) $f = 0.5\sqrt[3]{900N^2}$, and c) $f = 9\sqrt{N}$, respectively, and we compute the error ε as N, or equivalently f, increases. Figure 4.2 summarizes the dependence of the error ε on the wavenumber k in the three cases considered. We observe that when kh=const., the error grows with k, while it decays when k^2h is kept constant. When k^3h^2 is constant we note that the error tends to grow very weakly with k.

The outcome of the two experiments strengthens the case that the error behaves like k^3h^2 at least for the ranges of k and h that we experimented with.

Example 2 In this example our aim is to compare FENL with the standard coupled mode code COUPLE in a stratified environment with variable interface. (See [2] for a detailed comparison of FENL with other coupled mode methods.) Here, we shall present the outcome of an experiment where the interface is shaped like an underwater trench in the (r, z)-plane and is defined

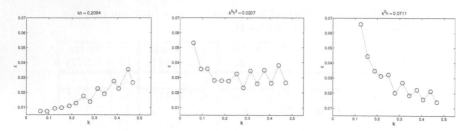

FIGURE 4.2: The dependence of the error ε on the wavenumber k for a) kh=const., b) k^3h^2=const., and c) k^2h=const.

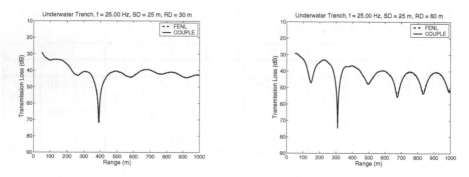

FIGURE 4.3: Trench. Comparison between FENL and COUPLE.

by

$$h(r) = \begin{cases} 65 + 25\cos\frac{2\pi(r-500)}{400} & \text{, for } 300 < r < 700, \\ 40 & \text{, elsewhere.} \end{cases}$$

(All distances are in meters.) The densities and sound speeds are taken constant in each layer, having values $\rho_w = 1.0$ g/cm^3, $c_w = 1500$ m/sec in the water, and $\rho_s = 1.5$ g/cm^3 and $c_s = 1700$ m/sec in the sediment. The source is located at $z = 25$ m and the hard horizontal bottom is placed at $H = 100$ m. As a near field condition on Γ_4 we specify $\Pi_h g$, defined as the piecewise linear interpolant of the acoustic field produced by COUPLE at $r = R_1$ on the grid induced by the triangulation $\mathcal{T}_h$ on Γ_4.

In underwater acoustics the field is usually expressed in terms of transmission loss (TL), defined as

$$\text{TL} := -20\log_{10}\left|\frac{p(r,z)}{p_0}\right|,$$

with decibels as units, where $p(r,z)$ is the pressure at the point (r,z) and p_0 is a reference pressure produced at a distance of 1 m from the same source in an

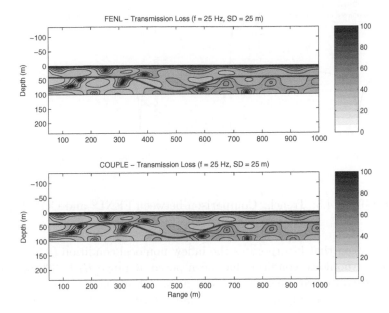

FIGURE 4.4: Trench. Comparison between FENL and COUPLE.

infinite homogeneous medium. The results are traditionally presented either as two-dimensional plots of the transmission loss, or as one-dimensional plots of the transmission loss vs. range for a fixed value of z called receiver depth (RD). Figures 4.3 and 4.4 present one- and two-dimensional transmission loss plots obtained by the two codes. In Figure 4.3 we show superimposed transmission loss vs. range curves at receiver depths RD = 30 and 60 m. The results are in excellent agreement. Indicatively we should like to report that for the FENL run we used a triangular mesh consisting of 53142 elements and 27087 nodes. The linear system solver (CPL of QMRPACK combined with the two sided SSOR preconditioner with parameter $\omega = 1.2$) required 1135 iterations to converge and 49 secs of CPU time on a Pentium IV PC with a RAM of 1 GB running under Linux at 2 GHz. For COUPLE we used 400 regions in range between $r = 300$ m and $r = 700$ m, and 15 contributing modes.

We would like to refer the reader in [2] for examples of cases in which the frequency of the harmonic point source is such that an eigenvalue of the local vertical problem remains small in magnitude and changes sign several times in the vicinity of the trench. Then the discrepancies between the results of the finite element and the coupled mode codes increase, but remain small in absolute terms.

Finally, in Figure 4.5 we show superimposed transmission loss plots at receiver depths RD = 30 and 60 m obtained by COUPLE and $FENL^2$, the

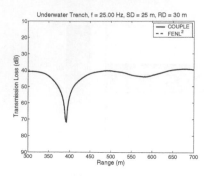 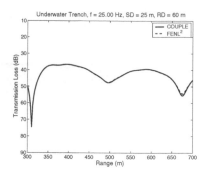

FIGURE 4.5: Trench. Comparison between FENL^2 and COUPLE.

FENL version that incorporates the inflow nonlocal condition $(4.9')$, see [27]. The inflow nonlocal condition has been posed at range $R_1 = 300$ m and the outflow condition at range $R_2 = 700$ m, i.e., the computational domain is confined in the vicinity of the trench, and is discretized by a triangulation of 50272 elements and 25537 nodes. The results are again in excellent agreement.

Example 3   With this last example we try to give the reader a flavor of the capability of the finite element method to handle problems in domains with complex interface and bottom topography, and with various boundary conditions. We consider the two-layer waveguide of Figure 4.6, for $0 \le r \le 200$ (all distances in meters). On the rectangular (ringlike) obstacle of 30 m height located at the bottom for $20 < r < 40$ we pose a homogeneous Neumann boundary condition, while a Dirichlet boundary condition is imposed on the rest of the bottom. (Of course the finite element formulation has to be appropriately altered in order to accommodate Dirichlet bottom b.c.'s, but this may be done in a straightforward manner.) An impermeable (axisymmetric) disk of diameter equal to 20 m is placed in the waveguide, with its centre located at the point $r = 100$, $z = 45$. The medium parameters have values $\rho_{\rm w} = 1.0$ g/cm^3, $c_{\rm w} = 1500$ m/sec in $\Omega^{\rm w}$, and $\rho_{\rm s} = 1.2$ g/cm^3, $c_{\rm s} = 1700$ m/sec in $\Omega^{\rm s}$. As an approximation of the point source at $r = 0$ we take the following combination of Gaussian functions

$$g(z) = \sqrt{\frac{k}{2}} \left({\rm e}^{-\frac{k^2}{4}(z-z_s)^2} - {\rm e}^{-\frac{k^2}{4}(z+z_s)^2} \right),$$

where $z_s = 20$ m, and the frequency is taken to be $f = 40$ Hz. (Note that the diameter of the disk is a little larger than half a wavelength for this frequency.)

In Figure 4.7 one may notice the distortions of the field brought upon by the presence of the disk. To obtain this figure we have used a discretization of 56256 elements for the case where the disk was absent and 49075 elements when the disk was present. Here, it is not possible to compare our results

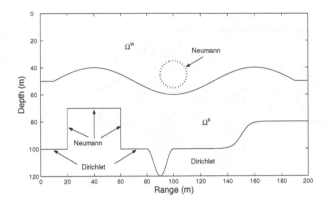

FIGURE 4.6: Configuration of a complex domain.

with a coupled mode code, so in order to check that numerical convergence was achieved, we ran our code with coarser discretizations of 36240 elements and 31408 elements, respectively, and we could not notice any discrepancy from Figure 4.7.

Acknowledgment

This work was partly supported by a "Pythagoras" EPEAEK II grant to the University of Athens, which was co-funded by the EU European Social Fund and national resources.

References

[1] G. A. Athanassoulis and K. A. Belibassakis. Rapidly-convergent local-mode representations for wave propagation and scattering in curved-boundary waveguides. in *Proc. 6th International Conference on Mathematical and Numerical Aspects of Wave Propagation (Waves 2003)*, G. C. Cohen. E. Heikkola, P. Joly, P. Neittaanmäki eds., pp. 451-456, Springer, 2003.

[2] G. A. Athanassoulis, K. A. Belibassakis, D. A. Mitsoudis, N. A. Kam-

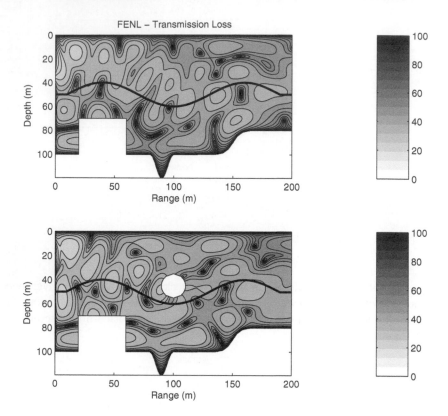

FIGURE 4.7: Transmission loss in a complex domain, with the absence (above) or the presence (below) of a rigid disk.

panis, and V. A. Dougalis. Coupled mode and finite element solutions of underwater sound propagation problems in stratified environments. *J. Comp. Acoustics*, to appear.

[3] A. K. Aziz, R. B. Kellogg, and A. B. Stephens. A two point boundary value problem with a rapidly oscillating solution. *Numer. Math.*, 53:107-121, 1988.

[4] I. Babuška and J. Osborn. "Eigenvalue problems", in *Handbook of numerical analysis, Vol. II*, eds. P. G. Ciarlet and J. L. Lions, pp. 641-787, Elsevier, 1991.

[5] A. Bayliss, C. I. Goldstein, and E. Turkel. On accuracy conditions for the numerical computation of waves. *J. Comp. Physics*, 59:396-404, 1985.

[6] A. Bayliss, C. I. Goldstein, and E. Turkel. The numerical solution of

the Helmholtz equation for wave propagation problems in underwater acoustics. *Comp. Maths. with Appls.*, 11:655-665, 1985.

[7] M. Bernadou, P. L. George, A. Hassim, P. Joly, P. Laug, A. Perronnet, E. Saltel, D. Steer, G. Vanderborck, and M. Vidrascu. *MODULEF: A Modular Library of Finite Elements*, INRIA, 1988.

[8] C. A. Boyles, *Acoustic Waveguides: Applications to Oceanic Science.* Wiley, 1984.

[9] M. J. Buckingham. Ocean-acoustic propagation models. *J. Acoustique*, 3:223-287, 1992.

[10] R. B. Evans. A coupled mode solution for the acoustic propagation in a waveguide with stepwise depth variations of a penetrable bottom. *J. Acoust. Soc. Am.*, 74:188-195, 1983.

[11] R. B. Evans. COUPLE: a user's manual, NORDA TN–332, 1986.

[12] J. A. Fawcett. A derivation of the differential equations of coupled mode propagation. *J. Acoust. Soc. Am.*, 97:290-295, 1992.

[13] G. J. Fix and S. P. Marin. Variational methods for underwater acoustic problems. *J. Comp. Physics*, 28:253-270, 1978.

[14] R. W. Freund and N. M. Nachtigal. QMR: a quasi–minimal residual method for non-Hermitian linear systems. *Numer. Math.*, 60:315-339, 1991.

[15] R. W. Freund and N. M. Nachtigal. An implementation of the QMR method based on coupled two-term recurrences. *SIAM J. Sci. Comput.*, 15:313-337, 1994.

[16] R. W. Freund and N. M. Nachtigal. QMRPACK: A package of QMR Algorithms. *ACM Trans. Math. Software*, 22:46-77, 1996.

[17] P. L. George. *Automatic mesh generation: Application to finite element methods.* Wiley, 1991.

[18] D. Givoli. *Numerical Methods for Problems in Infinite Domains.* Elsevier, 1992.

[19] C. I. Goldstein. A finite element method for solving Helmholtz type equations in waveguides and other unbounded domains. *Math. Comp.*, 39:309-324, 1982.

[20] I. Harari, M. Slavutin, and E. Turkel. Analytical and numerical studies of a finite element PML for the Helmholtz equation. *J. Comp. Acoustics*, 8:121-137, 2000.

[21] F. Ihlenburg and I. Babuška. Finite element solution of the Helmholtz equation with high wave number, Part I: The *h*-version of the FEM. *Computers Math. Applic.*, 30:9-37, 1995.

[22] F. Ihlenburg. *Finite Element Analysis of Acoustic Scattering.* Springer, 1998.

[23] F. B. Jensen, W. A. Kuperman, M. B. Porter, and H. Schmidt. *Computational Ocean Acoustics.* American Institute of Physics, 1994.

[24] N. A. Kampanis and V. A. Dougalis. A finite element code for the numerical solution of the Helmholtz equation in axially symmetric waveguides with interfaces. *J. Comp. Acoustics,* 7:83-110, 1999.

[25] J. B. Keller and D. Givoli. Exact non–reflecting boundary conditions. *J. Comp. Physics,* 82:172-192, 1989.

[26] D. A. Mitsoudis. *Finite element methods for axisymmetric, indefinite boundary-value problems and applications in underwater acoustics.* Ph.D. Thesis, University of Athens, Greece, 2003 (in Greek).

[27] D. A. Mitsoudis. Near- and far-field boundary conditions for a finite element method for the Helmholtz equation in axisymmetric problems of underwater acoustics. *Acta Acustica united with Acustica* (to appear).

[28] A. A. Oberai and P. M. Pinsky. A numerical comparison of finite element methods for the Helmholtz equation. *J. Comput. Acoustics,* 8:211-221, 2000.

[29] S. R. Rutherford and K. E. Hawker. Consistent coupled mode theory of sound propagation for a class of nonseparable problems. *J. Acoust. Soc. Am.,* 70:554-564, 1981.

[30] B. T. Smith, J. M. Boyle, B. S. Garbow, Y. Ikebe, V. C. Klema, and C. B. Moler. *Matrix eigensystem routines – EISPACK guide,* Springer, 1976.

[31] M. I. Taroudakis, G. A. Athanassoulis, and J. P. Ioannidis. A variational principle for underwater acoustic propagation in a three-dimensional ocean environment. *J. Acoust. Soc. Am.,* 88:1515-1521, 1990.

[32] L. L. Thompson. A review of finite-element methods for time-harmonic acoustics. *J. Acoust. Soc. Am.,* 119:1315-1330, 2005.

[33] L. N. Trefethen and D. Bau. *Numerical linear algebra.* SIAM, 1997.

[34] E. Turkel. Numerical difficulties solving time harmonic equations. in *Multiscale Computational Methods in Chemistry and Physics.* eds. A. Brandt, J. Bernholc, K. Binder, pages 319-337, IOS Press, Ohmsha, Tokyo, 2001.

[35] O. C. Zienkiewicz. Achievements and some unsolved problems of the finite element method. *Int. J. Numer. Meth. Engng.,* 47:9-28, 2000.

[36] *Partial differential equation toolbox user's guide (for use with MAT-LAB).* The MathWorks Inc., 1996.

Chapter 5

Parabolic Equation Techniques in Underwater Acoustics

D. J. Thomson, 733 Lomax Road, Victoria, BC, Canada V9C 4A4, drdjt@shaw.ca

G. H. Brooke, General Dynamics Canada, 3785 Richmond Road, Ottawa, ON, Canada K2H 5B7, Gary.Brooke@gdcanada.com

5.1 Introduction

For over 30 years, one-way wave equations that are derived from applying paraxial approximations to the reduced wave equation have been used to model underwater sound propagation in the ocean [73, 77]. Two criteria that underly the widespread use of parabolic approximations in this context are: (1) long-range sound propagation in the ocean waveguide is dominated by energy travelling at small angles to the horizontal, and (2) backscattered energy can be neglected compared to forward-scattered energy. When these criteria are met, the resulting one-way wave equations are first order in range and therefore admit accurate and efficient numerical solution by marching outward in range in a stepwise manner. Because of this range-stepping procedure, numerical solvers of one-way wave equations are faster and more memory efficient than numerical solvers that are applied directly to the reduced wave equation, e.g., the second order equation. In addition, parabolic equation (PE) models have the inherent capability for accommodating wave propagation in ocean waveguides where the material properties vary along the direction of propagation. Similar one-way propagation criteria are also relevant in other disciplines, e.g., in exploration seismology [15] and radar [57].

The traditional motivation for developing propagation models in underwater acoustics has been the requirement for computing single-frequency estimates of transmission loss (TL) for use in navy sonar performance prediction applications [89]. Here, TL is defined in decibels (dB) as $-10\log_{10}|p/p_0|^2$, where p is the excess pressure and p_0 is a reference pressure at a range of 1 m from the source. More recently, however, efficient and accurate propagation models for computing coherent values of p (both amplitude and phase), especially in shallow-water waveguides, have found widespread use in source localization methods based on matched-field processing (MFP) concepts [10, 69, 87]. Concurrently, there has been considerable activity in applying nonlinear, MFP-based, global optimization methods in geoacoustic inversion procedures to infer the material properties of the sea-bottom that significantly affect the propagation [12, 25, 88].

For stratified (range-independent) media, several standard representations for the acoustic field are available and each admits efficient numerical computation, e.g., wavenumber integral, normal mode, and multiple scattering [2, 50, 73]. The hierarchy of standard acoustic propagation models for layered media in relation to the parabolic equation model is shown in Fig. 5.1. The wavenumber integral representation makes use of the fast Fourier transform (FFT) algorithm to numerically invert the Hankel transform of the depth-dependent Green's function with respect to horizontal wavenumber [32,51,72]. Alternatively, the normal mode representation applies Cauchy's integral formula to replace the integral along the real wavenumber axis with a sum of residues at the poles (modal wavenumbers) of the depth-dependent Green's function plus contributions from any branch line integrals [66,91]. The multiple expansion representation relies on geometric ray concepts to partition the total field into various physical propagation components (refracted, surface reflected, and bottom bounce paths, etc.) which are then evaluated asymptotically [67,90]. Note, for each representation depicted in Fig. 5.1, an example of a well-developed, underwater acoustic propagation FORTRAN code is indicated in parentheses. These computer codes are freely available from the Ocean Acoustic Model Library web site that is supported by the US Office of Naval Research.

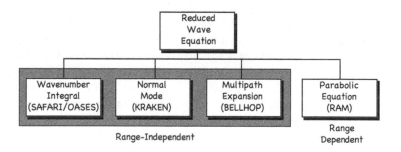

FIGURE 5.1: Hierarchy of standard acoustic propagation models. FORTRAN codes for each named model (in parentheses) are available at http://www.hlsresearch.com/oalib/.

While each of these range-independent models can be extended to treat range-dependent media, the resulting numerical codes can be computationally intensive, e.g., the coupled-mode code COUPLE [33]. In contrast, PE-based models are readily capable of accommodating range-varying media at little computational cost by simply updating the environmental variables and corresponding propagators during the marching algorithm. Although certain energy-conservation issues are associated with this simple updating procedure [68], several techniques have been proposed and implemented for ameliorating their effects in PE-type codes [24, 28]. Moreover, a one-way PE that is manifestly reciprocal in range- and depth-dependent acoustic waveguides has also recently been formulated and implemented numerically [42, 61].

At the time the parabolic equation approach was first applied to underwater sound propagation, the main focus of navy research was directed toward deep-water scenarios [35, 73, 76]. An early example of a standard PE transmission loss calculation (*circa* 1975) is presented in Fig. 5.2 for a 230-Hz source located at a depth of 25 m for a typical NE Pacific sound speed profile in summer conditions. The refracted arrivals exhibit a distinct convergence zone pattern that reveals an increasingly complicated interference structure with range. Note also the presence of sound energy that is trapped near the surface in the secondary subsurface sound channel. This calculation was carried out using a version of the split-step Fourier algorithm [50, 77, 82] that was developed at the Defence Research Establishment Pacific. [1]

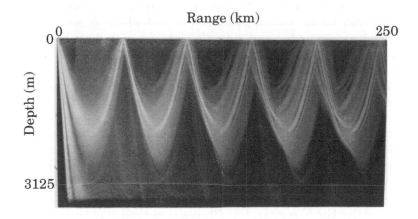

FIGURE 5.2: PE-based transmission loss calculation (*circa* 1975) for a summer NE Pacific Ocean sound speed profile. The 230-Hz source is at a depth of 25 m. The TL's vary between 60 dB (white) and 120 dB (black).

---

[1]This Canadian Laboratory was closed in March 1995. An informal history of DREP is documented in [11].

With the subsequent shift in emphasis to littoral (shallow-water) propagation in the 1980s, the development of PE algorithms centered mainly on finite-difference methods. Such methods are able to accommodate the large discontinuities in sound speed and density at the sea-bottom interface more effectively than methods based on the split-step Fourier solution procedure [23, 49, 80]. Over the past 15 years or so, numerical parabolic equation models based on high-order Padè-type finite-difference formulations have undergone extensive development. A selected survey of the literature reveals that many of these modern finite-difference PE codes are wide-angle accurate [3, 18], energy-conserving [24, 28, 42], and very efficient [22, 23]. Moreover, the latest models are capable of treating complicated waveguide effects such as elasticity [17, 19, 45, 92], backscatter [6, 21, 24], porosity [26, 27, 58], surface roughness [34, 53, 71, 81] and 3D propagation effects [8, 36, 55, 74, 75].

The full-field transmission-loss coverage diagram shown in Fig. 5.2 is a natural consequence of solving the one-way wave equation numerically on a two-dimensional computational grid in range r and depth z. This coverage byproduct merges naturally with modern source localization schemes such as matched-field processing [87] since they rely directly on correlating coherent acoustic predictions with measured array data (amplitude and phase) over a search region in the rz-plane. For example, in a typical PE application involving a vertical line array (VLA), the reciprocity of acoustic fields is invoked so that for a system of N sensors, N coherent coverage fields (replicas) that span the search grid due to sources positioned at each of the N sensor locations on the VLA are computed and stored. Then, for each node on the search grid, the set of N candidate replica fields, suitably normalized, are correlated against the measured VLA data to produce an ambiguity surface. The peak of the ambiguity surface can then be used to estimate the range and depth of the true source position.

As conventional matched field processing concepts were being developed, an alternate scheme of source localization, termed "acoustic retrogation" was introduced that uses a backpropagation algorithm based on the parabolic equation [64, 78, 84]. In addition to reciprocity, this procedure invokes the property of superposition by combining the measured data with vertically extended source fields at each sensor on the VLA to backpropagate a correlation surface directly. For N sensors, this PE-based method is N times faster than the standard approach that uses the PE to compute the replica fields. It is also more memory efficient since individual replica fields do not have to be computed and stored prior to being matched against the measured data.

A procedure that is closely related to the backpropagation method mentioned above is the spectral decomposition of PE fields. Because numerical solutions to one-way wave equations determine the total contribution to the acoustic field at each point on a computational grid, ancillary information on the propagation of individual spectral components (i.e., modes) is not immediately available. However, for some applications involving propagation in oceanic waveguides, it is instructive to be able to examine the propagation of each modal component, especially in range-varying situations where mode coupling may be important. Thus, in [52], a hybrid scheme was developed in which a PE field was numerically decomposed, at a given range, into its spectral components versus horizontal wavenumber. The decomposition was accomplished by regarding the PE field versus depth (at a given range) as a source field in a wavenumber integration code [51, 72]. In turn, this code was used to construct a "local" (range-independent) Green's function that correpsonds to the PE source-field excitation. This "local" Green's function is precisely the spectral decomposition in horizontal wavenumbers of the PE field versus depth at the given range in the waveguide.

In contrast to this hybrid approach, a modal decomposition scheme based on the "propagating beam method" had been developed previously for studying the propagation of light in optical fiber waveguides [37–41]. The propagating beam method in optics is equivalent to the PE method in underwater acoustics. This PE-based modal decomposition method was based on three related concepts: (1) an exact relationship exists between solutions to the standard PE and solutions to the Helmholtz equation, (2) an efficient solution procedure for numerically solving the standard PE, and (3) Fourier analysis of a suitably-defined PE amplitude correlation function along the axis of propagation. In underwater acoustics, the connection between the parabolic and exact fields in (1) formed the basis of a procedure to postprocess numerical solutions of the standard PE into solutions of the Helmholtz equation [30,85,86]. These concepts have also been applied in the context of source localization whereby standard PE modes were converted into normal modes for subsequent use in a matched mode processor [83].

In what follows, we begin with a section reviewing the development of the parabolic equation method in underwater acoustics. Subsequently, we describe a numerical solution procedure based on finite-differences that is particularly relevant for applications involving shallow-water propagation. The discussion includes several approximate techniques for treating profile interpolation, energy conservation, shear wave conversion, and the downgoing radiation condition, and concludes with a numerical example to illustrate sound propagation in a prototype shallow-water waveguide. In the next section, we introduce an acoustic source localization processor that uses a one-way wave equation to backpropagate a correlation function; the PE must be suitably initiated with the field versus depth at a given range. This PE-based correlation technique is compared to the standard Bartlett processor that is commonly used in matched field processing applications. The follow-on section contains the development of a PE-based method for numerically decomposing a given field as a function of depth into its spectral components. This modal decomposition method is based on the PE-based backpropagation procedure described previously. Numerical examples of both the backpropagation source localization and modal decomposition algorithms are illustrated for the same prototype shallow-water example introduced earlier. We conclude with a description of a PE-based matched mode localization scheme that incorporates both the backpropagation and the modal decomposition concepts.

5.2 Parabolic Approximations

5.2.1 Standard PE

We start by considering a stratified fluid half-space in cylindrical coordinates (r, φ, z), z positive downward, bounded above by a free surface at $z = 0$, with a sound-speed profile that supports long range propagation (as $r \to \infty$) in the upper part of the waveguide. For the topics treated in this paper, we also neglect azimuthal coupling by setting $\partial/\partial\varphi = 0$. Then, for a harmonic point source located at $(r, z) = (0, z_0)$, the spatial part of the 2D pressure field $p(r, z)e^{-i\omega t}$ in $r > 0$ satisfies the variable-density acoustic wave equation

$$\frac{1}{r}\frac{\partial}{\partial r}\left(r\frac{\partial p}{\partial r}\right) + \rho\frac{\partial}{\partial z}\left(\frac{1}{\rho}\frac{\partial p}{\partial z}\right) + k_0^2 N^2 p = 0. \tag{5.1}$$

Here, $k_0 = \omega/c_0$ is a suitable reference wavenumber, $\rho(z)$ is the density, $N(z) = n(z)[1 + i\alpha(z)]$ where $n(z) = c_0/c(z)$ is the refractive index, $c(z)$ is the sound speed, and $\alpha(z)$ is the absorption loss. It is common in low-frequency underwater acoustics to express the linear frequency-dependent attenuation in dB/$\lambda(z)$, where $\lambda(z)$ is the acoustic wavelength. In Eq. (5.1), α (in Nepers) is obtained from attenuation specified in dB/λ by dividing by $40\pi \log_{10} e$.

At low frequencies, the significant acoustic energy in the ocean waveguide propagates radially outward from the source. As a result, the acoustic field may be represented as an outgoing Hankel function (carrier wave) that is modulated by a slowly varying envelope function [50, 76],

$$p(r, z) = \psi(r, z)H_0^{(1)}(k_0 r) \sim \psi(r, z)\sqrt{(2/\pi i k_0 r)}\exp ik_0 r. \tag{5.2}$$

Substituting Eq. (5.2) into Eq. (5.1) yields

$$\frac{\partial^2\psi}{\partial r^2} + \left[\frac{2}{H_0^{(1)}}\frac{\partial H_0^{(1)}}{\partial r} + \frac{1}{r}\right]\frac{\partial\psi}{\partial r} + \rho\frac{\partial}{\partial z}\left(\frac{1}{\rho}\frac{\partial\psi}{\partial z}\right) + k_0^2\left(N^2 - 1\right)\psi = 0. \tag{5.3}$$

For far-field propagation ($k_0 r \gg 1$), the factor modifying $\partial\psi/\partial r$ tends to $2ik_0\left[1 + O(k_0 r)^{-2}\right]$. Neglecting the $O(k_0 r)^{-2}$ term in Eq. (5.3) leads to

$$\frac{\partial^2\psi}{\partial r^2} + 2ik_0\frac{\partial\psi}{\partial r} + \rho\frac{\partial}{\partial z}\left(\frac{1}{\rho}\frac{\partial\psi}{\partial z}\right) + k_0^2\left(N^2 - 1\right)\psi = 0. \tag{5.4}$$

At this point, the paraxial approximation involves neglecting the term $\partial^2\psi/\partial r^2$ compared to the term $2ik_0\partial\psi/\partial r$. The resulting one-way wave equation is known as the standard variable-density parabolic equation of underwater acoustics,

$$2ik_0 \frac{\partial\psi}{\partial r} + \rho \frac{\partial}{\partial z}\left(\frac{1}{\rho}\frac{\partial\psi}{\partial z}\right) + k_0^2\left(N^2 - 1\right)\psi = 0. \tag{5.5}$$

The narrow angle interpretation for invoking the paraxial approximation can be illustrated qualitatively by examining its effect in the case of a homogeneous ocean [31, pp. 34–35]. The fundamental solution to Eq. (5.5) when $N = 1$ is given by

$$\psi_0(r, z) = \frac{\exp(ik_0 z^2/2r)}{\sqrt{r}}. \tag{5.6}$$

Computing the ratio of the neglected term to the retained term yields

$$\frac{\partial^2\psi_0/\partial r^2}{2ik_0\partial\psi_0/\partial r} = \frac{iz^2}{4r^2} + \frac{i}{4k_0 r}\cdot\left[1 + 2\frac{1 + 2ik_0 z^2/r}{1 + ik_0 z^2/r}\right], \tag{5.7}$$

which is clearly small at long ranges provided the tangent of the grazing angle, $\tan\theta = z/r$, is small.

5.2.2   Exact PE

In this section we present a heuristic method for developing a generalized one-way evolution equation for layered media. It is convenient to rewrite Eq. (5.5) in the compact operator form

$$\frac{\partial\psi}{\partial r} = \tfrac{1}{2}\, ik_0 X\psi, \tag{5.8}$$

where we have introduced the notation

$$X = k_0^{-2}\rho\frac{\partial}{\partial z}\left(\rho^{-1}\frac{\partial}{\partial z}\right) + \left(N^2 - 1\right). \tag{5.9}$$

The basic idea [14] is to estimate the neglected $\partial^2\psi/\partial r^2$ term by first differentiating Eq. (5.8) with respect to r and then adding $(2ik_0)^{-1}$ times this estimated term back into the left hand side of Eq. (5.8). The result is readily put into the form

$$\frac{\partial\psi}{\partial r} = \tfrac{1}{2}\, ik_0 X\left[1 + \tfrac{1}{4} X\right]^{-1}\psi. \tag{5.10}$$

Formally iterating on this procedure yields the continued fraction expansion

$$\frac{\partial\psi}{\partial r} = \tfrac{1}{2}\, ik_0 X\left[1 + \frac{\tfrac{1}{4}X}{1+}\frac{\tfrac{1}{4}X}{1+}\frac{\tfrac{1}{4}X}{1+}\cdots\right]^{-1}\psi \doteq \tfrac{1}{2}\, ik_0 XF(X)\psi. \tag{5.11}$$

It is straightforward [1, p. 19] to terminate the continued fraction after a finite number of terms and expand the resulting rational approximation in a series to verify that

$$\tfrac{1}{2} XF(X) = \tfrac{1}{2} X - \tfrac{1}{8} X^2 + \tfrac{1}{16} X^3 - \tfrac{5}{128} X^4 + \cdots \tag{5.12}$$

which leads to a generalized one-way PE in the closed form

$$\frac{\partial \psi}{\partial r} = ik_0 \left\{ -1 + \sqrt{1 + X} \right\} \psi. \tag{5.13}$$

Next, we derive the generalized one-way parabolic equation using an operator formalism [29, 50]. We begin by rewriting the elliptic wave equation for ψ in Eq. (5.4) in terms of the notation given in Eq. (5.9), i.e.,

$$\frac{\partial^2 \psi}{\partial r^2} + 2ik_0 \frac{\partial \psi}{\partial r} + k_0^2 X \psi = 0. \tag{5.14}$$

This equation can be factored (exactly for layered media) according to

$$\left(\frac{\partial}{\partial r} + ik_0 - ik_0 \sqrt{1 + X} \right) \left(\frac{\partial}{\partial r} + ik_0 + ik_0 \sqrt{1 + X} \right) \psi = 0. \tag{5.15}$$

Selecting only the factor satisfied by the outgoing wave component leads directly to Eq. (5.13). It is worthwhile pointing out that for ocean waveguides in which the medium properties vary with range, the operator $\partial/\partial r$ does not commute with the operator $\sqrt{1 + X}$. In this case, the factorization of Eq. (5.14) needs to be modified by adding $-ik_0$ times the commutator term

$$\left[\frac{\partial}{\partial r}, \sqrt{1 + X} \right] \psi = \frac{\partial}{\partial r} \left(\sqrt{1 + X} \cdot \psi \right) - \sqrt{1 + X} \cdot \frac{\partial \psi}{\partial r}. \tag{5.16}$$

In most of the widely used PE's in underwater acoustics, the range-dependence of the ocean waveguide is assumed to be sufficiently weak that this commutator contribution may be neglected (but see [42, 61]).

5.2.3   Propagator Approximations

Since the exact PE in Eq. (5.13) is first order in the range variable, it can be integrated formally from r to $r + \Delta r$ to yield a marching algorithm that forms the basis for all PE methods, i.e.,

$$\psi(r + \Delta r, z) = \exp i\delta \left(-1 + \sqrt{1 + X} \right) \psi(r, z), \tag{5.17}$$

where we have set $\delta = k_0 \Delta r$. Due to the presence of the pseudo-differential square-root operator, the formal solution in Eq. (5.17) does not directly submit to numerical treatment. Different versions of PE's are obtained from different approximations to either the square-root operator and/or to the propagator itself. In the following, we will develop a PE that will be used to illustrate the topics that are described in the later sections of this paper.

Expanding the square-root operator as a finite sum of rational-linear fractions yields the approximation [3, 18]

$$-1 + \sqrt{1 + X} \approx \sum_{m=1}^{M} \frac{a_m X}{1 + b_m X}, \tag{5.18}$$

where the coefficients a_m and b_m are given in closed form as

$$a_m = \frac{2}{2M+1} \sin^2 \frac{m\pi}{2M+1}, \quad b_m = \cos^2 \frac{m\pi}{2M+1}. \tag{5.19}$$

When $M = 1$, Eq. (5.19) gives $a_1 = 1/2$ and $b_1 = 1/4$, which leads at once to Eq. (5.13). This single-term, rational-linear PE is originally due to Claerbout [14] and is sometimes referred to as a wide-angle PE in ocean acoustics [56]. On the other hand, a two-term Taylor expansion of $\sqrt{1+X}$ in Eq. (5.13) immediately recovers the standard PE of Eq. (5.8). In the context of a finite-difference implementation, Claerbout's wide-angle PE requires no more computational effort to solve than the standard PE, while accommodating a wider-angled capability.

The real-valued coefficients given in Eq. (5.19) result in a square-root approximation that maps the real axis onto itself. In the context of waveguide propagation, this leads to inappropriate treatment of evanescent (in range) energy (corresponding to $X < -1$), which should map onto the positive imaginary axis. This behaviour can be overcome by imposing certain stability conditions on the series in Eq. (5.18) that result in complex-valued coefficients that must be determined numerically [19,24]. Alternatively, we choose to make use of an analytic procedure for obtaining complex rational coefficients to dampen the evanescent energy. Rotating the branch cut an angle α^* into the lower half-plane determines the new rational-linear expansion [63]

$$\sqrt{1+X} \approx c_0^* + \sum_{m=1}^{M} \frac{a_m^* X}{1 + b_m^* X}, \tag{5.20}$$

where the complex-valued coefficients a_m^*, b_m^* and c_0^* are given by

$$a_m^* = \frac{a_m e^{-i\alpha^*/2}}{[1 + b_m(e^{-i\alpha^*} - 1)]^2}, \quad b_m^* = \frac{b_m e^{-i\alpha^*}}{1 + b_m(e^{-i\alpha^*} - 1)}, \tag{5.21}$$

and

$$c_0^* = e^{i\alpha^*/2} \left(1 + \sum_{m=1}^{M} \frac{a_m(e^{-i\alpha^*} - 1)}{1 + b_m(e^{-i\alpha^*} - 1)} \right). \tag{5.22}$$

Substituting Eq. (5.21) and Eq. (5.22) into Eq. (5.17) yields

$$\psi(r + \Delta r, z) = \exp i\delta \, (c_0^* - 1) \cdot \exp \left\{ \sum_{m=1}^{M} \frac{a_m^* X}{1 + b_m^* X} \right\} \cdot \psi(r, z),$$

$$= \exp i\delta \, (c_0^* - 1) \cdot \prod_{m=1}^{M} \exp \frac{i\delta a_m^* X}{1 + b_m^* X} \cdot \psi(r, z),$$

$$\approx \exp i\delta \, (c_0^* - 1) \cdot \prod_{m=1}^{M} \frac{1 + c_m^+ X}{1 + c_m^- X} \cdot \psi(r, z), \tag{5.23}$$

where $c_m^{\pm} = b_m^* \pm \frac{1}{2} i\delta a_m^*$. Here we have approximated each exponential in the product by its second-order accurate Cayley form [70, p. 844], which requires that δ be sufficiently small. It is evident that Eq. (5.23) admits a recursive solution in terms of M systems of equations of the form

$$\left(1 + c_m^- X\right) \psi_m(r, z) = \left(1 + c_m^+ X\right) \psi_{m-1}(r, z) \tag{5.24}$$

where $\psi_0(r, z) \equiv \psi(r, z)$ and $\exp i\delta \left(c_0^* - 1\right) \psi_M(r, z) \equiv \psi(r + \Delta r, z)$ for $m = 1, \ldots, M$. Alternatively, the product-to-sum expansion

$$\prod_{m=1}^{M} \frac{1 + c_m^+ X}{1 + c_m^- X} = 1 + \sum_{m=1}^{M} \frac{d_m X}{1 + e_m X} \tag{5.25}$$

can be used to transform Eq. (5.23) into the equivalent parallel form

$$\psi(r + \Delta r, z) = \exp i\delta \left(c_0^* - 1\right) \left\{ \psi(r, z) + \sum_{m=1}^{M} \psi_m^*(r, z) \right\} \tag{5.26}$$

where each partial component ψ_m^* satisfies

$$\left(1 + e_m X\right) \psi_m^*(r, z) = d_m X \psi(r, z). \tag{5.27}$$

Inspection of Eq. (5.25) indicates that $e_m = c_m^-$. The d_m can be found numerically by solving the linear system that results from evaluating the right-hand-side at the M zeros, $X = -(1/c_m^+)$. Because each ψ_m^* in Eq. (5.27) depends on the same total field ψ on the right-hand-side, the numerical solution for $m = 1, \ldots, M$ can proceed in parallel if M processors are available.

It should be pointed out that a parallel setup similar to that given in Eq. (5.26) and Eq. (5.27) was first proposed by Collins [22]. Moreover, instead of dealing with a rational-linear expansion of the square-root operator, he chose to work directly with a rational-linear expansion of the one-way propagator in Eq. (5.17). Consequently, his approximation was not restricted to the small-δ approximation that underlies Eq. (5.23), and allows much larger range steps to be used which lead to greater efficiency. In his case, however, the coefficients of the partial fraction expansion must be determined by a numerical iteration procedure. A similar approach has also been implemented elsewhere [8, 57, 61].

5.2.4   Finite-Difference Scheme

The numerical solution of Eq. (5.26) by finite-differences is carried out on a discrete computational grid in depth z and range r. At each node on the grid, values of sound speed c, density ρ, and attenuation α, need to be specified. These values are often derived by interpolation from a specified set of environmental profiles provided at irregularly spaced locations. Nominally, the grid is terminated above by a free (or rigid) surface at $z = 0$ and below by an absorbing layer overlying a free (or rigid) surface at $z = z_{\max}$. The absorbing layer is necessary to prevent any unphysical reflections at the base of the computational grid from returning to the water column.

For the discretization in depth, it is convenient to introduce the offset grid vector $\mathbf{z} = [z_1, z_2, \ldots, z_J]^T$, where $[\cdots]^T$ denotes vector transpose and $z_j = (j - 1/2)\Delta z$, $j = 1, \ldots, J$. The use of offset depths avoids the need to compute the field along the top and bottom of the computational domain where it is known to vanish if either boundary is a free surface. In addition, the implementation of either pressure-release or rigid boundary conditions can be effected in a symmetric way. In terms of the $\mathbf{z}$-grid, we apply a heterogeneous tridiagonal approximation for $X(\mathbf{z})\psi(\mathbf{z})$ in the form

$$X(\mathbf{z})\psi(\mathbf{z}) \approx L(\mathbf{z})\psi(\mathbf{z} - \Delta z) + D(\mathbf{z})\psi(\mathbf{z}) + U(\mathbf{z})\psi(\mathbf{z} + \Delta z), \qquad (5.28)$$

where the lower diagonal, upper diagonal and diagonal elements are given by

$$L(\mathbf{z}) = \gamma\rho^-(\mathbf{z}), \quad U(\mathbf{z}) = \gamma\rho^+(\mathbf{z}), \quad D(\mathbf{z}) = N^2(\mathbf{z}) - 1 - L(\mathbf{z}) - U(\mathbf{z}), \quad (5.29)$$

respectively, and we have introduced the z-step and density-ratio quantities

$$\gamma = 1/(k_0\Delta z)^2, \quad \rho^\pm(\mathbf{z}) = 2\rho(\mathbf{z})/[\rho(\mathbf{z}) + \rho(\mathbf{z} \pm \Delta z)]. \qquad (5.30)$$

The use of this heterogeneous form precludes the need for explicitly enforcing continuity of pressure, p, and the normal component of particle velocity, $(i\omega\rho)^{-1}\partial p/\partial z$, at any jump discontinuities in medium properties. In effect, once values of ρ, c, and α are specified on $\mathbf{z}$ there are no internal interfaces between material properties. From Eq. (5.30), it is seen that $\rho^\pm(\mathbf{z}) \to 1$ for constant density media and the weighted three-term approximation to the mixed derivative term for $X\psi$ in Eq. (5.28) reduces to the standard central-difference form for the second derivative of ψ.

Substituting Eq. (5.28) into Eq. (5.27) produces a tridiagonal matrix system for the split-step Padè algorithm whose jth row is given by

$$[1/e_m + X(z_j)]\,\psi_m(r + \Delta r, z_j) = (d_m/e_m)\,X(z_j)\psi(r, z_j). \qquad (5.31)$$

The diagonal matrix entries in the top ($j = 1$) and bottom ($j = J$) rows are modified by the boundary conditions imposed along $z = 0$ and $z = z_{\max}$, respectively. For a pressure-release surface, the antisymmetry of the field about $z = 0$ is ensured by setting $\psi_m(r, -z_1) = -\psi_m(r, z_1)$. This condition is implemented numerically by subtracting $L(z_1)$ from the diagonal entry $D(z_1)$. In a similar way, the bottom boundary condition corresponding to odd symmetry about $z = z_{\max}$ is handled by subtracting $L(z_J)$ from the diagonal entry $D(z_J)$.

5.2.5   Profile Interpolation

For the propagator approximation given in Eq. (5.26), typical grid spacings for shallow water low-frequency predictions are of the order of 1 m in depth and 5 m in range. Since environmental information rarely exists on this scale, the acoustic modeller is faced with the issue of interpolating coarse environmental information, usually in the form of depth profiles at specified ranges, onto the PE computational grid. Consider a sequence of coarse environmental profiles at the known ranges R_k km, $k = 1, \ldots, K$. For simplicity, each profile is restricted to have points at the same number of depths Z_ℓ^k m, $\ell = 1, \ldots, L$. The set of depth nodes will generally be different for each value of R_k. Here, we anticipate that the set of coarse profiles has been preprocessed to preserve significant oceanographic/geophysical features between profiles, such as the depth of the sound channel axis or surface duct and the sediment thickness beneath a sloping bathymetry. We assume that all environmental parameters (sound speed, density, absorption) vary linearly with depth between points on a given coarse profile. Moreover, between adjacent coarse profiles, we assume that each parameter varies linearly between depth points that have the same depth index. Any discontinuities in material properties, such as occur at the sea-bottom interface, are accommodated simply by including a pair of coarse profile points that are displaced in depth by a tiny amount, e.g., 1 cm. The resulting monotonic sequence of coarse profile depths can be safely processed by standard interpolation algorithms.

Let $E(Z_\ell^k)$ represent any of the material parameters at the depth of the ℓth point on the coarse profile at range R_k. For an intermediate range $R_k < R' < R_{k+1}$, the environment on the computational grid is determined using a two-step linear interpolation procedure. First a coarse profile at R' is obtained using

$$Z_\ell' = (1 - q)Z_\ell^k + qZ_\ell^{k+1}, \tag{5.32}$$

$$E(Z_\ell') = (1 - q)E(Z_\ell^k) + qE(Z_\ell^{k+1}), \tag{5.33}$$

where $q = (R' - R_k)/(R_{k+1} - R_k)$. Values for each material property on the computational grid associated with this intermediate coarse profile are then determined using linear interpolation between adjacent depths Z_ℓ'.

5.2.6   Energy Conservation

The derivation of the one-way propagation equation relies on factoring the Helmholtz equation into outgoing and incoming components and neglecting the coupling to the backscattered fields. Although this is formally exact only for range-independent waveguides, the algorithm is routinely applied to range-dependent cases in which the environment is modelled as a sequence of range-independent sections having different properties. That is, since the resulting propagators march the PE field step-by-step outward in range, the environment is simply updated after each range step and the coefficients implied by the terms in Eq. (5.27) are modified accordingly. Even for relatively benign range-dependent environments, however, this approach may not yield sufficiently accurate results [49]. The inaccuracy is related to the fact that, at abrupt changes in the environment between range sections, two boundary conditions need to be satisfied at the corresponding vertical interface in order to properly account for energy transfer along the waveguide. Of course, the one-way PE can only satisfy one boundary condition there [68]. Subsequent analysis [28] has shown, however, that if ψ is replaced by the scaled field $\tilde{\psi} \equiv \psi/\beta$ where $\beta = \sqrt{\rho(z)c(z)}$ in the PE algorithm, then a good approximation to the true energy-conserving condition is realized. To achieve this numerically, we replace Eq. (5.27) by the "energy-conserving" variant

$$\left(1 + e_m \tilde{X}\right) \tilde{\psi}_m^*(r, z) = d_m \tilde{X} \tilde{\psi}(r, z). \tag{5.34}$$

where the depth operator $\tilde{X}$ is defined by

$$\tilde{X} = N^2 - 1 + k_0^{-2} \sqrt{\frac{\rho}{\beta}} \frac{\partial}{\partial z} \left(\rho^{-1} \frac{\partial}{\partial z} \beta \right). \tag{5.35}$$

5.2.7 Equivalent Fluid

The current focus on shallow water propagation, where the physics of the bottom interaction must be taken into account, implies that PE models must be capable of handling the effects due to shear rigidity in the sediments. Although a fully elastic PE model could be employed for this purpose [50], it is desirable from an efficiency standpoint to have a less computationally intensive solution. Fortunately, for many problems, it is possible to represent the effects of shear on the propagation in the water column through the use of an "equivalent" fluid approximation [97]. The approach is based on choosing fluid parameters to match the reflection coefficient of the actual solid bottom. One way to achieve this in PE-based models is to convert the shear parameters into a complex density of the form

$$\rho_b' = \rho_b \left[\left(1 - 2/N_s^2\right)^2 + \frac{4i\gamma_s\gamma_b}{k_0^2 N_s^4} \right], \tag{5.36}$$

where ρ_b is the true value of density in the sediment, $N_s = (c_0/c_s)(1 + i\alpha_s)$, and c_s and α_s are the sediment shear speed and attenuation, respectively. In Eq. (5.36), the quantities $\gamma_s = k_0\sqrt{N_s^2 - 1}$ and $\gamma_b = ik_0\sqrt{1 - N_b^2}$ are the respective vertical wavenumbers of the shear and compressional waves in the sediments. Choosing the density to be complex in this way allows the plane-wave reflection coefficient of the fluid-elastic sub-bottom to be approximated by the reflection coefficient of an equivalent fluid for a range of angles that correspond to the propagating modes. Although the value of c_0 may require adjustment to optimize the matching (it is a free parameter), this does not significantly affect high-order PE algorithms which are inherently capable of modelling wide-angle propagation and, hence, insensitive to the value of c_0 that is chosen.

5.2.8   Initial Field

Before a PE algorithm can advance the solution of a one-way equation outward in range, an initial condition (starting field) is required at the range of the source. There are a number of simple starting fields that approximate asymptotically the field due to a point source embedded within a homogeneous half-space bounded by a pressure-release surface along $z = 0$ [50]. One source field that does account for all of the environmental information in the vicinity of a source located at $r = 0$, $z = z_0$ is the so-called self-starter [20]. It is sufficient for the calculations to be carried out in this paper to use a simple source field with good wide-angle properties that is given analytically in the form of a weighted Gaussian distribution [44],

$$S(0, z) = \sqrt{k_0} \left[1.4467 - 0.4201 k_0^2 \left(z - z_0\right)^2 \right] \exp \left[\frac{-k_0^2 \left(z - z_0\right)^2}{3.0512} \right]. \quad (5.37)$$

5.2.9   Perfectly Matched Absorber

Each one-way wave equation must be supplemented with relevant boundary conditions along the cross-range edges of the computational grid. For underwater sound propagation, the ocean surface is usually modelled simply as a pressure-release ($p = 0$) boundary. A more complicated procedure is required, however, to treat the interaction of the field beneath a penetrable sea-bottom where a downgoing radiation condition usually applies. Numerous absorbing boundary conditions of varying degrees of accuracy have been developed to remove this component of the acoustic field. The most direct procedure is to introduce a physical absorbing layer designed to remove a large fraction of the incident field by an attenuation mechanism without introducing spurious reflections from the base of the computational grid [35, 76, 80, 94]. This can be accomplished provided the absorber strength is increased gradually with depth. Generally, the absorber is terminated by imposing a pressure-release condition at the base of the layer. For some problems, the use of a physical absorber necessitates a large number of grid points, and can be computationally inefficient. Consequently, for our work we make use of an alternative absorbing mechanism (impedance matching) that is equivalent to introducing a fictitious imaginary component into the transverse coordinate, i.e., the z-direction. Such a procedure greatly reduces any unphysical (and unwanted) reflections from the boundary layer, independent of frequency, except at nearly grazing angle of incidence [5, 13, 16, 59, 96].

For an impedance-matched absorbing layer of thickness h, a z-dependent imaginary stretching of the grid point spacing is obtained by setting the complex distance between grid points z_l and z_{l+1} for $l = 1, \ldots, L$ to

$$\Delta z_l^* = [1 + ia\left(z_l\right)] \Delta z, \quad (5.38)$$

where the tapered profile $a(z)$, necessary for reducing discretization induced

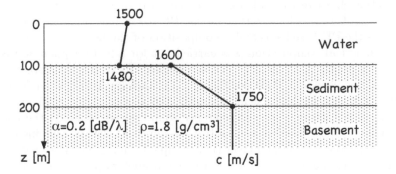

FIGURE 5.3: Nominal environmental parameters for '93 MFP Workshop shallow-water synthetic dataset [69].

reflections within the layer, is chosen to have the form

$$a(z) = \tfrac{1}{2}\, p\,\{1 - \cos\left[(z - z_J)\pi/h\right]\}^q\,.$$ (5.39)

Equation (5.39) is used in our subsequent finite-difference calculations incorporating perfectly matched layer absorbers.

Within the absorbing layer, the numerical implementation of the loss mechanisms differ for the two procedures. This is seen by inspection of the standard PE of Eq. (5.5) where we assume that the density is constant within the layer. For the physical absorber, the introduction of loss into the refractive index N affects the coefficient of ψ. In contrast, for the perfectly matched layer, the introduction of a stretched depth coordinate z^* affects the computation of the Laplacian. To implement the above z-dependent stretching of the imaginary part of the z-coordinate, the finite-difference approximation to the Laplacian operator is constructed from the standard three-point expression for non-equidistant grid points z_{l-1}, z_l, and z_{l+1}, namely,

$$\left.\frac{\partial^2\psi}{\partial z^{*2}}\right|_{z=z_l} \approx \frac{2/\Delta z^2}{2 + i a_l + i a_{l+1}}\left\{\frac{\psi_{l+1} - \psi_l}{1 + i a_{l+1}} - \frac{\psi_l - \psi_{l-1}}{1 + i a_l}\right\}.$$ (5.40)

5.2.10   Propagation Example

We conclude this section with a canonical numerical example of acoustic propagation in a shallow-water waveguide. A finite-difference PE code that incorporates the features given earlier in this section has been used to calculate the field. The environmental parameters that were used in the calculation are given in Fig. 5.3 and correspond to the nominal waveguide configuration that formed the basis of the synthetic datasets provided by the organizers of the 1993 workshop on benchmarking matched field processing methods [69]. The propagation characteristics of this waveguide will allow us in later sections to illustrate both source localization by the PE-based backpropagation scheme as well as the PE-based spectral decomposition of the field.

The numerical computation was carried out for a 250-Hz point source located at a depth of 82 m (cf., COLNOISE case (b) in [69]). Also, the PE computations used four Padè coefficients, a branch cut rotation angle of $10°$, a range step size of $\Delta r = 2$ m, a depth step size of $\Delta z = 0.25$ m, and a value of $c_0 = 1500$ m/s for the reference sound speed. Finally, the computational grid was extended vertically to a depth of 250 m and included a 20-point impedance-matched layer that occupied the region $245 < z < 250$ m. This layer was characterized by the taper parameters $p = 10$ and $q = 2$ in Eq. (5.39).

A full-field plot of the transmission losses out to a range of 10 km is shown in Fig. 5.4. The TL scale varies from 50 dB (white) to 100 dB (black). It is evident that the field structure exhibits a complicated interference pattern in the water column with a well-defined null in the vicinity of 6.5 km. Near the source, the field is seen to penetrate the sediments where it is absorbed by the perfectly matched layer. At longer ranges, the field is dominated by the propagating modes that are confined by the surficial sediments just below the sea-bottom interface.

5.3   PE-Based Matched Field Processing

Accurate predictions of the coherent sound field in a shallow-water waveguide have made possible the development of model-based signal processing schemes for the passive localization of a sound source (range and depth) or the determination of environmental properties that affect acoustic propagation [87]. Source localization by matched field processing, for example, is carried out by cross-correlating the fields measured on an array of hydrophones with hypothetical fields (replicas) predicted using a propagation model. These replica fields depend on known (or assumed) input parameters, such as the candidate ranges and depths for the source position, and the environmental parameters of the waveguide. The candidate source position showing the highest correlation between measured and replica fields is taken to correspond to the best estimate of the true location of the source.

5.3.1   Standard Processor

For tonal signals, the basic data that is input to the matched-field processor consists of the vector $\mathbf{d} = [d_1, d_2, \ldots, d_N]^T$ of complex-valued pressures obtained from an array of N hydrophones. In practice, a vertical line array

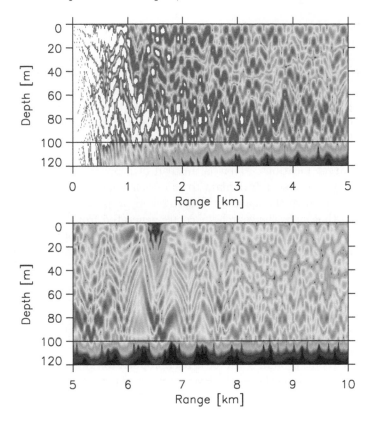

FIGURE 5.4: Full-field transmission losses at 250 Hz for the environmental parameters of Fig. 5.3. The source is at a depth of 82 m. The TL scale varies from 50 dB (white) to 100 dB (black).

(VLA) of sensors that span a significant portion of the water column is often used to acquire signals in littoral regions. The conventional (Bartlett) scheme compares these data to replica fields $\mathbf{p}(r, z) = [p_1, p_2, \ldots, p_N]^T$ computed for each location (r, z) on the search grid by forming the scalar product

$$h_1(r, z) = \frac{|\sum_n d_n^* p_n(r, z)|^2}{\sum_n |d_n|^2 \sum_n |p_n(r, z)|^2} = \frac{|\mathbf{d}^* \mathbf{p}(r, z)|^2}{|\mathbf{d}|^2 |\mathbf{p}(r, z)|^2} \equiv \frac{\mathbf{p}^*(r, z) \mathbf{C} \mathbf{p}(r, z)}{|\mathbf{d}|^2 |\mathbf{p}(r, z)|^2},$$
(5.41)

where $\mathbf{C} = \mathbf{d}\mathbf{d}^*$ is the cross spectral matrix and * denotes conjugate transpose. Here the vector $\mathbf{p}(r, z)$ is a prediction of the acoustic field on the array due to a point source located at range r and depth z. The evaluation of Eq. (5.41) for all candidate ranges and depths on the search domain produces an ambiguity surface that is typically multimodal, i.e., characterized by a peak value and several sidelobes. When several observations $\mathbf{d}_\ell$ of the field at successive times on the array are available, an improved estimate of the cross spectral matrix

$\mathbf{C}$ can be obtained by forming the average value $(1/L)\sum_{\ell=1}^{L} \mathbf{d}_\ell \mathbf{d}_\ell^*$, where L is the total number of time frames.

As indicated previously, there are a number of sophisticated propagation models that are available for computing coherent replica vectors $\mathbf{p}$ in range-independent waveguides. For non-adiabatic, range-dependent environments, however, only one-way models that are derived from the parabolic approximation satisfy the dual requirements of accuracy and computational efficiency. In practice, PE-based models make use of the reciprocity of acoustic fields to generate the replica fields on a specified grid that spans the search domain of potential target locations. That is, instead of propagating the field from each node (r, z) on the search domain to the N-elements on the VLA (which would typically involve thousands of step-by-step propagation runs), replica coverage fields that span the search space are generated from each sensor position on the VLA (which typically requires just tens of propagation runs). At each point on the search grid, N complex-valued replicas are available for matching against the N complex-valued measured data.

5.3.2   Backpropagated Processor

In contrast to the above methodology, a novel source localization scheme termed "acoustic retrogation" was introduced to the underwater acoustics community in 1985. It makes use of a backpropagation algorithm based on the parabolic equation method [64]. By combining the principle of reciprocity with linear superposition of acoustic fields, data from all N sensors of a VLA can be backpropagated simultaneously to generate a correlation surface in a single PE propagation run. The retrogation technique initiates the solution procedure by forming the scalar product $\mathbf{d}^*\mathbf{S}(0, z)$ at the array site for each point on the PE computational grid. Specifically, for each sensor centered at depth z_n, we weight the vertically extended PE source field $S(z)$ given in Eq. (5.37) by the complex conjugate of the signal received on the nth hydrophone and then add the weighted fields from all N sensors. This correlation measure satisfies the one-way PE and its boundary conditions, and allows the ambiguity surface to be propagated outward from the array toward the potential target locations spanning the search grid. Instead of the normalized correlation function given in Eq. (5.41), this backpropagation method yields the unnormalized correlation measure

$$h_2(r, z) = \frac{\sum_n |d_n^* p_n(r, z)|^2}{\sum_n |d_n|^2} = \frac{|\mathbf{d}^*\mathbf{p}(r, z)|^2}{|\mathbf{d}|^2}. \tag{5.42}$$

Calculations of h_2 based on Eq. (5.42) proceed N times faster than those for h_1 where the N individual replica fields must be generated and saved before the vector norm $||\mathbf{p}|| = \sqrt{\mathbf{p}^*\mathbf{p}}$ can be computed for use in Eq. (5.41). In practice, Eq. (5.42) can be partially normalized for cylindrical spreading in the far field by replacing $\mathbf{p}$ with $\mathbf{p}\sqrt{r}$. The high correlation levels in the vicinity of the VLA can be avoided by excluding this near field region from the search domain. As pointed out elsewhere [87], a potential drawback of the backpropagation processor in Eq. (5.42) compared to the forward modelling matched field processor in Eq. (5.41) is the difficulty of re-introducing attenuated energy in order to determine the correct signal level at the source location. This effect can be ameliorated somewhat by setting sediment absorption to zero during the backpropagation procedure. From an engineering perspective, there is an accuracy/runtime trade-off to be considered between the normalized accuracy of Eq. (5.41) and the N-fold speedup factor of the unnormalized measure of Eq. (5.42).

While Eq. (5.41) can deal with the averaged cross spectral matrix directly, for Eq. (5.42) there is the issue of how to obtain a good estimate of $\mathbf{d}$ from the averaged matrix $\mathbf{C}$. Suppose the observed data $\mathbf{d}_\ell$ consists of spatially-white, Gaussian noise w_ℓ added to a signal of the form $\mathbf{x}_\ell = A \exp(i\theta_\ell)\mathbf{u}$. Here $\mathbf{u}$ is an N-dimensional complex vector of unit norm ($\|\mathbf{u}\| = 1$), A is the amplitude and θ_ℓ is a random phase angle. Then, it can be shown that the maximum likelihood estimate for the value $\hat{\mathbf{d}}$ ($\|\hat{\mathbf{d}}\| = 1$) that maximizes the sum of correlations $|\mathbf{d}^*\mathbf{u}|$ over the L snapshots is given approximately by

$$\hat{\mathbf{d}} = \arg \max_{\|\mathbf{u}\|=1} \mathbf{u}^*\mathbf{C}\,\mathbf{u}. \tag{5.43}$$

The solution to Eq. (5.43) can be found from some standard results in matrix analysis. Moreover, the maximum eigenvalue ν_1 of the N (real) eigenvalues of $\mathbf{C}$ can be expressed as the maximum of the Rayleigh quotient [4]

$$\nu_1 = \max_{\mathbf{v} \neq 0} \frac{\mathbf{v}^*\mathbf{C}\,\mathbf{v}}{\|\mathbf{v}\|^2}. \tag{5.44}$$

This maximum is assumed when the (orthonormal) vector $\mathbf{v}$ is the eigenvector $\mathbf{v}_1$, i.e., $\nu_1 = \mathbf{v}_1\mathbf{C}\,\mathbf{v}_1$. It follows that $\hat{\mathbf{d}} = \mathbf{v}_1$ is the solution to Eq. (5.43). Numerically, then, we perform the eigen-decomposition of the averaged cross-spectral matrix $\mathbf{C}$, and use the eigenvector corresponding to the largest eigenvalue as the best estimate of the signal at the array hydrophones. This result is similar to one obtained elsewhere for a problem in bearing estimation [47].

5.3.3   Localization Example

To demonstrate the localization capability of the Bartlett and backpropagation processors in Eq. (5.41) and Eq. (5.42), we consider the canonical shallow water waveguide propagation example introduced in the previous section. The parameters of this waveguide are given in Fig. 5.3. Synthetic VLA data for this set of parameters, and perturbations from it, at a frequency of 250 Hz due to point sources at ranges within $5 < r < 10$ km from the vertical array are presented in [69]. The data signal vectors for a vertical array of 20 sensors at depths $5, 10, \ldots, 100$ m for several environmental scenarios were generated using the normal mode code KRAKEN to which 300 realizations of

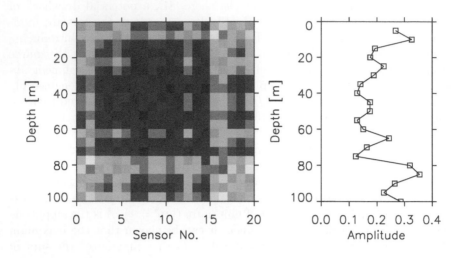

FIGURE 5.5: Cross spectral matrix for COLNOISE b case in [69] (left) and eigenvector $\mathbf{v}_1$ corresponding to its largest eigenvalue ν_1 (right).

Gaussian random noise were added to yield averaged cross spectral density matrices corresponding to single sensor signal-to-noise ratios of 20, 10, and -5 dB [69]. These noisy synthetic data sets are available from the Ocean Acoustic Modelling Library web site cited in the caption of Fig. 5.1.

The left panel of Fig. 5.5 shows the averaged cross spectral density matrix for the COLNOISE b case and corresponds to a signal-to-noise ratio of 10 dB. For this test case, the random noise that is most represenative of electrical noise on the array is replaced by a noise model that is most appropriate for surface noise due to breaking waves [69]. The right panel shows the eigenvector $\mathbf{v}_1$ that corresponds to the largest eigenvalue ν_1 of this averaged matrix.

Ambiguity surfaces for both processors are presented in Fig. 5.6. The top panel shows the output of the backpropagated processor and the bottom panel shows the output of the standard Bartlett processor. For display purposes, each ambiguity surface has been normalized to unity by its peak value. Both processors are seen to focus at the true source location of $r = 7.1$ km and $z = 82$ m for this dataset. The peak-to-sidelobe values are observed to differ between the two processors, however, due to the fact that the backpropagated processor is not normalized by the norm of the replica fields (no replicas are computed in this case). For this example, the backpropagated processor is 20 times faster and 20 times more memory efficient than the Bartlett processor. It is worthwhile remarking that for an environment that is not changing significantly over an observation interval, once the replicas have been pre-

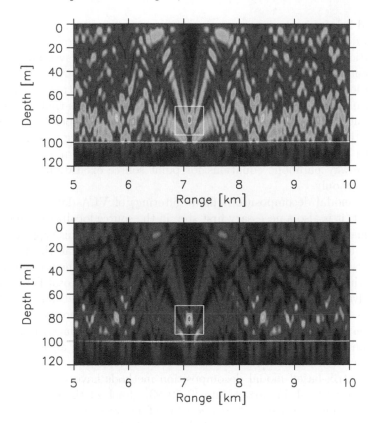

FIGURE 5.6: Ambiguity surfaces for the backpropagated (top) and the Bartlett processors (bottom) for the environment of Fig. 5.3 and the data of Fig. 5.5. The square box is centered on the true source location.

computed and saved, the Bartlett processor only involves matching successive data vectors against the replicas. Consequently, this straightforward matching could be carried out more efficiently than successively backpropagating a new ambiguity surface for each updated data vector that is required by the retrogation scheme. The replica storage requirements are still significantly more demanding for the Bartlett scheme however.

5.4   Modal Decomposition

The above matched-field processors require, as data, the frequency-domain complex pressure fields measured on an array of hydrophones. It is often convenient and helpful in the interpretation of complex propagation scenarios to be able to decompose these total field data into their spectral components. This is particularly true for propagation in range-dependent waveguides where mode coupling and/or mode cutoff effects may be important. Acoustic model representations that are based on normal modes or ray path trajectories provide this capability inherently. Additionally, because models that are based on wavenumber integration must first determine the depth-dependent Green's function with respect to horizontal wavenumber, they also are capable of exhibiting the modal structure of the field. In contrast, direct numerical solutions to one-way parabolic equations for point source excitation yield total pressure fields only.

Although modal decomposition (mode filtering) of VLA data is useful in its own right, it is also a necessary first step in the source localization method known as matched mode processing [87]. The second step of this type of signal processing, also known as modal beamforming, requires that the matched mode processor correlates the complex modal amplitudes of the normal mode replicas with those that are determined from the measured total fields. One possible advantage of the matched mode processor over the matched field processor is the opportunity of correlating a subset of the modes that make up the observed VLA field. Prior to processing, the modal data (and modal replicas) can be filtered to exclude those modes that are most affected by any mismatch between the modelled and true ocean envionments.

Several matrix-based modal decomposition methods have been developed for processing vertical line array data [9, 79, 93]. Each of these methods requires that the modal wavenumbers and shape functions that are appropriate for the environment in the vicinity of the VLA to be supplied by a normal mode model. Alternatively, a wavenumber integration scheme for numerically decomposing the field versus depth in a range-dependent waveguide was proposed and applied to PE-generated fields propagating in a wedge-shaped ocean [52]. By interpreting the PE field at a given range as a distributed source in the SAFARI code [72], the "local" range-independent Green's function corresponding to the given source field can be extracted. By carrying out this decomposition at different ranges in a range-dependent waveguide, it is possible to identify and study the propagation of individual modes in both adiabatic and coupled mode situations.

In contrast to the above spectral decomposition approaches, which require either normal mode or wavenumber integration codes, we present a PE-based decomposition scheme that is directly related to the backpropagation algorithm presented in the previous section. The basic tenets of this scheme were originally developed for analyzing the propagation of light in optical fibers [37–41]. This PE-based decomposition scheme proceeds from two related concepts: (1) An exact relationship between PE modes and normal modes; and (2) Spectral analysis of a suitably defined PE correlation function. Although both concepts hold true exactly for propagation in layered environments, the scheme can also be applied to fields propagating in range-varying waveguides. Moreover, by invoking the accurate multi-term Padè approximation to the one-way PE propagator given in Eq. (5.23), the need for exploiting the connection relation in (1) may be bypassed.

5.4.1 Relation between PE Modes and Normal Modes

For layered media, it can be shown that the acoustic pressure p in Eq. (5.1) can be recovered from the solution ψ to the standard parabolic equation in Eq. (5.5) by evaluating the *exact* integral transform [30]

$$p(r, z) = \frac{1}{\sqrt{2\pi i}} \int_0^\infty \psi(t, z) \exp\left[\frac{ik_0}{2t}\left(r^2 + t^2\right)\right] \frac{dt}{t}. \tag{5.45}$$

That is, numerical solutions to Eq. (5.5), which can be obtained efficiently by marching algorithms, can be postprocessed using Eq. (5.45) into solutions to Eq. (5.1). For numerical work, it is convenient to carry out this postprocessing in the appropriate wavenumber domains of p and ψ, where the nonlocal relationship between the fields in Eq. (5.45) becomes a local one [86]. The connections between the PE modes of Eq. (5.5) and the normal modes of Eq. (5.1) can be found by substituting the modal expansion of ψ, namely

$$\psi(t, z) = \sum_j a_j u_j(z) \exp(is_j r), \tag{5.46}$$

into Eq. (5.45), where the a_j are the PE modal amplitudes. It follows that the eigenfunctions u_j satisfy the eigenvalue equation [30]

$$\rho\frac{d}{dz}\left(\rho^{-1}\frac{du_j}{dz}\right) + k_0^2\gamma_j^2 u_j = 0, \tag{5.47}$$

where $\gamma_j^2 = N^2 - 1 + 2s_j/k_0$ is the square of the jth vertical wavenumber of the standard PE. If the PE eigenvalues (modal wavenumbers) s_j are expressed in terms of the new eigenvalues k_j by the mapping

$$s_j = \frac{k_j^2 - k_0^2}{2k_0}, \tag{5.48}$$

then Eq. (5.47) is recognized as the ordinary differential equation associated with the normal mode eigenfunctions of the acoustic wave equation Eq. (5.1) and the k_j are identified as their corresponding horizontal wavenumbers. The u_j can be normalized to form a complete orthonormal set. The usual boundary conditions imposed in underwater acoustics require each u_j to vanish at $z = 0$ and remain bounded as $z \to \infty$. As a result, Eq. (5.47) admits only a finite number of propagating modes.

Substituting Eq. (5.46) into Eq. (5.45) and using Eq. (5.48) yields the sequence

$$
p(r,z) = \frac{1}{\sqrt{2\pi i}} \sum_j a_j u_j(z) \int_0^\infty \exp\left[\frac{ik_j^2 t}{2k_0} + \frac{ik_0 r^2}{2t}\right] \frac{dt}{t}
$$

$$
= \sqrt{\frac{\pi i}{2}} \sum_j a_j u_j(z) H_0^{(1)}(k_j r) \approx \sum_j a_j u_j(z) \frac{\exp(ik_j r)}{\sqrt{k_j r}}, \qquad (5.49)
$$

where both the integral representation of the Hankel function $H_0^{(1)}(k_j r)$ [43, p. 956, No. 8] and the first term of its asymptotic expansion given in Eq. (5.2) were used. The right-hand side of Eq. (5.49) is recognized as the modal expansion of Eq. (5.1) into normal modes. In this case, the modal amplitudes have the known form [2, 50]

$$
a_j = A u_j(z_0), \qquad (5.50)
$$

where A is a complex constant. If the depth z_0 of a point source is located at a null of the jth mode, mode j will not be excited. Similarly, if a hydrophone is placed at a null of the jth mode, mode j will not contribute to the total field there. From Eq. (5.46), Eq. (5.48), and Eq. (5.49), it is seen that the amplitudes a_j' and wavenumbers k_j of the normal modes can be obtained *a posteriori* from the amplitudes a_j and wavenumbers s_j of the PE modes by the local mappings

$$
a_j' = \frac{a_j}{\sqrt{k_j}} \equiv \frac{A u_j(z_0)}{\sqrt{k_j}}, \qquad (5.51)
$$

and

$$
k_j = k_0 \sqrt{1 + \frac{2s_j}{k_0}} \approx k_0 + s_j. \qquad (5.52)
$$

For values of k_j near k_0, $s_j \approx 0$, and the nonlinear relation between the k_j of the wave equation and the s_j of the standard PE is given approximately by the linear shift k_0 that corresponds to the carrier wave connection between p and ψ expressed in Eq. (5.2).

5.4.2   Modal Excitations

From the preceding analysis, it is clear that the spectral properties of ψ lead directly to the spectral properties of p. An efficient algorithm for spectrally analyzing ψ, and thus p, that was previously developed for analyzing electromagnetic propagation in optical waveguides [37–41] was subsequently adapted to treat acoustic propagation [83, 85, 86]. Alternatively, instead of solving the standard PE for ψ and then postprocessing the ψ-field to yield the normal mode properties of p, it is also possible to spectrally analyze accurate solutions to the multi-term Padè one-way pressures p directly. In this case, the higher-angle approximations to the square-root operator and its associated propagator intrinsically yield accurate estimates of the normal mode wavenumbers k_j and shape functions u_j.

For processing measured data, it was pointed out in [83] that because it was not feasible to build a VLA with sensor spacings as small as the depth grid spacings required for accurate solutions to the one-way PE, an interpolation/extrapolation scheme was necessary for mapping the measured VLA data onto the PE computational grid. With the backpropagation approach described in the previous section, however, it is not necessary to carry out this step. Indeed, the unnormalized processor of Eq. (5.42), in which the measured VLA data are combined with vertically extended PE source fields, automatically determines the field at the resolution of the PE grid.

To estimate the modal wavenumbers and amplitudes associated with the pressure field observed on an N-element VLA, we first consider the modal expansion of the partially normalized backpropagated field $q(r, z) = \mathbf{d}^* \mathbf{p}(r, z) \sqrt{r}$. From Eq. (5.49) and Eq. (5.51), we find that

$$
\begin{aligned}
q(r, z) &= \sum_n d_n^* p_n(r, z) \sqrt{r} \\
&= \sum_n d_n^* \left\{ \sum_j \left(A / \sqrt{k_j} \right) u_j(z_n) u_j(z) \exp(i k_j r) \right\} \\
&= \sum_j g_j u_j(z) \exp(i k_j r),
\end{aligned}
\tag{5.53}
$$

where $g_j = (A / \sqrt{k_j}) \sum_n d_n^* u_j(z_n)$. Next, we introduce the acoustic PE backpropagated correlation function defined by

$$
\Gamma(r_0, r) = \int_0^\infty \rho^{-1}(z) q^*(r_0, z) q(r_0 + r, z) dz,
\tag{5.54}
$$

where $q(r_0 + r, z)$ is the (complex) one-way backpropagated field at range r from the VLA at range r_0 that is used to initiate the marching solution. The sequence of values $\Gamma(r_0, r_\ell)$ for $r_\ell = \ell \Delta r$, $\ell = 0, 1, \ldots, L - 1$, is readily computed during the step-by-step numerical solution of Eq. (5.5). On substituting the modal representation for q in Eq. (5.53) into Eq. (5.54), and noting that the u_j are orthonormal with respect to the weight function ρ^{-1}, it follows that the correlation function Γ assumes the analytical form

$$
\Gamma(r_0, r) = \sum_j |g_j|^2 \exp(i k_j r).
\tag{5.55}
$$

The Fourier transform of Eq. (5.55) yields

$$
\mathcal{F}\{\Gamma\} = \int_0^\infty \Gamma(r_0, r) \exp(-i k r) dr = \sum_j |g_j|^2 \delta(k - k_j),
\tag{5.56}
$$

which is seen to contain delta functions at the modal wavenumbers k_j of the normal modes. The strength of $\delta(k - k_j)$ corresponding to the jth mode is proportional to the power $|g_j|^2$ contained in that mode. In practice, Eq. (5.56) is implemented using a finite length discrete Fourier transform, and the k_j and $|g_j|^2$ are estimated from the finite peaks in the spectrum.

5.4.3   Modal Phases

It is seen from Eq. (5.55) that the correlation function Γ given in Eq. (5.54) does not contain the phase information that is accumulated by each mode as it propagates the distance r_0 between the source and receiving array. However, this range phase information for each mode, i.e., $\exp ik_j r_0$, is contained in the one-way backpropagated field q that is generated from the initial field data at r_0. The signal phase that is associated with each mode can be extracted by spectrally analyzing this propagated q-field. From the modal expansion in Eq. (5.53) we find

$$
\mathcal{F}\{q\} = \int_0^\infty q(r_0 + r, z) \exp(-ikr) dr
$$
$$
= \sum_j g_j u_j(z) \exp(ik_j r_0) \delta(k - k_j). \tag{5.57}
$$

Substituting for g_j in Eq. (5.53) into Eq. (5.56), the multiple-source, depth-dependent Green's function b_j for the jth normal mode is determined to be

$$
b_j(r_0, z) = \left(A/\sqrt{k_j}\right) \sum_n d_n^* u_j(z_n) u_j(z) \exp(ik_j r_0). \tag{5.58}
$$

It is seen that b_j depends on the source range r_0 and the hydrophone depths z_n as well as the depth z. Furthermore, the modal shapes (eigenfunctions) $|u_j|$ can be traced out by evaluating $|b_j|$ as a function of z at the appropriate horizontal wavenumbers determined from Eq. (5.56). When evaluated as a function of z, the b_j contain all of the information necessary for source localization by matched mode processing. This modal information is extracted from the backpropagated correlation processor without using any modal information obtained from a normal mode model. Instead, it is obtained by a suitable spectral analysis of weighted VLA data that is backpropagated using a high-order Padè PE algorithm of underwater acoustics.

5.4.4   Modal Decomposition Example

To illustrate the backpropagation-based method of modal decomposition, we reconsider the canonical shallow-water waveguide that was used in the earlier numerical examples of propagation and source localization. First, we combine the VLA data together with the environmental specification and PE

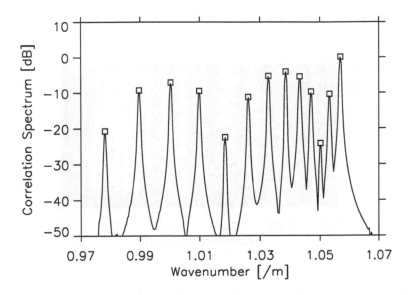

FIGURE 5.7: Partial mode spectrum of the backpropagated sequence Γ for the environment of Fig. 5.3 and the data of Fig. 5.5. The symbols ($\square$) denote peak estimates obtained using parabolic interpolation.

parameters for the COLNOISE case b. Second, we compute (and save) the back-propagated field q along with the correlation sequence Γ in Eq. (5.54) out to 16.384 km (8192 range steps) from the VLA. Finally, we filter the Γ-sequence using a Hann window to reduce edge effects and then apply an FFT to obtain its spectral estimates. Greater spectral resolution can be obtained simply by marching outwards to greater ranges. Alternatively, spectroscopic curve fit-ting techniques can be used to improve the estimates of the spectral peaks as described in the original fiber optic applications of the method [37–41]. For our work, we made use of a simple parabolic interpolator to refine the estimates of the modal wavenumbers from the FFT of Γ. The resulting nor-malized spectral estimates, expressed in dB via $10 \log_{10} \{|\mathcal{F}(\Gamma)| / \max |\mathcal{F}(\Gamma)|\}$ in the vicinity of the first thirteen propagating modes are shown in Fig. 5.7. The symbols ($\square$) denote the peak estimates obtained using parabolic interpo-lation. It is known from previous work [83] that the first thirteen modes are mostly confined to the water column while modes 13 to 26 exhibit significant penetration into the upward refracting sediment layer.

Spectral analysis of q in Eq. (5.58) as a function of depth z yield the mode shape functions versus wavenumber. The normalized spectral amplitudes ($10 \log_{10} |\mathcal{F}(q)| / \max |\mathcal{F}(q)|$) for the first 13 propagating modes are displayed in Fig. 5.8. The influence of the downward refracting sound speed profile is evident in the water column shapes of the lowest order modes.

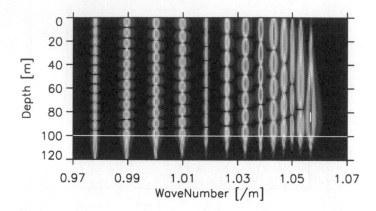

FIGURE 5.8: PE-derived mode amplitudes for modes 1 to 13 as a function of depth for the environment of Fig. 5.3 and the data of Fig. 5.5.

5.4.5   Modal Beamforming

The PE-based modal decomposition of the pressure field measured on a VLA can be used to carry out "range and depth beamforming" via a matched mode processor [83, 87, 93]. In contrast to matched field processing which is carried out in "phone space," the correlation in matched mode processing is carried out in "mode space." That is, the matched mode processor correlates the complex modal amplitudes that are inferred from the VLA pressure fields. One potential advantage of using a matched mode processor over the matched field processor is the option of including only a subset of the decomposed modes that make up the VLA field. This feature can be important since, prior to processing, the data can be filtered to exclude those modes that are most likely to be affected by any mismatch between the modelled and true ocean environment.

From these modal data, we define a PE-based matched mode processor by

$$h_3(r, z) = \left| \sum_j b_j^*(r_0, z) \exp(ik_j r) \right|^2 . \tag{5.59}$$

This range/depth estimator was originally introduced by Yang [93]. Unlike separate range and depth estimators that can be used whose peaks provide an estimate of the source range and source depth respectively, the processor in

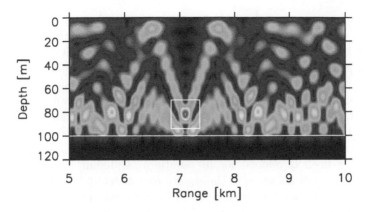

FIGURE 5.9: Matched mode ambiguity surface determined from PE-derived modal amplitudes 1 to 10 for the environment of Fig. 5.3 and the data of Fig. 5.5. The square box is centered on the true source location.

Eq. (5.58) does not contain any uncertainty in the polarity of u_j. Evaluation of Eq. (5.58) over a suitable search domain (r, z) will produce a peak at the true source coordinates (r_0, z_0).

To illustrate the implementation of matched mode processing, the complex amplitudes of the depth-dependent Green's function of Eq. (5.58) were evaluated at the discrete modal wavenumbers displayed in Fig. 5.7 for the first 10 modes of the COLNOISE b example. This was accomplished by computing the DFT of Eq. (5.57) at the interpolated values $k_j = 1, \ldots, 10$ at each node on the depth search grid. Then Eq. (5.59) was evaluated at each node on the range search grid to yield the matched mode ambiguity surface shown in Fig. 5.9. A comparison of Fig. 5.9 and Fig. 5.6 indicates that the focus of the ambiguity surface formed from modes 1 to 10 for the matched mode processor is not as sharp as the focus of the ambiguity surface for either matched field processor. It is evident, however, that the matched mode processor correctly peaks at the true source location of $r = 7.1$ km and $z = 82$ m for this example.

5.5   Summary

Acoustic propagation due to a point source in a shallow-water waveguide often exhibits two characteristics: (1) away from the source, the direction of propagation is nearly horizontal, and (2) backscattered energy is negligible compared to forward-scattered energy. In these situations, the propagation of underwater sound can be accurately described by a one-way wave equation as given in Eq. (5.13). This equation, or variants of it, are traditionally referred to as a parabolic equation (PE) in the underwater acoustic literature.

We began this chapter with a brief derivation of the standard PE, given in Eq. (5.5) and Eq. (5.8). This was followed by a heuristic derivation of an exact, one-way acoustic wave equation in Eq. (5.13) that was subsequently arrived at by factoring the reduced wave equation in a weakly range-dependent medium. The formal solution of this exact PE over from range r to range $r + \Delta r$, given in Eq. (5.17), forms the basis of all numerical marching methods that have been developed for its solution. Different PE's result from different approximations made to the square-root operator $\sqrt{1 + X}$ or its corresponding propagator. The numerical treatment used in this chapter makes use of a multi-term Padè-type rational-linear approximation to the square-root operator together with a rotation of its branch cut as given in Eq. (5.20) to Eq. (5.22). The subsequent second-order accurate approximation of each partial propagator gave rise to the product expansion in Eq. (5.23) and then to the sum expansion in Eq. (5.25), in which it was noted that accurate solutions required the marching procedure to take sufficiently small range steps Δr. The sum expansion led to the term-by-term partial PE formulation given in Eq. (5.27).

Next, we presented a heterogeneous finite-difference approach for solving Eq. (5.27) that made use of an offset depth grid where values of the material properties (sound speed, density and attenuation) were assumed given at each node. The resulting finite-difference representation was subsequently modified to incorporate profile interpolation in a range-dependent medium, to ameliorate energy conservation issues in the case of variable-depth bathymetry, to account for energy losses due to shear wave conversion at an elastic subbottom via an equivalent acoustic medium, and to efficiently suppress unphysical reflections from the base of the computational grid in the context of a perfectly matched layer formalism. A simple starting field for introducing a vertically extended source excitation was used to initiate the marching procedure. The finite-difference formulation and numerical solution procedure described above was demonstrated for modeling underwater sound propagation in a generic shallow water environment.

Given the availability of a high-order PE together with an efficient algorithm for numerically solving it, we then presented a method of source localization based on backpropagating the pressure field received on a vertical line array (VLA). This method, introduced over 20 years ago, was dubbed "acoustic retrogation." Basically, it involves weighting the measured pressure on each sensor of the array with a vertically extended PE starting field and summing over the sensors. This initializes a correlation field at each point on the PE depth grid. Then, invoking the principle of superposition, this sum of weighted source fields, which satisfies the PE, is backpropagated outwards from the array onto a range/depth grid that spans a given search domain. The result is an ambiguity surface calculated in a single PE run whose peak is taken as an estimate of the location of the point source that generated the data received on the VLA. This surface, although unnormalized, (i.e., in the same way that the standard Bartlett correlation surface is normalized), was demonstrated to peak at the same location as its Bartlett counterpart for synthetic data that was provided at a workshop on acoustic localization methods. For N sensors in the VLA, the backpropagation method proceeds N times faster than the Bartlett processor, which requires N propagation runs that must be stored in order to compute normalized replica fields prior to the correlation against the measured data.

Finally, we showed how the backpropgation method can be used to decompose the measured pressure field on a VLA into its modal components. That is, in addition to backpropagating the weighted sum of pressure fields, a correlation function was computed as a function of range, which, when analyzed spectrally via an FFT yields estimates of the modal wavenumbers. The backpropagated ambiguity surface was then evaluated at a subset of these modal wavenumbers by carrying out a DFT as a function of depth. The result is a set of (complex) mode shape functions that contain the necessary phase information required for matched mode processing. Using the same generic waveguide as before, the complex depth amplitude functions of the first 10 of the 26 propagating modes were matched against the range replicas to yield a source localization estimate that agreed with the previous source location estimates obtained with the matched field processors. One important advantage of the matched mode processor over the matched field processors is that those modes that suffer the most from any environmental mismatch issues can be removed from the localization processor.

We conclude this chapter by noting a useful aspect of PE modelling that has received considerable attention lately, namely, the use of a nonlocal boundary condition (NLBC) to replace the absorbing layer for terminating the PE computational depth grid. Such a condition can exactly transform the PE problem having a transverse radiation condition as $z \to \infty$ into an equivalent one in a bounded domain [7,54,65,85,95]. Of special note is the recent discrete NLBC that has been developed [62] for the manifestly reciprocal PE range-dependent formulation derived in [42] and numerically implemented in [61]. Moreover, the use of an NLBC for a single-term PE has recently been incorporated into a practical numerical scheme for geoacoustic inversion of waveguide properties via an adjoint PE-based approach [46, 48, 60] that involves backpropagation concepts similar to the one discussed herein.

References

[1] M. Abramowitz and I. A. Stegun. *Handbook of Mathematical Functions*. Dover, 1965.

[2] D. S. Ahluwalia and J. B. Keller. Exact and asymptotic representations of the sound field in a stratified ocean. *Wave Propagation and Underwater Acoustics*. J. B. Keller and J. S. Papadakis, eds., Springer, pages 14–85, 1977.

[3] A. Bamberger, B. Engquist, L. Halpern, and P. Joly. Higher order paraxial wave equation approximations in heterogeneous media. *SIAM J. Appl. Math.*, 48:129–154, 1988.

[4] R. Bellman. *Introduction to Matrix Analysis*, 2nd Ed, McGraw-Hill, 1970.

[5] J.-P. Bèrenger. A perfectly matched layer for the absorption of electromagnetic waves. *J. Comput. Phys.*, 114:185–200, 1994.

[6] G. H. Brooke and D. J. Thomson. A single-scatter formalism for improving PE calculations in range-dependent media. *PE Workshop II: Proceedings of the Second Parabolic Equation Workshop*, S. A. Chin-Bing, D. B. King, J. A. Davis and R. B. Evans, eds., pages 126–144, US Government Printing Office, 1993.

[7] G. H. Brooke and D. J. Thomson. Nonlocal boundary conditions for high-order parabolic equation algorithms. *Wave Motion*, 31:117–129, 2000.

[8] G. H. Brooke, D. J. Thomson, and G. R. Ebbeson. PECan: A Canadian parabolic equation model for underwater sound propagation. *J. Comput. Acoust.*, 9:69–100, 2001.

[9] J. R. Buck, J. C. Preisig, and K. E. Wage. A unified framework for mode filtering and the maximum a posteriori mode filter. *J. Acoust. Soc. Am.*, 103:1813–1824, 1998.

[10] H. P. Bucker. Use of calculated sound fields and matched field detection to locate sound sources in shallow water. *J. Acoust. Soc. Am.*, 59:368–373, 1976.

[11] R. P. Chapman. *Alpha and Omega, An Informal History of the Defence Research Establishment Pacific 1948–1995*. Defence Research Establishment Atlantic, Dartmouth, NS, 1998.

[12] N. R. Chapman, S. A. Chin-Bing, D. King, and R. B. Evans. Benchmarking geoacoustic inversion methods for range-dependent waveguides. *IEEE J. Oceanic Eng.*, 28:320–330, 2003.

[13] W. C. Chew and W. H. Weedon. A 3D perfectly matched medium from modified Maxwell's equations with stretched coordinates. *Microwave Opt. Technol. Lett.*, 7:599–604, 1994.

[14] J. F. Clearbout. Coarse grid calculations of waves in inhomogeneous media with application to the delineation of complicated seismic structures. *Geophysics*, 35:407–418, 1970.

[15] J. F. Claerbout. *Fundamentals of Geophysical Data Processing With Applications to Petroleum Prospecting*. Blackwell, 1985.

[16] F. Collino. Perfectly matched absorbing layers for the paraxial equations. *J. Comput. Phys.*, 131:164–180, 1997.

[17] M. D. Collins. A higher-order parabolic equation for wave propagation in an ocean overlying an elastic bottom. *J. Acoust. Soc. Am.*, 86:1459–1464, 1989.

[18] M. D. Collins. Benchmark calculations for higher-order parabolic equations. *J. Acoust. Soc. Am.*, 87:1535–1538, 1990.

[19] M. D. Collins. Higher-Order Padè approximations for accurate and stable elastic parabolic equations with application to interface wave propagation. *J. Acoust. Soc. Am.*, 89:1050–1057, 1991.

[20] M. D. Collins. A self starter for the parabolic equation method. *J. Acoust. Soc. Am.*, 92:2069–2074, 1992.

[21] M. D. Collins. Higher-order, energy-conserving, two-way, and elastic parabolic equations. *PE Workshop II: Proceedings of the Second Parabolic Equation Workshop*, S. A. Chin-Bing, D. B. King, J. A. Davis and R. B. Evans, eds., pages 145–168, US Government Printing Office, 1993.

[22] M. D. Collins. The split-step Padè solution for the parabolic equation method. *J. Acoust. Soc. Am.*, 93:1736–1742, 1993.

[23] M. D. Collins, R. J. Cederberg, D. B. King, and S. A. Chin-Bing. Comparison of algorithms for solving parabolic wave equations. *J. Acoust. Soc. Am.*, 100:178–182, 1996.

[24] M. D. Collins and R. B. Evans. A two-way parabolic equation for acoustic backscattering in the ocean. *J. Acoust. Soc. Am.*, 91:1357–1368, 1992.

[25] M. D. Collins and W. A. Kuperman. Inverse problems in ocean acoustics. *Inverse Problems*, 10:1023–1040, 1994.

[26] M. D. Collins, W. A. Kuperman, and W. L. Siegmann. A parabolic equation for poro-elastic media. *J. Acoust. Soc. Am.*, 98:1645–1656, 1995.

[27] M. D. Collins, J. F. Lingevitch, and W. L. Siegmann. Wave propagation in poro-acoustic media. *Wave Motion*, 25:265–272, 1997.

[28] M. D. Collins and E. K. Westwood. A higher-order energy-conserving parabolic equation for range-dependent ocean depth, sound speed, and density. *J. Acoust. Soc. Am.*, 89:1068–1075, 1991.

[29] J. A. Davis, D. White, and R. C. Cavanagh. *NORDA Parabolic Equation Workshop, 31 March–3 April, 1981*, Naval Ocean Research and Development Agency Tech. Note 143, 1982.

[30] J. A. DeSanto. Relation between the solutions of the Helmholtz and parabolic equations for sound propagation. *J. Acoust. Soc. Am.*, 62:295–297, 1977.

[31] J. A. DeSanto. Theoretical methods in ocean acoustics. *Ocean Acoustics*, J.A. DeSanto, Ed., pages 7–77, Springer, 1978.

[32] F. R. DiNapoli and R. L. Deavenport. Theoretical and numerical Green's function field solution in a plane multilayered medium. *J. Acoust. Soc. Am.*, 67:92–105, 1980.

[33] R. B. Evans. A coupled mode solution for acoustic propagation in a waveguide with stepwise depth variations of a penetrable bottom. *J. Acoust. Soc. Am.*, 74:188–195, 1983.

[34] R. B. Evans. The flattened surface parabolic equation. *J. Acoust. Soc. Am.*, 104:2167–2173, 1998.

[35] S. M. Flattè and F. D. Tappert. Calculation of the effect of internal waves on oceanic sound transmission. *J. Acoust. Soc. Am.*, 58:1151–1159, 1975.

[36] J. A. Fawcett. Modeling three-dimensional propagation in an oceanic wedge using parabolic equation methods. *J. Acoust. Soc. Am.*, 93:2627–2632, 1993.

[37] M. D. Feit and J. A. Fleck, Jr. Light propagation in graded-index optical fibers. *Appl. Opt.*, 17:3990–3998, 1978.

[38] M. D. Feit and J. A. Fleck, Jr. Computation of mode properties in optical fiber waveguides by a propagating beam method. *Appl. Optics*, 19:1154–1164, 1980.

[39] M. D. Feit and J. A. Fleck, Jr. Computation of mode eigenfunctions in gradex-index optical fibers by the propagating beam method, *Appl. Optics*, 19:2240–2246, 1980.

[40] M. D. Feit and J. A. Fleck, Jr. Mode properties of optical fibers with lossy components by the propagating beam method. *Appl. Optics*, 20:848–856, 1981.

[41] M. D. Feit, J. A. Fleck, Jr., and A. Steiger. Solution of the Schrodinger equation by a spectral method. *J. Comp. Phys.*, 47:412–433, 1982.

[42] O. A. Godin. Reciprocity and energy conservation within the parabolic approximation. *Wave Motion*, 29:175–194, 1999.

[43] I. S. Gradshteyn and I. M. Ryzhik. *Tables of Integrals, Series, and Products*, Academic Press, 1980.

[44] R. R. Greene. The rational approximation to the acoustic wave equation with bottom interaction. *J. Acoust. Soc. Am.*, 76:1764–1773, 1984.

[45] R. R. Greene. A high-angle one-way wave equation for seismic wave propagation along rough and sloping interfaces. *J. Acoust. Soc. Am.*, 77:1991–1998, 1985.

[46] J.-P. Hermand, M. Meyer, M. Asch, and M. Berrada. Adjoint-based acoustic inversion for the physical characterization of a shallow water environment. *J. Acoust. Soc. Am.*, 119:3860–3871, 2006.

[47] D. Hertz and I. Ziskind. Fast approximate maximum likelihood algorithm for single source localization. *IEE Proc.–Radar, Sonar Navig.*, 142:232–235, 1995.

[48] P. Hursky, M. B. Porter, W. S. Hodgkiss, and W. A. Kuperman. Adjoint modeling for acoustic inversion. *J. Acoust. Soc. Am.*, 115:607–619, 2004.

[49] F. B. Jensen and C. M. Ferla. Numerical solutions of range-dependent benchmark problems in ocean acoustics. *J. Acoust. Soc. Am.*, 87:1499–1510, 1990.

[50] F. B. Jensen, W. A. Kuperman, M. B. Porter, and H. Schmidt. *Computational Ocean Acoustics*. AIP Press, 1994.

[51] F. B. Jensen and H. Schmidt. Efficient numerical solution technique for wave propagation in horizontally stratified environments. *Comput. Math. Appl.*, 11:699–715, 1985.

[52] F. B. Jensen and H. Schmidt. Spectral decomposition of PE fields in a wedge-shaped ocean. *Progress in Underwater Acoustics*, H.M. Merklinger, Ed., pages 557–571, Plenum, 1987.

[53] N. A. Kampanis. Numerical simulation of low-frequency aeroacoustics over irregular terrain using a finite-element discretization of the parabolic equation. *J. Comput. Acoust.*, 10:97–111, 2002.

[54] N. A. Kampanis. A finite element method for the parabolic equation in aeroacoustics coupled with a nonlocal boundary condition for an inhomogeneous atmosphere. *J. Comput. Acoust.*, 13:569–584, 2005.

[55] D. Lee, G. Botseas, and W. L. Siegmann. Examination of three-dimensional effects using a propagation model with azimuthal coupling capability. *J. Acoust. Soc. Am.*, 91:3192–3202, 1992.

[56] D. Lee and S. T. McDaniel. Ocean acoustic propagation by finite-difference methods. *Comput. Math. Appl.*, 14:305–423, 1987.

[57] M. F. Levy. *Parabolic Equation Methods for Electromagnetic Wave Propagation*. Institution of Electrical Engineers, 2000.

[58] J. F. Lingevitch and M. D. Collins. Wave propagation in range-dependent poro-acoustic waveguides. *J. Acoust. Soc. Am.*, 104:783–790, 1998.

[59] Q. H. Liu and J. Tao. The perfectly matched layer for acoustic waves in absorptive media. *J. Acoust. Soc. Am.*, 102:2072–2081, 1997.

[60] M. Meyer and J.-P. Hermand. Optimal nonlocal boundary control of the wide-angle parabolic equation for inversion of a waveguide acoustic field. *J. Acoust. Soc. Am.*, 117:2937–2948, 2005.

[61] D. Mikhin. Energy-conserving and reciprocal solutions for higher-order parabolic equations. *J. Comput. Acoust.*, 9:183–203, 2001.

[62] D. Mikhin. Exact discrete nonlocal boundary conditions for high-order Padè parabolic equations. *J. Acoust. Soc. Am.*, 116:2864–2875, 2004.

[63] F. A. Milinazzo, C. A. Zala, and G. H. Brooke. Rational square-root approximations for parabolic equation algorithms. *J. Acoust. Soc. Am.*, 101:760–766, 1997.

[64] L. Nghiem-Phu, F. D. Tappert, and S. C. Daubin. *Systematic analysis of source localization using PE methods with back-propagation techniques*. Transcribed and Compiled by D. J. Thomson, Defence Research Establishment Atlantic Tech. Memo. TM 99-059, Dartmouth, NS, 1999.

[65] J. S. Papadakis. Exact, non-reflecting boundary conditions for parabolic-type approximations in underwater acoustics. *J. Comput. Acoust.*, 2:83–98, 1994.

[66] M. Porter. *The KRAKEN Normal Mode Program*. SACLANT Undersea Research Centre Memorandum SM–245, San Bartolomeo, Italy, 1991.

[67] M. B. Porter and H. P. Bucker. Gaussian beam tracing for computing ocean acoustic fields. *J. Acoust. Soc. Am.*, 82:1349–1359, 1987.

[68] M. B. Porter, F. B. Jensen, and C. M. Ferla. The problem of energy conservation in one-way models. *J. Acoust. Soc. Am.*, 89:1058–1067, 1991.

[69] M. B. Porter and A. Tolstoy. The matched field processing benchmark problems. *J. Comput. Acoust.*, 2:161–185, 1994.

[70] W. H. Press, S. A. Teukolsky, W. T. Vetterling, and B. P. Flannery. *Numerical Recipes in* FORTRAN*, 2nd Ed*, Cambridge, 1992.

[71] A. P. Rosenberg. A new rough surface parabolic equation program for computing low-frequency acoustic forward scattering from the ocean surface. *J. Acoust. Soc. Am.*, 105:144–153, 1999.

[72] H. Schmidt. *SAFARI, Seismo-Acoustic Fast field Algorithm for Range-Independent environments, User's Guide.* SACLANT Undersea Research Centre Report SR–113, San Bartolomeo, Italy, 1988.

[73] C. W. Spofford. *A synopsis of the AESD workshop on non-ray tracing techniques, 22-25 May 1973.* Acoustic Environmental Support Detachment TN–73–05, Arlington, VA (1973).

[74] F. Sturm. Examination of signal dispersion in a 3-D wedge-shaped waveguide using 3DWAPE. *Acta Acustica*, 88:714–717, 2002.

[75] F. Sturm and J. A. Fawcett. On the use of higher-order azimuthal schemes in 3-D PE modeling. *J. Acoust. Soc. Am.*, 113:3134–3145, 2003.

[76] F. D. Tappert. The parabolic approximation method. *Wave Propagation and Underwater Acoustics.* J. B. Keller and J. S. Papadakis, eds., pages 224–287, Springer, 1977.

[77] F. D. Tappert and R. H. Hardin. Computer simulation of long-range ocean acoustic propagation using the parabolic equation method, *Proceedings of the 8th International Conference on Acoustics*, Vol. 2, page 452, London, Goldcrest, 1974.

[78] F. D. Tappert, L. Nghiem-Phu and S. C. Daubin. Source localization using the PE method. *J. Acoust. Soc. Am. Suppl. 1*, 78:S30, 1985.

[79] C. T. Tindle, K. M. Guthrie, G. E. Bold, M. D. Johns, D. Jones, K. O. Dixon, and T. G. Birdsall. Measurements of the frequency dependence of normal modes. *J. Acoust. Soc. Am.*, 64:1178–1185, 1978.

[80] D. J. Thomson. Wide-angle parabolic equation solutions to two range-dependent benchmark problems. *J. Acoust. Soc. Am.*, 87:1514–1520, 1990.

[81] D. J. Thomson and G. H. Brooke. PE-based methods for treating forward scattering from a rough surface. *Theoretical and Computational Acoustics 2003*, A. Tolstoy, Yu-C. Teng, and E. C. Shang, eds., pages 390–402, World Scientific, 2004.

[82] D. J. Thomson and N. R. Chapman. A wide-angle split-step algorithm for the parabolic equation. *J. Acoust. Soc. Am.*, 74:1848–1854, 1983.

[83] D. J. Thomson and G. R. Ebbeson. A PE-based approach to modal decomposition and source localization. *J. Comput. Acoust.*, 2:231–250, 1994.

[84] D. J. Thomson, G. R. Ebbeson, and B. H. Maranda. A matched-field backpropagation algorithm for source localization. *Oceans '97 Proceedings (Halifax), IEEE/MTS*, pp. 602–607, 1997.

[85] D. J. Thomson and M. E. Mayfield. An exact radiation condition for use with the a posteriori PE method. *J. Comput. Acoust.*, 2:113–132, 1994.

[86] D. J. Thomson and D. H. Wood. A postprocessing method for removing phase errors in the parabolic equation. *J. Acoust. Soc. Am.*, 82:224–232, 1987.

[87] A. Tolstoy. *Matched Field Processing for Underwater Acoustics*. World Scientific, 1993.

[88] A. Tolstoy, N. R. Chapman, and G. Brooke. Workshop '97: Benchmarking for geoacoustic inversion in shallow water. *J. Comput. Acoust.*, 6:1–28, 1998.

[89] R. J. Urick. *Principles of Underwater Sound*. 3rd Ed., McGraw-Hill, 1983.

[90] H. Weinberg and R. E. Keenan. Gaussian ray bundles for modeling high-frequency loss under shallow-water conditions. *J. Acoust. Soc. Am.*, 100:1421–1431, 1996.

[91] E. K. Westwood, C. T. Tindle, and N. R. Chapman. A normal mode model for acousto-elastic ocean environments. *J. Acoust. Soc. Am.*, 100:3631–3645, 1996.

[92] B. T. R. Wetton and G. H. Brooke. One-way wave equations for seismoacoustic propagation in elastic waveguides. *J. Acoust. Soc. Am.*, 87:624–632, 1990.

[93] T. C. Yang. A method of range and depth estimation by modal decomposition. *J. Acoust. Soc. Am.*, 82:1736–1745, 1987.

[94] D. Yevick, J. Yu and Y. Yayon. Optimal absorbing boundary conditions. *J. Opt. Soc. Am. A*, 12:107–110, 1995.

[95] D. Yevick and D. J. Thomson. Nonlocal boundary conditions for finite-difference parabolic equation solvers. *J. Acoust. Soc. Am.*, 106:143–150, 1999.

[96] D. Yevick and D. J. Thomson. Impedance-matched absorbers for finite-difference parabolic equation algorithms. *J. Acoust. Soc. Am.*, 107:1226–1234, 2000.

[97] Z. Y. Zhang and C. T. Tindle. Improved equivalent fluid approximations for a low shear speed ocean bottom. *J. Acoust. Soc. Am.*, 98:3391–3396, 1995.

Chapter 6

Numerical Solution of the Parabolic Equation in Range–Dependent Waveguides

V. A. Dougalis, Mathematics Department, University of Athens, 15784 Zographou, Greece *and* Foundation for Research and Technology-Hellas, Institute of Applied and Computational Mathematics, 71110 Heraklion, Greece, doug@math.uoa.gr

N. A. Kampanis, Foundation for Research and Technology-Hellas, Institute of Applied and Computational Mathematics, 71110 Heraklion, Greece, kampanis@iacm.forth.gr

F. Sturm, Laboratoire de Mécanique des Fluides et d'Acoustique, UMR CNRS 5509, Ecole Centrale de Lyon, 36, avenue Guy de Collongue, 69134, Ecully Cedex, France, frederic.sturm@ec-lyon.fr

G. E. Zouraris, Mathematics Department, University of Crete, 71409, Heraklion, Greece *and* Foundation for Research and Technology-Hellas, Institute of Applied and Computational Mathematics, 71110 Heraklion, Greece, zouraris@math.uoc.gr

6.1 Introduction

The Helmholtz equation is often used to model sound propagation and scattering in a waveguide. Its solution represents the wave field (acoustic pressure) produced by a harmonic point source placed within the waveguide. Therefore, a central task in computational acoustics is the efficient numerical solution of the Helmholtz equation; we refer the reader to [18], [22], [79], [20] and Chapter 1 of Part II of the present book for more information and relevant references.

In a wide range of applications, for example in long–range underwater sound propagation, backscattering is limited and the dominant part of the wave consists of one–way propagating modes. This being the case, an approximation of the two–way wave field obtained by a direct numerical solution of a boundary value problem for the Helmholtz equation, a computationally very demanding task, may be avoided. It is well–known, [33], that an effective simulation of essentially one–way, long–range wave propagation may be provided by solving numerically *parabolic* (paraxial) approximations of the Helmholtz equation. The form of a parabolic approximation is directly related to the maximum angle of propagation of the wave field (relative to the horizontal direction); parabolic approximations of *standard* or *wide–angle* type describing propagation within narrow or wider angles, respectively, have been proposed and used; see, e.g., [22], [25], [26] and Chapter 5 (Part II) by Thomson and Brooke in this volume. In what follows we shall be concerned with the standard parabolic approximation, mainly in the context of underwater acoustic waveguides.

We begin by recalling the formal derivation of the standard parabolic approximation from the 3D Helmholtz equation. The reader is also referred to Chapter 5 in the present volume.

In a waveguide (consisting of a single fluid layer for simplicity), the wave field established by a harmonic point source of frequency f, may be represented by the solution of the Helmholtz equation expressed in cylindrical coordinates (z, r, θ). As is customary, the depth (z) axis is pointing downwards and the source is located on it. We shall refer to r as the range (horizontal distance from the source) and θ as the azimuthal angle. The complex-valued acoustic field $p(z, r, \theta)$ (acoustic pressure) satisfies

$$\nabla_\rho^2 p + k_0^2 n^2(z, r, \theta)p = -4\pi p_0 \delta(z, r), \tag{6.1}$$

where

$$\nabla_\rho^2 := \rho \left[\frac{1}{r} \frac{\partial}{\partial r} \left(\frac{r}{\rho} \frac{\partial}{\partial r} \right) + \frac{\partial}{\partial z} \left(\frac{1}{\rho} \frac{\partial}{\partial z} \right) + \frac{1}{r^2} \frac{\partial}{\partial \theta} \left(\frac{1}{\rho} \frac{\partial}{\partial \theta} \right) \right],$$

and

$$\delta(z, r) := \delta(z - z_s) \frac{\delta(r)}{2\pi r},$$

with z_S being the source depth, and with $\delta(z - z_s)$ and $\delta(r)$ Dirac delta functions. In (6.1) $k_0 = 2\pi f / c_0$ is a reference wave number associated with a reference sound speed c_0. The function $n^2 = n^2(z, r, \theta)$ is a complex–valued 'index of refraction' and is defined by the formula

$$n^2(z, r, \theta) := \left(\frac{c_0}{c(z, r, \theta)} \right)^2 (1 + i\alpha(z, r, \theta)),$$

where $c(z, r, \theta)$ is the speed of sound in the fluid and $\alpha(z, r, \theta)$ an empirically determined positive function of small magnitude, which models absorption (attenuation) of sound in the fluid. Finally, p_0 is a reference acoustic pressure characterizing the source, and $\rho = \rho(z, r, \theta)$ is the density of the fluid. In the sequel, we shall assume that ρ has negligible variation in r.

If we seek a separation of variables solution of (6.1) of the form $p(z, r, \theta) = h(r)u(z, r, \theta)$, the principal dependence on r being in $h(r)$, [33], it emerges that

$h(r)$ satisfies the zeroth–order Bessel equation. An outgoing solution of (6.1), therefore, representing the forward propagating acoustic field is expressed as $p(z, r, \theta) = H_0^{(1)}(k_0 r) u(z, r, \theta)$, where $H_0^{(1)}(k_0 r)$ is the first kind, zeroth–order Hankel function. For large values of r, so that $k_0 r \gg 1$, the far–field asymptotic formula

$$H_0^{(1)}(k_0 r) \approx \frac{e^{ik_0 r}}{\sqrt{k_0 r}}, \quad k_0 r \gg 1,$$

may be used and the equation satisfied by $u(z, r, \theta)$ becomes

$$\frac{\partial^2 u}{\partial r^2} + \rho \frac{\partial}{\partial z}\left(\frac{1}{\rho}\frac{\partial u}{\partial z}\right) + \rho \frac{1}{r^2}\frac{\partial}{\partial \theta}\left(\frac{1}{\rho}\frac{\partial u}{\partial \theta}\right)$$
$$+ 2ik_0 \frac{\partial u}{\partial r} + k_0^2(n^2 - 1)u = 0. \tag{6.2}$$

Since $u(z, r, \theta)$ varies slowly in r, the *paraxial approximation*

$$\left|\frac{\partial^2 u}{\partial r^2}\right| \ll \left|2ik_0 \frac{\partial u}{\partial r}\right|$$

applies in (6.2). Therefore the p.d.e. satisfied by $u(z, r, \theta)$, is

$$\frac{\partial u}{\partial r} = \frac{i\rho}{2k_0}\left[\frac{\partial}{\partial z}\left(\frac{1}{\rho}\frac{\partial u}{\partial z}\right) + \frac{1}{r^2}\frac{\partial}{\partial \theta}\left(\frac{1}{\rho}\frac{\partial u}{\partial \theta}\right)\right] + \frac{ik_0}{2}(n^2(z, r, \theta) - 1)u. \tag{6.3}$$

This is the (standard) Parabolic Equation (PE) in 3D. By our assumptions, it should describe the propagation of sound in weakly range–dependent fluid media outwards and far from the source. For constant ρ, (6.3) simplifies to the standard 3D PE of Tappert, [33].

For a physically and geometrically cylindrically symmetric waveguide the term $\frac{\partial}{\partial \theta}\left(\frac{1}{\rho}\frac{\partial u}{\partial \theta}\right)$ can be dropped from (6.3). The result is the (standard) 2D PE, which for constant ρ simplifies to

$$u_r = \frac{i}{2k_0}u_{zz} + \frac{ik_0}{2}(n^2(z, r) - 1)u. \tag{6.4}$$

Stratified media with sharp transitions in ρ are usually considered as multilayered media with constant density, separated by interfaces across which appropriate transmission conditions are valid.

To our knowledge, there is no complete, mathematically rigorous justification of the PE as an approximation to the Helmholtz equation in the appropriate parameter regime. (See however, [15], [16].) Nevertheless, there is ample experimental and computational evidence of its validation as a model for sound propagation under the assumptions made in its derivation. In particular these assumptions restrict its validity to modelling acoustic waves propagating within narrow angles both in depth and in azimuth, [33]. Various extensions of (6.3) capable of describing propagation of acoustic waves within larger angles have been proposed; these are referred to as *wide–angle* PE's, cf., e.g., [10], [11], [24], [25], [30] and Chapter 5 in this volume. In the present chapter we shall confine ourselves to the (standard) PE (6.3) and (6.4).

Although originally the PE was derived, in the context of underwater acoustics, to model propagation in single layer domains with horizontal bottom or multilayered media with horizontal interfaces, cf. [33], it has been extensively used for domains with bottom and interfaces that exhibit mild variations in range and azimuth so that backscattering effects are small. It is a natural issue, therefore, to study the well–posedness of initial–boundary value problems for the PE in the presence of range–dependent bottom and interfaces. We present briefly an introduction to some relevant mathematical and computational issues in the axially symmetric case in Section 6.2. The range–varying topography has often been approximated in practical computations by 'staircase' (piecewise horizontal) approximations. This raises the issue of boundary and interface conditions on the vertical part of the steps of the staircase. Moreover, it is well documented that staircase approximations lead to nonphysical energy losses or gains, cf., e.g., [21], [29]. We therefore use in this chapter range– (or range– and azimuth–dependent) change of the depth variable techniques that transform the problem into an equivalent one posed on horizontal strips. Of course this approach has its limitations and drawbacks, but it alleviates energy non–conservation problems. In Section 6.2 we present the change of variables in simple, axially symmetric, i.e., 2D, environments. In Section 6.3 a highly accurate finite element discretization of the problem is presented in the 2D case, while, in Section 6.4, the change of variable technique is extended in 3D and implemented within the finite element framework. Of great importance is the issue of what kind of approximation of the normal derivative of the field is used at range–dependent interfaces and bottoms.

In this chapter we have restricted ourselves to domains of finite depth. Semi–infinite domains are usually modelled by absorbing layers. We refer the reader to Chapter 4 (Part II) by Dawson, Brooke and Thomson, for a review of the nonlocal 'impedance' condition that should be properly used to model semi–infinite layers.

6.2   Initial–Boundary Value Problems in Axially Symmetric Range-Dependent Environments

In this section we shall first consider the PE (6.4) in a range-dependent axially symmetric waveguide consisting of a single layer of water of constant density over a bottom of varying topography, and discuss some issues con-

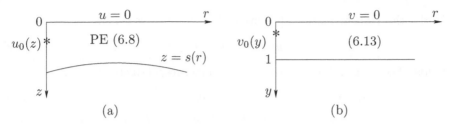

FIGURE 6.1: Variable bottom (a) and transformed problem (b).

cerning the well-posedness of two simple associated initial–boundary value problems (ibvp's). We shall then comment on some related mathematical issues that arise in the study of well–posedness of multilayered, variable interface problems.

Consider first the domain shown in Fig. 6.1(a). The water, of constant density, occupies the region $0 \leq z \leq s(r)$, $r \geq 0$, of the (z, r)–plane. Here, $z = 0$ is the surface and $z = s(r)$ is a positive smooth function representing the range–dependent bottom.

We consider the Helmholtz equation (6.1) in cylindrical coordinates in the presence of cylindrical symmetry and constant fluid density ρ, rewriting it here for notational convenience as

$$\Delta p + k_0^2 n^2(z, r) p = 0, \qquad 0 \leq z \leq s(r), \quad r > 0, \tag{6.5}$$

where

$$\Delta := \frac{\partial^2}{\partial r^2} + \frac{1}{r} \frac{\partial}{\partial r} + \frac{\partial^2}{\partial z^2}.$$

We supplement (6.5) with the surface pressure–release boundary condition $p(0, r) = 0$, $r \geq 0$, and consider two kinds of boundary conditions (b.c.'s) at the bottom: The 'soft' bottom (homogeneous Dirichlet) b.c.

$$p = 0 \quad \text{at} \quad z = s(r), \quad r \geq 0, \tag{6.6}$$

or the 'hard' bottom (homogenous Neumann) b.c.

$$p_z - s'(r) p_r = 0 \quad \text{at} \quad z = s(r), \quad r \geq 0, \tag{6.7}$$

where $s' = \frac{ds}{dr}$; Equation (6.7) expresses the fact that the normal derivative of p at the bottom vanishes.

As we saw in the previous section, the transformation

$$p(z, r) = H_0^{(1)}(k_0 r) u(z, r)$$

yields, under appropriate assumptions, the Parabolic Equation (6.4) that we rewrite here again as

$$u_r = \frac{i}{2k_0} u_{zz} + \frac{ik_0}{2} (n^2(z, r) - 1) u, \qquad 0 \leq z \leq s(r), \quad r \geq 0. \tag{6.8}$$

Under this transformation the surface b.c. is conserved, i.e.,

$$u(0, r) = 0, \quad r \geq 0, \tag{6.9}$$

holds. The bottom b.c.'s (6.6), (6.7) become, respectively,

$$u = 0 \quad \text{at} \quad z = s(r), \quad r \geq 0, \tag{6.10}$$

$$u_z - s'(r)\big(u_r + g(r)u\big) = 0 \quad \text{at} \quad z = s(r), \quad r \geq 0, \tag{6.11}$$

where $g(r) = (H_0^{(1)}(k_0 r))_r / (H_0^{(1)}(k_0 r))$, or simply $g(r) = ik_0$ if the long–range approximation of the Hankel function is used. The problem is also furnished with an initial condition

$$u(z, 0) = u_0(z), \quad 0 \leq z \leq s(0), \tag{6.12}$$

where u_0 is a complex function simulating the effect of the source at $r = 0$.

In order to study the ibvp's consisting of (6.8), (6.9), (6.12), and either (6.10) or (6.11), we transform them into equivalent ones posed on a horizontal strip of unit depth by the simple range–dependent change of depth variable

$$y = \frac{z}{s(r)}.$$

With respect to the new dependent variable y, $0 \leq y \leq 1$, we denote $v(y, r) := u(z, r)$, and (6.8) becomes

$$v_r = i\alpha(r)v_{yy} + y\mu(r)v_y + i\beta(y, r)v, \quad 0 \leq y \leq 1, \quad r \geq 0, \tag{6.13}$$

where $\alpha(r) := 1/(2k_0 s^2(r))$, $\mu(r) := s'(r)/s(r)$, $\beta(y, r) := k_0(n^2(ys(r), r) - 1)/2$. The auxiliary conditions (6.9)–(6.11), become, respectively,

$$v(0, r) = 0, \quad r \geq 0, \tag{6.14}$$

$$v(1, r) = 0, \quad r \geq 0, \tag{6.15}$$

$$v_y(1, r) = S_1(r)v_r(1, r) + S_2(r)v(1, r), \quad r \geq 0, \tag{6.16}$$

where

$$S_1(r) := \frac{s'(r)s(r)}{1 + (s'(r))^2}, \qquad S_2(r) := g(r)S_1(r),$$

and

$$v(y, 0) = v_0(y) := u_0(ys(0)), \quad 0 \leq y \leq 1. \tag{6.17}$$

We write the above equations as

$$v_r = i\alpha(r)v_{yy} + y\mu(r)v_y + i\beta(y, r)v, \quad 0 \leq y \leq 1, \quad r \geq 0, \tag{6.18}$$

$$v(0, r) = 0, \quad r \geq 0, \tag{6.19}$$

$$v(1, r) = 0, \quad r \geq 0, \tag{6.20}$$

$$v_y(1, r) = S_1(r)v_r(1, r) + S_2(r)v(1, r), \quad r \geq 0, \tag{6.21}$$

$$v(y, 0) = v_0(y), \quad 0 \leq y \leq 1. \tag{6.22}$$

In the above, as mentioned in Section 6.1, we shall allow the coefficient β to be complex–valued in order to model absorption phenomena in the water column. (Physically, this implies that $\mathrm{Im}\beta \geq 0$.)

The ibvp (6.18)–(6.20), (6.22), i.e., with the homogenous Dirichlet b.c. at the bottom, is straightforward to analyze. It is well–posed on any finite interval $0 \leq r \leq R$, [27], and can be easily solved numerically using e.g., second–order centered finite differences or continuous, piecewise polynomial finite elements for the discretization in y and the midpoint rule in r. The resulting Crank–Nicolson type fully discrete scheme is unconditionally stable and of second–order accuracy in y and r. Entirely analogous is the case when the Neumann b.c. is imposed on a horizontal bottom, i.e., when $s'(r) = 0$ and (6.21) is simply $v_y(1, r) = 0$.

The situation is much different, however, in the case of the Neumann problem when $s'(r)$ is not the zero function. The well–posedness of the resulting ibvp (6.18), (6.19), (6.21), (6.22), for $0 \leq r \leq R$, for any $R > 0$, has been studied by Abrahamsson and Kreiss, [1], [2]. They proved that the problem is well–posed in the case of a strictly monotone bottom, i.e., when $s'(r)$ is of one sign on $[0, R]$. They also provided numerical and analytical evidence suggesting that the solution of this problem may grow fast with R in some domains with downsloping bottom profiles. In [3] an unconditionally stable finite–difference Crank–Nicolson type scheme was constructed for this ibvp and was proved to be of second–order of accuracy in r and y when either $s'(r) > 0$ or $s'(r) \leq 0$, for all r in the interval of interest $[0, R]$. In [4], [5] a fully discrete finite element scheme was constructed and analyzed for the same problem, and was proved to be unconditionally stable and convergent at the optimal rates if $s'(r) \leq 0$, i.e., for upsloping bottoms. Indeed, numerical evidence in [4] suggests that the solution of the problem may develop a singularity in domains with downsloping bottom profiles with $s'(r) > 0$ for $r \neq r_*$ and $s'(r_*) = s''(r_*) = 0$.

In order to overcome the theoretical and computational problems that the b.c. (6.21) leads to, Abrahamsson and Kreiss proposed in [1] an alternative rigid bottom b.c., which, in the original variables u, r, z of the untransformed domain, is

$$u_z - ik_0 s'(r)u = 0 \quad \text{at} \quad z = s(r), \quad r \geq 0. \tag{6.23}$$

The new condition may be viewed as a parabolic (paraxial) approximation of (6.11) in the following sense: Using (6.8) we have for the term $s'(r)u_r$ in (6.11) that

$$s'(r)u_r = \frac{i}{2k_0}s'(r)u_{zz} + \frac{ik_0}{2}s'(r)(n^2(z, r) - 1)u.$$

Both terms in the right–hand side of this equation are small if $s'(r)$ and $n(z, r)$ vary slowly with r and the angle of propagation is small relative to the horizontal direction. Hence, if $s'(r)u_r$ is dropped from (6.8) and $g(r)$ is replaced by its long–range approximation ik_0, (6.23) follows.

When we apply the transformation $y = z/s(r)$ the Abrahamsson–Kreiss b.c. (6.23) becomes

$$v_y(1, r) = ik_0 s'(r) s(r) v(1, r). \tag{6.24}$$

The new ibvp consisting of (6.18), (6.19), (6.22) and (6.24) can be shown to be well–posed, cf. [1], [2], [3], with no restriction on the sign of $s'(r)$. In fact, if $\text{Im}\beta = 0$, multiplying both sides of (6.18) by $\bar{v}$ (the overbar denotes the complex conjugation), integrating by parts with respect to y, taking real parts, and using the b.c.'s (6.19) and (6.24), we obtain that

$$s(r) \int_0^1 |v(y, r)|^2 dy = s(0) \int_0^1 |v_0(y)|^2 dy,$$

i.e., that the L^2–norm with respect to z, $(\int_0^{s(r)} |u(z, r)|^2 dz)^{1/2}$, of the solution u of the original ibvp in the variable domain $0 \leq z \leq s(r)$, $r \geq 0$, is conserved under the Abrahamsson–Kreiss b.c., in the absence of attenuation.

The ibvp consisting of (6.18), (6.19), (6.22) and (6.24) may be solved numerically in a straightforward manner. For example, on a uniform mesh defined by $y_j := jh$, $0 \leq j \leq J + 1$, $h = 1/(J + 1)$ and $r^n := nk$, $n = 0, \ldots, N$, with $Nk = R$, let V_j^n approximate $v(y_j, r^n)$ and be defined by the finite difference scheme that follows, in which $r^{n+1/2} := r^n + k/2$, and $V_j^{n+1/2} := (V_j^n + V_j^{n+1})/2$,

$V_j^0 = v_0(y_j)$, $0 \leq j \leq J + 1$.

For $n = 0, \ldots, N - 1$:

$V_0^{n+1} = 0$,

$$\frac{V_j^{n+1} - V_j^n}{k} = i\alpha(r^{n+1/2}) \frac{V_{j+1}^{n+1/2} - 2V_j^{n+1/2} + V_{j-1}^{n+1/2}}{h^2}$$

$$+ y_j \mu(r^{n+1/2}) \frac{V_{j+1}^{n+1/2} - V_{j-1}^{n+1/2}}{2h} + i\beta(y_j, r^{n+1/2}) V_j^{n+1/2},$$

$$j = 1, \ldots, J,$$

$$\frac{V_{J+1}^{n+1} - V_{J+1}^n}{k} = i\alpha(r^{n+1/2}) \left[\frac{2}{h^2} (V_J^{n+1/2} - V_{J+1}^{n+1/2}) \right.$$

$$\left. + \frac{2i}{h} k_0 s'(r^{n+1/2}) s(r^{n+1/2}) V_{J+1}^{n+1/2} \right]$$

$$+ \mu(r^{n+1/2}) \frac{V_{J+1}^{n+1/2} - V_J^{n+1/2}}{h} + i\beta(1, r^{n+1/2}) V_{J+1}^{n+1/2}.$$

This scheme, whose solution requires solving a tridiagonal system of equations for each n, is unconditionally stable, has an error bound of $O(h^2 + k^2)$ in the discrete L^2–norm $\|V\|_h := (h \sum_{j=1}^J |V_j|^2 + \frac{h}{2} |V_{J+1}|^2)^{1/2}$ and, when $\text{Im}\beta = 0$, it is L^2–conservative in that it satisfies $\|V^{n+1}\|_h = \|V^n\|_h$ for all n, cf. [3].

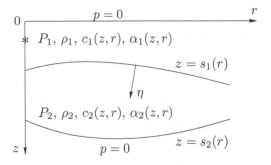

FIGURE 6.2: Two-layered medium with variable interface and bottom.

(See also [4], [5] for analysis and numerical experiments with an analogous fully discrete finite element scheme.)

Shown in [5] are also results of various numerical experiments aimed at comparing solutions of (6.18), (6.19), (6.22), and either of the two b.c.'s (6.21) and (6.24) on benchmark problems. In upslope problems, like, e.g., the ASA upsloping wedge, [21], with one fluid layer over a rigid bottom, numerical solutions of the two problems yield practically identical results. In downslope problems, there is good agreement for small range intervals $[0, R]$ but for larger R numerical discretizations of the exact Neumann b.c. in the form (6.21) apparently lose accuracy. If, however, one replaces in (6.21) the $v_r(1, r)$ term by the right–hand side of the PE (6.18) evaluated at $y = 1$, as done, e.g., in the program IFD, [23], [25], the discrete problem is stabilized and the numerical results are very close to those obtained by discretizing (6.23). We refer the reader to [5] for details.

We now comment on the well–posedness of PE problems in multilayered, axisymmetric fluid media separated by interfaces which, in general, may have variable topography. For simplicity we assume that the medium consists of two layers (cf. Fig. 6.2), the upper of which consists of water of constant density ρ_1 and speed of sound $c_1(z, r)$, and is bounded above by the pressure release free surface $z = 0$. The lower fluid medium has constant density $\rho_2 > \rho_1$, speed of sound $c_2(z, r)$, and is bounded above by the interface $z = s_1(r) > 0$ and below by the bottom surface $z = s_2(r)$ on which it is assumed that the acoustic pressure p is zero. We assume that $s_2(r) > s_1(r)$ for all r in the range of interest.

If we denote by $p_q(z, r)$ the acoustic pressure in the upper ($q = 1$) and the lower ($q = 2$) medium, we have (cf. Section 6.1)

$$\Delta p_1 + k_0^2 n_1^2(z, r) p_1 = 0, \quad 0 \le z \le s_1(r), \quad r > 0,$$
$$\Delta p_2 + k_0^2 n_2^2(z, r) p_2 = 0, \quad s_1(r) \le z \le s_2(r), \quad r > 0.$$

In these equations $k_0 = 2\pi f / c_0$ is the reference wavenumber, and $n_q^2(z, r) =$

$(c_0/c_q(z,r))^2(1 + i\alpha_q(z,r))$, $q = 1,2$, where $\alpha_q(z,r)$ are (small, nonnegative) absorption coefficients. The top and bottom b.c.'s are

$$p_1(0,r) = 0, \quad p_2(s_2(r),r) = 0, \quad r > 0.$$

Across the interface $z = s_1(r)$ we assume that the usual interface condition of acoustic, cf., e.g., [22], are valid, namely that the pressure is continuous

$$p_1(s_1(r),r) = p_2(s_1(r),r), \quad r > 0, \tag{6.25}$$

and that

$$\frac{1}{\rho_1}\frac{\partial p_1}{\partial \eta} = \frac{1}{\rho_2}\frac{\partial p_2}{\partial \eta} \quad \text{at} \quad z = s_1(r), \quad r > 0, \tag{6.26}$$

where η is the normal direction to the interface. The latter condition is, therefore, equivalent to the relation

$$\frac{\partial p_1}{\partial z}(s_1(r),r) - s_1'(r)\frac{\partial p_1}{\partial r}(s_1(r),r) =$$

$$\rho\left[\frac{\partial p_2}{\partial z}(s_1(r),r) - s_1'(r)\frac{\partial p_2}{\partial r}(s_1(r),r)\right], \quad r > 0, \tag{6.27}$$

where we put $\rho := \rho_1/\rho_2$.

The usual parabolic approximation procedure in each layer yields now the problem of determining a complex-valued function $u = u(z,r)$, *continuous* in $0 \le z \le s_2(r)$, $r \ge 0$, and satisfying

$$u_r = \frac{i}{2k_0}u_{zz} + i\beta_1(z,r)u, \quad 0 \le z \le s_1(r), \quad r \ge 0, \tag{6.28}$$

$$u_r = \frac{i}{2k_0}u_{zz} + i\beta_2(z,r)u, \quad s_1(r) \le z \le s_2(r), \quad r \ge 0, \tag{6.29}$$

with $\beta_q(z,r) := k_0(n_q^2(z,r) - 1)/2$, $q = 1,2$. In addition we have

$$u(0,r) = 0, \quad u(s_2(r),r) = 0, \quad r \ge 0, \tag{6.30}$$
$$u(z,0) = u_0(z), \quad 0 \le z \le s_2(0), \tag{6.31}$$

and

$$u_z(s_1^-(r),r)) - s_1'(r)\left(u_r(s_1^-(r),r) + ik_0u(s_1^-(r),r)\right) =$$

$$\rho\left[u_z(s_1^+(r),r) - s_1'(r)\left(u_r(s_1^+(r),r) + ik_0u(s_1^+(r),r)\right)\right], \quad r \ge 0, \tag{6.32}$$

where $\pm$ denote the directions ($-$ from above, $+$ from below) with respect to which one–sided derivatives of u are defined at the interface $z = s_1(r)$.

As in the case of the Neumann b.c. on the variable bottom of the single layer problem, we may replace the interface condition (6.32) by a paraxial approximation thereof. The new condition is

$$u_z(s_1^-(r),r) - ik_0s_1'(r)u(s_1^-(r),r) =$$

$$\rho\left[u_z(s_1^+(r),r) - ik_0s_1'(r)u(s_1^+(r),r)\right], \quad r \ge 0. \tag{6.33}$$

The resulting two interface ibvp's may be analyzed and solved numerically after a change of variables that transforms them into a system of two p.d.e.'s, on a horizontal strip of unit width, cf. [14]: Let $\delta_1(r) := s_1(r)$, $\delta_2(r) := s_2(r) - s_1(r)$ be the widths of the two layers. For $0 \le y \le 1$, $r \ge 0$, we let $w(y, r) := u(y\delta_1(r), r)$ and $v(y, r) := u(s_1(r) + y\delta_2(r), r)$. (Hence, w, v represent u in the upper and lower layer, respectively.) Then, the problem (6.28)–(6.32) is written, equivalently, as follow:

$$w_r = \frac{i}{2k_0(\delta_1(r))^2}w_{yy} + \frac{y\delta_1'(r)}{\delta_1(r)}w_y + i\gamma_1(y, r)w, \qquad 0 \le y \le 1, \qquad r \ge 0,$$

$$v_r = \frac{i}{2k_0(\delta_2(r))^2}v_{yy} + \frac{s_1'(r) + y\delta_2'(r)}{\delta_2(r)}v_y + i\gamma_2(y, r)v, \qquad 0 \le y \le 1, \qquad r \ge 0,$$

with $w(0, r) = 0$ and $v(1, r) = 0$, $r \ge 0$. In the above we have put $\gamma_1(y, r) := \beta_1(y\delta_1(r), r)$ and $\gamma_2(y, r) := \beta_2(s_1(r) + y\delta_2(r), r)$. In addition, for $r \ge 0$, w and v are coupled by the 'interface' conditions $w(1, r) = v(0, r)$, and

$$\frac{1}{\delta_1(r)}w_y(1, r) - \frac{s_1'(r)}{1 + (s_1'(r))^2}\left(w_r(1, r) + g(r)w(1, r)\right) =$$

$$\rho\left[\frac{1}{\delta_2(r)}v_y(0, r) - \frac{s_1'(r)}{1 + (s_1'(r))^2}\left(v_r(0, r) + g(r)v(0, r)\right)\right], \quad (6.34)$$

and for $r = 0$ are given in the terms of u_0 by the formula

$$w(y, 0) = u_0(y\delta_1(0)), \qquad v(y, 0) = u_0(s_1(0) + y\delta_2(0)), \qquad 0 \le y \le 1.$$

If, alternatively, the interface condition (6.33) holds, then (6.34) should be replaced by its 'paraxial' analog

$$\frac{1}{\delta_1(r)}w_y(1, r) - ik_0s_1'(r)w(1, r) = \rho\left[\frac{1}{\delta_2(r)}v_y(0, r) - ik_0s_1'(r)v(0, r)\right]. \quad (6.35)$$

In the analysis of [14], a key role is played by the condition

$$\frac{\delta_1'(r)}{2\delta_1(r)} + \text{Im}\beta_1(s_1(r), r) = \frac{\delta_2'(r)}{2\delta_2(r)} + \text{Im}\beta_2(s_1(r), r). \quad (6.36)$$

(If $\text{Im}\beta_1 = \text{Im}\beta_2$ for $z = s_1(r)$ and $r \ge 0$, (6.36) implies that $\delta_1(r)$ is a constant multiple of $\delta_2(r)$, i.e., that the two layers are *homothetic*.) If (6.36) is assumed to be valid, then, cf. [14], one may obtain *a priori* bounds in the Sobolev space $H^1(0, 1)$ for the solution (v, w) of the interface-ibvp with the 'exact' interface condition (6.34) provided $s_1'(r) \le 0$. When (6.34) is replaced by its 'paraxialization' (6.35), one may obtain an *a priori* L^2 bound for the solution without any additional hypothesis, and a H^1 bound under the condition (6.36) without assuming any sign condition on $s_1'(r)$. These bounds are basic steps for proving the well–posedness of the two interface-ibvp's.

One may solve the interface–ibvp's numerically by discretizing the (v, w) system but we will not go into details, referring instead to [14]. In Sections 6.3 and 6.4 that follow, we present two alternative variations of the change of variable technique for finite element, in 2D and 3D, discretizations using the analogs of interface condition (6.34) and (6.35), respectively.

We finally note that the rigorous analysis of the analogous ibvp's on variable domain for *wide–angle* extensions of the PE, cf., e.g., [10], [25], and Chapter 5 of Part II by D. Thomson and G. Brooke in this volume, also presents mathematical and numerical difficulties. For example, the well–posedness of the third–order 'Claerbout' wide–angle PE on a single layer domain with range–dependent bottom requires, in general, posing an additional boundary condition at the bottom; in this direction, cf., e.g., [13].

6.3 Finite Element Solution of the 2D PE in a General Stratified Waveguide

In this section, an initial–boundary value problem for the standard parabolic equation (6.4) is considered, in a fluid waveguide consisting of two layers, separated by an interface with mild variation in range, and an artificial absorbing layer, mimicking a semi-infinite outer fluid layer. The artificial layer is separated from the second layer by a horizontal interface. The variable interface is described for $r \geq 0$, by the curve $z = s(r)$, where

$$0 < s(r) < z_b < z_{\max}, \quad r \geq 0.$$

The upper layer is defined by $I_1 \times \{r \geq 0\}$, with $I_1 = (0, s(r))$, and has density ρ_1. The lower layer is defined by $I_2 \times \{r \geq 0\}$, with $I_2 = (s(r), z_b)$, and has density ρ_2. (In a marine waveguide the upper layer is the water and the second layer the sediment.) The absorbing layer occupies the strip $I_3 \times \{r \geq 0\}$, with $I_3 = (z_b, z_{\max})$; it has also density ρ_2 and a complex index of refraction incorporating large absorbtion coefficient to attenuate acoustics field in the vicinity of its outer boundary. The speed of sound $c(z, r)$ has possibly a jump discontinuity across the interfaces $z = s(r)$ and $z = z_b$. Here $I := I_1 \cup I_2 \cup I_3$, and $\overline{I_q}$, $1 \leq q \leq Q$, $\overline{I}$ denote the closure of intervals I_q, $1 \leq q \leq Q$, I.

In this section we rewrite the PE (6.4) as

$$u_r = i\alpha u_{zz} + i(\beta(z, r) + i\nu(z, r))u, \tag{6.37}$$

where $\alpha = 1/(2k_0)$ and $\beta + i\nu = k_0(n^2(z, r) - 1)/2$. Hence β and ν are the real and imaginary parts of the coefficient $k_0(n^2 - 1)/2$. (Recall that the index of refraction n is a complex-valued function.) We assume that $\nu \geq 0$. The

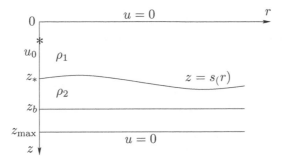

FIGURE 6.3: Variable interface problem.

transmission conditions for the PE at $z = s(r)$ have been derived in Section 6.2. They assert that

$$u(s^-(r), r) = u(s^+(r), r), \quad r \geq 0, \tag{6.38}$$

$$u_z(s^-(r), r) - s'(r)[ik_0 u(s^-(r), r) + u_r(s^-(r), r)] =$$
$$\rho \left(u_z(s^+(r), r) - s'(r)[ik_0 u(s^+(r), r) + u_r(s^+(r), r)] \right), \quad r \geq 0. \tag{6.39}$$

Note that (6.39) is deduced from (6.32) by taking $g(r) = ik_0$. In this section we shall implement the 'exact' interface conditions (6.38) and (6.39). Furthermore, we assume that $u(\cdot, r)$ is C^1 across $z = z_b$.

6.3.1 Horizontal Interface

For a horizontal interface, where $s(r) = z_* = $ constant, for $r \geq 0$, the transmission conditions simplify to

$$u(z_*^-, r) = u(z_*^+, r), \qquad u_z(z_*^-, r) = \rho u_z(z_*^+, r). \tag{6.40}$$

The boundary conditions

$$u(0, r) = u(z_{\max}, r) = 0, \quad r \geq 0, \tag{6.41}$$

are imposed. The harmonic point source is modelled by a function $u_0(z)$ which defines the initial field as

$$u(z, 0) = u_0(z), \quad 0 \leq z \leq z_{\max}. \tag{6.42}$$

We start by describing two high–order accurate finite element techniques and an implicit Runge–Kutta method, that we apply to discretize the problem in depth and in range, respectively.

The two finite element methods are illustrated on the o.d.e. two–point boundary value problem with interface given by

$$u''(z) = f(z), \qquad z \in I_1 \cup I_2 \cup I_3, \tag{6.43}$$

$$u(z_*^-) = u(z_*^+), \qquad u'(z_*^-) = \rho u'(z_*^+), \tag{6.44}$$

$$u(0) = u(z_{\max}) = 0, \tag{6.45}$$

where $u' = \frac{du}{dz}$, $u'' = \frac{d^2u}{dz^2}$. Here f is a given complex–valued function on $\bar{I}$, smooth on each I_q, with possible jump discontinuities at z_* and z_b. The solution u is assumed to be C^1 across $z = z_b$.

Let $H^1(I_q)$ denote the Sobolev spaces consisting of those complex–valued functions $v \in L^2(I_q)$ having a distributional derivative Dv of first order in $L^2(I_q)$. The subspace of $H^1(I)$ consisting of the functions that vanish at 0 and $z_{\max}$ in the sense of trace is denoted by $H_0^1(I)$. The $H^1(I_q)$ Sobolev space is equipped with the norm

$$\|v\|_{1,I_q} = \left(\int_{I_q} |v|^2 dz + \int_{I_q} |Dv|^2 dz \right)^{\frac{1}{2}}$$

and the $L^2(I_q)$ norm of a function on I_q will be denoted $\| \cdot \|_{0,I_q}$.

The weighted inner product on $L^2(I)$, appropriate for the problem at hand is defined for $u, v \in L^2(I)$ by

$$(u, v) := \int_{I_1} u\bar{v} dz + \rho \int_{I_2 \cup I_3} u\bar{v} dz.$$

The induced norm is $\|v\| = (v, v)^{1/2}$ (entirely equivalent to the $\| \cdot \|_{0,I}$ norm on $L^2(I)$).

In the associated weak formulation of (6.43)–(6.45), $u \in H_0^1(I)$ is sought satisfying

$$B(u, v) = -(f, v), \qquad \forall v \in H_0^1(I), \tag{6.46}$$

where for $u, v \in H_0^1(I)$,

$$B(u, v) := (u', v'). \tag{6.47}$$

Existence and uniqueness of the solution of (6.46) is studied in [12].

Two fourth–order accurate finite element methods are used to discretize (6.46). Let $\mathbf{P}_3(I)$ be the space of cubic complex–valued polynomials, defined on an interval I. The restriction on $e \subset \bar{I}$ of a function f is denoted by $f|_e$. For an integer M, $\mathcal{T}_h := \{z_0, z_1, \ldots, z_J\}$ is a nonuniform grid on $\bar{I}$, such that $z_0 = 0$, $z_J = z_{\max}$, and $z_{\ell_1} = z_*$ and $z_{\ell_2} = z_b$, for some $1 \le \ell_1 < \ell_2 < J$. Then, $\mathcal{T}_{h,1} = \{z_0, \ldots, z_{\ell_1}\}$, $\mathcal{T}_{h,2} = \{z_{\ell_1}, \ldots, z_{\ell_2}\}$ and $\mathcal{T}_{h,3} = \{z_{\ell_2}, \ldots, z_J\}$ are the induced grids on $\bar{I}_q$, $q = 1, 2, 3$, respectively. We let $h := \max_{1 \le j \le J}(z_j - z_{j-1})$ and $e_j := (z_{j-1}, z_j)$, $1 \le j \le J$.

The first method is the C^0–*polynomial finite element method.* Here the finite element space used is

$$Q_h = \{\chi : \chi \in C^0(\bar{I}) \text{ complex–valued, } \chi|_{e_j} \in \mathbf{P}_3(e_j),$$
$$1 \le j \le J, \text{ and } \chi(0) = \chi(z_{\max}) = 0\}. \tag{6.48}$$

For $v \in H_0^1(I)$ and piecewise (i.e., on each I_q, $q = 1, 2, 3$) suitably smooth, the approximation property

$$\inf_{\chi \in Q_h} \{\|v - \chi\|_{0,I} + h\|v - \chi\|_{1,I}\} = O(h^4), \tag{6.49}$$

holds for Q_h.

The discrete analog of (6.46) on Q_h is: Find $u_h \in Q_h$, satisfying

$$B(u_h, \chi) = -(f, \chi), \qquad \forall \chi \in Q_h, \tag{6.50}$$

which has a unique solution u_h approximating the solution u of (6.43)–(6.45) with fourth–order accuracy, [12],

$$\|u - u_h\|_{0,I} + h\|u - u_h\|_{1,I} = O(h^4). \tag{6.51}$$

The second method is the *nonstandard spline method* based on a nonstandard variational formulation of (6.43)–(6.45) using a discontinuous finite element space of cubic splines defined on each I_q, $q = 1, 2, 3$.

Let $H = \{v \in \prod_{q=1}^3 H^1(I_q), v(0) = v(z_{\max}) = 0\}$. For $v \in H$, we denote by $\tilde{D}v$, the function in $L^2(I)$ defined piecewise on I_q as the derivative of $v|_{I_q}$ on I_q, i.e. by $(\tilde{D}v)|_{I_q} = D(v|_{I_q})$, $q = 1, 2, 3$. The function space H is equipped with the h–dependent norm $\|\cdot\|_h$ defined by

$$\|v\|_h := \|v\| + \|\tilde{D}v\| + h^{\frac{1}{2}}|\tilde{D}v(z_*^-)| + h^{-\frac{1}{2}}|[v(z_*)]|$$
$$+ \rho\{h^{\frac{1}{2}}|\tilde{D}v(z_b^-)| + h^{-\frac{1}{2}}|[v(z_b)]|\}, \tag{6.52}$$

where $[\psi(z)] := \psi(z^+) - \psi(z^-)$.

A nonstandard weak formulation of (6.43)–(6.45) is to define the solution u so that

$$B_\gamma(u, v) = -(f, v), \qquad \forall v \in H, \tag{6.53}$$

where, for $w, v \in H$ and $\gamma > 0$,

$$B_\gamma(w, v) := (\tilde{D}w, \tilde{D}v) + \{\tilde{D}w(z_*^-)[\bar{v}(z_*)] + \tilde{D}\bar{v}(z_*^-)[w(z_*)]\}$$
$$+ \rho\{\tilde{D}w(z_b^-)[\bar{v}(z_b)] + \tilde{D}\bar{v}(z_b^-)[w(z_b)]\}$$
$$+ \frac{\gamma}{h}\{[w(z_*)][\bar{v}(z_*)] + \rho[w(z_b)][\bar{v}(z_b)]\}. \tag{6.54}$$

The parameter $\gamma > 0$ is chosen so that (6.53) is well posed.

A discontinuous finite element space $S_h \subset H$ is defined by

$$S_h = \{\chi : \chi|_{I_q} \in S_{h,q}, q = 1, 2, 3, \chi(0) = \chi(z_{\max}) = 0\}, \qquad (6.55)$$

with

$$S_{h,q} = \{\chi : \chi \in C^2(\bar{I}_q) \text{ complex–valued}, \chi|_{e_j} \in \mathbf{P}_3(e_j), \text{ for all } e_j \subset I_q\}.$$

For $v \in H$ and piecewise suitably smooth, S_h has the approximation property

$$\inf_{\chi \in S_h} \{\|v - \chi\|_{0,I} + h\|v - \chi\|_h\} = O(h^4), \qquad (6.56)$$

when the grid $\{\mathcal{T}_h\}$ is *quasi–uniform* in the sense that $\min_{1 \le j \le J} h_j/h \ge \xi$, for some constant $\xi > 0$.

The discrete analog of (6.53) is: Find $u_h \in S_h$, satisfying

$$B_\gamma(u_h, \chi) = -(f, \chi), \qquad \forall \chi \in S_h. \qquad (6.57)$$

For u and u_h the solutions of (6.53) and (6.57), respectively, the error estimate holds:

$$\|u - u_h\|_{0,I} + h\|u - u_h\|_h = O(h^4).$$

These finite element methods will be subsequently used to discretize (6.37) in the depth variable. A detailed study of both methods is found in [12].

A semidiscrete, i.e., continuous in range, approximation to the initial–boundary value problem (6.37), (6.40), (6.41), and (6.42), for which we recall that the interface is horizontal, is: Given a maximum range of propagation $R > 0$, find $u_h : [0, R] \to \mathcal{X}_h$, satisfying

$$(u_{hr}, \chi) + i\alpha\mathcal{B}(u_h, \chi) - i(\beta(r)u_h, \chi) + (\nu(r)u_h, \chi) = 0,$$
$$\forall \chi \in \mathcal{X}_h, \quad 0 \le r \le R, \qquad (6.58)$$
$$u_h(0) = u_h^0,$$

where the pair $(\mathcal{B}, \mathcal{X}_h)$ corresponds, as the case may be, either to the choice $\mathcal{B}(\cdot, \cdot) = B(\cdot, \cdot)$ and $\mathcal{X}_h = Q_h$ defined by (6.47) and (6.48), respectively, or to the choice $\mathcal{B}(\cdot, \cdot) = B_\gamma(\cdot, \cdot)$ and $\mathcal{X}_h = S_h$, defined by (6.54) and (6.55), respectively. Here u_h^0 is a suitable projection on $\mathcal{X}_h$ of the initial field u_0.

For $0 \le r \le R$, the linear operator $\mathcal{L}_h(r) : \mathcal{X}_h \to \mathcal{X}_h$ is defined by

$$(\mathcal{L}_h(r)\varphi, \chi) := -\alpha(r)\mathcal{B}(\varphi, \chi) + ((\beta(r) + i\nu(r))\varphi, \chi), \qquad \varphi, \chi \in \mathcal{X}_h,$$

and (6.58) is written compactly as

$$u_{hr} = i\mathcal{L}_h(r)u_h, \qquad 0 \le r \le R,$$
$$u_h(0) = u_h^0. \qquad (6.59)$$

The system of ordinary differential equations, (6.59), is discretized using the fourth–order accurate, two–stage implicit Runge–Kutta method of Gauss–Legendre type, determined by the set of parameters $a_{11} = a_{22} = 1/4, a_{12} =$

$1/4 - \sqrt{3}/6$, $a_{21} = 1/4 + \sqrt{3}/6$, $\tau_1 = 1/2 - \sqrt{3}/6$, $\tau_2 = 1/2 + \sqrt{3}/6$, and $b_1 = b_2 = 1/2$.

For $r^{n,1} := r^n + \tau_1 k$ and $r^{n,2} := r^n + \tau_2 k$, the fully discrete analog of (6.59) is: For $0 \le n \le N$, find $U^n \in X_h$, approximating $u_h(r^n)$, and $U^{n,1} \in X_h$, $U^{n,2} \in X_h$, that satisfy

$$U^0 = u_h^0,$$

for $n = 0, \ldots, N - 1$:

$$U^{n,1} = U^n + ik\left(a_{11}\mathcal{L}_h^{n,1}U^{n,1} + a_{12}\mathcal{L}_h^{n,2}U^{n,2}\right),$$

$$U^{n,2} = U^n + ik\left(a_{21}\mathcal{L}_h^{n,1}U^{n,1} + a_{22}\mathcal{L}_h^{n,2}U^{n,2}\right), \qquad (6.60)$$

$$U^{n+1} = U^n + ik\left(b_1\mathcal{L}_h^{n,1}U^{n,1} + b_2\mathcal{L}_h^{n,2}U^{n,2}\right),$$

where $\mathcal{L}_h^{n,1} = \mathcal{L}_h(r^{n,1})$ and $\mathcal{L}_h^{n,2} = \mathcal{L}_h(r^{n,2})$. The fully discrete scheme (6.60) produces a uniquely defined sequence of vectors $U^{n,1}, U^{n,2}$ and U^{n+1} in X_h, which is also unconditionally stable in the sense that $\max\{\|U^{n,1}\|, \|U^{n,2}\|\} \le c\|U^n\|$ and $\|U^{n+1}\| \le \|U^n\|$ (equality holds if $\nu = 0$ in (6.37)). Hence, the choice of this method for the range discretization ensures high order of accuracy in r, unconditional stability, and L^2–conservation in the absence of dissipation.

The following error estimate holds between the exact solution $u^n = u(z, r^n)$ and the fully discrete solution U^n: There exists a constant c independent of k and h such that

$$\max_{0 \le n \le N} \|U^n - u^n\|_{0,I} \le c(k^3 + h^4), \qquad (6.61)$$

which in the range–independent case $\beta = \beta(z)$ and $\nu = \nu(z)$ in (6.37), becomes

$$\max_{0 \le n \le N} \|U^n - u^n\|_{0,I} \le c(k^4 + h^4). \qquad (6.62)$$

We shall refer to the overall numerical scheme as the CCUB or the SPLN method when the C^0–*polynomial finite element method* (cf. (6.50)) or the *non-standard spline method* (cf. (6.57)) have been used, respectively, to discretize in depth. For a detailed analysis of the fully discrete scheme (6.60) the reader is addressed to [12]. A detailed description of an iterative scheme for solving the linear system determining $U^{n,1}$ and $U^{n,2}$ is also found in [12] along with comments on its parallelization and the optimal selection of the discretization parameters k and h regarding the minimization of the computational cost.

Both the CCUB and SPLN methods were applied to several problems of acoustic propagation in horizontally stratified marine waveguides with success, cf. [12] and the references therein. Two benchmark problems of acoustic propagation in sea waveguides with horizontal stratification are shown here.

- *The horizontal interface problem of Bucker.*
  For this problem, [25], let $z_* = 240$ m, $z_b = 512$ m, $z_{max} = 1200$ m, $\rho_1 =$

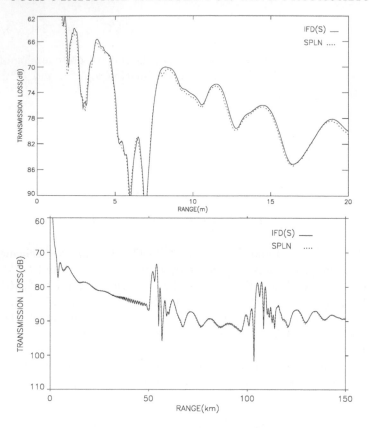

FIGURE 6.4: Comparison of SPLN and IFD(S) results for the horizontal interface problem (upper subplot) and the leaky surface duct (lower subplot).

1 and $\rho_2 = 2.1$ g/cm^3. No attenuation is present in $[0, z_b]$. The speed of sound is range–independent and linear in $[0,120]$ and $[120, z_*]$, with values $c = 1500$ m/sec at $z = 0$ and at $z = 240$ m, and $c = 1498$ m/sec at $z = 120$ m; it is constant with value $c = 1505$ m/sec in $[z_*, z_{max}]$. The reference sound speed is $c_0 = 1499$ m/sec. A source of frequency $f = 100$ Hz, described by a Gaussian function, is placed at a depth $z_S = 30$ m. In Fig. 6.4 (upper subplot), we present the transmission loss (transmission loss is defined by TL $= -20 \log_{10}(|u(z,r)|/\sqrt{r})$, at a given depth z, cf. also [22]) vs. range curve, for a receiver placed at 90 m, obtained with the SPLN method and the classical, second–order finite difference scheme IFD(S) of [25]. Both results are very close (and also very close to the results from CCUB method, not shown here). The large density jump across the interface, which is an inherent difficulty of this test case, was successfully handled by all methods.

- *The leaky surface duct problem.*
  For this problem, [28], let $z_* = 4000$ m, $z_b = 4100$ m, $z_{max} = 5125$ m, $\rho_1 = \rho_2 = 1$ g/cm^3. No attenuation is present in $[0, z_*]$, while within $[z_*, z_b]$ an attenuation coefficient of 0.1 dB/wavelength is added to ν of (6.37) (cf. [12] for the proper formula). The sound speed in $[0, z_*]$ is given in Fig. 1 of [28]. In $[z_*, z_b]$, $c = 1523.9$ m/sec, and $c_0 = 1500$ m/sec. The source, of frequency $f = 80$ Hz, modelled by a Gaussian function, is placed at a depth $z_S = 25$ m. In Fig. 6.4 (lower subplot) we present the corresponding transmission loss vs. range plot, for a receiver placed at 100 m, obtained by the SPLN and IFD(S) method. Both results, as well as those obtained by the CCUB method but not shown here, coincide.

In both cases, a strong dissipation is assumed within the absorbing layer $[z_b, z_{max}]$, expressed as in [12]. It is important to note that the SPLN method uses coarser discretizations (in z and r) than those used by IFD(S) to achieve the same accuracy, that despite its higher computational complexity it is more economical, [12].

6.3.2   Sloping Interface

We now proceed to the case of a stratified waveguide of two fluid layers, with a mildly range–dependent interface between them. A numerical solution will be provided based on the fully discrete schemes of the previous section. As in Section 6.2, a range–dependent change of the depth variable is used to transform the original physical domain to one with horizontal interfaces. This is done however at the expense of complicating the differential operator by adding low–order terms and variable coefficients in the parabolic equation and in the interface conditions. We now use the following change of the variables in the depth direction

$$y = \frac{zz_*}{s(r)}, \quad 0 \le z \le s(r),$$

$$y = z_b + \frac{(z_b - z)(z_* - z_b)}{z_b - s(r)}, \quad s(r) \le z \le z_b, \qquad (6.63)$$

$$y = z, \quad z_b \le z \le z_{max},$$

with $z_* = s(0)$. In the y coordinate the interface $z = s(r)$, is transformed to the horizontal interface $y = z_*$. The new dependent variable $v(y, r) := u(z, r)$ is defined for y and z related by the change of variable defined above. Equation (6.37) then transforms to

$$v_r = i\tilde{\alpha}(y, r)v_{yy} + i[\tilde{\beta}(y, r) + i\tilde{\nu}(y, r)]v + \tilde{\delta}(y, r)v_y,$$
$$y \in (0, z_*) \cup (z_*, z_b) \cup (z_b, z_{max}), \quad r \in [0, R], \qquad (6.64)$$

where

$$
\tilde{\alpha}(y,r) := \begin{cases} (\frac{z_*}{s(r)})^2\alpha, & y \in (0, z_*) \\ (\frac{z_* - z_b}{z_b - s(r)})^2\alpha, & y \in (z_*, z_b) \\ \alpha, & y \in (z_b, z_{\max}) \end{cases}
$$

$$
\tilde{\beta}(y,r) := \begin{cases} \beta(\frac{ys(r)}{z_*},r), & y \in (0, z_*) \\ \beta(z_b - \frac{(y - z_b)(z_b - s(r))}{z_* - z_b},r), & y \in (z_*, z_b) \\ \beta(y,r), & y \in (z_b, z_{\max}) \end{cases}
$$

$$
\tilde{\nu}(y,r) := \begin{cases} \nu(\frac{ys(r)}{z_*},r), & y \in (0, z_*) \\ \nu(z_b - \frac{(y - z_b)(z_b - s(r))}{z_* - z_b},r), & y \in (z_*, z_b) \\ \nu(y,r), & y \in (z_b, z_{\max}) \end{cases} \tag{6.65}
$$

$$
\tilde{\delta}(y,r) := \begin{cases} \frac{ys'(r)}{s(r)}, & y \in (0, z_*) \\ \frac{(z_b - y)s'(r)}{z_b - s(r)}, & y \in (z_*, z_b) \\ 0, & y \in (z_b, z_{\max}). \end{cases}
$$

In the new coordinates system, the transmission conditions (6.38) and (6.39) yield for the solution v of (6.64), that both v and v_r are continuous across z_* and z_b, and

$$
v_y(z_*^-,r) = \rho\frac{s(r)(z_b - z_*)}{z_*(z_b - s(r))}v_y(z_*^+,r)
$$

$$
+(1 - \rho)\frac{s'(r)s(r)}{z_*(1 + (s'(r))^2)}[v_r(z_*^+,r) + ik_0v(z_*^+,r)], \quad r \geq 0, \tag{6.66}
$$

$$
v_y(z_b^-,r) = \frac{z_b - s(r)}{z_b - z_*}v_y(z_b^+,r), \quad r \geq 0. \tag{6.67}
$$

(Recall that the third (absorbing) layer has the same density as the second layer.) Furthermore, the initial and boundary conditions (6.42) and (6.41) hold.

The problem (6.64)–(6.67) is discretized in depth using the standard C^0–polynomial finite element space defined by (6.48). Let $\tilde{\alpha}_q = \tilde{\alpha}_q(r)$, $q = 1, 2, 3$, be the restrictions of the function $\tilde{\alpha}(y,r)$ defined in (6.65), to $\tilde{I}_1 := (0, z_*)$, $\tilde{I}_2 := (z_*, z_b)$, $\tilde{I}_3 := (z_b, z_{\max})$, respectively. Similarly, let $\tilde{\delta}_q = \tilde{\delta}_q(y,r) := \tilde{\delta}(y,r)|_{y \in \tilde{I}_q}$, $q = 1, 2, 3$. Let $\tilde{I} := (0, z_*) \cup (z_*, z_b) \cup (z_b, z_{\max})$. Further, let

$$
d_1 = d_1(r) := \rho\frac{s(r)(z_b - z_*)}{z_*(z_b - s(r))}, \quad d_2 = d_2(r) := \frac{z_b - s(r)}{z_b - z_*},
$$

$$
\omega = \omega(r) := (1 - \rho)\frac{s'(r)s(r)}{z_*(1 + (s'(r))^2)}.
$$

For $w, v \in L^2(\tilde{I})$, $r \in [0, R]$, the r–dependent weighted L^2–inner product $\langle \cdot, \cdot \rangle$

is defined by

$$\langle w, v \rangle := \tilde{\alpha}_2 \tilde{\alpha}_3 \int_0^{z_*} w \bar{v} dy + d_1 \tilde{\alpha}_1 \tilde{\alpha}_3 \int_{z_*}^{z_b} w \bar{v} dy +$$
$$d_1 d_2 \tilde{\alpha}_1 \tilde{\alpha}_2 \int_{z_b}^{z_{\max}} w \bar{v} dy. \tag{6.68}$$

The weak formulation, that the solution v of (6.64)–(6.67) satisfies, is

$$\langle v_r, \psi \rangle = -i \langle \tilde{\alpha}(r) v_y, \psi_y \rangle + i \langle (\tilde{\beta}(r) + i \tilde{\nu}(r)) v, \psi \rangle + \langle \tilde{\delta}(r) v_y, \psi \rangle$$
$$+ i \tilde{\alpha}_1 \tilde{\alpha}_2 \tilde{\alpha}_3 \omega(r) [v_r(z_*, r) + i k_0 v(z_*, r)] \bar{\psi}(z_*),$$
$$\forall \psi \in H_0^1(I), \quad 0 \le r \le R, \tag{6.69}$$
$$v(0) = v^0, \quad y \in \tilde{I}.$$

Therefore, we seek $v_h : [0, R] \to Q_h$, where Q_h is the finite element space defined in (6.48), satisfying

$$\langle v_{hr}, \chi \rangle = -i \langle \tilde{\alpha}(r) v_{hy}, \chi_y \rangle + i \langle (\tilde{\beta}(r) + i \tilde{\nu}(r)) v_h, \chi \rangle + \langle \tilde{\delta}(r) v_{hy}, \chi \rangle$$
$$+ i \tilde{\alpha}_1 \tilde{\alpha}_2 \tilde{\alpha}_3 \omega(r) [v_{hr}(z_*, r) + i k_0 v_h(z_*, r)] \bar{\chi}(z_*), \tag{6.70}$$
$$\forall \chi \in Q_h, \quad 0 \le r \le R,$$
$$v_h(0) = v_h^0. \tag{6.71}$$

The system of ordinary differential equations (6.70) in r, which involves the coefficients of $v_h(y, r)$ with respect to a basis of Q_h, is discretized using the fourth–order accurate, two–stage implicit Runge–Kutta method (cf. (6.60)). The overall numerical method consists of combining this range–stepping scheme with continuous, piecewise cubic functions for the discretization in the depth variable.

In operator form (analog to that of (6.59)) equation (6.70) is written as

$$v_{hr} = i R_h(r) v_h, \quad 0 \le r \le R, \quad v_h(0) = v_h^0. \tag{6.72}$$

Here $R_h(r) := T_h^{-1}(r) L_h(r)$, with the operators $L_h(r), T_h(r) : Q_h \to Q_h$, for $0 \le r \le R$ and $\phi, \chi \in Q_h$, defined by

$$\langle L_h(r) \phi, \chi \rangle := -\langle \tilde{\alpha}(r) \phi_y, \chi_y \rangle + \langle (\tilde{\beta}(r) + i \tilde{\nu}(r)) \phi, \chi \rangle$$
$$- i \langle \tilde{\delta}(r) \phi_y, \chi \rangle + \tilde{\alpha}_1 \tilde{\alpha}_2 \tilde{\alpha}_3 \omega i k_0 \phi(z_*) \bar{\chi}(z_*),$$
$$\langle T_h(r) \phi, \chi \rangle := \langle \phi, \chi \rangle - i \tilde{\alpha}_1 \tilde{\alpha}_2 \tilde{\alpha}_3 \omega \phi(z_*) \bar{\chi}(z_*).$$

The fully discrete analog of (6.72) has the same form as (6.60), with $\mathcal{L}_h^{n,j}$ replaced by $R_h^{n,j}$ in the notation. The resulting method is an extension of the CCUB method of the previous subsection; it will be referred to in the sequel as the CCUB method again. The proposed method is fourth–order accurate and is analyzed in more details in [12]. In [12], the second–order accurate method CNP1, applicable for a general variable waveguide, is also

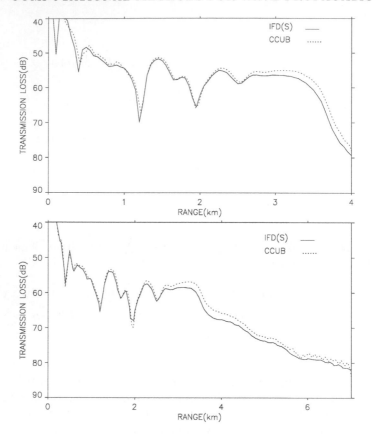

FIGURE 6.5: Comparison of CCUB and IFD(S) results for the wedge–shaped waveguide (upper subplot) and ridge–shaped waveguide (lower subplot).

presented and analyzed. The CNP1 method uses continuous, piecewise linear functions (using $\mathbf{P}_1(e_j)$ as local interpolation space in the definition of Q_h instead of $\mathbf{P}_3(e_j)$, cf. (6.48)) to discretize in the depth variable, coupled with the Crank–Nicolson scheme for stepping in range.

Both methods were applied successfully to several test problems of acoustic propagation in layered waveguides with range–dependent interfaces. Two examples are shown here.

- *The wedge–shaped waveguide problem.*
  For the wedge–shaped waveguide problem, [21], we take $s(r) = 200 - r/20$ m, $z_b = 1000$ m, $z_{\max} = 4000$ m, $\rho_1 = 1$ and $\rho_2 = 1.5$ g/cm^3. Within the second layer an attenuation coefficient of 0.5 dB/wavelength is added to ν of (6.37) (cf. [12]). The speed of sound is $c = 1500$ m/sec in $[0, s(r)]$, whereas $c = 1700$ m/sec in $[s(r), z_{\max}]$ and $c_0 = 1500$ m/sec. A source, of Greene type [19], with a frequency of $f = 25$ Hz

is placed at a depth of $z_S = 100$ m. In this problem backscattering is very weak, therefore, the acoustic field is essentially outgoing, [21], [29]. In Fig. 6.5 (upper subplot) we present the transmission loss vs. range plots, for a receiver placed at $z_R = 30$ m, obtained by the CCUB and IFD(S) codes. We observe that the CCUB transmission loss curve, in the region beyond the 3.5 km, remains about 2 dB above the IFD(S) curve. Therefore, the CCUB solution approximates more accurately the full reduced wave equation solution for this case, shown in [21].

- *The ridge–shaped waveguide problem.*
  Next, the upslope–downslope (ridge) problem of [9] (which is used to test energy conservation capabilities of PE models in a range–dependent waveguides) is shown. Here

$$s(r) = \begin{cases} 200 - r/20 \text{ m}, & 0 \le r \le 3500 \text{ m}, \\ r/20 - 150 \text{ m}, & 3500 \le r \le 7000 \text{ m}, \end{cases}$$

$z_b = 1000$ m, $z_{\max} = 4000$ m, $\rho_1 = 1$ and $\rho_2 = 1.5$ g/cm^3. An attenuation coefficient of 0.5 dB/wavelength is assumed for the second layer. The speed of sound is 1500 m/sec in $[0, s(r)]$ and 1700 m/sec in $[s(r), z_{\max}]$. c_0 is set to 1500 m/sec. A Greene type source, [19], of frequency $f = 25$ Hz is placed at a depth $z_S = 100$ m. In Fig. 6.5 (lower subplot) we present the transmission loss plots of the CCUB and IFD(S) results, for a receiver placed at $z_R = 20$ m. Again the CCUB curve is above the IFD(S) curve, therefore closer to the correct full reduced wave equation results, [9].

The superiority of the CCUB results is due to the fact that the 'exact' interface condition (6.39) is discretized in the finite element/ change of variable technique proposed. The IFD(S) method suffers the observed energy loss because a 'staircase' (piecewise horizontal) approximation of the variable bottom is used, [29]. This 'staircase' approximation of the variable bottom only consider partial derivatives with respect to z in the interface condition. (or (6.39)) actual interface condition. When the parabolized (6.35) interface condition was used by CCUB, the results obtained were almost identical to these of the IFD(S). In both cases, the CCUB method proved to be again less expensive computationally than the IFD(S) method. Further, the CNP1 results exhibit a complete agreement with the CCUB results, [12]; therefore the convergence of the change of variables technique is demonstrated.

6.4   Finite Element Solution of the 3D Standard PE in a General Stratified Waveguide

In this section an initial–boundary value problem for (6.3) in a general, stratified fluid waveguide will be established at first. This initial–boundary value problem will be solved in the sequel by a finite element technique coupled with an appropriate transformation of coordinates which simplify the geometry of the computational domain.

6.4.1   The Initial–Boundary Value Problem for the 3D Standard PE

A waveguide consisting in one water layer of constant density ρ_0, attenuation α_0, sound speed c_0, and Q fluid layers, each of constant density ρ_q, attenuation α_q, and sound speed c_q, for $1 \leq q \leq Q$, is considered. The geometry of the waveguide varies arbitrarily in 3D, with the only restriction of non mixing layers. No cylindrical symmetry is assumed on the waveguide geometry. Let $s(r, \theta)$ be the maximum depth of the waveguide and $z_q(r, \theta)$ denote the interface separating two adjacent layers. The physical assumption of non mixing layers is interpreted as

$$0 < z_1(r, \theta) < \ldots < z_Q(r, \theta) < s(r, \theta), \quad r \geq 0, \quad \theta \in [0, 2\pi].$$

Equation (6.3) holds therefore, for $z \in \cup_{q=0}^{Q}(z_q(r, \theta), z_{q+1}(r, \theta))$, $r \geq 0$ and $\theta \in [0, 2\pi]$, where $z_0(r, \theta) := 0$ and $z_{Q+1}(r, \theta) := s(r, \theta)$. The sound speed c, the density ρ and the dissipation α are defined by their restrictions on each fluid layer, respectively, i.e. for example $\rho = \rho_q$ if $z \in (z_q(r, \theta), z_{q+1}(r, \theta))$, for $0 \leq q \leq Q$. It is assumed that c, ρ and ϕ are continuous functions of θ and r, for $r \geq 0$, $0 \leq \theta \leq 2\pi$, $z \in \cup_{q=0}^{Q}(z_q(r, \theta), z_{q+1}(r, \theta))$, with possible jump discontinuities across the interfaces $z = z_q(r, \theta)$, $1 \leq q \leq Q$. For simplicity in the subsequent calculations, the notation $\beta(z, r, \theta) = n^2(z, r, \theta) - 1$ is used in (6.3). The solution u of (6.3) satisfies the pressure-release boundary conditions $u = 0$ at $z = 0$ and $z = s(r, \theta)$, and the 2π–periodicity condition in azimuth $u(z, r, 0) = u(z, r, 2\pi)$, for $r \geq 0$, $z \in \cup_{q=0}^{Q}(z_q(r, \theta), z_{q+1}(r, \theta))$. As for the initial condition, $u(z, 0, \theta) = u^0(z, \theta)$, where u^0 is the distribution of the acoustic pressure at $r = 0$ established by the harmonic point source. The transmission conditions for u along the interfaces $z = z_q(r, \theta)$, for $1 \leq q \leq Q$, emerge from those of the solution p of the Helmholtz equation (6.1). Indeed,

$$p(z_q^-(r, \theta), r, \theta) = p(z_q^+(r, \theta), r, \theta),$$
$$r \geq 0, \quad \theta \in [0, 2\pi], \quad 1 \leq q \leq Q. \tag{6.73}$$

and

$$\frac{1}{\rho_{q-1}} \frac{\partial p}{\partial \eta_q}(z_q^-(r,\theta), r, \theta) = \frac{1}{\rho_q} \frac{\partial p}{\partial \eta_q}(z_q^+(r,\theta), r, \theta),$$

$$r \geq 0, \quad \theta \in [0, 2\pi], \quad 1 \leq q \leq Q, \tag{6.74}$$

where $\partial/\partial \eta_q$ is the normal derivative along the q–th interface curve $z = z_q(r,\theta)$ and is expressed as

$$\frac{\partial}{\partial \eta_q} := \frac{\partial}{\partial z} - \left(\frac{\partial z_q}{\partial r}\right) \frac{\partial}{\partial r} - \frac{1}{r^2} \left(\frac{\partial z_q}{\partial \theta}\right) \frac{\partial}{\partial \theta}, \quad 1 \leq q \leq Q. \tag{6.75}$$

Let $R > 0$. The transmission condition for the solution u of (6.3), resulting from (6.73) is

$$u(z_q^-(r,\theta), r, \theta,) = u(z_q^+(r,\theta), r, \theta),$$

$$r \in [0, R], \quad \theta \in [0, 2\pi], \quad 1 \leq q \leq Q. \tag{6.76}$$

The exact transmission condition for u resulting from (6.74), is, however, approximated by the following *parabolized* transmission condition

$$\frac{1}{\rho_{q-1}} \mathcal{D}_q u(z_q^-(r,\theta), r, \theta) = \frac{1}{\rho_q} \mathcal{D}_q u(z_q^+(r,\theta), r, \theta),$$

$$r \in [0, R], \quad \theta \subset [0, 2\pi], \quad 1 \leq q \leq Q. \tag{6.77}$$

The differential operators $\mathcal{D}_q$, $1 \leq q \leq Q$, are defined, for $r \in [0, R]$ and $\theta \in [0, 2\pi]$, as

$$\mathcal{D}_q := \frac{\partial}{\partial z} - ik_0 \left(\frac{\partial z_q}{\partial r}\right) \mathcal{I} - \frac{1}{r^2} \left(\frac{\partial z_q}{\partial \theta}\right) \frac{\partial}{\partial \theta}, \tag{6.78}$$

with $\mathcal{I}$ the identity operator. The transmission condition (6.77) (with the differential operator $\mathcal{D}_q$ defined in (6.78)) represents a 3D analog of (6.35) of Section 2. For the derivation of (6.77) the range component of the normal derivative (6.75) has been replaced using the horizontal plane–wave impedance approximation

$$\frac{\partial p}{\partial r} \approx ik_0 p. \tag{6.79}$$

Note that the impedance approximation (6.79) underlying the derivation of (6.77) is compatible with the *paraxial* approximation used for the derivation of (6.3), [7]. It is expected, therefore, that the use of (6.77) will render stable for range marching, the initial–boundary value problem for (6.3) thus defined. In the following subsection, a stability condition ensuring the existence and uniqueness of the solution $u(z, r, \theta)$ of the initial–boundary value problem for (6.3) defined above (with (6.76) and (6.77) playing the role of transmission conditions) will be established. It will be a generalization of the stability condition of [12] for the 2D range–independent problem in a layered waveguide. On the contrary, using the complete normal derivative representation (6.75) in (6.74) to derive the exact analog of (6.77), which is mathematically correct for the Helmholtz equation, would lead to an ill–posed initial–boundary value problem for (6.3).

6.4.2   The Transformed Initial–Boundary Value Problem

In the sequel, the initial–boundary value problem of the previous subsection, posed on a physical domain of general topography, is transformed to one on a cylindrical domain, using an appropriate change of variables.

The azimuth and range dependent change of the depth variable z is defined, [31], by

$$y = z/s(r,\theta), \qquad z \in [0, s(r,\theta)], \tag{6.80}$$

which sends the azimuth and range dependent interval $[0, s(r,\theta)]$ onto the reference interval $[0,1]$.

The transformed domain, which is the actual computational domain, is described by the variables y, r, and θ. The range r and the azimuthal angle θ remain unchanged; they vary in the intervals $[0, R]$ and $[0, 2\pi]$, respectively. The transformed depth variable $y \in [0, 1]$, for all $r \in [0, R]$ and $\theta \in [0, 2\pi]$. The simplifying assumption that the layers are homothetic in the physical domain, in the sense that $z_q(r,\theta) = S_q s(r,\theta)$, for $1 \le q \le Q$, with the constants S_q, $q = 1, \ldots, Q$, satisfying

$$0 < S_1 < \cdots < S_Q < 1,$$

is also made. Then, the actual computational domain is bounded by the flat surfaces at $y = S_0 := 0$ and $y = S_{Q+1} := 1$, with interfaces placed at $y = S_q$, $1 \le q \le Q$. The dependent variable in the transformed coordinates is defined as $v(y, r, \theta) = u(z, r, \theta)$, with y and z related by (6.80).

Equation (6.3) then transforms to

$$v_r = \frac{i\rho}{2k_0} \left[\frac{1}{s^2(r,\theta)} \frac{\partial}{\partial y} \left(\frac{1}{\rho} \frac{\partial v}{\partial y} \right) + \frac{1}{r^2} \mathcal{L} \left(\frac{1}{\rho} \mathcal{L} v \right) \right]$$
$$+ \left(\frac{y}{s(r,\theta)} \frac{\partial s}{\partial r}(r,\theta) \right) \frac{\partial v}{\partial y} + \frac{ik_0}{2} \tilde{\beta}(y, r, \theta) v, \tag{6.81}$$
$$r \in [0, R], \quad \theta \in [0, 2\pi], \quad y \in \cup_{q=0}^{Q}(S_q, S_{q+1}),$$

Here $\tilde{\beta}(y, r, \theta) := \beta(ys(r, \theta), r, \theta)$, and the operator $\mathcal{L}$ is defined by

$$\mathcal{L} := \frac{\partial}{\partial \theta} - \left(\frac{y}{s(r,\theta)} \frac{\partial s}{\partial \theta}(r,\theta) \right) \frac{\partial}{\partial y}$$

accounting for horizontal refraction. It is evident that the effects of bottom geometry are now interpreted by the variable coefficients and the additional derivative terms of (6.81). The acoustic field v (in the new coordinates system) satisfies the initial condition $v(y, 0, \theta) = u^0(ys(0, \theta), \theta)$, the pressure release boundary conditions $v = 0$ at $y = 0$ and $y = 1$, the 2π–periodicity boundary condition in azimuth $v(y, r, 0) = v(y, r, 2\pi)$, and the transmission conditions

$$v(S_q^-, r, \theta) = v(S_q^+, r, \theta), \ r \in [0, R], \quad \theta \in [0, 2\pi], \tag{6.82}$$

$$\frac{1}{\rho_{q-1}}\tilde{\mathcal{D}}_q v(S_q^-, r, \theta) = \frac{1}{\rho_q}\tilde{\mathcal{D}}_q v(S_q^+, r, \theta), \quad r \in [0, R], \quad \theta \in [0, 2\pi], \qquad (6.83)$$

for all $1 \le q \le Q$. The differential operators $\tilde{\mathcal{D}}_q$, $1 \le q \le Q$, appearing in (6.83) are now defined, for $r \in [0, R]$ and $\theta \in [0, 2\pi]$, as

$$\tilde{\mathcal{D}}_q := \frac{1}{s(r, \theta)}\frac{\partial}{\partial y} - ik_0 S_q\frac{\partial s}{\partial r}(r, \theta)\mathcal{I} - \frac{S_q}{r^2}\frac{\partial s}{\partial \theta}(r, \theta)\mathcal{L}, \quad 1 \le q \le Q. \qquad (6.84)$$

Let $\Omega = \cup_{q=0}^{Q}(S_q, S_{q+1}) \times (0, 2\pi)$. For $r \in [0, R]$, the weighted, r–dependent inner product $\langle \cdot, \cdot \rangle_\rho$ is defined for $\phi, \varphi \in L^2(\Omega)$ by

$$\langle \phi, \varphi \rangle_\rho := \int_\Omega \phi \overline{\varphi}\frac{s(r, \theta)}{\rho}d\Omega, \qquad (6.85)$$

where, for the sake of brevity, we have used the following compact notation

$$\int_\Omega \phi \overline{\varphi}\frac{s(r, \theta)}{\rho}d\Omega = \sum_{q=0}^{Q}\int_0^{2\pi}\int_{S_q}^{S_{q+1}}\phi_q \overline{\varphi}_q\frac{s(r, \theta)}{\rho_q}dyd\theta.$$

The weighted norm on $L^2(\Omega)$, induced by the inner product $\langle \cdot, \cdot \rangle_\rho$ of (6.85), is denoted by $\| \cdot \|_\rho$.

Let ψ be an arbitrary smooth function defined on $[0, 1] \times [0, 2\pi]$, satisfying zero Dirichlet boundary conditions at $y = 0$ and $y = 1$, and a 2π-periodicity condition in azimuth. Multiplying both sides of Eq. (6.81) by $\overline{\psi}s/\rho$, and integrating by parts, results in

$$\int_\Omega \left(\frac{\partial v}{\partial r} + \frac{1}{2s}(\frac{\partial s}{\partial r})v\right)\overline{\psi}\frac{s}{\rho}d\Omega =$$

$$-\frac{i}{2k_0}\int_\Omega \left(\frac{1}{s^2}\frac{\partial v}{\partial y}\frac{\partial \overline{\psi}}{\partial y} + \frac{1}{r^2}(\mathcal{L}v)(\overline{\mathcal{L}\psi})\right)\frac{s}{\rho}d\Omega$$

$$+\frac{1}{2}\int_\Omega \left(\frac{y}{s}\frac{\partial s}{\partial r}\right)\left(\frac{\partial v}{\partial y}\overline{\psi} - v\frac{\partial \overline{\psi}}{\partial y}\right)\frac{s}{\rho}d\Omega$$

$$+\frac{ik_0}{2}\int_\Omega \tilde{\beta}v\overline{\psi}\frac{s}{\rho}d\Omega. \qquad (6.86)$$

Let V denotes in the sequel, the vector space of those functions belonging to $H^1(\Omega)$ and satisfying zero Dirichlet boundary conditions on $y = 0$ and $y = 1$, and a 2π-periodicity condition in azimuth. In terms of the inner product (6.85) the above weak formulation is expressed as seeking $v(r) \in V$ satisfying, for $r \in (0, R)$,

$$\langle \frac{\partial v}{\partial r} + (\frac{1}{2s}\frac{\partial s}{\partial r})v, \psi \rangle_\rho + ia(r; v, \psi) + \frac{k_0}{2}\langle \text{Im}(\tilde{\beta}(r))v, \psi \rangle_\rho = 0, \quad \forall \psi \in V, \quad (6.87)$$

and $v(0) = v^0$, $(y, \theta) \in \cup_{q=0}^{Q}(S_q, S_{q+1}) \times (0, 2\pi)$. In the arguments of the functions appearing above, both y and θ dependences are suppressed; Therefore $v(r) = v(\cdot, r, \cdot)$ and $\tilde{\beta}(r) = \tilde{\beta}(\cdot, r, \cdot)$. The r-dependent sesquilinear form

$a(r; \cdot, \cdot) : V \times V \longrightarrow \mathbb{C}$ in (6.87) is defined by

$$a(r; \phi, \psi) := \frac{1}{2k_0} \left(\langle \frac{1}{s^2} \frac{\partial \phi}{\partial y}, \frac{\partial \psi}{\partial y} \rangle_\rho + \frac{1}{r^2} \langle \mathcal{L}\phi, \mathcal{L}\psi \rangle_\rho \right)$$
$$+ \frac{i}{2} \left(\langle \frac{y}{s} \frac{\partial s}{\partial r} \frac{\partial \phi}{\partial y}, \psi \rangle_\rho - \langle \frac{y}{s} \frac{\partial s}{\partial r} \phi, \frac{\partial \psi}{\partial y} \rangle_\rho \right)$$
$$- \frac{k_0}{2} \langle \mathrm{Re}(\tilde{\beta}(r))\phi, \psi \rangle_\rho. \tag{6.88}$$

Letting $\psi = v(r)$ in the weak formulation (6.87), using the formula

$$\frac{\partial v}{\partial r} + \left(\frac{1}{2s} \frac{\partial s}{\partial r} \right) v = \frac{1}{\sqrt{s}} \frac{\partial(\sqrt{s}v)}{\partial r}$$

which gives the equality

$$\langle \frac{1}{\sqrt{s}} \frac{\partial(\sqrt{s}\,v)}{\partial r}, v \rangle_\rho = \frac{1}{2} \frac{\mathrm{d}}{\mathrm{d}r} \|v\|_\rho^2 + i\mathrm{Im} \left(\langle \frac{1}{\sqrt{s}} \frac{\partial(\sqrt{s}v)}{\partial r}, v \rangle_\rho \right),$$

and taking real parts, the following relation holds

$$\frac{1}{2} \frac{\mathrm{d}}{\mathrm{d}r} \|v(r)\|_\rho^2 = -\frac{k_0}{2} \langle \mathrm{Im}(\tilde{\beta}(r))v(r), v(r) \rangle_\rho,$$

which, since $\mathrm{Im}(\tilde{\beta}(r)) \geq 0$, yields that

$$\frac{\mathrm{d}}{\mathrm{d}r} \|v(r)\|_\rho \leq 0, \quad 0 \leq r \leq R. \tag{6.89}$$

Therefore, if $\mathrm{Im}(\tilde{\beta}(r)) \neq 0$, the following dissipation property holds

$$\|v(r_2)\|_\rho \leq \|v(r_1)\|_\rho, \quad 0 \leq r_1 \leq r_2 \leq R.$$

It is evident that if $\mathrm{Im}\tilde{\beta} = 0$, the above dissipation property is replaced by the conservation of the $\|\cdot\|_\rho$ norm.

6.4.3   The Numerical Scheme

The transformed initial–boundary value problem (6.81), (6.82), and (6.83) is discretized in y and θ by the standard Galerkin/finite-element method using piecewise linear continuous functions, while the conservative Crank-Nicolson marching scheme is used to discretize in range. Set $\theta_m := m\,h_\theta$, for $0 \leq m \leq M$, with $h_\theta = 2\pi/M$, and $y_j = j\,h_y$, for $0 \leq j \leq J+1$, with $h_y = 1/(J+1)$ such that $y_{\ell_q} = S_q$, $1 \leq q \leq Q$, for some $0 < \ell_1 < \ldots < \ell_q < J$. Let $P_{m,j} = (y_j, \theta_m)$ denote a typical node of the grid, and

$$\mathcal{K}_{m,j} = (y_{j-1}, y_j) \times (\theta_{m-1}, \theta_m), \quad 1 \leq m \leq M, \quad 1 \leq j \leq J+1,$$

where $\mathcal{K}_{m,j} \cap \mathcal{K}_{\tilde{m},\tilde{j}} = \emptyset$ if $(m,j) \neq (\tilde{m},\tilde{j})$. Then $\overline{\Omega} = \cup_{1 \leq m \leq M,\, 1 \leq j \leq J+1} \overline{\mathcal{K}}_{m,j}$.

Let now $\mathbf{Q}_1$ be the space of complex–valued polynomials of first degree in y and θ and define the finite dimensional vector space

$$V_h = \{\chi : \chi \in C^0(\overline{\Omega}) \text{ complex–valued, } \chi|_{\mathcal{K}_{m,j}} \in \mathbf{Q}_1(\mathcal{K}_{m,j}),$$
$$1 \leq m \leq M, \ 1 \leq j \leq J+1, \text{ and}$$
$$\chi(0,\theta) = \chi(1,\theta) = 0, \ \theta \in [0,2\pi], \ \chi(y,0) = \chi(y,2\pi), \ y \in [0,1]\}.$$

Then, V_h is a subspace of V, with dimension $\dim V_h = MJ$. For $k > 0$ so that $R = Nk$, define $r^n := nk$, for $0 \leq n \leq N$, and $r^{n+\frac{1}{2}} := r^n + k/2$, for $0 \leq n \leq N-1$. The fully discrete scheme for (6.81) is written as: Find $V^n \in V_h$, $1 \leq n \leq N$, satisfying, for $0 \leq n \leq N-1$,

$$\langle \frac{V^{n+1} - V^n}{k} + \left(\frac{1}{2s}\frac{\partial s}{\partial r}\right)^{n+\frac{1}{2}} \frac{V^{n+1} + V^n}{2}, \psi_h \rangle_\rho + ia(r^{n+\frac{1}{2}}; \frac{V^{n+1} + V^n}{2}, \psi_h)$$
$$+ \frac{k_0}{2} \langle \mathrm{Im}\left(\tilde{\beta}^{n+\frac{1}{2}}\right) \frac{V^{n+1} + V^n}{2}, \psi_h \rangle_\rho = 0, \tag{6.90}$$

for all $\psi_h \in V_h$, and $V^0 = v^0$. The stability and consistency of the fully discrete numerical scheme (6.90) have been investigated in [31]. It is shown that it preserves the energy properties of the continuous problem (6.81). The solution of a large, sparse linear system (of order $M \times J$) with a block-tridiagonal structure, is required at each range step of (6.90). These linear systems are solved using a non-stationary iterative algorithm equivalent to the preconditioned conjugate gradient iteration method for normal equations. The amount of storage required now depends linearly on the number of grid points. In addition, few vectors need to be stored. Hence the storage is much less than that required by any version of the Gaussian elimination. The efficiency of the solver highly depends on the preconditioning procedure. Adapting the preconditioning approach described in [6], a preconditioning is applied which uses the tridiagonal matrix derived from the M×2D associated model (M sections in the azimuthal direction). This preconditioning appears to be very efficient, except when propagating in the nearfield because of the singularity of the 3D PE (6.3) near the origin. Nevertheless, numerical results demonstrate that this preconditioning accelerates significantly the convergence of the iteration. Therefore, the additional operations inherent to the preconditioning procedure are negligible in the overall solution process when compared to the unpreconditioned iteration. For more details cf. [31], [32] and the references therein.

6.4.4 Numerical Examples

We consider the 3D conical seamount benchmark problem where an isovelocity water layer with sound speed 1500 m/s and density 1 g/cm^3 overlies a lossy, homogeneous semi–infinite sedimental layer with sound speed 1700 m/s,

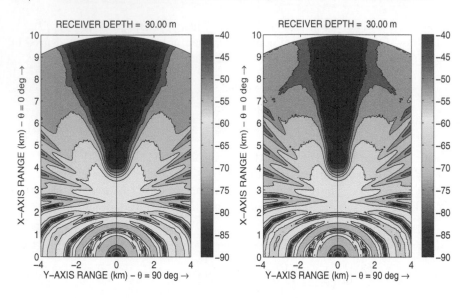

FIGURE 6.6: Transmission loss plots (horizontal slices at a constant depth of $z = 30$ m) corresponding to the full 3D solution (left figure) and to the pseudo 3D (or M×2D, i.e., azimuthally uncoupled) solution (right figure) for the 3D conical seamount test problem.

density 1.5 g/cm^3, and attenuation 0.5 dB per wavelength. There is no attenuation in the water column. The parametrization of the water/sediment interface is given by : $z_1(r, \theta) = \tilde{z}_1(x, y)$ with $x = r \cos \theta$, $y = r \sin \theta$, and

$$\tilde{z}_1(x, y) = \begin{cases} \dfrac{\sqrt{(x - 4000)^2 + y^2}}{20} & \text{if } \sqrt{(x - 4000)^2 + y^2} \leq 4000, \\ 200 & \text{elsewhere,} \end{cases}$$

and thus makes an angle of 2.86 deg with respect to the ocean surface at $\theta = 0$ deg (upslope direction) and leads to a zero–slope at both $\theta = 90$ deg and $\theta = 270$ deg (cross–slope directions). A source of frequency 25 Hz is placed at a depth of $z_S = 100$ m, at $r = 0$. The maximum computation range is $r_{max} = 10$ km and the reference sound speed is 1500 m/s. An accurate 3D computation requires the use of, at least, $M = 2880$ points in the azimuthal direction, i.e., $h_\theta = 0.125$ deg. In Fig. 6.6 (left subplot) is displayed the color–scaled transmission loss plot, where transmission loss is defined by $TL = -20 \log_{10}(|u(z, r, \theta)|/\sqrt{r}$, on a horizontal slice in the (r, θ)-plane at a depth of $z = 30$ m. The wave field is obtained using a full 3D calculation based on the fully discrete analog of (6.3) presented in (6.90). For comparison, the corresponding transmission loss plot obtained using a M×2D (i.e., azimuthally uncoupled) calculation is also shown in Fig. 6.6 (right subplot). Since the geometry of the waveguide, for this benchmark problem, is symmetric about the upslope direction, both M×2D and 3D solutions are symmetric about the same direction. For both computations, a shadow zone is observed in the rear of the seamount. The shadow zone is more pronounced in the fully 3D solution. This is due to the effect of the horizontal refraction on the energy in the downslope direction. Although the M×2D transmission loss plot is azimuthally dependent (because the bottom slope varies from one vertical plane to another), no 3D effects due to out–of–plane propagation are present. One should recall that M×2D models implicitly assume local cylindrical symmetry of the problem, and thus neglect the azimuthal diffraction term. On the contrary, in a 3D model, out-of-plane phenomena appear at various ranges, due to the azimuthal coupling term in the 3D parabolic equation (6.3), leading thus to a much more pronounced (and realistic) shadow zone.

Acknowledgement

This work was partly supported by the French–Greek research collaboration program PLATON, funded by EGIDE, France, and the General Secretariat of Research and Technology, Greece.

References

[1] L. Abrahamsson and H. O. Kreiss. Boundary conditions for the parabolic equation in a range dependent duct. *J. Acoust. Soc. Am.*, 87:2438-2441, 1990.

[2] L. Abrahamsson and H. O. Krciss. The initial–boundary value problem for the Schrödinger equation. *Math. Methods Appl. Sci.*, 13:385-390, 1990.

[3] G. D. Akrivis, V. A. Dougalis, and G. E. Zouraris. Finite difference schemes for the 'parabolic' equation in a variable depth environment with a rigid bottom boundary condition. *SIAM J. Numer. Anal.*, 39:539-565, 2001.

[4] D. C. Antonopoulou. Theory and Numerical Analysis of parabolic approximations. Ph.D. Thesis, *University of Athens*, 2006 (in Greek).

[5] D. C. Antonopoulou, V. A. Dougalis, and G. E. Zouraris. *Finite element*

discretizations of the 'parabolic' equation in an underwater variable bottom environment, preprint, 2007.

[6] A. Bayliss, C. I. Goldstein, and E. Turkel. An iterative method for the Helmholtz equation. *J. Comp. Physics*, 49:443-457, 1983.

[7] C. Bernardi and M.-C. Pèlissier. Spectral approximation of a Schrödinger type equation. *Math. Models and Methods in Appl. Sciences*, 4:49-88, 1994.

[8] M. J. Buckingham. Theory of three–dimensional acoustic propagation in a wedgelike ocean with a penetrable bottom. *J. Acoust. Soc. Am.*, 82:198-210, 1987.

[9] S. A. Chin–Bing, D. B. King, J. A. Davis, and R. B. Evans (Eds.) *PE Workshop II*, NRL, Stennis Space Center Mississippi, 1993.

[10] J. F. Claerbout. *Foundamentals of Geophysical Data Prospecting.* McGraw–Hill, New York, 1976.

[11] M. D. Collins. Applications and time–domain solution of higher–order parabolic equations in underwater acoustics. *J. Acoust. Soc. Am.*, 86:1097-1102, 1989.

[12] V. A. Dougalis and N. A. Kampanis. Finite element methods for the parabolic equation with interfaces. *J. Comp. Acoustics*, 4:55-88, 1996.

[13] V. A. Dougalis, F. Sturm, and G. E. Zouraris. Boundary conditions for the wide–angle PE at a slopping bottom. S. M. Jesus, O. C. Rodriguez (Eds.), *Proceedings of the 8th European Conference in Underwater Acoustics*, Vol. 1, p. 51-56, 2006.

[14] V. A. Dougalis and G. E. Zouraris. *Finite difference methods for the Parabolic Equation with interface condition* (to appear).

[15] S. L. Edelstein. Estimates and aymptotics of solution of two–dimensional boundary value problems for the Helmholtz equation with the coefficient weakly depending on horizontal coordinate. *Differential Equations*, 26:727-729, 1990.

[16] L. Fishman, A. K. Gautesen, and Z. Sun. Uniform high–frequency approximations of the square root Helmholtz operator symbol. *Wave Motion*, 26:127-161, 1997.

[17] D. Givoli. *Numerical Methods for Problems in Infinite Domains.* Elsevier, Amsterdam, 1992.

[18] C. I. Goldstein. A finite element method for solving Helmholtz type equations in waveguides and other unbounded domains. *Math. Comp.*, 39:309-324, 1982.

[19] R. R. Greene. The rational approximation to the acoustic wave equation with bottom interaction. *J. Acoust. Soc. Am.*, 76:1764–1773, 1984.

[20] F. Ihlenburg. Finite element analysis of acoustic scattering. *Applied Mathematical Sciences*, Volume 132, Springer–Verlag, New York, 1998.

[21] F. B. Jensen and M. C. Ferla. Numerical solutions of range–dependent benchmark problems. *J. Acoust. Soc. Amer.*, 87:1499–1510, 1990.

[22] F. B. Jensen, W. A. Kuperman, M. B. Porter, and H. Schmidt. *Computational Ocean Acoustics*. AIP Press, New York, 1994.

[23] D. Lee, G. Botseas, and J. S. Papadakis. Finite difference solution to the parabolic wave equation. *J. Acoust. Soc. Am.*, 70:795–800, 1981.

[24] D. Lee, G. Botseas, and W. L. Siegmann. Examination of three–dimensionnal effects using a propagation model with azimuth-coupling capability (FOR3D). *J. Acoust. Soc. Am.*, 91: 3192–3202, 1992.

[25] D. Lee and S. T. McDaniel. Ocean acoustic propagation by Finite Difference methods. *J. Comp. Maths. Applic.*, 14:305–423, 1987.

[26] D. Lee, A. D. Pierce, and E. C. Shang. Parabolic equation development in the twentieth century. *J. Comp. Acoustics*, 8:527-637, 2000.

[27] J. L. Lions and E. Magènes. *Problèmes aux Limites Non Homogènes et Applications*. Dunod, Paris, 1968.

[28] M. B. Porter and F. B. Jensen. Anomalous parabolic equation results for propagation in leaky surface ducts. *J. Acoust. Soc. Am.*, 94:1510–1516, 1993.

[29] M. B. Porter, F. B. Jensen, and C. M. Ferla. The problem of energy conservation in one–way models. *J. Acoust. Soc. Am.*, 89:1058–1067, 1991.

[30] W. L. Siegmann, G. A. Kriegsmann, and D. Lee. A wide-angle three–dimensional parabolic wave equation. *J. Acoust. Soc. Am.*, 78:659–664, 1985.

[31] F. Sturm. *Modélisation mathématique et numérique d'un problème de propagation en acoustique sous-marine: prise en compte d'un environnement variable tridimensionnel*. Ph.D. Thesis, Université de Toulon et du Var, 1997.

[32] F. Sturm. Numerical study of broadband sound pulse propagation in three-dimensional oceanic waveguides. *J. Acoust. Soc. Am.*, 117:1058–1079, 2005.

[33] F. D. Tappert. The parabolic approximation method. J. B. Keller and J. S. Papadakis (Eds.), *Wave Propagation and Underwater Acoustics, Lecture Notes in Physics*, Volume 70, p. 224–287, Springer–Verlag, 1977.

Chapter 7

Exact Boundary Conditions for Acoustic PE Modeling Over an N^2-Linear Half-Space

T. W. Dawson[1], Department of Electrical and Computer Engineering, University of Victoria, PO BOX 3055, STN CSC, Victoria, BC Canada V8V 3K3

G. H. Brooke, General Dynamics Canada, 3785 Richmond Road, Ottawa, ON Canada K2H 5B7, Gary.Brooke@gdcanada.com

D. J. Thomson[2], 733 Lomax Road, Victoria, BC Canada V9C 4A4, drdjt@shaw.ca

7.1 Introduction

Accurate and efficient propagation modeling is a topic of ongoing interest for a variety of underwater acoustic applications. A significant example is provided by Matched Field Processing, where a large number of acoustic replica fields must be computed in order to match a particular measurement or physical parameter of interest. Here the two numerical facets are particularly important–accuracy is required faithfully to represent the physical reality of the particular scenario being modelled, and speed is required to facilitate efficient generation of a sufficient number of replica fields to adequately sample the model parameter space.

A variety of parabolic equation (PE) approximations have provided a significant computational speed improvement over the more general, but also

---

[1]Deceased.

[2]This work was carried out while at Defence R&D Canada–Atlantic, PO Box 1012, Dartmouth, NS Canada B2Y 3Z7.

more computationally demanding, full-wave methods based on the Helmholtz equation. The improvement in computational speed arises, in part, from a neglect of back-scattered energy, which permits an efficient marching solution method along a preferred direction (range). Parabolic equation techniques are now routinely and successfully used to obtain full-field predictions in both underwater acoustic [5] and atmospheric radar [9] applications involving inhomogeneous environments.

Nevertheless, a variety of considerations must be addressed, particularly the issue of appropriate mesh truncation in the transverse (depth) direction. This is particularly so in shallow-water problems, where the computational domain may span many wavelengths in the transverse coordinate, even though the domain of interest (*i.e.*, the water column) may not. In many cases, much of the propagation domain (*i.e.*, the ocean bottom) has material properties with a relatively simple dependence on the transverse coordinate. In such cases, the relevant propagation equations may be tackled analytically. The resulting solution in the "analytical domain" can then be coupled across a boundary plane to a more general numerical solution in those portions of the waveguide with more complicated material parameter profiles ("numerical domain").

Subject to certain environmental constraints, the efficiency of one-way marching computations can in some cases be improved through the use of non-local boundary conditions (NLBCs). These are applied transversely to the direction of propagation at the interface between the "numerical" and "analytical" domains, and are based on analytic solutions of the field within the latter domain. They can improve computational efficiency by reducing the numerical solution domain to the regions of prime interest, such as the upper water column in the case of underwater sound propagation, or the lower atmosphere in the case of radar propagation. Previous successful applications of NLBCs to PE modeling involve integral convolutions between a known kernel and the field spatial-history up to the current range step. The convolution is evaluated numerically in conjunction with the PE marching algorithm.

To date, three types of half-space environments have been shown to be appropriate for application of this type of boundary condition. The first, of perhaps more relevance to the underwater acoustics case, is a homogeneous medium [10,14]. This scenario has been considered in the underwater acoustic case by several authors, involving both the standard (Tappert) [10,11,15] and simple wide-angle (Claerbout) [2, 8, 10] parabolic equation approximations. A similar application to parabolic equation modelling of atmospheric radar propagation over the surface of the Earth was also presented by Levy [7], where a uniform half-space was used to terminate the computational domain from above.

The next degree of complication, and also of physical realism, is provided by the "N^2-linear" medium involving a squared refractive index that is characterized by a constant gradient [4,8]. with respect to the transverse coordinate. Levy [7] considered an upward refracting half-space capping a numerical PE domain in atmospheric radar propagation. The analogous application to underwater acoustics, and associated extension to both upward and downward refracting profiles, was recently outlined by Dawson *et al.* [4] and Thomson and Brooke [14]. All of these applications involved only the standard narrow-angle parabolic equation approximation due to Tappert. The successful implementation of the non-local boundary condition under the Standard PE and Dirichlet–to–Neumann (DtN) formulation has been illustrated earlier [4,9] for both upward and downward refracting profiles.

A medium in which the refractivity index varies exponentially [8] provides a third example of a suitable scenario.

The implementation of a NLBC along the half-space interface requires that a representation of the vertical wavenumber for the field within the half-space be available. This vertical wavenumber depends on the choice of the particular Helmholtz equation factorization approximation. The NLBC readily simplifies analytically for only the Tappert and the Claerbout square-root approximations [5]. In the case of a N^2–linear medium, only the non-local boundary condition based on the so-called Tappert PE has been analyzed and tested [4,8]. Alternative methods for generating NLBCs for a uniform half-space for both Tappert and Claerbout PE's have also been considered [17].

In this paper, we extend the theory of non-local boundary conditions to include a higher-order PE representation based on the Claerbout approximation. Such a boundary condition enables steeper-angle interactions with the half-space to be accommodated more accurately.

We present an overview of the analysis for NLBCs with N^2–linear analytical domain for the underwater acoustic case. In addition to the earlier formulation in terms of the standard Dirichlet–to–Neumann map [10], an alternative formulation in terms of a Neumann–to–Dirichlet (NtD) map is also outlined. It is shown how attenuation in the analytical domain can be included, by allowing the sound speed to be complex. For the Claerbout case, the results are illustrated for several numerical domains, and the improvement over the simple Tappert approximation indicated.

Since the analysis is valid for any scalar PE problem, it is also applicable to the scalar radar case. Since N^2-linear profiles often provide a more accurate representation of the true profiles in both acoustical and atmospheric radar work, a Claerbout non-local boundary condition extension should find use in both fields.

Several final points can be regarding this work. Firstly, it is restricted to a low-order (Padé$(1,1)$) approximation. The results assume the Claerbout coefficients, but the present results are formulated for arbitrary Padé coefficients. Moreover, although superior PE algorithms, such as split–step and higher-order Padé have superseded the simple wide-angle PE algorithm, the hybrid numerical/analytical work presented here should prove useful in validation of purely numerical approaches, such as placing an absorbing material at depth and increasing the vertical numerical domain extent. The recent advent of perfectly matched layers (PMLs) for boundary conditions may in fact be more efficient than a NLBC. Nonetheless, the availability of alternative methods for validation is generally valuable.

It may be possible to extend the methods of this paper, but preliminary work by the authors (unpublished) has indicated that a Padé(2, 2) solution in the analytical domain shows at least one pole to the right of the imaginary axis in the Laplace transform domain, and the solution is unstable in range. Numerical experiments seem to support this. It is hoped that these difficulties may be surmounted, and the present NLBC find application coupled to more modern PE algorithms. However, such attempts are deferred to future research.

It is the authors' hope that the present first-order Padé development for N^2-linear media is a useful step in paving the way for future incorporation into a higher-order formulation, in the same way that was already done for a constant-N half-space and demonstrated in [2].

7.2   Theory

7.2.1   PE Theory

Consider an underwater acoustic waveguide in cylindrical coordinates (r, θ, z), with r denoting range and z depth increasing downwards from the surface $z = 0$. The ocean is modelled as a half-space $z > 0$, occupied by a horizontally stratified fluid of density $\rho(z)$ and sound speed $c(z)$. The waveguide is insonified by a time-harmonic azimuthally-symmetric acoustic source of angular frequency ω and situated at $z = 0$, which therefore generates a pressure field $p(r, z) \exp(-i\omega t)$. The waveguide has an associated attenuation profile $\alpha(z) = \operatorname{Im} \{\omega/c(z)\}$. With c_0 denoting a (constant) reference sound speed, the index of refraction of the medium is

$$N(z) \equiv c_0/c(z) \tag{7.1}$$

and

$$k_0 \equiv \omega/c_0 \tag{7.2}$$

is a reference wavenumber. The full pressure field is governed by Helmholtz's equation [5]. However, matters are considerably simplified if it is assumed that the far-field ($k_0 r \gg 1$) pressure envelope,

$$\psi(r, z) \equiv p(r, z) \sqrt{k_0 r} \exp(-i k_0 r), \tag{7.3}$$

is adequately described by the one-way outgoing parabolic equation [14]

$$\frac{\partial \psi(r, z)}{\partial r} = i k_0 \left\{ -1 + \sqrt{1 + X} \right\} \psi(r, z), \tag{7.4}$$

where X in general denotes the vertical differential operator

$$X \equiv \frac{\rho(z)}{k_0^2} \frac{\partial}{\partial z} \left(\frac{1}{\rho(z)} \frac{\partial}{\partial z} \right) + N^2(z) - 1. \tag{7.5}$$

This approximation will be valid if the waveguide is range-independent. The dependence of the material parameters on depth is allowed to be arbitrary over the finite vertical range $0 < z < d$. In this numerical domain, it is envisioned that the propagation equation Eq. (7.4) can be solved using one of a variety of standard numerical approaches [5]. Indeed, the formulation in the numerical domain also admits mild variation of waveguide properties with range, provided that the appropriate backscattered energy is negligible – essentially range-independence is assumed over each range-step in the solution procedure.

The focus of the present work, however, is the treatment of the problem in the analytical domain, namely the lower half-space $z > d$. Here, the medium is assumed to have a prescribed depth variation which allows the propagation equations to be treated analytically. Specifically, the present work is based on the assumptions (i) of uniform density, $\rho(z) \equiv \rho_0$, and (ii) a squared refractive index which varies linearly with depth,

$$N^2(z) \equiv 1 + \beta + \mu(z - d), \qquad (z > d). \tag{7.6}$$

In particular, the half-space is assumed to be independent of range. The refractive index has two adjustable parameters β and μ, both of which are assumed real for now. The former is a measure of the refractive index just inside the bottom half-space, *i.e.*,

$$\beta = N^2(d+) - 1. \tag{7.7}$$

The expression

$$c(z) = \frac{c_0}{\sqrt{1 + \beta + \mu(z - d)}}, \qquad (z > d), \tag{7.8}$$

for the half-space sound speed profile follows from Eqs. (7.1) and (7.6), and in turn leads to the expression

$$\frac{dc(z)}{dz} = -\left(\frac{\mu c_0}{2} \right) N^{-3/2}(z), \qquad (z > d), \tag{7.9}$$

for the sound speed gradient. Consequently, parameter μ is a measure of the refractive index gradient, with $\mu = 0$ corresponding to a homogeneous fluid half-space. This case has been explicitly dealt with elsewhere [15]. Positive values of μ are associated with a sound speed which decreases to 0 with increasing depth and hence a downward refracting medium. Similarly, negative values of μ indicate an upward refracting medium in which the sound speed

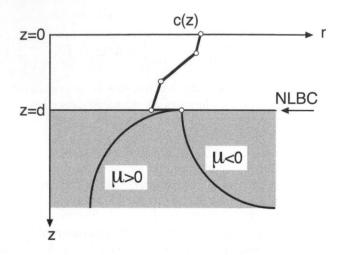

FIGURE 7.1: Sketch of the computational domain and the linear half space, with illustrative upward and downward refracting profiles.

increases with depth. In fact, in the latter case, the sound speed becomes infinite at the finite cutoff depth

$$z_c = d - \frac{1+\beta}{\mu} = d - \frac{N^2(d+)}{\mu} = d - \frac{c_0^2}{\mu\, c^2(d+)}. \tag{7.10}$$

Clearly, some restrictions on the two parameters are required in order that the half-space fall within the parabolic equation formulation. Figure 7.1 shows a schematic representation of the two inhomogeneous half-space sound speed profiles.

7.2.2   Solution in the Lower Half-Space

The task at hand is the solution of equation Eq. (7.4), with the simplified operator

$$X = X_+ \equiv \frac{1}{k_0^2} \frac{\partial^2}{\partial z^2} + \beta + \mu(z - d). \tag{7.11}$$

The first step in the solution is application of the Laplace transform pair [1]

$$\Psi(s, z) = \int_0^\infty \psi(r, z) e^{-sr}\, dr \tag{7.12}$$

and

$$\psi(r, z) = \frac{1}{2\pi i} \int_{\epsilon-i\infty}^{\epsilon+i\infty} \Psi(s, z) e^{+sr}\, ds, \tag{7.13}$$

where ϵ in Eq. (7.13) is chosen so that the contour of integration lies to the right of all singularities of the integrand.

On account of the material range-independence in the lower half space, equation Eq. (7.4) transforms to

$$\left[s - ik_0 \left(-1 + \sqrt{1 + X_+} \right) \right] \Psi(s, z) = \psi(0, z), \qquad (7.14)$$

independent of the particular parabolic equation approximation used. It remains to choose a particular parabolic equation approximation, *i.e.*, to choose an approximation for the square-root operator in Eq. (7.4). Attention will be restricted to a general first-order Padé approximation [5] of the form

$$\sqrt{1 + X_+} \approx 1 + \frac{aX_+}{1 + bX_+}, \qquad (7.15)$$

with the constants a and b remaining general for now. When this result is combined with the transformed equation Eq. (7.14) and the simplified vertical operator, Eq. (7.11), the resulting differential equation for the Laplace transform of the far-field pressure envelope can be arranged in the form

$$\left\{ \frac{\partial^2}{\partial z^2} + \mu k_0^2 \left(z - d \right) + \lambda^2(s) \right\} \Psi(s, z) =$$
$$\left(\frac{k_0^2}{sb - ik_0 a} \right) (1 + bX_+) \, \psi(0, z). \qquad (7.16)$$

The additional term on the left-hand side is

$$\lambda^2(s) \equiv k_0^2 \left(\beta + \frac{s}{sb - ik_0 a} \right). \qquad (7.17)$$

The main points are that this term is independent of z and, moreover, encapsulates all of the left-hand side dependence on the Padé parameters.

The right-hand side of Eq. (7.16) represents contributions from the starting field in the lower half-space. For a source located in $z < d$ and for emitted propagation angles which are aperture-limited with respect to the horizontal, it may be assumed (*e.g.*, the reference depth d could be chosen at a new depth $d' > d$) that the starting field vanishes in the half-space,

$$\psi(0, z) = 0, \qquad (z \geq d). \qquad (7.18)$$

Equation (7.16) then attains the simpler homogeneous form

$$\left\{ \frac{\partial^2}{\partial z^2} + \mu k_0^2 \left(z - d \right) + \lambda^2(s) \right\} \Psi(s, z) = 0, \qquad (z > d). \qquad (7.19)$$

One final simplification in notation is convenient. The additional definition,

$$\tilde{z} \equiv d - \lambda^2(s) / \left(\mu k_0^2 \right), \qquad (7.20)$$

reduces Eq. (7.19) to

$$\left\{ \frac{\partial^2}{\partial z^2} + \mu k_0^2 \left(z - \tilde{z} \right) \right\} \Psi(s, z) = 0, \qquad (z > d). \tag{7.21}$$

This equation is to be solved subject to appropriate boundary conditions as $z \to \infty$.

7.2.3   Solution Details

The differential equation Eq. (7.21) may be reduced to a canonical form by setting

$$\Psi(s, z) \equiv \text{const.} \times U\left[\zeta(s, z)\right], \tag{7.22}$$

where the argument on the right-hand side is

$$\zeta(s, z) \equiv \sigma \left(z - \tilde{z} \right). \tag{7.23}$$

The additional parameter σ is to be a solution of

$$\sigma^3 = -\mu k_0^2; \tag{7.24}$$

the particular branch choice to be specified later. Equation (7.21) then reduces to the form of a homogeneous Airy differential equation [1]

$$U''(\zeta) - \zeta U(\zeta) = 0, \tag{7.25}$$

where the prime denotes differentiation with respect to the argument ζ.

As noted earlier, the solution must be chosen to be either decaying as $z \to +\infty$ in the upward refracting case ($\mu < 0$), or else to take the form of downgoing waves in the downward refracting case ($\mu > 0$). It may be shown that the appropriate solution can be written simply in terms of the first Airy function $\text{Ai}(\zeta)$ [1] as

$$U(\zeta) = \text{Ai}(\zeta), \tag{7.26}$$

provided that the particular branch choice

$$\sigma = \begin{cases} \left(+\mu k_0^2\right)^{1/3} e^{-i\pi/3}, & (\mu > 0), \\ \left(-\mu k_0^2\right)^{1/3}, & (\mu < 0), \end{cases} \tag{7.27}$$

for σ is made. Under the present assumption of real nonzero μ, it is convenient to define the additional positive real parameter

$$\gamma \equiv \left(|\mu| k_0^2\right)^{1/3}, \tag{7.28}$$

which, like μ and k_0, has the dimensions of reciprocal length. Then for sufficiently large z,

$$\zeta(s, z) \sim \sigma z = \begin{cases} \gamma z e^{-i\pi/3} & (\mu > 0) \\ \gamma z & (\mu < 0) \end{cases}, \qquad (z \gg d). \tag{7.29}$$

Using the analysis presented in Appendix A, the asymptotic forms of the solution can be expressed as

$$\mathrm{Ai}\left[\zeta(s,z)\right] \sim \begin{cases} +\dfrac{i}{\pi}\sqrt{\dfrac{\gamma z}{3}}\,H_{1/3}^{(1)}\left(\dfrac{2}{3}(\gamma z)^{3/2}\right), & (\mu > 0)\,, \\[3mm] \dfrac{1}{\pi}\sqrt{\dfrac{\gamma z}{3}}\,K_{1/3}\left(\dfrac{2}{3}(\gamma z)^{3/2}\right), & (\mu < 0)\,, \end{cases} \tag{7.30}$$

in terms of Bessel functions [1]. Thus for positive μ, the presence of the first Hankel function $H_{1/3}^{(1)}(\bullet)$ in the solution, in combination with the assumed $e^{-i\omega t}$ time-dependence, ensures that the solution has the form of outgoing waves as $z \to +\infty$. Similarly, the modified Hankel function $K_{1/3}(\bullet)$ form for $\mu < 0$ ensures that the field decays exponentially into the bottom half-space in the case of an upward refracting profile.

7.2.4 Complex Half-Space Profile–Attenuation

As an aside, it may be noted that for the extension of μ to complex values, equation Eq. (7.27) is consistent with the prescription

$$\mu = \begin{cases} |\mu|, & (\mu > 0)\,, \\ |\mu|\,e^{+i\pi}, & (\mu < 0)\,. \end{cases} \tag{7.31}$$

This in turn is consistent with a requirement that $\mathrm{Im}\,\{\mu\} > 0$, and so from Eq. (7.8) that

$$\mathrm{Im}\,\{c(z)\} < 0, \qquad (z > d)\,. \tag{7.32}$$

This is the appropriate requirement for a complex sound speed profile in the lower half-space to be attenuating, in light of the assumed $e^{-i\omega t}$ time-dependence.

7.3 Non-Local Boundary Conditions

7.3.1 General Considerations

In light of Eq. (7.26), the solution for the Laplace transform of the far-field pressure envelope in the lower half-space ($z > d$) must therefore have the form

$$\Psi(s,z) = \frac{\mathrm{Ai}\left[\zeta(s,z)\right]}{\mathrm{Ai}\left[\zeta(s,d+)\right]}\,\Psi(s,d+). \tag{7.33}$$

Equation (7.33) is the starting point for development of the non-local boundary condition. The term $\Psi(s,d+)$ on the right-hand side represents the values of the far-field pressure envelope on the underside of the boundary $z = d$ between the numerical and analytical domains. The two solutions are to be coupled across this interface by boundary conditions of continuity of pressure and vertical particle velocity [5]. These translate to the conditions

$$\Psi(s, d+) = \Psi(s, d-) \tag{7.34}$$

and

$$\frac{1}{\rho(d+)} \frac{\partial}{\partial z} \Psi(s, d+) = \frac{1}{\rho(d-)} \frac{\partial}{\partial z} \Psi(s, d-). \tag{7.35}$$

The further relationship,

$$\partial_z \Psi(s, d+) = \sigma \frac{\mathrm{Ai}' \left[\zeta(s, d+) \right]}{\mathrm{Ai} \left[\zeta(s, d+) \right]} \Psi(s, d+), \tag{7.36}$$

between these two fields is obtained upon differentiation of both sides of Eq. (7.33) with respect to z, and subsequent evaluation of the result on the underside $z = d+$ of the interface. The factor σ arises from the derivative $d\zeta/dz$ applied to the definition Eq. (7.23).

Equation (7.36) can be arranged as an impedance-type equation

$$\partial_z \Psi(s, d+) = W(s) \left\{ s\Psi(s, d+) \right\}. \tag{7.37}$$

This constitutes the basis for the non-local boundary condition in standard, or Dirichlet-to-Neumann Map (DtN) [10], form. The kernel here is defined as

$$W(s) \equiv \left(\frac{\sigma}{s} \right) \mathrm{Yi} \left[\varsigma(s) \right], \tag{7.38}$$

in terms of the logarithmic derivative,

$$\mathrm{Yi}(\varsigma) \equiv \frac{\mathrm{Ai}'(\varsigma)}{\mathrm{Ai}(\varsigma)}, \tag{7.39}$$

of the Airy function. The factor s is introduced to keep the kernel integrable, and the Airy argument is

$$\varsigma(s) \equiv \zeta(s, d+) = \frac{\sigma \lambda^2(s)}{\mu k_0^2} = -\frac{\lambda^2(s)}{\sigma^2}, \tag{7.40}$$

with the last equality following from Eqs. (7.23) and (7.20).

With the help of the partial fraction expansion Eq. (7.104), the non-local boundary condition kernel may be expressed in the form

$$W(s) \equiv \left(\frac{\sigma}{s} \right) \left\{ \mathrm{Yi}(0) + \sum_{n=1}^{\infty} \left(\frac{1}{\varsigma(s) - a_n} + \frac{1}{a_n} \right) \right\}. \tag{7.41}$$

Further simplification, or reversion to the spatial domain, depends on the particular Padé approximation selected, *i.e.*, on the values of the parameters a and b in Eqs. (7.15) and (7.17).

In passing, note that the expression Eq. (7.37) can alternatively be arranged in the Neumann-to-Dirichlet Map (NtD) [10] form

$$\Psi(s, d+) = L(s)\left\{s\partial_z\Psi(s, d+)\right\}.$$

(7.42)

The kernel is now defined as

$$L(s) \equiv \left(\frac{1}{\sigma s}\right)\mathrm{Zi}\left[\varsigma(s)\right], \qquad \text{where} \qquad \mathrm{Zi}\left(\varsigma\right) \equiv \frac{\mathrm{Ai}\left(\varsigma\right)}{\mathrm{Ai}'\left(\varsigma\right)}.$$

(7.43)

7.3.2 Narrow Angle (Tappert) PE

This case corresponds to the standard parabolic equation and uses $a = 1/2$ and $b = 0$, corresponding to a simple linear Taylor approximation,

$$\sqrt{1 + X_+} \approx 1 + X_+/2,$$

(7.44)

for the square-root operator. Equation (7.17) reduces to

$$\lambda^2(s) \equiv k_0^2\left\{\beta + \frac{2is}{k_0}\right\},$$

(7.45)

which represents a linear conformal mapping from the s-plane to the λ^2-plane, reflective of the square root approximation Eq. (7.44).

The Airy function argument Eq. (7.40) can then be written in the form

$$\varsigma(s) \equiv \tau s + a_*,$$

(7.46)

in terms of the additional two parameters

$$\tau \equiv \frac{-2ik_0}{\sigma^2}$$

(7.47)

and

$$a_* \equiv -\frac{\beta k_0^2}{\sigma^2} = \frac{\beta\sigma}{\mu}.$$

(7.48)

Identity Eq. (7.106) then leads to the expression

$$\mathrm{Yi}\left(\tau s + a_*\right) = \mathrm{Yi}\left(a_*\right) + \sum_{n=1}^{\infty}\left\{\frac{1}{\tau s + a_* - a_n} - \frac{1}{a_* - a_n}\right\}.$$

(7.49)

This kernel therefore has poles in the s-plane at the points

$$s = s_n \equiv \frac{a_n - a_*}{\tau}.$$

(7.50)

The preceding results lead in turn to the expression

$$W(s) = \sigma\frac{\mathrm{Yi}\left(\tau s + a_*\right)}{s} = \sigma\left\{\frac{\mathrm{Yi}\left(a_*\right)}{s} + \frac{1}{\tau}\sum_{n=1}^{\infty}\frac{1}{s_n\left(s - s_n\right)}\right\}$$

(7.51)

for the kernel Eq. (7.41).

The location of these poles requires some investigation for Laplace-inversion purposes. Assuming that the reference wavenumber k_0 is real, Eq. (7.50) may be written as

$$\hat{s} \equiv \frac{2s_n}{k_0} = i\beta + i\left(\frac{\sigma}{k_0}\right)^2 a_n. \tag{7.52}$$

Equation (7.27) shows that

$$\frac{\sigma}{k_0} = \left(\frac{|\mu|}{k_0}\right)^{1/3} \times \begin{cases} e^{-i\pi/3}, & (\mu > 0), \\ 1, & (\mu < 0). \end{cases} \tag{7.53}$$

Since the Airy function zeros a_n lie on the negative real axis, it follows that

$$\eta \equiv -a_n\left(\frac{|\mu|}{k_0}\right)^{2/3}, \qquad (0 < \eta < \infty), \tag{7.54}$$

is a positive real parameter, and Eq. (7.52) can therefore be written as

$$\hat{s} = i\beta + \eta \times \begin{cases} e^{-7\pi i/6}, & (\mu > 0), \\ e^{-\pi i/2}, & (\mu < 0). \end{cases} \tag{7.55}$$

The s-plane poles therefore lie on the negative real axis when $\mu < 0$, and in the second quadrant of the s-plane when $\mu > 0$. The principal point is that they always lie in the left half s-plane, and ϵ in the Laplace inversion formula Eq. (7.13) can therefore be assigned any positive value.

The expression on the right-hand side of Eq. (7.51) may then readily be inverse Laplace-transformed [1], leading to the spatial-domain kernel representation

$$w(r) = \sigma \left\{ \text{Yi}(a_*) + \frac{1}{\tau} \sum_{n=1}^{\infty} \frac{e^{s_n r}}{s_n} \right\}. \tag{7.56}$$

This corresponds to previously published formulations [4, 14].

7.3.3   Simple Wide-Angle (Claerbout) PE

The other case of interest involves a first-order Padé approximation which is a true rational function, in the sense that the parameter b in Eq. (7.15) is non-zero. In this general case, definition Eq. (7.17) can be written in the form

$$\lambda^2(s) \equiv k_0^2 \left(\beta + 1/b\right) \left\{ \frac{s - (ik_0)\left(\frac{a}{b}\right)\left(\frac{\beta}{\beta+1/b}\right)}{s - (ik_0)\left(\frac{a}{b}\right)} \right\}, \tag{7.57}$$

which is a bilinear mapping from the s-plane to the λ^2-plane. Of particular interest to the present work is the simple first-order wide-angle parabolic equation of Claerbout [5]. Here the choices

$$a = 1/2, \qquad b = 1/4, \tag{7.58}$$

are made, corresponding to the square-root approximation

$$\sqrt{1 + X_+} \approx 1 + \frac{X_+/2}{1 + X_+/4}. \tag{7.59}$$

In this case, Eq. (7.57) reduces to

$$\lambda^2(s) \equiv k_0^2 \left(\beta + 4\right) \left\{ \frac{s - (2ik_0)\left(\frac{\beta}{\beta+4}\right)}{s - (2ik_0)} \right\}, \tag{7.60}$$

and the choice of a bilinear square-root approximation is reflected in a full bi-linear mapping of the s-plane. It should be noted, however, that other choices for a and b can be envisioned, for example, the parabolic approximation due to Greene [5].

The Airy function argument Eq. (7.40) in the present case may be written in the form

$$\varsigma(s) = f \left(\frac{s - \tau_u}{s - \tau_\ell} \right). \tag{7.61}$$

The new parameters here are

$$f \equiv -k_0^2 \left(\beta + 1/b\right) / \sigma^2 = -k_0^2 \left(\beta + 4\right) / \sigma^2, \tag{7.62}$$

$$\tau_\ell \equiv (ik_0) \left(\frac{a}{b}\right) \left(\frac{\beta}{\beta + 1/b}\right) = (2ik_0) \left(\frac{\beta}{\beta + 4}\right), \tag{7.63}$$

and

$$\tau_u \equiv (ik_0) \left(\frac{a}{b}\right) = 2ik_0. \tag{7.64}$$

The first equality in each of the definitions Eqs. (7.62), (7.63) and (7.64) is for the general first-order Padé approximation $b \neq 0$, while the second equality in each pair is for the particular case of the Claerbout coefficients Eq. (7.58).

The form Eq. (7.41) remains valid, but with the argument now being defined by Eq. (7.61). The nth term $[\varsigma(s) - a_n]^{-1}$ in the partial fraction expansion Eq. (7.41) has a pole in the s-plane at

$$s = s_n \equiv \frac{f \tau_u - a_n \tau_\ell}{f - a_n}, \tag{7.65}$$

with residue

$$t_n \equiv \frac{f \left(s_n - \tau_u\right)}{a_n \left(f - a_n\right)} = \frac{f \left(\tau_u - \tau_\ell\right)}{\left(f - a_n\right)^2}. \tag{7.66}$$

The locus of the s-plane poles again requires examination. The definition Eq. (7.65) can be arranged in the form

$$\hat{s} \equiv \frac{s_n}{2k_0} = i \left(\frac{1 + \chi\beta/(\beta + 4)}{1 + \chi} \right), \tag{7.67}$$

where

$$\chi \equiv -\frac{a_n \sigma^2}{(\beta + 4)k_0^2}. \tag{7.68}$$

Assuming that β is real and $\beta + 4 > 0$, and following arguments analogous to those surrounding Eqs. (7.54) and (7.53), it follows that

$$0 < |\chi| < \infty \tag{7.69}$$

and

$$\text{Arg}\,(\chi) = \begin{cases} e^{-2\pi i/3}, \ (\mu > 0), \\ 0, \qquad\quad (\mu < 0). \end{cases} \tag{7.70}$$

Equation (7.67) represents a mapping from a half-line defined by parameter χ to the complex $\hat{s}$-plane. The map begins at the point $\hat{s} = +i$ when $\chi = 0$, and terminates at the point $\hat{s} = i\beta/(\beta + 4)$ when $|\chi| \to \infty$. Equation (7.70) shows that the image is a segment of the imaginary $\hat{s}$-axis when $\mu < 0$. The case $\mu > 0$ is more complicated. In this case, the general properties of a bilinear mapping transform the half-line defined by χ to a portion of a circle in the $\hat{s}$-plane. It is readily shown that

$$\text{Re}\,\{\hat{s}\} = \frac{4\chi_i}{(\beta + 4)[(1 + \chi_r)^2 + \chi_i^2]}. \tag{7.71}$$

The denominator is positive provided that $\beta > 4$, and for $\mu > 0$, $\chi_i < 0$. Hence the s-poles all have non-positive real parts, and the inverse Laplace transform will be stable.

The identity Eq. (7.104) leads to the expression

$$\text{Yi}\,[\varsigma(s)] = \text{Yi}\,[\varsigma(0)] + \sum_{n=1}^{\infty} \left\{ \frac{1}{\varsigma(s) - a_n} - \frac{1}{\varsigma(0) - a_n} \right\}, \tag{7.72}$$

which may in turn be arranged in the form

$$\text{Yi}\,[\varsigma(s)] = \text{Yi}\,(\varsigma_0) + \sum_{n=1}^{\infty} t_n \left\{ \frac{1}{s - s_n} + \frac{1}{s_n} \right\}, \tag{7.73}$$

where the expression

$$\varsigma_0 \equiv \varsigma(0) = f\tau_u/\tau_\ell \tag{7.74}$$

follows immediately from Eq. (7.61).

The non-local boundary condition kernel Eq. (7.38) in the Laplace-transform domain may then be written

$$K(s) \equiv \sigma \left\{ \frac{\text{Yi}\,(\varsigma_0)}{s} + \sum_{n=1}^{\infty} \frac{t_n}{s_n\,(s - s_n)} \right\}. \tag{7.75}$$

This expression may readily be inverse Laplace-transformed to give the spatial domain kernel

$$w(r) \equiv \sigma \left\{ \mathrm{Yi}\,(\varsigma_0) + \sum_{n=1}^{\infty} \frac{t_n}{s_n} e^{s_n r} \right\}. \tag{7.76}$$

Equations (7.75) and (7.76) are the Claerbout counterparts, respectively, of the Tappert forms Eqs. (7.51) and (7.56). The differences lie in the altered form of the s-plane poles under the present square-root approximation, and in the presence of the residue term in the numerator.

7.4 Numerical Implementation

The numerical implementation of various PE algorithms in the numerical domain, and the coupling of those numerical techniques to non-local boundary conditions is adequately discussed elsewhere [10], [17], [2], [15]. In this section, one possible numerical implementation of the non-local boundary condition will be considered.

In order to simplify the discussion, it convenient to introduce the shorthand notations

$$F(r) = \psi(r, d + 0),$$
$$H(r) = \partial_r \psi(r, d + 0) \quad \text{and}$$
$$V(r) = \partial_z \psi(r, d + 0),$$

for the field and its horizontal and vertical gradients evaluated just inside the lower half-space. As noted earlier, these values are related to the numerical domain values on the opposite sides of the interface by the boundary conditions Eqs. (7.34) and (7.35).

The non-local boundary condition in the spatial domain is obtained from the inverse Laplace transform of the relationship Eq. (7.37). The result can be evaluated using the convolution theorem [1], and may be written in the form

$$V(r') = \int_0^{r'} H(r)\, w(r' - r)\, dr, \tag{7.77}$$

where the kernel is given by Eq. (7.76).

The remainder of this section is based on the assumption of a regular range grid,

$$r_n = nh, \qquad (n = 0, 1, \ldots), \tag{7.78}$$

with step $\delta_r \equiv h$. Abbreviations such as

$$F_n \equiv F(r_n) \tag{7.79}$$

will be used freely. Thus, with $r' = mh$, Eq. (7.77) can be written as

$$V_m = \sum_{\ell=1}^{m} \int_{(\ell-1)h}^{\ell h} H(r)\, w(mh - r)\, dr. \tag{7.80}$$

Further progress depends on the approximation for the horizontal derivative $H(r)$ under the integral.

The simplest possibility is a constant on each range step, based on a central difference formulation,

$$H(r) \approx \frac{F_\ell - F_{\ell-1}}{h}, \quad (r_{\ell-1} \le r \le r_\ell). \tag{7.81}$$

Then Eq. (7.80) leads to the result

$$V_m = \sum_{\ell=1}^{m} \left\{ \frac{F_\ell - F_{\ell-1}}{h} \right\} Q_{m-\ell+1}^{(0)}, \tag{7.82}$$

where the coefficients are

$$Q_j^{(0)} \equiv h \int_0^1 w[(j - u)h]\, du. \tag{7.83}$$

An alternative possibility is linear interpolation between values of the horizontal gradient at the range grid-points

$$H(r) \approx \left(\frac{r - r_{\ell-1}}{h} \right) H_\ell + \left(\frac{r_\ell - r}{h} \right) H_{\ell-1}, \quad (r_{\ell-1} \le r \le r_\ell). \tag{7.84}$$

This leads to a boundary condition of the form

$$V_m = \sum_{\ell=1}^{m} \left\{ C_{m-\ell+1}^{-} H_{\ell-1} + C_{m-\ell+1}^{+} H_\ell \right\}, \tag{7.85}$$

where

$$C_j^{-} \equiv Q_j^{(0)} - Q_j^{(1)} \quad \text{and} \quad C_j^{+} \equiv Q_j^{(1)}, \tag{7.86}$$

$Q_j^{(0)}$ is again defined by Eq. (7.83), and where

$$Q_j^{(1)} \equiv h \int_0^1 w[(j - u)h]\, u\, du. \tag{7.87}$$

When applied to the kernel Eq. (7.76), the requisite integrals Eqs. (7.83) and (7.87) can be written analytically,

$$Q_j^{(0)} = (\sigma h) \left\{ \mathrm{Yi}\,(\varsigma_0) + \left[\frac{S_2\,(r_j) - S_2\,(r_{j-1})}{h} \right] \right\} \tag{7.88}$$

and

$$Q_j^{(1)} = \left(\frac{\sigma h}{2}\right) \text{Yi}(\varsigma_0) + \sigma \left\{ \left[\frac{S_3(r_j) - S_3(r_{j-1})}{h}\right] - S_2(r_j) \right\}, \quad (7.89)$$

in terms of the series

$$S_k(r) \equiv \sum_{n=1}^{\infty} \frac{t_n}{s_n^k} \exp(s_n r), \quad (k = 2,3). \quad (7.90)$$

For the present Claerbout case, these series are poorly convergent. It follows from Eq. (7.65) that s_n terms have the asymptotic expansion

$$s_n \sim \tau_\ell, \quad (7.91)$$

i.e., they approach a constant for large value of the summation parameter n. Convergence is entirely provided by the numerator terms t_n, which have the asymptotic form

$$t_n \sim \frac{f(\tau_u - \tau_\ell)}{a_n^2}. \quad (7.92)$$

From the properties of the Airy function zeroes [1], it follows that

$$\frac{1}{a_n^2} \sim \left(\frac{2}{3\pi n}\right)^{4/3}, \quad (7.93)$$

which is sufficient to ensure convergence, albeit slow. Various mathematical techniques could be employed to accelerate convergence. However, this was not found necessary in practice, where the numerical results were insensitive to the sum truncation limit above several hundred terms.

It is of interest to note that the linear mapping provided by the simple Tappert PE ([4]) gives both better convergence properties, and easier numerical acceleration techniques than the present Claerbout bilinear mapping.

7.5 First-Order Claerbout Examples

The utility of the Claerbout NLBC is illustrated using several well-known refracting profiles.

7.5.1 Modified AESD [13] Case

This surface-duct example pertains to a 300-Hz, 15° aperture source at 91.44-m depth, with a receiver at 27.43-m depth. The upper numerical domain has a sound speed varying between 1536.5 m s^{-1} at the surface to 1539.24

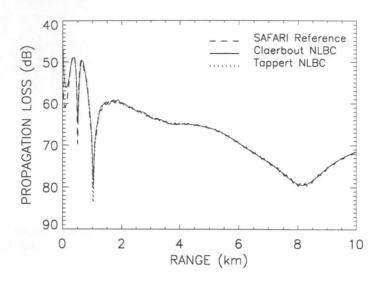

FIGURE 7.2: Transmission loss for the downward-refracting lower half-space.

m s^{-1} at the half-space boundary, which lies at 152.5-m depth. Two cases are considered. In one case, the lower half-space has an upward refracting gradient of $\mu = -199.36 \times 10^{-6}$ m^{-1}. In the other case, the profile has a downward-refracting gradient of $\mu = +199.36 \times 10^{-6}$ m^{-1}.

The transmission loss, computed using the new boundary condition, is compared to both a benchmark result computed using the standard SAFARI [12] propagation code, as well as to a Tappert PE implementation [4] of the non-local boundary condition for the present geometry. Results are presented below in Fig. 7.2 for the downward-refracting case and in Fig. 7.3 for the upward refracting case. In the downward refracting case, both the Tappert and the Claerbout NLBCs perform well when compared to SAFARI. Clearly this problem does not involve much high-angle energy that interacts with the boundary. In the upward refracting case there is a noticeable improvement in agreement between SAFARI and the Claerbout result (as compared to SAFARI and the Tappert result). Clearly, in this case, high-angle interaction with the lower waveguide boundary is very important and requires a high-order boundary condition in order to model it properly.

Full-field images comparing the results obtained using the Claerbout NLBC for the upward-refracting case are shown Fig. 7.4a. The upper 400 m of the corresponding SAFARI results are shown in Fig. 7.4b. This pair of images is quite remarkable in that it shows energy returning to the numerical domain (the interference that occurs around the 5 km mark in range) from the lower half-space. The upper figure shows the entire computational domain, and the entire effect of the lower half-space is completely contained in the non-

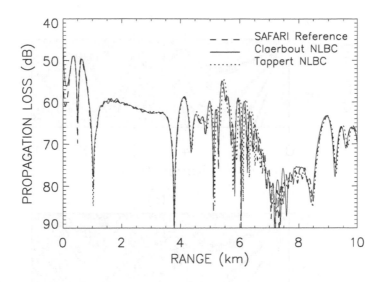

FIGURE 7.3: Transmission loss for the upward-refracting lower half-space.

local boundary condition. By contrast, the SAFARI run (or similarly a PE using a traditional absorbing layer) would require that a large portion of the lower-half space be included in the numerical domain.

Figure 7.4 is worthy of additional discussion. It is important to re-emphasize that, using the NLBC, the PE calculations were confined to a vertical grid that spanned the water column (numerical domain) above the half-space. This is clearly exhibited in Figure 7.4a, which depicts the entire PE+NLBC computation. Since the computational domain for SAFARI must be terminated by either a homogeneous half-space or a perfect reflector, a homogeneous half-space was chosen to lie below a depth of 5152.5 m. Given the choice of N^2 gradient, this depth is fractionally above the value at which the sound speed in the ocean becomes infinite in the upward refracting case (see Fig. 7.1). For convenience, the same depth was chosen for the downward refracting SAFARI results. Thus the SAFARI numerical domain spanned a depth exceeding 5 km. The image in Figure 7.4b only displays the upper 400 m of the complete SAFARI calculation. Finally, the limited source aperture of 15° was necessary in order to ensure that the narrow-angle accuracy restrictions were met for the PE calculations. The SAFARI results were spectrally limited to approximate the same 15° aperture, in order to closely match the PE calculations.

7.5.2 Modified Norda Test Cases

The NORDA Parabolic Equation Workshop [3] provided many useful and simple test cases designed to test the capabilities of acoustic propagation mod-

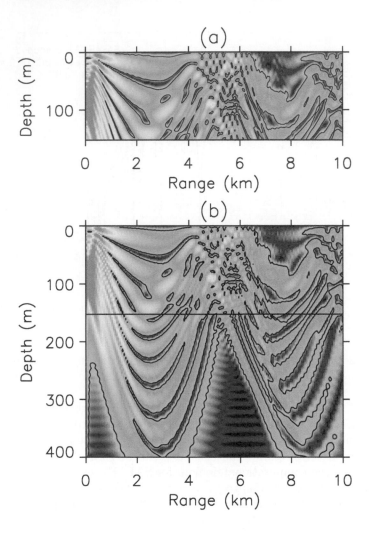

FIGURE 7.4: Full field images computed using the Claerbout NLBC and SAFARI methods. The 70-dB contour level is indicated.

els. The following examples are based on a modified version of what has become known as the NORDA Test Case 2. It consists of low-frequency (25 Hz) propagation in a range-independent bi-linear N^2 ocean environment (density equals 1 gm cm^{-3} and no attenuation) over a homogeneous half-space. The present modification removes the half-space representing the ocean bottom and treats the problem as if it were a bi-linear profile with the lowermost layer of infinite extent. Both source and receiver are placed at a 500 m depth. Three different cases, whose parameters are summarized in Table 7.1, are considered. The sound speeds at depths of 1500 m and 1000 m were used to compute the gradients used in the lower part of the bilinear profile. Note

TABLE 7.1:   Modified NORDA Case 2 Parameters.

Depth	Sound speeds (m/s)			Gradient
(m)	2a	2b	2c	$\times 10^{-3}/$m
0	1500	1500	1500	-0.108
1000	1520	1520	1520	-0.481
1500	1563	1744	1971	-0.811

that the gradient values represent the slope of $N^2 - 1$ and not those of the sound speed. Hence sound speeds between the tabulated depths have the form $c_0^2/c(z)^2 = az + b$, suitably fitted to the tabulated values.

Propagation loss computed for the three test cases are compared to both a benchmark result computed using the standard SAFARI [12] propagation code, as well as to a Tappert PE implementation [4] of the non-local boundary condition for the present geometry. The results for the modified NORDA 2a, 2b and 2c test cases are shown in Fig. 7.5, Fig. 7.6 and Fig. 7.7, respectively.

Panel (a) of each of the three Figures depicts comparison between the Claerbout NLBC implementation and the SAFARI reference, while panel (b) does the same for the Tappert NLBC.

Figure 7.5 shows that in the case a small N^2 gradient, there is little to choose between the Tappert and Claerbout boundary conditions. This is because the boundary interaction is relatively low-angle and the higher-order Claerbout NLBC offers little benefit.

For the NORDA 2b and 2c cases, however, the boundary interaction involves higher-angle energy, and the benefits of the more accurate Claerbout NLBC are readily apparent.

7.6   Closing Remarks

The preceding test cases show that the Claerbout non-local boundary condition is accurate, and clearly superior to the equivalent narrow-angle Tappert NLBC in cases involving higher propagation angles. In particular, the Claerbout NLBC improves the phase agreement in the computed results, and is in excellent agreement with the SAFARI results. The obvious remaining discrepancies are due to mismatches in the aperture-limitation process in the full-wave SAFARI results. The results of this section show that domain trun-

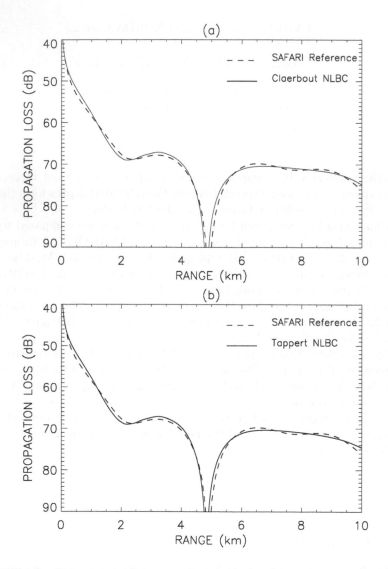

FIGURE 7.5: Comparisons between Claerbout (top) and Tappert (bottom) NLBC computations with the SAFARI reference results for the modified NORDA 2a profile.

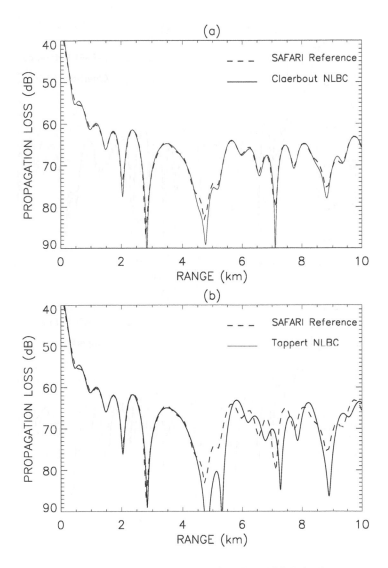

FIGURE 7.6: As Figure 7.5, but for the NORDA 2b case.

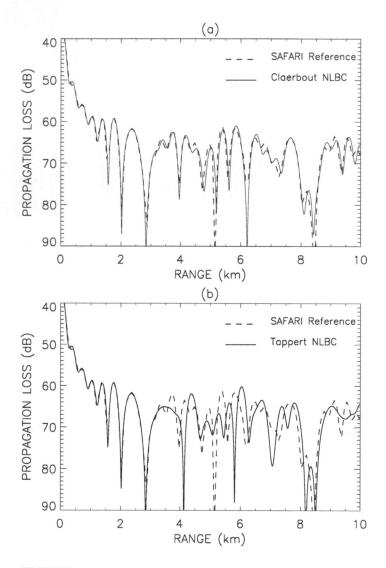

FIGURE 7.7: As Figure 7.5, but for the NORDA 2c case.

cation via NLBCs is a viable and useful addition to the underwater acoustic modelling tool kit.

The theory presented in this report provides several avenues for further research. Successful implementation of the theory has been demonstrated elsewhere [4] for the particular case of Standard PE with real value of the gradient parameter μ and in the DtN formulation. Remaining aspects to be considered with this formulation include better algorithms for faster computation of the non-local boundary condition kernel, particularly in the poorly convergent upward refracting case, which is probably the more interesting case for underwater acoustics. It also remains to implement and test the possibility of a complex gradient parameter μ, which would allow for non-zero attenuation in the lower half space in both the upward and downward refracting cases.

Although the theory for the non-local boundary condition associated with the wide-angle Claerbout PE with complex μ has been considered briefly here, the numerical implementation has not been attempted. The formulation appears to be stable, in that singularities are confined to the left-half plane in the Laplace transform domain. Furthermore, the kernel would appear to offer some advantages with regard to convergence acceleration in a numerical implementation, since the singularities have a finite limit-point. These aspects require further research.

Finally, the formulation in terms of a NtD mapping has been briefly indicated. However, neither an in-depth derivation nor the numerical implementation have been considered in this research, and any possible advantages are not clear.

Appendix

Airy Function Notes

Properties of the Kernel

The first Airy function Ai (ς) [1, §10.4] is defined to be the solution of

$$F''(\varsigma) - \varsigma F(\varsigma) = 0 \qquad (7.94)$$

which vanishes for Re $\{\varsigma\} \to \infty$ along the positive real ς-axis. It is an entire function over the whole complex ς-plane [6]. The zeros

$$\{a_n \mid n = 1, 2, \ldots\} \qquad (7.95)$$

lie along the negative real ς-axis, with increasing magnitude as n increases. The derivative Ai$'(\varsigma)$ likewise has a set of zeros along the negative real axis denoted

$$\{a'_n \mid n = 1, 2, \ldots\}. \tag{7.96}$$

Consequently the ratio,

$$\mathrm{Yi}\,(\varsigma) \equiv \frac{\mathrm{Ai}'\,(\varsigma)}{\mathrm{Ai}\,(\varsigma)}, \tag{7.97}$$

is a meromorphic function having a string of interlaced poles Eq. (7.95) and zeros Eq. (7.96) along the negative real axis. It follows from Eq. (7.94) that the ratio Eq. (7.97) satisfies the inhomogeneous nonlinear differential equation

$$\mathrm{Yi}'\,(\varsigma) + [\mathrm{Yi}\,(\varsigma)]^2 = \varsigma. \tag{7.98}$$

The Airy function and derivative have the nonzero values

$$\mathrm{Ai}\,(0) = 3^{-2/3}/\Gamma(2/3) \quad \text{and} \quad \mathrm{Ai}'\,(0) = 3^{-1/3}/\Gamma(1/3) \tag{7.99}$$

at the origin. All derivatives of order 2 and higher can be expressed in terms of the Airy function and its first derivative on account of the differential equation (7.94). The Airy function and its derivative have the asymptotic behaviors

$$\mathrm{Ai}\,(\varsigma) \sim +\frac{\varsigma^{-1/4}e^{-\chi}}{2\sqrt{\pi}}\left\{1 + O\left(\frac{1}{\chi}\right)\right\}, \qquad (|\arg\varsigma| < \pi) \tag{7.100}$$

and

$$\mathrm{Ai}'\,(\varsigma) \sim -\frac{\varsigma^{1/4}e^{-\chi}}{2\sqrt{\pi}}\left\{1 + O\left(\frac{1}{\chi}\right)\right\}, \qquad (|\arg\varsigma| < \pi), \tag{7.101}$$

respectively, where the auxiliary variable

$$\chi \equiv \frac{2}{3}\varsigma^{3/2}. \tag{7.102}$$

Consequently, the ratio Eq. (7.97) has the asymptotic expansion

$$\mathrm{Yi}\,(\varsigma) \sim -\sqrt{\varsigma}\left\{1 + O\left(\varsigma^{-3/2}\right)\right\}, \qquad |\arg\varsigma| < \pi. \tag{7.103}$$

This behavior is sufficient to allow for the partial fraction expansion [16]

$$\mathrm{Yi}\,(\varsigma) = \mathrm{Yi}\,(0) + \sum_{n=1}^{\infty}\left\{\frac{1}{\varsigma - a_n} + \frac{1}{a_n}\right\}, \tag{7.104}$$

or, equivalently,

$$\mathrm{Yi}\,(\varsigma) = \mathrm{Yi}\,(0) + \sum_{n=1}^{\infty}\frac{\varsigma}{a_n\,(\varsigma - a_n)}. \tag{7.105}$$

When an expansion of the form Eq. (7.104) is evaluated at each of the two points $\varsigma = \varsigma_1$ and $\varsigma = \varsigma_2$, the difference of the resulting expressions leads to the additional identity

$$\mathrm{Yi}\,(\varsigma_1) = \mathrm{Yi}\,(\varsigma_2) + \sum_{n=1}^{\infty} \left\{ \frac{\varsigma_2 - \varsigma_1}{(\varsigma_1 - a_n)\,(\varsigma_2 - a_n)} \right\}. \tag{7.106}$$

Related results in Titchmarsh [16] also lead to the similar expansion

$$\mathrm{Yi}\,(\varsigma) = \mathrm{Yi}\,(0) + \varsigma \mathrm{Yi}'\,(0) + \cdots + \frac{\varsigma^j \mathrm{Yi}^{(j)}\,(0)}{j!} +$$

$$+ \sum_{n=1}^{\infty} \frac{(\varsigma/a_n)^{j+1}}{\varsigma - a_n}, \quad (j = 0, 1, \ldots), \tag{7.107}$$

of which the $j = 1$ case,

$$\mathrm{Yi}'\,(\varsigma) = \mathrm{Yi}\,(0) + \varsigma \mathrm{Yi}'\,(0) + \sum_{n=1}^{\infty} \frac{1}{\varsigma - a_n} \left(\frac{\varsigma}{a_n} \right)^2, \tag{7.108}$$

is of particular interest. It may be noted that higher values of j in the expansion Eq. (7.107) lead to increasingly rapid convergence of the series for fixed argument ς, but have more difficult behavior for increasing ς. These expansions are particularly useful for numerical implementation of the non-local boundary condition kernel.

The partial fraction expansion Eq. (7.104) may be repeatedly differentiated term-by-term to obtain the results

$$\mathrm{Yi}^{(\ell)}\,(\varsigma) = \sum_{n=1}^{\infty} \frac{(-1)^{\ell}\ell!}{(\varsigma - a_n)^{\ell+1}}, \quad (\ell = 1, 2, \ldots). \tag{7.109}$$

By setting $\varsigma = 0$, this latter expression has a particularly useful role in summing reciprocal powers of the Airy function zeros, *e.g.*,

$$\sum_{n=1}^{\infty} \frac{1}{a_n^{\ell+1}} = -\frac{\mathrm{Yi}^{(\ell)}\,(0)}{\ell!} \quad (\ell = 1, 2, \ldots). \tag{7.110}$$

Equations (7.97), (7.98) and (7.99) may be used to develop analytical expressions for these sums by evaluating the right-hand side numerators in closed form. The first few such expressions are

$$
\begin{aligned}
\mathrm{Yi}\,(0) &= -3^{5/6}\left[\Gamma(2/3)\right]^2/(2\pi) \\
&= -0.72901113294722700
\end{aligned}
$$

$$
\begin{aligned}
\mathrm{Yi}^{(1)}\,(0) &= -\left[\mathrm{Yi}\,(0)\right]^2 \\
&= -0.53145723196099945
\end{aligned}
$$

$$
\begin{aligned}
\mathrm{Yi}^{(2)}\,(0) &= 1 + 2\left[\mathrm{Yi}\,(0)\right]^3 \\
&= +0.22512352243022916
\end{aligned}
$$

$$
\begin{aligned}
\mathrm{Yi}^{(3)}\,(0) &= -2\,\mathrm{Yi}\,(0) - 6\left[\mathrm{Yi}\,(0)\right]^4 \\
&= -0.23665847052743151
\end{aligned}
$$

(7.111)

$$
\begin{aligned}
\mathrm{Yi}^{(4)}\,(0) &= 10\left[\mathrm{Yi}\,(0)\right]^2 + 24\left[\mathrm{Yi}\,(0)\right]^5 \\
&= +0.37280782503895583
\end{aligned}
$$

$$
\begin{aligned}
\mathrm{Yi}^{(5)}\,(0) &= -6 - 60\left[\mathrm{Yi}\,(0)\right]^3 - 120\left[\mathrm{Yi}\,(0)\right]^6 \\
&= -0.76671233763494197
\end{aligned}
$$

These expressions have potential application in numerical convergence acceleration of the infinite series in the kernel Eq. (7.90). The terms in the infinite series have the asymptotic expansion

$$
\frac{t_n}{s_n}e^{s_n r} \sim e^{\tau_\ell r}\left\{\frac{A_2}{a_n^2} + \frac{A_3}{a_n^3} + \cdots + \frac{A_k}{a_n^k}\right\} \equiv \psi_{k,n}(r),
\tag{7.112}
$$

where the coefficients A_k depend on range and on the Claerbout and half-space parameters. The kernel can then be written in the modified form

$$
w(r) \equiv \sigma\left\{\mathrm{Yi}\,(\varsigma_0) + \Psi_k(r) + \sum_{n=1}^{\infty}\left[\frac{t_n}{s_n}e^{s_n r} - \psi_{k,n}(r)\right]\right\}.
\tag{7.113}
$$

The terms in square bracket have the accelerated convergence rate a_n^{-k-1} in n for each fixed range value. The additional term is

$$
\begin{aligned}
\Psi_k(r) &= \sum_{n=1}^{\infty}\psi_{k,n}(r) \\
&= e^{\tau_\ell r}\sum_{j=2}^{k}A_j\left(\sum_{n=1}^{\infty}\frac{1}{a_n^j}\right) \\
&= -e^{\tau_\ell r}\sum_{j=2}^{k}\frac{A_j}{(j-1)!}\mathrm{Yi}^{(j-1)}\,(0)
\end{aligned}
\tag{7.114}
$$

Together with Eq. (7.111), this is an analytic expression for the required function.

Similar acceleration techniques can be applied to the numerical sums Eq. (7.76) which appear in the numerical implementation.

Finally, we note that analogous expressions can be derived for the NtD kernel (7.43).

Relationships to Bessel Functions

The Airy function and its derivatives are closed related to Bessel functions of order $\pm 1/3$ and $\pm 2/3$. Of particular importance are (i) the relationship

$$\text{Ai}(\varsigma) = \frac{1}{\pi} \sqrt{\frac{\varsigma}{3}} K_{1/3}\left(\frac{2}{3}\varsigma^{3/2}\right) \tag{7.115}$$

between the Airy function and the modified Bessel function of the second kind and order $1/3$, and (ii), the relationships

$$\text{Ai}\left(\varsigma e^{-i\pi/3}\right) = +\frac{i}{\pi} \sqrt{\frac{\varsigma}{3}} H_{1/3}^{(1)}\left(\frac{2}{3}\varsigma^{3/2}\right) \tag{7.116}$$

and

$$\text{Ai}\left(\varsigma e^{+i\pi/3}\right) = -\frac{i}{\pi} \sqrt{\frac{\varsigma}{3}} H_{1/3}^{(2)}\left(\frac{2}{3}\varsigma^{3/2}\right) \tag{7.117}$$

between the Airy function and the two Hankel functions. These results are useful in choosing wavelike or decaying solutions of Airy's differential equation Eq. (7.94).

References

[1] M. Abramowitz and I. Stegun. *Handbook of Mathematical Functions.* Dover Publications, 1965.

[2] G. H. Brooke and D. J. Thomson. Non-local boundary conditions for high-order parabolic equation algorithms. *Wave Motion*, 31:117–129, 2000.

[3] J. A. Davis, D. White, and R. C. Cavanagh. *NORDA Parabolic Equation Workshop, 31 March–3 April, 1981.* Naval Ocean Research and Development Agency Tech. Note 143, 1982.

[4] T. W. Dawson, D. J. Thomson, and G. H. Brooke. Non-local boundary conditions for acoustic PE predictions involving inhomogeneous

layers. *Third European Conference on Underwater Acoustics*, Vol. 1, J. S. Papadakis, ed., pages 183–188, FORTH–IACM, 1996.

[5] F. B. Jensen, W. A. Kuperman, M. B. Porter, and H. Schmidt. *Computational Ocean Acoustics*. AIP Press, 1994.

[6] N. N. Lebedev. *Special Functions and Their Applications*. Dover Publications, 1972.

[7] M. F. Levy. Horizontal parabolic equation solution of radiowave propagation problems on large domains. *IEEE Trans. Antenn. Propag.*, 43:137–144, 1995.

[8] M. F. Levy. Transparent boundary conditions for parabolic equation solutions of radiowave propagation problems. *IEEE Trans. Antenn. Propag.*, 45:66–72, 1997.

[9] M. F. Levy. *Parabolic Equation Methods for Electromagnetic Wave Propagation*. IEE Press, 2000.

[10] J. S. Papadakis. Exact, nonreflecting boundary conditions for parabolic-type approximations in underwater acoustics. *J. Comput. Acoustics*, 2:83–98, 1994.

[11] J. S. Papadakis, M. I. Taroudakis, P. J. Papadakis, and B. Mayfield. A new method for a realistic treatment of the sea bottom in the parabolic approximation. *J. Acoust. Soc. Am.*, 92:2030–2038, 1992.

[12] H. Schmidt. SAFARI, *Seismo-Acoustic Fast field Algorithm for Range-Independent Environments, User's Guide*. SACLANT Undersea Research Centre Report SR–113, San Bartolomeo, Italy, 1988.

[13] C. W. Spofford. *A synopsis of the AESD workshop on non-ray tracing techniques, 22–25 May 1973*. Acoustic Environmental Support Detachment TN–73–05, Arlington, VA, 1973.

[14] D. J. Thomson and G. H. Brooke. Non-local boundary conditions for 1-way wave propagation. *Mathematical and Numerical Aspects of Wave Propagation*, J. A. DeSanto, ed., pages 348–352, SIAM, 1998.

[15] D. J. Thomson and M. E. Mayfield. An exact radiation condition for use with the *a posteriori* PE method. *J. Comput. Acoustics*, 2:113–132, 1994.

[16] E. C. Titchmarsh. *The Theory of Functions*. Oxford University Press, 1932.

[17] D. Yevick and D. J. Thomson. Nonlocal boundary conditions for finite-difference parabolic equation solvers. *J. Acoust. Soc. Am.*, 106:143–150, 1999.

Part III

Numerical Methods for Elastic Wave Propagation

Chapter 8

Introduction and Orientation

P. Joly, POEMS Project team, INRIA-Rocquencourt, France, Patrick.JOLY@inria.fr

There is a great need for numerical methods which treat time-dependent elastic wave propagation problems. Such problems appear in many applications, for example in geophysics or non-destructive testing. In particular, seismic wave propagation is one of the areas where intensive scientific computation has been developed and used since the beginning of the '70s, with the apparition of finite differences time domain methods (FDTD) [71, 127]. Although very old, these methods remain very popular and are widely used for the simulation of wave propagation phenomena or more generally for the numerical resolution of linear hyperbolic systems. They consist in obtaining discrete equations whose unknowns are generally field values at the points of a regular mesh with spatial step h and time step Δt. A prototype of these methods is the famous Yee's scheme [147] introduced in 1966 for Maxwell's equations. There are several reasons that explain the success of Yee-type schemes, among which their easy implementation and their efficiency which are related to the following properties:

- a uniform regular grid is used for the space discretization, so that there is a minimum of information to store and the data to be computed are structured: In other words, one avoids all the complications due to the use of non uniform meshes.

- an explicit time discretization is applied: No linear system has to be solved at each time step.

For instance, Yee's scheme is centered, of order two both in space and time, and completely explicit. The stability and accuracy properties of such a scheme are well known (at least in a homogeneous medium in which the classical Fourier analysis can be used). Due to its explicit nature, the scheme is stable under a C.F.L. condition (see for instance [57]), which says that the ratio $\Delta t/h$ must not exceed a given value. Hence the time step Δt cannot be too large, which is not restrictive in practice since a sufficient accuracy requires a small time step.

Of course, in thirty years of research (in the applied mathematics community as well as in the engineering community), considerable progress has been made in the development of more accurate, flexible, and efficient numerical methods for time-dependent wave propagation, beginning with higher order finite differences (see for instance [45] and the references therein). Attempting to give a complete overview of this abundant research would probably require thousands of pages and we had to make some omissions that we shall try to justify in this introduction. In particular, we have chosen not to talk about finite difference methods, even though they are still of practical and theoretical interest. Among the reasons of this choice, let us mention that the counterpart of the nice properties of FDTD schemes is a lack of "geometrical flexibility" which makes the use of such schemes awkward in the case of computational domains with complicated shape (consider for example the diffraction of waves by an obstacle). It may also be difficult (at least with a theoretical guaranty of stability) to treat boundary conditions and variable coefficients, to be able do mesh refinement and to extend the stability theory to higher order schemes. Let us describe in more detail the content of Part IV.

Our first goal (among two) in this part of the book is to present methods which aim at "going beyond finite differences." More precisely, we have chosen to deal with two categories of such methods, adopting a different point of view for each of them:

(i) **Methods that aim at emphasizing the accuracy and the broadness of application areas with a reasonable computation complexity.** This is the case of the methods based on the use of non-regular, unstructured computational grids whose prototypes are finite element methods [36, 40, 43, 68, 126]. The most tutorial part of our presentation, namely Chapter 10, will concern the presentation of numerical methods based on finite elements in space and finite differences in time. This chapter is conceived as an introductory course, even though some mathematical background is needed. We shall try to emphasize the main ideas and shall present the more instructive proofs. Of course, it will not be possible to give all details and proofs but bibliographical pointers will be provided to complete our presentation. We shall primarily emphasize energy methods for stability and convergence analyses, choosing to ignore other analytical tools such as dispersion analysis in the case of regular meshes, despite of its great importance from the practical point of view (in particular for the choice of values of discretization parameters). Finally, we shall include in our exposition some of the most recent developments of the research in this field, in particular those achieved by the ONDES Project at INRIA, on higher order (also called spectral) elements and mass lumping techniques.

The methods described above are not, of course, the only methods able to handle non- regular meshes. This is also feature of mixed finite-element methods [38] which are also variational methods. In our sense, they belong to the same class of methods as the usual finite element methods, even though they have their own interest and their analysis is more complicated. In Chapter 10, only a particular case of such methods will be mentionned in Section 10.6. However, we shall present another class of mixed methods in Chapter 11 (see point (ii)). Finite-volume methods [95, 108] and Discontinuous Galerkin methods [6, 90, 116] also belong to this category. Initially designed for computational fluid dynamics and conservation laws (see the first part of this book), they are becoming more popular in the world of linear wave propagation and would have surely deserved some place in this book. We chose not to cover them, first for a question of place, second because this class of methods is already covered in another part of this book, and finally because we feel that the degree of (at least our) theoretical understanding and practical experience on these methods is still less mature than for finite-element methods.

(ii) **The methods which aim at emphasizing the computational efficiency without sacrifying too much the accuracy**. These methods still work with regular grids like finite differences but try to handle complicated geometries or geometrical details with a higher accuracy. This will be the object of Chapters 11, 12 and 13. Our presentation will be closely related to the research work of the authors and, more than a course, it will be a review (with some details) of research articles. In particular, we do not claim any completeness: the methods that we shall deal with probably have a more limited field of application than those of Chapter 10. However, we think that they are based on interesting ideas and concepts that can be extended to many other problems, which gives them a pedagogical interest. The methods that we shall present have the following common features:

- The data of the problem remain (mostly) structured,
- The time discretization remains (essentially) explicit,
- The stability condition is not affected by the geometry of the computational domain or by the mesh refinement,
- They remain conservative (in the sense already mentionned of an underlying discrete energy).

The main two topics we shall address are the following

(a) **The treatment of boundary conditions along boundaries of complicated meshes.** We choose to present the so-called fictitious domain method (be careful, this denomination does not always mean the same thing in the literature) that we shall treat in Chapter 12. This method is of course not the only existing method that treats complicated boundaries with uniform meshes (let us cite for instance immersed or embedded boundary methods [148], [98], [106] or [111] for interfaces) but it is definitely the one which is the closest to the spirit of this book.

(b) **Non-conforming space-time mesh refinement methods.** It is clearly of practical importance to be able to deal with geometrical details in wave propagation problems. The need for a local time stepping is closely related to the use of explicit schemes and CFL conditions. Once again, this question has been the object of research in different directions (finite differences, finite volumes, domain decomposition,...) and we have chosen to restrict ourselves, in Chapter 13, to variational conservative methods, keeping the spirit of the rest of our presentation. Let us mention here that, even though our exposition is mostly oriented to the use of regular meshes, this is not a limitation and the ideas and principles are valid for unstructures meshes.

These two chapters have other common fundamental features:

- In each case, the method will require the introduction of a new (boundary or interface) unknown which can be interpreted as an auxiliary unknown (in fact a Lagrange multiplier) in order to satisfy, at least in some weak sense:
  - the physical boundary condition in case (a),
  - the artificial transmission conditions between two parts of the computational grid in case (b).

- We present them, even though this is not necessarily a required condition, using for the space discretization a particular **mixed finite element method with discontinuous displacement on regular meshes.**

This is the opportunity for us to present in Chapter 11 some of our recent research in this direction. This is also the occasion to treat the convergence analysis of mixed methods, as a complement to the more detailed analysis made in Chapter 10.

Finally, let us mention that

- a particular difficulty of the question of space-time refinement is the construction of a stable local time stepping procedure,

- one can combine the use of space-time refinement with the fictitious domain method (see [128]).

Our second and last objective is to present a state of the art in the fundamental transverse question in computational wave propagation: the transparent boundary conditions for bounding artificially a computational domain. This theme is related to the following question: how to reduce to a bounded domain the effective numerical calculation of the solution of a problem which is physically posed in an infinite (or at least very large) domain?

This is once again a vast problem which has retained considerable attention during the past thirty years, in particular from applied mathematicians. We shall focus ourselves to two attractive and competing techniques that have in common the feature of being local in space and time, contrary for instance to some exact methods [3,84,85] (that we shall not speak about here) no integral operator is involved:

- **Local absorbing boundary conditions.** This type of method has been developed since the late '70s [65]. The idea is to write on the boundary of the computational domain a boundary condition which is supposed to represent the effect of the presence of the exterior medium. Such a boundary is called local if this boundary condition can be expressed in terms of differential operators (which makes them compatible with traditional discretization techniques). Then, this boundary condition is necessarily not exact and implies some degree of approximation that needs to be quantified in some way. Moreover, the question of the stability of the coupling between the physical propagation interior model and the "unphysical" boundary conditions raises delicate and challenging mathematical questions.

- **Perfectly matched layers.** The idea of an absorbing layer is quite old: the principle is to surround the computational domain with a layer in which the waves are artificially damped. The concept of the perfectly matched layer is much more recent (it has appeared in the middle of the '90s in electromagnetics [31]). Roughly speaking, the new idea is to build a particular (non-physical) absorbing medium in such a way that no reflection occurs at the interface between the physical medium and the absorbing layer. This technique has generated a considerable literature and is now considered as a very attractive alternative to local absorbing boundary conditions.

We shall try, in Chapter 14, to give a kind of the state of the art about these two techniques with a particular emphasis on the application to elastic wave propagation, which is a cause of specific (sometimes still unsolved) difficulties. Let us mention in particular that the content of Section 14.2 will report on an original and unpublished research achieved in [145].

The five Chapters 10 to 14 will be preceded, in Chapter 9, by an introductory presentation of the mathematical model for elastic wave propagation (linear elastodynamics equations) and of its main properties. This will be the occasion to introduce basic material and notation.

This text is addressed to students (in particular in the field of applied mathematics), engineers, and researchers. Although we shall dedicate some place to some applied aspects of our subject, the overall style of our presentation will be definitely mathematically oriented, in the spirit of the French school of applied mathematics. In particular, some mathematical background in applied functional analysis is needed to be able to read the forthcoming chapters without too many difficulties. This material will not be provided in the present book and we refer the reader to classical textbooks [57, 64, 110, 119] for more information.

Note: The references for this chapter and for all chapters in Part III are given at the end of Chapter 14.

Chapter 9

The Mathematical Model for Elastic Wave Propagation

P. Joly, POEMS Project team, INRIA-Rocquencourt, France, Patrick.JOLY@inria.fr

9.1 Preliminary Notation

In what follows $d = 1, 2, 3$ will denote the space dimension. The euclidean scalar product in $\mathbb{R}^d$ will be denoted

$$u \cdot v = \sum_{i=1}^{d} u_i \cdot v_i, \quad \forall \, (u, v) \in \mathbb{R}^d \times \mathbb{R}^d. \tag{9.1}$$

We shall call a tensor a linear mapping from $\mathbb{R}^d$ into itself, i.e., a $d \times d$ matrix,

$$\varepsilon = ((\varepsilon_{ij})) \in \mathcal{L}(\mathbb{R}^d).$$

We equip $\mathcal{L}(\mathbb{R}^d)$ with the natural inner product

$$\sigma : \varepsilon = \sum_{i,j=1}^{d} \sigma_{ij} \, \varepsilon_{ij}, \tag{9.2}$$

and the associated norm will simply be denoted

$$|\varepsilon|^2 = \sum_{i,j=1}^{d} |\varepsilon_{ij}|^2.$$

We shall denote by $\mathcal{L}_s(\mathbb{R}^d)$ the subspace of $\mathcal{L}(\mathbb{R}^d)$ of symmetric tensors

$$\varepsilon \in \mathcal{L}_s(\mathbb{R}^d) \Leftrightarrow \varepsilon_{ij} = \varepsilon_{ji}, \forall \, 1 \leq i, j \leq d \Leftrightarrow \varepsilon u \cdot v = u \cdot \varepsilon v, \forall \, (u, v) \in \mathbb{R}^d \times \mathbb{R}^d.$$

The orthogonal supplement of $\mathcal{L}_s(\mathbb{R}^d)$ (for the inner product (9.2)) is the space $\mathcal{L}_a(\mathbb{R}^d)$ of antisymmetric tensors

$$\varepsilon \in \mathcal{L}_a(\mathbb{R}^d) \Leftrightarrow \varepsilon_{ij} = -\varepsilon_{ji}, \forall \, 1 \leq i, j \leq d \Leftrightarrow \varepsilon u \cdot v = -u \cdot \varepsilon v, \forall \, (u, v) \in \mathbb{R}^d \times \mathbb{R}^d.$$

The space of linear mappings from $\mathcal{L}(\mathbb{R}^d)$ into itself (also called fourth order tensors) will be denoted $\mathcal{L}^2(\mathbb{R}^d) \equiv \mathcal{L}(\mathcal{L}(\mathbb{R}^d), \mathcal{L}(\mathbb{R}^d))$

$$\mathbf{C} = ((\mathbf{C}_{ijkl})) \in \mathcal{L}^2(\mathbb{R}^d) \Longrightarrow (\mathbf{C}\varepsilon)_{ij} = \mathbf{C}_{ijkl}\,\varepsilon_{kl}, \quad \forall\, 1 \leq i,j \leq d,$$

where we have adopted the summation convention for repeated indices

$$\mathbf{C}_{ijkl}\,\varepsilon_{kl} = \sum_{k,l=1}^{d} \mathbf{C}_{ijkl}\,\varepsilon_{kl}.$$

We shall denote by $\mathcal{L}_s^2(\mathbb{R}^d)$ the subspace of $\mathcal{L}^2(\mathbb{R}^d)$ of symmetric fourth order tensors

$$\mathbf{C} \in \mathcal{L}_s^2(\mathbb{R}^d) \Leftrightarrow \mathbf{C}\varepsilon : \eta = \varepsilon : \mathbf{C}\eta, \quad \forall\, (\varepsilon, \eta) \in \mathcal{L}\mathbb{R}^d \times \mathcal{L}\mathbb{R}^d.$$

which is also equivalent to

$$\mathbf{C}_{ijkl} = \mathbf{C}_{klij}, \quad \forall\, 1 \leq i,j,k,l \leq d. \tag{9.3}$$

We shall be concerned by fourth order tensors $\mathbf{C}$ that map $\mathcal{L}(\mathbb{R}^d)$ into $\mathcal{L}_s(\mathbb{R}^d)$ which is equivalent to

$$\mathbf{C}_{ijkl} = \mathbf{C}_{jikl}, \quad \forall\, 1 \leq i,j,k,l \leq d. \tag{9.4}$$

By $\mathcal{L}_{ss}^2(\mathbb{R}^d)$ we shall denote, the class of tensors in $\mathcal{L}^2(\mathbb{R}^d)$ satisfying (9.3) and (9.4). Note that these two conditions automatically imply

$$\mathbf{C}_{ijkl} = \mathbf{C}_{ijlk}, \quad \forall\, 1 \leq i,j,k,l \leq d, \tag{9.5}$$

which, in other words, means that we have,

$$\forall\, \mathbf{C} \in \mathcal{L}_{ss}^2(\mathbb{R}^d), \quad \mathbf{C}\varepsilon = 0, \quad \forall \varepsilon \in \mathcal{L}_a(\mathbb{R}^d) \quad (\Leftrightarrow \mathcal{L}_a(\mathbb{R}^d) \subset \mathrm{Ker}\ \mathbf{C}). \tag{9.6}$$

That is why a tensor $\mathbf{C} \in \mathcal{L}_{ss}^2(\mathbb{R}^d)$ can be identified to its restriction to $\mathcal{L}_s(\mathbb{R}^d)$ as a linear mapping from $\mathcal{L}_s(\mathbb{R}^d)$ into $\mathcal{L}_s(\mathbb{R}^d)$. We shall say that $\mathbf{C} \in \mathcal{L}_{ss}^2(\mathbb{R}^d)$ is coercive if there exists $\alpha > 0$ such that

$$\mathbf{C}\varepsilon : \varepsilon \geq \alpha\, |\varepsilon|^2, \quad \forall\, \varepsilon \in \mathcal{L}_s(\mathbb{R}^d). \tag{9.7}$$

This is equivalent to saying that the smallest eigenvalue of $\mathbf{C}$, as a symmetric operator in $\mathcal{L}_s(\mathbb{R}^d)$, is strictly positive. As a consequence, as an element of $\mathcal{L}(\mathcal{L}_s(\mathbb{R}^d), \mathcal{L}_s(\mathbb{R}^d))$, $\mathbf{C}$ is invertible, namely there exists $\mathbf{A} \in \mathcal{L}_{ss}^2(\mathbb{R}^d)$, also coercive, such that

$$\forall(\varepsilon, \sigma) \in \mathcal{L}_s(\mathbb{R}^d) \times \mathcal{L}_s(\mathbb{R}^d), \quad \sigma = \mathbf{C}\varepsilon \iff \varepsilon = \mathbf{A}\sigma. \tag{9.8}$$

As it is done classically, in dimension 2, the symmetries (9.3), (9.4) of the tensor $\mathbf{C}$ allow us to simplify its representation into a 3×3 matrix C_{pq}, still denoted by $\mathbf{C}$ for simplicity, such that

$$\mathbf{C}_{ijkl} = C_{p(i,j)p(k,l)},$$

where the function p is defined by

$$p(1,1) = 1, \quad p(2,2) = 2, \quad p(1,2) = p(2,1) = 3.$$

REMARK 9.1 *We give two examples of tensors $\mathcal{L}^2_{ss}(\mathbb{R}^d)$ that arise in linear elasticity*

- *Given (λ, μ) two strictly positive numbers, we set*

$$\mathbf{C}\varepsilon = \lambda \, (tr \, \varepsilon) \, I + \mu \, (\varepsilon + \varepsilon^t), \quad \forall \varepsilon \in \mathcal{L}(\mathbb{R}^d), \tag{9.9}$$

  where I is the identity matrix (of order d), $tr \, \varepsilon$ denotes the trace of ε and ε^t the transpose of ε. It is easy to check that $\mathbf{C} \in \mathcal{L}^2_{ss}(\mathbb{R}^d)$ from the formula

$$\mathbf{C}\varepsilon = \lambda \, (tr \, \varepsilon) \, I + 2\mu \, \varepsilon, \quad \forall \varepsilon \in \mathcal{L}_s(\mathbb{R}^d), \tag{9.10}$$

  which yields the identity

$$\mathbf{C}\varepsilon : \eta = \lambda \, (tr \, \varepsilon)(tr \, \eta) + 2\mu \, \varepsilon : \eta, \quad \forall (\varepsilon, \eta) \in \mathcal{L}_s(\mathbb{R}^d)^2,$$

  This shows that $\mathbf{C}$ is coercive with $\alpha = 2\mu$. Such a tensor occurs in the description of isotropic elastic media [2].

- *In an orthotropic medium in 2D whose principal axes coincides with the coordinate axes, we have $c_{13} = c_{23} = 0$ so that the elasticity matrix can be written as,*

$$\mathbf{C} = \begin{pmatrix} c_{11} & c_{12} & 0 \\ c_{12} & c_{22} & 0 \\ 0 & 0 & c_{33} \end{pmatrix}. \tag{9.11}$$

  In this case the coercivity of $\mathbf{C}$ implies,

$$c_{11} > 0, \quad c_{22} > 0, \quad c_{33} > 0, \quad c_{11}c_{22} - c_{12}^2 > 0. \tag{9.12}$$

9.2 The Equations of Linear Elastodynamics

We present the equations for a very standard model problem. Of course our goal is simply to briefly describe the partial differential equations that are currently used for modelling elastic wave propagation. It is not our purpose to make a course in continuum mechanics and, for more details and physical explanations, we refer the reader to standard monographs such as [2, 11, 66, 67, 114].

In the sequel, $x \in \mathbb{R}^d$ will denote the space variable and $t \geq 0$ will denote the time. The domain Ω represents an open set in $\mathbb{R}^d$ (not necessarily bounded) filled by an elastic medium at time $t = 0$. The boundary of Ω, $\Gamma = \partial\Omega$, is supposed to be made of the union of two parts, $\Gamma = \Gamma_0 \cup \Gamma_1$ such that

- The elastic medium is clamped along Γ_0, *i.e.*, the corresponding material particles do not move.

- The part Γ_1 is subject to a density of external surface forces. If these forces vanish, we say that Γ_1 is a free boundary.

9.2.1   The Unknowns of the Problem

Elastic waves correspond to the movement and deformations of the elastic body due to external (volume or surface) forces or initial deformations of the medium. We shall make the assumptions of small displacements and small deformations, in such a way that

- At any time, the domain of the elastic body can be identified to the initial configuration Ω.

- The equations can be linearized.

Under these assumptions, the mathematical quantities that describe the state of the elastic body are

- The displacement field $u(x,t) : \Omega \times \mathbb{R}^+ \mapsto \mathbb{R}^d$. $u(x,t)$ represents the displacement at time t of the material particle that occupies the position x in the medium Ω at rest.

- The stress tensor field $\sigma(x,t) : \Omega \times \mathbb{R}^+ \mapsto \mathcal{L}_s(\mathbb{R}^d)$ that describes the internal efforts inside the material.

9.2.2   Useful Differential Operators and Green's Formulas

For writing the equations, it is useful to introduce the (linearized) deformation (or strain) tensor, more precisely a tensor field that is directly connected to the displacement u

$$
\left|
\begin{aligned}
&\varepsilon(u)(x,t) : \Omega \times \mathbb{R}^+ \mapsto \mathcal{L}_s(\mathbb{R}^d), \\
&\varepsilon(u)(x,t) = \frac{1}{2}(\nabla u + \nabla u^t)(x,t),
\end{aligned}
\right.
\tag{9.13}
$$

where $\nabla u(x,t)$ is the Jacobian matrix of the vector field $x \mapsto u(x,t)$. In other words

$$
\varepsilon_{ij}(u)(x,t) = \frac{1}{2}\left(\frac{\partial u_i}{\partial x_j} + \frac{\partial u_j}{\partial x_i}\right), \quad \forall\, 1 \le i,j \le d.
\tag{9.14}
$$

It will be also useful to introduce the divergence of a symmetric tensor field $\sigma : \Omega \mapsto \mathcal{L}_s(\mathbb{R}^d)$, where Ω is a (sufficiently regular) domain of $\mathbb{R}^d$ as the vector field

$$
\operatorname{div} \sigma : \Omega \mapsto \mathbb{R}^d,
$$

defined by

$$(\operatorname{div} \sigma)_i = \sum_{j=1}^{d} \frac{\partial \sigma_{ij}}{\partial x_j}, \quad \forall\, 1 \le i \le d. \tag{9.15}$$

We recall below the Green's formula associated to this operator. In the formula below, σ denotes a (sufficiently smooth) tensor field in Ω and v a (sufficiently smooth) vector field.

$$-\int_{\Omega} \operatorname{div} \sigma \cdot v \, dx = \int_{\Omega} \sigma : \nabla v \, dx - \int_{\Gamma} \sigma n \cdot v \, ds, \tag{9.16}$$

where Γ is the boundary of Ω, n the unit outgoing normal vector to Γ and ds the surface measure along Γ.

If moreover σ is symmetric (almost everywhere in Ω), we have

$$\int_{\Omega} \sigma : \nabla v \, dx = \int_{\Omega} \sigma^t : \nabla v \, dx = \int_{\Omega} \sigma : \nabla v^t \, dx = \frac{1}{2} \int_{\Omega} \sigma : (\nabla v + \nabla v^t) \, dx,$$

where σ^t denotes the transpose of σ, and formula (9.16) yields in this case

$$-\int_{\Omega} \operatorname{div} \sigma \cdot v \, dx = \int_{\Omega} \sigma : \varepsilon(v) \, dx - \int_{\Gamma} \sigma n \cdot v \, ds. \tag{9.17}$$

9.2.3 The Equations of the Problem

The equations for elastic wave propagation result from the fundamental law of mechanics (equilibrium equations), whose validity is somewhat universal, and the constitutive equations that characterize the mechanical behavior on the material and, of course, depend on the nature of this material.

The equilibrium equations are

$$\rho \frac{\partial^2 u}{\partial t^2} - \operatorname{div} \sigma = f, \quad x \in \Omega, \quad t > 0, \tag{9.18}$$

or equivalently

$$\rho \frac{\partial^2 u_i}{\partial t^2} - \sum_{j=1}^{d} \frac{\partial \sigma_{ij}}{\partial x_j} = f_i, \quad x \in \Omega, \quad t > 0, \quad i = 1, ..., d. \tag{9.19}$$

The constitutive law expresses the relationship between the deformation tensor ε and the stress tensor σ, providing an answer to the question: for a given material, what is the distribution of stresses σ produced by a deformation field ε? We shall restrict our case to purely elastic materials. This means that

- The tensor σ depends linearly on ε (contrary, for instance, to elasto-plastic materials which obey to a nonlinear constitutive law). This is consistent with the assumption of small deformations: the linear relation can be seen as a first order approximation, for small ε, of a more general relationship.

- The relationship is instantaneous: the stresses at time t only depend on the deformation at time t (contrary, for instance, to viscoelastic materials, for which the stresses at time t depend on the whole history of the deformation of the materials for times $s \leq t$).

- The relationship is local in space: stresses at point x only depend on the deformations at point x.

Under these assumptions, the deformation-stress relationship is given by the Hooke's law

$$\sigma(x,t) = \mathbf{C}(x)[\varepsilon(u)(x,t)], \quad x \in \Omega, \quad t > 0, \tag{9.20}$$

where $\mathbf{C}(x) \in \mathcal{L}_{ss}^2(\mathbb{R}^d)$ is by definition the elasticity tensor of the material at point x (the dependence of $\mathbf{C}$ with respect to x reflects the possible heterogeneity of the material). In addition to the symmetry conditions (9.4) and (9.5), this tensor has to satisfy a uniform coercivity assumption

$$\forall \varepsilon \in \mathcal{L}_s(\mathbb{R}^d), \quad \forall x \in \Omega, \quad \mathbf{C}(x)\,\varepsilon : \varepsilon \geq \alpha\,|\varepsilon|^2, \tag{9.21}$$

where $\alpha > 0$ is uniform with respect to x.

The equations of linear elastodynamics are made of (9.18) and (9.20). To define a well-posed problem, one must provide initial conditions, which means that the initial displacement field and the initial velocity field are known

$$u(x,0) = u_0(x), \quad \frac{\partial u}{\partial t}(x,0) = u_1(x). \tag{9.22}$$

One must also prescribe boundary conditions along Γ. For our problem, these conditions are

$$\begin{cases} u(x,t) = 0, & x \in \Gamma_0, \quad t > 0, \\[2mm] \sigma(x,t)\,n = g(x,t), & x \in \Gamma_1, \quad t > 0. \end{cases} \tag{9.23}$$

The initial boundary value problem to be solved is made of (9.18), (9.20), (9.22) and (9.23). This constitutes what we shall call the stress-displacement formulation of the problem.

Of course, we can eliminate the unknown σ and obtain a pure displacement formulation in the unknown u. The problem to be solved is then

$$
\begin{cases}
\rho \dfrac{\partial^2 u}{\partial t^2} - \operatorname{div} \mathbf{C}\varepsilon(u) = f, & x \in \Omega, \quad t > 0, \\[2mm]
u(x,0) = u_0(x), \quad \dfrac{\partial u}{\partial t}(x,0) = u_1(x), & x \in \Omega, \\[2mm]
u(x,t) = 0, & x \in \Gamma_0, \quad t > 0, \\[2mm]
\mathbf{C}\varepsilon(u)(x,t)\, n = g(x,t), & x \in \Gamma_1, \quad t > 0.
\end{cases}
\tag{9.24}
$$

This is the problem we shall consider in the following section. Note that this problem belongs to the general class of second-order hyperbolic systems with variable (in space) coefficients.

9.3 Variational Formulation and Weak Solutions

We derive in this section the variational formulation in space of problem (9.24). This is one of the possible weak formulations of the problem. It is adapted to both the mathematical analysis of the problem and its space discretization by finite elements.

Let us assume that u is a sufficiently smooth solution of (9.24) (we avoid in our presentation the details of functional analysis) and consider v a test function which is a function (also smooth) of x only. We have

$$
\int_\Omega \rho \frac{\partial^2 u}{\partial t^2} \cdot v \, dx - \int_\Omega \operatorname{div} \mathbf{C}\varepsilon(u) \cdot v \, dx = \int_\Omega f \cdot v \, dx.
$$

Using Green's formula (9.16), or more precisely (9.17) since $\mathbf{C}\varepsilon(u)$ is symmetric, this can be rewritten

$$
\frac{d^2}{dt^2} \int_\Omega \rho\, u \cdot v \, dx + \int_\Omega \mathbf{C}\varepsilon(u) : \varepsilon(v) \cdot v \, dx - \int_\Gamma \mathbf{C}\varepsilon(u) n \cdot v \, ds = \int_\Omega f \cdot v \, dx.
$$

Next, we assume that v vanishes along Γ_0

$$
v \in V = H^1_{0,\Gamma_0}(\Omega)^d = \{v \in H^1(\Omega)^d \,/\, v|_{\Gamma_0} = 0\}.
\tag{9.25}
$$

Thus, using the boundary condition on Γ_1, we have

$$
\int_\Gamma \mathbf{C}\varepsilon(u) n \cdot v \, d\sigma = \int_{\Gamma_1} g \cdot v \, ds, \quad \forall v \in V.
$$

We have proven, at least formally, that the function $t \in \mathbb{R}^+ \mapsto u(t) \in V$ (we take into account in this way the boundary condition on Γ_0) defined by

$$[u(t)](x) = u(x,t),$$

is solution of the problem,

$$\begin{cases} \dfrac{d^2}{dt^2} \big(u(t), v\big)_H + a\big(u(t), v\big) \; = \; < L(t), v >, \quad \forall \, v \in V, \\[2mm] u(0) = u_0, \quad \dfrac{du}{dt}(0) = u_1 \end{cases} \tag{9.26}$$

where $(\cdot, \cdot)_H$ denotes the scalar product in $H = L^2(\Omega)^d$ defined by

$$(u, v)_H = \int_\Omega \rho \, u \cdot v \, dx, \quad \forall \, (u, v) \in H \times H, \tag{9.27}$$

$a(\cdot, \cdot)$ the bilinear form in V

$$a(u, v) = \int_\Omega \mathbf{C}\varepsilon(u) : \varepsilon(v) \, dx \quad \forall \, (u, v) \in V \times V, \tag{9.28}$$

and $L(t)$ is the linear form in V

$$< L(t), v > = \int_\Omega f(\cdot, t) \cdot v \, dx + \int_{\Gamma_1} g(\cdot, t) \cdot v \, ds, \quad \forall \, v \in V. \tag{9.29}$$

The formulation (9.26) is called the variational formulation (or weak formulation) in space of (9.24). It allows us to define more precisely what we mean by solution of (9.24).

DEFINITION 9.1 *A weak solution of (9.24) is a function*

$$u \in C^1(\mathbb{R}^+; H) \cap C^0(\mathbb{R}^+; V) \tag{9.30}$$

that satisfies (9.26) (the first equation being taken in the sense of distributions).

Problem (9.26) enters the framework of a general class of abstract evolution problems whose theory is well known. More precisely, assume that V and H are two Hilbert spaces with $V \subset H$, H being identified to its dual so that,

$$V \subset H \subset V', \tag{9.31}$$

where V' denotes the dual of V. We denote by $\| \cdot \|$ the norm of V and we assume moreover that

- $(u, v) \rightarrow (u, v)_H$ is the scalar product in H for the Hilbert space structure.

- $(u, v) \rightarrow a(u, v)$ is a continuous symmetric bilinear form on V.

- There exist $\alpha > 0$ and $\nu \geq 0$ such that

$$a(u, u) + \nu \, |u|_H^2 \geq \alpha \, \|u\|^2, \quad \forall \, u \in V. \tag{9.32}$$

THEOREM 9.1 *Assuming that the above assumptions hold and that*

$$(u_0, u_1) \in V \times H, \quad L \, (\equiv f + g) \in L_{loc}^1(\mathbb{R}^+; H) + W_{loc}^{1,1}(\mathbb{R}^+; V'),$$

then, the problem (9.26) admits a unique solution

$$u \in C^1(\mathbb{R}^+; H) \cap C^0(\mathbb{R}^+; V).$$

PROOF This theorem is essentially an adaptation of well known results that can be found for instance in [110]. ⬚

COROLLARY 9.1 *Under the assumptions*

$$(u_0, u_1) \in H^1_{0,\Gamma_0}(\Omega)^d \times L^2(\Omega)^d, \quad f \in L_{loc}^1(\mathbb{R}^+; H), \quad g \in W_{loc}^{1,1}(\mathbb{R}^+; H^{-\frac{1}{2}}(\Gamma_1)^d)$$

problem (9.24) admits a unique solution

$$u \in C^1(\mathbb{R}^+; L^2(\Omega)^d) \cap C^0(\mathbb{R}^+; H^1_{0,\Gamma_0}(\Omega)^d).$$

PROOF It is a simple exercise to check that the abstract theorem can be applied to (9.24). Let us simply mention that the the coercivity inequality (9.32) relies on Korn's inequality [64, 72, 121], which requires some regularity to the boundary Γ_1 [76, 104].

There exists a constant $C(\Omega) > 0$ such that

$$\forall \, u \in H^1(\Omega)^d, \quad \int_\Omega |\nabla u|^2 \, dx \leq C(\Omega) \left[\int_\Omega |u|^2 \, dx + \int_\Omega |\varepsilon(u)|^2 \, dx \right] \tag{9.33}$$

⬚

Weak solutions in the sense of definition 9.1 are often called finite energy solutions, in the sense that the energy

$$E(t) = \frac{1}{2} \left| \frac{du}{dt}(t) \right|_H^2 + \frac{1}{2} \, a\big(u(t), u(t)\big), \tag{9.34}$$

is well defined as a continuous function of time. In the application to elasto-dynamics, this energy is

$$E(t) = \frac{1}{2} \int_\Omega \rho \, \left| \frac{\partial u}{\partial t} \right|^2 \, dx + \frac{1}{2} \int_\Omega \mathbf{C}\varepsilon(u) : \varepsilon(u) \, dx. \tag{9.35}$$

The interest of this quantity lies in a fundamental energy identity

THEOREM 9.2   *Under the assumptions of theorem 9.1, the solution u of problem (9.26) satisfies, for any $T \geq 0$ ($< \cdot, \cdot >$ denoting the duality product between V' and V)*

$$\frac{1}{2} \left| \frac{du}{dt}(T) \right|_H^2 + \frac{1}{2} \, a\big(u(T), u(T)\big) = \frac{1}{2} \, |u_1|_H^2 + \frac{1}{2} \, a(u_0, u_0)$$

$$+ \int_0^T < f(t), \frac{du}{dt} > \, dt - \int_0^T < \frac{dg}{dt}, u > \, dt$$

$$+ < g(T), u(T) > - < g(0), u_0 > .$$

$$(9.36)$$

PROOF   We restrict ourselves to give a simple proof when the solution is assumed more regular, namely when

$$u \in C^2(\mathbb{R}^+; V).$$

In that case, we can write, for any time t

$$\big(\frac{d^2 u}{dt^2}(t), v\big)_H + a\big(u(t), v\big) = < L(t), v >, \quad \forall \, v \in V.$$

Choosing $v = \dfrac{du}{dt}(t) \in V$, we get

$$\big(\frac{d^2 u}{dt^2}(t), \frac{du}{dt}(t)\big)_H + a\big(u(t), \frac{du}{dt}(t)\big) = < L(t), \frac{du}{dt}(t) > .$$

Next we observe that (this is where the symmetry of $a(\cdot, \cdot)$ is important)

$$\big(\frac{d^2 u}{dt^2}(t), \frac{du}{dt}(t)\big)_H = \frac{d}{dt}\left[\frac{1}{2} \left| \frac{du}{dt}(t) \right|_H^2 \right], \quad a\big(u(t), \frac{du}{dt}(t)\big) = \frac{d}{dt}\left[\frac{1}{2} a\big(u(t), u(t)\big) \right].$$

Therefore, $E(t)$ being defined by (9.34), we have

$$\frac{dE}{dt}(t) = < L(t), \frac{du}{dt}(t) > = \big(f(t), \frac{du}{dt}(t)\big)_H + < g(t), \frac{du}{dt}(t) > .$$

One concludes by integrating the above equality in time between 0 and T. □

The identity (9.36) is sometimes called the energy conservation identity. Indeed, if $0 < t_1 < t_2$ are such that the source terms f and g vanish in the interval $[t_1, t_2]$, one deduces from (9.36) that

$$E(t) \text{ is constant in } [t_1, t_2].$$

More generally, thanks to standard techniques (Gronwall's lemma), it is easy to deduce from (9.36) a priori estimates that constitute stability results for the problem. Such estimates are the basis of an existence proof (see for instance [110]).

9.4 Plane Wave Propagation in Homogeneous Media

We recall in this section classical results about the propagation of plane waves in homogeneous media. In this paragraph, we assume that the domain of propagation Ω is the whole space $\mathbb{R}^d$ and suppose that the density ρ and the tensor $\mathbf{C}$ are constant. Plane waves are particular solutions of elastodynamics equations of the form,

$$u(x,t) = \mathbf{d} \exp \imath (\mathbf{k} \cdot x - \omega t) \tag{9.37}$$

where by definition,

- $\mathbf{d} \in \mathbb{R}^d$ is the displacement (or polarization) vector.

- $\mathbf{k} \in \mathbb{R}^d$ is the wave vector.

- $\omega \in \mathbb{R}$ is the pulsation.

Physically, (9.37) represents a plane wave (at a given time t, the sets of $\mathbb{R}^d$ along which u is constant are hyperplanes orthogonal to $\mathbf{k}$) that propagates in the direction $\nu = \mathbf{k}/|\mathbf{k}|$ at the velocity (called phase velocity)

$$V = \frac{\omega}{|\mathbf{k}|}, \quad (\vec{V} = \frac{\omega}{|\mathbf{k}|} \nu),$$

$|\mathbf{k}|$ being the wave number and $\lambda = 2\pi/|\mathbf{k}|$ the wavelength. It is straightforward to see that u defined by (9.37) is solution of

$$\rho \frac{\partial^2 u}{\partial t^2} - \operatorname{div} \mathbf{C}\varepsilon(u) = 0, \tag{9.38}$$

if and only if $\mathbf{d}, \mathbf{k}$ and ω are related by

$$\mathbf{C}(\mathbf{k})\, \mathbf{d} = \rho\, \omega^2\, \mathbf{d}, \tag{9.39}$$

where $\mathbf{C}(\mathbf{k})$ is the Christoffel tensor, namely the $d \times d$ symmetric matrix defined by

$$\mathbf{C}_{ij}(\mathbf{k}) = \sum_{p,q=1}^{d} \mathbf{C}_{ipjq}\, k_p\, k_q. \tag{9.40}$$

The symmetry of $\mathbf{C}(\mathbf{k})$ follows from the symmetry property (9.4) of $\mathbf{C}$. Moreover this matrix is positive definite. In fact, introducing $\mathbf{k} \otimes \xi \in \mathcal{L}(\mathbb{R}^d)$ such that

$$(\mathbf{k} \otimes \xi)_{ij} = \mathbf{k}_i \, \xi_j, \quad \forall \, 1 \leq i, j \leq d, \tag{9.41}$$

one easily computes

$$\mathbf{C}(\mathbf{k})\xi \cdot \xi = \mathbf{C} \, \mathbf{k} \otimes \xi : \mathbf{k} \otimes \xi, \quad \forall \, \xi \in \mathbb{R}^d. \tag{9.42}$$

Therefore, using (9.7), one obtains

$$\mathbf{C}(\mathbf{k})\xi \cdot \xi \geq \alpha \, |\mathbf{k} \otimes \xi|^2 = \alpha \, |\mathbf{k}|^2 \, |\xi|^2, \quad \forall \, \xi \in \mathbb{R}^d. \tag{9.43}$$

As a consequence, the matrix $\mathbf{C}(\mathbf{k})$ has only real positive eigenvalues (which also proves the hyperbolicity of system (9.24)) and is diagonalizable in an orthonormal basis of $\mathbb{R}^d$. Since $\mathbf{C}(\mathbf{k})$ is a homogeneous polynomial of degree 2 with respect to $\mathbf{k}$, its eigenvalues share the same homogeneity. In other words, there exist d smooth functions defined on the unit sphere of $\mathbb{R}^d$,

$$V_j(\nu) : S^{d-1} \to \mathbb{R}^+, \quad S^{d-1} = \{\nu \in \mathbb{R}^d \, / \, |\nu| = 1\},$$

such that, if $\mathbf{k} = |\mathbf{k}| \, \nu$, the spectrum of $\mathbf{C}(\mathbf{k})$ is given by,

$$\mathrm{sp} \, \mathbf{C}(\mathbf{k}) = \{ \, \rho \, |\mathbf{k}|^2 \, V_j(\nu)^2, 1 \leq j \leq d \, \}. \tag{9.44}$$

Thus, for each direction $\nu \in S^{d-1}$, there exists an orthonormal basis $\{p_j(\nu), 1 \leq j \leq d\}$ of $\mathbb{R}^d$ (the polarization vectors) such that

$$\mathbf{C}(\mathbf{k}) \, p_j(\nu) = \rho \, |\mathbf{k}|^2 \, V_j(\nu)^2 \, p_j(\nu), \quad \nu = \frac{k}{|k|}. \tag{9.45}$$

The relationship (9.39) means that $\mathbf{d}$ is an eigenvector of $\mathbf{C}(\mathbf{k})$ associated to the eigenvalue $\rho \, \omega^2$. In other words, the function u given by (9.37) is a non trivial solution of (9.38) if and only if, there exists $1 \leq j \leq d$ such that

$$\mathbf{k} = |\mathbf{k}| \, \nu \Longrightarrow \omega = \pm \, \omega_j(\mathbf{k}) = |\mathbf{k}| \, V_j(\nu), \quad \text{and } \mathbf{d} \text{ is colinear to } p_j(\nu). \tag{9.46}$$

Let $D(\omega, \mathbf{k}) = \det [\mathbf{C}(\mathbf{k}) - \rho \, \omega^2 \, I]$. By definition, the dispersion relation of the elastic medium is

$$D(\omega, \mathbf{k}) \equiv \rho^d \, \prod_{j=1}^{d} (\omega_j(\mathbf{k})^2 - \omega^2) = 0. \tag{9.47}$$

and the dispersion relation of the jth plane wave is given by $\omega = \omega_j(\mathbf{k})$.

REMARK 9.2 *To be more rigorous, (9.46) is valid in the case where $\rho \, |\mathbf{k}|^2 \, V_j(\nu)^2$ is a simple eigenvalue of $\mathbf{C}(\mathbf{k})$. In the case of a multiple eigenvalue, the sentence "$\mathbf{d}$ is colinear to $p_j(\nu)$" must be replaced by "$\mathbf{d}$ belongs to the eigenspace of $\mathbf{C}(\mathbf{k})$ associated to $\rho \, |\mathbf{k}|^2 \, V_j(\nu)^2$".*

The discussion above means that, up to the change of the sense of propagation (the choice of the $\pm$ sign), in each direction, there exist d plane waves propagating in the direction ν, with respective phase velocities $V_j(\nu)$ and polarizations $p_j(\nu)$. Since the phase velocities only depend on ν but not on $|\mathbf{k}|$, one says that elastic media may be anisotropic but are non dispersive.

Of course, a particular case of an anisotropic media is an isotropic medium ! This corresponds to the case where $\mathbf{C}$ is given by (9.9). The positive coefficients λ and μ are the Lamé coefficients of the medium. In this case, it is easy to see that equation (9.38) can be rewritten

$$\rho \frac{\partial^2 u}{\partial t^2} - (\lambda + \mu) \nabla(\mathrm{div}\ u) - \mu\, \Delta u = 0, \tag{9.48}$$

and that the Christoffel tensor $\mathbf{C}(\mathbf{k})$ is given by

$$\mathbf{C}(\mathbf{k})\, \mathbf{d} = (\lambda + \mu)\, (\mathbf{k} \cdot \mathbf{d})\, \mathbf{k} + \mu\, |\mathbf{k}|^2\, \mathbf{d}, \quad \forall\, \mathbf{d} \in \mathbb{R}^d. \tag{9.49}$$

In this case, $\mathbf{C}(\mathbf{k})$ has only two eigenvalues. One is simple and the second one of multiplicity $d - 1$. In fact, one sees immediately that

$$\begin{cases} \mathbf{C}(\mathbf{k})\, \mathbf{k} = (\lambda + 2\mu)\, |\mathbf{k}|^2\, \mathbf{k} = \rho\, V_P^2\, |\mathbf{k}|^2\, \mathbf{k} \quad \text{where } V_P = \left(\frac{\lambda + 2\mu}{\rho}\right)^{\frac{1}{2}}, \\[2mm] \mathbf{k} \cdot \mathbf{d} = 0 \Longrightarrow \mathbf{C}(\mathbf{k})\, \mathbf{d} = \mu\, |\mathbf{k}|^2\, \mathbf{d} = \rho\, V_S^2\, |\mathbf{k}|^2\, u \quad \text{where } V_S = \left(\frac{\mu}{\rho}\right)^{\frac{1}{2}}. \end{cases} \tag{9.50}$$

This means that in a homogeneous isotropic medium, there are two types of plane waves

- The P waves or pressure waves, which are the fastest ones. They propagate at velocity V_P and their polarization is colinear to the propagation direction $\nu = k/|k|$.

- The S waves or shear waves, which are the slowest ones. They propagate at velocity V_S and their polarization is orthogonal to the propagation direction ν.

Note that, this time, the phase velocities, namely V_P and V_S, are independent of the propagation direction ν, which justifies the name of isotropic media.

Another important notion associated to plane wave propagation is the notion of group velocity. More precisely, if the plane wave obeys the dispersion relation

$$\omega = \omega_j(\mathbf{k}),$$

the corresponding group velocity is given by

$$V_j^g(\mathbf{k}) = \nabla \omega_j(\mathbf{k}). \tag{9.51}$$

Since $\omega_j(\mathbf{k})$ is homogeneous of degree 1,

$$\mathbf{k} = |\mathbf{k}| \, \nu \Longrightarrow \vec{V}_j^g(\mathbf{k}) = \vec{V}_j^g(\nu). \tag{9.52}$$

The notions of phase velocity $\vec{V}_j(\nu) = V_j(\nu) \, \nu$ and group velocity $\vec{V}_j^g(\nu)$ coincide only in isotropic media. They do not for a general anisotropic media. Let us illustrate this in 2D with the notion of slowness curves and wavefronts.

Slowness curves and wavefronts.   By homogeneity, if we set

$$\widehat{D}(\mathbf{s}) = D(1, \mathbf{s}) \equiv \rho^d \prod_{j=1}^{d} (\, \omega_j(\mathbf{s})^2 - 1),$$

for $\omega \neq 0$, (9.47) is equivalent to

$$\widehat{D}(\mathbf{s}) = 0 \quad \text{where } \mathbf{s} = \frac{\mathbf{k}}{\omega} \text{ is the so-called slowness vector.} \tag{9.53}$$

The *slowness curves* are defined in the s-plane by equation (9.53). Of course, this corresponds to the union of two curves of respective equations

$$\mathcal{C}_j := \{ \mathbf{s} \in \mathbb{R}^2 \, / \quad \omega_j(\mathbf{s}) = 1 \}, \quad j = 1, 2.$$

Obviously, since $\omega_j(|\mathbf{s}| \, \nu) = |\mathbf{s}| \, V_j(\nu)$, $\mathcal{C}_j$ admits a polar representation: More precisely, the intersection of $\mathcal{C}_j$ with the half line $\mathbf{s} = \rho \, \nu, \rho > 0$ is the slowness vector $\mathbf{s} = \nu / V_j(\nu)$. Clearly, the curve $\mathcal{C}_j$ is a closed curve, symmetric with respect to the origin, and the set,

$$\mathcal{S}_j := \{ \mathbf{s} \in \mathbb{R}^2 \, / \quad \omega_j(\mathbf{s}) \leq 1 \},$$

is the connected component of $\mathbb{R}^2 \setminus \mathcal{C}_j$ which contains the origin (see Figure 9.1 for a general representation of a slowness curve, the slowness vector and the group velocity in an anisotropic medium).

Since the gradient of a function is normal to the level lines of this function, the orientation of the group velocity of the j-th plane wave propagating in direction ν is given by the normal vector to $(\mathcal{C}_j)$ at the point $\mathbf{s} = \nu / V_j(\nu)$, outgoing with respect to $(\mathcal{S}_j)$.

We shall define the *wave fronts* (we do not necessarily use here a standard definition, for instance in [11] these are called *ray curves*) as the curves $\mathcal{C}_j^g$ described by the extremities of the group velocity vectors $V_j^g(\nu)$ when ν describes the unit circle. There exists a standard geometrical construction of the so-defined wavefronts from the slowness curves: the wavefront is the polar reciprocal curve of the slowness curve (see remark 9.3 and [16] for more details).

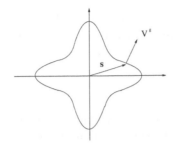

FIGURE 9.1: Slowness curve, slowness vector and group velocity in an anisotropic medium.

REMARK 9.3 *The construction of the polar reciprocal of a closed curve $\mathcal{C}$ in the plane $\mathbf{s} \in \mathbb{R}^2$ is as follows. For any $\nu \in S^1$, we consider the set:*

$$\mathcal{E}(\nu, \mathcal{C}) = \{\mathbf{s} \in \mathbb{R}^2 \ / \ \text{the tangent to } \mathcal{C} \text{ at point } \mathbf{s} \text{ is orthogonal to } \nu.\}$$

Let $\mathcal{P}(\nu, \mathcal{C}) = \{\rho_j(\nu) \ \nu\}$ be the finite set of the orthogonal projections of $\mathcal{E}(\nu, \mathcal{C})$ on the straight line directed by ν. The polar reciprocal $\mathcal{C}^*$ is the curves described by the points in $\mathcal{P}^*(\nu, \mathcal{C}) = \{\rho_j(\nu)^{-1} \ \nu\}$ when ν describes the unit circle.

The reader may check that when $\mathcal{C}$ is a closed curve delimiting a bounded strictly convex set, each $\mathcal{P}(\nu, \mathcal{C})$ is made of a unique point and $\mathcal{C}^*$ is also a closed curve delimiting a bounded strictly convex set.

In general, $\mathcal{P}(\nu, \mathcal{C})$ may contain several points. As a consequence, even if $\mathcal{C}$ admits a polar representation. it is not necessarily the case of $\mathcal{C}^*$.

The concept of wave fronts is of interest since it allows us predict the localization of the singularities of solutions (which often corresponds to the localization of most of the energy), in particular of solutions emitted from a point source (of the fundamental solution in particular). It would be difficult to give a very precise mathematical statement of what we mean here. This would require mathematical developments on microlocal analysis and pseudodifferential operators that are far beyond the scope of this book (see for instance [136] or [96]). Let us however give a very simple theorem in this direction, that can be found using this kind of technique:

THEOREM 9.1 *Let (u_0, u_1) be the (compactly supported) Cauchy data for the Cauchy problem (i.e., the domain Ω is the whole space $\mathbb{R}^d$) associated with (9.38). We assume that (u_0, u_1) are C^∞ except at the origin. Then the function $x \to u(x, t)$ is also of class C^∞ except (maybe) at the points of the curves $t \ C_j^g$.*

More generally, if S_0 is the set of points where (u_0, u_1) are not C^∞, the set where $x \to u(x,t)$ is (a priori) not C^∞ is the union of the $S_0 + t \, C_j^g$.

In isotropic media, the two slowness curves are two circles of respective radius V_P^{-1} and V_S^{-1} and the corresponding waves fronts are the two circles of respective radius V_P and V_S (see Figure 9.2).

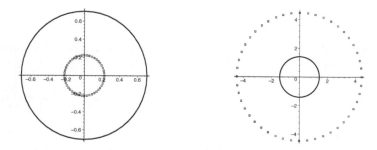

FIGURE 9.2: Slowness diagrams and the waves fronts in an isotropic medium.

In general, in anisotropic media, phase and group velocities associated to a same direction ν are not colinear and the wavefronts do not necessarily have a polar representation. Let us illustrate this with orthotropic media in 2D, *i.e.*, the particular class of anisotropic media associated to elasticity tensors of the form (9.11) (cf. remark (9.1)). We illustrate in Figure 9.3 an example of slowness diagrams and waves fronts for an othotropic medium ($c_{11} = 4$, $c_{22} = 20$, $c_{33} = 2$, $c_{12} = 7.5$).

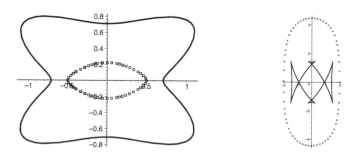

FIGURE 9.3: Slowness diagrams and waves fronts in othotropic medium.

9.5   Finite Propagation Velocity

One of the main characteristic properties of hyperbolic systems is that their solutions propagate with finite velocity in a sense which is close to the physical intuition of propagation. Let us state a more precise result in the case of elastic waves. We assume in this section that the domain of propagation is the whole space $\Omega = \mathbb{R}^d$ but allow the medium to be heterogeneous: ρ and $\mathbf{C}$ depend on x.

To each point $x \in \mathbb{R}^d$ and $\nu \in S^{d-1}$, we can associate the (local) Christoffel tensor $\mathbf{C}(x, \nu)$, defined as in an homogeneous media whose elastic parameters would correspond to the ones of the material at point x,

$$\mathbf{C}_{ij}(x, \nu) = \sum_{p,q=1}^{d} \mathbf{C}_{ipjq}(x) \, \nu_p \, \nu_q. \tag{9.54}$$

Then, we define the quantities (the notation $V_P(\cdot, \cdot)$ is chosen here by analogy with the P-wave velocity in isotropic media),

$$V_p(x, \nu) = \left(\sup_{v \neq 0} \frac{\mathbf{C}(x, \nu)v.v}{\rho \, |v|^2} \right)^{1/2}, \quad V_p^+(\nu) = \sup_{x \in \mathbb{R}^d} \{ V_p(x, \nu) \}. \tag{9.55}$$

We consider the mobile half-space with velocity $V \in \mathbb{R}_+^*$,

$$\mathcal{H}_\nu(V, t) = \{ x \in \mathbb{R}^d \ / \ x.\nu \leq Vt \}. \tag{9.56}$$

Finite propagation velocity means in particular that if the data of the problem have compact support in space, then the support of the solution remains compact for any time. The more precise result is the following,

THEOREM 9.3   *Let K be a compact subset or $\mathbb{R}^d$ such that,*

$$supp \ u_0 \cup supp \ u_1 \cup supp \ \sigma_0 \cup supp \ f(., t) \subset K,$$

then, for any time $t > 0$,

$$supp \ u(., t) \subset K + \mathcal{B}(t), \quad \forall \, 0 \leq t \leq T,$$

where $\mathcal{B}(t) = \bigcap_{\nu \in S^{d-1}} \mathcal{H}_\nu(V_p^+(\nu), t)$.

PROOF   For simplicity, we restrict ourselves to give the proof in the case of sufficiently smooth data (in which case the solution u is itself smooth enough to justify the forthcoming manipulations) and where $K = B(0, R)$ the ball of center 0 and radius R. Let $\nu \in S^{d-1}$, we introduce the mobile half-space

$$\Omega_\nu^t = \big(B(0, R) + E_\nu(V, t)\big)^c = \{x \in \mathbb{R}^d \ / \ x.\nu > R + Vt\},$$

with $\Gamma_\nu^t = \partial \Omega_\nu^t$ the boundary of Ω_ν^t (see Figure 9.4). Note that by construction

$$\begin{cases} u_0(x) = u_1(x) = 0, \ \sigma_0(x) = 0, \ \forall x \in \Omega_\nu^0, \\ f(x, t) = 0, \ \forall t > 0, \ \forall x \in \Omega_\nu^t. \end{cases} \tag{9.57}$$

The idea of the proof is to determine V (large enough) so that the solution

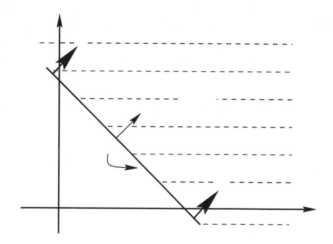

FIGURE 9.4: The mobile half-space.

$u(., t)$ remains 0 in Ω_ν^t for any $t > 0$.

We define the energy of the solution contained in the domain Ω_ν^t at time t as the quantity

$$E_\nu(t) = \int_{\Omega_\nu^t} e(x, t) \, dx,$$

where $e(x, t)$ is the energy density defined by

$$e(x, t) = \frac{1}{2} \Big[\rho |\partial_t u|^2 + \mathbf{C} \, \varepsilon(u) : \varepsilon(u) \Big]. \tag{9.58}$$

We multiply (in the sense of the inner product of $\mathbb{R}^d$) the equation

$$\rho \, \partial_{tt}^2 u - \mathrm{div} \, \big(\mathbf{C} \, \varepsilon(u)\big) = f, \tag{9.59}$$

by $\partial_t u$ and integrate over Ω_ν^t. Using Green's formula, one obtains, since $f(\cdot, t)$ vanishes in Ω_ν^t,

$$\int_{\Omega_\nu^t} \partial_t e \, dx + \int_{\Gamma_\nu^t} (\mathbf{C}\,\varepsilon(u))\nu \cdot \partial_t u \, d\sigma = 0,$$

Using the identity

$$\frac{d}{dt} \int_{\Omega_\nu^t} e(x, t) \, dx = \int_{\Omega_\nu^t} \frac{\partial e}{\partial t}(x, t) \, dx - V \int_{\Gamma_\nu^t} e(x, t) \, dx, \qquad (9.60)$$

we deduce that

$$\frac{dE_\nu}{dt}(t) + \frac{1}{2} \int \phi_\nu(x, t) \, d\sigma = 0,$$

where

$$\phi_\nu(x, t) = V\left[\rho(x(\,|\partial_t u|^2 + \mathbf{C}(x)\,\varepsilon(u) : \varepsilon(u)\,] + 2(\mathbf{C}(x)\,\varepsilon(u))\nu \cdot \partial_t u. \quad (9.61)$$

Next, we prove the inequality, for any $v \in \mathbb{R}^d$ and $\varepsilon \in \mathcal{L}_s(\mathbb{R}^d)$:

$$(\mathbf{C}(x)\varepsilon)\,\nu.v \le (\mathbf{C}(x)\,\varepsilon : \varepsilon)^{\frac{1}{2}} (\mathbf{C}(x, \nu)\,v.v)^{\frac{1}{2}}, \qquad (9.62)$$

For this, we write

$$(\mathbf{C}(x)\varepsilon)\,\nu \cdot v = \mathbf{C}_{ijkl}(x)\,\varepsilon_{kl}\,\nu_j\,v_i = \mathbf{C}_{ijkl}(x)\,\varepsilon_{kl}\,(\nu \otimes v)_{ij} = \mathbf{C}(x)\,\varepsilon : \nu \otimes v.$$

As $(\sigma, \tau) \mapsto \mathbf{C}(x)\,\sigma : \tau$ defines a scalar product in $\mathcal{L}_s(\mathbb{R}^d)$, we can use Cauchy-Schwarz inequality and obtain

$$\mathbf{C}(x)\,\varepsilon : \nu \otimes v \le (\mathbf{C}(x)\,\varepsilon : \varepsilon)^{\frac{1}{2}} (\mathbf{C}(x)\,\nu \otimes v : \nu \otimes v)^{\frac{1}{2}}.$$

To get (9.62), it suffices to remark that

$$\begin{vmatrix} \mathbf{C}(x)\,\nu \otimes v : \nu \otimes v = \sum_{i,j,k,l=1}^{d} \mathbf{C}_{ijkl}(x)\,(\nu \otimes v)_{kl}\,(\nu \otimes v)_{ij} \\[2mm]
= \sum_{i,j,k,l=1}^{d} \mathbf{C}_{ijkl}(x)\,\nu_k\,v_l\,\nu_i\,v_j \\[2mm]
= \sum_{i,j,k,l=1}^{d} (\mathbf{C}_{ijkl}(x)\,\nu_i\,\nu_k)\,v_l\,v_j \\[2mm]
= \sum_{i,j,k,l=1}^{d} (\mathbf{C}_{lkji}(x)\,\nu_i\,\nu_k)\,v_j\,v_l \\[2mm]
= \sum_{j,l=1}^{d} \mathbf{C}_{lj}(x, \nu)\,v_j\,v_l = \mathbf{C}(x, \nu)\,v \cdot v. \end{vmatrix}$$

Using (9.62) in (9.61), we get

$$\phi_\nu(x,t) \geq \rho(x) V |\partial_t u|^2 + V \left(\mathbf{C}(x) \, \varepsilon : \varepsilon\right) - 2\big(\mathbf{C}(x) \, \varepsilon : \varepsilon\big)^{\frac{1}{2}} \left(\mathbf{C}(x,\nu)\partial_t u \cdot \partial_t u\right)^{\frac{1}{2}},$$

which implies, thanks to (9.55)

$$\phi_\nu(x,t) \geq \rho(x) V |\partial_t u|^2 + V \left(\mathbf{C}(x) \, \varepsilon : \varepsilon\right) - 2 \, \rho(x)^{\frac{1}{2}} \, V_P^+(\nu) \left(\mathbf{C}(x) \, \varepsilon : \varepsilon\right)^{\frac{1}{2}} |\partial_t u|.$$

The right hand side of the above inequality is a quadratic form in

$$|\partial_t u| \quad \text{and} \quad \left(\mathbf{C}(x) \, \varepsilon : \varepsilon\right)^{\frac{1}{2}}$$

whose discriminant is given by

$$D(x) = \rho(x) \left(\, V_P^+(\nu)^2 - V^2 \, \right).$$

Therefore, if we choose $V = V_P^+(\nu)$. the quantity $\phi_\nu \geq 0$ is positive. In particular

$$E_\nu(t) \leq E_\nu(0) = 0, \quad \text{(by construction)}.$$

Thus, $u(x,t) = 0$ in $\Omega_\nu^t \implies supp \, u(.,t) \subset \{\Omega_\nu^t\}^c = B(0,R) + \mathcal{H}_\nu(V_p^+(\nu),t).$

Since this is valid for any $\nu \in S^{d-1}$, we have finally

$$supp \, u(.,t) \subset \bigcap_{\nu \in S^{d-1}} \{\Omega_\nu^t\}^c = B(0,R) + \mathcal{B}(t).$$

$\square$

REMARK 9.4 *In dimension 2 and in a homogeneous medium, there is a clear geometrical link between the set* $\mathcal{B}(t) = t \, \mathcal{B}(0)$ *and the two wavefronts* $\mathcal{C}^*1$ *and* $\mathcal{C}^*2$*. More precisely, B(0) is the convex envelope of* $\mathcal{C}^*1 \cap \mathcal{C}^*2$*.*

Note: The references for this chapter and for all chapters in Part III are given at the end of Chapter 14.

Chapter 10

Finite Element Methods with Continuous Displacement

P. Joly, POEMS Project team, INRIA-Rocquencourt, France,
Patrick.JOLY@inria.fr

We develop in this chapter the very well-known basic principles of the standard Lagrange finite element approximation of elastodynamics equations in their form (9.24), i.e., as a second order hyperbolic system in the displacement field. In fact, this method applies to any abstract second order variational evolution problem of the form (9.26) whose a particular case is the variational (or weak) formulation of the second order elastodynamics system (9.24). That is why we choose a general abstract presentation in the next paragraphs, assuming that the general assumptions (9.31) to (9.32) hold. Of course, the space V is supposed to be of infinite dimension, as it is the case of the space V defined by (9.25). For the simplicity of the exposition, we shall use some stronger hypotheses that are needed (see theorem 9.1), namely,

$$\begin{cases} L(t) \in C^1(\mathbb{R}^+; V') \\ \\ (9.32) \text{ holds with } \nu = 0, \quad \text{i.e., } a \text{ is coercive.} \end{cases} \qquad (10.1)$$

10.1 Galerkin Approximation of Abstract Second Order Variational Evolution Problems

Principle of the method. This method is linked to the existence of $\{V_h, h > 0\}$ finite dimensional subspaces of V where h is an approximation parameter devoted to tend to zero. From the theoretical point of view, it suffices that h describes a subset of $\mathbb{R}^+$ having 0 as accumulation point. In practice, for finite element methods, h will be the stepsize of a spatial mesh of the computational domain Ω. The important property from the theoretical point of view is a *consistency* condition meaning that V_h approaches V as h goes to 0.

$$\forall\, u \in V, \quad \lim_{h \to 0} \inf_{v_h \in V_h} \|u - v_h\| = 0. \tag{10.2}$$

This implies of course that the dimension of V_h goes to infinity when h goes to 0,

$$N_h = \dim V_h \to +\infty \quad \text{when } h \to 0. \tag{10.3}$$

In practice, such a property will appear as a consequence of a stronger assumption,

$$
\begin{vmatrix}
\exists\, W_\ell \text{ (Hilbert spaces)}, \ 0 \le \ell \le L, \text{ and } \delta_\ell(h) : \mathbb{R}^+ \to \mathbb{R}^+ \ 0 \le \ell \le L \ / \\[2mm]
W_0 = V, \quad \forall\, \ell \le L - 1, \quad W_{\ell+1} \subset W_\ell, \qquad W_\ell \text{ is dense in } V, \\[2mm]
\delta_0(h) = 1, \quad \delta_{\ell+1}(h) = o\big(\delta_\ell(h)\big) \text{ when } h \to 0 \quad \text{and for any } \ell \ge 1, \\[2mm]
\forall\, u \in W_\ell, \quad \inf_{v_h \in V_h} \|u - v_h\| \le \delta_\ell(h)\, \|u\|_{W_\ell}
\end{vmatrix}
\tag{10.4}
$$

REMARK 10.1 *In practice, when V is a functional space of H^1 type, the W_ℓ are functional spaces of more and more regular functions (typically higher order Sobolev spaces). A typical example of functions δ_ℓ are $\delta_\ell(h) = C_\ell\, h^\ell$.*

The semi-discrete problem (semi-discretization refers to the fact that the time remains continuous) is,

$$
\begin{cases}
\dfrac{d^2}{dt^2}\big(u_h(t), v_h\big) + a\big(u_h(t), v_h\big) \ = \ < L(t), v_h >, \quad \forall\, v_h \in V_h, \\[4mm]
u_h(0) = u_{0,h}, \quad \dfrac{du}{dt}(0) = u_{1,h}.
\end{cases}
\tag{10.5}
$$

From the practical point of view, this results into an ordinary differential system. Introducing a basis $\{w_j, 1 \le j \le N_h\}$ of V_h and decomposing the solution $u_h(t)$ as,

$$u_h(t) = \sum_{j=1}^{N_h} u_j(t)\, w_j, \tag{10.6}$$

one obtains a differential system where the unknown is the so-called vector of degrees of freedom,

$$t \longrightarrow U_h(t) = \big(u_j(t)\big)_{1 \le j \le N_h} \in \mathbb{R}^{N_h}. \tag{10.7}$$

$$\begin{cases} \mathbf{M}_h \dfrac{d^2 U_h}{dt^2}(t) + \mathbf{A}_h U_h(t) = F_h(t), \\[2mm] U_h(0) = U_{0,h}, \quad \dfrac{dU_h}{dt}(0) = U_{1,h}, \end{cases} \tag{10.8}$$

where $\mathbf{M}_h$, $\mathbf{A}_h$ are respectively the *mass matrix* and the *stifness matrix*,

$$\mathbf{M}_{i,j} = (w_j, w_i), \quad \mathbf{A}_{i,j} = a(w_j, w_i). \tag{10.9}$$

The right hand side is given by,

$$F_j(t) = < L(t), w_j > \in C^1(0, T), \tag{10.10}$$

while $U_{0,h}$ and $U_{1,h}$ are the vectors of the degrees of freedom of $u_{0,h}$ and $u_{1,h}$.

From the assumptions about the bilinear form $a(\cdot, \cdot)$ (in particular from (9.32) with $\nu = 0$) we infer that,

The matrices $\mathbf{M}_h$ and $\mathbf{A}_h$ are positive definite symmetric. $\qquad (10.11)$

Using classical theorems for ordinary differential equations it is then easy to prove the following existence and uniqueness theorem,

THEOREM 10.1 *The problem (10.5) admits a unique solution,*

$$u_h \in C^1(\mathbb{R}^+; V_h). \tag{10.12}$$

REMARK 10.2 *Although the solution $U_h(t)$ and the meaning of the degrees of freedom depend on the choice of the basis $\{w_j, 1 \le j \le N_h\}$, $u_h(t)$ does not. In the same way the more precise structure of $\mathbf{M}_h$ and $\mathbf{A}_h$ depends on $\{w_j, 1 \le j \le N_h\}$. We shall come back later on this point.*

Stability and convergence analysis. A first fundamental remark is that the stability of Galerkin approximations for the space discretization of wave-like equations is an automatic consequence of the variational nature of the approximation process. Proving stability means being able to establish *a priori* estimates, in appropriate norms, of the approximate solution which are independent of the approximation parameter h. This is of course a necessary condition for proving convergence: convergence implies in particular boundedness.

Galerkin approximations automatically provide stability in the energy sense. Indeed, exploiting the symmetry of $a(\cdot, \cdot)$, one can prove the following result (the discrete energy identity),

THEOREM 10.2 *The solution u_h of (10.5) satisfies the energy identity,*

$$\frac{d}{dt} E_h(t) = < L(t), \frac{du_h}{dt}(t) >, \tag{10.13}$$

with

$$E_h(t) = \frac{1}{2}\{ \ |\frac{du_h}{dt}(t)|^2 + a\big(u_h(t), u_h(t)\big) \ \}. \tag{10.14}$$

PROOF   We simply use the fact that, for any v_h in V_h and any $t \geq 0$,

$$\big(\frac{d^2 u_h}{dt^2}(t), v_h\big)_H + a\big(u_h(t), v_h\big) \ = \ < L(t), v_h > .$$

Then, we choose $v_h = u_h(t)$. The conclusion follows immediately.   ☐

A priori estimates easily follow from standard Gronwall-like techniques. For example we demonstrate the following,

COROLLARY 10.1   *Assume that,*

$$L(t) \in C^1(\mathbb{R}^+; H), \tag{10.15}$$

then, one has the a priori estimates, for any positive T,

$$\|\frac{du_h}{dt}\|_{L^\infty(0,T;H)} \leq \big(\ |u_{1,h}|^2 + a(u_{0,h}, u_{0,h}) \ \big)^{\frac{1}{2}} + \|L\|_{L^1(0,T;H)},$$

$$\|u_h\|_{L^\infty(0,T;V)} \leq \frac{1}{\sqrt{\alpha}} \left\{ \ \big(\ |u_{1,h}|^2 + a(u_{0,h}, u_{0,h}) \ \big)^{\frac{1}{2}} + \|L\|_{L^1(0,T;H)} \ \right\}. \tag{10.16}$$

PROOF   Thanks to the assumption (10.15) and the positivity of $a(\cdot, \cdot)$, we can write,

$$| < L(t), \frac{du_h}{dt}(t) > | \leq |L(t)| \ |\frac{du_h}{dt}(t)| \leq \sqrt{2E_h(t)} \ |L(t)|.$$

Therefore, we deduce from the differential inequation,

$$\frac{d}{dt} E_h(t) \leq \sqrt{2E_h(t)} \ |L(t)|,$$

which leads after time integration to,

$$\sqrt{E_h(t)} \leq \sqrt{E_h(0)} + \int_0^t |L(t)| \ ds.$$

The *a priori* estimates easily follow from the coercivity property, which yields,

$$\sqrt{E_h(t)} \ \geq \ \frac{\sqrt{2}}{2} |\frac{du_h}{dt}(t)| \quad \text{and} \quad \sqrt{E_h(t)} \ \geq \ \frac{\sqrt{2\alpha}}{2} \|u_h(t)\|.$$

☐

The *a priori* estimates can be exploited, using standard compacity methods, to show the weak convergence (for instance in the $H^1(0,T;H) \cap L^2(0,T;V)$ topology) of u_h to u, provided that the approximation property (10.2) holds.

We explain below how to obtain strong convergence and error estimates. A classical approach consists in combining the use of an appropriate "elliptic projection" (which allows one to apply standard approximation results for elliptic problems) with energy estimates (see for instance [62]). We introduce the elliptic projection operator,

$$\Pi_h : u \in V \longrightarrow \Pi_h u \in V_h,$$

which is nothing but the orthogonal projection on V_h for the scalar product $a(\cdot, \cdot)$, namely,

$$a(u - \Pi_h u, v_h) = 0, \quad \forall v_h \in V_h. \tag{10.17}$$

Of course, $\Pi_h u$ tends to u when h goes to 0, and we have the more precise estimate (simply notice that $a(u - \Pi_h u, u - \Pi_h u) \le a(u - v_h, u - v_h)$ for any v_h in V_h)

$$\forall u \in V, \quad \|u - \Pi_h u\| \le \left(\frac{M}{\alpha}\right)^{\frac{1}{2}} \inf_{v_h \in V_h} \|u - v_h\|. \tag{10.18}$$

In particular, for $\ell \ge 1$, using (10.4),

$$\forall u \in W_\ell, \quad \|u - \Pi_h u\| \le \left(\frac{M}{\alpha}\right)^{\frac{1}{2}} \delta_\ell(h) \|u\|_{W_l}. \tag{10.19}$$

Moreover, one is able to get better convergence in the H norm using so-called duality techniques. This relies upon a property of the following map,

$$\Phi : H \to V, \tag{10.20}$$

where, for g given in H, $\Phi(g)$ is the unique solution of the problem ($\Phi(g)$ is well defined thanks to Lax-Milgram's lemma),

$$a(\Phi(g), v) = (g, v), \quad \forall v \in V. \tag{10.21}$$

If we make the following assumption (see also (10.54) in the particular context of elastodynamics),

$$\forall g \in H, \quad \Phi(g) \in W_1, \quad \text{and} \quad \|\Phi(g)\|_{W_1} \le |g|. \tag{10.22}$$

Then, using Aubin-Nitsche's trick [43], one proves that,

$$\forall u \in W_\ell, \quad |u - \Pi_h u| \le \left(\frac{M}{\alpha}\right)^{\frac{1}{2}} \delta_\ell(h) \, \delta_1(h) \|u\|_{W_l}. \tag{10.23}$$

Next, the idea is to split the error $e_h(t) = u_h(t) - u(t)$ into two parts,

$$e_h(t) = \eta_h(t) - \varepsilon_h(t), \quad \eta_h(t) = u_h(t) - \Pi_h u(t), \quad \varepsilon_h(t) = u(t) - \Pi_h u(t).$$
$$(10.24)$$

The convergence of ε_h to 0 results from (10.18). It remains to look at η_h, which satisfies, for any t (note that Π_h commutes with d/dt),

$$\left(\frac{d^2\eta_h}{dt^2}(t), v_h\right) + a(\eta_h(t), v_h) = \left(\frac{d^2\varepsilon_h}{dt^2}(t), v_h\right), \quad \forall \, v_h \in V_h.$$

Choosing $v_h = \dfrac{d\eta_h}{dt}(t)$ and setting,

$$\mathcal{E}_h(t) = \frac{1}{2}\left\{ \left\|\frac{d\eta_h}{dt}(t)\right\|^2 + a(\eta_h(t), \eta_h(t)) \right\},$$

we get the identity,

$$\frac{d}{dt}\mathcal{E}_h(t) = \left(\frac{d^2\varepsilon_h}{dt^2}(t), \frac{d\eta_h}{dt}(t)\right) \le \sqrt{2} \left\|\frac{d^2\varepsilon_h}{dt^2}(t)\right\| \times \mathcal{E}_h^{\frac{1}{2}}(t).$$

After integration in time, we obtain the estimate,

$$\mathcal{E}_h(t)^{\frac{1}{2}} \le \mathcal{E}_h(0)^{\frac{1}{2}} + \int_0^t \left\|\frac{d^2\varepsilon_h}{dt^2}(s)\right\| ds,$$

which yields, as in corollary 10.1, to

$$\left|
\begin{array}{l}
\left\|\frac{d\eta_h}{dt}\right\|_{L^\infty(0,T;H)} \le \left(|u_1 - u_{1,h}|^2 + a(u_0 - u_{0,h}, u_0 - u_{0,h}) \right)^{\frac{1}{2}} \\
\qquad\qquad + \|\frac{d^2\varepsilon_h}{dt^2}\|_{L^1(0,T;H)}, \\[2mm]
\|\eta_h\|_{L^\infty(0,T;V)} \le \frac{1}{\sqrt{\alpha}} \left\{ \left(|u_1 - u_{1,h}|^2 + a(u_0 - u_{0,h}, u_0 - u_{0,h}) \right)^{\frac{1}{2}} \right. \\
\qquad\qquad \left. + \|\frac{d^2\varepsilon_h}{dt^2}\|_{L^1(0,T;H)} \right\}.
\end{array}
\right.$$

Using (10.24) and the triangular inequality, we prove the following lemma.

LEMMA 10.1 *The approximate solution u_h of (10.5) satisfies the following error estimates:*

$$\left\|\frac{du_h}{dt} - \frac{du}{dt}\right\|_{L^\infty(0,T;H)} \le \left(|u_1 - u_{1,h}|^2 + a(u_0 - u_{0,h}, u_0 - u_{0,h}) \right)^{\frac{1}{2}}$$

$$+ \left\|\frac{d^2\varepsilon_h}{dt^2}\right\|_{L^1(0,T;H)} + \left\|\frac{d\varepsilon_h}{dt}\right\|_{L^\infty(0,T;H)},$$

$$\|u_h - u\|_{L^\infty(0,T;V)} \le \frac{1}{\sqrt{\alpha}} \left(|u_1 - u_{1,h}|^2 + a(u_0 - u_{0,h}, u_0 - u_{0,h}) \right)^{\frac{1}{2}}$$

$$+ \frac{1}{\sqrt{\alpha}} \left\|\frac{d^2\varepsilon_h}{dt^2}\right\|_{L^1(0,T;H)} + \left\|\frac{d\varepsilon_h}{dt}\right\|_{L^\infty(0,T;V)}.$$
$$(10.25)$$

These a priori estimates can be converted into a convergence result with error estimates provided that the exact solution satisfies some regularity assumptions.

THEOREM 10.3 *We assume that the property (10.22) holds and that the solution u of the continuous problem satisfies the regularity assumption,*

$$u \in W_{loc}^{2,1}(\mathbb{R}^+; W_\ell) \cap W_{loc}^{1,\infty}(\mathbb{R}^+; W_m), \quad 1 \le \ell \le m \le L, \qquad (10.26)$$

for any $T > 0$, u_h strongly converges to u in $W^{1,\infty}(0,T;H) \cap L^\infty(0,T;V)$ and one has the error estimates (C denotes a positive constant that does not depend on h),

$$
\begin{aligned}
\left\| \frac{du_h}{dt} - \frac{du}{dt} \right\|_{L^\infty(0,T;H)} &\le C \left(|u_1 - u_{1,h}| + \|u_0 - u_{0,h}\| \right) \\
&+ C\, \delta_1(h) \left(\delta_\ell(h)\, \left\| \frac{d^2 u}{dt^2} \right\|_{L^1(0,T;W_\ell)} + \delta_m(h)\, \left\| \frac{du}{dt} \right\|_{L^\infty(0,T;W_m)} \right), \\[2mm]
\|u_h - u\|_{L^\infty(0,T;V)} &\le C \left(|u_1 - u_{1,h}| + \|u_0 - u_{0,h}\| \right) \\
&+ C \left(\delta_1(h)\, \delta_\ell(h)\, \left\| \frac{d^2 u}{dt^2} \right\|_{L^1(0,T;W_\ell)} + \delta_m(h) \left\| \frac{du}{dt} \right\|_{L^\infty(0,T;W_m)} \right), \\[2mm]
\|u_h - u\|_{L^\infty(0,T;H)} &\le C\, |u_0 - u_{0,h}| \\
&+ C\, \delta_1(h) \left(\delta_\ell(h)\, \left\| \frac{du}{dt} \right\|_{L^1(0,T;W_\ell)} + \delta_m(h)\, \|u\|_{L^\infty(0,T;W_m)} \right).
\end{aligned}
$$

$$(10.27)$$

PROOF The first two inequalities of (10.27) result from Lemma 10.1 and the use of the inequalities (10.19) and (10.23). For the last inequality, we simply observe that if,

$$U(t) = \int_0^t u(s)\, ds \quad \text{and} \quad U_h(t) = \int_0^t u_h(s)\, ds,$$

then one has

$$
\begin{cases}
\dfrac{d^2}{dt^2}\big(U(t), v\big) + a\big(U(t), v\big) = \left\langle \displaystyle\int_0^t L(s)ds, v \right\rangle, \quad \forall\, v \in V, \\[4mm]
\dfrac{d^2}{dt^2}\big(U_h(t), v_h\big) + a\big(U_h(t), v_h\big) = \left\langle \displaystyle\int_0^t L(s)ds, v_h \right\rangle, \quad \forall\, v_h \in V_h,
\end{cases}
$$

with the initial conditions

$$\begin{cases} U(0) = 0, & \dfrac{dU}{dt}(0) = u_0, \\[3mm] U_h(0) = 0, & \dfrac{dU_h}{dt}(0) = u_{0,h}. \end{cases}$$

Then applying the equivalent of the first inequality of Theorem 10.3 to the difference $U - U_h$ leads to the desired result. □

REMARK 10.3 *The error estimates clearly distinguish the two sources of errors,*

- *The approximation of the initial data (the first line in the right hand side of each inequality),*

- *The error due to the approximation of the equation (the second line in the right hand side of each inequality).*

10.2 Space Approximation of Elastodynamics Equations with Lagrange Finite Elements

In this section, we come back to the concrete evolution problem (9.24) through its variational formulation (9.26) with (9.27), (9.28) and (9.29). In particular, we remind that

$$V = \mathcal{V}^d, \quad \mathcal{V} = H^1_{0,\Gamma_0}(\Omega) = \{v \in H^1(\Omega) \, / \, v|_{\Gamma_0} = 0\}. \tag{10.28}$$

Note that the conditions (10.1) are satisfied under time regularity assumptions on f and g for the first one, and as soon as

$$\Omega \text{ is connected} \quad \text{and} \quad meas(\Gamma_0) > 0,$$

for the second one (as a consequence of Korn and Poincare's inequalities).

We consider successively the Galerkin approximation of this problem with P_k and Q_k Lagrange finite elements. According to (10.28), the finite element space V_h will be of the form

$$V_h = \mathcal{V}_h^d, \tag{10.29}$$

where $\mathcal{V}_h$ is a finite element approximation subspace of $\mathcal{V}$.

REMARK 10.4 *Using the two letters V and $\mathcal{V}$, we have chosen to distinguish a space of vector fiels from a space of scalar functions. Consistently,* *we should use a different notation for a element of V (a vector field) and an element of $\mathcal{V}$, a scalar function. For the sake of simplicity in the notation, we have chosen not to do the disticntion and will use the same letter v. In principle, the context will be sufficient to help the reader in making the difference.*

P_k **Lagrange finite elements.** For simplicity, we assume that $d = 2$ or 3 and that Ω is a polygon in 2D, a polyhedron in 3D. The construction of the approximation spaces $\mathcal{V}_h$ is based on a family $\mathcal{T}_h$ of conforming triangulations (triangular meshes in 2D, tetrahedral meshes in 3D) of Ω, of stepsize $h > 0$ (h is supposed to describe a set $\mathcal{H}$ of real positive numbers admitting 0 as point of accumulation). To define this more precisely, we introduce $\mathcal{T}$ the set of triangles of $\mathbb{R}^2$ in $d = 2$, the set of tetrahedra of $\mathbb{R}^3$ in $d = 3$,

$$\mathcal{T}_h \text{ is a finite subset of } \mathcal{T}, \quad \overline{\Omega} = \bigcup_{K \in \mathcal{T}_h} K, \quad h = \sup_{K \in \mathcal{T}_h} \{diam(K)\},$$

$$\forall (K, L) \in \mathcal{T}_h^2, \quad K \cap L = \begin{cases} \emptyset, \\ \text{a common vertex,} \\ \text{a common edge,} \\ \text{a common face (in 3D),} \end{cases} \tag{10.30}$$

where $diam(K)$ denotes the diamctcr of K. We assume moreover that

$$\forall h, \quad \Gamma_0 \text{ is a finite union of edges (in 2D) or faces (in 3D) of } \mathcal{T}_h. \tag{10.31}$$

REMARK 10.5 *A triangle in 2D, a tetrahedron in 3D and more generally a $d-$simplex is usually defined as the convex envelope of a non degenerate set $\{M_1 \cdots, M_{d+1}\}$ of $d + 1$ points (non degeneracy means that there does not exist any subset of m points embedded in an affine space or dimension $m-1$),*

$$x \in K \iff \exists (\lambda_1(x), \cdots, \lambda_{d+1}(x)) \in \mathbb{R}^{d+1} \ / \ x = \sum_{\ell=1}^{d+1} \lambda_\ell(x) \, M_\ell. \tag{10.32}$$

The points M_ℓ are the vertices of the $d-$simplex and the functions $\lambda_\ell(x)$ (polynomials of degree 1 in x) are called the barycentric coordinates of x.

A more analytical point of view consists in defining a $d-$simplex K as the image by an invertible affine map of the so-called reference element $\widehat{K}$,

$$\begin{cases} K = F_K(\widehat{K}), \quad \widehat{K} = \{\widehat{x} \in \mathbb{R} \ / \ \widehat{x}_\ell \geq 0, \ \widehat{x}_1 + \cdots + \widehat{x}_{d+1} \leq 1 \}, \\ F_K(\widehat{K}) = B_K \, \widehat{K} + b_K, \quad B_K \in \mathcal{L}(\mathbb{R}^d), \quad \text{invertible}, \quad b_K \in \mathbb{R}^d. \end{cases} \tag{10.33}$$

This trivial remark is not purely anecdotic: this point of view plays a major role from both theoretical (for the interpolation theory) and practical (programming of the computation of the mass and stifness matrices) points of view [43, 150].

Let $k \geq 1$ be a positive integer (the order of the finite element), the approximation space $\mathcal{V}_h$ is defined as,

$$\mathcal{V}_h = \{v_h \in C^0(\overline{\Omega}) \,/\, \forall\, K \in \mathcal{T}_h, v_K = v_h|_K \in P_k \text{ and } v_h|_{\Gamma_0} = 0\}, \quad (10.34)$$

where P_k is the space of polynomials in $\mathbb{R}$ of total degree less or equal than k

$$P_k = span[x_1^{\alpha_1} \cdots x_d^{\alpha_d}], \quad \alpha \in \mathcal{M}_k = \{\alpha \in \mathbb{N}^d, \; |\alpha| = \alpha_1 + \cdots + \alpha_d = k\}.$$

Thanks to classical properties of $H^1(\Omega)$, one easily sees that V_h is a subspace of V [43]. The standard description of V_h consists in constructing a particular basis of vector fields with small supports. Let us introduce some notation. For each K and $\alpha \in \mathcal{M}_k$, we denote by M_α^K the point of K whose barycentric coordinates are $\{\alpha_l/k, \; l = 1, \cdots, d\}$ (we shall call this set the canonical P_k frame of K) . The important property of the points $M_\alpha^K, \alpha \in \mathcal{M}_k$ is that (this is the so-called *unisolvence* property [43]),

$$\left\{ \begin{array}{l} \text{- The number of points } M_\alpha^K \text{ coincides with the dimension of } P_k, \\[2mm] \text{- A polynomial of } P_k \text{ is entirely determined by its values at the } M_\alpha^K\text{'s,} \\[2mm] \text{- The restriction of any polynomial in } P_k \text{ to an edge or a face (in 3D)} \\ \quad \text{is completely determined by its values at the points } M_\alpha^K \\ \quad \text{that belong to this edge or this face.} \end{array} \right.$$

$$(10.35)$$

REMARK 10.6 *The last property of is essential for ensuring the continuity of v_h from one element to the other.*

REMARK 10.7 *The reader will easily check that all what follows in this paragraph remains true if one replaces, for each K, $\{M_\alpha^K, \alpha \in \mathcal{M}_k\}$ by any set of points satisfying the above properties. By definition, such a set $\{M_\alpha^K, \alpha \in \mathcal{S}_k\}$ is a P_k unisolvent set of points in K.*

REMARK 10.8 *A standard choice for a P_k unisolvent set of points is a set of points $\{M_\alpha^K, \alpha \in \mathcal{S}_k\}$ that have fixed barycentric coordinates (as the canonical frame of K). In other words we have,*

$$M_\alpha^K = F_K(\widehat{M_\alpha}), \quad \forall\, \alpha \in \mathcal{S}_k, \quad (10.36)$$

where the $\widehat{M_\alpha}$'s form a P_k unisolvent set in the reference element $\widehat{K}$.

We set,

$$\mathcal{N}_h^k = \bigcup_{K \in \mathcal{T}_h} \{M_\alpha^K, \alpha \in \mathcal{M}_k\},$$

that we call the set of the nodes of the finite element space V_h. For the sequel, we assume that the nodes are numbered in such a way that,

$$\mathcal{N}_{h,0}^k = \{M \in \mathcal{N}_h^k \, / \, M \notin \Gamma_0\} = \{M_j, \, j \leq N_{0,h}\}.$$

REMARK 10.9 *For $k = 1$, $\mathcal{N}_h^1$ is nothing but the set of the vertices of the triangulation. For larger k one can distinguish,*

- *the vertices of the nodes,*

- *the edge nodes which are on an edge, but are not a vertex (exist for $k \geq 2$),*

- *the face nodes in 3D (exist for $k \geq 3$),*

- *the interior nodes, interior to an element, (exist for $k \geq 3$ in 2D, for $k \geq 4$ in 3D).*

To each $j \leq N_{0,h}$ we associate the unique scalar function w_j defined by,

$$w_j \in C^0(\overline{\Omega}), \quad w_j|_K \in P_k, \, \forall \, K \in \mathcal{T}_h, \quad w_j(M_l) = \delta_{jl}, \quad \forall \, l \leq N_{0,h}. \quad (10.37)$$

One easily sees that the support of w_j is the union of elements (triangles or tetrahedra) of $\mathcal{T}_h$ admitting M_j as a common node. This support thus depends on the nature of this node,

- In dimension 2:

  - for a vertex M_j, the support is the union of triangles (in variable number) admitting M_j as a common vertex,

  - for an edge node, the support is the union of the two triangles (with an exception for edges along the boundary Γ_1) that share this edge,

  - for an interior node, this support is made of one single triangle.

- In dimension 3:

  - for a vertex M_j, the support is the union of tetrahedra (in variable number) admitting M_j as a common vertex,

  - for an edge node, the support is the union of tetrahedra (in variable number) that share this edge,

  - for a face node, the support is the union of the two tetrahedra (with an exception for faces of the boundary Γ_1) that share this face,

  - for an interior node, this support is made of one single tetrahedron.

We can now describe a basis of the subspace V_h of V defined as when $\mathcal{V}_h$ is given by (10.34). Let $\{e_1, \cdots, e_d\}$ be the canonical basis of $\mathbb{R}^d$, for $j \leq N_{0,h}$ and $\ell \in \{1, \cdots, d\}$, we set,

$$w_j^\ell = w_j \, e_\ell \quad \in V_h. \tag{10.38}$$

It is straightforward to prove that,

$$\{ \, w_j^\ell, \, j \leq N_{0,h}, \, 1 \leq \ell \leq d \, \} \tag{10.39}$$

is a basis of V_h, which shows in particular that,

$$N_h = d \, N_{0,h}. \tag{10.40}$$

More precisely, if $v_h = (v_{\ell,h})_{1 \leq \ell \leq d}$ belongs to V_h, we have,

$$v_h = \sum_{\ell=1}^{d} \sum_{j=1}^{N_{0,h}} v_{\ell,h}(M_j) \, w_j^\ell, \quad (\equiv \sum_{j=1}^{N_{0,h}} v_h(M_j) \, w_j). \tag{10.41}$$

In other words, for this basis, the degrees of freedom of the finite element space V_h, defined as the coefficients of the expansion of any element v_h of V_h in this basis, are nothing but the values of the components of v_h at the nodes: this caracterizes Lagrange finite elements.

REMARK 10.10 *More generally a Lagrange finite element space is a space of the form (10.34), (except that the elements K may also have different shapes, as we shall see later, as quadrangles or hexaedra) but locally, the space P_k is replaced by a finite dimensional space $\mathbf{V}_K$ of smooth vector valued functions (at least continuous) defined in K,*

$$V_h = \{v_h \in C^0(\overline{\Omega}) \, / \, \forall \, K \in \mathcal{T}_h, v_K = v_h|_K \in \mathbf{V}_K \text{ and } v_h|_{\Gamma_0} = 0 \, \}. \tag{10.42}$$

Up to these changes, one can still define the notion of $\mathbf{V}_K$ unisolvent set of points $\{M_\ell^K, 1 \leq \ell \leq \dim \mathbf{V}_K\}$ (i.e., (10.35) where one replaces P_k by $\mathbf{V}_K$, α by ℓ, ...). Of course, one can define the related notion of associated Lagrange basis functions and interpolation operator [43].

The matrix reformulation of the semi-discrete problem follows the presentation we made in the abstract case. Simply remark that, because of the notation we have adopted, the indices $i \in \{1, \cdots, N_h\}$ of the previous section have to be replaced by a double index (i, ℓ), $i \in \{1, \cdots, N_h\}$, $\ell \in \{1, \cdots, d\}$. The vector $U_h(t)$ is the vector with components,

$$\{u_i^\ell(t) = u_{h,\ell}(M_i, t), \, 1 \leq i \leq N_{0,h}, \, 1 \leq \ell \leq d \, \} \tag{10.43}$$

and the matrices $\mathbf{M}_h$ and $\mathbf{A}_h$ are of the form (with obvious notation),

$$\mathbf{M}_h = ((\mathbf{M}_{ij}^{\ell m})), \quad \mathbf{A}_h = ((\mathbf{A}_{ij}^{\ell m})). \tag{10.44}$$

REMARK 10.11 *With an appropriate (and obvious) ordering of the indices* (i, ℓ), *the matrices* $\mathbf{M}_h$ *and* $\mathbf{A}_h$ *can be written in a* $d \times d$ *block form*

$$\mathbf{M}_h = ((\mathbf{M}^{\ell m})) \, 1 \leq \ell, m \leq d, \quad \mathbf{A}_h = ((\mathbf{A}^{\ell m})) \, 1 \leq \ell, m \leq d,$$

where the blocks $\mathbf{M}^{\ell m}$ *and* $\mathbf{A}^{\ell m}$ *are* $N_{0,h} \times N_{0,h}$ *matrices. Note that the off diagonal blocks of* $\mathbf{M}_h$, $\mathbf{M}^{\ell m}$ *for* $m \neq \ell$ *are 0, due to the orthogonality of* e_ℓ *and* e_m. *Moreover, the diagonal blocks are identical to the same* $N_{0,h} \times N_{0,h}$ *matrix* $\mathbf{M}_h^s$ *with:*

$$\mathbf{M}_{ij}^s = \int_\Omega \rho \, w_i \, w_j \, dx \tag{10.45}$$

A pleasant consequence of having chosen basis functions with small support is that the matrices $\mathbf{M}_h$ and $\mathbf{A}_h$ are both very sparse. This can be expressed as follows.

- For each node M_j of $\mathcal{N}_{h,0}$, we can defined the neighbors of M_j as the nodes M_l such that the support of w_l intersects the one of w_j. Of course, the number of neighbors depends on the nature of the node: it is minimal for interior nodes and maximal for vertices.

- Each line of the differential system can be associated to a degree of freedom u_j^ℓ. In this equation the unknown $u_j^\ell(t)$ is linked to the $u_l^m(t)$'s corresponding to the nodes M_l that are neighbours of M_j.

If one roughly defines the connectivity of the method as "the" number of non zero elements per line of the matrices $\mathbf{M}_h$ and $\mathbf{A}_h$, we observe that the connectivity increases with the order k of the finite element method (roughly as k^d).

We conclude by checking that the theoretical requirements are well satisfied. For this, we need to introduce the notion of *regular* family of triangulations. First, for any element K of $\mathcal{T}$, we define ρ_K as the radius of the largest disk (in 2D) or sphere (in 3D) which is included in K and we set

$$\sigma_K = \frac{diam(K)}{\rho_K}. \tag{10.46}$$

By definition, one says that $\{\mathcal{T}_h, h \in \mathcal{H}\}$ is a regular family of triangulations if and only if there exists $\sigma > 0$, independent of h, such that,

$$\forall \, h \in \mathcal{H}, \quad \forall \, K \in \mathcal{T}_h, \quad \sigma_K \leq \sigma. \tag{10.47}$$

Clearly, this property means that, when the mesh is refined (i.e., when h goes to 0), the geometry of the elements does not degenerate. For instance, in 2D, this is equivalent to saying that all the angles of the triangles remain larger than a fixed angle θ_0 [43].

REMARK 10.12 *A systematic way to construct a regular family of triangu- lations is as follows. Starting from an initial triangulation T_{h_0}, one constructs recursively the triangulations,*

$$\{T_{h_n}, \; h_n = h_0/2^n\},$$

where $T_{h_{n+1}}$ is constructed from T_{h_n} by dividing, in 2D (resp. 3D) each trian- gle (resp. tetrahedron) of T_{h_n} into 4 (resp. 8) equal triangles (resp. tetrahedra) by introducing the middles of the edges of the original element as new vertices.

Approximation theory.   The approximation theory relies on the interpo- lation theory in Sobolev spaces. We shall restrict ourselves here to recall the main useful results of this theory.

First, for each $h \in \mathcal{H}$, we can define the interpolation operator,

$$\mathcal{I}_h : \left(C^0(\overline{\Omega})\right)^d \to V_h, \tag{10.48}$$

such that for any $v \left(\equiv (v_\ell) \right) \in C^0(\overline{\Omega})^d$,

$$I_h v = \sum_{\ell=1}^{d} \sum_{j=1}^{N_{0,h}} v_\ell(M_j) \, w_j^\ell, \quad \left(\equiv \sum_{j=1}^{N_{0,h}} v(M_j) \, w_j\right). \tag{10.49}$$

Note that since $\Omega \subset \mathbb{R}^d$ with $d \leq 3$, one has,

$$\left(H^2(\Omega)\right)^d \subset \left(C^0(\overline{\Omega})\right)^d,$$

so that I_h is well defined in $H^2(\Omega)^d$. The main result of interpolation theory is the following [43],

THEOREM 10.4   *Assume that assumption (10.47) holds. Then, for each $1 \leq \ell \leq k$, there exists a constant $C_\ell > 0$ such that,*

$$\forall \, v \in W^l = \left(H^{1+\ell}(\Omega)\right)^d \cap V, \quad \|v - \mathcal{I}_h v\| \; \leq \; C_\ell \, h^\ell \, \|v\|_{W^l}, \tag{10.50}$$

i. e. (10.4) holds with $L = k$, $W_\ell = H^{1+\ell}(\Omega) \cap V$ and $\delta_\ell(h) = C_\ell \, h^\ell$.

REMARK 10.13 *Assume the existence of open sets $\Omega_m, 1 \leq m \leq M$ such that,*

$$\overline{\Omega} = \bigcup_{m=1}^{M} \overline{\Omega}_m, \quad \Omega_m \cap \Omega_l = \emptyset \; for \; l \neq m. \tag{10.51}$$

We assume moreover that the triangulation T_h verifies the compatibility con- dition,

For each m, for any $h \in \mathcal{H}$, $\overline{\Omega}_m$ is a finite union of elements of T_h. (10.52)

Then, (10.50) and thus (10.4) still hold if we replace $W_\ell = \left(H^{1+\ell}(\Omega)\right)^d \cap V$ by

$$W_\ell = \{\, v \in V \,/\, \forall\, 1 \le m \le M, \; v|_{\Omega_m} \in \left(H^{1+\ell}(\Omega_m)\right)^d \,\}$$

equipped with the norm

$$\|v\|^2_{W_\ell} = \sum_{m=1}^{M} \|v\|^2_{H^{1+\ell}(\Omega)} \,.$$

REMARK 10.14 *The interpolation inequalities (10.50) or (10.4) are in fact a consequence of a general local interpolation result valid for $0 \le \ell < m \le k$,*

$$\forall\, K \in \mathcal{T}_h, \quad \forall\, v \in H^m(K), \quad |v - \mathcal{I}_h v|_{\ell,K} \le C_{\ell,m}\, h^{m-\ell}\, \|v\|_{m,K}. \quad (10.53)$$

Moreover, it is easy to see that the interpolation results (10.53) are valid as soon as the interpolation operator $\mathcal{I}_h$ is constructed with interpolation points having the property (10.36).

We explain now the meaning of the property (10.22). We first remark that, in the context of elastodynamics equations, for any $g \in H = \left(L^2(\Omega)\right)^d$ the vector field $\mathbf{u} = \Phi(g)$ is the solution of the boundary value problem (a static elasticity problem),

$$\begin{cases} -\operatorname{div} C\varepsilon(\mathbf{u}) = g, & x \in \Omega, \quad t > 0, \\[2mm] \mathbf{u}(x) = 0, & x \in \Gamma_0, \quad t > 0, \\[2mm] C\varepsilon(\mathbf{u})(x)\, n = 0, & x \in \Gamma_1, \quad t > 0. \end{cases} \quad (10.54)$$

Then, the property (10.22) is simply,

$$\forall\, g \in \left(L^2(\Omega)\right)^d, \quad \mathbf{u} \in H^2(\Omega) \text{ and } \|\mathbf{u}\|_{2,\Omega} \le C\, \|g\|_{0,\Omega}. \quad (10.55)$$

REMARK 10.15 *If one is under the assumptions of remark 10.14, in (10.55), $H^2(\Omega)$ can be replaced by W^2.*

REMARK 10.16 *A property of this type relies on the delicate regularity theory for the solution of elliptic boundary value problems. Clearly, since one has a second order problem the H^2 (or W^2) regularity is the best possible regularity. Such a result requires smoothness properties on Γ and C but also depends on the geometry of Γ_0 and Γ_1. An example is [82],*

$$\Omega \text{ is convex}, \quad C \in C^1(\overline{\Omega}, \mathcal{L}_s^2(\mathbb{R}^d)) \quad \text{and} \quad \Gamma_1 = \emptyset. \quad (10.56)$$

If such assumptions are not satisfied, one can nevertheless obtain intermediate regularity results of the type $H^{1+\sigma}$ with $0 < \sigma < 1$. Of course, the error estimates that we shall give in theorem 10.5 can be modified accordingly.

Applying theorem 10.3 we can thus derive error estimates for the semi-discrete P_k finite element semi-discretization of the problem (9.24).

THEOREM 10.5   *Assume that (10.56) and (10.52) hold. Assume also that the regularity property (10.55) holds. Then, provided that the solution u of (9.24) has the regularity,*

$$u \in W_{loc}^{2,1}(\mathbb{R}^+; W_\ell) \cap W_{loc}^{1,\infty}(\mathbb{R}^+; W_m), \quad 1 \le \ell, m \le k, \tag{10.57}$$

for any $T > 0$, one has the error estimates (C denotes a positive constant independent of h),

$$
\left\|\frac{du_h}{dt} - \frac{du}{dt}\right\|_{L^\infty(0,T;L^2(\Omega)^d)} \le C \left(|u_1 - u_{1,h}| + \|u_0 - u_{0,h}\| \right)
$$

$$
+ C \left(h^{\ell+1} \left\|\frac{d^2u}{dt^2}\right\|_{L^1(0,T;W_\ell)} + h^{m+1} \left\|\frac{du}{dt}\right\|_{L^\infty(0,T;W_m)} \right),
$$

$$
\|u_h - u\|_{L^\infty(0,T;H^1(\Omega)^d)} \le C \left(|u_1 - u_{1,h}| + \|u_0 - u_{0,h}\| \right)
$$

$$
+ C \left(h^{\ell+1} \left\|\frac{d^2u}{dt^2}\right\|_{L^1(0,T;W_\ell)} + h^m \left\|\frac{du}{dt}\right\|_{L^\infty(0,T;W_m)} \right), \tag{10.58}
$$

$$
\|u_h - u\|_{L^\infty(0,T;L^2(\Omega)^d)} \le C \, |u_1 - u_{1,h}|_{L^2(\Omega)^d}
$$

$$
+ C \left(h^{\ell+1} \left\|\frac{du}{dt}\right\|_{L^1(0,T;W_\ell)} + h^{m+1} \|u\|_{L^\infty(0,T;W_m)} \right).
$$

It is worthwhile noticing that the optimal order of convergence is obtained in the following sufficient conditions,

- $u_0 \in H^{k+1}(\Omega)$, $u_1 \in H^{k+1}(\Omega)$, $u_{0,h} = \mathcal{I}_h \, u_0$ and $u_{1,h} = \mathcal{I}_h \, u_1$,

- (10.57) holds with $\ell = m = k$.

In such a situation one has,

$$
\begin{aligned}
\left\|\frac{du_h}{dt} - \frac{du}{dt}\right\|_{L^\infty(0,T;L^2(\Omega)^d)} &\le C\, h^k \left(|u_1|_{k,\Omega} + |u_0|_{k+1,\Omega} \right.\\
&\left. + h\left\|\frac{du}{dt}\right\|_{L^\infty(0,T;W_k)} + h\left\|\frac{d^2u}{dt^2}\right\|_{L^1(0,T;W_k)} \right),\\[2mm]
\|u_h - u\|_{L^\infty(0,T;H^1(\Omega)^d)} &\le C\, h^k \left(|u_1|_{k+1,\Omega} + |u_0|_{k+1,\Omega} \right.\\
&\left. + \left\|\frac{du}{dt}\right\|_{L^\infty(0,T;W_k)} + h\left\|\frac{d^2u}{dt^2}\right\|_{L^1(0,T;W_k)} \right),\\[2mm]
\|u_h - u\|_{L^\infty(0,T;L^2(\Omega)^d)} &\le C\, h^{k+1} \left(|u_1|_{k+1,\Omega} + \|u\|_{L^\infty(0,T;W_k)} \right.\\
&\left. + \left\|\frac{du}{dt}\right\|_{L^1(0,T;W_k)} \right).
\end{aligned}
\tag{10.59}
$$

Q_k Lagrange finite elements. This approximation process is linked to meshes of the computational domain Ω using quadrilaterals in 2D or hexahedra in 3D. We shall denote by $\mathcal{Q}$ the set of quadrilaterals in 2D or hexahedra in 3D. To define more precisely $\mathcal{Q}$, it is first useful to introduce the space Q_1 as the space of real valued functions in $\mathbb{R}^d$ that are polynomials of degree 1 with respect to each variable, namely,

$$
Q_1 = span[x_1^{\alpha_1} \cdots x_d^{\alpha_d}], \quad \alpha \in \mathcal{S}_1 = \{\alpha \in \mathbb{N}^d, \quad max\, \alpha_\ell \le 1\}. \tag{10.60}
$$

We shall define the admissible Q_1 transforms from $\mathbb{R}$ into itself as,

$$
Q_{1,ad}^d = \{F \in Q_1^d \,/\, F|_{\widehat{K}} \text{ is injective }\}, \tag{10.61}
$$

where $\widehat{K}$ is the reference unit square (in 2D) or cube (in 3D) : $\widehat{K} = [0,1]^d$.

REMARK 10.17 *In general, a map F in Q_1^d will not be injective on $\mathbb{R}^d$. It suffices to realize that its Jacobian $J(\widehat{x})$ ($= det\, DF(\widehat{x})$, where $DF(\widehat{x})$ is the differential (or Jacobian matrix) of F at point $\widehat{x}$) is a polynomial (of degree 1 if $d=2$, of degree 4 if $d=3$). The important point is to check that the set $\{\widehat{x} \in \mathbb{R}^d \,/\, J(\widehat{x}) = 0\}$ (in principle a curve in 2D, a surface in 3D) does not intersects $\widehat{K}$.*

A natural question is: given 2^d points in $\mathbb{R}$

$$
\{x_\alpha, \; \alpha \in \mathcal{S}_1\} \text{ where } \mathcal{S}_k = \{\alpha \in \mathbb{N}^d, \quad max\, \alpha_\ell \le k\},
$$

does there exist an admissible transform F such that the vertices of $K = F(\widehat{K})$ are the x_α's?

It is straightforward to see that such a transform is necessarily given by (up to a preliminary permutation of the x_α's),

$$F(\widehat{x}) = \sum_{\alpha \in \mathcal{S}_1} x_\alpha \, \varphi_\alpha(\widehat{x}),$$

where φ_α is the Q^1 function such that

$$\varphi_\alpha(\widehat{x}_\beta) = \delta_{\alpha\beta},$$

the points $\widehat{x}_\alpha$ denoting the vertices of $\widehat{K}$, with coordinates $(\alpha_1, \cdots, \alpha_{d+1})$.

The difficult question is the admissibility. In 2D, one can show that this means that the four points x_α form a convex non degenerate quadrilateral in the usual sense. In 3D, this is a much more delicate question for which we are not aware of any satisfactory answer.

We define an element K of $\mathcal{Q}$ as the image by an admissible Q_1 transformation of $\widehat{K}$.

$$K \in \mathcal{Q} \Longleftrightarrow \exists \, F_K \in Q_{1,ad}^d \; / \; K = F_K(\widehat{K}) \tag{10.62}$$

By definition, a vertex of K is the image by F_K of a vertex of $\widehat{K}$, an edge of K is the image by F_K of an edge of $\widehat{K}$ and, in 3D, a face of K is the image by F_K of a face of $\widehat{K}$. In 2D, any $K \in \mathcal{Q}$ is a quadrilateral in the usual sense, i.e., with straight edges. In 3D, if each edge of K is a segment, a face of K is in general not plane but curved. In that sense, K is not an hexahedron in the usual sense (i.e., with plane faces).

A family $\{\mathcal{T}_h, h \in \mathcal{H}\}$ of conforming quadrangulations of Ω is such that, for each $h \in \mathcal{H}$

$$\mathcal{T}_h \text{ is a finite subset of } \mathcal{Q}, \quad \overline{\Omega} = \bigcup_{K \in \mathcal{T}_h} K, \quad h = \sup_{K \in \mathcal{T}_h} diam(K),$$

$$\forall \, (K, L) \in \mathcal{T}_h^2, \quad K \cap L = \begin{cases} \emptyset \\ \text{a common vertex,} \\ \text{a common edge,} \\ \text{a common face (in 3D),} \end{cases} \tag{10.63}$$

Let $k \geq 1$ be a positive integer (the order of the finite element), the approximation space V_h is defined as,

$$V_h = \{v_h \in C^0(\overline{\Omega}) / \forall K \in \mathcal{T}_h, v_K = v_h|_K \to v_K \circ F_K \in Q_k, v_h|_{\Gamma_0} = 0\}, \tag{10.64}$$

where Q_k is the space of polynomials in $\mathbb{R}$ of degree less or equal than k in each variable

$$Q_k = span[x_1^{\alpha_1} \cdots x_d^{\alpha_d}], \quad \alpha \in \mathcal{S}_k = \{\alpha \in \mathbb{N}^d, \quad max \, \alpha_\ell \leq k\}.$$

REMARK 10.18 *It is worthwhile to notice that, contrary to P_k elements, the restriction to an element of $v_h \in V_h$ may not be a polynomial. Indeed the condition $v_K \circ F_K \in Q_k$ means that there exists a function $\widehat{v}_K \in Q_k$ such that,*

$$v_K = \widehat{v}_K \circ F_K^{-1}. \tag{10.65}$$

In other words, the space V_h enters the general definition of Lagrange finite element spaces of remark (10.10) with,

$$\mathbf{V}_K = \{\widehat{q} \circ F_K^{-1}, \widehat{q} \in Q_k\}.$$

If K is a parallelogram (in 2D) or a parallelepiped (in 3D), the transformation F_K is linear. So is F_K^{-1} and as a consequence $v_K \circ F_K \in Q_k$. For a general quadrilateral or tetrahedron, this is no longer the case. Indeed, the expression of F_K^{-1} will result from the inversion of a non linear (bilinear in 2D, trilinear in 3D) system, which is hard to do analytically. It is possible in 2D, where after ellimination of one variable, the problem is reduced to the resolution of a quadratic equation (as a consequence the explicit expression of the components of F_K^{-1} can be written as rational fractions and square roots of polynomials in x and y). In 3D, it is not clear that there exists any closed form for F_K^{-1}.

As for P_k elements, one can describe V_h through the construction of a particular basis. For each $\alpha \in \mathcal{S}_k$ we denote by $\widehat{M}_\alpha$ the point of $\widehat{K}$ whose coordinates are $\{\alpha_1/k, \cdots, \alpha_d/k\}$. These points satisfy the following unisolvence properties,

- The number of points $\widehat{M}_\alpha$ coincides with the dimension of Q_k,

- A polynomial of Q_k is entirely determined by its values at the $\widehat{M}_\alpha$'s,

- The restriction of any polynomial in Q_k to an edge or a face (in 3D) of $\widehat{K}$ is completely determined by its values at the points $\widehat{M}_\alpha$ that belong to this edge or this face.

REMARK 10.19 *As for P_k elements, the set $\{M_\alpha^K, \alpha \in \mathcal{S}_k\}$ could be replaced by any Q_k unisolvent set of points, i.e., any set of points in K satisfying the above properties.*

Then we define for each element K the $\{M_\alpha^K, \alpha \in \mathcal{M}_k\}$ where,

$$M_\alpha^K = F_K(\widehat{M}_\alpha). \tag{10.66}$$

and as in the previous section, we introduce the sets of nodes,

$$\left| \begin{array}{l} \mathcal{N}_h^k = \displaystyle\bigcup_{K \in \mathcal{T}_h} \{M_\alpha^K, \alpha \in \mathcal{M}_k\} \\[2em] \mathcal{N}_{h,0}^k = \{M \in \mathcal{N}_h^k \,/\, M \notin \Gamma_0\} = \{M_j, \, j \leq N_{0,h}\} \end{array} \right. \tag{10.67}$$

REMARK 10.20 *As for P_k elements (see remark 10.9), the nodes can be splitted into the vertices, edge nodes, face nodes (in 3D) and interior nodes.*

To each $j \leq N_{0,h}$ and $\ell \in \{1, \cdots, d\}$, we associate the unique scalar function w_j defined by,

$$w_j \in C^0(\overline{\Omega}), \quad w_j|_K \in Q_k, \, \forall \, K \in \mathcal{T}_h, \quad w_j(M_l) = \delta_{jl}, \forall \, l \leq N_{0,h}. \quad (10.68)$$

Obviously the support of w_j is the union of elements (quadrilaterals or hexahedra) of $\mathcal{T}_h$ admitting M_j as a common node. One can then construct a basis $\{ w_j^\ell, \, j \leq N_{0,h}, \, 1 \leq \ell \leq d \}$ via the formula (10.38) as for P_k elements and the representation formula (10.41) still holds: the degrees of freedom are the values of the components of the vector fields at the nodes.

The convergence theory of Q_k elements is very similar to the theory of P_k elements. One needs to introduce the notion of regular family of quadrangulations, which is once again a condition of non degeneracy of the geometry of the elements when the mesh is refined. This condition is however slightly more difficult to express than for triangulations [5]. It is still written as:

$$\forall \, h \in \mathcal{H}, \quad \forall \, K \in \mathcal{T}_h, \quad \sigma_K \leq \sigma. \quad (10.69)$$

where σ_K is still defined by K but the definition of ρ_K has to be changed:

- In dimension 2, let T_K^i be the triangle whose vertices are the vertices of K except vertex $n°i$ then:

$$\rho_K = \inf_{i=1}^{4} \rho_K^i, \quad (10.70)$$

where ρ_K^i is the radius of the largest disk included in T_K^i.

- In dimension 3, let we define a non degenerate sub-tetrahedron of K as a tetrahedron whose vertices are the image by F_K of four non coplanar vertices of the reference unit cube $\widehat{K}$. Let $\mathcal{T}_K$ be the set of non-degenerate sub-tetrahedra of K, then:

$$\rho_K = \inf_{T \in \mathcal{T}_K} \rho_T, \quad (10.71)$$

where ρ_T is the radius of the largest ball included in T.

Then, if we define the Q_k interpolation operator $\mathcal{I}_h$ with formula (10.49) (only the definition of the functions w_j has changed), one has the following approximation theorem,

THEOREM 10.6 *Assume that assumption (10.69) holds. Then, for each $1 \leq \ell \leq k$, there exists a constant $C_\ell > 0$ such that,*

$$\forall \, v \in \left(H^{1+\ell}(\Omega) \right)^d \cap V, \quad \|v - \mathcal{I}_h v\| \leq C_\ell \, h^\ell \, \|v\|_{H^{1+\ell}(\Omega)}. \quad (10.72)$$

Using this result it is easy to write a convergence theorem for Q_k elements, similar to the theorem 10.5 for P_k elements, by application of the abstract result of theorem 10.3. The details are left to the reader.

10.3 On the Use of Quadrature Formulas

The general problematic. One of the practical problems posed by the implementation of the Lagrange finite element method is the effective numerical computations of the matrices $\mathbf{M}_h$ and $\mathbf{A}_h$, which amounts, at the elementary level to compute the following integrals,

$$\begin{cases} \displaystyle\int_K \rho \, w_i^\ell \cdot w_j^\ell \, dx, & \text{for the mass matrix,} \\[2mm] \displaystyle\int_K C \, \varepsilon(w_i^\ell) : \varepsilon(w_j^m) \, dx, & \text{for the stiffness matrix.} \end{cases} \qquad (10.73)$$

In the expressions above, ℓ and m vary in $\{1, \cdots, d\}$, K describes $\mathcal{T}_h$ while (i, j) vary in the subset of $\{1, \cdots, N_{h,0}\}^2$ for which the nodes M_i and M_j are neighbours, in the sense of the previous section.

REMARK 10.21 *Note that, since ρ is scalar, the mass matrix does not couple w_i^ℓ and w_i^m for $m \neq \ell$ which means that the mass matrix is naturally block diagonal wih $N_{h,0} \times N_{h,0}$ blocks.*

In principle, these integrals should be computed exactly. Clearly, this is not possible in general (in particular for arbitrary variations in space of ρ or C). A particular case is the case of piecewice constant ρ and C and compatible triangulations (in the sense that ρ and C are constant inside each element). Then, with P_k elements, we observe that the functions,

$$x \to \rho(x) \, w_i^\ell(x) \cdot w_j^\ell(x) \quad \text{and} \quad x \to C(x) \, \varepsilon(w_i^\ell)(x) : \varepsilon(w_j^m)(x)$$

are polynomials (of respective degrees $2k$ and $2k - 2$). This remains true, of course, for piecewise polynomials ρ and C. In such a situation the integrals (10.73) can be analytically computed in closed (but quite complicated) form. However, this property is lost immediately with Q_k elements except if K is a parallelogram or a parallepiped (see remark 10.18).

In any case, the standard practical solution to the computation of the matrices consists in computing (exactly in some particular cases, approximately most of the time) the integrals (10.73) using so called quadrature formulas.

A quadrature formula can be seen as a way of approximating the integral of a (sufficiently smooth - at least continuous) function f in a domain K of sufficiently simple shape as an appropriate combination of the values of the function at some particular points called quadrature points (or quadrature nodes). Typically [45, 134],

$$\int_K f\, dx \sim \int_K^{\mathcal{Q}_K} f\, dx = meas\ K \sum_{\mathcal{M}_l \in \mathcal{Q}_K} \omega_l^K\, f(\mathcal{M}_l^K), \qquad (10.74)$$

- the $\mathcal{M}_l^K$'s ($\in K$) are the quadrature nodes,

- the ω_l^K's ($\in \mathbb{R}$) are the corresponding quadrature weights.

In practice, for the implementation, admitting that different quadrature rules can be used for the mass and rigidity matrices, one makes the substitutions

$$\begin{cases} \displaystyle\int_K \rho\, w_i^\ell \cdot w_j^\ell\, dx & \rightarrow & \displaystyle\int_K^{\mathcal{Q}_K^m} \rho\, w_i^\ell \cdot w_j^\ell\, dx, \\[2mm] \displaystyle\int_K C\, \varepsilon(w_i^\ell) : \varepsilon(w_j^m)\, dx & \rightarrow & \displaystyle\int_K^{\mathcal{Q}_K^s} C\, \varepsilon(w_i^\ell) : \varepsilon(w_j^m)\, dx, \end{cases} \qquad (10.75)$$

so that, in the semi discrete problem,

- (u_h, v_h) is replaced by $\quad (u_h, v_h)_h = \displaystyle\sum_{K \in \mathcal{T}_h} \int_K^{\mathcal{Q}_K^m} \rho\, u_h\, v_h\, dx,$

- $a(u_h, v_h)$ is replaced by $\quad a_h(u_h, v_h) = \displaystyle\sum_{K \in \mathcal{T}_h} \int_K^{\mathcal{Q}_K^s} C\, \varepsilon(u_h) : \varepsilon(v_h)\, dx.$

Because of this (in most cases, the quadrature formulas will not be exact, i.e., $(u_h, v_h)_h \neq (u_h, v_h)$ and $a_h(u_h, v_h) \neq a(u_h, v_h)$), the corresponding method enters the class of non conforming Galerkin methods. Therefore, the convergence theory presented in section 10.2 can not be directly applied any longer and the new approximation needs to be analyzed. In particuler, the quadrature formulas should be chosen in such a way that one does not lose any order of accuracy. We shall come back later (see the paragraph on the analysis of the approximation of the mass matrix) on this point.

Let us now consider separately the case of P_k and Q_k elements.

Quadrature formulas in triangles or tetrahedra. Usual quadrature formulas on triangles or tetrahedra are constructed in order to be exact for some polynomials, typically exact in P_q for some integer q which is then called the order of the quadrature formula.

Since any d-simplex K can be seen as the image of the reference element $\widehat{K}$

by an affine transformation $F_K(x) = B_K x + b_K$, it is natural to construct a quadrature formula in K from a quadrature formula in $\widehat{K}$. This simply follows from the remark that f_K being defined in K, if we set,

$$\widehat{f}_K = f_K \circ F_K \ : \widehat{K} \to \mathbb{R}, \tag{10.76}$$

one has the well known formula

$$\int_K f_K \, dx = J_K \int_{\widehat{K}} \widehat{f}_K \, d\widehat{x}, \tag{10.77}$$

where

$$J_K = |\det B_K| \quad (\equiv \frac{meas\ K}{(d+1)!}). \tag{10.78}$$

Let us consider a quadrature formula in $\widehat{K}$, of order q (note that $meas\ \widehat{K} = (d+1)!^{-1}$),

$$\int_{\widehat{K}}^{\widehat{Q}} \widehat{f} \, d\widehat{x} = \frac{1}{(d+1)!} \sum_{\widehat{M}_l \in \widehat{Q}} \omega_l \, \widehat{f}(\widehat{M}_l) \quad (\text{exact in } P_q), \tag{10.79}$$

according to formula (10.77) we construct a quadrature formula of the form (10.74) where simply the quadrature nodes are "transported" by affine transformation,

$$M_l^K = F_K(\widehat{M}_l), \tag{10.80}$$

and where the weights do not depend on K : $\omega_l^K = \omega_l$.

Note that, as the set P_q is invariant by affine transformation (in other words $\widehat{f}_K$ describes P_q when f_K describes P_q) this quadrature formula on K is also of order q. The construction of quadrature formulas in $\widehat{K}$ is not a such an easy task and we refer the reader to reference monographies on this subject. We restrict ourselves to give the most well known low order quadrature formulas. These formulas are symmetric as they respect the natural "symmetries" of triangles or tetrahedra, in the sense defined for instance in [47]

- **Quadrature formulas in a triangle.** We indicate below the quadrature points on the left (with their 3 barycentric coordinates) and the associated weights on the right.

  - Formula of order 1 with 1 point:

$$(\frac{1}{3}, \frac{1}{3}, \frac{1}{3}) \qquad\qquad\qquad 1$$

– Formula of order 1 with 3 points:

$$(1,0,0) \quad (0,1,0) \quad (0,0,1) \qquad \frac{1}{3}$$

– Formula of order 2 with 3 points:

$$(\frac{1}{2},\frac{1}{2},0) \quad (\frac{1}{2},1,\frac{1}{2}) \quad (0,\frac{1}{2},\frac{1}{2}) \qquad \frac{1}{3}$$

– Formula of order 3 with 7 points:

$$
\begin{array}{ll}
(\frac{1}{3},\frac{1}{3},\frac{1}{3}) & \frac{9}{20} \\[2mm]
(\frac{1}{2},\frac{1}{2},0) \quad (\frac{1}{2},1,\frac{1}{2}) \quad (0,\frac{1}{2},\frac{1}{2}) & \frac{2}{15} \\[2mm]
(1,0,0) \quad (0,1,0) \quad (0,0,1) & \frac{1}{20}
\end{array}
$$

– Formula of order 5 with 7 points $(\alpha = \dfrac{6-\sqrt{15}}{21}, \beta = \dfrac{6+\sqrt{15}}{21})$:

$$
\begin{array}{ll}
(\frac{1}{3},\frac{1}{3},\frac{1}{3}) & \frac{9}{20} \\[2mm]
(\alpha,\alpha,1-2\alpha) \quad (\alpha,1-2\alpha,\alpha) \quad (1-2\alpha,\alpha,\alpha) & \dfrac{155-\sqrt{15}}{1200} \\[2mm]
(\beta,\beta,1-2\beta) \quad (\beta,1-2\beta,\beta) \quad (1-2\beta,\beta,\alpha) & \dfrac{155+\sqrt{15}}{1200}
\end{array}
$$

- **Quadrature formulas in a tetrahedron.** We indicate below the quadrature points on the left (with their 4 barycentric coordinates) and the associated weights on the right.

  – Formula of order 1 with 1 point:

$$(\frac{1}{4},\frac{1}{4},\frac{1}{4},\frac{1}{4}) \qquad\qquad 1$$

– Formula of order 1 with 4 points:

$$(1,0,0,0) \quad (0,1,0,0) \quad (0,0,1,0) \quad (0,0,0,1) \qquad \frac{1}{4}$$

– Formula of order 2 with 4 points $\left(\alpha = \dfrac{5 - \sqrt{5}}{20}\right)$:

$$(\alpha, \alpha, \alpha, 1 - 2\alpha) \quad (\alpha, \alpha, 1 - 2\alpha, \alpha) \quad (\alpha, 1 - 2\alpha, \alpha, \alpha)$$

$$(1 - 2\alpha, \alpha, \alpha, \alpha) \quad \frac{1}{4}$$

– Formula of order 3 with 5 points:

$$\left(\frac{1}{4}, \frac{1}{4}, \frac{1}{4}, \frac{1}{4}\right) \qquad\qquad -\frac{4}{5}$$

$$\left(\frac{1}{6}, \frac{1}{6}, \frac{1}{6}, \frac{1}{2}\right) \quad \left(\frac{1}{6}, \frac{1}{6}, \frac{1}{2}, \frac{1}{6}\right) \quad \left(\frac{1}{6}, \frac{1}{2}, \frac{1}{6}, \frac{1}{6}\right) \quad \left(\frac{1}{2}, \frac{1}{6}, \frac{1}{6}, \frac{1}{6}\right) \qquad -\frac{9}{20}$$

Quadrature formulas in quadrilaterals or hexahedra. Let us recall that any quadrilateral or hexahedral K can be seen as the image of the reference element $\widehat{K}$ (unit square or unit cube) by a Q_1 transformation F_K. Once again, it is natural to construct a quadrature formula in K from a quadrature formula in $\widehat{K}$ thanks to the following formula,

$$\int_K f_K(x)\,dx = \int_{\widehat{K}} J_K(\widehat{x})\,\widehat{f}_K(\widehat{x})\,d\widehat{x} \tag{10.81}$$

where we have set,

$$\widehat{f}_K = f_K \circ F_K : \widehat{K} \to \mathbb{R}, \quad J_K = \det DF_K. \tag{10.82}$$

The novelty with respect to formula (10.77) is that the Jacobian J_K remains inside the integral, because it is not constant.

It remains to construct quadrature formulas in $\widehat{K}$. Such formulas will be constructed in order to be exact in some Q_q for some integer q. The basic remark is that, since $\widehat{K} = [0,1]^d$

$$\int_{\widehat{K}} \prod_{j=1}^d \widehat{f}_j(\widehat{x}_j)\,d\widehat{x} = \prod_{j=1}^d \int_0^1 \widehat{f}_j(\widehat{x}_j)\,d\widehat{x}_j. \tag{10.83}$$

As a consequence, let us consider a 1D quadrature formula,

$$\int_{[0,1]}^{\nu,\mu} f(t)dt = \sum_{\ell=1}^{q} \mu_\ell \, f(\nu_\ell), \tag{10.84}$$

where the ν_ℓ's are quadrature points in $[0,1]$ and the μ_ℓ's corresponding weights. A natural idea for constructing a quadrature formula in $\widehat{K}$ is to consider,

$$\begin{cases} \widehat{\mathcal{Q}} = \{\widehat{\mathcal{M}}_\alpha = (\nu_{\alpha_1}, \cdots, \nu_{\alpha_1}), \alpha \in S_q = \{1, \cdots, q\}^d \} & \text{as quadrature points,} \\ \{\omega_\alpha = \mu_{\alpha_1} \cdots \mu_{\alpha_l}, \ \alpha \in S_q \} & \text{as the corresponding quadrature weights.} \end{cases} \tag{10.85}$$

The "natural" quadrature formula is thus,

$$\int_{\widehat{K}}^{Q} \widehat{f}(\widehat{x}) \, d\widehat{x} = \sum_{\alpha \in S_q} \omega_\alpha \, \widehat{f}(\widehat{\mathcal{M}}_\alpha), \tag{10.86}$$

in such a way that, if $\widehat{f}(\widehat{x}) = \prod_{j=1}^{d} \widehat{f}_j(\widehat{x}_j)$ then,

$$\int_{\widehat{K}}^{Q} \prod_{j=1}^{d} \widehat{f}_j(\widehat{x}_j) \, d\widehat{x} = \prod_{j=1}^{d} \int_{[0,1]}^{\nu,\mu} \widehat{f}_j(\widehat{x}_j) \, d\widehat{x}_j. \tag{10.87}$$

Thus , if the 1D formula is exact for each $\widehat{f}_j$, the formula (10.86) is exact for

$$\widehat{f} = \prod_{j=1}^{d} \widehat{f}_j.$$

As a consequence, as Q_q is generated by polynomials of the form

$$\widehat{f}(\widehat{x}) = \prod_{j=1}^{d} \widehat{f}_j(\widehat{x}_j)$$

where each $\widehat{f}_j$ is a 1D-polynomial of degree less than q, a necessary and sufficient condition for (10.86) to be exact in Q_q is that the $1D$ formula (10.84) integrates exactly in $[0,1]$ polynomials of degree less or equal than q.

Following (10.81), the quadrature formula on K corresponding to (10.86) will be,

$$\int_{K}^{Q_K} f(x) \, dx = meas \, K \sum_{\alpha \in S_q} \omega_\alpha^K \, f(\mathcal{M}_\alpha). \tag{10.88}$$

$$\mathcal{M}_\alpha^K = F_K(\widehat{\mathcal{M}}_\alpha), \quad \omega_\alpha^K \, meas \, K = \omega_\alpha \, J_K(\widehat{\mathcal{M}}_\alpha). \tag{10.89}$$

REMARK 10.22 *For the construction of the mass and stiffness matrices, at least in the case of piecewise coefficients ρ and C, the functions to be integrated in K will be known analytically in $\widehat{K}$ (products of Q_k basis functions or of their derivatives) so that only the Jacobian of F_K at the quadrature points needs to be computed, which can be done analytically.*

REMARK 10.23 *In general, even if the quadrature formula (10.86) is exact in Q_q, the formula (10.88) is not, due to the fact that J_K is not constant. However, we shall still say that this formula is of order q.*

To be complete, is suffices to recall that there exists a very complete and very general theory for 1D quadrature based on the theory of orthogonal polynomials. This allows us to construct quadrature formulas in quadrilaterals or hexahedrals of arbitrary high order. The most well known (and most accurate) formulas are the Gauss-Legendre, that we describe below, based on Legendre polynomials. We shall see in the next section that another class of quadrature formulas is very useful for the purpose of mass lumping : the Gauss-Lobatto formulas.

In what follows we denote by $P_k(\mathbb{R})$ the set of polynomials in 1 variable whose degree is less or equal than k. Legendre polynomials in $[0,1]$ are characterized by the following lemma (see for instance [74]),

LEMMA 10.2 *There exists a unique family of polynomials $\{\mathbf{p}_n, n \in \mathbb{N}\}$ such that (by convention $\mathbf{p}_0 = 1$),*

- $\mathbf{p}_n \in P_n(\mathbb{R})$ *and* $\mathbf{p}_n(\xi) - \xi^n \in P_{n-1}(\mathbb{R})$,

- $\mathbf{p}_n$ *is orthogonal to* $P_{n-1}(\mathbb{R})$: $\displaystyle\int_0^1 \mathbf{p}_n(\xi)\, q(\xi)\, d\xi = 0, \quad \forall\, q \in P_{n-1}(\mathbb{R}).$

These polynoms are known explicitly,

$$\mathbf{p}_n(\xi) = \frac{n!}{(2n)!}\, \frac{d^n}{d\xi^n}\, \xi^n(1-\xi)^n$$

and satisfy, as any family of orthogonal polynomials, a two level recurrence relation. The first three of them are,

$$\mathbf{p}_1(\xi) = \xi - \frac{1}{2}, \quad \mathbf{p}_2(\xi) = \left(\xi - \frac{1}{2}\right)^2 - \frac{1}{3}, \quad \mathbf{p}_3(\xi) = \left(\xi - \frac{1}{2}\right)\left[\left(\xi - \frac{1}{2}\right)^2 - \frac{1}{3}\right].$$

The important property for us is the following

LEMMA 10.3 *For each $n \geq 1$ the n zeros of $\mathbf{p}_n$ are real, distinct, included in $]0,1[$ and symmetrically distributed in this interval,*

$$0 < \xi_{n,\ell} < \cdots < \xi_{n,n} < 1, \; for \; 1 \leq \ell \leq n, \quad \xi_{n,\ell} + \xi_{n,n-\ell} = 1.$$

For $n = 1, 2, 3$, we have,

$$
\begin{cases}
\xi_{1,1} = \dfrac{1}{2} \\[2ex]
\xi_{2,1} = \dfrac{1}{2} - \dfrac{\sqrt{3}}{3}, \quad \xi_{2,2} = \dfrac{1}{2} + \dfrac{\sqrt{3}}{3}, \\[2ex]
\xi_{3,1} = \dfrac{1}{2} - \sqrt{\dfrac{3}{5}}, \quad \xi_{3,2} = \dfrac{1}{2}, \qquad \xi_{3,3} = \dfrac{1}{2} + \sqrt{\dfrac{3}{5}}.
\end{cases}
\tag{10.90}
$$

We are now in position to answer the following question:

Given an integer $n \geq 1$, what is the maximal value of m for which there exists a quadrature formula that integrates exactly polynomials in $P_m(\mathbb{R})$?

The answer to this question is $m = 2n - 1$. The more precise answer is given by the following theorem,

THEOREM 10.7   *Given $n \geq 1$, there exists a unique quadrature formula of the form (10.84) with n points that integrates exactly on $[0, 1]$ polynomials in $P_{2n-1}(\mathbb{R})$. This formula is such that,*

- *The quadrature points are the zeros of $\mathbf{p}_n$: $\nu_\ell = \xi_{n,\ell}, \quad \ell = 1, \cdots, n$,*

- *The weights ω_ℓ are strictly positive and given by:*

$$
\omega_\ell = \int_0^1 \left[\frac{\mathbf{p}_n(\xi)}{\xi - \xi_{n,\ell}} \right]^2 d\xi.
$$

Moreover, this formula is not exact in $P_{2n}(\mathbb{R})$.

Analysis of the approximation of the mass matrix.   We consider here the case where the stiffness matric is computed exactly and only the mass matrix is computed with quadrature formulas (see however the last paragraph of this section). We present here the analysis in the case of P_k elements and refer to remark 10.32 for the comments concerning the more complicated case of the Q_k elements. We consider here that the problem consists in finding $u_h(t) : \mathbb{R} \to V_h$ such that,

$$
\begin{cases}
\dfrac{d^2}{dt^2}\big(u_h(t), v_h\big)_h + a\big(u_h(t), v_h\big) \ = \ <L(t), v_h>, \quad \forall\, v \in V_h, \\[2ex]
u_h(0) = u_{0,h}, \quad \dfrac{du}{dt}(0) = u_{1,h}
\end{cases}
\tag{10.91}
$$

where V_h is defined by (10.34) and where

$$
(u_h, v_h)_h = \sum_{K \in \mathcal{T}_h} \int_K^{Q_K^m} \rho\, u_h\, v_h\, dx,
\tag{10.92}
$$

the quadrature form $\int_K^{Q_K^m} f \, dx$ being of the form (10.74) with (10.79) and (10.89). We assume, for simplicity, that the same quadrature formula is applied in each element. Keeping the notation of section 10.1, (10.91) is equivalent to the ordinary differential linear system,

$$
\begin{cases}
\widetilde{\mathbf{M}}_h \dfrac{d^2 U_h}{dt^2}(t) + \mathbf{A}_h U_h(t) = F_h(t), \\[2mm]
U_h(0) = U_{0,h}, \quad \dfrac{dU_h}{dt}(0) = U_{1,h},
\end{cases}
\tag{10.93}
$$

where $\widetilde{\mathbf{M}}_h$ is the modified mass matrix defined by,

$$
\widetilde{\mathbf{M}}_{i,j} = (w_j, w_i)_h.
\tag{10.94}
$$

The first step of the analysis amounts to establishing an existence and stability result for (10.91). Such a result is guaranteed if the symmetric matrix $\widetilde{\mathbf{M}}_h$ is positive definite, uniformly with respect to h. This is where the following assumptions play a fundamental role,

($\mathcal{H}1$)   The weights ω_ℓ in (10.74) are strictly positive.

($\mathcal{H}2$)
$$
\begin{cases}
\text{The quadrature points } \mathcal{M}_l \in \widehat{\mathcal{Q}} \text{ in (10.74) are such that:} \\[2mm]
\widehat{v} \in P_k \quad \text{and} \quad \widehat{v}(\widehat{\mathcal{M}}_\ell) = 0, \quad \text{for all } \mathcal{M}_l \in \widehat{\mathcal{Q}} \implies \widehat{v} = 0.
\end{cases}
$$

REMARK 10.24 *The assumption ($\mathcal{H}2$) implies that the number of quadrature points is larger or equal than the dimension of P_k. However, this does not mean that the set $\{\mathcal{M}_l \in \widehat{\mathcal{Q}}\}$ is P_k unisolvent: the number of quadrature points can be strictly larger.*

Under such assumptions, we can show the following

LEMMA 10.4 *Suppose that assumptions ($\mathcal{H}1$) and ($\mathcal{H}2$) hold. There exist a strictly positive constant γ such that,*

$$
\forall \, v_h \in V_h, \quad (v_h, v_h)_h \geq \gamma \, |v_h|^2.
\tag{10.95}
$$

PROOF   We first prove that there exists $\widehat{\gamma} > 0$ such that,

$$
\forall \, \widehat{v} \in P_k, \quad \int_{\widehat{K}}^{\widehat{Q}} |\widehat{v}|^2 \, d\widehat{x} \geq \widehat{\gamma} \int_{\widehat{K}} |\widehat{v}|^2 \, d\widehat{x}.
$$

Indeed

$$
\int_{\widehat{K}}^{\widehat{Q}} |\widehat{v}|^2 \, d\widehat{x} = \sum_{\widehat{M}_l \in \widehat{Q}} \omega_l \, |\widehat{v}(\widehat{M}_l)|^2.
$$

Using successively ($\mathcal{H}1$) and ($\mathcal{H}2$), we thus have, if $\hat{v} \in P_k$

$$\int_{\widehat{K}}^{\widehat{Q}} |\hat{v}|^2 \, d\hat{x} = 0 \quad \Longrightarrow \quad \hat{v}(\widehat{M}_l) = 0, \quad \forall \, \widehat{M}_l \in \widehat{Q} \quad \Longrightarrow \quad \hat{v} = 0.$$

As a consequence $\hat{v} \to \left(\int_{\widehat{K}}^{\widehat{Q}} |\hat{v}|^2 \, d\hat{x} \right)^{\frac{1}{2}}$ is a norm in P_k.

The existence of $\hat{\gamma} > 0$ follows from the fact that, in a finite dimensional space, here P_k, all norms are equivalent.

By scaling, using the change of variable $x = F_K(\hat{x})$, we get

$$\forall \, v_K \in P_k, \quad \int_K^{Q_K} |v_K|^2 \, dx \geq \hat{\gamma} \int_K |v_K|^2 \, dx.$$

Now, for any $v_h \in V_h$

$$\left| (v_h, v_h)_h \right. \geq \rho_- \sum_{K \in \mathcal{T}_h} \int_K^{Q_K} |v_K|^2 \, dx$$

$$\geq \hat{\gamma} \, \rho_- \sum_{K \in \mathcal{T}_h} \int_K |v_K|^2 \, dx$$

$$\geq \gamma \int_\Omega \rho \, |v_h|^2 \, dx,$$

with $\gamma = \hat{\gamma} \, (\rho_-/\rho_+)$. $\qquad\qquad\qquad\qquad\qquad\qquad$ ☐

The direct consequence of this result is the stability of the method (10.91). Indeed, the same energy techniques as in section (see theorem 10.2 and corollary 10.1) permit us to control the *modified* discrete energy,

$$E_h^m(t) = \frac{1}{2} \{ \, |\frac{du_h}{dt}(t)|_h^2 + a(u_h(t), u_h(t)) \}, \qquad (10.96)$$

which itself controls the discrete energy $E_h(t)$ defined by (10.14) using (10.95).

The next step is a consistency error estimate, i.e., the evaluation of the quadrature error committed when replacing (u_h, v_h) by $(u_h, v_h)_h$, namely,

$$\mathcal{E}_h(u_h, v_h) = (u_h, v_h) - (u_h, v_h)_h, \quad \forall \, (u_h, v_h) \in V_h \times V_h. \qquad (10.97)$$

Getting the optimal order of approximation, requires appropriate regularity assumptions on the function $\rho(x)$. More precisely, if $\{\Omega_m, m \leq M\}$ is a partition of Ω in the sense of (10.51)

$$\forall \, 1 \leq m \leq M, \quad \rho \, |_{\Omega_m} \in W^{k,\infty}(\overline{\Omega}_m). \qquad (10.98)$$

The following result applies to $k \geq 2$. For the case $k = 1$, see remark 10.25.

LEMMA 10.5 *Assume that* $(\mathcal{H}1)$ *and* $(\mathcal{H}2)$ *hold, that the regularity assumption (10.98) is satisfied and that the triangulations* $\mathcal{T}_h$ *satisfy the compatibility condition (10.52). Assume that* $k \geq 2$ *and suppose that the quadrature formula is of order* $2k - 2$,

$$\int_{\widehat{K}} \widehat{f} \, d\widehat{x} = \int_{\widehat{K}}^{\widehat{Q}} \widehat{f} \, d\widehat{x}, \quad \forall \, \widehat{f} \in P_{2k-2}. \tag{10.99}$$

There exists a constant $C > 0$ *independent of* h *such that,*

$$\forall \, (u_h, v_h) \in V_h \times V_h, \quad |\mathcal{E}_h(u_h, v_h)| \leq C \, h^k \left(\sum_{K \in \mathcal{T}_h} |u_h|_{k-1,K}^2 \right)^{\frac{1}{2}} \|v_h\| \tag{10.100}$$

PROOF For simplicity, we shall give the proof when $\rho = 1$. The case of smoothly varying ρ only adds technical difficulties and in such a case the constant C in (10.100) depends on the norms $\|\rho\|_{W^{k,\infty}(\overline{\Omega}_m)}$.

The proof is decomposed in three steps.

Step 1. We introduce the quadrature error in $\widehat{K}$,

$$\widehat{E}(\widehat{f}) = \int_{\widehat{K}} \widehat{f} \, d\widehat{x} - \int_{\widehat{K}}^{\widehat{Q}} \widehat{f} \, d\widehat{x}, \quad \forall \, \widehat{f} \in C^0(\widehat{K}).$$

Let us prove that there exists $\widehat{C} > 0$ such that,

$$\forall (\widehat{u}, \widehat{v}) \in P_k \times P_k, \quad |\widehat{E}(\widehat{u} \, \widehat{v})| \leq \widehat{C} \, |\widehat{u}|_{k-1,\widehat{K}} \, |\widehat{v}|_{1,\widehat{K}}. \tag{10.101}$$

Indeed, let q be an integer and $\mathbf{\Pi}_{q-1}$ be the L^2-orthogonal projection from $L^2(\widehat{K})$ in P_{q-1}. A well known consequence of Bramble-Hilbert's lemma is,

$$\forall \, \widehat{v} \in H^q(\widehat{K}), \quad |\widehat{v} - \mathbf{\Pi}_{q-1} \, \widehat{v}|_{0,\widehat{K}} \leq \widehat{C}_q \, |\widehat{v}|_{q,\widehat{K}}. \tag{10.102}$$

Next, we remark that if $(\widehat{u}, \widehat{v}) \in P_k \times P_k$,

$$\mathbf{\Pi}_{k-2} \, \widehat{u} \cdot \widehat{v} \in P_{2k-2} \quad \text{and} \quad (\widehat{u} - \mathbf{\Pi}_{k-2} \, \widehat{u}) \cdot \mathbf{\Pi}_0 \, \widehat{v} \in P_k \subset P_{2k-2}.$$

Consequently, the quadrature formula being of order $2k - 2$,

$$\widehat{E}(\widehat{u} \, \widehat{v}) = \widehat{E}\big((\widehat{u} - \mathbf{\Pi}_{k-2} \, \widehat{u}) \, \widehat{v} \big) = \widehat{E}\big((\widehat{u} - \mathbf{\Pi}_{k-2} \, \widehat{u}) (\widehat{v} - \mathbf{\Pi}_0 \, \widehat{v}) \big).$$

Using that, in a finite dimensional space, a bilinear form is necessarily continuous and all the norms are equivalent, we can write

$$|\widehat{E}(\widehat{u}\,\widehat{v})| \leq \widetilde{C}_k \, \|\widehat{u} - \mathbf{\Pi}_{k-2}\widehat{u}\|_{0,\widehat{K}} \, \|\widehat{v} - \mathbf{\Pi}_0\widehat{v}\|_{0,\widehat{K}}. \tag{10.103}$$

Using (10.102) with $q = 1$ and $q = k - 1$ leads to (10.101) with $\widehat{C} = \widetilde{C}_k \, \widehat{C}_{k-1} \, \widehat{C}_0$.

Step 2. Let $E_K(\cdot)$ be the quadrature error in K,

$$E_K(f) = \int_K f \, dx - \int_K^{Q_m} f \, dx, \quad \forall \, f \in C^0(K).$$

Let $(u_K, v_K) \in P_k \times P_k$ and $(\widehat{u}_K, \widehat{v}_K) = (u_K \circ F_K^{-1}, v_K \circ F_K^{-1}) \in P_k \times P_k$, by definition,

$$E_K(u_K \, v_K) = J_K \, \widehat{E}(\widehat{u}_K \, \widehat{v}_K).$$

Therefore, by step 1,

$$|E_K(u_K \, v_K)| \leq J_K \, \widehat{C} \, |\widehat{u}_K|_{k-1,\widehat{K}} \, |\widehat{v}_K|_{1,\widehat{K}}.$$

Using the change of variable $x = F_K(\widehat{x})$, one easily sees that

$$J_K^{\frac{1}{2}} \, |\widehat{u}_K|_{k-1,\widehat{K}} \leq h_K^{k-1} \, |u_K|_{k-1,K}, \quad J_K^{\frac{1}{2}} \, |\widehat{v}_K|_{1,\widehat{K}} \leq h_K \, |u_K|_{1,K}.$$

Step 3. This is a simple computation. If $(u_h \, v_h) \in V_h$, $u_K = u_h|_K$ and $v_K = v_h|_K$

$$|\mathcal{E}_h(u_h \, v_h)| = |\sum_{K \in \mathcal{T}_h} E_K(u_K \, v_K)| \leq C \, h^k \sum_{K \in \mathcal{T}_h} |u_K|_{k-1,K} \, |v_K|_{1,K}.$$

One concludes with Cauchy-Schwartz inequality. □

REMARK 10.25 *It is easy to see how to modify the proof and the result for* $k = 1$ *and if the quadrature formula is assumed to be of order 1. Indeed in that case, if* $(\widehat{u}, \widehat{v}) \in P_1 \times P_1$,

$$\mathbf{\Pi}_0 \, \widehat{u} \cdot \widehat{v} \in P_1 \quad \text{and} \quad (\widehat{u} - \mathbf{\Pi}_0 \, \widehat{u}) \cdot \mathbf{\Pi}_0 \, \widehat{v} \in P_1.$$

Consequently, the quadrature formula being of order 1, (10.103) becomes,

$$|\widehat{E}(\widehat{u}\,\widehat{v})| \leq \widetilde{C}_k \, \|\widehat{u} - \mathbf{\Pi}_0 \, \widehat{u}\|_{0,\widehat{K}} \, \|\widehat{v} - \mathbf{\Pi}_0 \, \widehat{v}\|_{0,\widehat{K}}, \tag{10.104}$$

and (10.101) becomes,

$$\forall (\widehat{u}, \widehat{v}) \in P_1 \times P_1, \quad |\widehat{E}(\widehat{u}\,\widehat{v})| \leq \widehat{C} \, |\widehat{u}|_{1,\widehat{K}} \, |\widehat{v}|_{1,\widehat{K}}. \tag{10.105}$$

Finally, the estimate (10.100) of lemma 10.5 is replaced by,

$$\forall \, (u_h, v_h) \in V_h \times V_h, \quad |\mathcal{E}_h(u_h, v_h)| \leq C \, h^2 \left(\sum_{K \in \mathcal{T}_h} |u_h|_{1,K}^2 \right)^{\frac{1}{2}} \|v_h\|. \tag{10.106}$$

REMARK 10.26 *We have treated u_h and v_h differently. This is because our objective was to make appear the norm $\|v_h\|$ in the right hand side of (10.100), and not higher order norms. The reason will appear clearly in the proof of theorem 10.8 (see also the remark 10.29).*

Of course, applying a symmetric treatment would have led to

$$|\mathcal{E}_h(u_h, v_h)| \leq C\, h^{2k} \left(\sum_{K \in \mathcal{T}_h} |u_h|^2_{k-1,K} \right)^{\frac{1}{2}} \left(\sum_{K \in \mathcal{T}_h} |v_h|^2_{k-1,K} \right)^{\frac{1}{2}}. \qquad (10.107)$$

REMARK 10.27 *If we replace the standard P_k finite element space V_h by a more general Lagrange finite element space V_h of the form (10.42) such that each local space $\mathbf{V}_K$ satisfies the double inclusion:*

$$P_k \subset \mathbf{V}_K \subset P_{k'}\,, \qquad \text{with } k' > k, \qquad (10.108)$$

it is rather easy to prove (with a slight modification of the proof of lemma 10.5) that the conclusion (10.100) still holds provided that:

$$\text{The quadrature formula is exact in } P_{k+k'-2}\,. \qquad (10.109)$$

Next, we show why the consistency estimate leads to optimal error estimates, i.e., in $O(h^k)$ for H^1-norms in space and $O(h^{k+1})$ for L^2-norms in space, provided adequate regularity for the exact solution u. This is a very technical (although not very hard) piece of work. In order to present a rather short proof we shall restrict ourselves to the obtention of the $O(h^k)$ H^1-estimates. Moreover, we do not pay attention to obtaining the optimal order of convergence for the minimal regularity. Our proof follows in its principle the proof of [13], except that we avoid the use of an elliptic projection. The duality technique presented also in [13] leads to optimal L^2-estimates. We state below the result for $k = 2$. For $k = 1$, this result has to be modified according to remark 10.25. The details are left to the reader.

For simplicity, we assume that

$$u_{0,h} = \mathcal{I}_h u_0 \quad \text{and} \quad u_{1,h} = \mathcal{I}_h u_1, \qquad (10.110)$$

where $\mathcal{I}_h$ denotes an interpolation operator. This is not restrictive since it coincides with the usual choice in practice.

THEOREM 10.8 *Let $k \geq 2$. Assume that (10.52), (10.56) and (10.110) hold. Assume also that ρ satisfies (10.98). Then, provided that the solution u of (9.24) satisfies the regularity assumption,*

$$u \in W^{3,1}_{loc}(\mathbb{R}^+; W_{k-1}) \cap W^{2,1}_{loc}(\mathbb{R}^+; W_k) \cap W^{1,1}_{loc}(\mathbb{R}^+; W_{k+1}), \qquad (10.111)$$

for any $T > 0$, one has the error estimates (C denotes a positive constant independent of h),

$$\|u_h - u\|_{L^\infty(0,T;H^1(\Omega)^d)} \leq C\, h^k \sum_{j=0}^{2} \|\frac{d^j u}{dt^j}\|_{L^\infty(0,T;W_{k+1-j})}$$
$$+ C\, h^k \sum_{j=1}^{3} \|\frac{d^j u}{dt^j}\|_{L^1(0,T;W_{k+2-j})} .$$

(10.112)

PROOF The idea is to split the error into two parts, the difference with what we did in section 10.2 being that we use here the interpolation operator $\mathcal{I}_h$ instead of the elliptic projection Π_h (cf remark 10.31),

$$e_h(t) = \eta_h(t) - \varepsilon_h(t), \quad \eta_h(t) = u_h(t) - \mathcal{I}_h u(t), \quad \varepsilon_h = u(t) - \mathcal{I}_h u(t). \quad (10.113)$$

We already know, via interpolation results, how to estimate ε_h. It remains to bound η_h and to conclude by the triangular inequality.

In order to do so, we first establish an evolution equation for η_h where the right hand side contains terms in ε_h.

Substituting u_h by $\eta_h + \mathcal{I}_h u$ in (10.91) leads to (for any $v_h \in V_h$),

$$(\frac{d^2 \eta_h(t)}{dt^2}, v_h)_h + a(\eta_h(t), v_h) + (\frac{d^2 \mathcal{I}_h u(t)}{dt^2}, v_h)_h + a(\mathcal{I}_h u(t), v_h) = <L(t), v_h> .$$

Replacing $(\cdot, \cdot)_h$ by $(\cdot, \cdot) - \mathcal{E}_h(\cdot, \cdot)$ in the third term of the left hand side of the previous equality leads to,

$$\left| (\frac{d^2 \eta_h}{dt^2}(t), v_h)_h + a(\eta_h(t), v_h) + (\frac{d^2 \mathcal{I}_h u}{dt^2}(t), v_h) + a(\mathcal{I}_h u(t), v_h) = \right.$$
$$= <L(t), v_h> + \mathcal{E}_h(\frac{d^2 \mathcal{I}_h u(t)}{dt^2}, v_h).$$

Substracting to the above equality the identity,

$$(\frac{d^2 u}{dt^2}(t), v_h) + a(u(t), v_h) = <L(t), v_h>$$

we finally obtain,

$$(\frac{d^2 \eta_h}{dt^2}(t), v_h)_h + a(\eta_h(t), v_h) = (\frac{d^2 \varepsilon_h}{dt^2}(t), v_h) + a(\varepsilon_h(t), v_h) + \mathcal{E}_h(\frac{d^2 \mathcal{I}_h u}{dt^2}(t), v_h).$$

We now proceed via energy estimates. We take $v_h = \dfrac{d\eta_h}{dt}(t)$ and get,

$$\frac{d}{dt}\left\{ \frac{1}{2} |\frac{d\eta_h}{dt}|_h^2 + \frac{1}{2} a(\eta_h, \eta_h) \right\} = \mathcal{E}_h(\mathcal{I}_h \frac{d^2 u}{dt^2}, \frac{d\eta_h}{dt}) + (\frac{d^2 \varepsilon_h}{dt^2}, \frac{d\eta_h}{dt}) + a(\varepsilon_h, \frac{d\eta_h}{dt}).$$

Here we notice that a brute estimate of the right hand side using the continuity of $a(\cdot, \cdot)$ and the estimate (10.100) for $\mathcal{E}_h(\cdot, \cdot)$ would make appear the norm,

$$\|\frac{d\eta_h}{dt}\|$$

that cannot be controlled by the left hand side. The only authorized norms are

$$|\frac{d\eta_h}{dt}| \quad \text{or} \quad \|\eta_h\|.$$

That is why we are led to some manipulations. First, we remark that the previous equality can be rewritten as

$$\left|\frac{d}{dt}\left\{\frac{1}{2}|\frac{d\eta_h}{dt}|_h^2 + \frac{1}{2}a(\eta_h,\eta_h)\right\} = \frac{d}{dt}\left(\mathcal{E}_h(\mathcal{I}_h\frac{d^2u}{dt^2},\eta_h) + a(\varepsilon_h,\eta_h)\right)\right.$$

$$- \mathcal{E}_h(\mathcal{I}_h\frac{d^3u}{dt^3},\eta_h) - a(\frac{d\varepsilon_h}{dt},\eta_h) + (\frac{d^2\varepsilon_h}{dt^2},\frac{d\eta_h}{dt}).$$

Thanks to (10.110), $\eta_h(0) = \dfrac{d\eta_h}{dt}(0) = 0$. Thus, after time integration,

$$\frac{1}{2}|\frac{d\eta_h}{dt}|_h^2 + \frac{1}{2}a(\eta_h,\eta_h) = \mathcal{E}_h(\mathcal{I}_h\frac{d^2u}{dt^2},\eta_h) + a(\varepsilon_h,\eta_h)$$

$$- \int_0^t \mathcal{E}_h(\mathcal{I}_h\frac{d^3u}{dt^3},\eta_h)\,ds - \int_0^t a(\frac{d\varepsilon_h}{dt},\eta_h)\,ds + \int_0^t (\frac{d^2\varepsilon_h}{dt^2},\frac{d\eta_h}{dt})\,ds.$$

which yields, using the uniform coercivity estimate (10.4)

$$\left|\frac{\gamma}{2}|\frac{d\eta_h}{dt}|_h^2 + \frac{1}{2}a(\eta_h,\eta_h) \le \left|\mathcal{E}_h(\mathcal{I}_h\frac{d^2u}{dt^2},\eta_h)\right| + |a(\varepsilon_h,\eta_h)|\right.$$

$$+ \left|\int_0^t \mathcal{E}_h(\mathcal{I}_h\frac{d^3u}{dt^3},\eta_h)\,ds\right|$$

$$+ \left|\int_0^t a(\frac{d\varepsilon_h}{dt},\eta_h)\,ds\right|$$

$$+ \left|\int_0^t (\frac{d^2\varepsilon_h}{dt^2},\frac{d\eta_h}{dt})\,ds\right|.$$

(10.114)

Now we estimate the five terms of the right hand side of (10.114). In the inequalities below, C denotes a positive constant which may vary from one line to another but is always independent of h.

For the first term, we use the estimate (10.100) of lemma 10.5, together with the interpolation estimate (10.53) (see also remark 10.31), to get

$$\left| \mathcal{E}_h \left(\mathcal{I}_h \frac{d^2 u}{dt^2}, \eta_h \right) \right| \le C\, h^k \left(\sum_{K \in \mathcal{T}_h} |\mathcal{I}_h \frac{d^2 u}{dt^2}|^2_{k-1, K} \right)^{\frac{1}{2}} \|\eta_h\| \,,$$

$$\le C\, h^k \, |\frac{d^2 u}{dt^2}|_{W_k} \|\eta_h\| \,.$$

For any $\delta > 0$, we have (this is Young's inequality)

$$C\, h^k \, |\frac{d^2 u}{dt^2}|_{W_{k-1}} \|\eta_h\| = C\, h^k \, \delta^{-\frac{1}{2}} |\frac{d^2 u}{dt^2}|_{W_{k-1}} \cdot \delta^{\frac{1}{2}} \|\eta_h\|$$

$$\le \frac{\delta}{2} \|\eta_h\|^2 + \frac{C^2\, h^{2k}}{2\delta} |\frac{d^2 u}{dt^2}|^2_{W_{k-1}} \,.$$

Using the coercivity of $a(.,.)$, we thus get:

$$C\, h^k \, |\frac{d^2 u}{dt^2}|_{W_k} \|\eta_h\| \le \frac{\delta}{2\alpha}\, a(\eta_h, \eta_h) + \frac{C^2\, h^{2k}}{2\delta} |\frac{d^2 u}{dt^2}|^2_{W_{k-1}}$$

If we choose $\delta / 2\alpha = 1/8$, we finally obtain

$$\left| \mathcal{E}_h \left(\mathcal{I}_h \frac{d^2 u}{dt^2}, \eta_h \right) \right| \le \frac{1}{8}\, a(\eta_h, \eta_h) + C\, h^{2k} |\frac{d^2 u}{dt^2}|^2_{W_{k-1}} \,. \tag{10.115}$$

With the same kind of trick, we get for the second term,

$$| a(\varepsilon_h, \eta_h) | \le \frac{1}{8}\, a(\eta_h, \eta_h) + C\, a(\varepsilon_h, \varepsilon_h). \tag{10.116}$$

REMARK 10.28 *The choice of $1/8$ in (10.115) and (10.116) is somewhat arbitrary. In each of the estimates (10.115) and (10.116), we could replace $1/8$ by $\theta\gamma$ with $\theta > 0$. However, for which follows, it is important that $\theta < 1/4$.*

For the third term, using Cauchy-Schwartz inequality and estimate (10.100) we get,

$$\left| \int_0^t \mathcal{E}_h \left(\mathcal{I}_h \frac{d^3 u}{dt^3}, \eta_h \right) ds \right| \le C\, h^k \int_0^t |\frac{d^3 u_h}{dt^3}|_{W_{k-1}} \|\eta_h\|\, ds, \tag{10.117}$$

while for the last two terms, we have,

$$\begin{cases} \left| \int_0^t a\left(\frac{d\varepsilon_h}{dt}, \eta_h \right) ds \right| \le \int_0^t a\left(\frac{d\varepsilon_h}{dt}, \frac{d\varepsilon_h}{dt} \right)^{\frac{1}{2}} a(\eta_h, \eta_h)^{\frac{1}{2}}\, ds \\[2mm] \left| \int_0^t \left(\frac{d^2 \varepsilon_h}{dt^2}, \frac{d\eta_h}{dt} \right) ds \right| \le \int_0^t |\frac{d^2 \varepsilon_h}{dt^2}| |\frac{d\eta_h}{dt}|\, ds; . \end{cases} \tag{10.118}$$

Let us set

$$\Phi_h = \|\frac{d\eta_h}{dt}\|^2 + a(\eta_h, \eta_h). \tag{10.119}$$

Substituting (10.115) to (10.118) into (10.114), it is easy to get an inequality of the form (this is where the importance of the 1/8 factors appears)

$$\Phi_h(t) \leq \psi_h(t) + \int_0^t m_h(s) \, \Phi_h(s)^{\frac{1}{2}} \, ds, \tag{10.120}$$

where, C designing appropriate constants,

$$\left| \begin{array}{l} \psi_h = C \left(h^{2k} \left| \dfrac{d^2 u}{dt^2} \right|^2_{W_{k-1}} + \|\varepsilon_h\|^2 \right), \\[3mm] m_h = C \left(\left| \dfrac{d^2 \varepsilon_h}{dt^2} \right| + \left\| \dfrac{d\varepsilon_h}{dt} \right\| + h^k \left| \dfrac{d^3 u}{dt^3} \right|_{W_{k-1}} \right). \end{array} \right.$$

Applying an appropriate Gronwall's lemma (see for instance the proof of corollary 10.1) to (10.120), we finally get,

$$\sup_{t \in [0,T]} \Phi_h(t)^{\frac{1}{2}} \leq \sup_{t \in [0,T]} \psi_h(t)^{\frac{1}{2}} + \int_0^T m_h(s) \, ds. \tag{10.121}$$

Recalling the interpolation estimates,

$$\|\varepsilon_h(t)\| \leq C \, h^k \, |u(t)|_{W_{k+1}}, \quad \left\| \dfrac{d\varepsilon_h}{dt}(t) \right\| \leq C \, h^k \, \left| \dfrac{du}{dt}(t) \right|_{W_{k+1}},$$

and

$$\left| \dfrac{d^2 \varepsilon_h}{dt^2}(t) \right| \leq C \, h^k \, \left| \dfrac{d^2 u}{dt^2}(t) \right|_{W_k},$$

one sees that,

$$\left| \begin{array}{l} \sup_{t \in [0,T]} \psi_h(t)^{\frac{1}{2}} \leq C \, h^k \left\{ \left\| \dfrac{d^2 u}{dt^2} \right\|_{L^\infty(0,T;W_{k-1})} + \|u\|_{L^\infty(0,T;W_{k+1})} \right\}, \\[5mm] \int_0^T m_h(s) \, ds \leq C \, h^k \sum_{j=1}^3 \left\| \dfrac{d^j u}{dt^j} \right\|_{L^1(0,T;W_{k+2-j})}. \end{array} \right. \tag{10.122}$$

Since, by definition of Φ_h (10.119), using the coercivity of $a(\cdot, \cdot)$

$$\|\eta_h(t)\| \leq C \, \Phi_h(t)^{\frac{1}{2}}$$

Sustituting the inequalities 10.122 in (10.121) yields in particular

$$\|\eta\|_{L^\infty(0,T;H^1)} \leq C \, h^k \left[\sum_{j=0,2} \left\| \dfrac{d^j u}{dt^j} \right\|_{L^\infty(0,T;W_{k+1-j})} + \sum_{j=1}^3 \left\| \dfrac{d^j u}{dt^j} \right\|_{L^1(0,T;W_{k+2-j})} \right],$$

while we already know that

$$\|\varepsilon_h\|_{L^\infty(0,T;H^1)} \leq C \, h^k \, \|u\|_{L^\infty(0,T;W_{k+1})} \, .$$

It is then straightforward to conclude since

$$\|u - u_h\|_{L^\infty(0,T;H^1)} \leq \|\eta_h\|_{L^\infty(0,T;H^1)} + \|\varepsilon_h\|_{L^\infty(0,T;H^1)}$$

$$\square$$

REMARK 10.29 *It is clear why one can only use (10.100) and not (10.107). Indeed, when estimating the left hand side of (10.115), one wants to make appear $\|\eta_h\|$, i.e., not higher Sobolev norms than the 1-norm (otherwise, Gronwall's lemma cannot be applied).*

REMARK 10.30 *According to remark 10.27, it is clear that the optimal $O(h^k)$ H^1-estimates are still valid for spaces of approximation of the form (10.42) with (10.108) as soon as the accuracy property (10.109) is satisfied by the quadrature formula.*

REMARK 10.31 *The reader may wonder why we use the interpolation operator $\mathcal{I}_h$ instead of the elliptic projection operator Π_h which would have permitted us to get rid of $a(\varepsilon_h, v_h)$. The explanation is the inequality (10.115) for which we have used the property:*

$$\|\mathcal{I}_h v\|_{k-1,K} \leq \|v\|_{k-1,K} + \|v - \mathcal{I}_h v\|_{k-1,K} \leq \|v\|_{k-1,K} + C\, h\, \|v\|_{k,K} \leq C\, \|v\|_{k,K}$$

where we have used the interpolation inequality (10.53) with $m = k$ and $\ell = k - 1$. The problem is that a priori we can not use such an estimate with Π_h instead of $\mathcal{I}_h$. A notable exception is the case where $k = 1$ or $k = 2$ for which we can use directly the inequality

$$\|\Pi_h v\|_{H^1(\Omega)} \leq C\, \|v\|_{H^1(\Omega)},$$

which is a straightforward consequence of the definition of Π_h.

REMARK 10.32 *The equivalent approximation theory for Q_k elements is very similar to the P_k one in the special case when the transformations F_K are linear. In this case, one can show that the optimal order of accuracy is preserved when the quadrature formula in the reference element is exact in Q_{2k-2} for $k \leq 2$ and in Q_1 for $k = 1$.*

The general case is more delicate due to the fact that the Jacobian of the transform F_K is no longer constant over $\widehat{K}$. Additional technical difficulties appear, similar to the ones that appear in the case of a non constant ρ. One can find results in this direction in [83] where it is shown that having a quadrature formula in the reference element exact in Q_{2k-2} still preserves the optimal accuracy at least for large k's.

About the approximation of the stiffness matrix.   Of course, the use of quadrature formulas for the approximation of the stiffness matrix also requires to be justified. In any case, the positivity of the weights is required for stability reasons. In the case of P_k elements, it can be shown that, in order to preserve an optimal order of accuracy under reasonable assumptions on $C(x)$, it is sufficient (and probably necessary) that the quadrature formula be exact in P_{2k-2} (this is shown in [43], for instance). The reader will note that, in the case of a piecewise constant $C(x)$, this means that the stiffness matrix is computed exactly since the scalar product of the gradients of two P_k polynomials is in P_{2k-2}.

For the case of Q_k elements, the theory is, to our knowledge much less complete. One of the additional difficulties if the fact that the Jacobian J_K is not constant. One can find preliminary results in these directions in [123] or [83]. However, even in the case of affine transformations F_K, in particular for regular meshes, it appears that this time it is not needed to have a formula which would be exact in the case of a piecewise constant $C(x)$. Numerical simulations [63] show that it is sufficient to have a formula exact in Q_{2k-1}. In the case of general meshes, with the same type of accuracy, it appears that the order of approximation is preserved in 2D but deteriorated (by one order) in 3D. In 3D, the optimal accuracy is restored with formulas exact in Q_k. All these numerical observations still need to be clarified from the theoretical point of view.

10.4   The Mass Lumping Technique

The general problematic.   One drawback of the finite element approach for evolution hyperbolic problems is the presence of the mass matrix $\mathcal{M}_h$: even with an explicit time discretization (we anticipate here section 10.6), this matrix has to be inverted at each time step. With finite difference or finite volume methods for instance, this matrix is by construction diagonal (or at least block diagonal, the dimension of the blocks — typically 2 or 3 for anisotropic vectorial problems — being independent of h) which leads to "really" explicit schemes. This is coherent to the completely local nature of the continuous operator which is approximated by the mass matrix (typically the identity operator). With the finite element method, due to the fact that it is in general impossible to construct a basis of functions with disjoint supports, this matrix is no longer diagonal (although very sparse). To invert it, as direct methods are in practice prohibited for reasons of size, one uses in general an iterative algorithm, typically the conjugate gradient method, that will converge with very few iterations (typically between 10 and 50, depending on the complexity of the problem and the desired accuracy, which should a priori increase with the order of the finite element method, see the remark below). Nevertheless, this has an important cost: simply realize that one step of the iterative algorithm corresponds to one step of an explicit scheme with a diagonal mass matrix.

REMARK 10.33 *The iterative methods converge faster due to the good conditionning of the mass matrix. Note however that the condition number increases with the space dimension, with the size and the quality of the mesh and in the case of very heterogeneous media.*

An alternative approach is to apply the so-called mass lumping procedure. This consists in replacing the exact mass matrix by an approximation, the lumped mass matrix, which is diagonal. This should be done without losing any (order of) accuracy.

In the case of first order P_1 or Q_1 elements, the method is well known by engineers but is often presented differently. It corresponds to consider the diagonal matrix in which each diagonal element is the sum of all the elements of the same line in the original mass matrix!

The heuristic justification of this method is easy. For simplicity, let us consider the case where $\rho = 1$ and $\Gamma_0 = \emptyset$. Let us consider as the equation for the unknown u_i^ℓ the equation of the differential system obtained by taking in the variational formulation (10.5) the test function $v_h = w_i^\ell$. This equation is of the form,

$$\sum_{j \in \mathcal{V}(i)} \mathbf{M}_{ij}^s \frac{d^2 u_j^\ell}{dt^2} + \text{non differential terms} = \text{r.h.s},$$

where by definition the set $\mathcal{V}(i)$ is such that $\{M_j,\ j \in \mathcal{V}(i)\}$ is the set of nodes which are neighbors of M_i and where:

$$\mathbf{M}_{ij}^s = \int_\Omega w_i\, w_j\, dx.$$

For a sufficiently fine mesh, one can reasonably consider that

$$u_j^\ell \sim u_i^\ell, \quad \forall j \in \mathcal{V}(i).$$

As a consequence

$$\sum_{j \in \mathcal{V}(i)} \mathbf{M}_{ij}^s \frac{d^2 u_j^\ell}{dt^2} \sim \left(\sum_{j \in \mathcal{V}(i)} \mathbf{M}_{ij}^s \right) \frac{d^2 u_i^\ell}{dt^2}.$$

If one makes this substitution in the differential equations for all (i, ℓ) one obtains a differential system with a diagonal modified mass matrix. In fact this can simply be reinterpreted as the result of the application of a simple quadrature formula using the vertices of the elements as the quadrature nodes. Let us consider for example the case of the P_1 elements in triangular or tetrahedric meshes (our argumentation also works for Q_1 elements, as the reader will easily realize !). If $\mathcal{V}(K)$ is the set of the $d+1$ vertices of K, the appropriate quadrature formula is,

$$\int_K f \, dx \sim \frac{meas\ K}{d+1} \sum_{M \in \mathcal{V}(K)} f(M), \qquad (10.123)$$

which can easily be checked to be exact in P_1. Using this formula we have,

$$\mathbf{M}_{ij}^s \sim \sum_{K \in \mathcal{T}_h} \frac{meas\ K}{d+1} \sum_{M \in \mathcal{V}(K)} w_i(M)\, w_j(M).$$

As the points M are the vertices of the elements,

$$i \neq j \quad \Longrightarrow \quad \forall\, K \in \mathcal{T}_h, \quad \forall\, M \in \mathcal{V}(K), \quad w_i(M)\, w_j(M) = 0.$$

Therefore $\mathbf{M}_{ij}^s = 0$ for $j \neq i$ and the approximate mass matrix is diagonal. For $j = i$ we have,

$$\mathbf{M}_{ii}^s \sim \sum_{K \in \mathcal{T}_h} \frac{meas\ K}{d+1} \sum_{M \in \mathcal{V}(K)} w_i(M)^2.$$

Since $w_i(M) = 0$ or 1, at each vertex $w_i(M)^2 = w_i(M)$ and,

$$\mathbf{M}_{ii}^s \sim \sum_{K \in \mathcal{T}_h} \frac{meas\ K}{d+1} \sum_{M \in \mathcal{V}(K)} w_i(M) = \sum_{K \in \mathcal{T}_h} \int_K w_i \, dx = \int_\Omega w_i \, dx,$$

since $w_i|_K \in P_1$ and the quadrature formula is exact in P_1.

Noticing that $\sum_j w_j = 1$ we can write,

$$\mathbf{M}_{ii}^s \sim \int_\Omega w_i \left(\sum_j w_j\right) dx - \sum_j \int_\Omega w_i\, w_j \, dx = \sum_j \mathbf{M}_{ij}^s,$$

which is precisely what we announced.

If we use the same idea for higher order P_k or Q_k elements, we will still preserve the consistency of the scheme. However, the order of accuracy will be deteriorated to the $O(h^2)$ (in L^2) or $O(h)$ (in H^1) given by the lowest order elements.

In order to generalize this approach to higher orders the challenge is to find a quadrature formula of higher accuracy leading to a diagonal mass matrix. This relies upon the following remark whose easy proof relies on the same arguments than for the P_1 case and is left to the reader.

LEMMA 10.6   *Assume that V_h is a Lagrange finite element approximation space for V in the sense of remark (10.10) and assume that for each $K \in T_h$ the ($\mathbf{V}_K$ - unisolvent) set of nodes M_ℓ^K of the finite element space coïncides with the set of quadrature points $\mathcal{M}_\ell^K$ of the quadrature formula used for approximating the mass matrix. Then the corresponding approximate (mass lumped) mass matrix is diagonal.*

As a consequence, the challenge is to find in each element K a quadrature formula such that,

(i) The quadrature nodes (which coincide with the locations of the degrees of freedom) form a $\mathbf{V}_K$ unisolvent set of points.

(ii) The quadrature weights are strictly positive in order to ensure the stability result via lemma 10.24.

(iii) The quadrature formula is sufficiently precise in order to preserve the order of accuracy obtained with the exact mass matrix.

The precise consequence of the last criterion (iii) relies upon the error analysis and depends more on the nature of the element.

The case of Q_k elements.   We shall see that, in order to get mass lumping with preserved accuracy, it is sufficient to change the location of the degrees of freedom (with respect to their canonical location, i.e., when they are uniformly distributed in the reference element).

To exploit the "tensorial" nature of Q_k elements, we look for a quadrature formula of the form (10.85) with $q = k + 1$ since $dim\, Q_k = (k+1)^d$. In order to satisfy the criterion (i), we need to keep $k + 1$ points along each edge of $\widehat{K}$. Coming back to (10.85), this imposes,

$$\nu_1 = 0, \quad \nu_{k+1} = 1. \tag{10.124}$$

The theory (see remark 10.32) says that, in order to satisfy (iii), the quadrature formula in the reference element must be exact in Q_{2k-2} which amounts to saying that the corresponding 1D formula in $[0,1]$ (10.84) must be exact in $P_{2k-2}(\mathbb{R})$. In fact there exists a unique possible quadrature: the Gauss-Lobatto formula. To describe this formula, we introduce for each n, the zeros $\{\xi'_{n,1}, \cdots, \xi'_{n,n-1}\}$ of $\mathbf{p}'_n$ where $\mathbf{p}_n$ is the nth Legendre polynomial which satisfy obviously,

$$0 < \xi_{n,1} < \xi'_{n,1} < \xi_{n,1} < \cdots < \xi'_{n,n-1} < \xi_{n,n} < 1,$$

$$\xi'_{n,\ell} + \xi'_{n,n-\ell} = 1, \quad \forall\, 1 \leq \ell \leq n - 1.$$

LEMMA 10.7 *There exists a unique quadrature formula of the form (10.84) satisfying (10.124) and exact in $P_{2k-2}(\mathbb{R})$.*

- *The interior quadrature points are the zeros of $\mathbf{p}'_k$,*

$$\nu_{\ell+1} = \xi'_{k,\ell}, \quad \forall\, 1 \le \ell \le k-1. \tag{10.125}$$

- *The quadrature weights are strictly positive and given by,*

$$\omega_\ell = \int_0^1 \left[\frac{\mathbf{p}'_n(\xi)}{\xi - \xi'_{n,\ell}}\right]^2 \xi(1-\xi)\, d\xi. \tag{10.126}$$

Moreover, this formula is exact in $P_{2k-1}(\mathbb{R})$.

REMARK 10.34 *The uniqueness of a quadrature formula satisfying (10.124) exact for $P_{2k-1}(\mathbb{R})$ is expected: integrating exactly $P_{2k-1}(\mathbb{R})$ gives $2k$ equations while we have a priori $2k$ degrees of freedom, namely the $k-1$ interior nodes and the $k+1$ quadrature weights.*

REMARK 10.35 *The theory of Gauss-Lobatto formulas is based, as for Gauss-Legendre formulas, on the theory of orthogonal polynomials [74] except that here one considers a L^2 weighted inner product,*

$$\int_0^1 f(\xi)\, g(\xi)\, \xi(1-\xi)\, d\xi.$$

We give below the quadrature points and quadrature weights of the Gauss-Lobatto formula for $k = 1, 2, 3, 4$ (for $k = 1$ the formula is known as the trapezoidal rule and for $k = 2$ as the Simpson's formula).

$$k = 1 \quad \begin{cases} \nu_1 = 0,\ \nu_2 = 1, \\[2mm] \omega_1 = \omega_2 = \dfrac{1}{2}. \end{cases}$$

$$k = 2 \quad \begin{cases} \nu_1 = 0,\ \nu_2 = \dfrac{1}{2},\ \nu_3 = 1, \\[2mm] \omega_1 = \omega_3 = \dfrac{1}{6},\ \omega_2 = \dfrac{2}{3}. \end{cases}$$

$$k = 3 \quad \begin{cases} \nu_1 = 0,\ \nu_2 = \dfrac{5-\sqrt{5}}{10},\ \nu_3 = \dfrac{5+\sqrt{5}}{10},\ \nu_4 = 1, \\[2mm] \omega_1 = \omega_4 = \dfrac{1}{12},\ \omega_2 = \omega_3 = \dfrac{5}{12}. \end{cases}$$

$$k = 4 \quad \begin{cases} \nu_1 = 0,\ \nu_2 = \dfrac{7-\sqrt{21}}{14},\ \nu_3 = \dfrac{1}{2},\ \nu_4 = \dfrac{7+\sqrt{21}}{14},\ \nu_5 = 1, \\[2mm] \omega_1 = \omega_5 = \dfrac{1}{20},\ \omega_2 = \omega_4 = \dfrac{49}{180},\ \omega_3 = \dfrac{16}{45}. \end{cases}$$

$$\tag{10.127}$$

The case of P_k elements.   In the case of triangles or tetrahedra, the solution is more complicated in particular because 1D theory can not be used. Let us examine the difficulties that occur with triangles for $k = 2$ and $k = 3$.

In what follows the equality $M = (\lambda_1, \lambda_2, \lambda_3)$ means that M is the point in the triangle with barycentric coordinates $(\lambda_1, \lambda_2, \lambda_3)$. In what follows, we shall restrict ourselves to *symmetric* quadrature formulas in the triangle in the sense defined in [47] for instance,

- Applying any permutation of the barycentric coordinates leaves the set of quadrature points invariant.

- The quadrature weights associated to points having the same barycentric coordinates up to a permutation are the same.

Such formulas are optimal in the sense that they preserve the symmetry properties of the exact integral (see [47] for more details).

For P_2, the natural quadrature points are the three summits and the three middles of the edges (this is even the only symmetric P_2 unisolvent set of points since one needs 3 points along each edge),

$$\begin{cases} S_1 = (1,0,0), \quad S_2 = (0,1,0), \quad S_3 = (0,0,1), \\ M_1 = (0, \frac{1}{2}, \frac{1}{2}), \quad M_2 = (\frac{1}{2}, 0, \frac{1}{2}), \quad M_3 = (\frac{1}{2}, \frac{1}{2}, 0). \end{cases} \tag{10.128}$$

The quadrature formula must be completed by quadrature weights ω_S (associated to the summits) and ω_M (associated to the middle of the edges). According to the error analysis, one needs a quadrature exact in P_2 ($2k - 2 = 2$). There does exists a unique quadrature formula of this type corresponding to,

$$\omega_S = 0, \quad \omega_M = \frac{1}{3}.$$

The trouble is that, the weight associated to the summits being 0, the resulting mass lumped mass matrix has 0 entries and is thus non invertible.

If we now look at $k = 3$ we need a quadrature formula exact in P_4 ($2k - 2 = 4$). One easily sees [47] that it is not possible to find such a formula if one takes as quadrature points the usual canonical frame for P_3. However, there exists a one parameter family of symmetric P_3 unisolvent sets of points in which we simply authorize the 6 boundary points that are strictly interior to the edges to move (symetrically) along the boundary. This corresponds to (β denoting a real parameter in $]0, \frac{1}{2}[)$,

$$
\begin{cases}
S_1 = (1,0,0), \quad S_2 = (0,1,0), \quad S_3 = (0,0,1), \\[1ex]
M_{12}^\beta = (0, 1-\beta, \beta), \quad M_{21}^\beta = (0, \beta, 1-\beta), \\[1ex]
M_{13}^\beta = (1-\beta, 0, \beta), \quad M_{31}^\beta = (\beta, 0, 1-\beta), \\[1ex]
M_{23}^\beta = (1-\beta, \beta, 0), \quad M_{32}^\beta = (\beta, 1-\beta, 0), \\[1ex]
G = (\dfrac{1}{3}, \dfrac{1}{3}, \dfrac{1}{3}).
\end{cases}
\tag{10.129}
$$

One can easily show that there exists a unique value of β and quadrature weights ω_S, ω_M and ω_G so that the corresponding quadrature formula is exact in P_4. More precisely,

$$
\beta = \quad \text{et} \quad \omega_S = -\frac{1}{120}, \quad \omega_M = \frac{9}{40}, \quad \omega_G = \frac{1}{20}.
$$

This time the trouble is that, the weight associated to the vertices is strictly negative. As a consequence, the resulting mass lumped mass matrix, although invertible, is not positive definite and the resulting scheme is unstable !

As a consequence, for P_k elements, the solution to the mass lumping problem for $k \geq 2$ can not be found only by changing the location of the degrees of freedom, as we did for Q_k-elements. One solution can be found by also changing (in fact enriching) the space of functions locally in each triangle.

For $k = 2$, in the definition (10.42) of V_h we take (in a vector space, the notation $[v_1, \cdots, v_M]$ represents the subspace spanned by the vectors $(v_1, \cdots, v_M)$),

$$
\mathbf{V}_K = P_2 + [b], \quad b = \lambda_1 \lambda_2 \lambda_3.
\tag{10.130}
$$

Due to the fact that b vanishes along the boundary of the triangle, it is called a bubble function. Of course we need to introduce one additional node per triangle, namely the center of gravity. Then we look for a quadrature formula with quadrature points,

$$
\begin{cases}
S_1 = (1,0,0), \quad S_2 = (0,1,0), \quad S_3 = (0,0,1), \\[1ex]
M_1 = (0, \dfrac{1}{2}, \dfrac{1}{2}), \quad M_2 = (\dfrac{1}{2}, 0, \dfrac{1}{2}), \quad M_3 = (\dfrac{1}{2}, \dfrac{1}{2}, 0), \\[1ex]
G = (\dfrac{1}{3}, \dfrac{1}{3}, \dfrac{1}{3}).
\end{cases}
\tag{10.131}
$$

Since $P_2 \subset \mathbf{V}_K \subset P_3$, the theory (see remark 10.27) says that, in order to preserve the P_2 accuracy, the quadrature formula must be exact in P_3. One can show that there exists a unique choice of quadrature weights associated to (10.131) satisfying this property,

$$\omega_S = \frac{1}{40}, \quad \omega_M = \frac{1}{15}, \quad \omega_G = \frac{9}{40}. \tag{10.132}$$

As all these weights are positive and therefore the resulting scheme is stable.

For P_3 elements, we need to add more bubble functions. More precisely, we take,

$$\mathbf{V}_K = P_3 + b\, P_1. \tag{10.133}$$

As $P_3 \cap b\, P_1 = [b]$, *dim* $\mathbf{V}_K = 12$ and we need two additional nodes in the triangles with respect to P_3. The idea is to "explode" symmetrically the center of gravity into 3 interior points. In other words, we replace G by the following three points (α denoting a real parameter in $]0, \frac{1}{3}[$),

$$G_1^\alpha = (1 - \alpha, \frac{\alpha}{2}, \frac{\alpha}{2}), \quad G_2^\alpha = (\frac{\alpha}{2}, 1 - \alpha, \frac{\alpha}{2}), \quad G_3^\alpha = (\frac{\alpha}{2}, \frac{\alpha}{2}, 1 - \alpha) \tag{10.134}$$

Since

$$P_3 \subset \mathbf{V}_K \subset P_4$$

the theory says that, to preserve the P_3 accuracy, one needs a quadrature formula exact in P_5 (we apply the criterion of remark 10.27 with $k = 3$ and $k' = 4$). One finds (see [47]) that there exists a unique choice of values of α and β and of quadrature weights ω_S (associated to the S_i's), ω_M (associated to the M_{ij}^β's) and ω_G (associated to the G_i^α's) leading to such a formula, namely

$$\begin{cases} \beta = \dfrac{21 - \sqrt{84\sqrt{7} - 147}}{42}, \quad \alpha = \dfrac{7 + 2\sqrt{7}}{21}, \\[3mm] \omega_S = \dfrac{8 - \sqrt{7}}{720}, \quad \omega_M = \dfrac{7 + 4\sqrt{7}}{720}, \quad \omega_G = \dfrac{7(14 - \sqrt{7})}{720}. \end{cases} \tag{10.135}$$

This process can be continued for higher orders but there is no known general theory. To our knowledge the solution is known up to $k = 7$ (see [42] and [118]) and still consists in enriching the local approximation space with more and more bubble-like basis functions. Note in particular that the degrees of the traces of the functions in V_h along an edge of the mesh does not change (it remains equal to k) and, as a consequence, the additional degrees of freedom are purely interior (which automatically limits the increase of connectivity of the stiffness matrix). The corresponding method has been tested successfully and appears to be efficient in comparison with the non mass lumped case. This is due to the fact that the price to pay for obtaining mass lumping is only a few additional degrees of freedom: 1 (for 6) with $k = 2$, 2 (for 10) with $k = 3$, 3 (for 15) with $k = 4$, 9 (for 21) with $k = 5$, 18 (for 28) with $k = 6$ and

16 (for 35) with $k = 7$!

In dimension 3, one meets the same kind of trouble and one can use the same approach for overcoming the difficulty. Of course, the machinery and the computations are much more complicated to implement and the solution is known up to $k = 3$ (see [42]). Moreover, the interest is much less clear than in the 2D case because mass lumping requires a large increase in the number of degrees of freedom. Moreover the (still polynomial) enrichment of the local approximation space, although it still does not increase the degree of the functions along the edges of the mesh, leads to an increase of the degree of the traces on the faces of the mesh. As a consequence the additional degrees of freedom are not purely interior: there are also additional face nodes

For instance, for $k = 2$ (*dim* $P_2 = 10$ in 3D) the new space $\mathbf{V}_K$ satisfies,

$$P_2 \subset \mathbf{V}_K \subset P_4, \quad dim \, \mathbf{V}_K = 23, \tag{10.136}$$

and there are 13 additional degrees of freedom: 12 face nodes and one interior node.

10.5   Time Discretization by Finite Differences

In this section, we are interested in the time discretization of the semi-discrete problems obtained after finite element approximation in space. We shall restrict ourseles to the finite difference method. The conservative nature (cf. the conservation of the energy) of the abstract wave equation (9.26) can be seen as a consequence of the time reversibility of this equation. That is why we shall privilege centered finite difference schemes which preserve such a property at the discrete level. In this section, we essentially come back to the notation of the section relative to abstract second order evolution problems.

The second order leap-frog scheme.   Let us consider a time step $\Delta t > 0$ and denote by $u_h^n \in V_h$ an approximation of $u_h(t^n)$ where $t^n = n\Delta t, n \in \mathbb{N}$. The simplest finite difference scheme for the approximation of (10.5) is the so-called leap-frog scheme,

$$\left(\frac{u_h^{n+1} - 2u_h^n + u_h^{n-1}}{\Delta t^2}, v_h\right) + a(u_h^n, v_h) = (L(t^n), v_h), \quad \forall \, v_h \in V_h, \tag{10.137}$$

or if one uses an approximation for the calculation of (u_h, v_h) $(\rightarrow (u_h, v_h)_h)$ and $a(u_h, v_h)$ $(\rightarrow a_h(u_h, v_h))$ (using for instance quadrature formulas for calculating the integrals)

$$(\frac{u_h^{n+1} - 2u_h^n + u_h^{n-1}}{\Delta t^2}, v_h)_h + a_h(u_h^n, v_h) = (L(t^n), v_h)_h, \quad \forall \, v_h \in V_h, \quad (10.138)$$

If one introduces the operator $\mathcal{A}_h \in \mathcal{L}(V_h)$ defined by,

$$(\mathcal{A}_h u_h, v_h) = a(u_h, v_h) \quad \text{for (10.137)}, \quad (\mathcal{A}_h u_h, v_h)_h = a_h(u_h, v_h) \quad \text{for (10.138)},$$

the scheme can be rewritten as,

$$\frac{u_h^{n+1} - 2u_h^n + u_h^{n-1}}{\Delta t^2} + \mathcal{A}_h u_h^n = f_h^n, \quad (10.139)$$

where $f_h^n \in V_h$ is defined by,

$$(f_h^n, v_h) = (L(t^n), v_h) \quad \text{for (10.137)}, \quad (f_h^n, v_h)_h = (L(t^n), v_h)_h \quad \text{for (10.138)},$$

After introducing a basis $\{w_j\}$ of V_h, we can define U_h^n as the vector of the degrees of freedom of u_h^n to rewrite (10.137)

$$\mathbf{M}_h \frac{U_h^{n+1} - 2U_h^n + U_h^{n-1}}{\Delta t^2} + \mathbf{A}_h \, U_h^n = 0. \quad (10.140)$$

where $\mathbf{M}_h$ and $\mathbf{A}_h$ are respectively the mass and stiffness matrices. Note that the operator $\mathcal{A}_h$ is nothing but the operator whose matrix in the basis $\{w_j\}$ is $\mathbf{M}_h^{-1}\mathbf{A}_h$. By construction, (10.139) is explicit. If the mass matrix $\mathbf{M}_h$ is diagonal, which is the case with finite elements if one applies a mass lumping procedure (see section 10.4), it is "really" explicit since it allows us to compute U_h^{n+1} from the solution at the two previous time steps through a simple matrix-vector product,

$$U_h^{n+1} = 2U_h^n - U_h^{n-1} + \Delta t^2 \, \mathbf{M}_h^{-1}\mathbf{A}_h \, U_h^n. \quad (10.141)$$

Of course, (10.140) (or (10.137)) must be completed by a start-up procedure using the initial conditions to compute u_h^0 and u_h^1. We omit this here for simplicity.

By construction, this scheme is second order accurate in time. We next show how to improve the accuracy of this scheme.

Higher order schemes. It is possible to construct higher order schemes (which is *a priori* natural when using higher order finite elements in space) which remain explicit and centered. In particular, all the machinery of Runge-Kutta methods for ordinary differential equations is available. Let us concentrate here on a classical approach, the so-called modified equation approach.

For instance, to construct a fourth order scheme, we start by looking at the truncation error of (10.139),

$$\frac{u_h(t^{n+1}) - 2u_h(t^n) + u_h(t^{n-1})}{\Delta t^2} = \frac{d^2 u_h}{dt^2}(t^n) + \frac{\Delta t^2}{12}\frac{d^4 u_h}{dt^4}(t^n) + O(\Delta t^4).$$

Using the equation satisfied by u_h, we get the identity,

$$\frac{u_h(t^{n+1}) - 2u_h(t^n) + u_h(t^{n-1})}{\Delta t^2} = \frac{d^2 u_h}{dt^2}(t^n) + \frac{\Delta t^2}{12}\mathcal{A}_h^2 u_h(t^n) + O(\Delta t^4).$$

which leads to the following fourth order scheme in time,

$$\frac{u_h^{n+1} - 2u_h^n + u_h^{n-1}}{\Delta t^2} + \mathcal{A}_h u_h^n - \frac{\Delta t^2}{12}\mathcal{A}_h^2 u_h^n = 0, \qquad (10.142)$$

that can be implemented in such a way that each time step involes only 2 applications of the operator $\mathcal{A}_h$, using Horner's rule,

$$u_h^{n+1} = 2u_h^n - u_h^{n-1} - \Delta t^2 \mathcal{A}_h \left(I - \frac{\Delta t^2}{12}\mathcal{A}_h\right) u_h^n.$$

More generally, an explicit centered scheme of order $2m$ in time is given by,

$$\frac{u_h^{n+1} - 2u_h^n + u_h^{n-1}}{\Delta t^2} + \mathcal{A}_h(\Delta t)\, u_h^n = 0, \qquad \mathcal{A}_h(\Delta t) = \mathcal{A}_h\, P_m(\mathcal{A}_h \Delta t), \quad (10.143)$$

where the polynomial $P_m(x)$ is defined by,

$$P_m(x) = 1 + 2\sum_{l=1}^{m-1}(-1)^l \frac{x^l}{(2l+2)!}. \qquad (10.144)$$

Using again Horner's rule for the representation of the polynomial P_m permits to reduce the calculation of u_h^{n+1} to m successive applications of the operator $\mathcal{A}_h$. In other words, the computational cost for one time step of the scheme of order $2m$ is only m times larger than the computational cost for one time step of the scheme of order 2.

Stability analysis.   We present below the energy technique for analyzing the L^2 (or H^1) stability of (10.137) or (10.139). Here, stability means being able to obtain *uniform* estimates of the type,

$$\sup_{0 \le t^n \le T} |\varepsilon_h^n| \le C \quad \text{or} \quad \sup_{0 \le t^n \le T} \|\varepsilon_h^n\| \le C, \qquad (10.145)$$

where the constant C depends on T and on the data of the continuous problem but not on Δt and h. With explicit schemes such as (10.139), it is not possible to obtain such estimates for any values of Δt and h. In fact the time step must be constrained by a stability condition that depends on h. This is what we are going to show.

REMARK 10.36 *When we use finite elements with mass lumping for elastodynamics equations, such a condition can be intuitively understood. Indeed, one can show a discrete finite velocity propagation property for the solution u_h^n (an equivalent of the property for the continuous problem). More precisely, $\mathbf{M}_h$ being diagonal, $\mathbf{M}_h^{-1}\mathbf{A}_h$ has the same connectivity properties (i.e., same non zero entries) as $\mathbf{A}_h$. As a consequence at times t^n and t^{n-1}, the support of u_h^n and u_h^{n-1} are included in some set,*

$$K_h^n : \text{ a finite union of elements of } \mathcal{T}_h,$$

the support of u_h^{n+1} satisfies

$$K_h^{n+1} = K_h^n \cap \{K \in \mathcal{T}_h \ / \ K \cap K_h^n \neq \emptyset\}.$$

In other words, at each time step, the support of the solution progresses at each time step by one layer of elements of the mesh. Roughly speaking the diameter of the support of the solution increases by $2h$.

A priori, to guarantee the convergence of the method, one needs to prove that the support of the approximate solution contains the support of the exact one: the propagation velocity of the discrete solution (of the order of h/dt) must be larger than the one of the exact solution. This already provides an upper bound for $\Delta t/h$.

For the sake of simplicity, we shall consider the case where the initial data are the only data of the problem ($L(t) = 0$), in which case the solution of the semi-discrete problem (10.5) satisfies an energy conservation property: the energy defined by (10.14) is constant in time.

The idea is first to determine a discrete (in time) equivalent of this energy conservation property. The principle consists in taking a test function v_h in (10.137) which is a discrete equivalent of the time derivative of $u_h(t)$ at time t^n, namely,

$$v_h = \frac{u_h^{n+1} - u_h^{n-1}}{2\Delta t} .$$

We observe that (we use in particular the symmetry of $a(\cdot, \cdot)$),

$$\begin{cases} (\dfrac{u_h^{n+1} - 2u_h^n + u_h^{n-1}}{2\Delta t}, \dfrac{u_h^{n+1} - u_h^{n-1}}{2\Delta t}) = \dfrac{1}{2\Delta t}\left\{|\dfrac{u_h^{n+1} - u_h^n}{\Delta t}|^2 - |\dfrac{u_h^n - u_h^{n-1}}{\Delta t}|^2\right\}, \\[4mm] a(u_h^n, \dfrac{u_h^{n+1} - u_h^{n-1}}{2\Delta t}) = \dfrac{1}{2\Delta t}\left\{a(u_h^{n+1}, u_h^n) - a(u_h^n, u_h^{n-1})\right\}. \end{cases}$$

After summation, these two equalities lead to the discrete conservation property,

$$\frac{E_h^{n+\frac{1}{2}} - E_h^{n-\frac{1}{2}}}{\Delta t} = 0,$$

where

$$E_h^{n+\frac{1}{2}} = \frac{1}{2} \left|\frac{u_h^{n+1} - u_h^n}{\Delta t}\right|^2 + \frac{1}{2} a(u_h^{n+1}, u_h^n). \qquad (10.146)$$

To get a stability result, it is necessary to prove that $E_h^{n+\frac{1}{2}}$ is a positive energy which is not obvious since the second term in (10.146) has *a priori* no sign. However one can expect that, if Δt is small enough, u_h^{n+1} will be close to u_h^n and $a(u_h^{n+1}, u_h^n)$ will be "almost positive." More precisely, if we define

$$u_h^{n+1/2} = \frac{1}{2} \left(u_h^{n+1} + u_h^n\right),$$

we can write

$$\begin{aligned} E_h^{n+\frac{1}{2}} = \frac{1}{2} \left|\frac{u_h^{n+1} - u_h^n}{\Delta t}\right|^2 + \frac{1}{2} a(u_h^{n+1/2}, u_h^{n+1/2}) \\ - \frac{\Delta t^2}{8} a(\frac{u_h^{n+1} - u_h^n}{\Delta t}, \frac{u_h^{n+1} - u_h^n}{\Delta t}). \end{aligned} \qquad (10.147)$$

Introducing the norm of the operator $\mathcal{A}_h$ defined by,

$$\|\mathcal{A}_h\| - \sup_{u_h \in V_h, u_h \neq 0} \frac{a(u_h, u_h)}{|u_h|^2}, \qquad (10.148)$$

we get a lower bound for $E_h^{n+\frac{1}{2}}$,

$$E_h^{n+\frac{1}{2}} \geq \frac{1}{2} \left(1 - \frac{\Delta t^2}{4} \|\mathcal{A}_h\|\right) \left|\frac{u_h^{n+1} - u_h^n}{\Delta t}\right|^2 + \frac{1}{2} a(u_h^{n+1/2}, u_h^{n+1/2}). \quad (10.149)$$

This is the basic estimate for proving the following stability result.

THEOREM 10.9 *A sufficient condition for the stability of (10.137) is,*

$$\frac{\Delta t^2}{4} \|\mathcal{A}_h\| \leq 1. \qquad (10.150)$$

REMARK 10.37 *The inequality (10.149) proves the stability of the scheme under the stronger condition,*

$$\frac{\Delta t^2}{4} \|\mathcal{A}_h\| \leq \alpha, \quad for \ \alpha < 1.$$

Proving the stability result for $\alpha = 1$ requires some additional effort (see [100]).

REMARK 10.38 *This stability result appeals some comments: The condition (10.150) is a priori a sufficient stability condition. However, in this simple case, due to the fact that $\mathcal{A}_h$ can be diagonalized, the Von Neumann*

*analysis can be applied to prove that this condition is also necessary. It suffices
to look at solutions of the form*

$$u_h^n = a^n \, \mathbf{w}_h, \quad a^n \in \mathbb{R},$$

where $\mathbf{w}_h$ is the eigenvector of $\mathcal{A}_h$ associated to its greatest eigenvalue $\lambda = \|\mathcal{A}_h\|$.

REMARK 10.39 *The condition (10.150) appears as an abstract CFL condition. In the applications to concrete equations, in particuler for the finite element approximation of elastodynamics equations, it is possible to get a bound for $\|\mathcal{A}_h\|$ of the form,*

$$\|\mathcal{A}_h\| \leq \frac{4c_+^2}{h^2},$$

where c_+ is a positive constant which is homogeneous to a velocity and only depends on the continuous problem: it is typically related to the maximum wave velocity for the continuous problem. Therefore, a (weaker) sufficient stability condition takes the form,

$$\frac{c_+ \, \Delta t}{h} \leq 1.$$

Under an assumption of uniform regularity (see [43] for a definition) of the computational mesh, it is also possible to show that (where $c_- \leq c_+$ is also homogeneous to a velocity),

$$\|\mathcal{A}_h\| \geq \frac{4c_-^2}{h^2},$$

so that a necessary stability condition is,

$$\frac{c_- \, \Delta t}{h} \leq 1.$$

REMARK 10.40 *To illustrate the "quasi-optimal" nature of the necessary and sufficient conditions of remark (10.39), it suffices to remark that, for instance when the domain Ω has a rectangular or parallelepipedic shape, the medium is homogeneous and one uses a uniform mesh of stepsize h, one can show, using Fourier techniques (or more precisely by computing analytically the eigenvalues of $\mathcal{A}_h$), that*

$$\|\mathcal{A}_h\| \sim \frac{4c^2}{h^2}, \quad (h \to 0).$$

where the constant $c > 0$ is once again homogeneous to a velocity. We refer the reader to the work by Bamberger-Chavent-Lailly [14] in he case of P_1 and Q_1 finite elements.

REMARK 10.41 *The above stability analysis also applies to the higher order scheme (10.143) but it is complicated by the fact that one must verify that the operator $\mathcal{A}_h(\Delta t)$ is positive, which already imposes an upper bound on Δt. Using Von Neumann analysis, one can show that a sufficient (probably necessary) stability condition is given by,*

$$\frac{\Delta t^2}{4}\,\|\mathcal{A}_h\| \le \alpha_m, \tag{10.151}$$

where we have defined,

$$\alpha_m = \sup\{\,\alpha\;/\;\forall\,x \in [0,\alpha],\;0 \le x\,P_m(x) \le 4\,\}. \tag{10.152}$$

It is of course difficult to compute explicitly α_m except for the first values of m. One has in particular (for the exact — but very complicated — expression of α_4, we refer to [140]),

$$\alpha_2 = 3, \quad \alpha_3 = \frac{5 + 5^{\frac{1}{3}} - 5^{\frac{2}{3}}}{2} \simeq 1.893, \quad \alpha_4 = 5.3703, \cdots \tag{10.153}$$

It is particularly interesting to note than for the fourth order scheme, one is allowed to take a time step which is 1.732 ($= \sqrt{3}$) times larger than for the second order scheme, which almost balances the fact that the cost of one time step is twice larger. In the same way, with the scheme of order 8, one can take a time step 2.31739 ($= \sqrt{5.3703}$) times larger (while each time step costs four times more). Surprisingly, the scheme of order 6 seems less interesting: the stability condition is more constraining than that for the fourth order scheme.

Convergence analysis. The classical convergence theory relies on energy estimates. We give only a flavor of the proof (a rigorous proof would need tedious details) in the case where we assume that the exact solution u is smooth enough in time. Let us introduce the error,

$$e_h^n = u_h^n - u_h(t^n), \qquad \text{(where } u_h(t) \text{ is the exact solution of (10.5)).} \tag{10.154}$$

We have immediately,

$$\frac{e_h^{n+1} - 2e_h^n + e_h^{n-1}}{\Delta t^2} + \mathcal{A}_h e_h^n = \varepsilon_h^n, \tag{10.155}$$

where the truncation error ε_h^n defined by,

$$\varepsilon_h^n = \frac{u_h(t^{n+1}) - 2u_h(t^n) + u_h(t^{n-1})}{\Delta t^2} + \mathcal{A}_h u_h(t^n), \tag{10.156}$$

tends to 0 with Δt (the precise estimate uses Taylor expansions and uniform estimates of the semi-discrete solution). A typical estimate is,

$$\sup_{0 \leq t^n \leq T} |\varepsilon_h^n| \leq C \, \Delta t^2 \sup_{0 \leq t \leq T} \|\frac{d^4 u_h(t^n)}{dt^4}(t)\| \qquad (\leq C(u,T) \, \Delta t^2). \qquad (10.157)$$

Let us introduce the energy of the error,

$$\mathcal{E}_h^{n+\frac{1}{2}} = \frac{1}{2} |\frac{e_h^{n+1} - e_h^n}{\Delta t}|^2 + \frac{1}{2} a(e_h^{n+1}, e_h^n). \qquad (10.158)$$

From (10.155), we easily deduce the identity,

$$\frac{\mathcal{E}_h^{n+\frac{1}{2}} - \mathcal{E}_h^{n-\frac{1}{2}}}{\Delta t} = (\varepsilon_h^n, \frac{e_h^{n+1} - e_h^{n-1}}{2\Delta t}). \qquad (10.159)$$

Assume that,

$$\frac{\Delta t^2}{4} \|\mathcal{A}_h\| \leq \alpha^2 < 1.$$

From (10.149), we deduce in particular that,

$$|\frac{e_h^{n+1} - e_h^n}{\Delta t}| \leq (1 - \alpha^2)^{-\frac{1}{2}} \sqrt{2\mathcal{E}_h^{n+\frac{1}{2}}},$$

and therefore that,

$$|\frac{e_h^{n+1} - e_h^{n-1}}{2\Delta t}| \leq (1 - \alpha^2)^{-\frac{1}{2}} \left(\sqrt{2\mathcal{E}_h^{n-\frac{1}{2}}} + \sqrt{2\mathcal{E}_h^{n+\frac{1}{2}}} \right).$$

Using this inequality in (10.159) leads to

$$\frac{\sqrt{\mathcal{E}_h^{n+\frac{1}{2}}} - \sqrt{\mathcal{E}_h^{n-\frac{1}{2}}}}{\Delta t} \leq \sqrt{2} \, (1 - \alpha^2)^{-\frac{1}{2}} |\varepsilon_h^n|.$$

After summation over n, we finally get an error estimate in terms of the energy of the error,

$$\sqrt{\mathcal{E}_h^{n+\frac{1}{2}}} \leq \sqrt{\mathcal{E}_h^{\frac{1}{2}}} + \sqrt{2} \, (1 - \alpha^2)^{-\frac{1}{2}} \sum_{k=1}^{n} |\varepsilon_h^k| \, \Delta t, \qquad (10.160)$$

where the reader will notice that,

- $\sqrt{\mathcal{E}_h^{\frac{1}{2}}}$ represents the error due to the approximation of the initial conditions,

- The term $\sum_{k=1}^{n} |\varepsilon_h^k| \, \Delta t$ is a discrete $L^1(0, t^n; L^2)$ norm of the truncation error, in $O(\Delta t^2)$.

From (10.160), the derivation of error estimates in appropriate norms is only a question of technical details that we do not want to treat in the framework of this exposition.

REMARK 10.42 *The estimate (10.160) blows up when α tends to 1. This is not an optimal result which can be in fact improved with some work (cf [100]).*

10.6   Computational Issues

10.6.1   General Considerations

The formula (10.141) shows that, at each time step the expensive part of the algorithm is the evaluation of

$$\mathbf{M}_h^{-1} \mathbf{A}_h \, U_h^n. \tag{10.161}$$

Of course, this involves the computation of the matrix-vector product $\mathbf{A}_h \, U_h^n$ and the inversion of a system with the mass matrix $\mathbf{M}_h$. Let us concentrate ourselves on the case where mass lumping is applied, in which case the inversion has no cost and the main cost at each time step is the evaluation of a matrix-vector product,

$$V_h = \mathbf{A}_h \, U_h.$$

The most natural (or naive) way to implement this is to form and store the matrix $\mathbf{A}_h$. It appears that this is efficient for low orders (Q_1 or P_1 typically) but not for higher orders. We describe in the next section a general algorithm for computing the matrix-vector product that appears to be efficient for high order finite elements [63, 69].

REMARK 10.43 *The efficiency of the algorithm presented in the next section has been evaluated in particular for the case of Q_k spectral elements which correspond to the case where the same Gauss-Lobatto formulas are used for the evaluation of the mass and the stiffness matrices [63, 69]. With respect to the notation of the following section, the dimension M of the local finite element space for scalar functions and the number L of quadrature points in each element are given by*

$$M = L = (k+1)^2.$$

One additional advantage of this choice is that the matrices $\widehat{\mathbf{G}}_{lm}^p$ that appear in the computations (see lemma 10.8) have more zero entries than a priori expected, due to the tensorial nature of the local basis functions $\widehat{w}_m$ which permits to reduce the computational cost in consequence.

10.6.2   An Efficient Algorithm for General Lagrange Elements

We consider a general Lagrange finite element approximation space (in the sense of the remark 10.10) constructed with a mesh $\mathcal{T}_h$ of $N_{T,h}$ elements. We assume that in the definition (10.42) the space $\mathbf{V}_K$ is of the form,

$$\mathbf{V}_K = \{v_K = \widehat{v} \circ F_K^{-1}, \ \widehat{v} \in \widehat{\mathbf{V}} = \widehat{\mathbf{P}}^{\,d}\},$$

where $\widehat{\mathbf{P}}$ is a space of scalar smooth functions in $\widehat{K}$ with $dim\ \widehat{\mathbf{P}} = M$. We consider a $\widehat{\mathbf{P}}$ unisolvent set of points $\{\widehat{M}_m, 1 \le m \le M\}$ and denote by $\{\widehat{w}_m\}$ the related Lagrange basis,

$$\widehat{w}_m \in \widehat{P}, \quad \widehat{w}_m(\widehat{M}_m) = \delta_{mn}.$$

The set of nodes of the mesh $\mathcal{T}_h$ is

$$\{M_j, 1 \le j \le N_h\} = \bigcup_{K \in \mathcal{T}_h} \{F_K(\widehat{M}_m), 1 \le m \le M\}.$$

We assume that, for the calculation of the stiffness matrix, we first use the symmetry property of $C(x)$ (which yields $C\varepsilon(u) = C\nabla u$) to rewrite the bilinear form $a(u, v)$ as,

$$a(u, v) = \int_\Omega C\,\nabla u : \nabla v\,dx, \tag{10.162}$$

and we use a quadrature formula in $\widehat{K}$ with quadrature points $\widehat{Q} = \{\widehat{\mathcal{M}}_l, 1 \le l \le L\}$ and weights $\{\omega_l, 1 \le l \le L\}$. In other words

$$a_h(u_h, v_h) = \sum_{K \in \mathcal{T}_h} a_K(u_K, v_K),$$

where, by definition

$$a_K(u_K, v_K) = \int_{\widehat{K}}^{\widehat{Q}} \left(C\,(DF_K^*)^{-1} \nabla \widehat{u}_K : (DF_K^*)^{-1} \nabla \widehat{v}_K \right) \widehat{J}_K\,d\widehat{x}$$

$$= \sum_{l=1}^{L} \omega_l\,\widehat{J}_K(\widehat{M}_l) \times$$

$$\left[C(\widehat{M}_l)\,(DF_K^*)^{-1}(\widehat{M}_l)\,\nabla \widehat{u}_K(\widehat{M}_l) : (DF_K^*)^{-1}(\widehat{M}_l)\,\nabla \widehat{v}_K(\widehat{M}_l) \right]. \tag{10.163}$$

For the implementation, it is useful to move to more algebraic and matrix notation. This is not so straightforward and requires some formalism. We start the expansion of the local fields $\widehat{u}_K$ and $\widehat{v}_K$ as

$$\widehat{u}_K = \sum_{p=1}^{d} \sum_{m=1}^{M} U_{K,m}^p\,\widehat{w}_m\,\mathbf{e}^p, \quad \widehat{v}_K = \sum_{q=1}^{d} \sum_{n=1}^{M} V_{K,n}^q\,\widehat{w}_n\,\mathbf{e}^q, \tag{10.164}$$

and we introduce the vector U_K (and the same for V_K) of the local degrees of freedom as an element of $(\mathbb{R}^d)^M$ with the block decomposition,

$$U_K = \begin{pmatrix} U_{K,1} \\ \vdots \\ U_{K,M} \end{pmatrix} \quad \text{where} \quad U_{K,m} = \begin{pmatrix} U_{K,m}^1 \\ \vdots \\ U_{K,m}^d \end{pmatrix} \in \mathbb{R}^d. \tag{10.165}$$

We shall equip $(\mathbb{R}^d)^M$ with its natural inner product

$$(U, V)_{(\mathbb{R}^d)^M} = \sum_{m=1}^{M} U_m \cdot V_m. \tag{10.166}$$

On the other hand it will be useful to work with $\mathcal{L}(\mathbb{R}^d)^L$,

$$S = (S_l)_{1 \leq l \leq L}, \quad S_l \in \mathcal{L}(\mathbb{R}^d), \quad \forall\, 1 \leq l \leq L,$$

that we equip with the inner product

$$(S, T) = \sum_{l=1}^{L} S_l : T_l. \tag{10.167}$$

We shall introduce also linear mappings from $(\mathbb{R}^d)^M$ in $\mathcal{L}(\mathbb{R}^d)^L$,

$$G \in \mathcal{L}\Big((\mathbb{R}^d)^M, \mathcal{L}(\mathbb{R}^d)^L\Big).$$

Any operator of this type can be represented by a family of $M \times L \times d$ square $d \times d$ matrices,

$$G_{lm}^p, \quad 1 \leq p \leq d, \quad 1 \leq l \leq L, \quad 1 \leq m \leq M. \tag{10.168}$$

Indeed by linearity

$$U = (U_m)_{1 \leq m \leq M} \in (\mathbb{R}^d)^M \implies GU = \big((GU)_l\big)_{1 \leq l \leq L} \in \mathcal{L}(\mathbb{R}^d)^L,$$

is such that,

$$(GU)_l = \sum_{m=1}^{M} G_{lm}\, U_m, \quad G_{lm} \in \mathcal{L}\Big(\mathbb{R}^d, \mathcal{L}(\mathbb{R}^d)\Big), \tag{10.169}$$

and, as for any element of $\mathcal{L}\Big(\mathbb{R}^d, \mathcal{L}(\mathbb{R}^d)\Big)$ we can write, for each (l, m) (if $U_m = (U_m^p)_{1 \leq p \leq d}$),

$$G_{lm}\, U_m = \sum_{p=1}^{d} G_{lm}^p\, U_m^p, \quad G_{lm}^p \in \mathcal{L}(\mathbb{R}^d), \quad \forall\, 1 \leq p \leq d. \tag{10.170}$$

Note that the ajoint of such an operator,

$$G^* \in \mathcal{L}\left(\mathcal{L}(\mathbb{R}^d)^L, (\mathbb{R}^d)^M\right),$$

is defined by,

$$G^* S = \left((G^* S)_m\right) \in (\mathbb{R}^d)^M, \quad (G^* S)_m \mathbb{R}^d \text{ and } (G^* S)_m^p = \sum_{l=1}^{L} G_{lm}^p : S_l.$$

$$(10.171)$$

Indeed, using (10.169) and (10.170),

$$(GU, S)_{\mathcal{L}(\mathbb{R}^d)^L} = \sum_{l=1}^{L} (GU)_l : S_l = \sum_{m=1}^{M} \sum_{p=1}^{d} U_m^p \left(\sum_{l=1}^{L} G_{lm}^p : S_l\right).$$

Next we introduce, for each $l \in \{1, \cdots, L\}$ the $d \times d$ matrix

$$\mathbf{C}_K^l = \omega_l \, \widehat{J}_K(\widehat{M_l}) \, (DF_K)^{-1} (\widehat{M_l}) \, C(\widehat{M_l}) \, (DF_K^*)^{-1} (\widehat{M_l}). \qquad (10.172)$$

REMARK 10.44 *We can write $\mathbf{C}_K^l$ as the product of two $d \times d$ matrices, namely*

$$\mathbf{C}_K^l = \widehat{J}_K(\widehat{M_l}) \, (DF_K^*)^{-1} (\widehat{M_l}) \cdot B_K^l,$$

where $B_K^l = C(\widehat{M_l}) \, (DF_K^*)^{-1} (\widehat{M_l})$ is the result of the application of the linear mapping $C(\widehat{M_l}) \in \mathcal{L}^2(\mathbb{R}^d)$ to the matrix $(DF_K^*)^{-1} (\widehat{M_l}) \in \mathcal{L}(\mathbb{R}^d)$.

Then we introduce the block diagonal $dM \times dM$ matrix (with $d \times d$ diagonal blocks)

$$\mathbf{C}_K = \begin{pmatrix} \mathbf{C}_K^1 & & 0 \\ & \ddots & \\ 0 & & \mathbf{C}_K^L \end{pmatrix},$$

that we shall consider as a block diagonal application from $\mathcal{L}(\mathbb{R}^d)^L$ into itself,

$$\mathbf{C}_K \in \mathcal{L}\left(\mathcal{L}(\mathbb{R}^d)^L, \mathcal{L}(\mathbb{R}^d)^L\right),$$

in the sense that,

$$M = (M_l)_{1 \le l \le L} \quad \in \mathcal{L}(\mathbb{R}^d)^L \quad \Longrightarrow \quad \mathbf{C}_K M = (\mathbf{C}_K^l M_l)_{1 \le l \le L} \quad \in \mathcal{L}(\mathbb{R}^d)^L.$$

The main result is thus the following,

LEMMA   10.8 *Let $\widehat{G}$ be the (independent of K) operator in $\mathcal{L}((\mathbb{R}^d)^M, \mathcal{L}(\mathbb{R}^d)^L)$ characterized (in the sense of (10.168), (10.169), (10.170)) by the $d \times d$ matrices,*

$$\widehat{G}_{lm}^p = \nabla \left[\widehat{w}_m \mathbf{e}_p\right](\widehat{M_l}) \qquad (10.173)$$

One has the identity,

$$a_K(u_K, v_K) = \left(\left[\widehat{G}^* \, \mathbf{C}_K \, \widehat{G} \right] U_K, V_K \right)_{(\mathbb{R}^d)^M}. \qquad (10.174)$$

PROOF By definition of the matrices $\mathbf{C}_K^l$, we deduce from (10.163) that

$$a_K(u_K, v_K) = \sum_{l=1}^{L} \mathbf{C}_K^l \, \nabla \widehat{u}_K(\widehat{M_l}) : \nabla \widehat{v}_K(\widehat{M_l}),$$

where we have used the fact that, if $(\mathbf{A}, \mathbf{B}, \mathbf{C}) \in \mathcal{L}(\mathbb{R}^d)^3$,

$$\mathbf{A} : \mathbf{B} \, \mathbf{C} = \mathbf{C}^* \, \mathbf{A} : \mathbf{B}.$$

Subsituting (10.164) in the above formula, we get

$$a_K(u_K, v_K) = \sum_{p=1}^{d} \sum_{m=1}^{M} \sum_{q=1}^{d} \sum_{n=1}^{M} \sum_{l=1}^{L} U_{K,m}^p \, V_{K,n}^q \times \left(\mathbf{C}_K^l \, \nabla \left[\widehat{w}_m(\widehat{M_l}) \mathbf{e}_p \right] : \nabla \left[\widehat{w}_n(\widehat{M_l}) \mathbf{e}_q \right] \right).$$

On the other hand,

$$(\widehat{G}^* \, \mathbf{C}_K \, \widehat{G} \, U_K, V_K)_{(\mathbb{R}^d)^m} = (\mathbf{C}_K \, \widehat{G} \, U_K, \widehat{G} \, V_K)_{\mathcal{L}(\mathbb{R}^d)^L}$$

(by (10.167))
$$= \sum_{l=1}^{L} (\mathbf{C}_K \, \widehat{G} \, U_K)_l : (\widehat{G} \, V_K)_l$$

(by (10.169))
$$= \sum_{l=1}^{L} (\sum_{m=1}^{L} \mathbf{C}_K^l \, \widehat{G}_{lm} \, U_{K,m}) : (\sum_{n=1}^{L} \widehat{G}_{ln} \, V_{K,n})$$

(by (10.170))
$$= \sum_{l=1}^{L} (\sum_{m=1}^{L} \sum_{p=1}^{d} \mathbf{C}_K^l \, \widehat{G}_{ln}^p \, U_{K,m}^p) : (\sum_{n=1}^{L} \sum_{q=1}^{d} \widehat{G}_{lm}^q \, V_{K,u}^q))$$

$$= \sum_{l=1}^{L} \sum_{m=1}^{L} \sum_{p=1}^{d} \sum_{n=1}^{L} \sum_{q=1}^{d} (\mathbf{C}_K^l \, \widehat{G}_{ln}^p : \widehat{G}_{lm}^q) \, U_{K,m}^p \, V_{K,u}^q.$$

One easily concludes by identification with the expression of $a_K(u_K, v_K)$. $\square$

REMARK 10.45 *The reader will note that each $d \times d$ matrix has only one column of non zero elements, namely the pth one. More precisely,*

$$(\widehat{G}_{lm}^p)_{p,q} = \frac{\partial \widehat{w}_m}{\partial x_q}(M_l), \quad (\widehat{G}_{lm}^p)_{r,q} = 0, \quad \text{for } r \neq p. \qquad (10.175)$$

We now explain how to implement the matrix-vector product $W_h = A_h\, U_h$, where $(U_h, W_h) \in \mathbb{R}^{N_h} \times \mathbb{R}^{N_h}$. We remind that, for each K, there exists a renumbering function,

$$j_K : \{1, \cdots, M\} \to \{1, \cdots, N_h\},$$

such that one can relate the global scalar Lagrange basis functions w_j to the local ones $\widehat{w}_m$ (note that the union over all the elements K of the sets $\{j_K(m), 1 \le m \le M\}$ fills the set $\{1, \cdots, N_h\}$), using the formula

$$w_{j_K(m)}|_K = \widehat{w}_m \circ F_K^{-1}.$$

The relationship between the vector $U_h = U_{j,p} \in \mathbb{R}^{N_h}$ and the vectors $U_K \in (\mathbb{R}^d)^M$ is,

$$U_h = \sum_{K \in \mathcal{T}_h} L_K\, U_K \quad \Longleftrightarrow \quad U_K = L_K^*\, U_h,$$

where $\widetilde{U}_h = L_K\, U_K \in \mathbb{R}^{N_h}$ is such that,

$$\widetilde{U}_{j_K(m)}^p = U_{K,m}^p, \quad \widetilde{U}_j^p = 0 \quad \text{if } j \notin \{j_K(m), 1 \le m \le M\}, \tag{10.176}$$

while $\widetilde{U}_K = L_K^*\, U_h$ is such that,

$$(\widetilde{U}_K)_m^p = U_{j_K(m)}^p. \tag{10.177}$$

The algorithm for the product $W_h = A_h\, U_h$ is as follows

- For $K \in \mathcal{T}_h$,

1. Determine $U_K = L_K^*\, U_h$,

2. Compute $\Gamma_K \in \mathcal{L}(\mathbb{R}^d)^L$ by $\Gamma_K = \widehat{G}\, U_K \quad \Longleftrightarrow \quad (\Gamma_K)_l = \sum_{p=1}^{d} \sum_{m=1}^{M} \widehat{G}_{lm}^p U_{K,m}^p,$

3. Compute $\Sigma_K \in \mathcal{L}(\mathbb{R}^d)^L$ by $\Sigma_K = \mathbf{C}_K \Gamma_K \quad \Longleftrightarrow \quad (\Sigma_K)_l = \mathbf{C}_K^l\, (\Gamma_K)_l,$

4. Compute $W_K = \widehat{G}^*\, \Sigma_K \in (\mathbb{R}^d)^M \quad \Longleftrightarrow \quad (W_K)_m^p = \sum_{l=1}^{L} \widehat{G}_{lm}^p : (\Sigma_K)_l,$

5. Increment $W_h = W_h + L_K\, W_K$.

One sees that the needed memory storage is,

- $L \times N_{h,T}$ square $d \times d$ symmetric matrices $\mathbf{C}_K^l$,

- $M \times L \times d$ (number independent of h) square $d \times d$ matrices $\widehat{G}_{lm}^p$.

while the needed computational cost is, for each $K \in \mathcal{T}_h$ (we take into account the particular sparse structure of the matrices $\widehat{G}_{lm}^p$),

- $2\,d^2\,L\,M$ operations for step 2.

- $2\,d^3\,L$ operations for step 3.

- $2\,d^2\,L\,M$ operations for step 4.

- $d\,M$ operations for step 5.

10.6.3   A Link with Mixed Finite Element Methods

In fact one can reinterpret the result of lemma 10.8 as a local decomposition of the operator A_h, which represents a discrete version of $A = -div(C\nabla u)$, into the product of three operators

- $\widehat{G}$ as an approximation of the operator ∇,

- $\widehat{G}^*$ as an approximation of the operator $-div$ (the formal adjoint of ∇),

- $\mathbf{C}_K^l$ as an approximation of the operator C.

In the same way that $C(x)$ contains the information of the elastic (continuous) medium, the matrices $\mathbf{C}_K^l$ contains the information about the discrete medium which include both the geometry of the mesh (through J_K and DF_K) and the physical characteristics of the medium (via C). The independence of $\widehat{G}$ with respect to K can then be interpreted as the consequence of the intrinsic character of the operators ∇ and div. Moreover

- the vector Γ_K is a local discrete version of ∇u,

- the vector Σ_K is a local discrete version of the stress tensor σ,

- $\Sigma_K = \mathbf{C}_K^l \Gamma_K$ is a local discrete version of the constitutive law.

As it was observed for the first time in the works by Cohen-Fauqueux (cf. [46]), the interpretation we give here appears more naturally if one sees the pure displacement finite element approximation of the elastodynamics equations as the displacement version of a mixed finite element approximation of the elastodynamics system in a first order (in space form), namely (we consider here the source free problem for simplicity),

$$
\begin{cases}
\rho \, \dfrac{\partial^2 u}{\partial t^2} - \operatorname{div} \sigma = 0, \\[2mm]
\sigma = C\,\gamma, \\[2mm]
\gamma = \nabla u.
\end{cases}
\tag{10.178}
$$

where the new unknowns γ and σ are tensors. Note that we do not prescribe a priori the symmetry of σ: this is a consequence of $\sigma = C\,\gamma$. Without entering in the details, obtaining a mixed variational formulation consists in multiplying the three equations of (10.178) by test fields of the space variables only, respectively v, τ and δ, and integrating over Ω. Then one chooses to integrate by parts one of the two equations involving space derivatives. Here, to be consistent with the pure displacement formulation, we choose to keep the regularity on the unknown u but relax the one of σ. Then, we integrate by parts the last equation of (10.178). This influences of course the choice of the functional spaces in which leave v, τ and δ. One ends up with the following formulation (that is considered in [46], the primal-dual formulation by opposition with the dual-primal formulation that we shall present in the next chapter - see also [60] for more details),

Find $(u,\gamma,\sigma) : \mathbb{R}^+ \to V \times S \times S$, such that

$$
\begin{cases}
\dfrac{d^2}{dt^2}(u,v) + \displaystyle\int_\Omega \sigma : \nabla v \, dx = 0, & \forall v \in V, \\[2mm]
\displaystyle\int_\Omega \sigma : \tau \, dx = \int_\Omega C\,\gamma : \tau \, dx, & \forall \tau \in S, \\[2mm]
\displaystyle\int_\Omega \gamma : \delta \, dx - \int_\Omega \nabla u : \delta \, dx = 0, & \forall \delta \in S.
\end{cases}
\tag{10.179}
$$

where V is the same space as in the displacement formulation (see (9.25)) while S is simply,

$$
S = L^2(\Omega; \mathcal{L}(\mathbb{R}^d)). \tag{10.180}
$$

For the discrete problem, we approach V by the same Lagrange finite element space V_h and S by another finite dimensional subspace $S_h \subset S$, typically a set of discontinuous tensor fields which are piecewise smooth (polynomial Lagrange elements) in each element K of the mesh $\mathcal{T}_h$. After decomposition on appropriate bases (we keep in particular the basis $\{w_j\}$ of V_h) we get the matrix formulation (with obvious notation - we assume that the integrals can be computed approximately via quadrature formulas),

$$
\begin{cases}
\mathbf{M}_h \dfrac{\partial^2 U_h}{\partial t^2} + \mathbf{G}_h^* \, \Sigma_h = 0, \\[2mm]
\Sigma_h = \mathbf{C}_h \, \Gamma_h, \\[2mm]
\Gamma_h = \mathbf{G}_h \, U_h.
\end{cases}
\tag{10.181}
$$

After elimination of Σ_h and Γ_h, we get,

$$
\mathbf{M}_h \frac{\partial^2 U_h}{\partial t^2} + \mathbf{A}_h U_h = 0, \quad \text{with} \quad \mathbf{A}_h = \mathbf{G}_h\,\mathbf{C}_h\,\mathbf{G}_h^* \tag{10.182}
$$

which is of course formally identical to the ordinary differential system that we obtain with the pure displacement formulation. Moreover the factorization $\mathbf{A}_h = \mathbf{G}_h\, \mathbf{C}_h \mathbf{G}_h^*$ is very similar to the local factorization (10.174). It is easy to realize that such a local factorization exists since, as S_h are made of discontinuous functions, the matrices $\mathbf{G}_h$ and $\mathbf{C}_h$ have naturally a "block diagonal" structure, each block being associated to the degrees of freedom (for γ and σ) of one element K. Finally, one easily realizes the "intrinsic" nature of the blocks composing $\mathbf{G}_h$ if the space S_h is constructed from a single reference space $\widehat{S}$ through the use of the so called Piola transform (this transform usually appears for the finite element approximation of $H(\mathrm{div})$ spaces, see [126], [38]),

$$
\begin{vmatrix}
S_h = \{\tau_h \in S \,/\, \forall K \in \mathcal{T}_h,\ \tau_h|_K \in \mathbf{S}_K\}, \\[2mm]
S_K = \{\tau_K = J_K\, DF_K^*\, \widehat{\tau} \circ F_K^{-1}, \widehat{\tau} \in \widehat{S}\}.
\end{vmatrix}
\tag{10.183}
$$

The key point is to remark that, if $u_K = \widehat{u} \circ F_K^{-1}$ then, one has the identity,

$$
\int_K \tau_K : \nabla u_K \, dx = \int_{\widehat{K}} \widehat{\tau} : \nabla \widehat{u} \, d\widehat{x}.
\tag{10.184}
$$

Finally, one can show (we shall omit the general proof here and we refer the reader to [46], [123] or [83] for various special cases) that, provided that,

- The set $\widehat{Q}$ is $\widehat{S}$ unisolvent: $dim\ \widehat{S} = L$ and any $\widehat{\tau}$ in $\widehat{S}$ is completely determined by its values at the points M_l (here, no requirement on the boundary),

- The corresponding local Lagrange interpolation functions are used to construct the global basis of the space S_h,

- The quadrature formula associated to $\widehat{Q}$ is used for approximating the integrals,

$$
\int_\Omega \sigma : \nabla v \, dx, \int_\Omega \sigma : \tau \, dx, \text{ and } \int_\Omega C\, \gamma : \tau \, dx,
$$

then the mixed method we have just described and the "classical" pure displacement finite element method are equivalent in the sense that the matrix $\mathbf{A}_h$ and the computed solution U_h are the same as in the classical method. Conversely the vectors Γ_K and Σ_K appearing in the algorithm we gave for the classical method, are nothing but the vectors of the local degrees of freedom for γ and σ, respectively.

Note: The references for this chapter and for all chapters in Part III are given at the end of Chapter 14.

Chapter 11

Finite Element Methods with Discontinuous Displacement

P. Joly, POEMS Project team, INRIA-Rocquencourt, France,
Patrick.JOLY@inria.fr

C. Tsogka, Dept. of Mathematics, University of Chicago, USA,
tsogka@tem.uoc.gr

11.1 Introduction

We consider in this chapter the first order in time velocity-stress formulation of elastodynamics. In comparison with the second order formulation that we studied in the previous chapter, the first order system has the following two advantages: (i) it can be naturally coupled with the fictitious domain method for taking into account the free surface boundary condition on the stress tensor, (ii) as for any first order hyperbolic problem it is possible for this formulation to derive a perfectly matched absorbing layer model which permits modelling elastic wave propagation in infinite domains.

The fictitious domain method and its application to elastodynamics will be discribed in Chapter 12. Let us recall here briefly the principle of this approach for computing the scattered field by an object with a free surface boundary condition (i.e., the normal stress is zero). The fictitious domain method consists in extending artificially the solution of the problem inside the object so that the new domain of computation has a very simple shape (typically a rectangle in 2D). To account for the boundary condition, a new auxiliary unknown, defined only at the boundary of the object, is introduced. The main advantage of the method is that the mesh for the solution on the enlarged domain can be chosen independently of the geometry of the object. In particular, one can use regular grids or structured meshes which allow for simple and efficient computations.

Motivated by the use of the fictitious domain method we will consider in this chapter finite element methods for structured (regular) grids. In particular we are concerned with the space discretization of the mixed velocity-stress formulation of elastodynamics. For efficiency reasons (to reduce the computation time and memory requests) we want to use mass lumping techniques (as the ones described in Chapter 10) in order to get a really explicit method after time discretization.

Several families of mixed finite elements have been proposed in the literature for static plane elasticity. The main difficulty appearing in this problem is finding a way to take into account the symmetry of the stress tensor. Namely, taking into account the symmetry in the approximation space is not easy and can lead to numerical locking (this is what happens with Raviart-Thomas elements). The approach which is usually followed is the relaxed symmetry approach: the symmetry is imposed in a weak sense via the introduction of a Lagrange multiplier [4, 7, 117, 131, 132]. Another approach using spaces of symmetric stress tensors, based on composite elements, was introduced in [99]. None of these, however, are adapted to mass lumping.

We present here the mixed finite elements proposed in [24, 26] which are inspired by Nédelec's second family [120]. These elements have two basic characteristics: they allow mass lumping and they use spaces of symmetric stress tensors.

11.2   Mixed Variational Formulation

Let $\mathbf{A}(x) \in \mathcal{L}_{ss}^2(\mathbb{R}^d)$ be the inverse of the elasticity tensor defined by (9.8) in §9.2 and satisfying the uniform inequalities (α and M are strictly positive real numbers)

$$0 < \alpha |\sigma|^2 \le A(x)\sigma : \sigma \le M |\sigma|^2 , \ \forall \sigma \in \mathcal{L}_s(\mathbb{R}^d), \ \forall x \in \Omega. \qquad (11.1)$$

The elastodynamic problem in Ω can be written as a first order hyperbolic system, the velocity-stress system, with the unknowns: $v = \partial u / \partial t$, the velocity vector and $\sigma = \sigma(u)$ the stress tensor,

$$\begin{cases} \varrho \dfrac{\partial v}{\partial t} - \operatorname{div} \sigma = f, \\[2mm] A \dfrac{\partial \sigma}{\partial t} - \varepsilon(v) = 0, \end{cases} \qquad (11.2)$$

subject to the initial conditions $v(0) = 0$, $\sigma(0) = 0$ and the boundary condition

$$v = 0 \text{ on } \partial\Omega. \qquad (11.3)$$

Let us now introduce the Hilbert spaces

$$M = (L^2(\Omega; \mathbb{R}^2))^2, \ H = L^2(\Omega; \mathcal{L}(\mathbb{R}^2)) \text{ and } X = \{\sigma \in H \ / \ \operatorname{div} \sigma \in M\}.$$

We denote $X^{sym} \subset X$ the subspace of symmetric tensors in X

$$X^{sym} = \{\sigma \in X \ / \ \mathrm{as}(\sigma) = 0\} \, .$$

The dual-primal mixed variational formulation of system (11.2, 11.3) can be written in the form

$$\left\{ \begin{array}{l} \text{Find } (\sigma, v) : [0, T] \longmapsto X^{sym} \times M \text{ such that} \\[2ex] \dfrac{d}{dt} \, a(\sigma(t), \tau) + b(v(t), \tau) = 0 \, , \qquad \forall \tau \in X^{sym}, \\[2ex] \dfrac{d}{dt} \, c(v(t), w) - b(w, \sigma(t)) = (f, w) \, , \ \forall w \in M, \end{array} \right. \qquad (11.4)$$

where

$$\left\{ \begin{array}{ll} a(\sigma, \tau) = \displaystyle\int_\Omega A\sigma : \tau dx, & \forall \, (\sigma, \tau) \in H \times H, \\[2ex] c(v, w) = \displaystyle\int_\Omega \varrho v \cdot w dx, & \forall \, (v, w) \in M \times M, \\[2ex] b(w, \tau) = \displaystyle\int_\Omega \mathrm{div} \, \tau \cdot w dx, & \forall \, (w, \tau) \in X \times M. \end{array} \right. \qquad (11.5)$$

The bilinear form $a(\cdot, \cdot)$ (resp. $b(\cdot, \cdot)$) is continuous on $H \times H$ (resp. on $X \times M$), thus we can define linear continuous operators $\mathcal{A} : H \to H'$ and $B : X \to M'$ by

$$\langle \mathcal{A}\sigma, \tau \rangle_{H' \times H} = a(\sigma, \tau) \, ; \ \langle B\tau, w \rangle_{M' \times M} = b(w, \tau),$$

where H' (resp. M') holds for the dual space of H (resp. M). We set $B^{sym} = B_{|X^{sym}} \in \mathcal{L}(X^{sym}, M')$. The following properties are satisfied, e.g., [38]

(i)   The continuous inf-sup condition

$$\exists \, \beta > 0 \, / \, \forall \, w \in M, \ \exists \, \tau \in X^{sym}, \tau \neq 0 \, / \, b(w, \tau) \geq \beta \, \|w\|_M \, \|\tau\|_X \, .$$

(ii)   The coercivity of the form $a(\cdot, \cdot)$ on Ker $B^{sym} = $ Ker $B \cap X^{sym}$

$$\exists \, \alpha > 0 \, / \, \forall \, \sigma \in \mathrm{Ker} \, B^{sym}, \ a(\sigma, \sigma) \geq \alpha \, \|\sigma\|_X^2 \, ,$$

with Ker $B^{sym} = \{\tau \in X^{sym} \ / \ b(w, \tau) = 0, \ \forall \, w \in M\}$.

$$(11.6)$$

The mixed formulation (11.4) is the one we shall work with for the numerical approximation. Note that, it is crucial to work in the space X^{sym} of symmetric tensors because the operator $-\varepsilon$ is not the adjoint of div if one works in X.

Let us note that the dual-primal formulation (11.4) is not the only mixed formulation of the elastodynamic problem. One could also consider the following primal-dual formulation,

$$
\begin{cases}
\text{Find } (v, \sigma) : [0, T] \rightarrow V \times S, \text{such that} \\[2mm]
\dfrac{d}{dt} \displaystyle\int_{\Omega} \sigma : \tau \, dx - \int_{\Omega} C \, \nabla v : \tau \, dx = 0, \quad \forall \tau \in S, \\[4mm]
\dfrac{d}{dt} \displaystyle\int_{\Omega} \varrho v \cdot w \, dx + \int_{\Omega} \sigma : \nabla w \, dx = 0, \quad \forall w \in V,
\end{cases}
\tag{11.7}
$$

with

$$
V = (H_0^1(\Omega))^2, \quad S = L^2(\Omega; \mathcal{L}(\mathbb{R}^2)).
$$

Remark that to obtain (11.7) we integrated by parts the first equation of system (11.2), whereas (11.4) was obtained by integrating by parts the second equation of (11.2). As a consequence, the velocity in (11.7) is sought in a more regular space then the stress tensor. As we have already mentioned in section 10.6.3, formulation (11.7) can be descretized with the finite elements proposed by Cohen-Fauqueux [46] when written in the equivalent form of system (10.179). In [60] both the primal-dual and the dual-primal formulations for elastodynamics were considered and coupled with the acoustic equation for modelling fluid-structure interaction problems. In this chapter we will focus our attention in the dual-primal formulation (11.4).

11.3   Space Discretization

We suppose now that Ω is a union of rectangles in such a way that we can consider a regular mesh $(\mathcal{T}_h)$ with square elements (K) of edge length $h > 0$. To obtain the finite element spaces, we shall adopt a constructive approach which aims, in particular, at exploiting the geometry of the mesh. We look for approximation spaces

$$
\begin{cases}
M_h = \{v_h \in M \ / \ \forall K \in \mathcal{T}_h, \ v_h \,|_K \in \mathcal{P}\}, \\[2mm]
X_h = \{\sigma_h \in X \ / \ \forall K \in \mathcal{T}_h, \ \sigma_h \,|_K \in \mathcal{Q}\}, \\[2mm]
X_h^{sym} = \{\sigma_h \in X_h \ / \ \text{as}(\sigma_h) = 0\},
\end{cases}
\tag{11.8}
$$

where $\mathcal{P}$ and $\mathcal{Q}$ are finite dimensional spaces of C^∞ functions, for instance polynomials, that are to be determined. We can remark that X_h^{sym} is sought as a subspace of X^{sym}: we want to take into account the symmetry condition in the strong sense. Then the approximate problem associated to the mixed velocity-stress system for elastodynamics can be written in the following form

$$
\begin{cases}
\text{Find } (\sigma_h, v_h) : [0, T] \longmapsto X_h^{sym} \times M_h \text{ such that} \\[2mm]
\dfrac{d}{dt} a(\sigma_h, \tau_h) + b(v_h, \tau_h) = 0, \qquad \forall \tau_h \in X_h^{sym}, \\[2mm]
\dfrac{d}{dt} c(v_h, w_h) - b(w_h, \sigma_h) = (f, w_h), \ \forall w_h \in M_h,
\end{cases}
\tag{11.9}
$$

with initial conditions $\sigma_h(0) = 0$, $v_h(0) = 0$.

We will explain in the next section the construction of the lowest order element of two families of mixed finite elements introduced in [24, 26] (see also [21, 26] where the dispersion analyses for these elements are presented). In what follows, we shall use the following standard notation for spaces of polynomials of two variables. We denote by P_k the space of polynomials of degree at most k, and define

$$
P_{k,l} = \left\{ p(x_1, x_2) \mid p(x_1, x_2) = \sum_{i \le k, j \le l} a_{ij} x_1^i x_2^j \right\},
$$

and $Q_k = P_{k,k}$.

11.3.1 Choice of the Approximation Space for the Stress Tensor

A possible choice for X_h is the lowest order Raviart-Thomas element $RT_{[0]}$

$$
X_h^{RT} = \left\{ \sigma_h \in X \ / \ \forall K \in \mathcal{T}_h, \ (\sigma_1, \sigma_2)|_K \in (RT_{[0]})^2 \right\},
$$

where $RT_{[0]} = P_{1,0} \times P_{0,1}$. However, this choice is not satisfactory, since the space $X_h^{RT} \cap X^{sym}$ is too small and thus cannot be considered as a good approximation space for X^{sym}: indeed, if σ_h is a symmetric tensor in X_h^{RT}, then it is easy to see that σ_{12} is necessarily constant in Ω. This is usually called *numerical locking*.

The construction of X_h^{sym} as a subspace of X^{sym}, is based on the following observation, which is true only for meshes with squares whose edges are parallel to the coordinate axis (the proof, which is trivial, is omitted).

THEOREM 11.1 *For all $\sigma_h \in X_h$, where X_h is given by (11.8), we have*

$$
\sigma_h \in X_h^{sym} \iff \begin{pmatrix} \sigma_{11} \\ \sigma_{22} \end{pmatrix} \in H(\mathrm{div}, \Omega) \quad \text{and} \quad \sigma_{12} = \sigma_{21} \in H^1(\Omega).
$$

Theorem 11.1 shows that σ_{12} must belong to an approximation space of $H^1(\Omega)$. Thus, in order to define the lowest order element, it is natural to choose

$$\sigma_{12} \in H^1(\Omega)/\forall K \in \mathcal{T}_h, \ \sigma_{12}|_K \in Q_1.$$

It remains now to define the approximation space of $H(\mathrm{div}, \Omega)$ in which the vector $(\sigma_{11}, \sigma_{22})^t$ belongs. Once again, a natural choice is the lowest order Raviart-Thomas element $RT_{[0]}$. We denote

$$\widetilde{X}_h^{sym} = \left\{ \sigma_h \in X^{sym}/\forall K \in \mathcal{T}_h, \ \sigma_{12}|_K \in Q_1 \text{ and } (\sigma_{11}, \sigma_{22})|_K \in RT_{[0]} \right\}.$$

However, it can be easily shown that this choice does not allow for an explicit time discretization scheme. This was explained in [22] for the anisotropic wave equation. The reason is that the degrees of freedom for the stress tensor are associated either to a vertex of an element K (for σ_{12}) or to an edge (for $(\sigma_{11}, \sigma_{22})$). To obtain an explicit time discretization scheme, we want to use a mass lumping technique for the approximation of the mass matrix associated to the bilinear form $a(\sigma_h, \tau_h)$ (the reader can verify that the matrix associated to $c(v_h, w_h)$ is already diagonal in the usual basis of M_h). Thus, we are led to approximate the mass matrix $a(\sigma_h, \tau_h)$ by

$$a_h(\sigma_h, \tau_h) = \sum_{K \in \mathcal{T}_h} \int_K^{\mathcal{Q}_K} A\sigma_h : \tau_h \ dx,$$

where $\int_K^{\mathcal{Q}_K} \cdot dx$ is some quadrature formula defined in general by (10.74), that is to be determined. The key point for choosing the approximation space of $H(\mathrm{div}, \Omega)$ is to regroup all the degrees of freedom at the nodes of the quadrature formula, namely the nodes of the mesh. Under this condition the adequate choice for $(\sigma_{11}, \sigma_{22})$ is the lowest order element of the second family of mixed finite elements proposed by Nédélec in [120], that is

$$\left\{ q_h \in H(\mathrm{div}, \Omega) \text{ such that } q_h|_K \in (Q_1(K))^2, \quad \forall K \in \mathcal{T}_h \right\},$$

and thus the choice for the space X_h^{sym} can be written as

$$X_h^{sym} = \left\{ \sigma_{12} \in H^1(\Omega)/\sigma_{12}|_K \in Q_1, \ \forall K \in \mathcal{T}_h \text{ and} \right.$$
$$\left. (\sigma_{11}, \sigma_{22}) \in H(\mathrm{div}, \Omega)/(\sigma_{11}, \sigma_{22})|_K \in (Q_1)^2, \ \forall K \in \mathcal{T}_h \right\}. \tag{11.10}$$

With this choice, the degrees of freedom for the stress tensor are all associated with the vertices of an element K as we can see in Figure 11.1 (we choose point values instead of moments as in [120], see also section 11.3.3 for a complete description of the degrees of freedom). In this case the approximation of $a(\sigma_h, \tau_h)$ using the following quadrature formula in K

$$\int_k f dx \approx \int_K^{\mathcal{Q}_K} f dx = \frac{h^2}{4} \sum_{M \in \{\text{vertices of } K\}} f(M), \quad \forall f \in C^o(K),$$

leads to a block diagonal mass matrix (this is nothing but the Gauss-Lobatto quadrature formula of order 1, cf. section 10.4). Each block is associated to a node of the mesh and its dimension is equal to the number of degrees of freedom at this point (that is 5, see Figure 11.1).

11.3.2 Choice of the Approximation Space for the Velocity

To construct the lowest order space M_h, it is natural to approximate the velocity with piecewise constants

$$M_h = \left\{ v_h \in M \ / \ \forall K \in T_h, \ v_h|_K \in (Q_0)^2 \right\}. \tag{11.11}$$

The finite element defined by (11.10)-(11.11) is called the $Q_1^{div} - Q_0$ element and its degrees of freedom are illustrated in Figure 11.1.

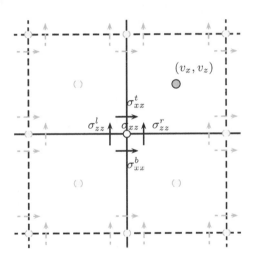

FIGURE 11.1: Degrees of freedom for the $Q_1^{div} - Q_0$ element.

The convergence analysis for this finite element will be presented in the next section and, as we will see, it is quite technical because the finite element spaces defined by (11.10)-(11.11) do not fit the Brezzi-Fortan approximation theory [38, 78]. More precisely, introducing the discrete operator (the continuous operator B is defined by (11.5)),

$$(B_h \tau_h, w_h) = b(w_h, \tau_h) = \int_\Omega \operatorname{div} \tau_h \cdot w_h dx, \forall (w_h, \tau_h) \in X_h^{sym} \times M_h, \tag{11.12}$$

it is easy to verify that the inclusion

$$\operatorname{Ker}(B_h) \subset \operatorname{Ker}(B), \tag{11.13}$$

is not satisfied and that furthermore the bilinear form $a(\cdot, \cdot)$ is not coercive on $\text{Ker}(B_h)$ (even if it is on $\text{Ker}(B)$).

Although it is possible to overcome these difficulties and show that the method converges, the same technique cannot be applied when the method is coupled with fictitious domains (denoted FDM, cf. Chapter 12). Moreover, numerical results illustrate that the method does not always converge when coupled with the FDM (numerically, convergence is obtained only for very particular geometries).

In order to avoid this problem, another choice can be made for the space M_h (cf. [25, 26]), namely,

$$M_h = \left\{ w_h \in M \, / \, \forall \, K \in \mathcal{T}_h, \; w_{h|K} \in (P_1(K))^2 \right\}. \qquad (11.14)$$

With this enrichment of the approximation space, we will have six degrees of freedom per element on the unknown v_h as shown in Figure 11.2. The

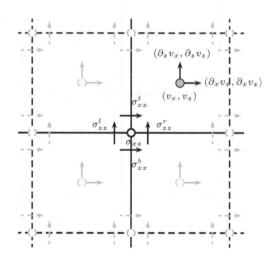

FIGURE 11.2: Degrees of freedom for the $Q_1^{div} - P_1$ element.

space M_h defined by (11.14) was chosen so that for the mixed finite element (11.10)-(11.14), called the $Q_1^{div} - P_1$ element, we have

$$\text{div}\,(X_h^{sym}) \; \subset \; M_h, \qquad (11.15)$$

which easily implies (11.13) and therefore one can prove that the bilinear form $a(\cdot, \cdot)$ is now coercive on $\text{Ker}(B_h)$.

11.3.3 Extension to Higher Orders and Mass Lumping

The natural generalization of the lowest order elements presented in §11.3.1 consists in taking

$$
\begin{cases}
X_h = \left\{ \tau_h \in X \ / \ \forall K \in \mathcal{T}_h, \ \tau_h|_K \in (Q_{k+1})^4 \right\}, \\[2mm]
X_h^{sym} = X_h \cap X^{sym}, \\[2mm]
M_h = \left\{ w_h \in M \ / \ \forall K \in \mathcal{T}_h, \ w_h|_K \in (Q_k)^2 \right\}.
\end{cases}
\tag{11.16}
$$

This will be referred to as the $Q_{k+1}^{div} - Q_k$ element. For the second family of mixed finite elements denoted $Q_{k+1}^{div} - P_{k+1}$ the space M_h is defined by

$$
M_h \ = \ = \ \left\{ w_h \in M \ / \ \forall \, K \in \mathcal{T}_h, \ w_{h|K} \in (P_{k+1}(K))^2 \right\}.
\tag{11.17}
$$

The locations of the degrees of freedom for these elements correspond to tensor products of 1D quadrature points associated with Gauss-Lobatto (for σ_h) or Gauss-Legendre (for v_h) quadrature formulas [58]. As an illustration, we represent the degrees of freedom of the element corresponding to $k = 1$ in Figure 11.3 and we also indicate the number of degrees of freedom per node. We can notice -and this is a general remark for all the higher order elements- that, for σ_h, there are three kind of nodes:

1. The nodes located at a vertex of an element, that are associated to 5 degrees of freedom: the value of σ_{12}, the two values (up and down) of σ_{11}, the two values (left and right) of σ_{22}.

2. The nodes located on an edge, associated to 4 degrees of freedom:

   - for a vertical edge: the two values (left and right) of σ_{22}, the value of σ_{11}, the value of σ_{12},

   - for a horizontal edge: the two values (up and down) of σ_{11}, the value of σ_{22}, the value of σ_{12}.

3. The interior nodes located inside one element, associated to 3 degrees of freedom: the value of σ_{12}, the value of σ_{11}, the value of σ_{22}.

Mass lumping can then be achieved by approximating the integrals in the mass matrices $a(\sigma_h, \tau_h)$ and $c(v_h, w_h)$ by adequate quadrature formulas. More precisely by using the Gauss-Lobatto quadrature formulas to compute $a(\sigma_h, \tau_h)$, one obtains a block diagonal matrix. Each block is associated to one quadrature point and its dimension is equal to the number of degrees of freedom at this point: from 3×3 for the interior nodes to 5×5 for the vertices of the mesh. Note that $c(v_h, w_h)$ leads naturally to a block diagonal matrix. However, using the Gauss-Legendre quadrature formula and appropriate degrees of freedom leads to a diagonal matrix (see also section 10.4 for more details on the quadrature formulas).

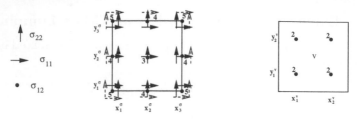

FIGURE 11.3: Degrees of freedom in the $Q_2^{div} - Q_1$ element.

11.4   Theoretical Issues

11.4.1   The $Q_{k+1}^{div} - P_{k+1}$ Element

The analysis of the $Q_{k+1}^{div} - P_{k+1}$ element can be carried out using the abstract theory presented in section 10.2. Indeed, using (11.15) and assuming that the density is constant on each element we have

$$\tau_h \in X_h \implies w_h := \frac{1}{\rho}\mathrm{div}(\tau_h) \in M_h.$$

Introducing this particular test function in the second equation of (11.9) we obtain

$$\frac{d}{dt}\int v_h \cdot \mathrm{div}(\tau_h)dx - \int \frac{1}{\rho}\mathrm{div}(\sigma_h) \cdot \mathrm{div}(\tau_h)dx = \int \frac{1}{\rho}\, f \, \cdot \mathrm{div}(\tau_h)dx.$$

Deriving now the first equation of (11.9) with respect to time and using the last expression we deduce that the variational formulation (11.9) implies the following second order formulation

$$\begin{cases} \text{Find } \sigma_h : [0,T] \longmapsto X_h^{sym} \text{ such that} \\ \dfrac{d}{dt}a(\sigma_h,\tau_h) + \displaystyle\int \frac{1}{\rho}\mathrm{div}(\sigma_h) \cdot \mathrm{div}(\tau_h)dx = -\int \frac{1}{\rho}f \cdot \mathrm{div}(\tau_h)dx, \forall \tau_h \in X_h^{sym} \end{cases}$$

This second order in time formulation can be analyzed following the theory presented in section 10.2. This analysis does not present any particular difficulty and is left to the reader. The only technical point concerns the non-classical approximation properties of the space X_h^{sym} that will be presented in section 11.4.2.2.

11.4.2   The $Q_{k+1}^{div} - Q_k$ Element

We first present in this section some important theoretical properties of the spaces X_h and M_h that will be useful for the convergence analysis. Let us introduce a scalar version of those spaces, the spaces X_h and M_h (remark that we use the same *abusive* notation for X_h-M_h and their scalar version) defined by

$$\begin{cases} X_h = \{q_h \in X = H(\text{div}; \Omega) \,/\, \forall K \in \mathcal{T}_h, \; q_h|_K \in X_k\}, \\[2mm] X_k = Q_{k+1} \times Q_{k+1}, \\[2mm] M_h = \{w_h \in L^2(\Omega) \,/\, \forall K \in \mathcal{T}_h, \; w_h|_K \in Q_k\}. \end{cases} \tag{11.18}$$

It was shown in [22] that the scalar version of X_h admits the following orthogonal (in L^2) decomposition

$$\left| \begin{array}{l} X_h = X_h^s \oplus X_h^r, \\[2mm] X_h^s = \{p_h \in H(\text{div}; \Omega) \,/\, \forall K \in \mathcal{T}_h, \; p_h|_K \in RT_{[k]}\}, \\[2mm] X_h^r = \{p_h \in H(\text{div}; \Omega) \,/\, \forall K \in \mathcal{T}_h, \; p_h|_K \in \Psi_k\}, \end{array} \right. \tag{11.19}$$

where Ψ_k is defined as the orthogonal complement in X_k of $RT_{[k]}$ (for the inner product of $L^2(K)$)

$$\Psi_k(K) = \left\{ \psi \in X_k \,/\, \int_K \psi\,\phi\,dx = 0, \; \forall\phi \in RT_{[k]} \right\}. \tag{11.20}$$

We recall here the well known property of Raviart-Thomas elements:

$$\forall p_h^s \in X_h^s, \quad \text{div} p_h^s \in M_h.$$

Thus, it is straightforward that the tensor space X_h admits the orthogonal decomposition

$$X_h = X_h^s \oplus X_h^r, \tag{11.21}$$

and we have

$$\forall \tau_h^s \in X_h^s, \quad \text{div} \tau_h^s \in M_h. \tag{11.22}$$

Moreover, one can prove easily the following fundamental properties of the tensor space X_h^r (see [22])

$$\begin{cases} (i) \quad \forall \eta_h \in M_h, \forall \tau_h^r \in X_h^r, \; ((\tau_h^r)_{ij}, \eta_h) = 0, \quad \forall i,j = 1,2, \\[3mm] (ii) \quad \forall K \in \mathcal{T}_h, \forall T_j, \; j = 1,\ldots,4, \; \int_{T_j} \tau_h^r n \cdot q \, d\gamma = 0, \quad \forall q \in (P_k(T_j))^2, \\ \qquad \text{where } T_j \text{ are the edges of } K, \text{ i.e., } \partial K = T_1 \cup T_2 \cup T_3 \cup T_4, \\[3mm] (iii) \; \forall w_h \in M_h, \forall \tau_h^r \in X_h^r, \; (\text{div} \tau_h^r, w_h) = 0, \end{cases} \tag{11.23}$$

where property (iii) directly follows from properties (i) and (ii).

Finally, we give another characterization of the space X_h^{sym} defined in (11.16) that follows from Theorem 11.1:

$$X_h^{sym} = \left\{ \sigma \in X^{sym} / \ \sigma_{12} = \sigma_{12} \in Q_h^{k+1}, \ (\sigma_{11}, \sigma_{22}) \in X_h \right\}, \qquad (11.24)$$

where Q_h^{k+1} is the space of continuous functions, locally Q_{k+1}:

$$Q_h^{k+1} = \left\{ \phi \in C^0(\overline{\Omega}), \ \phi|_K \in Q_{k+1}, \ \forall K \in \mathcal{T}_h \right\}. \qquad (11.25)$$

11.4.2.1   Error Estimates for the Evolution Problem

To study the error between the solution (σ, v) of (11.4) and the solution (σ_h, v_h) of (11.9), we follow the same technique as in [75, 112]. This classical approach consists of two main steps. The first step relates, thanks to energy estimates, error estimates for the evolution problem to the estimation of the difference between the exact solution and its elliptic projection (that has to be cleverly defined). The second step, which amounts to analyzing the elliptic projection error, directly follows from the analysis of the approximation of the stationary problem associated to the evolution problem (11.9).

REMARK 11.1 *Although the family of mixed finite elements is compatible with mass lumping, we present here the error analysis for the discrete problem without mass lumping. Of course, when doing mass lumping, one should add to this error the quadrature error due to the numerical integration (see section 10.3 and references [13, 139, 140]).*

In this section, we shall use the notation ∂_t for the time derivative and introduce the spaces (r and m are integers, and T is positive and fixed)

$$C^{r,m} = C^r(0, T; H^m(\Omega)), \ C^{r,(m,m+1)} = C^r(0, T; H^{m,m+1}(\Omega)),$$
$$C^{r,(m+1,m)} = C^r(0, T; H^{m+1,m}(\Omega)), \qquad (11.26)$$

and we introduce

$$\|u\|_{C^{r,m}} = \sup_{t \in [0,T]} \sup_{q \leq r} |\partial_t^q u|_m \ \text{ and } \ \|u\|_{C^{r,(m,m+1)}} = \sup_{t \in [0,T]} \sup_{q \leq r} |\partial_t^q u|_{m,m+1}.$$

In the following we shall make regularity assumptions on the solution (σ, v) of (11.4) that are sufficient in order to obtain the optimal rate of convergence when using elements of order k:

$$\sigma_{11} \in C^{1,(k+2,k+1)}, \ \sigma_{22} \in C^{1,(k+1,k+2)}, \sigma_{12} \in C^{1,k+2}, v \in (C^{1,k+2})^2. \quad (11.27)$$

Following [75, 112] we introduce $\widehat{\sigma}_h(t)$, $\widehat{v}_h(t)$ s.t.,

$$(\widehat{\sigma}_h(t), \widehat{v}_h(t)) \in X_h^{sym} \times M_h, \ \ \widehat{\sigma}_h(0) = 0,$$

and for each $t > 0$, $(\partial_t \widehat{\sigma}_h(t), \widehat{v}_h(t))$ is the solution of

$$\begin{cases} a(\partial_t \widehat{\sigma}_h, \tau_h) + b(\widehat{v}_h, \tau_h) = 0, & \forall \tau_h \in X_h{}^{sym}, \\ b(w_h, \partial_t \widehat{\sigma}_h) = b(w_h, \partial_t \sigma), & \forall w_h \in M_h, \end{cases} \tag{11.28}$$

or equivalently

$$\begin{cases} \dfrac{d}{dt} a(\widehat{\sigma}_h, \tau_h) + b(\widehat{v}_h, \tau_h) = 0, & \forall \tau_h \in X_h{}^{sym}, \\ b(w_h, \widehat{\sigma}_h) = b(w_h, \sigma), & \forall w_h \in M_h. \end{cases} \tag{11.29}$$

We can estimate the elliptic projection error and prove the following result

LEMMA 11.1 *We have the following error estimates*

$$|(\partial_t \sigma - \partial_t \widehat{\sigma}_h)(t)|_H + \|(v - \widehat{v}_h)(t)\|_M \leq C h^{k+1} \left(|\partial_t \sigma_{11}(t)|_{k+2,k+1} \right.$$

$$\left. + |\partial_t \sigma_{22}(t)|_{k+1,k+2} + |\partial_t \sigma_{12}(t)|_{k+2} + |v(t)|_{k+1} + |\nabla v(t)|_{k+1} \right).$$

LEMMA 11.2 *Setting*

$$C(v,\sigma) = |\sigma_{11}|_{C^{1,(k+2,k+1)}} + |\sigma_{22}|_{C^{1,(k+1,k+2)}} + |\sigma_{12}|_{C^{1,k+2}} + |v|_{C^{1,k+1}} + |\nabla v|_{C^{1,k+1}},$$

we have

$$\|v - \widehat{v}_h\|_{C^0(0,T,M)} \leq C(v,\sigma) h^{k+1},$$

$$|\sigma - \widehat{\sigma}_h|_{C(0,T;H)} \leq C(v,\sigma) T h^{k+1}.$$

REMARK 11.2 *Obviously, the same estimates hold for $\partial_t^p(v - \widehat{v}_h)$ and $\partial_t^p(\sigma - \widehat{\sigma}_h)$ if one replaces $C(v,\sigma)$ with $C(\partial_t^p v, \partial_t^p \sigma)$.*

Before proving these lemmas let us remark that they permit us to obtain the following global error estimates.

LEMMA 11.3 *$\exists C_1$, independent of h such that $\forall t \in [0,T]$*

$$|(\widehat{\sigma}_h - \sigma_h)(t)|_H + \|(\widehat{v}_h - v_h)(t)\|_M \leq C_1 \int_0^t \|\partial_t(v - \widehat{v}_h)(s)\|_M ds. \tag{11.30}$$

PROOF By difference between (11.9) and (11.29) and using the second equation of (11.4) we observe that

$$(i) \quad a(\partial_t(\widehat{\sigma}_h - \sigma_h), \tau_h) + b(\widehat{v}_h - v_h, \tau_h) = 0, \quad \forall \tau_h \in X_h{}^{sym},$$

$$(ii) \quad b(w_h, \widehat{\sigma}_h - \sigma_h) = c(\partial_t(v - v_h), w_h), \quad \forall w_h \in M_h.$$

$$\tag{11.31}$$

Taking $\tau_h = \widehat{\sigma}_h - \sigma_h$ and $w_h = \widehat{v}_h - v_h$, we get

$$a(\partial_t(\widehat{\sigma}_h - \sigma_h), \widehat{\sigma}_h - \sigma_h) = c(\partial_t(v_h - v), \widehat{v}_h - v_h),$$

or equivalently, setting $2E_h = a(\widehat{\sigma}_h - \sigma_h, \widehat{\sigma}_h - \sigma_h) + c(\widehat{v}_h - v_h, \widehat{v}_h - v_h)$,

$$\frac{dE_h}{dt} = c(\partial_t(\widehat{v}_h - v), \widehat{v}_h - v_h),$$

and Gronwall's lemma leads to (11.30). □

Joining Lemmas 11.2, 11.3 and Remark 11.2, we can easily prove our final result which is summarized in the following theorem.

THEOREM 11.2 *Let (σ, v) be the solution of (11.4) and (σ_h, v_h) the solution of the approximate problem (11.9), we have the following error estimates*

$$\|v - v_h\|_{C^0(0,T;M)} \leq (C(v, \sigma) + T\, C(\partial_t v, \partial_t \sigma))\; h^{k+1},$$

$$|\sigma - \sigma_h|_{C(0,T;H))} \leq T\; (C(v, \sigma) + C(\partial_t v, \partial_t \sigma))\; h^{k+1}.$$

REMARK 11.3 *We will prove in the sequel Lemma 11.1 by studying the elliptic projection error. We would like to notify the reader that the following section is quite technical because the finite element spaces (11.16) do not satisfy the hypotheses of the classical theory [12, 37] and we need to combine several non-trivial techniques (such as macroelements) to obtain Lemma 11.1.*

The elliptic projection error. We define the elliptict projection $(\widetilde{\sigma}_h, \widetilde{v}_h) = \Pi_h(\sigma, v) \in X_h^{sym} \times M_h$ as the solution of the following problem,

$$\begin{cases} a(\widetilde{\sigma}_h - \sigma, \tau_h) + b(\widetilde{v}_h - v, \tau_h) = 0, \forall \tau_h \in X_h^{sym}, \\[2mm] b(w_h, \widetilde{\sigma}_h - \sigma) = 0, \forall w_h \in M_h. \end{cases} \quad (11.32)$$

REMARK 11.4 *Remark that the elliptic projection in Lemma 11.1 is defined so that $(\partial_t \widetilde{\sigma}_h, \widetilde{v}_h) = \Pi_h(\partial_t \sigma, v)$ with (σ, v) the solution of the continuous problem (11.4).*

The main difficulty in the convergence analysis of problem (11.32) is that the finite element spaces (11.16) do not satisfy the hypotheses of the classical theory [12, 37]. In particular, one can easily prove that the uniform discrete coercivity condition is not satisfied, because $\mathrm{div}(X_h{}^{sym}) \not\subset M_h$. Thus, we will present here a different convergence analysis in the L^2 framework for both unknowns σ and v. For this, we only use the coercivity of the form $a(\cdot, \cdot)$ on the space H

$$\exists \alpha > 0 \, / \, \forall\, \tau \in H,\; a(\tau, \tau) \geq \alpha\, |\tau|_H^2. \quad (11.33)$$

It is less obvious to obtain estimates of $\text{div}(\sigma - \sigma_h)$. Let us introduce

$$V_h(\sigma) = \{\tau_h \in X_h^{sym} \, / \, b(w_h, \sigma - \tau_h) = 0, \; \forall w_h \in M_h\},$$

$$V_h = V_h(0) \equiv \text{Ker} B_h,$$

with B_h defined by (11.12). For proving error estimates, we shall use the following lemma.

LEMMA 11.4   *The following discrete uniform inf-sup condition holds*

$$\left| \begin{array}{l} \exists \beta > 0 \, / \, \forall h, \forall w_h \in M_h, \; \exists \, \tau_h \in X_h^{sym}, \; \tau_h \neq 0 \, / \\[2mm] b(w_h, \tau_h) \geq \beta \, \|w_h\|_M \, \|\tau_h\|_X \, . \end{array} \right. \tag{11.34}$$

The proof of this lemma is based on a macroelement technique and is quite technical, we will present its proof in section 11.4.2.1.1.

Notation.   For any integer $m \geq 0$, we introduce the space

$$H^{m,m+1}(\Omega) = \{\psi \in H^m(\Omega), / \, \partial_2 \psi \in H^m(\Omega)\}, \tag{11.35}$$

and similarly for $H^{m+1,m}(\Omega)$, and we set

$$|\psi|_{m,m+1} = |\psi|_m + |\partial_2 \psi|_m, \quad |\psi|_{m+1,m} = |\psi|_m + |\partial_1 \psi|_m, \tag{11.36}$$

where $|\cdot|_m$ denotes the semi-norm on $H^m(\Omega)$.

As a consequence of (11.33) and of the continuity of the bilinear form $a(\cdot, \cdot)$ on $H \times H$, we deduce the following result (see [38], Prop. 2.13 and Rem. 2.14):

THEOREM 11.3   *Problem (11.32) admits a unique solution* $(\tilde{\sigma}_h, \tilde{v}_h) \in X_h^{sym} \times M_h$, *which satisfies*

$$|\sigma - \tilde{\sigma}_h|_H + \|v - \tilde{v}_h\|_M \leq C \left\{ \inf_{\tau_h \in V_h(\sigma)} |\sigma - \tau_h|_H + \inf_{w_h \in M_h} \|v - w_h\|_M + \right.$$
$$\left. + \inf_{w_h \in M_h} \sup_{\tau_h \in V_h} \frac{b(v - w_h, \tau_h)}{\alpha \, |\tau_h|_H} \right\}. \tag{11.37}$$

This allows us to establish the following convergence theorem:

THEOREM 11.4   *Let* $(\tilde{\sigma}_h, \tilde{v}_h)$ *be the unique solution of (11.32). If* $(\sigma_{11}, \sigma_{22}) \in H^{1,0}(\Omega) \times H^{0,1}(\Omega)$ *and* $\sigma_{12} \in H^1(\Omega)$ *then*

$$|\sigma - \tilde{\sigma}_h|_H + \|v - \tilde{v}_h\|_M \longrightarrow 0 \quad \text{when} \;\; h \to 0. \tag{11.38}$$

Furthermore, if $(\sigma_{11}, \sigma_{22}) \in H^{k+2,k+1}(\Omega) \times H^{k+1,k+2}(\Omega)$, $\sigma_{12} \in H^{k+2}(\Omega)$,
and $v \in (H^{k+2}(\Omega))^2$, *then, one has the error estimates*

$$|\sigma - \tilde{\sigma}_h|_H + \|v - \tilde{v}_h\|_M \leq Ch^{k+1} \left(|\sigma_{11}|_{k+2,k+1} + |\sigma_{22}|_{k+1,k+2} + |\sigma_{12}|_{k+2} \right.$$

$$\left. + |v|_{k+1} + |\nabla v|_{k+1} \right).$$

(11.39)

The proof of this theorem uses the following technical lemma which will be proved in section 11.4.2.1.2:

LEMMA 11.5   *For any* $w \in (H_0^1(\Omega))^2$,

$$\lim_{h \to 0} \inf_{w_h \in M_h} \sup_{\tau_h \in V_h} \frac{b(w - w_h, \tau_h)}{\alpha |\tau_h|_H} = 0.$$

(11.40)

Furthermore for all $w \in (H_0^1(\Omega))^2 \cap (H^{k+2}(\Omega))^2$, *there exists a* $w_h \in M_h$ *such that*

$$\sup_{\tau_h \in V_h} \frac{b(w - w_h, \tau_h)}{\alpha |\tau_h|_H} \leq Ch^{k+1} |w|_{k+2,\Omega}.$$

(11.41)

Proof of Theorem 11.4: The proof consists in estimating each term of the right hand side of (11.37).

• Using the inf-sup condition (11.34), we can bound the first term by (see [38]):

$$\inf_{\tau_h \in V_h(\sigma)} |\sigma - \tau_h|_H \leq \inf_{\tau_h \in V_h(\sigma)} \|\sigma - \tau_h\|_X \leq C \inf_{\tau_h \in X_h^{sym}} \|\sigma - \tau_h\|_X.$$

Then from specific properties of the Raviart-Thomas approximation space X_h^s on regular meshes, combined with classical approximations properties of Q_h^{k+1}, one gets, if $(\tau_{11}, \tau_{22}) \in H^{1,0} \times H^{0,1}$ and $\tau_{12} \in H^1$

$$\lim_{h \to 0} \inf_{\tau_h \in X_h^{sym}} \|\tau - \tau_h\|_X = 0.$$

(11.42)

Furthermore, if $(\tau_{11}, \tau_{22}) \in H^{k+2,k+1} \times H^{k+1,k+2}$ and $\tau_{12} \in H^{k+2}$, we have

$$\inf_{\tau_h \in X_h^{sym}} \|\tau - \tau_h\|_X \leq Ch^{k+1} (|\tau_{11}|_{k+2,k+1} + |\tau_{22}|_{k+1,k+2} + |\tau_{12}|_{k+2}).$$

(11.43)

• The standard approximation property in M_h gives:

$$\lim_{h \to 0} \inf_{w_h \in M_h} \|v - w_h\|_M = 0, \ \forall v \in M.$$

Furthermore, for all $w \in (H^{k+1}(\Omega))^2$, we have

$$\inf_{w_h \in M_h} \|w - w_h\|_M \leq C h^{k+1} |w|_{k+1}.$$

(11.44)

• Finally the last term is estimated thanks to Lemma 11.5.   ∎

11.4.2.1.1   Proof of Lemma 11.4 (the inf-sup condition).   To prove
(11.34) we first construct a particular $\tau \in X^{sym}$ satisfying the continuous
inf-sup condition

LEMMA 11.6   *For all $v \in M$ there exists a $\tau \in X^{sym}$ such that*

$$(i) \quad \tau \text{ is diagonal (thus } as(\tau) = 0),$$

$$(ii) \quad \text{div } \tau = v,$$

$$(iii) \quad \|\tau\|_X \leq C_1 \|v\|_M.$$

PROOF   In the case of a rectangular domain Ω it is easy to prove that for all
$v = (v_1, v_2) \in (L^2(\Omega))^2 (= M)$, there exists a $\tau = (\tau_1, \tau_2) \in H(\text{div}, \Omega)(= X)$
with

$$\tau_1 = \begin{pmatrix} \tau_{11} \\ 0 \end{pmatrix}, \quad \tau_2 = \begin{pmatrix} 0 \\ \tau_{22} \end{pmatrix},$$

where τ_1, τ_2 verify

$$\text{div}\tau_1 = \frac{\partial \tau_{11}}{\partial x_1} = v_1 \text{ and div}\tau_2 = \frac{\partial \tau_{22}}{\partial x_2} = v_2.$$

Indeed, we define $\tilde{v}_1(\text{resp. } \tilde{v}_2)$ the extension by zero of $v_1(\text{resp. } v_2)$ at the
exterior of Ω. We can then define τ_{11} (resp. τ_{22}) as a primitive of $\tilde{v}_1$ (resp.
$\tilde{v}_2$)

$$\tau_{11}(x_1, x_2) = \int_0^{x_1} \tilde{v}_1(s, x_2)ds, \quad \tau_{22}(x_1, x_2) = \int_0^{x_1} \tilde{v}_2(x_1, s)ds.$$

It is clear then, that $\tau = (\tau_1, \tau_2)$ satisfy (i), (ii) and (iii) of Lemma 11.6.   ∎

Then, by using the well known properties of the Raviart-Thomas mixed
finite element space, we are able to show the following lemma, that implies
immediately Lemma 11.4.

LEMMA 11.7   *For all $v_h \in M_h$, there exists $\tau_h \in X_h^{sym}$ such that*

$$(i) \quad (\text{div}\tau_h, w_h) = (v_h, w_h), \quad \forall w_h \in M_h,$$

$$(ii) \quad \|\tau_h\|_X \leq C_1 \|v_h\|_M,$$

with C_1 a positive constant independent of h.

PROOF   Take v_h any element of M_h. By Lemma 11.6, we can construct

$\tau^h \in X^{sym}$ such that

$$(a1)\ \tau^h \text{ is diagonal} \ \Rightarrow \text{as}(\tau^h) = 0,$$

$$(a2)\ \text{div}\tau^h = v_h,$$

$$(a3)\ \|\tau^h\|_X \le C_1\|v_h\|_M.$$

(11.45)

Let Π_h^s be the usual interpolation operator on the Raviart-Thomas space X_h^s. It is well known (see [38]) that

$$(b1)\ (\text{div}(\tau - \Pi_h^s\tau), w_h) = 0,\ \forall(\tau, w_h) \in X \times M_h,$$

$$(b2)\ \|\Pi_h^s\tau\|_X \le C_2\|\tau\|_X,\ \forall\tau \in X.$$

(11.46)

Moreover, because of the particular mesh we work with, we have

$$\tau \text{ is diagonal} \ \Rightarrow \Pi_h^s\tau \text{ is diagonal too and thus } \Pi_h^s\tau \in X_h^{sym}. \quad (11.47)$$

Let us now take $\tau_h = \Pi_h^s\tau^h$ and check that it satisfies properties (i) and (ii) of Lemma 11.7. Indeed,

(i) is a consequence of (11.45)-(a2) and (11.46)-(b1).

(ii) is a consequence of (11.45)-(a3) and (11.46)-(b2).   □

11.4.2.1.2   Proof of Lemma 11.5.

A Macro-Element Partitioning

To prove Lemma 11.5, we will use a macro-element technique (cf. [133]). This technique will permit us to obtain a global estimate by simply adding together analogous local estimates. We need to introduce some notations. We define $\{\tau^j,\ j = 1, N_\tau\}$ the basis function of X_h^{sym} ($N_\tau = \dim X_h^{sym}$) associated to the degrees of freedom defined in section 11.3.3. It is easy to show that these basis functions satisfy

(a) $\left\|\tau^j\right\|_{L^\infty} = 1,$

(b) $\exists\, C > 0,\ \ C \text{ independent of } h,\ \ \forall j,\ \ \left\|\text{div}\tau^j\right\|_{L^2} \le C,$

$\exists\, C_1 > 0 \text{ and } C_2 > 0,\ C_1 \text{ and } C_2 \text{ independent of } h,$

(c) $\left| \forall\, \tau_h = \sum_j \alpha_j \tau_j,\ \ C_1\, h \left(\sum_j \alpha_j^2\right)^{1/2} \le |\tau_h|_H \le C_2\, h \left(\sum_j \alpha_j^2\right)^{1/2}. \right.$

(11.48)

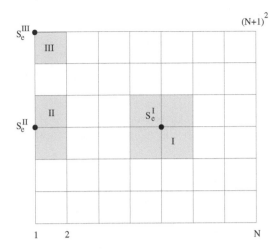

FIGURE 11.4: Definition of the macro-element - I: for an interior node, II: for a node on the boundary (distinct of a vertex), III: for a vertex.

We shall define a macro-element M_e associated with each vertex of the mesh T_h, S_e $(1 \leq e \leq Ne = (N+1)^2)$, as the union of the elements having S_e as a common node, see Figure 11.4. We have, obviously:

$$\Omega = \bigcup_{e=1,...,N_e} M_e, \tag{11.49}$$

as well as the finite overlapping property (that is essential for property (c)):

Each element $K \in T_h$ is included in at most 4 macro-elements. (11.50)

For each macro-element M_e $(e = 1, ..., N_e)$, we define the basis functions τ^j, $j \in \mathcal{J}(e)$ of the macro-element as the τ^j which have a support included in M_e.

The number of basis functions in the macro-elements is bounded, the bound depending only on the order of the element:

$$\exists J_k > 0 \text{ independent of } h \text{ such that } \text{card}(\mathcal{J}(e)) \leq J_k. \tag{11.51}$$

One can notice that each basis function belongs to at least one macro-element and at most four macro-elements. That is why we define weighting coefficients λ_{je} for each basis function τ^j, $j = 1, ..., N_\tau$, associated to a degree of freedom located at a node R_j (the nodes are the locations of the degrees of freedom, see section 11.3.3):

- If R_j is a vertex of the mesh, there exists a unique $e(j) \in 1, .., N_e$ such that $R_j = S_{e(j)}$,

$$\lambda_{jf} = \delta_{fe(j)}, \quad f = 1, ..., N_e.$$

- If R_j is located on an edge of the mesh, there exists a unique $(e_1(j), e_2(j))$ such that $R_j \in [S_{e_1(j)}, S_{e_2(j)}]$,

$$\lambda_{jf} = \frac{1}{2}\delta_{fe_1(j)} + \frac{1}{2}\delta_{fe_2(j)}, \quad f = 1, ..., N_e.$$

- If R_j is an interior node, there exists a unique $(e_1(j), e_2(j), e_3(j), e_4(j))$ such that R_j belongs to the element of the mesh whose vertices are $S_{e_m(j)}, m = 1, ..., 4$

$$\lambda_{jf} = \frac{1}{4}\delta_{fe_1(j)} + \frac{1}{4}\delta_{fe_2(j)} + \frac{1}{4}\delta_{fe_3(j)} + \frac{1}{4}\delta_{fe_4(j)}, \quad f = 1, ..., N_e.$$

By construction, the λ_{je}'s have the property:

$$\sum_{e=1}^{N_e} \lambda_{je} = 1, \quad \forall j = 1, .., N_\tau. \tag{11.52}$$

DEFINITION 11.1 *For each macro-element M_e ($e = 1, ..., N_e$), and for any function $w \in M$, we define the polynomial function w_e as follows: $w_e = 0$ outside M_e and inside M_e, it is defined as:*

- *If S_e is an interior node, w_e is the $L^2(M_e)$ projection of w on $(P_{k+1}(M_e))^2$, the space of functions which are polynomial of degree $k+1$*

$$P_{k+1}(M_e) = \left\{ p(x_1, x_2) \, / \, p(x_1, x_2) = \sum_{i+j \leq k+1} a_{ij} x_1^i x_2^j \right\}.$$

- *If S_e is a boundary node, distinct of a vertex, w_e is the $L^2(M_e)$ projection of w on to the subspace $(P_{k+1}^t(M_e))^2$ of $(P_{k+1}(M_e))^2$, defined by functions which are polynomial of degree k in the variable tangent to the boundary. For instance, in the case of a vertical boundary (the tangential direction being $t = x_2$), the polynomial space $(P_{k+1}^t(M_e))^2$ is defined by :*

$$P_{k+1}^t(M_e) = P_{k+1}^{x_2}(M_e) = \left\{ p(x_1, x_2) \, / \, p(x_1, x_2) = \sum_{\substack{i+j \leq k+1, \\ j \leq k}} a_{ij} x_1^i x_2^j \right\}.$$

- *If S_e is a vertex of the domain, w_e is zero.*

We can then prove the following lemmas,

LEMMA 11.8 *For all $w \in H_0^1(\Omega) \cap (H^{k+2}(\Omega))^2$, we have*

$$\|w - w_e\|_{0,M_e} \leq Ch^{k+2} |w|_{k+2,M_e}. \qquad (11.53)$$

PROOF If S_e is an interior node, the result is standard and can be obtained using Bramble-Hilbert's lemma.

For a node S_e located on the boundary, we use the fact that w vanishes on a part of the boundary of M_e. We can first prove (for the proof see [24]) that for a reference macro-element $\widehat{M}_e$, if $\widehat{w} \in (H^1(\widehat{M}_e))^2 \cap (H^{k+2}(\widehat{M}_e))^2$, and $\widehat{w}$ vanishes on a vertical edge of $\widehat{M}_e$ for instance, then there exists a constant $C > 0$ such that

$$\|\widehat{w} - \widehat{w}_e\|_{0,\widehat{M}_e} \leq C |\widehat{w}|_{k+2,\widehat{M}_e}, \qquad (11.54)$$

$\widehat{w}_e$ being the $L^2(\widehat{M}_e)$ projection of $\widehat{w}$ onto $(P_{k+1}^t(\widehat{M}_e))^2$. Coming back on the current macro-element M_e gives the adequate power of h in estimate (11.53).
□

The key point of the analysis is the following nice result.

LEMMA 11.9 *For all $w \in M$, for all macro-element M_e and for every basis function of the macro-element τ^j, $j \in \mathcal{J}(e)$,*

$$\int_{M_e} (P_{M_h} w_e - w_e) \, \mathrm{div} \, \tau^j dx = 0, \qquad (11.55)$$

where $P_{M_h} : M \to M_h$ denotes the orthogonal projection onto M_h and w_e is defined from w by Definition 11.1.

PROOF For each $e = 1, ..., N_e$, and for all $j \in \mathcal{J}(e)$, we set

$$I = \int_{M_e} (P_{M_h} w_e - w_e) \, \mathrm{div} \, \tau^j dx.$$

For proving this result, we use the orthogonal decomposition of space X_h given in (11.21), which shows that

$$\mathrm{div} \, \tau^j = \mathrm{div} \, (\tau^j)^s + \mathrm{div} \, (\tau^j)^r, \quad \text{with } (\tau^j)^s \in X_h^s \text{ and } (\tau^j)^r \in X_h^r.$$

From property (11.22), we deduce that $\mathrm{div}(\tau^j)^s \in M_h$, therefore, from the definition of P_{M_h}, we get

$$\int_{M_e} (P_{M_h} w_e - w_e) \, \mathrm{div} \, (\tau^j)^s dx = 0.$$

Thus

$$I = \int_{M_e} (P_{M_h} w_e - w_e) \, \mathrm{div} \, (\tau^j)^r dx.$$

Furthermore property (11.23)-(iii) implies

$$\int_{M_e} P_{M_h} w_e \ \text{div} \ (\tau^j)^r dx = 0.$$

Finally, by Green's formula, we have

$$I = -\int_{M_e} w_e \ \text{div} \ (\tau^j)^r dx = \int_{M_e} \nabla w_e \ : \ (\tau^j)^r dx - \int_{\partial M_e} (\tau^j)^r n.w_e \ d\gamma.$$

The vector function w_e belongs to $(P_{k+1}(M_e))^2$, so that ∇w_e belongs to M_h and using (11.23)-(i) the first term of the right hand side vanishes.
The term on the boundary also vanishes, if S_e is an interior node, since $(\tau^j)^r \in H(\text{div}, \Omega)$ is supported in M_e. If S_e is a vertex it also vanishes since $w_e = 0$. For a node S_e on the boundary, distinct of a vertex of Ω, for instance on a vertical boundary, the normal component of $(\tau^j)^r$ does not vanish any more on $\partial\Omega \cap \partial M_e$. However, as $w_e \in P_{k+1}^{x_2}$, its restriction to $\partial\Omega \cap \partial M_e$ is a polynomial of degree k in x_2, and property (11.23)-(ii) permits us to conclude.   ☐

We can now prove the following lemma which will imply in particular the estimate (11.41) of Lemma 11.5.

LEMMA 11.10   *For all* $w \in (H_0^1(\Omega))^2 \cap (H^{k+2}(\Omega))^2$ *and for all* $\tau_h \in X_h^{sym}$, *we have*

$$b(w - P_{M_h} w, \tau_h) \le C h^{k+1} |\tau_h|_H |w|_{k+2,\Omega}, \tag{11.56}$$

where P_{M_h} *is defined as in Lemma 11.9, and* $C > 0$ *is a constant independent of* h.

PROOF   We can decompose τ_h with respect to the basis, $\tau_h = \sum_{j=1}^{N_\tau} \alpha_j \tau^j$, which can be written using property (11.52) as:

$$\tau_h = \sum_{j=1}^{N_\tau} \sum_{e=1}^{N_e} \lambda_{je} \alpha_j \tau^j = \sum_{e=1}^{N_e} \sum_{j=1}^{N_\tau} \lambda_{je} \alpha_j \tau^j = \sum_{e=1}^{N_e} \sum_{j \in \mathcal{J}(e)} \lambda_{je} \alpha_j \tau^j.$$

We then obtain

$$b(w - P_{M_h} w, \tau_h) = \sum_{e=1}^{N_e} \sum_{j \in \mathcal{J}(e)} \lambda_{je} \alpha_j \int_{M_e} (w - P_{M_h} w) \ \text{div}\tau^j dx.$$

From (11.55) in Lemma 11.9, this can be written as

$$b(w - P_{M_h}w, \tau_h) = \sum_{e=1}^{N_e} \sum_{j \in \mathcal{J}(e)} \lambda_{je} \alpha_j \int_{M_e} (w - w_e + P_{M_h}(w_e - w)) \, \mathrm{div}\tau^j \, dx$$

$$\leq \sum_{e=1}^{N_e} \sum_{j \in \mathcal{J}(e)} \lambda_{je} |\alpha_j| (\|w - w_e\|_{0,M_e} + \|P_{M_h}(w - w_e)\|_{0,M_e}) \|\mathrm{div}\tau^j\|_{0,M_e}.$$

Using that P_{M_h} is a projection operator, property (11.48)-(b) of the basis functions and estimate (11.53) of Lemma 11.8, we deduce that

$$b(w - P_{M_h}w, \tau_h) \leq Ch^{k+2} \sum_{e=1}^{N_e} \sum_{j \in \mathcal{J}(e)} \lambda_{je} |\alpha_j| |w|_{k+2,M_e}$$

$$\leq Ch^{k+2} \left(\sum_{e=1}^{N_e} \sum_{j \in \mathcal{J}(e)} \lambda_{je} |\alpha_j|^2 \right)^{1/2} \left(\sum_{e=1}^{N_e} \sum_{j \in \mathcal{J}(e)} \lambda_{je} |w|_{k+2,M_e}^2 \right)^{1/2}$$

$$\leq Ch^{k+2} \left(\sum_{j=1}^{N_\tau} \left(\sum_{e=1}^{N_e} \lambda_{je} \right) |\alpha_j|^2 \right)^{1/2} \left(\sum_{e=1}^{N_e} \left(\sum_{j \in \mathcal{J}(e)} \lambda_{je} \right) |w|_{k+2,M_e}^2 \right)^{1/2}.$$

Using properties (11.51) and (11.52) we obtain:

$$b(w - P_{M_h}w, \tau_h) \leq J_k \, h^{k+2} \left(\sum_{j=1}^{N_\tau} |\alpha_j|^2 \right)^{1/2} \left(\sum_{e=1}^{N_e} |w|_{k+2,M_e}^2 \right)^{1/2}.$$

We conclude thanks to the finite overlapping property (11.50) and to inequality (11.48)-(c)

$$b(w - P_{M_h}w, \tau_h) \leq J_k/C_1 \, h^{k+1} |\tau_h|_H |w|_{k+2,\Omega}.$$

$$\square$$

End of the proof of Lemma 11.5. It remains to prove (11.40) by the density of $(H_0^1(\Omega))^2 \cap (H^{k+2}(\Omega))^2$ in $(H_0^1(\Omega))^2$: for any $w \in (H_0^1(\Omega))^2$, and for any $\varepsilon > 0$, there exists $w_\varepsilon \in (H_0^1(\Omega))^2 \cap (H^{k+2}(\Omega))^2$ such that:

$$|w - w_\varepsilon|_1 \leq \varepsilon/2.$$

We then decompose,

$$\frac{b(w, \tau_h)}{|\tau_h|_H} = \frac{b(w - w_\varepsilon, \tau_h)}{|\tau_h|_H} + \frac{b(w_\varepsilon, \tau_h)}{|\tau_h|_H}.$$

Since $w_\varepsilon \in (H_0^1(\Omega))^2 \cap (H^{k+2}(\Omega))^2$ we can apply Lemma 11.10, thus for all $\tau_h \in V_h$ $(b(P_{M_h}w_\varepsilon, \tau_h) = 0)$, we have

$$\frac{b(w_\varepsilon, \tau_h)}{|\tau_h|_H} \leq Ch^{k+1} |w_\varepsilon|_{k+2}.$$

For the first term, since $w - w_\varepsilon \in (H_0^1(\Omega))^2$, by an integration by parts, we have:

$$b(w - w_\varepsilon, \tau_h) = -(\tau_h, \nabla(w - w_\varepsilon)) \leq |\tau_h|_H |w - w_\varepsilon|_1, \quad \forall \tau_h \in X_h^{sym}.$$

Thus, for all $\tau_h \in V_h$,

$$\frac{b(w, \tau_h)}{|\tau_h|_H} \leq |w - w_\varepsilon|_1 + Ch^{k+1} |w_\varepsilon|_{k+2} \leq \varepsilon/2 + Ch^{k+1} |w_\varepsilon|_{k+2}.$$

Then choosing h_0 such that $Ch_0^{k+1} |w_\varepsilon|_{k+2} \leq \varepsilon/2$ we get

$$\forall h \leq h_0, \quad \frac{b(w, \tau_h)}{|\tau_h|_H} \leq \varepsilon,$$

which implies (11.40). ∎

11.4.2.2   Approximation properties in the space X_h^{sym}

To get the approximation property (11.42) we use a specific property of the Raviart-Thomas elements on a rectangular mesh, whose proof is easy and left to the reader:

LEMMA 11.11   *Let $\mathcal{T}_h$ be a regular mesh composed by rectangular elements K. For any K, if $q = (q_1, q_2) \in H^{1,0}(\Omega) \times H^{0,1}(\Omega)$, one can define the local Raviart-Thomas interpolant $\Pi_K^{RT_k} = (\Pi_1^{RT_k}, \Pi_2^{RT_k})$ as in [38]. Moreover, one has:*

$$\left(\frac{\partial}{\partial x_1} \Pi_1^{RT_k} q, \frac{\partial}{\partial x_2} \Pi_2^{RT_k} q \right) = \left(\Pi_K^{Q_k} \frac{\partial q_1}{\partial x_1}, \Pi_K^{Q_k} \frac{\partial q_2}{\partial x_2} \right), \tag{11.57}$$

where $\Pi_K^{Q_k}$ is the orthogonal projection on $Q_k(K)$.

As a consequence, if $\Pi_h^s = (\Pi_1^s, \Pi_2^s)$ denotes the standard Raviart-Thomas global interpolant [38], one has the following property, specific to our regular mesh (the proof is left to the reader):

LEMMA 11.12   *If $q \in H^{1,0}(\Omega) \times H^{0,1}(\Omega)$ then*

$$\left(\frac{\partial}{\partial x_1} \Pi_1^s q, \frac{\partial}{\partial x_2} \Pi_2^s q \right) = \left(\Pi_h^{Q_k} \frac{\partial q_1}{\partial x_1}, \Pi_h^{Q_k} \frac{\partial q_2}{\partial x_2} \right), \tag{11.58}$$

where $\Pi_h^{Q_k}$ is the orthogonal projection on Q_h^k defined in (11.25).

Using classical estimates for the orthogonal projection on Q_h^k, it is then straightforward to obtain:

LEMMA 11.13 *On a rectangular mesh, if $q \in (H^{1,0}(\Omega) \times H^{0,1}(\Omega))$, then*

$$\|q - \Pi_h^s q\|_{L^2} + \left\|\frac{\partial}{\partial x_1}(q_1 - \Pi_1^s q)\right\|_{L^2} + \left\|\frac{\partial}{\partial x_2}(q_2 - \Pi_2^s q)\right\|_{L^2} \to 0, \ as \ h \to 0.$$

Furthermore if $q \in H^{k+2,k+1}(\Omega) \times H^{k+1,k+2}(\Omega)$ then

$$\|q - \Pi_h^s q\|_{L^2} \le Ch^{k+1} |q|_{k+1},$$

$$\|\partial_1(q_1 - \Pi_1^s q)\|_{L^2} + \|\partial_2(q_2 - \Pi_2^s q)\|_{L^2} \le \quad\quad (11.59)$$

$$Ch^{k+1}\left(\left|\frac{\partial q_1}{\partial x_1}\right|_{k+1} + \left|\frac{\partial q_2}{\partial x_2}\right|_{k+1}\right).$$

In order to establish approximation properties for the space X_h^{sym}, we will use the characterization (11.24) and the fact that $X_h \supset X_h^s$, where X_h^s is the Raviart-Thomas space (see (11.19)). The idea is then to obtain an approximation of (τ_{11}, τ_{22}) in X_h^s and an approximation of $\tau_{12} = \tau_{21}$ in Q_h^{k+1}. To do so, it is necessary to require regularity not only on the divergence of the tensor τ but also on each derivative of the tensor: we assume that $\tau \in X^{sym}$ satisfies $(\tau_{11}, \tau_{22}) \in H^{1,0} \times H^{0,1}$, and $\tau_{12} = \tau_{21} \in H^1$. Observing that

$$\|\tau - \tau_h\|_X \le \|\tau_{12} - \tau_{12}^h\|_{H^1} + \|\tau_{11} - \tau_{11}^h\|_{L^2} + \|\tau_{22} - \tau_{22}^h\|_{L^2}$$

$$+ \|\partial_1(\tau_{11} - \tau_{11}^h)\|_{L^2} + \|\partial_2(\tau_{22} - \tau_{22}^h)\|_{L^2},$$

it is clear from Lemma 11.13 and from classical approximation results of H^1 with Q_h^{k+1} that if $(\tau_{11}, \tau_{22}) \in H^{k+2,k+1} \times H^{k+1,k+2}$ and $\tau_{12} \in H^{k+2}$ then

$$\inf_{\substack{(\tau_{11}^h, \tau_{22}^h) \in X_h^s, \\ \tau_{12}^h = \tau_{21}^h \in Q_h^{k+1}}} \|\tau - \tau_h\|_X \le Ch^{k+1}(|\tau_{11}|_{H^{k+2,k+1}} + |\tau_{22}|_{H^{k+1,k+2}} + |\tau_{12}|_{H^{k+2}}),$$

and by standard density arguments that if $(\tau_{11}, \tau_{22}) \in H^{1,0} \times H^{0,1}$ and $\tau_{12} \in H^1$

$$\lim_{h \to 0} \inf_{(\tau_{11}^h, \tau_{22}^h) \in X_h^s, \tau_{12}^h = \tau_{21}^h \in Q_h^{k+1}} \|\tau - \tau_h\|_X = 0.$$

These results imply (11.42) and (11.43) since $X_h^s \subset X_h$.

11.5 Time Discretization

To construct the equivalent of the leap-frog scheme that was presented in section 10.5 in the case of the first order system, we must use centered finite difference approximations in time. This naturally leads to use a staggered grid approximation. More precisely

- v_h will be computed at times $t^n = n\Delta t$; v_h^n.

- σ_h will be computed at times $t^{n+1/2} = \left(n + \dfrac{1}{2}\right)\Delta t$; $\sigma_h^{n+1/2}$.

The fully discrete scheme is simply

$$
\begin{cases}
a\left(\dfrac{\sigma_h^{n+1/2} - \sigma_h^{n-1/2}}{\Delta t}, \tau_h\right) + b(v_h^n, \tau_h) = 0, \quad \forall\, \tau_h \in X_h^{sym}, \\[4mm]
c\left(\dfrac{v_h^{n+1} - v_h^n}{\Delta t}, w_h\right) - b(w_h, \sigma_h^{n+1/2}) = (f^{n+1/2}, w_h), \quad \forall\, w_h \in M_h,
\end{cases}
\tag{11.60}
$$

By introducing $B_{N_1} = \{\tau_i\}_{i=1}^{N_1}$ and $B_{N_2} = \{w_i\}_{i=1}^{N_2}$ the bases of X_h and M_h respectively ($N_1 = \dim X_h$ and $N_2 = \dim M_h$), $[\Sigma_h] = (\Sigma_1, ..., \Sigma_{N_1})$ and $[V_h] = (V_1, ..., V_{N_2})$ are the coordinates of σ_h and v_h in the bases B_{N_1} and B_{N_2}. In these bases, (11.60) can be written in the following matrix form,

$$
\begin{cases}
\text{Find } (\Sigma_h^{n+1/2}, V_h^{n+1}) \in \mathbb{R}^{N_1} \times \mathbb{R}^{N_2} \text{ such that:} \\[3mm]
\mathcal{M}_h^{\sigma}\dfrac{\Sigma_h^{n+1/2} - \Sigma_h^{n-1/2}}{\Delta t} + \mathcal{B}_h^{\star}V_h^n \quad = 0, \\[4mm]
\mathcal{M}_h^{v}\dfrac{V_h^{n+1} - V_h^n}{\Delta t} \quad - \mathcal{B}_h\Sigma_h^{n+1/2} = F_h^{n+1/2},
\end{cases}
\tag{11.61}
$$

This is the scheme that one implements in practice.

To prove that scheme (11.60) is stable, we can use the same energy techniques as in section 11.63. More precisely, let us define the discrete energy in the following way,

$$
\mathcal{E}_h^n = \frac{1}{2}c(v_h^n, v_h^n) + a(\sigma_h^{n+1/2}, \sigma_h^{n-1/2}).
$$

We will first show that this energy is conserved. Indeed, by taking the first equation of system (11.60) at times n and $n+1$ and by adding the two, we obtain, for $\tau_h = \sigma_h^{n+1/2}$,

$$
a(\sigma_h^{n+3/2} - \sigma_h^{n-1/2}, \sigma_h^{n+1/2}) = -\Delta t\, b(v_h^n + v_h^{n+1}, \sigma_h^{n+1/2}).
$$

Then by multiplying the last equation by $1/2$ and adding it to the second equation of system (11.60), for $w_h = \dfrac{v_h^n + v_h^{n+1}}{2}$ and zero source terms, we get the energy conservation,

$$
\frac{\mathcal{E}_h^{n+1} - \mathcal{E}_h^n}{\Delta t} = 0.
$$

To get the stability condition of the scheme, let us observe that, setting $\sigma_h^n = \frac{1}{2}(\sigma_h^{n+1/2} + \sigma_h^{n-1/2})$, the energy can be written in the following form,

$$\mathcal{E}_h^n = \frac{1}{2}c(v_h^n, v_h^n) + a(\sigma_h^n, \sigma_h^n) - \frac{\Delta t^2}{4}a\left(\frac{\sigma_h^{n+1/2} - \sigma_h^{n-1/2}}{\Delta t}, \frac{\sigma_h^{n+1/2} - \sigma_h^{n-1/2}}{\Delta t}\right).$$

Or, by going to the matricial form and using the first equation of system (11.61), we get

$$\mathcal{E}_h^n = \frac{1}{2}(\mathcal{M}_h^v V_h^n, V_h^n) + (\mathcal{M}_h^\sigma \Sigma_h^n, \Sigma_h^n) - \frac{\Delta t^2}{4}(\mathcal{M}_h^{\sigma,-1}\mathcal{B}_h^\star V_h^n, \mathcal{B}_h^\star V_h^n).$$

It is now clear that the positivity of $\mathcal{E}_h^n$ is guaranteed under the stability condition,

$$\frac{\Delta t^2}{4}\|\mathcal{A}_h\| \leq 1, \quad \|\mathcal{A}_h\| = \sup_{V_h \in \mathbb{R}^{N_2}} \frac{(\mathcal{M}_h^{\sigma,-1}\mathcal{B}_h^\star V_h, \mathcal{B}_h^\star V_h)}{(\mathcal{M}_h^v V_h, V_h)}. \qquad (11.62)$$

REMARK 11.5 *Remark that by using mass lumping we obtain diagonal (or block diagonal) mass matrices $\mathcal{M}_h^\sigma$ and $\mathcal{M}_h^v$ and, therefore, the computational scheme (11.61) is explicit. Moreover, in this case we can eliminate Σ_h and implement the following scheme,*

$$\begin{cases} \text{Find } V_h^{n+1} \in \mathbb{R}^{N_2} \text{ such that:} \\ \mathcal{M}_h^v \dfrac{V_h^{n+1} - 2V_h^n + V_h^{n-1}}{\Delta t^2} + \mathcal{B}_h \mathcal{M}_h^{\sigma,-1}\mathcal{B}_h^\star V_h^n = \dfrac{F_h^{n+1/2} - F_h^{n-1/2}}{\Delta t}. \end{cases}$$
$$(11.63)$$

This is a second order scheme in time that can be analyzed with the same techniques as the ones presented in section 10.5. In particular, the generalization of the stability analysis to this case is straightforward and the stability condition for this scheme can be obtained directly by applying theorem 10.9. The resulting CFL condition coincides with (11.62). We can also use a higher order scheme for the time discretization of the problem by applying the modified equation technique presented in section 10.5.

Note: The references for this chapter and for all chapters in Part III are given at the end of Chapter 14.

Chapter 12

Fictitious Domains Methods for Wave Diffraction

P. Joly, POEMS Project team, INRIA-Rocquencourt, France,
Patric.JOLY@inria.fr

C. Tsogka, Dept. of Mathematics, University of Chicago, USA,
tsogka@tem.uoc.gr

12.1 Introduction

Let us consider as a model problem the propagation of waves in an exterior domain, defined as the complement of a bounded object. Solving this problem with a finite difference method requires a staircase approximation of the boundary of the object (see Figure 12.1). The great disadvantage of this approach is that spurious numerical diffractions are introduced in the solution from the inadequate geometric approximation. To avoid this drawback one can use a finite element method with a mesh that follows exactly the boundary of the object (see Figure 12.1). To have a method which is comparable in cost with the finite defferences method one should use finite elements with mass-lumping (see section 10.4 and cf. [48]).

However, other drawbacks are introduced. First of all, the numerical implementation is much more difficult and the efficiency of the computations is decreased by the unstructured nature of the data. Furthermore, for complicated three dimensional geometries it is quite challenging to create meshes of the whole computational domain. Last but not least, the time step has to be chosen in accordance with the grid mesh size (CFL condition), which leads sometimes to small time steps (in the presence of small elements in the mesh).

In this chapter, we present an alternative method for handling the scattering problem, namely, the fictitious domain method (denoted FDM). Special interest has been given to this approach as it has been shown to lead to efficient numerical methods for solving complicated problems [8–10, 70, 80, 113] particularly in the stationary case.

Recently, in [81] the use of the FDM was also proposed for time dependent problems. As for the stationary case the FDM is very well adapted for problems dealing with complex geometries. In the case of the exterior wave propagation problem, the FDM, also called the domain embedding method, consists in extending artificially the solution inside the object so that the new domain of computation has a very simple shape (typically a rectangle in 2D). To account for the boundary condition a new auxilary unknown, defined only at the boundary of the object, is introduced. The solution of this extended problem has now a singularity across the boundary of the object which can be related to the new unknown.

This idea will be developed in more detail in section 12.2, in the case of elastic waves. The main point is that the mesh for the solution on the enlarged domain can be chosen independently of the geometry of the object. In particular, the use of regular grids or structured meshes allows for simple and efficient computations.

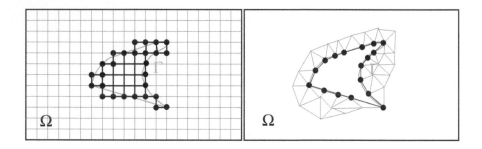

FIGURE 12.1: Schematic of the finite difference (left) and finite element (right) meshes.

In comparison with the finite difference or the finite element method there is an additional cost for the FDM which is due to the determination of the new boundary unknown. However, the final numerical scheme appears to be a slight perturbation of the scheme for the problem without an object and thus this cost may be considered as marginal.

Theoretically, the convergence of the method is linked to a uniform inf-sup condition which leads to a compatibility condition between the boundary mesh and the uniform mesh: this implies that the two mesh grids cannot be chosen completely independently, but this is not an important constraint for the applications. Another important point is that the stability condition of the resulting scheme is the same as the one of the finite difference scheme.

12.2 The Continuous Formulation

In the case of elastic waves motivated by applications such as seismic wave propagation and non-destructive testing we use the fictitious domain method to model boundary conditions on the stress tensor. The simplest condition is the free surface boundary condition (normal stress is zero) which is the natural condition on the topography, *i.e.*, the interface between the (solid) earth and the air. The same condition can be used in the context of non-destructive testing on the boundary of an open crack [23]. Also, the FDM can be used for the unilateral contact condition which is non-linear [20].

Our model problem is seismic wave propagation with complex topography (see Figure 12.2-left). The solution is governed by the elastic wave equation in Ω together with a free surface boundary condition on the topography Γ,

$$\begin{cases} \varrho\dfrac{\partial^2 u}{\partial t^2} - \operatorname{div} \sigma(u) = g & \text{in } \Omega, \\ u = 0 & \text{on } \Gamma_D, \\ \sigma \cdot n = 0 & \text{on } \Gamma, \end{cases} \tag{12.1}$$

with some initial conditions at time $t = 0$ that we will systematically omit in the following.

To obtain a finite computational domain, Ω can be bounded using the method of absorbing boundary conditions or absorbing layers (PML) on the boundary Γ_D. For the sake of simplicity, a Dirichlet condition is assumed here on Γ_D. The question of modeling wave propagation in unbounded domains will be addressed in chapter 14.

The main idea of the FDM consists in extending the solution of problem (12.1) to a larger domain of simple geometry, denoted C, (see Figure 12.2) and in taking into account the boundary condition in a weak way, thanks to the introduction of a Lagrange multiplier. The key point of the approach is that it can be applied to essential type boundary conditions, *i.e.*, conditions that can be considered as an equality constraint in the functional space in which the solution is searched.

To do so with the free surface condition, σ has to be one of the unknowns. This can be done by considering for example the mixed formulation (11.4) with σ sought in the subspace of symmetric tensors of $(H(\operatorname{div}, C))^2$. In this case the Neumann boundary condition $\sigma \cdot n = 0$ can be considered as an equality constraint in the functional space, i.e., we seek for σ in

$$X_0^{sym} = \left\{ \sigma \in (H(\operatorname{div}, C))^2, \sigma \cdot n = 0 \right\}.$$

Let us consider, as in chapter 11, the first order velocity stress system for elastodynamics,

$$\begin{cases} A\dfrac{\partial \sigma}{\partial t} - \varepsilon(v) = 0 \text{ in } \Omega & (i) \\[2mm] \varrho\dfrac{\partial v}{\partial t} - \text{div } \sigma = f \text{ in } \Omega & (ii) \\[2mm] v \qquad\qquad\quad = 0 \text{ on } \Gamma_D & (iii) \\[2mm] \sigma \cdot n \qquad\quad = 0 \text{ on } \Gamma & (iv) \end{cases} \qquad (12.2)$$

where we recall that $v = \dfrac{\partial u}{\partial t}$ is the velocity in Ω and $f = \dfrac{\partial g}{\partial t}$. Following the

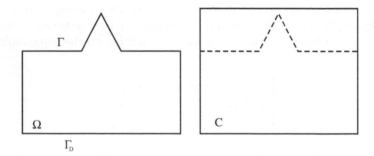

FIGURE 12.2: Left: the original domain. Right: the enlarged domain.

FDM we extend the solution, still denoted (σ, v) for simplicity, to the domain C,

$$\begin{cases} A\dfrac{\partial \sigma}{\partial t} - \varepsilon(v) = \Lambda\delta_\Gamma & \text{in } C \\[2mm] \varrho\dfrac{\partial v}{\partial t} - \text{div } \sigma = f + ([\sigma \cdot n]_\Gamma \equiv 0) \text{ in } C \\[2mm] v \qquad\qquad = 0 & \text{on } \partial C \\[2mm] \sigma \cdot n \qquad = 0 & \text{on } \Gamma \end{cases} \qquad (12.3)$$

where Λ is the tensor of components $\Lambda_{ij} = \lambda_i n_j = [v_i]n_j$, λ being the new unknown.

The variational formulation of (12.3) can be written,

$$\begin{cases} \text{Find } (\sigma, v, \lambda) \in X^{sym} \times M \times G \text{ such that:} \\[2mm] \dfrac{d}{dt}a(\sigma, \tau) + b(\tau, v) - b_\Gamma(\tau, \lambda) = 0, & \forall \tau \in X^{sym} \ (i) \\[2mm] \dfrac{d}{dt}c(v, w) - b(\sigma, w) = (f, w), & \forall w \in M \ (ii) \\[2mm] b_\Gamma(\sigma, \mu) = 0, & \forall \mu \in G \ (iii) \end{cases} \qquad (12.4)$$

where

$$\begin{cases} X = \{\tau \in (H(\text{div}, C))^2\} \quad ; \quad X^{sym} = \{\tau \in X, \tau \text{ symmetric}\}, \\ M = (L^2(C))^2 \quad ; \quad G = (H^{1/2}(\Gamma))^2, \end{cases} \tag{12.5}$$

and the bilinear forms $a(\cdot, \cdot), b(\cdot, \cdot), c(\cdot, \cdot)$ and $b_\Gamma(\cdot, \cdot)$ are defined by

$$a(\sigma, \tau) = \int_C A\sigma : \tau \, dx, \quad \forall(\sigma, \tau) \in X^{sym} \times X^{sym},$$

$$c(v, w) = \int_C \varrho v \cdot w dx, \quad \forall(v, w) \in M \times M,$$

$$b(\tau, w) = \int_C \text{div}(\tau) w \, dx, \forall(\tau, v) \in X^{sym} \times M,$$

$$b_\Gamma(\tau, \lambda) = \langle \tau \vec{n}, \lambda \rangle_\Gamma \qquad \forall(\tau, \lambda) \in X^{sym} \times G.$$

$$\tag{12.6}$$

Obviously, the restriction of (σ, v) to Ω still satisfies (12.2). Moreover, we can remark that the restriction of the solution to $\overline{\Omega}^c$ (where $\overline{\Omega}^c$ denotes the complementary of $\overline{\Omega}$ in C) also satisfies (12.2) where Ω is replaced by $\overline{\Omega}^c$ (and Γ_D denotes $\partial C \cap \partial \overline{\Omega}^c$). Now if we multiply (12.2)-(i) with a function $\tau \in X^{sym}$ (X^{sym} being defined by (12.5)), and integrate in $\Omega \cup \overline{\Omega}^c$, an integration by parts of the second term gives

$$-\int_{\Omega \cup \Omega^c} \varepsilon(v)\tau dx = \int_{\Omega \cup \Omega^c} v \cdot \text{div}\tau dx - \langle \tau \vec{n}, v^+ - v^- \rangle_\Gamma \equiv b(\tau, v) - b_\Gamma(\tau, [v]_\Gamma),$$

which yields (12.4-(i)), if we set $\lambda = [v]_\Gamma$. Since the free surface boundary condition on Γ is not taken into account any more in the new definition of space X^{sym}, it has to be imposed in the formulation which is done with (12.4-(iii)). This establishes an analogy between the FDM and the integral equations for scattering problems [30]. Indeed, in this kind of method λ is typically the quantity that is chosen as the unknown. Nevertheless let us point out a very important difference between our approach and these methods. Integral equations are known to lead, after discretization, to the solution of full linear systems in λ, as will be shown later, this will not be the case for the FDM.

12.3 Finite Element Approximation and Time Discretization

Space discretization. Consider now finite dimension spaces $X_h^{sym} \subset X^{sym}$, $M_h \subset M$ and $\mathcal{G}_H \subset G$. Here $h > 0$ and $H > 0$ represent two approximation parameters (*a priori* independent) devoted to tend to 0. In practice, they will be the stepsizes of a (regular) volume mesh of C (h) and a surface mesh of Γ (H). We approximate the variational problem (12.4) by,

$$\begin{cases} \text{Find } (\sigma_h, v_h, \lambda_H) \in X_h{}^{sym} \times M_h \times \mathcal{G}_H \text{ such that :} \\[2mm] \dfrac{d}{dt} a(\sigma_h, \tau_h) + b(\tau_h, v_h) - b_\Gamma(\tau_h, \lambda_H) = 0, \qquad \forall \tau_h \in X_h{}^{sym}, \\[2mm] \dfrac{d}{dt}(v_h, w_h) - b(\sigma_h, w_h) = (f, w_h), \qquad \forall w_h \in M_h, \\[2mm] b_\Gamma(\sigma_h, \mu_H) = 0, \qquad \forall \mu_H \in \mathcal{G}_H. \end{cases} \quad (12.7)$$

The spaces $X_h{}^{sym} - M_h$ and $\mathcal{G}_H$ will be assumed to satisfy the usual approximation properties. Typically, $X_h{}^{sym} - M_h$ can be the finite element spaces introduced in 11.3. For $\mathcal{G}_H$, which is a subspace of $(H^{1/2}(\Gamma))^2$, one can use continuous functions, for instance piecewise linear functions.

Let us introduce now $B_{N_1} = \{\tau_i\}_{i=1}^{N_1}$, $B_{N_2} = \{w_i\}_{i=1}^{N_2}$ and $B_{N_3} = \{\mu_i\}_{i=1}^{N_3}$ the bases of X_h, M_h and $\mathcal{G}_H$ respectively ($N_1 = \dim X_h$, $N_2 = \dim M_h$ and $N_3 = \dim \mathcal{G}_H$), $[\Sigma_h] = (\Sigma_1, ..., \Sigma_{N_1})$, $[V_h] = (V_1, ..., V_{N_2})$ and $[\lambda_H] = (\Lambda_1, ..., \Lambda_{N_3})$ the coordinates of σ_h, v_h and λ_H in the bases B_{N_1}, B_{N_2} and B_{N_3}. In these bases, (12.7) can be written in the following matrix form: Find $(\Sigma_h, V_h, \lambda_H) \in L^2(0, T; \mathbb{R}^{N_1}) \times L^2(0, T; \mathbb{R}^{N_2}) \times L^2(0, T; \mathbb{R}^{N_3})$ such that,

$$\begin{cases} \mathcal{M}_h^\sigma \dfrac{d\Sigma_h}{dt} + \mathcal{B}_h^\star V_h - \mathcal{B}_H^\Gamma \lambda_H = 0, \\[2mm] \mathcal{M}_h^v \dfrac{dV_h}{dt} - \mathcal{B}_h \Sigma_h \qquad\qquad = F_h, \\[2mm] \mathcal{B}_H^{\Gamma,\star} \Sigma_h \qquad\qquad\qquad = 0, \end{cases} \quad (12.8)$$

where $\mathcal{M}^\star$ denotes the transpose of the matrix $\mathcal{M}$. In practice, and this is the interesting point in the fictitious domain method, we introduce two meshes: the volume unknowns (Σ_h, V_h) are defined on a regular grid, $\mathcal{T}_h$ made of squares of size h while the surface unknown λ_H is computed on a non-uniform mesh on Γ $\mathcal{T}_H$, made of segments of size H_j, $H = \sup_j H_j$ (see Figure 12.3). From the theoretical point of view, the well-posedness of (12.8) and the convergence of the method is linked to the verification of a uniform inf-sup condition which leads to a compatibility condition between the boundary mesh and the uniform mesh of the form $H \geq Ch$ [77].

At this point we can see the importance of mass-lumping. Indeed, when the matrices $\mathcal{M}_h^v$ and $\mathcal{M}_h^\sigma$ are diagonal or block diagonal, we can eliminate the unknown Σ_h (which implies important savings in memory requirements especially in the 3D case) and write system (12.8) as the following second order system in time: Find $(V_h, \lambda_H) \in L^2(0, T; \mathbb{R}^{N_2}) \times L^2(0, T; \mathbb{R}^{N_3})$ such that,

$$\begin{cases} \mathcal{M}_h^v \dfrac{d^2 V_h}{dt^2} + \mathcal{B}_h \mathcal{M}_h^{\sigma,-1} \mathcal{B}_h^\star V_h - \mathcal{B}_h \mathcal{M}_h^{\sigma,-1} \mathcal{B}_H^\Gamma \lambda_H = \dfrac{dF_h}{dt}, \\[2mm] \mathcal{B}_H^{\Gamma,\star} \mathcal{M}_h^{\sigma,-1} \mathcal{B}_H^\Gamma \lambda_H \qquad\qquad\qquad = \mathcal{B}_H^{\Gamma,\star} \mathcal{M}_h^{\sigma,-1} \mathcal{B}_h^\star V_h. \end{cases} \quad (12.9)$$

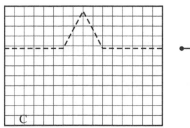 

FIGURE 12.3: Left: the regular mesh of C. Right: the irregular mesh on Γ.

REMARK 12.1 *The matrix $\mathcal{B}_H$, which represents a discrete trace operator on Γ, depends on the two approximation parameters h and H.*

Problem (12.9) appears as a system of ordinary differential equations with an algebraic constraint, *i.e.*, the second equation directly relates λ_H and V_h,

$$\mathcal{Q}_H \lambda_H = \mathcal{B}_H^{\Gamma,\star} \mathcal{M}_h^{\sigma,-1} \mathcal{B}_h^{\star} V_h, \quad \mathcal{Q}_H = \mathcal{B}_H^{\Gamma,\star} \mathcal{M}_h^{\sigma,-1} \mathcal{B}_H^{\Gamma}. \tag{12.10}$$

If $\mathcal{Q}_H$, which is by construction symmetric and positive, is invertible (this issue will be discussed in section 12.4), we see that V_h is the solution of the ordinary differential system,

$$\mathcal{M}_h^v \frac{d^2 V_h}{dt^2} + \left(I - \mathcal{B}_h \mathcal{M}_h^{\sigma,-1} \mathcal{B}_H^{\Gamma} \mathcal{Q}_H^{-1} \right) \mathcal{B}_h \mathcal{M}_h^{\sigma,-1} \mathcal{B}_h^{\star} V_h = \frac{dF_h}{dt}, \tag{12.11}$$

which proves the existence of the discrete solution (V_h and λ_H being now known, Σ_h can be obtained by solving the ordinary differential system corresponding to the first equation (12.8)).

Moreover, for the semi-discrete system (12.7) it is easy to prove the following energy conservation result (for $f = 0$, simply take $\tau_h = \sigma_h$, $w_h = v_h$ and $\mu_H = \lambda_H$ and add the three equations of (12.7)),

$$\frac{d}{dt} \left\{ \frac{1}{2} \left(a(\sigma_h, \sigma_h) + (v_h, v_h) \right) \right\} = 0.$$

Time discretization. According to section 11.5, we shall apply the standard leap frog procedure. With obvious notation, this leads to the following problem,

$$\begin{cases} \text{Find } (\Sigma_h^{n+1/2}, V_h^{n+1}, \lambda_H^n) \in \mathbb{R}^{N_1} \times \mathbb{R}^{N_2} \times \mathbb{R}^{N_3} \text{ such that:} \\[2mm] \mathcal{M}_h^{\sigma} \dfrac{\Sigma_h^{n+1/2} - \Sigma_h^{n-1/2}}{\Delta t} + \mathcal{B}_h^{\star} V_h^n - \mathcal{B}_H^{\Gamma,\star} \Lambda_H^n = 0, \\[3mm] \mathcal{M}_h^{v} \dfrac{V_h^{n+1} - V_h^n}{\Delta t} - \mathcal{B}_h \Sigma_h^{n+1/2} = F_h^{n+1/2}, \\[3mm] \mathcal{B}_H^{\Gamma} \Sigma_h^{n+1/2} = 0, \end{cases} \tag{12.12}$$

or,

$$\begin{cases} \text{Find } (\Sigma_h^{n+1/2}, V_h^{n+1}, \lambda_H^n) \in \mathbb{R}^{N_1} \times \mathbb{R}^{N_2} \times \mathbb{R}^{N_3} \text{ such that:} \\[2mm] \mathcal{M}_h^\sigma \dfrac{\Sigma_h^{n+1/2} - \Sigma_h^{n-1/2}}{\Delta t} + \mathcal{B}_h^\star V_h^n \quad - \mathcal{B}_H^{\Gamma,\star} \Lambda_H^n \qquad\quad = 0, \\[3mm] \mathcal{M}_h^v \dfrac{V_h^{n+1} - V_h^n}{\Delta t} \quad - \mathcal{B}_h \Sigma_h^{n+1/2} \qquad\qquad = F_h^{n+1/2}, \\[3mm] \qquad\qquad \mathcal{Q}_H \lambda_H^n \quad = \mathcal{B}_H^{\Gamma,\star} \mathcal{M}_h^{\sigma,-1} \mathcal{B}_h^\star V_h^n \end{cases}$$

$$(12.13)$$

REMARK 12.2 *We can show that (12.12) and (12.13) are equivalent if $\mathcal{B}_H \Sigma_h^0 = 0$, which expresses in a discrete way a compatibility condition between the initial data and the boundary condition.*

For system (12.12), assuming that the solution is known up to time t^n, the algorithm to compute the solution at t^{n+1} has three steps:

- Compute $\Sigma_h^{n+1/2}$ via the first equation of (12.13) (explicit step).

- Compute V_h^{n+1} via the second equation of (12.13) (explicit step).

- Solve $\mathcal{Q}_H \lambda_H^{n+1} = \mathcal{B}_H^{\Gamma,\star} \mathcal{M}_h^{\sigma,-1} \mathcal{B}_h^\star V_h^{n+1}$ in order to compute λ_H^{n+1} (the invertibility of $\mathcal{Q}_H$ is thus the only condition for the existence of the discrete solution).

As for the semi-discrete problem, we can eliminate Σ_h to obtain,

$$\begin{cases} \text{Find } (V_h^{n+1}, \lambda_H^n) \in \mathbb{R}^{N_2} \times \mathbb{R}^{N_3} \text{ such that:} \\[2mm] \mathcal{M}_h^v \dfrac{V_h^{n+1} - 2V_h^n + V_h^{n-1}}{\Delta t^2} + \mathcal{B}_h \mathcal{M}_h^{\sigma,-1} \mathcal{B}_h^\star V_h^n - \mathcal{B}_h \mathcal{M}_h^{\sigma,-1} \mathcal{B}_H^\Gamma \lambda_H^n = \\[3mm] \qquad\qquad\qquad\qquad\qquad\qquad\qquad \dfrac{F_h^{n+1/2} - F_h^{n-1/2}}{\Delta t} \\[3mm] \mathcal{Q}_H \lambda_H^n = \mathcal{B}_H^{\Gamma,\star} \mathcal{M}_h^{\sigma,-1} \mathcal{B}_h^\star V_h^n \end{cases}$$

$$(12.14)$$

In (12.14), the terms coming from the coupling with the fictitious domain method are the ones containing the unknown λ_H^n. This means that without topography, we would have to solve

$$\mathcal{M}_h^v \frac{V_h^{n+1} - 2V_h^n + V_h^{n-1}}{\Delta t^2} + \mathcal{B}_h \mathcal{M}_h^{\sigma,-1} \mathcal{B}_h^\star V_h^n = \frac{F_h^{n+1/2} - F_h^{n-1/2}}{\Delta t}, \quad (12.15)$$

which can be reinterpreted as a finite difference scheme and is comparable from the computational point of view to the standard FDTD scheme. Therefore, the additional cost for taking into account the topography thanks to the fictitious domain method, compared to the standard FDTD scheme, is due to the second equation of system (12.14). This cost is marginal: the matrix $\mathcal{Q}_H$ is of small size (number of degrees of freedom on Γ) and independent of the step n, so that it can be factorized once and at each time step only a forward backward solve is performed. Note that the term λ_H^n in the first equation of system (12.14)-(i) can be interpreted as an additional source term located on the boundary.

In fact, from the computational point of view, the difficult step in the implementation lies in the construction of the matrix $\mathcal{B}_H$ (and then of $\mathcal{Q}_H$): in 3D for instance, it involves the determination of the intersections between a cubic mesh and a surfacic mesh (typically with triangular plane facets, cf. [73]),

REMARK 12.3 *System (12.9) has the advantage of being a second-order system in time : it is easier to get higher-order discretization in time, using the modified equation technique (see section 10.5), than for the first-order system.*

REMARK 12.4 *The invertibility of the matrix $\mathcal{Q}_H$ of system (12.14)-(ii) (and thus the well posedness of (12.14) is assured by the inf-sup condition already mentioned before.*

12.4 Existence of the Discrete Solution and Stability

Properties of the matrix $\mathcal{Q}_H$. The following properties of the matrix $\mathcal{Q}_H = \mathcal{B}_H^{\Gamma,*} \mathcal{M}_h^{\sigma,-1} \mathcal{B}_H^\Gamma$ are general but we shall illustrate them in the case $d = 2$ when we make the following choice:

- (σ_h, v_h) are appoximated in $X_h{}^{sym} \times M_h$,

$$
\begin{cases}
M_h = \left\{ v_h \in M \ / \ \forall K \in \mathcal{T}_h, \ v_h\,|_K \in (P_1(K))^2 \right\}, \\[2mm]
X_h = \left\{ \sigma_h \in X \ / \ \forall K \in \mathcal{T}_h, \ \sigma_h\,|_K \in (Q_1(K))^4 \right\}, \qquad (12.16) \\[2mm]
X_h{}^{sym} = \left\{ \sigma_h \in X_h; \, / \ as(\sigma_h) = 0 \right\},
\end{cases}
$$

- $\mathcal{G}_H$ is the space of P_1 piecewise linear continuous finite elements,

$$
\mathcal{G}_H = \left\{ \mu_H \in (C^0(\Gamma))^2 \, / \ \forall T \in \Gamma_h, \ \mu_H|_T \in (P^1(T))^2 \right\}.
$$

The entries of the matrix $\mathcal{B}_H$ are:

$$
b_\Gamma(\tau_i, \mu_j) = \int_\Gamma \tau_i \cdot n \, \mu_j \, ds. \qquad (12.17)
$$

To emphasize that $\mathcal{Q}_H$ should be easy to invert, let us observe that

- $\mathcal{Q}_H$ is symmetric and positive (by construction!)

- $\mathcal{Q}_H$ is a "small" matrix: its dimension is exactly (N_3, N_3), to be compared with the one of $\mathcal{B}_h$ which is (N_1, N_3) with $N_1 >> N_3$.

- $\mathcal{Q}_H$ is a sparse matrix with narrow bandwidth. This is due to the sparsity of the matrix $\mathcal{B}_H$: indeed the coefficient $b_\Gamma(\tau_i, \mu_j)$ vanishes as soon as the supports of the two basis functions (τ_i, μ_j) do not intersect.

To illustrate the sparsity of the matrix $\mathcal{Q}_H$, let us consider Λ_I the degree of freedom associated to point M_I on the mesh $\mathcal{T}_H$ of Γ (see Figure 12.4-left). The basis function μ_I is the hat function whose value is one on M_I and zero

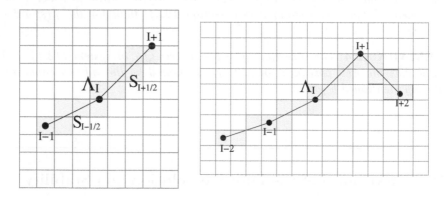

FIGURE 12.4: The coupling between the two meshes.

on the points M_{I+1} and M_{I-1}. The support of this function is thus limitted to the two segments $S_{I-1/2}$, $S_{I+1/2}$. It is then easy to see that the only non-zero elements $(\mathcal{B}_H)_{I,J}$ are the ones that correspond to the indices J of the volume mesh that belong to an element that intersects the segments $S_{I-1/2}$, $S_{I+1/2}$ (see Figure 12.4-left). Concerning the matrix $\mathcal{Q}_H$, we can easily deduce that the summit M_I being coupled with its four neighbors the size of the band of the matrix at this point is 10 (Λ being a two-dimensional vector here), see Figure 12.4-right.

The (crucial) remaining question is the definiteness of $\mathcal{Q}_H$ which corresponds to the fact that the kernel of the matrix is equal to zero, or equivalently that $\mathcal{B}_H$ is surjective from X_h^{sym} onto $\mathcal{G}_H$. This suggests that the space $\mathcal{G}_H$ must not be too large, or in other words that one must not impose too many "boundary"constraints to the discrete solution. As for any mixed method, there is a compatibility condition with the two spaces X_h^{sym} and $\mathcal{G}_H$ that can be reduced to a compatibility relation between the two meshes of C and Γ: the volume mesh cannot be too large with respect to the boundary mesh or, roughly speaking, the ratio H/h must be large enough and thus in this sense the two meshes cannot be completely independent.

Let us illustrate this in the $X_h{}^{sym} - \mathcal{G}_H$ choice. The invertibility of $\mathcal{Q}_H$ can be reformulated as,

$$b_\Gamma(\tau_h, \mu_H) = 0, \quad \forall\, \tau_h \in X_h{}^{sym} \implies \mu_H = 0. \tag{12.18}$$

Let us consider the following opposite cases:

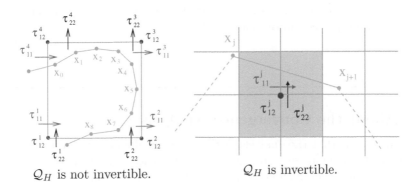

$\mathcal{Q}_H$ is not invertible. $\mathcal{Q}_H$ is invertible.

FIGURE 12.5: The two opposite cases oncerning the invertbility of $\mathcal{Q}_H$.

- **A case where the volume mesh is (locally) too fine.** Assume that one square K of the 2D mesh contains 8 segments of the 1D mesh (cf. Figure 12.4). Let $\{\tau_i, 1 \leq i \leq 12\}$ be the twelve basis functions associated to the vertices of K (cf. Figure 12.4) and $\{\mu_j, 1 \leq j \leq 14\}$ the fourteen basis functions associated to the seven grid points on Γ. Obviously we can find $\mu_H \in \mathcal{G}_H$ of the form:

$$\mu_H = \sum_{l=1}^{7} \mu_l \nu_l, \text{ such that} \mu_H \neq 0 \text{ and } b_\Gamma(\tau_i, \mu_H) = 0, \quad \forall\, 1 \leq i \leq 12.$$

For any $\tau_h \in X_h{}^{sym}$, $b_\Gamma(\tau_h, \mu_H)$ is a linear combination of the $b(\tau_j, \mu_H)$ and therefore vanishes, which contradicts (12.18).

- **A case where the volume mesh is large enough.** Assume that to each segment $I_j = [x_j, x_{j+1}]$ we can associate a basis function τ_{12}^j such that

$$\begin{cases} (i) \;\; I_j \cap \operatorname{Supp}\tau_{12}^j \neq \emptyset, \\ (ii) \;\; \operatorname{Supp}\tau_{12}^j \cap \Gamma \subset I_j. \end{cases}$$

Then there also exist τ_{11}^j and τ_{22}^j satisfying (i) and (ii). In such a case, we can easily prove that (12.18) is satisfied.

Stability analysis.   One very interesting property of the fictitious domain method is that the numerical scheme is stable under the same CFL condition as for the problem without the object. In other words, the stability condition is independent of the geometry of the problem. To prove that, we first show that the the discrete energy,

$$E_h^{n+\frac{1}{2}} = \frac{1}{2} \left(c(v_h^{n+1}, v_h^n) + a(\sigma_h^{n+1/2}, \sigma_h^{n+1/2}) \right),$$

is conserved and then conclude as in section 10.5. The key point being here that the discrete energy is the same as for the case without the object and therefore the stability condition remains the same.

12.5   About the Convergence Analysis

Let us point out that the analysis of the fictitious domain method that we presented in the previous section is not complete at least from the theoretical point of view. Moreover, results obtained with numerical simulations indicate that the FDM does not converge with the $Q_1^{div} - Q_0$ element while it converges when the $Q_1^{div} - P_1$ element is used. Although, theoretical convergence results were recently reported in [25] for the scalar acoustic case, their generalization to elastodynamics is not straightforward.

We present in what follows some numerical results that illustrate the difficulties related with the convergence of the FDM.

12.5.1   FDM with the $Q_{k+1}^{div} - Q_k$ Element

We will present in this section some numerical results that illustrate that the fictitious domain method with the $Q_{k+1}^{div} - Q_k$ element does not always converge.

We consider a computational domain that is the square $[0, 10] \times [0, 10]$ mm^2 composed by a homogeneous isotropic material with density and Lame coefficients given by

$$\rho = 1000 \text{ Kgr/m}^3, \lambda = 3.45 \times 10^9 \text{ Pa}, \mu = 2.04 \times 10^9 \text{ Pa}. \tag{12.19}$$

We introduce an initial condition on the velocity field centered at $(x_c, z_c) = (5, 5)$mm,

$$v((x, z), t = 0) = 0.1 \ F \left(\frac{r}{r_0} \right) \frac{\mathbf{r}}{r},$$

where $F(r)$ is supported in $[0, 1]$ and given by (for $r \in [0, 1]$)

$$F(r) = A_0 - A_1 \cos(2\pi r) + A_2 \cos(3\pi r) - A_3 \cos(6\pi r), \tag{12.20}$$

with $\mathbf{r} = (x - x_c, z - z_c)^t$, $r = \|\mathbf{r}\|$, $r_0 = 1.5$mm and

$$A_0 = 0.35875, \quad A_1 = 0.48829, \quad A_2 = 0.14128, \quad A_3 = 0.01168.$$

We consider a uniform mesh of squares using a discretization step $h = 0.025$mm. The time discretization step Δt is chosen in such a way that the ratio $\Delta t/h$ is equal to the maximal value that ensures the stability. Perfectly matched layers (see Chapter 14 and cf. [31,56]) are used to simulate a non-bounded domain at all the boundaries.

For the discretization of σ and v we use the lowest order element, i.e., the $Q_1^{div} - Q_0$ element and λ is discretized by piecewise continuous functions, i.e.,

$$G_H = \left\{ \nu_h \in G \ / \ \forall S_j, j = 1, ..., N \ ; \ \nu_{H|S_j} \in (P_1)^2 \right\}. \tag{12.21}$$

Horizontal obstacle. In the first experiment we consider a plane horizontal crack

$$(x, z) = (5 + 2\sqrt{2}(2t - 1), 5 - 2\sqrt{2})\text{mm}, \qquad t \in [0, 1], \tag{12.22}$$

that we discretize using a uniform mesh of step $H = Rh$. The method converges and we obtain good results for reasonable values of the parameter R (in the interval $[0.75, 3]$). In Figure 12.6-(a) we show a snapshot of the modulus of the velocity field for $R = 1.2$. We can see the incident pressure wave that has reached the boundary of the computational domain, the reflected waves and the scattered waves created by the two tips of the crack.

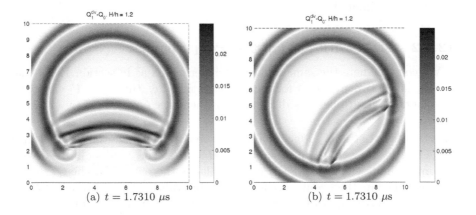

(a) $t = 1.7310 \ \mu s$ (b) $t = 1.7310 \ \mu s$

FIGURE 12.6: $Q_1^{div} - Q_0$. Modulus of the velocity field. $H/h = 1.2$.

Diagonal obstacle. In the second experiment we treat a plane diagonal defect given by

$$(x, z) = (5 + 4t, 1 + 4t)\text{mm}, \qquad t \in [0, 1], \tag{12.23}$$

that is, the same obstacle considered in the previous paragraph rotated by $\pi/4$ radians with respect to (x_c, z_c), the center of the initial condition. We point out that the initial condition satisfies

$$v((x, z), t = 0) \;=\; Q \, v((\tilde{x}, \tilde{z}), t = 0),$$

with

$$(\tilde{x}, \tilde{z})^t = Q^t \, (x, z)^t \;+\; (x_c, z_c)^t, \qquad Q = \begin{pmatrix} \cos(\frac{\pi}{4}) & \sin(\frac{\pi}{4}) \\ -\sin(\frac{\pi}{4}) & \cos(\frac{\pi}{4}) \end{pmatrix}.$$

As the domain is isotropic, the velocity field obtained with the diagonal crack, denoted by v_{diag}, and the one obtained with the horizontal one, denoted by v_{hor}, satisfy

$$v_{hor}(x, z) \;=\; Q \, v_{diag}(\tilde{x}, \tilde{z}), \quad \text{and } |v_{hor}|(x, z) \;=\; |v_{diag}|(\tilde{x}, \tilde{z}).$$

In the same way as in the horizontal case, we discretize the Lagrange multiplier using a uniform grid of step $H = Rh$ with several values for the parameter R. However, this time, the approximated solution does not seem to converge towards the physical solution (see, for instance, Figure 12.6-(b) for $R = 1.2$). The incident wave is not completely reflected but also transmitted through the interface. The scattered waves created by the tips of the crack are also poorly approximated.

12.5.2 FDM with the $Q^{div}_{k+1} - P_{k+1}$ Element

We consider here the same numerical experiments as in section 12.5.1 and use the lowest order element for the discretization (i.e., the $Q^{div}_1 - P_1$ element). The lagrange multiplier λ is discretized in the space G_H defined by (12.21).

Horizontal obstacle. Once again we discretize the horizontal crack defined by (12.22) using a uniform mesh of step $H = Rh$. The results obtained with the mixed finite element $Q^{div}_1 - P_1$ are similar to those given by the $Q^{div}_1 - Q_0$ element. The method converges for reasonable values of the parameter R (in the interval $[0.75, 3]$). In Figure 12.7-(a) we can see the results for $R = 1.2$.

Diagonal obstacle. We now consider the diagonal crack defined by the expression (12.23). We recall that the continuous problem is a rotation of $\pi/4$ radians with respect to the point $(x_c, z_c) = (5, 5)$. The Lagrange multiplier is again discretized using an uniform mesh of step $H = Rh$. Contrary to the results obtained with the element $Q_1^{div} - Q_0$, the ones given by the element $Q_1^{div} - P_1$ converge towards the physical solution when choosing reasonable values for the ratio H/h. In Figure 12.7-(b), this time the incident wave is almost completely reflected by the obstacle. The scattered waves created by the tips of the crack are well approached.

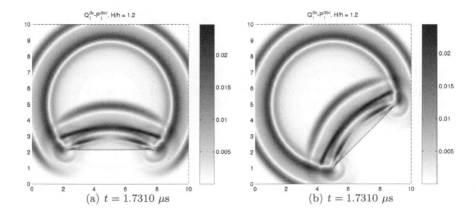

(a) $t = 1.7310 \; \mu s$ (b) $t = 1.7310 \; \mu s$

FIGURE 12.7: $Q_1^{div} - P_1$. Modulus of the velocity field. $H/h = 1.2$.

Numerical error estimates. In this section we are interested in estimating numerically the order of convergence of the method. To do so, we consider solving the elastic wave equations on a disk $\Omega \subset I\!R^2$ with homogeneous Neumann boundary conditions on its boundary $\Gamma = \partial\Omega$. The geometry of the problem is presented in Figure 12.8 with $R = 4$ mm. To compute the solution we extend the unknowns in the domain of simple geometry C (see Figure 12.8) and use the fictitious domain formulation (12.4) with a zero force term $f = 0$ and the initial conditions

$$v((x, z), t = 0) = 0.1 \; F\left(\frac{r}{r_0}\right) \frac{\mathbf{r} + \mathbf{r}^\perp}{r},$$

where $F(r)$ is supported in $[0, 1]$ and is given by (12.20). The center of the initial condition, $(x_c, z_c) = (5, 5)$mm, coincides with the center of the domain Ω. The material in Ω is a homogeneous isotropic solid with density and Lame coefficients given by (12.19). We consider the final time equal to $T = 5\mu s$ when both the pressure and shear waves, have reached the boundary.

The fact that we have a problem that is rotationally invariant allows us to compute a reference solution solving an one-dimensional problem. More

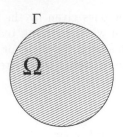
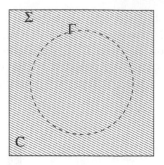

FIGURE 12.8: The geometry of the problem. On the left the initial domain of propagation Ω and on the right the extended domain C introduced by the fictitious domain formulation of the problem.

precisely, rewriting equations (12.2) in polar coordinates

$$
\left|
\begin{aligned}
\rho\frac{\partial v_r}{\partial t} &= \frac{\partial \sigma_{rr}}{\partial r} + \frac{1}{r}\frac{\partial \sigma_{r\theta}}{\partial \theta} + \frac{1}{r}(\sigma_{rr} - \sigma_{\theta\theta}), && \text{in } [0,R] \times [0,2\pi], \\
\rho\frac{\partial v_\theta}{\partial t} &= \frac{\partial \sigma_{r\theta}}{\partial r} + \frac{1}{r}\frac{\partial \sigma_{\theta\theta}}{\partial \theta} + \frac{2}{r}\sigma_{r\theta}, && \text{in } [0,R] \times [0,2\pi], \\
\frac{\partial \sigma_{rr}}{\partial t} &= (2\mu+\lambda)\frac{\partial v_r}{\partial r} + \lambda\frac{v_r}{r} + \lambda\frac{1}{r}\frac{\partial v_\theta}{\partial \theta}, && \text{in } [0,R] \times [0,2\pi], \\
\frac{\partial \sigma_{\theta\theta}}{\partial t} &= (2\mu+\lambda)\left(\frac{1}{r}\frac{\partial v_\theta}{\partial \theta} + \frac{v_r}{r}\right) + \lambda\frac{\partial v_r}{\partial r}, && \text{in } [0,R] \times [0,2\pi], \\
\frac{\partial \sigma_{r\theta}}{\partial t} &= \mu\left(\frac{1}{r}\frac{\partial v_r}{\partial \theta} + \frac{\partial v_\theta}{\partial r} - \frac{v_\theta}{r}\right), && \text{in } [0,R] \times [0,2\pi], \\
\sigma_{rr} &= 0, \quad \sigma_{r\theta} = 0, && \text{in } R \times [0,2\pi],
\end{aligned}
\right. \tag{12.24}
$$

with the initial conditions

$$
v_s((r,\theta), t=0) = 0.1 F\left(\frac{r}{r_0}\right), \quad s \in \{r, \theta\}, \tag{12.25}
$$

we remark that the solution depends only on r and the former equations are equivalent to the two following decoupled one-dimensional problems,

$$
\left|
\begin{aligned}
\rho\frac{\partial v_r}{\partial t} &= \frac{\partial \sigma_{rr}}{\partial r} + \frac{1}{r}(\sigma_{rr} - \sigma_{\theta\theta}), && \text{in } [0,R], \\
\frac{\partial \sigma_{rr}}{\partial t} &= (2\mu+\lambda)\frac{\partial v_r}{\partial r} + \lambda\frac{v_r}{r}, && \text{in } [0,R], \\
\frac{\partial \sigma_{\theta\theta}}{\partial t} &= (2\mu+\lambda)\frac{v_r}{r} + \lambda\frac{\partial v_r}{\partial r}, && \text{in } [0,R], \\
\sigma_{rr} &= 0, && \text{in } R,
\end{aligned}
\right. \tag{12.26}
$$

$$\left| \begin{aligned} \rho \frac{\partial v_\theta}{\partial t} &= \frac{\partial \sigma_{r\theta}}{\partial r} + \frac{2}{r}\sigma_{r\theta}, && \text{in } [0, R] \times [0, 2\pi], \\ \frac{\partial \sigma_{r\theta}}{\partial t} &= \mu\left(\frac{\partial v_\theta}{\partial r} - \frac{v_\theta}{r}\right), && \text{in } [0, R] \times [0, 2\pi], \\ \sigma_{r\theta} &= 0, && \text{in } R \times [0, 2\pi]. \end{aligned} \right. \qquad (12.27)$$

We solve numerically these systems using piecewise constant functions for the discretization of the velocity field and continuous linear functions for the stress tensor. For the time discretization we use a leap frog scheme. The reference solution is obtained using a very fine mesh ($h_{1d} \approx 1/800$mm). The two dimensional problem is solved using four different meshes with $h_z = h_z = 1/10, 1/20, 1/40$ and $1/80$mm. We use the largest time step Δt authorized by the CFL condition. The mesh for the object is uniform and with a discretization step H such that $H/h \approx 4$ for each mesh. For each numerical experiment we compute the difference between the approximated solution and the reference solution. In Figure 12.9 we display the logarithm of the error on the stress tensor, the velocity field and the Lagrange multiplier versus the logarithm of the discretization step. The rate of convergence is thus given by the slope of the lines. We observe that the order of convergence for σ in the $L^\infty(0, T, (H(div, C))^2)$ norm and for v in the $L^\infty(0, T, (L^2(C))^2)$ norm are near the values we could expect (the theory would give $1/2 - \epsilon$). Finally, notice that the convergence rate on λ (approximately 1) is computed in the $L^2(\Gamma)$ norm and therefore we recover the expected convergence rate $(1/2)$ in $H^{1/2}(\Gamma)$.

In Figure 12.10 we display the same results but with the norm of the error now computed in $\tilde{C} = C/B_b(\Gamma)$, *i.e.*, the domain C restricted from $B_b(\Gamma)$, defined by

$$B_b(\Gamma) = \left\{ \mathbf{x} \in C \text{ s.t. } \min_{\mathbf{y} \in \Gamma} |\mathbf{x} - \mathbf{y}| \leq b \right\}, \qquad (12.28)$$

with $b = 0.15$mm. We remarked numerically that $b = h$ is the critical value, *i.e.*, when a layer of this width is removed the convergence rate of the method is higher, while it does not change for bigger values of b and it decreases for $b < h$. This agrees with our intuition in the sense that the elements that we need to remove are the ones in which the solution has less regularity, *i.e.*, the elements that have non-zero intersection with the boundary Γ.

REMARK 12.5 *As for the case without the FDM, using (11.15) and assuming that the density is constant on each element we have that*

$$\tau_h \in X_h \implies w_h := \frac{1}{\rho}\text{div}(\tau_h) \in M_h.$$

Introducing this particular test function in the second equation of (12.7) we obtain

$$\frac{d}{dt}\int v_h \cdot \text{div}(\tau_h)dx - \int \frac{1}{\rho}\text{div}(\sigma_h) \cdot \text{div}(\tau_h)dx = \int \frac{1}{\rho} f \cdot \text{div}(\tau_h)dx.$$

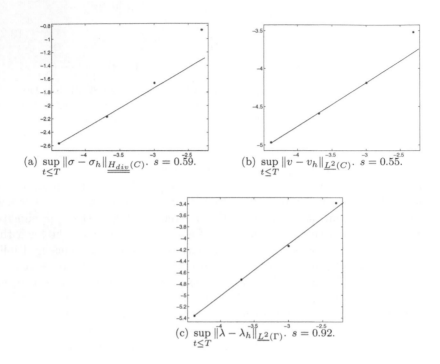

FIGURE 12.9: Numerical error on σ, v and λ versus the discretization step.

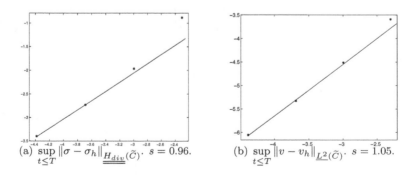

FIGURE 12.10: Numerical error σ, v versus the discretization step. Here we compute the norm of the error in $\tilde{C}$.

Deriving with respect to time the first and third equations of (11.9) and using the last expression we deduce that our variational formulation implies the following second order formulation

$$
\begin{cases}
\text{Find } (\sigma_h(t), \tilde{\lambda}_H(t)) : [0,T] \longmapsto X_h^{sym} \times G_H \text{ such that} \\
\dfrac{d}{dt}a(\sigma_h, \tau_h) + \displaystyle\int \dfrac{1}{\rho}\text{div}(\sigma_h) \cdot \text{div}(\tau_h)dx - b_\Gamma(\tau_h, \tilde{\lambda}_H) = \\
\qquad\qquad\qquad\qquad\qquad -\displaystyle\int \dfrac{1}{\rho}f \cdot \text{div}(\tau_h)dx, \forall \tau_h \in X_h^{sym} \\
b_\Gamma(\sigma_h, \mu_H) = 0, \forall \mu_H \in G_H
\end{cases}
$$

where $\tilde{\lambda}_H = \dfrac{\partial \lambda_H}{\partial t}$. *The nature of this problem is close to the ones analyzed in [77, 101]. We will present in a more general framework the theory that permits to analyze this problem in the next section.*

12.5.3   An Abstract Result

Consider (u, λ) the exact solution of problem,

$$
\begin{cases}
\dfrac{d^2}{dt^2}(u, v) + a(u, v) = b(v, \lambda), \ \forall v \in U, \\
b(u, \mu) = 0, \qquad\qquad\qquad \forall \mu \in L,
\end{cases}
\tag{12.29}
$$

and (u_h, λ_h) the solution of the semi-discrete problem,

$$
\begin{cases}
\dfrac{d^2}{dt^2}(u_h, v_h) + a(u_h, v_h) = b(v_h, \lambda_H), \ \forall v_h \in U_h, \\
b(u_h, \mu_H) = 0, \qquad\qquad\qquad\qquad \forall \mu_H \in L_H,
\end{cases}
\tag{12.30}
$$

assuming the following properties:

The continuous inf-sup condition

$$
\inf_{\mu \in L} \sup_{v \in U} \frac{b(v, \mu)}{\|\mu\|_L \|v\|_U} \geq \alpha > 0,
\tag{12.31}
$$

The coercivity of the form $a(\cdot, \cdot)$ **on Ker** B

$$
\exists \, \alpha > 0 \, / \, \forall \, u \in \text{Ker } B, \ a(u, u) \geq \alpha \|u\|_U^2,
\tag{12.32}
$$

with Ker $B = \{u \in U \, / \, b(\mu, u) = 0, \ \forall \mu \in L\}$.

The uniform discrete inf-sup condition

$$
\inf_{\mu_H \in L_H} \sup_{\tau_h \in X_h^{sym}} \frac{b(\tau_h, \mu_H)}{\|\mu_H\|_L \|v\|_U} \geq \alpha > 0.
\tag{12.33}
$$

The coercivity of the form $a(\cdot, \cdot)$ on Ker B_h

$$\exists\, \alpha > 0 \,/\, \forall\, u_h \in \text{Ker } B_h, \ a(u_h, u_h) \geq \alpha \|u_h\|_U^2\,,$$

$$\text{with Ker } B_h = \{u_h \in U_h \,/\, b(\mu_H, u_h) = 0, \ \forall \mu_H \in L_H\}.$$

(12.34)

Approximation properties

$$\lim_{h \to 0} \inf_{v_h \in U_h} \|u - v_h\|_U = 0, \quad \forall u \in U. \tag{12.35}$$

$$\lim_{H \to 0} \inf_{\mu_H \in L_H} \|\mu - \mu_H\|_L = 0, \quad \forall \mu \in L. \tag{12.36}$$

Existence and uniqueness for both continuous and semi-discrete solutions follows from the classical mixed theory [38].

We want here to estimate the error between the approximate solution (u_h, λ_h) and the exact solution (u, λ) provided that this solution is regular enough. We simply indicate here the main steps of the proof and refer the reader to [53] for more details. Thanks to (12.31) and the coercivity of $a(\cdot, \cdot)$ in Ker$B$, we can introduce the elliptic projection,

$$\left| \begin{array}{l} U \times L \to U_h \times L_H \\[2mm] (u, \lambda) \ \to (\Pi_h u, \Pi_H \lambda) \end{array} \right. \tag{12.37}$$

where $(\Pi_h u, \Pi_H \lambda)$ is defined by,

$$\left\{ \begin{array}{ll} a(u - \Pi_h u, v_h) = b(v_h, \lambda - \Pi_H \lambda), & \forall v_h \in U_h, \\[2mm] b(u - \Pi_h u, \mu_H) = 0, & \forall \mu_H \in L_H. \end{array} \right. \tag{12.38}$$

Moreover, we have the classical result (cf. [38]),

$$\|u - \Pi_h u\|_U + \|\lambda - \Pi_H \lambda\|_L \leq C \times$$

$$\left(\inf_{v_h \in U_h} \|u - v_h\|_U + \inf_{\mu_H \in L_H} \|\lambda - \mu_H\|_L \right). \tag{12.39}$$

Let us write,

$$\left| \begin{array}{ll} u_h - u = \eta_h - \varepsilon_h, & \eta_h = u_h - \Pi_h u, \quad \varepsilon_h = u - \Pi_h u, \\[2mm] \lambda_H - \lambda = \tau_H - \theta_H, & \tau_H = \lambda_H - \Pi_H \lambda, \quad \theta_H = \lambda - \Pi_H \lambda. \end{array} \right. \tag{12.40}$$

As ε_h and θ_H tend to 0, thanks to (12.39), it suffices to estimate η_h and τ_H. We easily see that,

$$\left\{ \begin{array}{ll} \left(\dfrac{d^2\eta_h}{dt^2}, v_h \right) + a(\eta_h, v_h) - b(v_h, \tau_H) = \left(\dfrac{d^2\varepsilon_h}{dt^2}, v_h \right), & \forall v_h \in U_h, \\[4mm] b(\eta_h, \mu_H) = 0, & \forall \mu_H \in L_H. \end{array} \right. \tag{12.41}$$

From the second equation, we deduce (we first differentiate in time this equation for fixed μ_H and then take $\mu_H = \tau_H$),

$$b(\frac{d\eta_h}{dt}, \tau_H) = 0.$$

Then taking $v_h = \dfrac{d\eta_h}{dt}$ in the first equation of (12.41) leads to,

$$\frac{d}{dt}\mathcal{E}_h = (\frac{d^2\varepsilon_h}{dt^2}, \frac{d\eta_h}{dt}) \text{ where } \mathcal{E}_h(t) = \frac{1}{2}\{ \|\frac{d\eta_h}{dt}(t)\|^2 + a(\eta_h(t), \eta_h(t)) \}. \quad (12.42)$$

After integration in time, we obtain the estimate,

$$\mathcal{E}_h(t)^{\frac{1}{2}} \leq \mathcal{E}_h(0)^{\frac{1}{2}} + \int_0^t \|\frac{d^2\varepsilon_h}{dt^2}(s)\| \, dt,$$

where $\mathcal{E}_h(0)$ refers to the approximation of the initial data. The integral term can be shown to tend to 0 as h tends to 0, thanks to (12.35), provided some regularity assumptions (in time) on the exact solution u. We thus get an estimate on η_h.

Finally, to get an estimate for τ_H, we use the uniform inf-sup condition,

$$\|\tau_H\|_L \leq \alpha^{-1} \sup_{v_h \in U_h} \frac{b(v_h, \tau_H)}{\|v_h\|_U} \quad (12.43)$$

From the first equation of (12.41) we deduce that,

$$|b(v_h, \tau_H)| \leq C \left\{ \|\frac{d^2\eta_h}{dt^2}\| + \|\eta_h\| + \|\varepsilon_h\| \right\} \|v_h\|_U,$$

so that finally,

$$\|\tau_H\|_L \leq C \left\{ \|\frac{d^2\eta_h}{dt^2}\| + \|\eta_h\| + \|\varepsilon_h\| \right\}, \quad (12.44)$$

which means essentially that one controls τ_H in terms of quantities which have been already estimated (η_h and ϵ_h).

Accuracy of the fictitious domain method. The counterpart to the good properties of the fictitious domain method, in terms of simplicity and robustness, is its limited accuracy. As a matter of fact, the convergence proof shows that the accuracy is essentially driven by the "interpolation error" associated with the exact solution (u, λ) of the continuous problem,

$$\inf_{v_h \in U_h} \|u - v_h\|_U + \inf_{\mu_H \in L_H} \|\lambda - \mu_H\|_L$$

(and the same quantity where (u, λ) are replaced by successive time derivatives). The limitation of the accuracy is then due to the fact that the regularity of the exact solution in C – or inside an element K of the volume mesh – is limited, independently of the smoothness of the data of the problem. Indeed, consider the acoustic case, for which $U = H(\text{div}, C)$, $L = H^{1/2}(\Gamma)$ and the bilinear forms are defined by,

$$a(u, v) = (\text{div}(u), \text{div}(v)), \quad \forall (u, v) \in U \times U$$
$$b(\mu, v) = < v \cdot n, \mu >_{-1/2, 1/2}, \forall (v, \mu) \in U \times L.$$

In the elements of the volume mesh that intersect Γ, the tangential derivative of u presents a jump (across Γ). As a consequence, the maximal space regularity for $u(\cdot, t), \lambda(\cdot, t)$ is (it is the same for time derivatives),

$$u(\cdot, t) \in H^{\frac{1}{2} - \varepsilon}(\text{div}, C), \quad \forall \varepsilon > 0,$$
$$\lambda(\cdot, t) \in H^{1 - \varepsilon}(\Gamma), \quad \forall \varepsilon > 0,$$

This implies in particular that the fictitious domain method is essentially of order 1/2. That estimate holds for a Neumann type condition, while in the case of a Dirichlet type (in which case we consider the primal formulation) the FDM is of order 1.

About the uniform inf-sup condition.   It is natural to look for what kind of compatibility relation between the two meshes would imply the uniform inf-sup condition. We already know that the inversibility of $\mathcal{Q}_H$ requires a sufficiently fine mesh. It appears that the same kind of condition, is needed here, namely,

$$H \geq C\, h \tag{12.45}$$

is sufficient for the uniform inf-sup condition. We will present in what follows a very general scheme of proof of this result (inspired in particular by the work of Babuska [12]). The objective is to show that for some $\alpha > 0$, and for a small enough ratio h/H, we have

$$\forall \mu_H \in L_H, \exists\, v_h \in U_h \,/\, b_\Gamma(\mu_H, v_h) \geq \alpha \, \|\mu_H\|_L \, \|v_h\|_U. \tag{12.46}$$

To do so, it is sufficient to make the following two assumptions,

1. There exists an operator $\mathcal{R} \in \mathcal{L}(L, U)$ such that,

$$\forall \mu \in L, \quad b_\Gamma(\mu, \mathcal{R}\mu) \geq \|\mathcal{R}\mu\|_U, \quad \|\mathcal{R}\mu\|_X \geq \nu \, \|\mu\|_L \tag{12.47}$$

   which obviously provides the continuous inf-sup condition.

2. There exists an "interpolation" operator $\Pi_h \in \mathcal{L}(U, U_h)$ with the property,

$$\forall \mu_H \in L_H, \quad \|\mathcal{R}\mu_H - \Pi_h \mathcal{R}\mu_H\|_U \leq o(\frac{h}{H}) \, \|\mu_H\|_L^2. \tag{12.48}$$

Indeed, let us assume that these two hypotheses are fullfilled. To obtain (12.46), for any $\mu_H \in L_H$, we choose $v_h = \Pi_h \mathcal{R} \mu_H$, to get,

$$\left|\begin{aligned} b_\Gamma(\mu_H, v_h) &= b_\Gamma(\mu_H, \Pi_h \mathcal{R} \mu_H), \\[2mm] &= b_\Gamma(\mu_H, \mathcal{R} \mu_H) + b_\Gamma(\mu_H, \mu_H - \Pi_h \mathcal{R} \mu_H), \\[2mm] &\geq \|\mathcal{R} \mu_H\|_U^2 - o(\frac{h}{H}) \, \|\mu_H\|_L^2. \end{aligned}\right.$$

We then use the inequalities,

$$\left|\begin{aligned} \|\mathcal{R} \mu_H\|_U^2 &\geq \nu \, \|\mu_H\|_L \, \|\mathcal{R} \mu_H\|_U, \\[2mm] \|\mu_H\|_L^2 &\leq \frac{1}{\mu} \, \|\mu_H\|_L \, \|\mathcal{R} \mu_H\|_U, \end{aligned}\right.$$

to conclude that,

$$b_\Gamma(\mu_H, v_h) \geq \{\nu - \frac{1}{\mu} \, o(\frac{h}{H})\} \, \|\mu_H\|_L \, \|v_h\|_U,$$

which proves the inf-sup condition for $\dfrac{h}{H}$ small enough.

Application of the abstract theory to the case of the scalar wave equation with a Dirichlet condition. The main drawback of the "very general" type of proof that we presented in the previous section is that it does not provide an explicit numerical value for the constant C in the inequality (12.45). In [77], Girault and Glowinski have obtained an explicit value for the constant in the case of the following 2D elliptic problem

$$a(u,v) = b(v, \lambda), \, \forall v \in U,$$
$$b(u, \mu) = 0, \qquad \forall \mu \in L,$$

with $U = H^1(C)$, $L = H^{-1/2}(\Gamma)$ and

$$a(u,v) = (\nabla u, \nabla v), \quad \forall (u,v) \in U \times U,$$
$$b(u, \mu) = < u \cdot n, \mu >_\Gamma, \, \forall (u, \mu) \in U \times L.$$

Their proof is technically difficult and relies on the construction of a so called Fortin's operator. This construction uses the Clement's interpolation operator [44], that is known to exist for Lagrange finite elements. That is why this proof is not directly generalizable to other equations (such as Maxwell's equations or elasticity).

Application of the abstract theory to the case of the scalar wave equation with a Neumann condition. In [26] the authors consider the mixed velocity-pressure formulation for the scalar acoustic waves coupled with the fictitious domain method for taking into account a Neumann boundary condition on a complex geometry. They prove the convergence of the method when using, for the discretization, the scalar version of the $Q_1 - P_1$ element and piecewise continuous functions for the Lagrange multiplier. Their proof is a straightforward application of the general theory presented in the previous section. Namely, it is based in eliminating the velocity and obtaining a second order in time system, using the technique proposed in Remark 12.5.

Application of the abstract theory to the case of the elastic wave equation with a Neumann condition. The application of the abstract theory previously presented to the elastic case is non-trivial. The main difficulty comes from the non-standard regularity required to obtain the approximation properties for $X_h{}^{sym}$ (see section 11.4.2.2). We recall here the approximation properties for the space $X_h{}^{sym}$. Let $\tau \in X^{sym}$ with $(\tau_{11}, \tau_{22}) \in H^{1,0} \times H^{0,1}$ (these spaces are defined by (11.35)) and $\tau_{12} = \tau_{21} \in H^1$. Then

$$\lim_{h \to 0} \inf_{\tau_h X_h{}^{sym}} \|\tau - \tau_h\|_X = 0.$$

Moreover, if $(\tau_{11}, \tau_{22}) \in H^{2,1} \times H^{1,2}$ and $\tau_{12} \in H^2$ then

$$\inf_{\tau_h \in X_h{}^{sym}} \|\tau - \tau_h\|_X \leq Ch(|\tau_{11}|_{H^{2,1}} + |\tau_{22}|_{H^{1,2}} + |\tau_{12}|_{H^2}). \tag{12.49}$$

The maximal regularity (in space) of the stress tensor in the FDM case is $\sigma \in (H^{\frac{1}{2}-\varepsilon}(\text{div}, C))^2$, and σ is symmetric. This regularity is not sufficient to obtain (12.49) and thus we cannot conclude.

From the numerical point of view we do observe that the method converges under a compatibility condition of the form (12.45) between the two discretization meshes. The results shown in section 12.5.2 illustrate numerically the convergence of the method.

12.6   Illustration of the Efficiency of FDM

In this section, we show a numerical experiment devoted to demonstrate that the fictitious domain method provides a better accuracy than a staircase approximation of the boundary. To do so we borrow an example from [52] where the scattering of an electromagnetic wave by a perfect conductor is modelled with the FDM. The geometry of the problem is given in Figure 12.11. The incident wave is a plane wave, with gaussian pulse, coming from the left

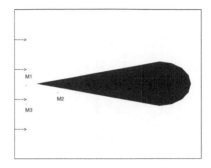

FIGURE 12.11: Geometry of the obstacle.

of the picture and propagating to the right. We represent two snapshots of the solution of the problem in Figure 12.12. We clearly distinguish in these pictures the incident wave and the scattered field. Next we want to compare

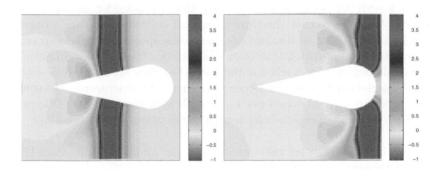

FIGURE 12.12: The solution at two successive instants.

the convergence of the fictitious domain method with the one of the pure FDTD approach combined with a staircase approximation of the boundary γ. To do that, we look at the solution at point M_1 as a function of time for $1 \leq t \leq 3$, which corresponds to the passage of the scattered field. In each case we have computed the discrete solutions for three values of the stepsize h which correspond approximately to 10, 20 and 40 points per wavelength (of the incident wave). The ratio h/H is kept constant.

The three curves corresponding to the fictitious domain calculations are depicted in Figure 12.13-left. It is difficult to distinguish one curve from the other, which expresses the fact that the fictitious domain method has converged. In Figure 12.13-right, we superpose the three curves obtained

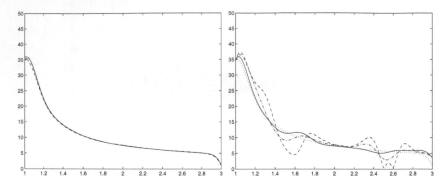

FIGURE 12.13: Left: Solution at point M_1 for $h/\lambda = 1/10, 1/20$ and $1/40$ using the fictitious domain method. Right: we superpose the solution at point M_1 for $h/\lambda = 1/10, 1/20$ and $1/40$ computed with the FDTD, with the curve computed with the FDM using 40 points per wavelength (dotted line).

with the FDTD approach with the curve computed with the fictitious domain approach with 40 points per wavelength that we consider as the reference solution. We observe that the three numerical solutions oscillate around the exact solution. These oscillations are due to the spurious diffractions related to the staircase approximation of the boundary. Clearly, the amplitude of these oscillation decreases when h diminishes but they are still visible with the FDTD calculation with 40 points per wavelength. In other words, the solution is less accurate with 40 points per wavelength with the pure FDTD approach than with 10 points per wavelength with the fictitious domain method.

Note: The references for this chapter and for all chapters in Part III are given at the end of Chapter 14.

Chapter 13

Space Time Mesh Refinement Methods

G. Derveaux, MACS Project team, INRIA-Rocquencourt, France,
Derveaux@inria.fr

P. Joly, POEMS Project team, INRIA-Rocquencourt, France,
Patrick.JOLY@inria.fr

J. Rodríguez, POEM Project team, ENSTA, France

13.1 Introduction

In practical applications, one may have to treat complex geometries or geometrical details in diffraction problems. In such situations, it is desirable to use local mesh refinements. Moreover, it is highly desirable to be able to treat non matching grids (this is even needed if one wants to use FDTD like scheme in each grid). A first idea consists in using only spatial refinement (see [15] for acoustic waves, [28] and [115] for Maxwell's equations). However, when a uniform time step is used, it is the finest mesh that will impose the time step because of the stability condition. There are two problems with this: (1) the computational costs will be increased and (2) the ratio $c\Delta t/h$ on the coarse grid will be much smaller than its optimal value, which will generate dispersion errors. A way to avoid these problems is to use a local time step Δt, related to h in order to keep the ratio $c\Delta t/h$ constant. This solution however raised other practical and theoretical problems that are much more intricate than in the case of a simple spatial refinement, in particular in terms of stability.

The question of mesh refinement was intially studied for electromagnetic problems. The solutions suggested in the associated literature are primarily based on interpolation techniques (in time and/or in space) especially designed to guarantee the consistency of the scheme at the coarse grid-fine grid interface [107], [103], [125] and [41]. Unfortunately, the resulting schemes appear to be very difficult to analyze and may suffer from some instability phenomena.

Recently, we developped alternative solutions to these interpolation procedure. Our purpose in this chapter is to describe these new methods in the case of elastodynamics's equations. However, these methods are applicable to a large class of problems including acoustics, electromagnetics, fluid-structure interaction, etc. They are based on the principle of domain decomposition techniques and consist essentially in ensuring a priori the numerical stability of the scheme via the conservation of a discrete energy.

We first present the method for the case of two spatial grids with one that is twice finer than the other. We will show that the criterion of energy conservation leads to necessary conditions for the discrete transmission equations between the two grids. We can then derive the appropriate transmission equations from consistency and accuracy considerations. We shall also give some elements about the convergence analysis of such methods. This convergence analysis is completed by a Fourier type analysis of the reflexion transmission problem at the interface of the two different grids. This analysis exhibits highly oscillating modes which pollute the solution. An averaging method is proposed to reduce the effect of those spurious oscillations. The method is finally generalized to any refinement rate. Numerical results illustrate the performance of the methods presented in this chapter.

13.2   The Domain Decomposition Approach

13.2.1   Velocity Stress Formulation

Let Ω be a bounded domain whose boundary is denoted Γ^0. Since we want to use the mixed finite elements method which is presented in Chapter 11, we consider the velocity-stress formulation of the elastodynamics problem in Ω:

$$\begin{cases} \rho\dfrac{\partial v}{\partial t} - \operatorname{div} \sigma = f, & \text{in } \Omega, \\[2mm] \mathbf{A}\dfrac{\partial \sigma}{\partial t} - \varepsilon(v) = 0, & \text{in } \Omega, \\[2mm] v = 0, & \text{on } \Gamma^0, \end{cases} \tag{13.1}$$

with some initial conditions that will be omitted in the sequel. ρ is the material density and $\mathbf{A} = \mathbf{C}^{-1}$ is the compliance tensor (see Chapter 9 for notations). For the presentation, we have considered a fixed boundary condition on Γ^0, but we could have also considered Neumann type conditions or propagation in free space. In the latter case, the computational domain can be bounded with artificial boundaries or artificial layers (see Chapter 14).

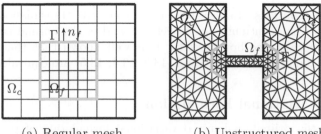

(a) Regular mesh   (b) Unstructured mesh

FIGURE 13.1: Two possible configuration for the decomposition of the domain Ω into two subdomains Ω_c and Ω_f.

13.2.2 Transmission Problem

Our aim is to solve numerically (13.1) using a locally refined mesh. Therefore, we decompose the domain Ω into two non-overlapping subdomains Ω_c and Ω_f (see Figure 13.1):

$$\overline{\Omega} = \overline{\Omega}_c \cup \overline{\Omega}_f, \qquad \Omega_c \cap \Omega_f = \emptyset. \tag{13.2}$$

The geometrical details or singularities of the solutions are located inside the *fine* subdomain Ω_f. Ω_c is the *coarse* sub-domain. The common boundary of Ω_c and Ω_f is denoted Γ (*i.e.*, $\Gamma = \overline{\Omega}_c \cap \overline{\Omega}_f$). Γ_c^0 and Γ_f^0 denote the outer boundary on Ω_c and Ω_f respectively, so that the boundary of Ω is $\Gamma^0 = \Gamma_c^0 \cup \Gamma_f^0$. The unit outgoing normal vector to Ω_c (resp. Ω_f) is denoted n_c (resp. n_f). For any function g of the space variable x, defined on the domain Ω, we denote by g_c and g_f its restriction to Ω_c and Ω_f respectively.

The basic idea of the method is to reformulate problem (13.1) as a transmission problem between Ω_c and Ω_f:

$$\begin{cases} \rho_c \dfrac{\partial v_c}{\partial t} - \text{div}\, \sigma_c = f_c, & \text{in } \Omega_c \quad (a) \\[2mm] \mathbf{A_c} \dfrac{\partial \sigma_c}{\partial t} - \varepsilon(v_c) = 0, & \text{in } \Omega_c \quad (b) \\[2mm] v_c = 0, & \text{on } \Gamma_c^0 \quad (c) \end{cases} \tag{13.3}$$

$$\begin{cases} \rho_f \dfrac{\partial v_f}{\partial t} - \text{div}\, \sigma_f = f_f, & \text{in } \Omega_f \quad (a) \\[2mm] \mathbf{A_f} \dfrac{\partial \sigma_f}{\partial t} - \varepsilon(v_f) = 0, & \text{in } \Omega_f \quad (b) \\[2mm] v_f = 0, & \text{on } \Gamma_f^0 \quad (c) \end{cases} \tag{13.4}$$

$$\begin{cases} \sigma_c \cdot n_c = -\sigma_f \cdot n_f, & \text{on } \Gamma \quad (a) \\[2mm] v_c = v_f, & \text{on } \Gamma \quad (b) \end{cases} \tag{13.5}$$

The two equations (13.5-a) and (13.5-b) are the transmission conditions which ensure the continuity of the velocity and of the normal component of the stress accross the boundary Γ. It is well known that thanks to those conditions the problems (13.1) and (13.3, 13.4, 13.5) are equivalent.

13.2.3   Variational Formulation

We use the primal-dual mixed variational formulation of problem (13.3, 13.4, 13.5). This formulation is similar to the fictitious domain method presented in Chapter 12. That means in particular that the regularity is put on the stress tensors σ_c and σ_f. More precisely v_c (resp. v_f) is sought in $L^2(\Omega_c)$ (resp. $L^2(\Omega_f)$) and σ_c (resp. σ_f) is sought in $H(\text{div}, \Omega_c)$ (resp. $H(\text{div}, \Omega_f)$). This choice allows to use the mixed finite elements introduced in chapter 11 which lead to mass lumping.

After multiplication by test fields and integration in space, only equations (13.3-b) and (13.4-b) are integrated by parts. This makes appear the traces of v_c and v_f on the boundary Γ. In order to give a sense to those traces, which appear also in the second transmission condition (13.5-b), we introduce the new unknowns $j_c \equiv v_{c|\Gamma}$ and $j_f \equiv v_{f|\Gamma}$. The variational formulation of problem (13.3, 13.4, 13.5) is thus written

$$\text{Find} \quad (\sigma_c, v_c, j_c) : \mathbb{R}_+ \to\in X_c^{sym} \times M_c \times G,$$

$$\text{and} \quad (\sigma_f, v_f, j_f) : \mathbb{R}_+ \to\in X_f^{sym} \times M_f \times G, \text{ such that}$$

$$\begin{cases} \dfrac{d}{dt} c_c(v_c, \tilde{v}_c) - b_c(\sigma_c, \tilde{v}_c) = (f_c, \tilde{v}_c), & \forall \tilde{v}_c \in M_c, \\[2mm] \dfrac{d}{dt} a_c(\sigma_c, \tilde{\sigma}_c) + b_c(\tilde{\sigma}_c, v_c) - b_c^{\Gamma}(\tilde{\sigma}_c, j_c) = 0, & \forall \tilde{\sigma}_c \in X_c^{sym}, \end{cases} \tag{13.6}$$

$$\begin{cases} \dfrac{d}{dt} c_f(v_f, \tilde{v}_f) - b_f(\sigma_f, \tilde{v}_f) = (f_f, \tilde{v}_f), & \forall \tilde{v}_f \in M_f, \\[2mm] \dfrac{d}{dt} a_f(\sigma_f, \tilde{\sigma}_f) + b_f(\tilde{\sigma}_f, v_f) - b_f^{\Gamma}(\tilde{\sigma}_f, j_f) = 0, & \forall \tilde{\sigma}_f \in X_f^{sym}, \end{cases} \tag{13.7}$$

$$\begin{cases} b_c^{\Gamma}(\sigma_c, \tilde{\jmath}) + b_f^{\Gamma}(\sigma_f, \tilde{\jmath}) = 0, & \forall \tilde{\jmath} \in G, \\[2mm] j_c = j_f, \end{cases} \tag{13.8}$$

where we have introduced the following functional spaces

$$\begin{vmatrix} X_c^{sym} = \left\{ \sigma_c \in (H(\text{div}, \Omega_c))^2, \ \sigma_c \text{ is symmetric} \right\}, \\[2mm] X_f^{sym} = \left\{ \sigma_f \in (H(\text{div}, \Omega_f))^2, \ \sigma_f \text{ is symmetric} \right\}, \\[2mm] M_c = (L^2(\Omega_c))^2 \quad \text{and} \quad M_f = (L^2(\Omega_f))^2, \\[2mm] G = \left(H^{\frac{1}{2}}(\Gamma) \right)^2, \end{vmatrix} \tag{13.9}$$

The bilinear forms appearing in (13.6), (13.7) and (13.8) are defined by

$$
\begin{cases}
a_c(\sigma_c, \tilde{\sigma}_c) = \displaystyle\int_{\Omega_c} \mathbf{A}_c \sigma_c : \tilde{\sigma}_c \, dx, & a_f(\sigma_f, \tilde{\sigma}_f) = \displaystyle\int_{\Omega_f} \mathbf{A}_f \sigma_f : \tilde{\sigma}_f \, dx, \\[2mm]
c_c(v_c, w_c) = \displaystyle\int_{\Omega_c} \rho_c v_c \cdot w_c dx, & c_f(v_f, w_f) = \displaystyle\int_{\Omega_f} \rho_f v_f \cdot w_f dx, \\[2mm]
b_c(\tilde{\sigma}_c, w_c) = \displaystyle\int_{\Omega_c} \mathrm{div}(\tilde{\sigma}_c) w_c \, dx, & b_f(\tilde{\sigma}_f, w_f) = \displaystyle\int_{\Omega_f} \mathrm{div}(\tilde{\sigma}_f) w_f \, dx, \\[2mm]
b_c^\Gamma(\tilde{\sigma}_c, j_c) = \langle \pi_c \tilde{\sigma}_c, j_c \rangle_\Gamma, & b_f^\Gamma(\tilde{\sigma}_f, j_f) = \langle \pi_f \tilde{\sigma}_f, j_f \rangle_\Gamma.
\end{cases}
$$

$$(13.10)$$

The operators π_c and π_f denote the normal trace map defined by

$$
\pi_c \tilde{\sigma}_c = \tilde{\sigma}_c \, n_c \qquad \text{and} \qquad \pi_f \tilde{\sigma}_f = \tilde{\sigma}_f \, n_f, \tag{13.11}
$$

which map continuously X_c^{sym} and X_f^{sym} respectively onto the dual space of G, $G' = H^{-\frac{1}{2}}(\Gamma)$. $< \, . \, >$ denotes the duality product between G and G'.

REMARK 13.1 *One can show that the trace maps π_c and π_f are surjective onto G' [22, 24] which ensures that the bilinear forms b_c^Γ and b_f^Γ fullfill the inf-sup condition associated to the mixed problem (13.6) to (13.8).*

REMARK 13.2 *It would also be possible to use the primal mixed variational formulation with $v_c \in H^1(\Omega_c)$ and $\sigma_c \in L^2(\Omega_c)$, but this method is not adressed here.*

Also another formulation which avoids the introduction of a Lagrange multiplier is possible. The idea is to use the primal-dual formulation on one grid and the primal mixed formulation on the other one (see for example [100]).

Notice that the Dirichlet boundary conditions on Γ_c^0 and Γ_f^0 are verified as a natural condition in the formulation. The two transmission conditions are treated in a different way. The continuity of the normal component of the stress (condition (13.5-a)) has been taken into account in a weak way in (13.8). This process is similar to the one used in the Mortar elements method where a Lagrance multiplier is used in order to ensure the continuity of the traces of the solution [27]. The continuity of the velocity field accross Γ (condition (13.5-b)) is imposed in a strong way. In principle, thanks to this continuity condition, we could eliminate equation (13.8-b) and define only one Lagrange multiplier $j \equiv j_c = j_f$. The interest of the introduction of a Lagrange multiplier associated to each subdomain will be justified at the time-discretization level.

REMARK 13.3 (Energy identity) *The total energy of the system is de-*

fined by

$$E(t) = \frac{1}{2} \left(\int_{\Omega_c} \mathbf{A_c} \sigma_c : \sigma_c \; dx + \int_{\Omega_c} \rho_c v_c \cdot v_c dx \right)$$

$$+ \frac{1}{2} \left(\int_{\Omega_f} \mathbf{A_f} \sigma_f : \sigma_f \; dx + \int_{\Omega_f} \rho_f v_f \cdot v_f dx \right). \quad (13.12)$$

Taking $(\tilde{\sigma}_c, \tilde{v}_c) = (\sigma_c, v_c)$ and $(\tilde{\sigma}_f, \tilde{v}_f) = (\sigma_f, v_f)$, one gets after adding (13.6) and (13.7) the following result,

$$\frac{dE}{dt}(t) = b_c^\Gamma(\sigma_c, j_c) + b_f^\Gamma(\sigma_f, j_f) + \int_{\Omega_c} f_c \cdot v_c \, dx + \int_{\Omega_f} f_f \cdot v_f \, dx, \quad (13.13)$$

and thanks to equations (13.5) we see that the total energy fulfills the following identity

$$\frac{dE}{dt}(t) = \int_{\Omega_c} f_c \cdot v_c \, dx + \int_{\Omega_f} f_f \cdot v_f \, dx. \quad (13.14)$$

In particular, the energy is conserved when external forces are zero. The energy conservation is a consequence of the conservation of the energetic flux between the two domains, namely $b_c^\Gamma(\sigma_c, j) + b_f^\Gamma(\sigma_f, j) = 0$. This important property will be exploited at the discrete level to ensure the stability of the numerical scheme.

13.3   Space Discretization

13.3.1   Semi–Discretized Variational Formulation

For the space discretization of the variational formulation (13.6, 13.7, 13.8), we will use finite element spaces. We introduce a mesh of the coarse subdomain Ω_c with stepsize h_c, denoted $\mathcal{T}_{h_c}(\Omega_c)$, a mesh of the fine subdomain Ω_f with stepsize h_f, denoted $\mathcal{T}_{h_f}(\Omega_f)$, and a mesh of the surface Γ with stepsize H, denoted $\mathcal{T}_H(\Gamma)$. Two possible configurations are reprensented on Figure 13.1, one with regular meshes, the other with unstructured triangular meshes. By construction, the meshes $\mathcal{T}_{h_c}(\Omega_c)$ and $\mathcal{T}_{h_f}(\Omega_f)$ do not have to be confirmingly connected, as the coupling between them will be ensured through the Lagrange multiplier j. The mesh of the surface Γ can be the trace on Γ of either volumic mesh, but it is not necessary. Based on those meshes, we construct finite dimensional spaces

$$X_{h_c}^{sym} \subset X_c^{sym}, M_{h_c} \subset M_c, X_{h_f}^{sym} \subset X_f^{sym}, M_{h_f} \subset M_f \text{ and } G_H \subset G.$$

The semi-discretized in space problem is written

$$\text{Find } (\sigma_c^{h_c}, v_c^{h_c}, j_c^{h_c}) : \mathbb{R}_+ \to X_{h_c}^{sym} \times M_{h_c} \times G_H$$

$$\text{and } (\sigma_f^{h_f}, v_f^{h_f}, j_f^{h_f}) : \mathbb{R}_+ \to X_{h_f}^{sym} \times M_{h_f} \times G_H \text{ such that}$$

$$\begin{cases} \dfrac{d}{dt} c_c(v_c^{h_c}, \tilde{v}_c^{h_c}) - b_c(\sigma_c^{h_c}, \tilde{v}_c^{h_c}) = (f_c^{h_c}, \tilde{v}_c^{h_c}), & \forall \tilde{v}_c^{h_c} \in M_{h_c}, \\[2mm] \dfrac{d}{dt} a_c(\sigma_c^{h_c}, \tilde{\sigma}_c^{h_c}) + b_c(\tilde{\sigma}_c^{h_c}, v_c^{h_c}) - b_c^\Gamma(\tilde{\sigma}_c^{h_c}, j_c^H) = 0, & \forall \tilde{\sigma}_c^{h_c} \in X_{h_c}^{sym}, \end{cases} \tag{13.15}$$

$$\begin{cases} \dfrac{d}{dt} c_f(v_f^{h_f}, \tilde{v}_f^{h_f}) - b_f(\sigma_f^{h_f}, \tilde{v}_f^{h_f}) = (f_f^{h_f}, \tilde{v}_f^{h_f}), & \forall \tilde{v}_f^{h_f} \in M_{h_f}, \\[2mm] \dfrac{d}{dt} a_f(\sigma_f^{h_f}, \tilde{\sigma}_f^{h_f}) + b_f(\tilde{\sigma}_f^{h_f}, v_f^{h_f}) - b_f^\Gamma(\tilde{\sigma}_f^{h_f}, j_f^H) = 0, & \forall \tilde{\sigma}_f^{h_f} \in X_{h_f}^{sym}, \end{cases} \tag{13.16}$$

$$\begin{cases} b_c^\Gamma(\sigma_c^{h_c}, \tilde{j}^H) + b_f^\Gamma(\sigma_f^{h_f}, \tilde{j}^H) = 0, & \forall \tilde{j}^H \in G_H, \\[2mm] j_c^H = j_f^H. \end{cases} \tag{13.17}$$

13.3.2 Matrix Formulation

Let N_σ^c (resp. N_σ^f) be the dimension of $X_{h_c}^{sym}$ (resp. $X_{h_f}^{sym}$), N_v^c (resp. N_v^f), be the dimension of M_{h_c} (resp. M_{h_f}) and N_j be the dimension of Γ_H. A basis of each of the finite dimensional spaces defined above is introduced. Σ_c, V_c, J_c and Σ_f, V_f, J_f denote the coordinate vectors of the unknowns $\sigma_c^{h_c}, v_c^{h_c}, j_c^H$ and $\sigma_f^{h_f}, v_f^{h_f}, j_f^H$ respectively. For the sake of simplicity, the parameters h_c, h_f and H are omitted in the notation of the coordinate vectors. In those bases, the semi-discretized problem (13.15, 13.16, 13.17) leads to the following matricial differential system

$$\text{Find } (\Sigma_c, V_c, J_c) : \mathbb{R}_+ \to \mathbb{R}^{N_\sigma^c} \times \mathbb{R}^{N_v^c} \times \mathbb{R}^{N_j},$$

$$\text{and } (\Sigma_f, V_f, J_f) : \mathbb{R}_+ \to \mathbb{R}^{N_\sigma^f} \times \mathbb{R}^{N_v^f} \times \mathbb{R}^{N_j}, \text{ such that}$$

$$\begin{cases} M_{v,c} \dfrac{d}{dt} V_c - B_c \Sigma_c = F_c, & (a) \\[2mm] M_{\sigma,c} \dfrac{d}{dt} \Sigma_c + B_c^* V_c - C_c^* J_c = 0, & (b) \\[2mm] M_{v,f} \dfrac{d}{dt} V_f - B_f \Sigma_f = F_f, & (c) \\[2mm] M_{\sigma,f} \dfrac{d}{dt} \Sigma_f + B_f^* V_f - C_f^* J_f = 0, & (d) \\[2mm] C_c \Sigma_c + C_f \Sigma_f = 0, & (e) \\[2mm] J_c = J_f (\equiv J). & (f) \end{cases} \tag{13.18}$$

where A^* denotes the transpose of a matrix A. $M_{v,c}, M_{v,f}, M_{\sigma,c}$ and $M_{\sigma,f}$ denote positive definite mass matrices. The matrices B_c and B_f represent a discrete divergence and their transpose a discrete version of the operator ε. The matrices C_c and C_f represent discrete trace operators on the surface Γ from the coarse subdomain and the fine subdomain respectively. All those matrices obviously depend on the discretization parameters h_c, h_f and H.

Well posedness of the semi-discrete problem. It is possible to eliminate the Lagrange multiplier J from system (13.18). In this way the latter can be reduced to a standard ordinary differential system composed of two sets of independant systems of ODE's on each subdomain. Indeed let us multiply (13.18-b) by $C_c M_{\sigma,c}^{-1}$ and (13.18-d) by $C_f M_{\sigma,f}^{-1}$ and add the result. We obtain the following stationnary relation, thanks to the transmission conditions (13.18-e) and (13.18-f)

$$\left(C_c M_{\sigma,c}^{-1} C_c^* + C_f M_{\sigma,f}^{-1} C_f^* \right) J = C_c M_{\sigma,c}^{-1} B_c^* V_c + C_f M_{\sigma,f}^{-1} B_f^* V_f. \qquad (13.19)$$

This shows that the well-posedness of (13.18) is conditionned by the invertibility of the symmetric and positive matrix

$$\mathcal{M} = C_c M_{\sigma,c}^{-1} C_c^* + C_f M_{\sigma,f}^{-1} C_f^* \qquad (13.20)$$

which is also equivalent to the injectivity of either matrix C_c^* or C_f^* that is

$$\ker C_f^* \cap \ker C_c^* = 0. \qquad (13.21)$$

REMARK 13.4 *This condition can be expressed as a discrete inf-sup condition associated to the space discretized mixed problem ((13.15), (13.16), (13.17)). It can be rewritten as*

$$\forall\, j^H \in G_H(\Gamma), \quad \left[\begin{array}{ll} b_c^\Gamma(\sigma_c^{h_c}, j^H) = 0, & \forall \sigma_c^{h_c} \in X_{h_c}^{sym} \\[2mm] b_f^\Gamma(\sigma_f^{h_f}, j^H) = 0, & \forall \sigma_f^{h_f} \in X_{h_f}^{sym} \end{array} \right] \implies j^H = 0. \quad (13.22)$$

Thus it represents a compatibility condition between the spaces $X_{h_c}^{sym}$, $X_{h_f}^{sym}$ and G_H. This condition implies that the space G_H must not be too large. In practice, we shall often use as the boundary mesh for Γ the trace of the coarse grid mesh for Ω_c (see also remark 13.14).

A well known particular case which ensures (13.21) is to choose an approximation space for the Lagrange multiplier which satisfies the inclusion [27, 38],

$$G^H \subset \pi_c(X_{h_c}^{sym}) \qquad (13.23)$$

REMARK 13.5 (Semi-discrete energy identity) *Since we have considered a conforming approximation of the variational formulation, it is straightforward to prove a semi-discrete energy identity result which is similar to the* *continuous one (see remark 13.3). It can be written in the following matricial form, when external forces are zero*

$$\frac{d}{dt}\left[\frac{1}{2}\left((\Sigma_c,\Sigma_c)_{M_{\sigma,c}}+(V_c,V_c)_{M_{v,c}}\right)+\frac{1}{2}\left((\Sigma_f,\Sigma_f)_{M_{\sigma,f}}+(V_f,V_f)_{M_{v,f}}\right)\right]=0,$$

where $(.,.)_M$ *denotes the scalar product* $(X,Y)_M = X^* M Y$ *for any positive definite matrix* M.

13.3.3 Choice for the Approximation Spaces

In the sequel, for the approximation of the velocity and the stress tensor, we will use the mixed finite elements introduced and analysed in chapter 11, the so-called $Q_1^{div} \times Q_0$ elements. This element is based on regular meshes of the subdomains Ω_c and Ω_f composed of rectangular elements. So we consider here the configuration represented on Figure 13.1-a. It has been designed in order to obtain mass lumping which consists in computing the mass matrices with an appropriate quadrature formula without loss of accuracy. This technique reduces the mass matrices $M_{v,c}, M_{v,f}, M_{\sigma,c}$ and $M_{\sigma,f}$ to diagonal and block diagonal matrices. It allows therefore for inverting them easily (see section 10.4). The second main property of the $Q_1^{div} \times Q_0$ elements is to take into acount the symmetry of the stress tensor in a strong way, as can be seen on the definition of $X_{h_l}^{sym}$ in (13.3.3) below. The approximation spaces are defined for $l \in \{c, f\}$ by (see section 11.3 or [24] for more details)[1]

$$\left| \begin{array}{l} X_{h_l}^{sym} = \left\{ \sigma_l^{h_l} \in X_l \ / \ \forall \ K \in \mathcal{T}_{h_l}(\Omega_l), (\sigma_{l1}^{h_l}, \sigma_{l2}^{h_l})_{|K} \in Q_1^2 \times Q_1^2 \right\} \cap X^{sym}, \\[2mm] M_{h_l} = \left\{ v_l^{h_l} \in M_l \ / \ \forall \ K \in \mathcal{T}_{h_l}(\Omega_l), v_l^{h_l}{}_{|K} \in Q_0^2 \right\}. \end{array} \right.$$

The choice of the approximation space of the Lagrange multiplier, space G, must be done carrefully in order to verify the inf-sup condition (13.22). A natural choice is to take a conforming approximation which satisfies the inclusion $G^H \subset \pi_c(X_{h_c}^{sym})$ as mentioned in remark 13.4. We consider thus the mesh of Γ which is defined by the trace of the coarse subdomain volumic mesh $\mathcal{T}_{h_c}(\Omega_c)$, and we will denote it $\mathcal{T}_{h_c}(\Gamma)$. Then we introduce the piecewise linear continuous functions based on this mesh for the approximation of $(H^{\frac{1}{2}}(\Gamma))^2$ that is

$$G_{h_c}^1 = \left\{ j^{h_c} \in C^0(\Gamma) \ / \ \forall \ S \in \mathcal{T}_{h_c}(\Gamma), j^{h_c}{}_{|S} \in P_1^2 \right\}. \tag{13.24}$$

REMARK 13.6 *We have also investigated the approximation of the space G with discontinuous P_0 finite elements, that is*

$$G_{h_c}^0 = \left\{ j^{h_c} \in L^2(\Gamma) \ / \ \forall \ S \in \mathcal{T}_{h_c}(\Gamma), j^{h_c}{}_{|S} \in P_0^2 \right\}. \tag{13.25}$$

---

[1]$(\sigma_{l1}^{h_l})$ and $(\sigma_{l2}^{h_l})$ denote the two row vectors of the tensor $\sigma_l^{h_l}$, $l \in \{c, f\}$.

Since this space is not included in $G = (H^{\frac{1}{2}}(\Gamma))$ this approximation is not conforming and the convergence analysis is more complicated. However, $b(\sigma_l, j_{h_l})$ still makes sense for $j_{h_l} \in G_{h_l}^0$ and one can show that in this case also the inf-sup condition (13.22) is verified. Namely $\ker C_c^* = \{0\}$, which ensures the well posedness of the semi-discretized scheme (see [128]). In practice, a numerical study has shown that this choice gives more accurate results. Therefore, this is the approximation space which is used for numerical experiments presented in section 13.8.

REMARK 13.7 *A general approach in the Mortar elements literature is to choose*

$$G^H \subset \pi_c(X_{h_c}^{sym}) \quad or \quad G^H \subset \pi_f(X_{h_f}^{sym}).$$

This choice presents the advantage to be the largest space which verifies (13.23) and thus ensures the well posedness of the semi-discrete problem. It can lead to conforming approximations [27] or non conforming approximations of G [35].

REMARK 13.8 (About error estimates) *The variational problem (13.6, 13.7, 13.8) and its discrete form (13.15, 13.16, 13.17) is a double mixed problem: one for the primal dual mixed stress-velocity formulation of the elastodynamic problem and one for the transmission problem. This kind of problem is by itself complicated to analysize. Further, as explained in section 11.3.2, the convergence analysis for the finite element introduced above is quite technical because it does not fit the classical Babuska-Brezzi theory. However, it is possible to derive error estimates of the semi-discretized problem (13.15, 13.16, 13.17) for another choice of the approximation spaces M_{h_c} and M_{h_f}, namely*

$$M_{h_c} = \left\{ v_c^{h_c} \in M_c \ / \ \forall \ K \in \mathcal{T}_{h_c}, \ v_c^{h_c}{}_{|K} \in (P_1)^2 \right\}, \tag{13.26}$$

and a similar definition for M_{h_f}. This choice has been made so that

$$div \left(X_{h_c}^{sym} \right) \subset M_{h_c} \quad and \quad div \left(X_{h_f}^{sym} \right) \subset M_{h_f},$$

and therefore one can eliminate the velocity from problem (13.15, 13.16, 13.17). The latter reduces to a standard evolution mixed problem in the variables Σ_c, Σ_f and J. If in addition the approximation space for the Lagrange multiplier verifies the inclusion

$$G^H \subset \pi_c(X_{h_c}^{sym}),$$

then it is easy to derive error estimates for the elliptic projector naturally and therefore for the associated evolution problem. Those estimates rely on the classical mixed theory in the case of a conforming approximation of G and on the second Strang lemma when the approximation is not conforming (see [128] for details).

13.4 Time Discretization: the 1-2 Refinement

13.4.1 A Conservative Time Scheme

For the time discretization of (13.15, 13.16, 13.17) a constant time step Δt is chosen and a second order centered finite difference scheme in time is used. We suppose here that the mesh of the fine subdomain is twice finer than the mesh of the coarse one, so that one has $h_c = 2h_f$. Since we want to keep the same ratio between the time and space steps in both subdomains, we use a time step $\Delta t_c = 2\Delta t$ in the coarse subdomain Ω_c and a time step $\Delta t_f = \Delta t$ in the fine subdomain Ω_f.

The scheme is constructed in such a way that almost all computations are explicit. Therefore, the discrete velocity and the discrete stress tensor are computed on a staggered time grid. Namely, with obvious notations, Σ_f is computed at time t^n, V_f and J_f are computed at time $t^{n+\frac{1}{2}}$, Σ_c is computed at time t^{2n}, V_c and J_c are computed at time t^{2n+1}. With such a choice, the two computational grids only meet at "even instants" t^{2n}, at which the two stress tensors Σ_c and Σ_f will be computed simultaneously. The time distribution of the unknowns is represented on Figure 13.2.

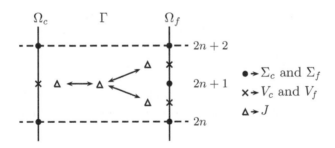

FIGURE 13.2: Time distribution of the unknowns for the (1, 2) refinement.

In addition this scheme is chosen in order to be able to derive a totally discrete energy identity which ensures its stability. In this aim a conservative time approximation of the transmission conditions (13.18-e) and (13.18-f) is proposed. This choice is explained below in the proof of 13.1, but it is not the only possible one (see remark 13.10). The following totally discretized scheme is finally obtained

Find $(\Sigma_c^{2n}, V_c^{2n+1}, J_c^{2n+1}) : \mathbb{R}_+ \to \mathbb{R}^{N_\sigma^c} \times \mathbb{R}^{N_v^c} \times \mathbb{R}^{N_j}$,

and $(\Sigma_f^{2n}, V_f^{2n+\frac{1}{2}}, J_f^{2n+\frac{1}{2}}) : \mathbb{R}_+ \to \mathbb{R}^{N_\sigma^f} \times \mathbb{R}^{N_v^f} \times \mathbb{R}^{N_j}$,

and $(\Sigma_f^{2n+1}, V_f^{2n+\frac{3}{2}}, J_f^{2n+\frac{3}{2}}) : \mathbb{R}_+ \to \mathbb{R}^{N_\sigma^f} \times \mathbb{R}^{N_v^f} \times \mathbb{R}^{N_j}$, such that

$$
\begin{cases}
M_{v,c} \dfrac{V_c^{2n+1} - V_c^{2n-1}}{2\Delta t} - B_c \Sigma_c^{2n} = F_c^{2n}, & (a) \\[4mm]
M_{\sigma,c} \dfrac{\Sigma_c^{2n+2} - \Sigma_c^{2n}}{2\Delta t} + B_c^* V_c^{2n+1} - C_c^* J_c^{2n+1} = 0, & (b)
\end{cases}
\tag{13.27}
$$

$$
\begin{cases}
M_{v,f} \dfrac{V_f^{2n+\frac{1}{2}} - V_f^{2n-\frac{1}{2}}}{\Delta t} - B_f\, \Sigma_f^{2n} = F_f^{2n}, & (a) \\[4mm]
M_{\sigma,f} \dfrac{\Sigma_f^{2n+1} - \Sigma_f^{2n}}{\Delta t} + B_f^* V_f^{2n+\frac{1}{2}} - C_f^*\, J_f^{2n+\frac{1}{2}} = 0, & (b) \\[4mm]
M_{v,f} \dfrac{V_f^{2n+\frac{3}{2}} - V_f^{2n+\frac{1}{2}}}{\Delta t} - B_f\, \Sigma_f^{2n+1} = F_f^{2n+1}, & (c) \\[4mm]
M_{\sigma,f} \dfrac{\Sigma_f^{2n+2} - \Sigma_f^{2n+1}}{\Delta t} + B_f^* V_f^{2n+\frac{3}{2}} - C_f^*\, J_f^{2n+\frac{3}{2}} = 0, & (d)
\end{cases}
\tag{13.28}
$$

$$
\begin{cases}
C_c \dfrac{\Sigma_c^{2n+2} + \Sigma_c^{2n}}{2} + C_f \dfrac{\Sigma_f^{2n+2} + 2\Sigma_f^{2n+1} + \Sigma_f^{2n}}{4} = 0, & (a) \\[4mm]
J_c^{2n+1} = J_f^{2n+\frac{1}{2}} = J_f^{2n+\frac{3}{2}} (\quad \equiv J^{2n+1}). & (b)
\end{cases}
\tag{13.29}
$$

In the sequel, thanks to (13.29-b) we will denote by J^{2n+1} the value of the discrete Lagrange multiplier wich is the same on both grids at times $t^{2n+\frac{1}{2}}, t^{2n+1}$ and $t^{2n+\frac{3}{2}}$.

REMARK 13.9 *The time discretization of the second transmission condition (13.29-b) is consistant with (13.18-f), but it is not centered, and therefore it is only first order accurate in time. All other equations of the scheme are centered and thus second order accurate.*

13.4.2   Stability Analysis

The stability of the numerical scheme ((13.27), (13.28), (13.29)) is ensured through a totally discrete energy identity. One has:

THEOREM 13.1 Energy conservation of the numerical scheme
The discrete energies of the scheme ((13.27), (13.28), (13.29)) in the coarse

and fine subdmain are respectively defined by

$$\begin{cases} \mathbf{E}_c^{2n} = \frac{1}{2}\left((\Sigma_c^{2n}, \Sigma_c^{2n})_{M_{\sigma,c}} + (V_c^{2n+1}, V_c^{2n-1})_{M_{v,c}}\right), \\ \mathbf{E}_f^n = \frac{1}{2}\left((\Sigma_f^{2n}, \Sigma_f^{2n})_{M_{\sigma,f}} + (V_f^{2n+\frac{1}{2}}, V_f^{2n-\frac{1}{2}})_{M_{v,f}}\right). \end{cases} \tag{13.30}$$

The total energy can only be defined at even instants

$$\mathbf{E}^{2n} = \mathbf{E}_c^{2n} + \mathbf{E}_f^{2n}. \tag{13.31}$$

When external forces are zero, the total energy is conserved, that is

$$\mathbf{E}^{2n+2} = \mathbf{E}^{2n}. \tag{13.32}$$

This is why we will call this scheme the conservative scheme.

PROOF We assume that external forces are zero. Multiplying equation (13.27-b) by $(\Sigma_c^{2n+2} + \Sigma_c^{2n})/2$ yields

$$\frac{1}{2}\left(\frac{\left(\Sigma_c^{2n+2}, \Sigma_c^{2n+2}\right)_{M_{\sigma,c}} - \left(\Sigma_c^{2n}, \Sigma_c^{2n}\right)_{M_{\sigma,c}}}{2\Delta t}\right)$$
$$= -\left(B_c^* V_c^{2n+1}, \frac{\Sigma_c^{2n+2} + \Sigma_c^{2n}}{2}\right) + \left(C_c^* J_c^{2n+1}, \frac{\Sigma_c^{2n+2} + \Sigma_c^{2n}}{2}\right). \tag{13.33}$$

The mean value of (13.27-a) written at time steps t^{2n} and t^{2n+2} is multiplied by V_c^{2n+1} which gives

$$\frac{1}{2}\left(\frac{\left(V_c^{2n+3}, V_c^{2n+1}\right)_{M_{v,c}} - \left(V_c^{2n+1}, V_c^{2n-1}\right)_{M_{v,c}}}{2\Delta t}\right)$$
$$= \left(B_c^* V_c^{2n+1}, \frac{\Sigma_c^{2n+2} + \Sigma_c^{2n}}{2}\right). \tag{13.34}$$

The sum of equations (13.33) and (13.34) gives the variations of the discrete energy on the coarse subdomain between two even time steps:

$$\frac{\mathbf{E}_c^{2n+2} - \mathbf{E}_c^{2n}}{2\Delta t} = \left(C_c^* J_c^{2n+1}, \frac{\Sigma_c^{2n+2} + \Sigma_c^{2n}}{2}\right). \tag{13.35}$$

Similar operations on the fine subdomain give

$$\begin{cases} \dfrac{\mathbf{E}_f^{2n+1} - \mathbf{E}_f^{2n}}{\Delta t} = \left(C_f^* J_f^{2n+\frac{1}{2}}, \dfrac{\Sigma_f^{2n+1} + \Sigma_f^{2n}}{2}\right) \\ \dfrac{\mathbf{E}_f^{2n+2} - \mathbf{E}_f^{2n+1}}{\Delta t} = \left(C_f^* J_f^{2n+\frac{3}{2}}, \dfrac{\Sigma_f^{2n+2} + \Sigma_f^{2n+1}}{2}\right). \end{cases} \tag{13.36}$$

Combining equalities (13.35) and (13.36) we derive the following energy identity

$$
\frac{\mathbf{E}^{2n+2} - \mathbf{E}^{2n}}{2\Delta t} = \left(C_f \frac{\Sigma_f^{2n+2} + \Sigma_f^{2n+1}}{4}, J_f^{2n+\frac{3}{2}} \right)
$$
$$
+ \left(C_f \frac{\Sigma_f^{2n+1} + \Sigma_f^{2n}}{4}, J_f^{2n+\frac{1}{2}} \right) + \left(C_c \frac{\Sigma_c^{2n+2} + \Sigma_c^{2n}}{2}, J_c^{2n+1} \right). \quad (13.37)
$$

The time discretization of the transmission conditions (13.29) has been chosen in order to be consistant with (13.18-e) and (13.18-f) and in order to ensure the discrete flux conservation between the two space-time grids:

$$
\left(C_f \frac{\Sigma_f^{2n+2} + \Sigma_f^{2n+1}}{4}, J_f^{2n+\frac{3}{2}} \right) + \left(C_f \frac{\Sigma_f^{2n+1} + \Sigma_f^{2n}}{4}, J_f^{2n+\frac{1}{2}} \right)
$$
$$
+ \left(C_c \frac{\Sigma_c^{2n+2} + \Sigma_c^{2n}}{2}, J_c^{2n+1} \right) = 0. \quad (13.38)
$$

$\square$

REMARK 13.10 *The choice of a time aproximation of the transmission conditions is not unique. For instance, another choice, which somehow is dual to the choice made in (13.29) is given by*

$$
\left|
\begin{aligned}
C_c \frac{\Sigma_c^{2n+2} + \Sigma_c^{2n}}{2} &= -C_f \frac{\Sigma_f^{2n+2} + \Sigma_f^{2n+1}}{2}, \\
C_c \frac{\Sigma_c^{2n+2} + \Sigma_c^{2n}}{2} &= -C_f \frac{\Sigma_f^{2n+1} + \Sigma_f^{2n}}{2}, \\
J_c^{2n+1} &= \frac{J_f^{2n+\frac{3}{2}} + J_f^{2n+\frac{1}{2}}}{2}.
\end{aligned}
\right. \quad (13.39)
$$

At first guess, both schemes seem to be first order accurate in time. However a deeper analysis shows that this is not the case. The scheme (13.29) is more accurate than scheme (13.39) (see section 13.5 below and [51, 102]). That is why we shall restrict this presentation to (13.29).

Using the energy identity 13.1 and with a similar proof as in section 11.5, the L^2 stability of the scheme will be ensured under the double strict CFL stability condition:

$$
\frac{\|b_{c,h_c}\|\Delta t_c}{2} < 1 \quad \text{and} \quad \frac{\|b_{f,h_f}\|\Delta t_f}{2} < 1, \quad (13.40)
$$

where we have set

$$\|b_{l,h_l}\| = \sup_{\sigma_l^{h_l} \in X_{h_l}^{sym}, v_l^{h_l} M_{h_l}} \frac{b_l(\sigma_l^{h_l}, v_l^{h_l})}{\|\sigma_l^{h_l}\|_{A_l} \|v_l^{h_l}\|_{\rho_l}}, \quad l \in \{c, f\}. \tag{13.41}$$

REMARK 13.11 *The definition of the operator norm (13.41) can be rewritten equivalently in the matricial form [17]*

$$\|b_{l,h_l}\| = \sup_{\Sigma_l \in \mathbb{R}^{N_\sigma^l}} \frac{\|B_l \Sigma_l\|_{(M_{v,l})^{-1}}}{\|\Sigma_l\|_{M_{\sigma,l}}}. \tag{13.42}$$

REMARK 13.12 *If we use the mixed finite element $Q_1^{div} \times Q_0$ presented above for the approximation of the velocity and stress tensor, it can be shown for a homogeneous infinite medium and a regular mesh composed of squares of size h_c and h_f that conditions (13.40) become (see section 10.5 and 11.5)*

$$\alpha = V_p \frac{\Delta t_c}{h_c} = V_p \frac{\Delta t_f}{h_f} < \frac{\sqrt{2}}{2} = \alpha_{opt}, \tag{13.43}$$

where $V_p = \sqrt{\lambda + 2\mu/\rho}$ is the pressure wave velocity (see section 9.4). By construction, the Courant number α is the same in both grids.

REMARK 13.13 *Each of the two strict inequalities (13.40) corresponds to the stability condition of the same scheme in each grid separately. One interest of the method is thus to get optimal stability conditions in each subdomain, without being affected by the coupling between the two grids. Similarly, as already mentionned, the ability to chose optimaly the time step on each grid reduces dispersion errors.*

In practice, since the space step will be twice larger in the coarse grid, we shall have:

$$\|b_{f,h_f}\| = 2 \|b_{c,h_c}\|,$$

so that (13.40) reduces to one single inequality.

13.4.3   Practical Computations

The practical resolution of the scheme (13.27, 13.28, 13.29) has to be explained. Assume that all unknowns up to time t^{2n} have been computed. We need to compute the unknowns up to the time t^{2n+2}. In other words, we seek for V_c^{2n+1}, Σ_c^{2n+2}, $V_f^{2n+\frac{1}{2}}$, Σ_f^{2n+1}, $V_c^{2n+\frac{3}{2}}$, Σ_f^{2n+2}, J^{2n+1}, all other terms being known at this step of the computations.

For clarity, we suppose that external forces are zero. First, since the mass matrices $M_{v,c}$ and $M_{v,f}$ are diagonal, equations (13.27-a) and (13.28-a) give explicitly V_c^{2n+1} and $V_f^{2n+\frac{1}{2}}$. Now we note that if the Lagrange multiplier

J^{2n+1} is known, then all other remaining unknowns can be computed explicitly, because the mass matrices $M_{\sigma,c}$ and $M_{\sigma,f}$ are block diagonal. Equation (13.29-a) will be used to derive a linear system verified by J^{2n+1}. In this aim, we first compute the difference between (13.28-d) and (13.29-b):

$$M_{\sigma,f} \frac{\Sigma_f^{2n+2} - 2\Sigma_f^{2n+1} + \Sigma_f^{2n}}{\Delta t^2} = -B_f^* \frac{V_f^{2n+\frac{3}{2}} - V_f^{2n+\frac{1}{2}}}{\Delta t} + C_f^* \frac{J_f^{2n+\frac{3}{2}} - J_f^{2n+\frac{1}{2}}}{\Delta t}.$$

Using (13.28-c) and the transmission condition (13.29-b), we obtain

$$M_{\sigma,f} \frac{\Sigma_f^{2n+2} - 2\Sigma_f^{2n+1} + \Sigma_f^{2n}}{\Delta t^2} = \left[-B_f^* M_{v,f}^{-1} B_f \right] \Sigma_f^{2n+1},$$

so we have

$$C_f \frac{\Sigma_f^{2n+2} + 2\Sigma_f^{2n+1} + \Sigma_f^{2n}}{4} = \left[C_f M_{\sigma,f}^{-1} N_f \right] \Sigma_f^{2n+1}, \qquad (13.44)$$

where we have introduced

$$N_f = M_{\sigma,f} - \frac{\Delta t^2}{4} B_f^* M_{v,f}^{-1} B_f. \qquad (13.45)$$

Using (13.28-b), equation (13.44) becomes

$$C_f \frac{\Sigma_f^{2n+2} + 2\Sigma_f^{2n+1} + \Sigma_f^{2n}}{4} = \left[\Delta t\, C_f\, M_{\sigma,f}^{-1} N_f\, M_{\sigma,f}^{-1} C_f^* \right] J^{2n+1} + \tilde{J}_f, \qquad (13.46)$$

where $\tilde{J}_f$ is known at this step of the computations. On the other hand, using (13.27-b) we compute

$$C_c \frac{\Sigma_c^{2n+2} + \Sigma_c^{2n}}{2} = \left[\Delta t\, C_c\, M_{\sigma,c}^{-1} C_c^* \right] J^{2n+1} + \tilde{J}_f, \qquad (13.47)$$

where once again $\tilde{J}_f$ is known at this step of computations. Finally adding (13.46) and (13.47) and using (13.29-a) leads to

$$\mathcal{M}_{\delta t} J^{2n+1} = \tilde{J}_c + \tilde{J}_f, \qquad (13.48)$$

where the matrix $\mathcal{M}_{\delta t}$ is defined by

$$\mathcal{M}_{\delta t} = \Delta t\, C_c\, M_{\sigma,c}^{-1} C_c^* + \Delta t\, C_f\, M_{\sigma,f}^{-1} N_f\, M_{\sigma,f}^{-1} C_f^*. \qquad (13.49)$$

Well-posedness of the discrete solution.   The existence and uniqueness of the discrete solution is guaranteed by the invertibility of the linear system (13.48). The matrix $\mathcal{M}_{\delta t}$ is obviously symmetric. It is easy to see that this matrix is positive definite as soon as the two following conditions are fullfilled:

- The matrix N_f is positive definite. This property is equivalent to the matrix inequality

$$\frac{\Delta t^2}{4} B_f^*(M_{v,f})^{-1} B_f < M_{\sigma,f}, \tag{13.50}$$

which is nothing but the (strict) CFL condition in the fine grid (13.40) (see remark 13.11).

- The compatibility condition (13.22) is verified.

Note that the matrix $\mathcal{M}_{\delta t}$ is small compared to the dimensions of the volumic unknowns (its dimension is N_j, the number of degrees of freedom of the Lagrange multiplier J^{2n+1}, which lives on the the surface Γ). Moreover, $\mathcal{M}_{\delta t}$ is independent of the time step so that it can be factorized preliminary to the computations. The linear system (13.48) requires thus only a forward-backward resolution at each time step.

REMARK 13.14 *Note that it is preferable that C_c^* be injective rather than C_f^*. Indeed, if C_f^* only is injective, then the matrix $\Delta t\, C_c\, M_{\sigma,c}^{-1}\, C_c^*$ is not invertible and the invertibility of $\mathcal{M}_{\delta t}$ relies on the invertibility of the matrix $\mathcal{N}_f$ defined by (13.45). But if the courant number α is close to α_{opt} (see remark 13.12) then $\mathcal{N}_f$ becomes singular and consquently $\mathcal{M}_{\delta t}$ becomes also singular.*

Resolution algorithm. Finally, the structure of the global algorithm is:

- Preliminary computation of the matrix $\mathcal{M}_{\delta t}$ (defined by (13.49)) and its Cholesky factorization,

- At each time step, compute successively:

  1. V_c^{2n+1} and $V_f^{2n+\frac{1}{2}}$ with (13.27-a) and (13.28-a) respectively,

  2. J^{2n+1} by solving the system (13.48),

  3. Σ_c^{2n+2}, Σ_f^{2n+1}, $V_f^{2n+\frac{3}{2}}$ and Σ_f^{2n+2} with (13.27-b), (13.28-b), (13.28-c) and (13.28-d) respectively.

13.5 About the Error Analysis

Up to now, there are very few complete results about the error analysis of the mesh refinement methods that we presented in the previous section. That is why we shall restrict ourselves to give some insights about what such an analysis could be and indicate some directions of research. In fact, it is natural to distinguish two sources of errors:

- Space discretization: change of space step between Ω_c and Ω_f,

- Time stepping: change of time step between Ω_c and Ω_f.

To analyze the error due to the space discretization, it suffices to look at the semi-discrete problems of section 13.2. We have a "non conforming" method at the interface in the sense that one of the two continuity conditions across Γ is an essential one and appears in the continuous spaces for the variational formulation. The introduction of the Lagrange multiplier allows us to take into account in a weak way this continuity condition but also introduces a non conformity at the interface of the numerical method. However, in this case all the methodology developed for the analysis of the mortar element method, which relies either on the second Strang's lemma (in this case the Lagrange multiplier does no longer appear in the analysis, [34], [28]) or the mixed finite element technology [27], is applicable (see remark 13.8). This remains however an open question. In particular, the role of the choice of the space G_H for the Lagrange multiplier (which is of obvious importance at least for the uniqueness of the discrete solution through the discrete inf-sup conditions) on the quality of the treatment of the numerical interface should be clarified.

The error due to the time discretization appears to be a more original issue. It has already been investigated in the case of the 1D wave equation for which the question of the space approximation does not introduce any difficulty. Two different kinds of analysis can be made: L^2 error analysis and Fourier analysis of the reflection-transmission problem at the interface between the two grids. After presenting the notations for the 1D case, we summarize the results of both analysis in the this section.

13.5.1   The Case of the 1D Wave Equation

We consider in this section a simple 1D model problem on the real line:

$$\frac{\partial u}{\partial t} + \frac{\partial v}{\partial x} = 0, \qquad \frac{\partial v}{\partial t} + \frac{\partial u}{\partial x} = 0, \qquad x \in \mathbb{R}, \quad t > 0, \tag{13.51}$$

The two domains Ω_c and Ω_f are respectively the two half spaces $\{x < 0\}$ and $\{x > 0\}$. The 1D equivalent of the transmission problem of section 13.2 is

$$\begin{cases} \dfrac{\partial u_c}{\partial t} + \dfrac{\partial v_c}{\partial x} = 0, \\[2mm] \dfrac{\partial v_c}{\partial t} + \dfrac{\partial u_c}{\partial x} = 0, \end{cases} \text{in } \Omega_c, \qquad \begin{cases} \dfrac{\partial u_f}{\partial t} + \dfrac{\partial v_f}{\partial x} = 0, \\[2mm] \dfrac{\partial v_f}{\partial t} + \dfrac{\partial u_f}{\partial x} = 0, \end{cases} \text{in } \Omega_f, \tag{13.52}$$

with the (artificial) interface conditions

$$u_c(0,t) = u_f(0,t), \qquad v_c(0,t) = v_f(0,t). \tag{13.53}$$

We apply the following procedure: in each grid, we use the primal-dual formulation with u in H^1 and v in L^2. The Lagrange multiplier j is simply a real number. For the space discretization, we use two regular meshes of respective sizes h in Ω_f and $2h$ in Ω_c, continuous P_1 finite elements for u and piecewise constant elements for v. For the time discretization we use the staggered grid procedure of section 13.4 with time step Δt in Ω_f and $2\Delta t$ in Ω_c. With obvious notations, the unknowns of the scheme are simply:

$$
\begin{cases}
u_{c,2j}^{2n}, & v_{c,2j-1}^{2n+1}, \quad j \leq 0, \quad n \geq 0, \quad \text{in} \quad \Omega_c, \\[2mm]
u_{f,j}^{n}, & v_{f,j+1/2}^{n+1/2}, \quad j \geq 0, \quad n \geq 0, \quad \text{in} \quad \Omega_f.
\end{cases}
\tag{13.54}
$$

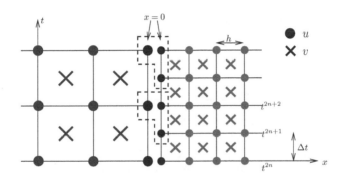

FIGURE 13.3: The new unknowns: large (resp. small) crosses correspond to v_c (resp. v_f), large (resp. small) circles correspond to u_c (resp. u_f).

In order to understand the scheme, look at Figure 13.3: in each interval of time $[t^{2n}, t^{2n+2}]$, the interior leap-frog scheme allows for computing the discrete solution everywhere except the three "interface values" at $x = 0$ (inside the small boxes in Figure 13.3), namely:

$$
u_{c,0}^{2n+2}, u_{f,0}^{2n+1}, u_{f,0}^{2n+2}.
\tag{13.55}
$$

We will denote by scheme (A) the time discretization of the transmission conditions given by (13.29) and by scheme (B) the alternative scheme given by (13.39). The use of the transmission scheme (A) provides the three missing equations, after elimination of the Lagrange multiplier j^{2n+1}

$$
\begin{cases}
v_{c,-1}^{2n+1} - \dfrac{h}{2\Delta t}\left(u_{c,0}^{2n+2} - u_{c,0}^{2n}\right) = v_{f,1/2}^{2n+1/2} + \dfrac{h}{2\Delta t}(u_{f,0}^{2n+1} - u_{f,0}^{2n}), \\[3mm]
v_{c,-1}^{2n+1} - \dfrac{h}{2\Delta t}\left(u_{c,0}^{2n+2} - u_{c,0}^{2n}\right) = v_{f,1/2}^{2n+\frac{3}{2}} + \dfrac{h}{2\Delta t}(u_{f,0}^{2n+2} - u_{f,0}^{2n+1}), \\[3mm]
\dfrac{u_{f,0}^{2n+2} + 2u_{f,0}^{2n+1} + u_{f,0}^{2n}}{4} = \dfrac{\left(u_{c,0}^{2n+2} + u_{c,0}^{2n}\right)}{2}.
\end{cases}
\qquad (13.56)
$$

If one uses scheme (B) one obtains

$$
\begin{cases}
v_{c,-1}^{2n+1} - \dfrac{h}{2\Delta t}\left(u_{c,0}^{2n+2} - u_{c,0}^{2n}\right) = \dfrac{v_{f,1/2}^{2n+1/2} + v_{f,1/2}^{2n+\frac{3}{2}}}{2} + \dfrac{h}{2\Delta t}(u_{f,0}^{2n+2} - u_{f,0}^{2n}), \\[3mm]
\dfrac{\left(u_{c,0}^{2n+2} + u_{c,0}^{2n}\right)}{2} = \dfrac{u_{f,0}^{2n+2} + u_{f,0}^{2n+1}}{2}, \\[3mm]
\dfrac{\left(u_{c,0}^{2n+2} + u_{c,0}^{2n}\right)}{2} = \dfrac{u_{f,0}^{2n+1} + u_{f,0}^{2n}}{2}.
\end{cases}
$$

$$(13.57)$$

13.5.2 L^2 Error Estimates

One can perform the L^2 error analysis of both schemes (cf. [102]). We shall assume that the Courant number $\alpha = \Delta t/h$ is constant, strictly smaller than 1 (which corresponds to the L^2 stability condition (13.40)). To state our result, let us introduce the discrete L^2 norm of the error committed on the variable u at time t^{2n}:

$$
\begin{cases}
\|u(.,t^{2n}) - u_h^{2n}\|_h^2 = \|u_c(.,t^{2n}) - u_{c,h}^{2n}\|_h^2 + \|u_f(.,t^{2n}) - u_{f,h}^{2n}\|_h^2, \\[3mm]
\|u_c(.,t^{2n}) - u_{c,h}^{2n}\|_h^2 = \displaystyle\sum_{j\le -1} |(u_c)_{2j}|^2\, 2h + |(u_c)_0|^2\, h, \\[3mm]
\|u_f(.,t^{2n}) - u_{f,h}^{2n}\|_h^2 = \displaystyle\sum_{j\ge 1} |(u_f)_j|^2\, h + |(u_f)_0|^2\, \dfrac{h}{2}.
\end{cases}
\qquad (13.58)
$$

We shall measure the error in some discrete $L^\infty(0,T,L^2)$ norm, namely:

$$
\|u - u_h\|_{h,T} = \sup_{t^{2n}\le T} \|u(.,t^{2n}) - u_h^{2n}\|_h . \qquad (13.59)
$$

Analogously, we can define a discrete $L^\infty(0,T,L^2)$ norm for the error $v - v_h$.

THEOREM 13.1 *Assume that the exact solution (u,v) has the regularity:*

$$
(u,v) \in C^3(0,T;L^2), \qquad (13.60)
$$

then with scheme (B), one has the error estimate:

$$\|u - u_h\|_{h,T} + \|v - v_h\|_{h,T} \leq C \, (1 - \alpha^2)^{-1/2} \, h^{1/2} \, \|(u,v)\|_{C^3(0,T;L^2)}. \quad (13.61)$$

Assume that the exact solution (u,v) has the regularity:

$$(u,v) \in C^{3+k}(0,T;L^2), \quad k \geq 0, \quad (13.62)$$

then with scheme (A), one has the error estimate:

$$\|u - u_h\|_{h,T} + \|v - v_h\|_{h,T} \leq C_k \, (1 - \alpha^2)^{-1/2} \, h^{(\frac{3}{2} - \frac{1}{2k})} \, \|(u,v)\|_{C^3(0,T;L^2)}. \quad (13.63)$$

The result of this theorem expresses essentially that the method is of order $1/2$ with scheme (13.57) and of order $\frac{3}{2}$ (take k arbitrarily large in (13.63)) with scheme (13.56). In both cases, one loses some accuracy with respect to the interior scheme which is second order accurate. A challenging - still open - question is to find more precise conservative transmission schemes.

REMARK 13.15 *The constants appearing in the error estimates (13.61) and (13.63) blow up when α tends to 1, similarly to what we obtain with the estimate (10.160) (see section 10.5 and remark 10.42). Moreover, the numerical experiments indicate that the method is not strongly convergent when $\alpha = 1$ (however, we conjecture that, in this case, the discrete solution converges weakly in L^2 to the continuous one).*

13.5.3   Fourier Analysis Results

This different analysis of schemes (13.56) and (13.57) completes and confirms the results of theorem 13.1 We give here a brief recap of the results given in [51]. This study is based on the behavior of plane wave solutions in the presence of a space-time mesh refinement. For example, a discrete plane wave on the fine grid is a solution of the form

$$(u^f)^n_j = u_0 e^{i(k_f x_j - \omega t^n)}, \quad (v^f)^{n+\frac{1}{2}}_{j+\frac{1}{2}} = v_0 e^{i(k_f x_{j+\frac{1}{2}} - \omega t^{n+\frac{1}{2}})}, \quad (13.64)$$

where the wave number k is related to the frequency ω by the discrete dispersion relation

$$\frac{4}{\Delta t^2} \sin^2\left(\frac{\omega \Delta t}{2}\right) = \frac{4}{h^2} \sin^2\left(\frac{kh}{2}\right), \quad (13.65)$$

which is the discrete analogous of the classical continuous dispersion relation of plane wave solutions of (13.51) $\omega^2 = k^2$. For a given ω, (13.65) is considerd as an equation in the wave number k, which is thus written $k(\omega)$. Note that the frequency ω is defined modulo $\frac{2\pi}{\Delta t}$. This is the so-called aliasing phenomenon. One can show that if $\omega \in [-\omega^*, \omega^*] \mod [2\pi/\Delta t]$, where

$$\omega^* = \frac{2}{\Delta t} \arcsin \alpha$$

then $k(\omega)$ is real and the plane wave is *propagative*. On the other hand, if $\omega \notin [-\omega^*, \omega^*] \mod [2\pi/\Delta t]$ then $k(\omega)$ has a non zero imaginary part and the wave is *evanescent* with a penetration length $l(h, \alpha)$ which is proportional to h and a highly oscillatory behavior (precisely $\mathrm{Re}(k) = \pi/h$; see [51] for details).

For the analysis of schemes A and B, we study how an incident harmonic wave in the coarse grid Ω_c of amplitude 1 and frequency ω is reflected and transmited through the artificial interface between the coarse and the fine grids. This incident wave gives rise to

- a reflected wave of frequency ω and amplitude R_c in the coarse grid.

- a transmitted wave of frequency ω and amplitude T_f in the fine grid.

- a parasitic transmitted wave of frequency $\omega + \frac{\pi}{\Delta t}$ in the fine grid. The existence of this high frequency parasitic wave is due to the aliasing phenomenon: the two frequencies ω are $\omega + \frac{\pi}{\Delta t}$ are "equal" for the coarse grid but distinct for the fine grid.

REMARK 13.16 *It is easy to verify that the set of frequencies for which the parasitic transmitted wave is evanescent* $[\frac{-2 \arccos \alpha}{\Delta t}, \frac{-2 \arccos \alpha}{\Delta t}]$ *decreases when α incereases and tends to 0 when α tends to 1. In other words for $\alpha = 1$, the parasitic transmitted wave is always propagative.*

The particular solution one looks for is thus given by

$$\begin{cases} (u_c)_{2j}^{2n} = e^{i(k_c x_{2j} - \omega t^{2n})} + R_c \, e^{i(-k_c x_{2j} - \omega t^{2n})}, & j \leq 0, \\[2mm] (u_f)_j^n = T_f \, e^{i(k_f x_j - \omega t^n)} + T_f^p \, (-1)^n \, e^{i(k_f^p x_j - \omega t^n)} \, e^{-\frac{x_j}{l(h,\alpha)}}, & j \geq 0. \end{cases}$$

and similar expressions for v. The unknown coefficients R_c, T_f, and T_f^p are determined from the coupling equations (13.56) or (13.57). As the interface $x = 0$ is purely artificial, if we consider the continuous case, we should find the *physical values* of the parameters

$$R_c = 0, \quad T_f = 1, \quad T_f^p = 0,$$

In the discrete case the coefficients R_c, T_f, and T_f^p depend only on ωh and α. For fixed $0 < \alpha < 1$, their Taylor expansions for small ωh are given by (see [51])

$$R_c = \mathcal{O}(\omega^2 h^2), \ T_f = 1 - \mathcal{O}(\omega^2 h^2), \ T_f^p = \mathcal{O}(\omega h), \ \text{with scheme A,}$$
$$R_c = \mathcal{O}(\omega h), \ T_f = 1 - \mathcal{O}(\omega h), \ T_f^p = \mathcal{O}(1), \qquad \text{with scheme B.}$$

(13.66)

This illustrates once again the superiority of scheme (13.56) (or scheme (A)) with respect to scheme (13.57) (or scheme (B)). It also confirms in some sense the optimality of the estimates (13.61) and (13.63). Roughly speaking, the amplitude of the parasitic wave is a function of the form (γ is a given positive constant):

$$T_f^p \exp(-\gamma \frac{x}{h}), \quad x > 0,$$

and the L^2 norm of such a function is equal to $|T_f^p| \sqrt{h/\gamma}$ which gives $O(h^1/2)$ with scheme (13.57) and $O(h^{\frac{3}{2}})$ with scheme (13.56).

REMARK 13.17 *When one considers the limit case* $\alpha = 1$, *the parasitic wave becomes propagative and we have for both schemes (see [51])*

$$R_c = 0, \quad T_f = 1 + O(\omega^2 h^2), \quad T_f^p = -1 + O(\omega^2 h^2).$$

13.5.4 Numerical Illustration in the 1D Case

We illustrate the results of the present section by a numerical example. We consider the 1D wave equation (13.51) with the initial condition

$$\begin{cases} u(x, t = 0) = u_0(\frac{x-x_0}{L}), \, x \in \mathbb{R}, \\ v(x, t = 0) = 0, \quad\quad\quad x \in \mathbb{R}. \end{cases} \quad (13.67)$$

where $x_0 = -0.25$, $L = 0.25$, and

$$u_0(x) = \begin{cases} 256(x - 1/2)^4(x + 1/2)^4 & \text{if } x \in [-1/2, 1/2], \\ 0 & \text{otherwise.} \end{cases}$$

The exact solution of the problem is given by d'Alembert's formula. The computational domain for the numerical resolution of the equations is the interval $\Omega = [-0.5, 0.5]$. We use transparent boundary conditions to simulate the unbounded domain. A space step of size h is used in $\Omega_c = [-0.5, 0]$ and of size $h/2$ in $\Omega_f = [0, 0.5]$. We recall that both schemes also depend on the parameter $\alpha = \Delta t/h$ that we must choose in the interval $(0, 1)$ to ensure the stability of the method. In practice, it is interesting to choose α to be as large as possible to reduce the computational costs. The problem is that all the error and stability estimates given in section 13.5.2 blow up when α tends to 1. In Figure 13.4 we can notice this phenomenon as well. For $\alpha = 1$ both schemes give similar results. A high frequency wave appears when the waves cross the artificial boundary (see Figures 13.4(a) and 13.4(d)). Even so, the method seems to be L^2 stable. Taking $\alpha < 1$ most of oscillatory parasitic waves become evanescent and we obtain a good solution if we remove the behavior near $x = 0$. The penetration depth of the transmitted parasitic wave increases as α goes to the limit value 1. As we can see in Figures 13.4(c) and 13.4(f), $\alpha = 0.95$ is sufficient to obtain a good result. We can also see that the amplitude of the parasitic wave is higher for scheme B than for scheme A. In particular, scheme B does not converge in the L^∞ norm.

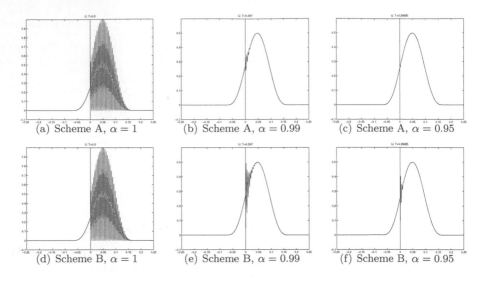

FIGURE 13.4: Dependence on α of u^h. $T \approx 0.3$.

13.6   A Post-Processed Scheme

Motivation. The Fourier analysis presented in the previous section shows parasitic transmitted waves of high frequency (both in time and space) which generate the most important part of the error. This phenomenon is illustrated by the numerical results of section 13.5.4 in the 1D case and in section 13.8 for the 2D case. On this later example, one can observe that this spurious parasitic phenomenon gets worse with higher refinement rates. A first solution consists in taking a smaller time step, which ensures that all parasitic waves will be evanescent and hence have their influence localized around the interface. However this choice limitates the interests of the space-time refinement method. Our aim in this section is to present an amelioration of the method. It is based on the fact that the parasitic waves are highly oscillating in time. It is thus possible to reduce their effect by averaging the solution over several time steps.

We denote by $(u_f^p)_j^n$ the parasitic wave introduced in the previous section. One has:

$$(u_f^p)_j^n = T_f^p \, (-1)^n \, e^{i(k_f^p x_j - \omega t^n)}. \tag{13.68}$$

This parasitic wave is highly oscillating in time so we consider the following averaging over three different time steps, which is the averaging formula used in the fine domain for the conservative discretization of the transmission condition:

$$(\overline{u_f^p})_j^n = \frac{(u_f^p)_j^{n-1} + 2(u_f^p)_j^n + (u_f^p)_j^{n+1}}{4}. \tag{13.69}$$

It is very simple to verify that for a given α and for small ωh one has

$$(\overline{u_f^p})_j^n = T_f^p \, \sin^2\left(\frac{\omega \Delta t}{2}\right)(-1)^n \, e^{i(k_f^p x_j - \omega t^n)}. \tag{13.70}$$

So, using (13.66), the amplitude of the averaged parasitic wave verifies, with scheme A, $\overline{T}_f^p = \mathcal{O}(\omega^3 h^3)$. On the other hand since this averaging is centered, one easily derives that the same averaging performed on the physical transmitted waves remains a second order approximation. Similarly for the wave in the coarse grid, we consider the averaging

$$(\overline{u_c})_{2j}^{2n+1} = \frac{(u_c)_{2j}^{2n+2} + (u_c)_{2j}^{2n}}{2}, \tag{13.71}$$

and again, since this averaging is centered, the discrete reflection waves remains a second order approximation.

Finally the treatment given by (13.71) on the coarse grid and (13.69) on the fine grid leads to the following estimates of the averaged reflection and transmission coefficients, for scheme A:

$$\overline{R}_c = \mathcal{O}(\omega^2 h^2), \quad \overline{T}_f = 1 - \mathcal{O}(\omega^2 h^2), \quad \overline{T}_f^p = \mathcal{O}(\omega^3 h^3), \tag{13.72}$$

which shows that the amplitude of the parasitic wave becomes small compared to the physical reflected and transmitted waves.

New equations for the post-processed scheme. Based on the previous observation, we introduce the following post-processed velocity and stress tensor vectors

$$\begin{cases} \overline{\Sigma}_c^{2n+1} = \dfrac{\Sigma_c^{2n+2} + \Sigma_c^{2n}}{2}, & \overline{\Sigma}_f^n = \dfrac{\Sigma_f^{n+1} + 2\Sigma_f^{n+1} + \Sigma_f^n}{4}, \\[2mm] \overline{V}_c^{2n} = \dfrac{\overline{V}_c^{2n+1} + \overline{V}_c^{2n-1}}{2}, & \overline{V}_f^{n+\frac{1}{2}} = \dfrac{\overline{V}_f^{n+\frac{3}{2}} + 2\overline{V}_f^{n+\frac{1}{2}} + \overline{V}_f^{n-\frac{1}{2}}}{4}. \end{cases} \tag{13.73}$$

It is then possible to write directly the system of equations verified by this new set of unknowns by averaging the scheme (13.27, 13.28,13.29) over 2 time steps in the coarse grid and 3 time steps in the fine grid. We obtain

Find $(\overline{\Sigma}_c^{2n+1}, \overline{V}_c^{2n}, \overline{J}_c^{2n}) : \mathbb{R}_+ \to \mathbb{R}^{N_\sigma^c} \times \mathbb{R}^{N_v^c} \times \mathbb{R}^{N_j}$,

and $(\overline{\Sigma}_f^{2n}, \overline{V}_f^{2n-\frac{1}{2}}, \overline{J}_f^{2n-\frac{1}{2}}) : \mathbb{R}_+ \to \mathbb{R}^{N_\sigma^f} \times \mathbb{R}^{N_v^f} \times \mathbb{R}^{N_j}$,

and $(\overline{\Sigma}_f^{2n+1}, \overline{V}_f^{2n+\frac{1}{2}}, \overline{J}_f^{2n+\frac{1}{2}}) : \mathbb{R}_+ \to \mathbb{R}^{N_\sigma^f} \times \mathbb{R}^{N_v^f} \times \mathbb{R}^{N_j}$, such that

$$\begin{cases} M_{v,c}\dfrac{\overline{V}_c^{2n} - \overline{V}_c^{2n-2}}{2\Delta t} - B_c\,\overline{\Sigma}_c^{2n-1} = \overline{F}_c^{2n-1}, & (a) \\[2mm] M_{\sigma,c}\dfrac{\overline{\Sigma}_c^{2n+1} - \overline{\Sigma}_c^{2n-1}}{2\Delta t} + B_c^*\overline{V}_c^{2n} - C_c^*\,\overline{J}_c^{2n} = 0, & (b) \end{cases} \tag{13.74}$$

$$\begin{cases} M_{v,f}\dfrac{\overline{V}_f^{2n-\frac{1}{2}} - \overline{V}_f^{2n-\frac{3}{2}}}{\Delta t} - B_f\,\overline{\Sigma}_f^{2n-1} = \overline{F}_f^{2n-1}, & (a) \\[2mm] M_{\sigma,f}\dfrac{\overline{\Sigma}_f^{2n} - \overline{\Sigma}_f^{2n-1}}{\Delta t} + B_f^*\,\overline{V}_f^{2n-\frac{1}{2}} - C_f^*\,\overline{J}_f^{2n-\frac{1}{2}} = 0, & (b) \\[2mm] M_{v,f}\dfrac{\overline{V}_f^{2n+\frac{1}{2}} - \overline{V}_f^{2n-\frac{1}{2}}}{\Delta t} - B_f\,\overline{\Sigma}_f^{2n} = \overline{F}_f^{2n}, & (c) \\[2mm] M_{\sigma,f}\dfrac{\overline{\Sigma}_f^{2n+1} - \overline{\Sigma}_f^{2n}}{\Delta t} + B_f^*\overline{V}_f^{2n+\frac{1}{2}} - C_f^*\,\overline{J}_f^{2n+\frac{1}{2}} = 0, & (d) \end{cases} \tag{13.75}$$

$$\begin{cases} C_c\overline{\Sigma}_c^{2n+1} + C_f\overline{\Sigma}_f^{2n+1} = 0, & (a) \\[2mm] \overline{J}_c^{2n} = \dfrac{J^{2n+1} + J^{2n-1}}{2}, & (b1) \\[2mm] \overline{J}_f^{2n-\frac{1}{2}} = \dfrac{3J^{2n-1} + J^{2n+1}}{4}, & (b2) \\[2mm] \overline{J}_f^{2n+\frac{1}{2}} = \dfrac{J^{2n-1} + 3J^{2n+1}}{4}. & (b3) \end{cases} \tag{13.76}$$

This new scheme is obviously equivalent to the conservative scheme (13.27, 13.28,13.29) (up to some condition on the initial time steps). Consequently it is stable. Note that it would not be easy to prove directly its stability by an energy technique since it is not a conservative scheme as is (13.27, 13.28,13.29).

The fundamental difference between the two schemes is that the 3 transmission conditions (13.76-b1) (13.76-b2) and (13.76-b3) are centered and second order accurate in time with 13.17b while (13.29-b) is decentered and only first order accurate (see remark 13.9). Here, the Lagrange multiplier J_c is evaluated at time t^{2n} and J_f is evaluated at time $t^{2n-\frac{1}{2}}$ and $t^{2n+\frac{1}{2}}$ by computing a weighted average of J_f at times t^{2n-1} and t^{2n} which is second order consistant in time (this is illustrated on figure Figure 13.5). Since the averaging is centered, all other equations remain centered and second order acurate in time.

REMARK 13.18 *Note that the Lagrange multiplier J^{2n+1} is the same in both schemes. The pratical resolution of the linear system (13.74, 13.75,13.76) is very similar to the one of (13.27, 13.28,13.29). Assuming that all unknowns have been computed up to time t^{2n-1}, we first compute $\overline{V}_c^{2n}$ and $\overline{V}_f^{2n-\frac{1}{2}}$. Then the Lagrange multiplier J^{2n+1} is obtained by solving the linear system (13.48)*

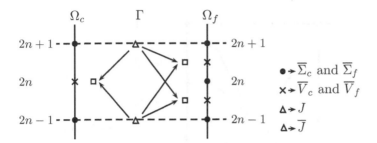

FIGURE 13.5: Time distribution of the unknowns for the (1, 2) refinement with the post-processed scheme.

(but in the current case, the right hand side will be computed in a slighlty diffent way). Then one can compute the 3 versions of the Lagrange multiplier on time interval $[t^{2n-1}, t^{2n+1}]$ $\overline{J}_c^{2n}, \overline{J}_f^{2n-\frac{1}{2}}$ *and* $\overline{J}_f^{2n+\frac{1}{2}}$ *and finally compute all remaining unknowns by explicit forward resolution in time.*

13.7 Generalization to Any Refinement Rate

The refinement method presented in section 13.4 allows to make higher rates of refinement by constructing successive included refinements. But this technique is limited to refinement rates that are powers of 2. It is complicated to implement and requires an implicit scheme at each artificial boundary. To circumvent this limitation, we present in this section a generalization of the method to any (rational) space-time refinement rate between the coarse and the fine grid.

13.7.1 Construction of the (q_c, q_f) Refinement

Suppose that the space discretization of the coarse subdomain is based on a mesh with space step $h_c = h/q_c$ and that the space discretization on the fine subdomain is based on a mesh with space step $h_f = h/q_f$. h is a given reference space step and q_c and q_f are two positive integers. The starting point is thus the matrix formulation of the spatial semi-discretization given by (13.18).

For the time discretization, we will use the second order centered scheme presented in section 13.4. In a homogeneous medium, we remind that this scheme is stable under the CFL condition $\alpha_l < 1$. α_l is the Courant number

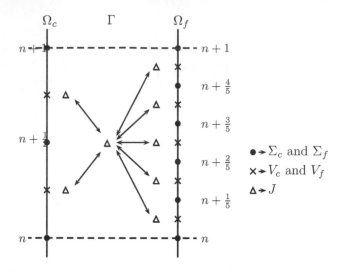

FIGURE 13.6: Time distribution of the unknowns - (2,5) refinement.

defined by $\alpha_l = V_p \Delta t_l / h_l$, $l \in \{c, f\}$, where V_p is the velocity of the pressure wave in the medium (see remark 13.12). In order to keep the same Courant number α in each subdomain, we use the time discretization step $\Delta t_c = \Delta t / q_c$ in the coarse subdomain and $\Delta t_f = \Delta t / q_f$ in the fine sub-domain, where the reference time step is $\Delta t = \alpha h / V_p$. With this choice, the two time grids will meet only at 'integer' instants $t^n = n\Delta t$. Between two such instants, there will be q_c time subcycles in the coarse grid and q_f subcycles in the fine grid.

As in the $(1, 2)$ refinement case, the stress tensor and velocity are computed on a staggered time grid, in order to obtain a scheme in which almost all computations are explicit. More precisely V_c is computed at time $t^{n + \frac{k}{q_c} + \frac{1}{2q_c}}$ and Σ_c at time $t^{n + \frac{k}{q_c}}, 0 \le k < q_c - 1$. V_f is computed at time $t^{n + \frac{k}{q_f} + \frac{1}{2q_f}}$ and Σ_f at time $t^{n + \frac{k}{q_f}}, 0 \le k < q_f - 1$. An example of the time distribution of the unknowns is presented on Figure 13.6 for the $(2, 5)$ refinement case.

The transmission conditions are discretized in a conservative way so that the total energy of the system is conserved when external forces are zero, which ensures the stability of the scheme. We obtain the following conservative scheme

Find for $l \in \{c, f\}$,

$$(\Sigma_l^{n + \frac{k+1}{q_l}}, V_l^{n + \frac{2k+1}{2q_l}}, J_l^{n + \frac{2k+1}{2q_l}}) : \mathbb{R}_+ \to \mathbb{R}^{N_v^l} \times \mathbb{R}^{N_\sigma^l} \times \mathbb{R}^{N_j},$$

such that for all $k \in \{0, \dots, q_l - 1\}$:

$$\begin{cases} M_{v,l}\dfrac{V_l^{n+\frac{2k+1}{2q_l}} - V_l^{n+\frac{2k-1}{2q_l}}}{\Delta t_l} - B_l\,\Sigma_l^{n+\frac{k}{q_l}} = F_l^{n+\frac{k}{q_l}}, & (a) \\[2em] M_{\sigma,l}\dfrac{\Sigma_l^{n+\frac{k+1}{q_l}} - \Sigma_l^{n+\frac{k}{q_l}}}{\Delta t_l} + B_l^*\,V_l^{n+\frac{2k+1}{2q_l}} = C_l^*\,J_l^{n+\frac{2k+1}{2q_l}}, & (b) \end{cases} \qquad (13.77)$$

$$\begin{cases} \displaystyle\sum_{k=0}^{q_c-1} C_c\dfrac{\Sigma_c^{n+\frac{k+1}{q_c}} + \Sigma_c^{n+\frac{k}{q_c}}}{2q_c} + \sum_{k=0}^{q_f-1} C_f\dfrac{\Sigma_f^{n+\frac{k+1}{q_f}} + \Sigma_f^{n+\frac{k}{q_f}}}{2q_f} = 0, & (a) \\[2em] J_l^{n+\frac{2k+1}{2q_l}} = J^{n+\frac{1}{2}}, l \in \{c,f\}, \forall k\{0\ldots q_l - 1\}, & (b) \end{cases} \qquad (13.78)$$

Equation (13.78-b) is centered and second order accurate with the transmission condition (13.18-e). On the other hand, all the Lagrange multipliers have the same value denoted $J^{n+\frac{1}{2}}$ on the time interval $[t^n, t^{n+1}]$. So the approximation of the second transmission condition is not centered and it is only one order accurate in time. As for the $(1,2)$ refinement case, we can fix this problem by introducing adaquate average variables, and obtain a second order scheme (see section 13.7.2 below).

Energy conservation and stability. The discrete energies in the two subdomains are defined by

$$\mathbf{E}_l^{n+\frac{k}{q_l}} = \frac{1}{2}\left\{ \left(M_{\sigma,l}\Sigma_l^{n+\frac{k}{q_l}}, \Sigma_l^{n+\frac{k}{q_l}}\right) + \left(M_{v,l}V_l^{n+\frac{2k+1}{2q_l}}, V_l^{n+\frac{2k-1}{2q_l}}\right) \right\}, \qquad (13.79)$$

for $0 \le k < q_l$, $l \in \{c, f\}$. The total energy can only be defined at integer instants:

$$\mathbf{E}^n = \mathbf{E}_c^n + \mathbf{E}_f^n \qquad (13.80)$$

When external forces are zero, the scheme (13.77,13.78) is conservative, i.e., it verifies the equality $\mathbf{E}^{n+1} = \mathbf{E}^n$. Indeed following the same technique as for the $(1,2)$ refinement case, one can show the following energy identity when external forces are zero:

$$\frac{1}{\Delta t}\left(\mathbf{E}^{n+1} - \mathbf{E}^n\right) = \sum_{l\in\{c,f\}}\sum_{k=0}^{q_l-1}\left(C_l\dfrac{\Sigma_l^{n+\frac{k+1}{q_l}} + \Sigma_l^{n+\frac{k}{q_l}}}{2q_l} \cdot J_l^{n+\frac{2k+1}{2q_l}}\right). \qquad (13.81)$$

The transmission conditions (13.78) have been discretized so that the right hand side of (13.81) is zero and hence the total energy (13.80) is conserved. This time discretization of the transmission condition generalizes the choice made for the $(1,2)$ refinement case.

This energy identity ensures the L^2 stability of the scheme under the double strict CFL conditions (13.40).

Practical computations. Assuming that all unknowns have been computed up to time t^n, we need to compute the unknowns up to time t^{n+1}. Following the same line as in section 13.4, the computational algorithm is given by

1. compute $V_c^{n+\frac{1}{2q_c}}$ and $V_f^{n+\frac{1}{2q_f}}$ with the equation (13.77-a),

2. compute the Lagrange multiplier $J^{n+\frac{1}{2}}$ by solving a small linear system of the form

$$\mathcal{M}_{\delta t} J^{n+\frac{1}{2}} = \tilde{J}^{n+\frac{1}{2}}, \tag{13.82}$$

   where the right hand side $\tilde{J}^{n+\frac{1}{2}}$ is known at this step of computation. The expression of the symmetric matrix $\mathcal{M}_{\delta t}$ is given below and its invertibility is discussed.

3. Once the Lagrange multiplier is known, all the remaining unknowns can be computed explicitly by solving successively equations (13.77-a) and (13.77-b) as time goes forward.

Expression of $\mathcal{M}_{\delta t}$ and well posedness of the discrete solution. As for the (1,2) refinement case, the transmission condition (13.78-a) is used to derive the linear system (13.82) verified by $J^{n+\frac{1}{2}}$. The principle of the computation is the following. We suppose that $J^{n+\frac{1}{2}}$ is known. Therefore the stress tensors ($\Sigma_l^{n+\frac{k}{2q_l}}, l \in \{c, f\}, 0 \leq k < q_l$) can be computed explicitly in each subdomain by forward resolution as a linear function of the lagrange multiplier $J^{n+\frac{1}{2}}$. Injecting the result into the transmission condition (13.78-a) gives the linear system (13.82).

Let us consider the subcycles in only one of the subdomains. In order to clarify the presentation, the indices l and n are omitted and we introduce the following notation changes:

$$\Delta t\, M_v^{-1} D \to D \qquad \Delta t\, M_\sigma^{-1} D^* \to D^* \qquad \Delta t\, M_\sigma^{-1} C^*. \tag{13.83}$$

We suppose also that external forces are zero. We are finally led to study the following scheme in each subdomain (note that J is a constant)

$$\begin{cases} V^{k+\frac{1}{2}} = V^{k-\frac{1}{2}} + D\,\Sigma^k, \\ \Sigma^{k+1} = \Sigma^k - D^*\, V^{k+\frac{1}{2}} + C^*J, \end{cases} \tag{13.84}$$

subjected to the initial conditions[2] Σ^0 and $V^{\frac{1}{2}}$. We have

---

[2]Note that, in principle, the initial condition should be $V^{-\frac{1}{2}}$, but taking $k = 0$ in the first equation of (13.84) gives explicitly $V^{n+\frac{1}{2}}$.

LEMMA 13.1 *With initial conditions Σ^0 and $V^{\frac{1}{2}}$, the solution of scheme (13.84) verifies*

$$\forall q \in \mathbb{N}^*, \quad \sum_{k=0}^{q-1} \left(\Sigma^{k+1} + \Sigma^k \right) = Q_q(N) \, C^* J + P_q(N) \, \Sigma^0 - Q_q(N) \, D^* V^{\frac{1}{2}},$$

(13.85)

where $N = I - \frac{DD^*}{4}$ and the polynoms P_q and Q_q are defined by induction by

$$\begin{cases} P_1(x) = 2, \qquad Q_1(x) = 1, \\ P_{q+1}(x) = P_q(x) + 4(x-1)Q_q(x) + 2, \\ Q_{q+1}(x) = P_q(x) + (4x-3)Q_q(x) + 1, \end{cases}$$

(13.86)

PROOF Using (13.84) with $k = 0$, we first compute,

$$\Sigma^1 = \Sigma^0 - D^* V^{\frac{1}{2}} + C^* J,$$

(13.87)

so that one has

$$\Sigma^0 + \Sigma^1 = C^* J + 2\Sigma^0 - D^* V^{\frac{1}{2}},$$

(13.88)

which corresponds to (13.85) with $q = 1$. By induction, we now suppose that (13.85) is true for some $\tilde{q} > 0$ (Induction Hypothesis). We show that it is also true for $\tilde{q}+1$. The induction hypothesis for initial conditions Σ^1 and $V^{\frac{3}{2}}$ writes

$$\sum_{k=1}^{\tilde{q}} \left(\Sigma^{k+1} + \Sigma^k \right) = Q_{\tilde{q}}(N) \, C^* J + P_{\tilde{q}}(N) \, \Sigma^1 - Q_{\tilde{q}}(N) \, D^* V^{\frac{3}{2}}.$$

(13.89)

Now, using (13.87), (13.88) and the first equation of (13.84) for $k = 1$ (and also that $DD^* = 4N - 4I$), one easily verifies that (13.85) holds for $q = \tilde{q}+1$ with the induction formula given by (13.86). □

We apply lemma 13.1 in the coarse subdomain and in the fine subdomain. Then, using (13.78-a) and comming back to the initial notations (13.83), it is straightforward to show that the matrix $\mathcal{M}_{\delta t}$ is given by

$$\begin{vmatrix} \mathcal{M}_{\delta t} = \mathcal{M}_{\delta t}^c + \mathcal{M}_{\delta t}^l, \\ \mathcal{M}_{\delta t}^l = \frac{\Delta t_l}{2q_l} C_l Q_{q_l}(\tilde{N}_l) M_{\sigma,c}^{-1} C_l^*, \qquad l \in \{c, f\}, \end{vmatrix}$$

(13.90)

where the matrices $\tilde{N}_l$, $l \in \{c, f\}$ are defined by

$$\tilde{N}_l = (M_{\sigma,l})^{-1} N_l, \qquad N_l = M_{\sigma,l} - \frac{\Delta t_l^2}{4} B_l^* (M_{\sigma,l})^{-1} B_l.$$

(13.91)

We remind that the CFL conditions (13.40) are equivalent to the positive definiteness of the matrices N_c and N_f.

In order to show the existence and uniqueness of a solution to the scheme (13.77,13.78), we proove that the matrix $\mathcal{M}_{\delta t}$ defined by (13.90) is symmetric positive definite under a CFL type condition. This result is based on the computations of the roots of the polynoms $Q_q(x)$. First, eliminating $P_k(x)$ in (13.86), we see that the polynoms $Q_k(x)$ defined by (13.86) are caracterized by

$$\begin{cases} Q_0(x) = 0, \quad Q_1(x) = 1, \\ Q_{k+1}(x) = (4x - 2)Q_k(x) - Q_{k-1}(x) + 2. \end{cases} \qquad (13.92)$$

Based on the observation for that any $x \in \mathbb{R}$, the sequence $(Q_k(x))$ is a two step linear recursive sequence, it is possible to derive an analytical expression for $Q_k(x)$. Indeed, the set of solutions is an affine space of dimension 2. It is natural to look for solutions of (13.92) of the form $[(e^{i\theta_x})^k + (\text{particular solution})]$. One can show (see [128] for the detailed proof)

LEMMA 13.2 *The polynomes $Q_k(x)$ defined by (13.92) are given by*

$$Q_k(x) = \frac{1 - \cos(k \arccos(2x - 1))}{2(1 - x)}, \qquad \forall x \in (0,1), \quad \forall k \geq 0. \qquad (13.93)$$

Their degree is $k - 1$ for all $k \in \mathbb{N}^*$. Their root decomposition is given by

$$Q_k(x) = 4^{k-1} \prod_{l=1}^{k-1} \left(x - \cos^2\left(\frac{\pi l}{k}\right) \right). \qquad (13.94)$$

Using this result and the fact that $\cos^2\left(\frac{\pi l}{k}\right) = \cos^2\left(\frac{\pi(k-l)}{k}\right), \forall l \in \{1 \ldots k-1\}$. one easily verifies that

$$Q(x) \geq 0, \qquad \forall x \geq 0 \qquad (13.95)$$

Hence one has

THEOREM 13.2 Positive definiteness of $\mathcal{M}_{\delta t}$ *We suppose that:*

(h1) The matrices N_c and N_f defined by (13.91) are positive definite. We remind that this condition is equivalent to the CFL condition (13.40).

(h2) For at least one $l_0 \in \{c, f\}$, the matrix

$$\sin^2\left(\frac{\pi}{\max\{2, q_{l_0}\}}\right) M_{\sigma, l_0} - \frac{\Delta t_{l_0}^2}{4} D_{l_0}^* M_{v, l_0}^{-1} D_{l_0}, \qquad (13.96)$$

is positive definite. This condition is more restrictive than the usual CFL condition on Ω_{l_0} as soon as $q_{l_0} > 2$.

(h3) $\ker C_{l_0}^* = \{0\}$, i.e. $C_{l_0}^*$ is injective,

then the matrix $\mathcal{M}_{\delta t}$ is positive definite.

PROOF We remind that $\mathcal{M}_{\delta t} = \mathcal{M}_{\delta t}^c + \mathcal{M}_{\delta t}^f$. We show (1) that the matrices $\mathcal{M}_{\delta t}^c$ qnd $\mathcal{M}_{\delta t}^l$ are symmetric positive, and (2) that $\mathcal{M}_{\delta t}^{l_0}$ is positive definite, which implies that $\mathcal{M}_{\delta t}$ is itself positive definite.

(1) For any positive definite symmetric matrix M, we denote $(.,.)_M$ the scalar product defined by $(X, Y)_M \equiv X^* M Y$ and by $(.,.)$ the canonical scalar product $(X, Y) \equiv X^* Y$. It is straightfroward to verify that a matrix N is positive (definite) symmetric for the cannonical scalar product $(.,.)$ if and only if the matrix $M^{-1}N$ is positive (definite) symmetric for the scalar product $(.,.)_M$. Similarly, Q is a positive (definite) symmetric matrix for the scalar product $(.,.)_M$ if and only if QM^{-1} is positive (definite) symmetric for the canonical scalar product.

The matrix $N_l, l \in \{c, f\}$ defined by (13.91) is positive symmetric for $(.,.)$, so is $\tilde{N}_l$ thus, thanks to (13.95) and the remark above, the matrix $Q_q(\tilde{N}_l)(M_{\sigma,l})^{-1}$ is positive symmetric for the canonical scalar product. Consequently the matrices $\mathcal{M}_{\delta t}^c$ and $\mathcal{M}_{\delta t}^f$ are symmetric positive matrices.

(2) Further, thanks to condition (h2) the matrices

$$N_{l_0} - \cos^2\left(\frac{\pi k}{q_{l_0}}\right) M_{\sigma,l_0} = \sin^2\left(\frac{\pi k}{q_{l_0}}\right) M_{\sigma,l_0} - \frac{\Delta t_{l_0}^2}{4} D_{l_0}^* M_{v,l_0}^{-1} D_{l_0}, \quad (13.97)$$

are positive definite for the canonical scalar product for all $k \in \{0 \ldots q_{l_0} - 1\}$. Hence, using (13.94) one infers that the matrix

$$Q_q(\tilde{N}_{l_0})(M_{\sigma,l_0})^{-1},$$

is positive definite. Using this result and condition (h3) the matrix $\mathcal{M}_{\delta t}^{l_0}$ is positive definite, and hence the result follows.

$\Box$

REMARK 13.19 *The CFL condition (13.96) also appears in the Fourier analysis presented in section 13.5.3. Indeed, one can see that if (13.96) is satisfied, then for a sufficiently small time step Δt all parasitic waves will be evanescent (see remark 13.16).*

REMARK 13.20 *The assumptions of theorem (13.2) are sufficient conditions to ensure the well posedness of the totally deiscretized scheme. However we have observed in practice that for some cases $\mathcal{M}_{\delta t}$ can be invertible even if (h2) is not verified.*

Computation of $\mathcal{M}_{\delta t}$. In practice we do not use expression (13.90) to compute the matrix $\mathcal{M}_{\delta t}$, which requires too many matrices multiplications. We use instead the following numerical algorithm, based on the inductive construction of the matrix $\mathcal{M}_{\delta t}$. This technique proposed by Francis Collino allows for computing in an elegant way the polynomial evaluations $Q_{q_c}(\tilde{N}_c)$ and $Q_{q_f}(\tilde{N}_f)$. We denote by $(J^i), 1 \leq i \leq N_j$ the canonical basis of the vector space $\mathbb{R}^{N_j}$. J^i is thus the vector whose i^{th} coordinate is 1 and is zero elsewhere. We note $\mathcal{M}^i_{\delta t}$ the ith column of the matrix $\mathcal{M}_{\delta t}$. By definition of a matrix, we have $\mathcal{M}^i_{\delta t} = \mathcal{M}_{\delta t} J^i$.

Taking $J = J^i$, we solve separately the explicit scheme (13.84) in the coarse subdomain and in the fine subdomain during q_c time steps and q_f time steps respectively, with zero initial conditions. We have thus computed $\Sigma_c^{n+\frac{k_c}{q_c}}$, $k_c \in \{0 \ldots q_c - 1\}$ and $\Sigma_f^{n+\frac{k_f}{q_f}}$, $k_f \in \{0 \ldots q_f - 1\}$. Then, one has by construction (see the definition of $\mathcal{M}_{\delta t}$ (13.90) and lemma 13.1)

$$\mathcal{M}^i_{\delta t} = C_c \sum_{k_c=0}^{q_c-1} \left(\Sigma_c^{k_c+1} + \Sigma_c^{k_c} \right) + C_f \sum_{k_f=0}^{q_f-1} \left(\Sigma_f^{k_f+1} + \Sigma_f^{k_f} \right). \qquad (13.98)$$

The procedure is repeated for each $i \in \{1 \ldots N_j\}$ in order to obtain all columns of the matrix $\mathcal{M}_{\delta t}$.

13.7.2   A Post-Processed Scheme

Following the same idea as for the $(1, 2)$ refinement case one can reduce the error due to the parasitic waves by computing adequate averages of the solution in each subdomain (see section 13.6). We introduce the following post-processed velocity and stress tensors vectors in each subdomain Ω_l, $l \in \{c, f\}$:

$$\begin{cases} \bar{\Sigma}_l^{\frac{n}{q_l}+\frac{1}{2}} = \displaystyle\sum_{m=0}^{q_f-1} \frac{\Sigma_l^{\frac{n+m+1}{q_l}} + \Sigma_l^{\frac{n+m}{q_l}}}{2q_l}, & \forall k \in \{0 \ldots q_l - 1\}, \\[4mm] \bar{V}_l^{\frac{2n+1}{2q_l}+\frac{1}{2}} = \displaystyle\sum_{m=0}^{q_l-1} \frac{V_l^{\frac{2n+2m+3}{2q_l}} + V_l^{\frac{2n+2m+1}{2q_l}}}{2q_l}, & \forall k \in \{0 \ldots q_l - 1\}. \end{cases} \qquad (13.99)$$

And one verifies that they are solution of the following scheme by averaging the equations of the scheme (13.77,13.78)

Find for $l \in \{c, f\}$,

$$(\Sigma_l^{n+\frac{k+1}{q_l}}, V_l^{n+\frac{2k+1}{2q_l}}, J_l^{n+\frac{2k+1}{2q_l}}) : [0, T] \rightarrow \mathbb{R}^{N_v^l} \times \mathbb{R}^{N_\sigma^l} \times \mathbb{R}^{N_j},$$

such that for all $k \in \{0, \ldots, q_l - 1\}$:

$$\begin{cases} M_{v,l} \dfrac{\bar{V}_l^{n+\frac{2k+1}{2q_l}-\frac{1}{2}} - \bar{V}_l^{n+\frac{2k-1}{2q_l}-\frac{1}{2}}}{\Delta t_l} - B_l\, \bar{\Sigma}_l^{n+\frac{k}{q_l}-\frac{1}{2}} = \bar{F}_l^{n+\frac{k}{q_l}-\frac{1}{2}}, \\[4mm] M_{\sigma,l} \dfrac{\bar{\Sigma}_l^{n+\frac{k+1}{q_l}-\frac{1}{2}} - \bar{\Sigma}_l^{n+\frac{k}{q_l}-\frac{1}{2}}}{\Delta t_l} + B_l^*\, \bar{V}_l^{n+\frac{2k+1}{2q_l}-\frac{1}{2}} = C_l^*\, \bar{J}_l^{n+\frac{2k+1}{2q_l}-\frac{1}{2}}, \end{cases}$$

$$\tag{13.100}$$

$$\begin{cases} C_f \bar{\Sigma}_f^{n+\frac{1}{2}} + C_c \bar{\Sigma}_c^{n+\frac{1}{2}} = 0, \\[3mm] \bar{J}_l^{n+\frac{2k+1}{2q_l}-\frac{1}{2}} = \left(1 - \dfrac{2k+1}{2q_l}\right) J^{n-\frac{1}{2}} + \dfrac{2k+1}{2q_l} J^{n+\frac{1}{2}}. \end{cases}$$

$$\tag{13.101}$$

By construction, this scheme is stable under the CFL condition (13.40). Contrary to the scheme (13.77,13.78) the discretization of the second transmission condition is centered and second order accurate in time. Since the averaging is centered, all other equations remain centered and second order acurate. Consequently the scheme (13.100, 13.101) is centered, second order accurate in time. The Lagrange multiplier J_l is evaluated at time $t^{n+\frac{2k+1}{2q_l}-\frac{1}{2}}$ by computing a weighted average of J_l at times $t^{n-\frac{1}{2}}$ and $t^{n+\frac{1}{2}}$ which is second order consistant in time.

13.8   Numerical Results

In order to show the performance of this space-time refinement method, we present a numerical experiment with different values of the refinement rate. We consider the square computational domain $\Omega =]0,10\times]0,10[$ made with a homogeneous isotropic material with density $\rho = 1$ and Lame's coefficients $\lambda = 3.45$ and $\mu = 2.04$. Perfectly matched layers are used on the boundaries of the domain to simulate the propagation in the free space. The medium is excited by an initial condition on the velocity field centered at the middle of the square $\underline{r}_0 = (5,5)$ which is given by

$$\underline{v}(\underline{r}, t=0) = 0.1\, H(|\underline{r} - \underline{r}_0|/\gamma_0)(\underline{e}_r + \underline{e}_\tau), \tag{13.102}$$

where $H(r)$ is given by

$$H(r) = \begin{cases} A_0 - A_1 \cos(2\pi r) + A_2 \cos(3\pi r) - A_3 \cos(6\pi r) & \text{if } r \in [0,1], \\ 0 & \text{else,} \end{cases}$$

where $\underline{e}_r = (\underline{r} - \underline{r}_0)/|\underline{r} - \underline{r}_0|$, $(\underline{e}_r, \underline{e}_\tau)$ is a direct orthonormal basis of $\mathbb{R}^2$, and $A_0 = 0.35875$, $A_1 = 0.48829$, $A_2 = 0.14128$, $A_3 = 0.01168$.

We consider four space-time mesh refinements $(q_c, q_f) = (1, q)$ with different values of q in order to compare the results obtained with respect to the refinement rate. The refined boxes B_1 to B_4 are defined by (see Figure 13.7)

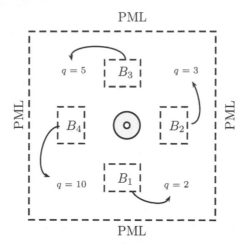

FIGURE 13.7: Geometry of the numerical experiment with four refinement boxes $(q_c, q_f) = (1, q)$.

- $B_1 = [4, 6] \times [1.5, 3]$ with a refinement rate $q = 2$,

- $B_2 = [7, 8.5] \times [4, 6]$ with a refinement rate $q = 3$,

- $B_3 = [4, 6] \times [7, 8.5]$ with a refinement rate $q = 5$,

- $B_4 = [1.5, 3] \times [4, 6]$ with a refinement rate $q = 10$.

The space step on the coarse grid $\Omega_c = \Omega \setminus \{\overline{B_1} \cup \overline{B_2} \cup \overline{B_3} \cup \overline{B_4}\}$ is $h = 1/15$. The time step on the coarse grid Δt is chosen so that the Courant number $\alpha = V_p \Delta t$ verifies $\alpha = 0.95\alpha_{opt}$ which ensures the stability of the scheme (see remark 13.12). The space and time steps inside a box with refinement $(1, q)$ are thus given by h/q and $\Delta t/q$ respectively.

Solution provided by the conservative scheme (13.77,13.78) and parasitic phenomena. We have represented on Figure 13.8 snapshots of the modulus of the velocity at two different time steps, computed with the scheme (13.77,13.78). In addition, the velocity at the center of each box versus time is represented on Figure 13.9.

Since the medium is homogeneous, the refinement boxes are artificial. The elastic wave generated by the initial condition should propagate freely through them, as if they were not there. Consequently once the wave front is away, the solution inside all boxes should be almost 0. It is the case for the $(1, 2)$ refinement, which shows that the method is performant for this refinement rate and this value of the Courant number. However, it is not the case for higer refinement rates. One can see that spurious waves remain trapped inside the boxes B_2, B_3 and B_4. This is due to the parasitic waves, that are transmitted inside the finer grids as the incident wave impiges on the box. Those parasitic waves are highly oscillating as can be seen on Figure 13.9.

Solution provided by the post-processed scheme (13.100,13.101). Now we compute the solution with the second order "post-processed" scheme (13.100,13.101) whose purpose is to reduce the effect of the parasitic waves by averaging. We have represented on Figure 13.10 snapshots of the modulus of the velocity at two different time steps, and the velocity at the center of each box versus time is represented on Figure 13.11.

The results are clearly better than with the former scheme. The parasitic phenomena have almost disappeared. After the wave has passed, the solution is almost zero in all the boxes, even for a high refinement rate. The velocity at the center of each box is not polluted by highly oscillating parasitic waves.

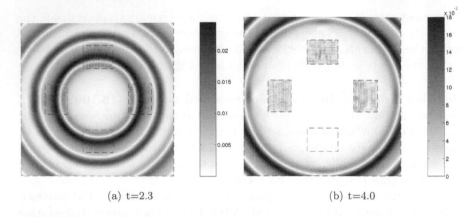

(a) t=2.3   (b) t=4.0

FIGURE 13.8: Snapshots of the modulus of the velocity at two different times. The solution is computed with the **conservative scheme (13.77,13.78)**.

Note: The references for this chapter and for all chapters in Part III are given at the end of Chapter 14.

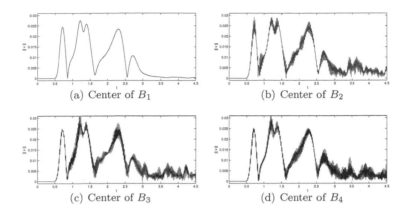

(a) Center of B_1 (b) Center of B_2

(c) Center of B_3 (d) Center of B_4

FIGURE 13.9: Modulus of the velocity at the center of each refined box versus time. The solution is computed with the **conservative scheme** (**13.77,13.78**).

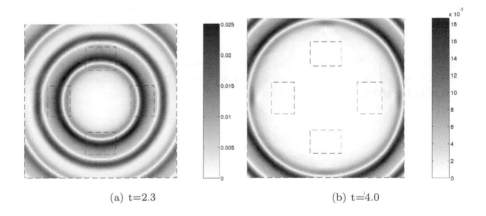

(a) t=2.3   (b) t=4.0

FIGURE 13.10: Snapshots of the modulus of the velocity at two different times. The solution is computed with the **post-processed scheme** (**13.100,13.101**).

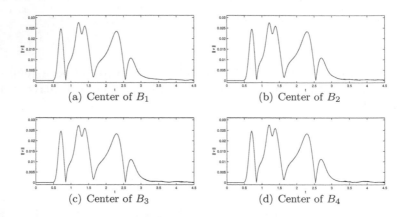

(a) Center of B_1

(b) Center of B_2

(c) Center of B_3

(d) Center of B_4

FIGURE 13.11: Modulus of the velocity at the center of each refined box versus time. The solution is computed with the **post-processed scheme (13.100,13.101)**.

Chapter 14

Numerical Methods for Treating Unbounded Media

P. Joly, POEMS Project team, INRIA-Rocquencourt, France, Patrick.JOLY@inria.fr

C. Tsogka, Dept. of Mathematics, University of Chicago, USA, tsogka@tem.uoc.gr

---

14.1 Introduction

In many important scientific areas such as geophysics, non-destructive testing, mine detection, etc., we deal with waves propagating in unbounded media. When using finite differences or finite element methods for simulating wave propagation phenomena as they appear in such applications we need to employ a special treatment on the boundaries of the necessarily truncated computational domain. Note that such treatement is not necessary for boundary integral methods because the behaviour of the solution at infinity is taken into account through the Green's function of the problem.

The importance and difficulty of this problem attracted the attention of numerous researchers first in the engineering community and then in the applied mathematics one. We can classify the proposed solutions in two families: the absorbing boundary conditions (ABC's) and the absorbing layers. ABC's were inititally introduced in the late 70's [65,109]. They consist in introducing some suitable local boundary conditions that simulate the outgoing nature of the waves impinging on the boundaries. This method works particularly well for absorbing waves nearly normally incident to the artificial boundaries. For waves traveling obliquely to the boundary, higher order ABC must be used to achieve acceptable accuracy. Since then there has been a significant progress in this domain with the introduction of higher order local [?, ?, ?, 49] and non-local [84,85] absorbing conditions. These are only a few references on the subject and a more complete list can be found in the review papers [?, 86, 87].

For elastic waves, the situation is slightly more complex. First, the transparent condition, *i.e.*, the exact condition relating normal stress and displacement on a line for outgoing waves, is no longer a scalar but a matrix integro-differential relation. Its approximation by partial differential equations, which is the usual way to construct ABC's, leads to a very complex system of equations, especially for higher order methods. The stability of the coupled problem composed of the elastodynamic system completed by these artificial conditions is then very difficult to analyze and the situation is even more intricate when discretization is considered [39]. To overcome these difficulties Higdon [93, 94], proposed to combine several first order boundary conditions designed for the wave equation, each of them associated with either the pressure wave velocity or the shear wave velocity. These conditions are theoretically stable, [92], relatively easy to implement, and efficient for waves traveling in a direction close to the normal of the artificial boundaries. They can be adapted to the case of surface waves too, [130]. However, for other directions of propagation, important spurious reflections may occur. Some authors, [122] for instance, have proposed to optimize the coefficients of Higdon's method in order to make these reflections decrease. However, numerical experiments still show relatively strong spurious reflections in some situations and stability problems when higher order numerical schemes are used, [129].

Following the approach of local ABC's for the elastic wave equations we propose the design and analysis of higher order absorbing boundary conditions that can be naturally coupled with a variational formulation. The construction and analysis of these ABC's will be presented in section 14.2. The main disavantage of ABC's is that they are not so easy to implement, especially at higher orders which are necessary for obtaining an accurate solution. Their generalization to corners can be quite tricky [49]. Last but not least ABC's are usually restricted to homogeneous isotropic media and very few works exist in the case of variable coefficients (cf. the review paper [87] and references there in) and anisotropic media.

Layer models are an alternative to ABC's. The idea is to surround the domain of interest by some artificial absorbing layers in which waves are trapped and attenuated. This idea seems quite natural and was first introduced by engineers in the 70's. For elastic waves, several models have been proposed. For instance, Sochacki et al., [130] suggest adding inside the layers some attenuation term, proportional to the first time derivative of the displacement to the elastodynamic equations. This technique, inspired by physics, is quite delicate in practice. The main difficulty is that, when entering the layers, the wave "sees" the change in impedance of the medium and then is reflected artificially into the domain of interest. The use of smooth and not too high attenuation profiles allows us to weaken the difficulty but requires the use of thick layers, [97].

The "revolutionary" idea in this domain came by J.P Bérenger in '94 [31]. He invented the perfectly matched layer (PML) that possesses the property of generating no reflection at the interface between the free medium and the artificial absorbing medium. The initial PML was designed for Maxwell's equations (electromagnetic waves) [31–33] and is now one of the most widely used methods for modeling wave propagation in unbounded domains [86, 87, 137,146,149]. Besides the fact that the PML is particularly easy to implement, it works very efficiently for complex problems (even when theorical proofs are lacking) and that is why it became very popular. Although theoretical questions concernig the well-posedness and stability of the problem where adressed in [1, 18, 19, 124, 149] there are still some open questions, such as the design of stable PMLs for anisotropic elastic media.

In section 14.3 we will present the construction and analyis of the PML for the velocity stress formulation of elastodynamics [56]. Another way to construct the PML for elastic waves was presented in [89] where the authors propose a PML for the compressional and shear potentials formulation. Note that the latter can be only used in homogeneous isotropic media.

14.2   Local Absorbing Boundary Condition

Following the approach of local ABC's for the elastic wave equations we propose in this section the design and analysis of higher order absorbing boundary conditions. Our main objective is to derive ABC's that can be naturally coupled with a variational formulation so that they can be easily incorporated in any finite element discretization scheme. Note that this is not so obvious, and, in particular, it is not satisfied, for example, by the higher order boundary conditions proposed by Higdon [93, 94].

14.2.1   The Model Problem and the Dirichlet to Neumann Map

We consider the elastodynamic problem in a homogeneous, isotropic material and assume that the initial conditions have a compact support in the half-space $\mathbb{R}^2_-(=\mathbb{R}\times\mathbb{R}_-)$. Our initial problem is,

$$\begin{cases} \varrho\dfrac{\partial^2 u}{\partial t^2} - \operatorname{div}(\mathbf{C}\varepsilon(u)) = 0, \text{ in } \mathbb{R}^2\times]0,T[, \\[2ex] u(x,y,0) = u_0(x,y), \qquad \text{in } \mathbb{R}^2, \\[2ex] \dfrac{\partial u}{\partial t}(x,y,0) = u_1(x,y), \quad \text{in } \mathbb{R}^2, \end{cases} \qquad (14.1)$$

where we recall that $u = (u_x, u_y)$ is the displacement, $\varepsilon(u)$ the strain tensor, ϱ the density and $\mathbf{C}$ the elasticity tensor.

Our aim is to compute an approximation of the solution u by solving, instead, a problem in the half space $\mathbb{R}^2_-$,

$$\begin{cases} \varrho \dfrac{\partial^2 v}{\partial t^2} - \mathrm{div}(\mathbf{C}\varepsilon(v)) = 0, & \text{in } \mathbb{R}^2_- \times]0, T[, \\[2mm] \mathcal{T}v = 0, & \text{on } x = 0, \\[2mm] v(x,y,0) = u_0(x,y), & \text{in } \mathbb{R}^2_-, \\[2mm] \dfrac{\partial v}{\partial t}(x,y,0) = u_1(x,y), & \text{in } \mathbb{R}^2_-. \end{cases} \tag{14.2}$$

The condition

$$\mathcal{T}v = 0, \quad \text{on } x = 0. \tag{14.3}$$

is said to be a transparent (exact) condition if, for all initial data u_0 and u_1, we get

$$v = u\big|_{\mathbb{R}^2_-}.$$

To design ABC's that can be easily incorporated in a variational formulation, it seems natural to seek the transparent condition as a relation between the normal stress and the displacement, relation which is also called the Dirichlet to Neumann map. This is a well-known approach initially proposed in [88] that we briefly recall in what follows. Let us first define $\widehat{u}(k,y,\omega)$,

$$\widehat{u}(k,y,\omega) = \int_0^{+\infty} \int_{-\infty}^{+\infty} u(x,y,t)e^{i(\omega t - kx)}\,dx\,dt, \tag{14.4}$$

the Fourier transform of $u(x,y,t)$ with respect to the variables (x,t). We know that in the half-space $\mathbb{R}^+$, $\widehat{u}(k,y,\omega)$ admits a plane wave decomposition, or, in other words, it can be written in the following form,

$$\widehat{u}(k,y,\omega) = \widehat{A}_p(k,\omega)\begin{pmatrix} ik \\ i\dfrac{\xi_p}{V_P} \end{pmatrix} e^{i\frac{\xi_p}{V_P}y} + \widehat{A}_s(k,\omega)\begin{pmatrix} i\dfrac{\xi_s}{V_S} \\ -ik \end{pmatrix} e^{i\frac{\xi_s}{V_S}y}, \tag{14.5}$$

with,

$$\left(i\frac{\xi_p}{V_P}\right)^2 = k^2 - \frac{\omega^2}{V_P{}^2},$$

$$\left(i\frac{\xi_s}{V_S}\right)^2 = k^2 - \frac{\omega^2}{V_S{}^2}. \tag{14.6}$$

Here V_P, V_S are the pressure and shear wave velocities defined by (9.50), whereas ξ_p, ξ_s are

$$\xi_p = \sqrt{\omega^2 - k^2 V_P^2},$$

$$\xi_s = \sqrt{\omega^2 - k^2 V_S^2},$$

with the square root $z \to \sqrt{z}$ being defined as the one that has non-negative imaginary part (this expresses the fact that the physical solution of (14.1) cannot increase exponentially in y). In the following, to simplify the notation, we omit the dependence of $\widehat{A}_p$ and $\widehat{A}_s$ on (k, ω). On the artificial boundary $y = 0$ we get,

$$\widehat{u}_x(k, 0, \omega) = ik\widehat{A}_p + i\frac{\xi_s}{V_S}\widehat{A}_s,$$

$$\widehat{u}_y(k, 0, \omega) = i\frac{\xi_p}{V_P}\widehat{A}_p - ik\widehat{A}_s,$$

(14.7)

$$\frac{d\widehat{u}_x}{dy}(k, 0, \omega) = (ik)\left(i\frac{\xi_p}{V_P}\right)\widehat{A}_p + \left(i\frac{\xi_s}{V_S}\right)^2\widehat{A}_s,$$

$$\frac{d\widehat{u}_y}{dy}(k, 0, \omega) = \left(i\frac{\xi_p}{V_P}\right)^2\widehat{A}_p - (ik)\left(i\frac{\xi_s}{V_S}\right)\widehat{A}_s,$$

(14.8)

and

$$\widehat{\sigma}_{yx}(k, 0, \omega) = \mu\left\{(ik)\left(i\frac{\xi_p}{V_P}\right)\widehat{A}_p + \left(k^2 - \frac{\omega^2}{V_S^2}\right)\widehat{A}_s + ik\widehat{u}_y(k, \omega)\right\},$$

$$\widehat{\sigma}_{yy}(k, 0, \omega) = (\lambda + 2\mu)\left\{\left(k^2 - \frac{\omega^2}{V_P^2}\right)\widehat{A}_p - (ik)\left(i\frac{\xi_s}{V_S}\right)\widehat{A}_s\right\}$$
$$+ \lambda(ik)\widehat{u}_x(k, \omega).$$

(14.9)

Usually the transparent condition on $y = 0$ is obtained in the following way: first using (14.7) we compute $\widehat{A}_p$ and $\widehat{A}_s$ as a function of $\widehat{u}_x$ and $\widehat{u}_y$. Then the obtained expressions are replaced in (14.9) and we deduce two equations that relate the normal stress $(\widehat{\sigma}_{yx}, \widehat{\sigma}_{yy})$ to the displacement. In this final expression the function $k^2 + \dfrac{\xi_p \xi_s}{V_S V_P}$ appears, which is difficult to approximate. We chose instead to keep the functions $\widehat{A}_p$ and $\widehat{A}_s$ as unknowns to the problem and write the transparent condition as a system of four equations with four unknowns. The main advantage is that now $k^2 + \dfrac{\xi_p \xi_s}{V_S V_P}$ does not appear in (14.7)–(14.9) and the derivation of ABC's from the approximation of the transparent condition becomes less complicated.

Thus by introducing $T_p(x, t)$ and $T_s(x, t)$ the pseudo-differential operators defined by

$$\widehat{T_p A_p} = i\xi_p \widehat{A}_p, \quad \widehat{T_s A_s} = i\xi_s \widehat{A}_s,$$

we can re-write the system (14.7)-(14.9), as

$$
\begin{cases}
\sigma_{yx} = \mu \left\{ \dfrac{1}{V_S^2} \dfrac{\partial^2 A_s}{\partial t^2} - \dfrac{\partial^2 A_s}{\partial x^2} + \dfrac{1}{V_P} \dfrac{\partial [T_p A_p]}{\partial x} + \dfrac{\partial u_y}{\partial x} \right\}, \\[2ex]
\sigma_{yy} = (\lambda + 2\mu) \left\{ \dfrac{1}{V_P^2} \dfrac{\partial^2 A_p}{\partial t^2} - \dfrac{\partial^2 A_p}{\partial x^2} - \dfrac{1}{V_S} \dfrac{\partial [T_s A_s]}{\partial x} \right\} + \lambda \dfrac{\partial u_x}{\partial x}, \\[2ex]
u_x = \dfrac{\partial A_p}{\partial x} + \dfrac{1}{V_S} [T_s A_s], \\[2ex]
u_y = \dfrac{1}{V_P} [T_p A_p] - \dfrac{\partial A_s}{\partial x},
\end{cases}
\tag{14.10}
$$

or after some manipulation

$$
\begin{cases}
\sigma_{yx} = \mu \left\{ \dfrac{1}{V_S^2} \dfrac{\partial^2 A_s}{\partial t^2} + 2 \dfrac{\partial u_y}{\partial x} \right\}, \\[2ex]
\sigma_{yy} = (\lambda + 2\mu) \dfrac{1}{V_P^2} \dfrac{\partial^2 A_p}{\partial t^2} - 2\mu \dfrac{\partial u_x}{\partial x}, \\[2ex]
u_x = \dfrac{\partial A_p}{\partial x} + \dfrac{1}{V_S} [T_s A_s], \\[2ex]
u_y = \dfrac{1}{V_P} [T_p A_p] - \dfrac{\partial A_s}{\partial x}.
\end{cases}
\tag{14.11}
$$

System (14.11) seems simpler as the non-local operators T_p and T_s appear only in the last two equations.

It is easy to show that the solution of problem (14.2) with (14.11) as boundary conditions on $y = 0$ (with v instead of u), coincides with the restriction of u in $\mathbb{R}^2_-$, u being the solution of the initial problem (14.1). This means that the artificial boundary at $y = 0$ does not introduce any reflection, *i.e.*, condition (14.11) is a transparent condition.

The operators T_p and T_s that appear in (14.11) are not local in space and time and therefore they are very difficult to approximate numerically. The principle of absorbing boundary conditions is based on an approximation of the symbols of these operators. The resulting approximate conditions usually involve operators that are local in space and time but they are no longer transparent (*i.e.*, there is a reflection introduced by the artificial boundary). We introduce a family of such absorbing boundary conditions in the next section.

14.2.2   Construction of the Approximate Conditions

The absorbing boundary conditions can be written in the following general form,

$$
\begin{cases}
\sigma_{yx} = \mu \left\{ \dfrac{1}{V_S{}^2} \dfrac{\partial^2 A_s}{\partial t^2} + 2\dfrac{\partial u_y}{\partial x} \right\}, \\[2ex]
\sigma_{yy} = (\lambda + 2\mu) \dfrac{1}{V_P{}^2} \dfrac{\partial^2 A_p}{\partial t^2} - 2\mu \dfrac{\partial u_x}{\partial x}, \\[2ex]
u_x = \dfrac{\partial A_p}{\partial x} + \dfrac{1}{V_S} \left[T_s^N A_s \right], \\[2ex]
u_y = \dfrac{1}{V_P} \left[T_p^N A_p \right] - \dfrac{\partial A_s}{\partial x},
\end{cases}
\tag{14.12}
$$

where T_p^N and T_s^N are approximations of the pseudo-differential operators T_p and T_s. In order to obtain local operators, T_p^N and T_s^N are constructed by approximating the symbols of T_p and T_s with a polynomial or a rational function. In particular, we consider (k, ω) such that,

$$
\left| \frac{kV_P}{\omega} \right| < 1 \; ; \; \left| \frac{kV_S}{\omega} \right| < 1,
$$

which corresponds to the propagating waves and we are seeking an approximation of the square roots

$$
\sqrt{\omega^2 - k^2 V_P^2} \text{ and } \sqrt{\omega^2 - k^2 V_S^2}.
$$

We remark that for $\mathcal{I}m(\omega) > 0$, we have

$$
\mathcal{I}m \left(\sqrt{\omega^2 - k^2 V_i^2} \right) > 0 \Leftrightarrow \omega \sqrt{1 - x_i^2}, \quad i = p, s,
$$

with

$$
x_p = \frac{k}{\omega V_P}, \; |x_p| < 1 \text{ and } x_s = \frac{k}{\omega V_S}, \; |x_s| < 1.
$$

A polynomial approximation of the square root $\sqrt{1 - x^2}$ can be obtained for example using a Taylor series expansion. This is not very satisfactory however, as it leads to ill-posed problems (for any order larger than two) [65]. An alternative consists in considering a rational approximation of the form,

$$
\sqrt{1 - x^2} \approx 1 - \sum_{l=1}^{L} \beta_l \frac{x^2}{1 - \alpha_l x^2}.
$$

The well-posedness of problem (14.2) with (14.12) as boundary conditions on $y = 0$ depends on the choice of the coefficients α_l and β_l [29, 141]. More precisely, for the scalar acoustic problem, L. Halpern and L. Trefethen [141] showed, using the Kreiss criterion, that the problem is well-posed if and only if α_l and β_l are such that,

$$\begin{cases} 0 \leq \alpha_1 \leq \alpha_2 \leq ... \leq \alpha_L < 1, \\[2mm] \displaystyle\sum_{l=1}^{L} \frac{\beta_l}{1-\alpha_l} < 1 \; ; \; \beta_l \geq 0. \end{cases}$$

Also, T. Ha Duong and P. Joly studied the well-posedness of the same problem with energy techniques in [61], where they established a link between the Kreiss criterion and the energy.

For the elastodynamic problem we propose to approximate T_p (resp. T_s) by T_p^N (resp. T_s^N) with ξ_p^N and ξ_s^N, defined by

$$\begin{cases} \xi_p^N = \omega \left(1 - \displaystyle\sum_{l=1}^{N} \frac{\beta_l k^2 V_P^2}{\omega^2 - \alpha_l V_P^2 k^2} \right), \\[4mm] \xi_s^N = \omega \left(1 - \displaystyle\sum_{l=1}^{N} \frac{\beta_l k^2 V_S^2}{\omega^2 - \alpha_l V_S^2 k^2} \right). \end{cases}$$

Here α_l and β_l are defined by the Padé approximation, *i.e.*,

$$\begin{aligned} \alpha_l &= \cos^2 \left(\frac{l\pi}{2L+1} \right), \; l = 1, .., L, \\[2mm] \beta_l &= \frac{2}{2L+1} \sin^2 \left(\frac{l\pi}{2L+1} \right), \; l = 1, .., L. \end{aligned} \tag{14.13}$$

By replacing the expressions for ξ_p^N and ξ_s^N in the last two equations of (14.12) we get,

$$\begin{aligned} \widehat{u}_x &= (ik)\widehat{A}_p + \frac{1}{V_S}\widehat{A}_s \left(i\omega - i\omega \sum_{i=1}^{N} \frac{\beta_l k^2 V_S^2}{\omega^2 - \alpha_l V_S^2 k^2} \right), \\[2mm] \widehat{u}_y &= -(ik)\widehat{A}_s + \frac{1}{V_P}\widehat{A}_p \left(i\omega - i\omega \sum_{i=1}^{N} \frac{\beta_l k^2 V_P^2}{\omega^2 - \alpha_l V_P^2 k^2} \right). \end{aligned} \tag{14.14}$$

We introduce now the auxiliary functions $\phi_l^p(x,t)$ and $\phi_l^s(x,t)$ (with Fourier transforms $\widehat{\phi}_l^p(k,\omega)$ and $\widehat{\phi}_l^s(k,\omega)$) defined by

$$\begin{aligned} (\omega^2 - \alpha_l V_P^2 k^2)\widehat{\phi}_l^p &= k^2 V_P^2 \widehat{A}_p, \; \forall l = 1, .., N, \\[2mm] (\omega^2 - \alpha_l V_S^2 k^2)\widehat{\phi}_l^s &= k^2 V_S^2 \widehat{A}_s, \; \forall l = 1, .., N. \end{aligned} \tag{14.15}$$

Remark that $\phi_l^p(x,t)$ and $\widehat{\phi}_l^p(x,t)$ are solutions of 1D wave equations,

$$\frac{\partial^2 \phi_l^p}{\partial t^2} - \alpha_l V_P^2 \frac{\partial^2 \phi_l^p}{\partial x^2} = V_P^2 \frac{\partial^2 A_p}{\partial x^2}, \quad \forall l = 1, .., N,$$

$$\frac{\partial^2 \phi_l^s}{\partial t^2} - \alpha_l V_S^2 \frac{\partial^2 \phi_l^s}{\partial x^2} = V_S^2 \frac{\partial^2 A_s}{\partial x^2}, \quad \forall l = 1, .., N.$$

(14.16)

By combining (14.14) and (14.15) we obtain,

$$\widehat{u}_x = (ik)\widehat{A}_p + \frac{1}{V_S} i\omega \widehat{A}_s - i\omega \sum_{i=1}^{N} \frac{\beta_l}{V_S} \widehat{\phi}_l^s,$$

$$\widehat{u}_y = -(ik)\widehat{A}_s + \frac{1}{V_P} i\omega \widehat{A}_p - i\omega \sum_{i=1}^{N} \frac{\beta_l}{V_P} \widehat{\phi}_l^p,$$

or, in the time domain,

$$u_x = \frac{\partial A_p}{\partial x} - \frac{1}{V_S}\frac{\partial A_s}{\partial t} + \sum_{l=1}^{N} \frac{\beta_l}{V_S}\frac{\partial \phi_l^s}{\partial t},$$

$$u_y = -\frac{\partial A_s}{\partial x} - \frac{1}{V_P}\frac{\partial A_p}{\partial t} + \sum_{l=1}^{N} \frac{\beta_l}{V_P}\frac{\partial \phi_l^p}{\partial t}.$$

(14.17)

The final system of absorbing boundary conditions is obtained from (14.12) using (14.17) and (14.16),

$$\begin{cases} \sigma_{21} = \mu \left\{ \frac{1}{V_S^2}\frac{\partial^2 A_s}{\partial t^2} + 2\frac{\partial u_y}{\partial x} \right\}, \\[2mm] \sigma_{22} = (\lambda + 2\mu)\frac{1}{V_P^2}\frac{\partial^2 A_p}{\partial t^2} - 2\mu\frac{\partial u_x}{\partial x}, \\[2mm] u_x = \frac{\partial A_p}{\partial x} - \frac{1}{V_S}\frac{\partial A_s}{\partial t} + \sum_{l=1}^{N} \frac{\beta_l}{V_S}\frac{\partial \phi_l^s}{\partial t}, \\[2mm] u_y = -\frac{\partial A_s}{\partial x} - \frac{1}{V_P}\frac{\partial A_p}{\partial t} + \sum_{l=1}^{N} \frac{\beta_l}{V_P}\frac{\partial \phi_l^p}{\partial t}, \\[2mm] \frac{\partial^2 \phi_l^p}{\partial t^2} - \alpha_l V_P^2 \frac{\partial^2 \phi_l^p}{\partial x^2} - V_P^2 \frac{\partial^2 A_p}{\partial x^2} = 0, \\[2mm] \frac{\partial^2 \phi_l^s}{\partial t^2} - \alpha_l V_S^2 \frac{\partial^2 \phi_l^s}{\partial x^2} - V_S^2 \frac{\partial^2 A_s}{\partial x^2} = 0. \end{cases}$$

(14.18)

It is interesting to remark that we can write these conditions as a perturbation of the well known first order boundary conditions,

$$
\begin{cases}
\sigma_{21} + \dfrac{\mu}{V_S}\dfrac{\partial u_x}{\partial t} = 2\mu\dfrac{\partial u_y}{\partial x} + \dfrac{\mu}{V_S}\dfrac{\partial^2 A_p}{\partial x\partial t} + \mu\displaystyle\sum_{l=1}^{N}\dfrac{\beta_l}{V_S^2}\dfrac{\partial^2 \phi_l^s}{\partial t^2}, \\[4mm]
\sigma_{22} + \dfrac{\lambda+2\mu}{V_P}\dfrac{\partial u_y}{\partial t} = -2\mu\dfrac{\partial u_x}{\partial x} - \dfrac{\lambda+2\mu}{V_P}\dfrac{\partial^2 A_s}{\partial x\partial t} + (\lambda+2\mu)\displaystyle\sum_{l=1}^{N}\dfrac{\beta_l}{V_P^2}\dfrac{\partial^2 \phi_l^p}{\partial t^2}, \\[4mm]
\dfrac{1}{V_S}\dfrac{\partial A_s}{\partial t} - \dfrac{\partial A_p}{\partial x} + u_x = \displaystyle\sum_{l=1}^{N}\dfrac{\beta_l}{V_S}\dfrac{\partial \phi_l^s}{\partial t}, \\[4mm]
\dfrac{1}{V_P}\dfrac{\partial A_p}{\partial t} + \dfrac{\partial A_s}{\partial x} + u_y = \displaystyle\sum_{l=1}^{N}\dfrac{\beta_l}{V_P}\dfrac{\partial \phi_l^p}{\partial t}, \\[4mm]
\dfrac{\partial^2 \phi_l^p}{\partial t^2} - \alpha_l V_P^2\dfrac{\partial^2 \phi_l^p}{\partial x^2} - V_P^2\dfrac{\partial^2 A_p}{\partial x^2} = 0, \quad l = 1, ..., N, \\[4mm]
\dfrac{\partial^2 \phi_l^s}{\partial t^2} - \alpha_l V_S^2\dfrac{\partial^2 \phi_l^s}{\partial x^2} - V_S^2\dfrac{\partial^2 A_s}{\partial x^2} = 0, \quad l = 1, ..., N.
\end{cases}
$$

$$(14.19)$$

14.2.3   Stability Analysis

Using the Kreiss criterion, we will prove in this section that problem (14.2) combined with (14.12) as boundary conditions on $y = 0$ is strongly well-posed. More precisely we have the following result.

THEOREM 14.1   *Under the conditions*

$$
\begin{cases}
\bullet \quad 0 \le \alpha_l < 1, \quad \beta_l \ge 0, \quad 1 \le l \le N, \\[4mm]
\bullet \quad 1 - \displaystyle\sum_{l=1}^{N}\dfrac{\beta_l}{1-\alpha_l} > 0,
\end{cases}
$$

$$(14.20)$$

the system (14.2)–(14.12) is strongly well-posed in the sense of Kreiss.

PROOF   Let us consider solutions of system (14.2)–(14.12) in the following form,

$$
\widehat{u}(k,\omega,x_2) =
\begin{cases}
\widehat{\alpha}_p \begin{pmatrix} ik \\ -i\dfrac{\xi_p}{V_P} \end{pmatrix} e^{-\left(i\frac{\xi_p}{V_P}x_2\right)} \\[2em]
+\widehat{\alpha}_s \begin{pmatrix} -i\dfrac{\xi_s}{V_S} \\ -ik \end{pmatrix} e^{-\left(i\frac{\xi_s}{V_S}x_2\right)}
\end{cases} e^{i(kx_1-\omega t)},
$$

$$
A_p(x_1,t) = \widehat{A}_p\, e^{i(kx_1-\omega t)}, \qquad A_s(x_1,t) = \widehat{A}_s\, e^{i(kx_1-\omega t)}.
$$
(14.21)

We would like to prove that there are no solutions of this form that satisfy,

$$
\begin{cases}
\bullet \quad k \in \mathbb{R}, \ k \neq 0, \\[0.5em]
\bullet \quad Im(\omega) \le 0, \ \omega \neq 0.
\end{cases}
$$
(14.22)

This is a consequence of the following result which is implied by the analysis carried out in the case of the scalar wave equation (cf. [141]). $\forall k \in \mathbb{R}$ the equations,

$$
\begin{cases}
\xi_p + \xi_p^N = 0, \\[0.5em]
\xi_s + \xi_s^N = 0,
\end{cases}
$$
(14.23)

do not admit any solution ω in the half space $Im(\omega) \le 0$.

To prove the theorem let us remark that by construction (14.21) satisfy (14.2). Thus we just need to verify that the boundary conditions (14.12) are also satisfied. By introducing (14.21) in (14.12) we obtain a linear system in $(\widehat{\alpha}_p, \widehat{\alpha}_s, \widehat{A}_p$ and $\widehat{A}_s)$,

$$
\begin{cases}
\mu\left[(ik)\left(-i\dfrac{\xi_p}{V_P}\right)\widehat{\alpha}_p + \left(k^2 - \dfrac{\omega^2}{V_S{}^2}\right)\widehat{\alpha}_s + (ik)\left(-i\dfrac{\xi_p}{V_P}\widehat{\alpha}_p - ik\widehat{\alpha}_s\right)\right] \\[1em]
\qquad = -\dfrac{\mu}{V_S{}^2}\omega^2\widehat{A}_s + 2\mu(ik)\left(-i\dfrac{\xi_p}{V_P}\widehat{\alpha}_p - ik\widehat{\alpha}_s\right), \\[1.5em]
(\lambda+2\mu)\left[\left(k^2 - \dfrac{\omega^2}{V_P{}^2}\right)\widehat{\alpha}_p + (ik)\left(i\dfrac{\xi_s}{V_S}\right)\widehat{\alpha}_s\right] + \lambda(ik)\left(ik\widehat{\alpha}_p - i\dfrac{\xi_s}{V_S}\widehat{\alpha}_s\right) \\[1em]
\qquad = -\dfrac{\lambda+2\mu}{V_P{}^2}\omega^2\widehat{A}_p - 2\mu(ik)\left(ik\widehat{\alpha}_p - i\dfrac{\xi_s}{V_S}\widehat{\alpha}_s\right), \\[1.5em]
ik\widehat{\alpha}_p - i\dfrac{\xi_s}{V_S}\widehat{\alpha}_s = ik\widehat{A}_p + i\dfrac{\xi_s^N}{V_S}\widehat{A}_s, \\[1em]
-i\dfrac{\xi_p}{V_P}\widehat{\alpha}_p - ik\widehat{\alpha}_s = -ik\widehat{A}_s + i\dfrac{\xi_p^N}{V_P}\widehat{A}_p,
\end{cases}
$$
(14.24)

for which we would like to show that it admits only the trivial (zero) condition. From the first equation of (14.24) we get,

$$\left(k^2 - \frac{\omega^2}{V_S^2}\right)\widehat{\alpha}_s + k^2\widehat{\alpha}_s = -\frac{\omega^2}{V_S^2}\widehat{A}_s + 2k^2\widehat{\alpha}_s,$$

or equivalently,

$$\frac{\omega^2}{V_S^2}\left(\widehat{\alpha}_s - \widehat{A}_s\right) = 0,$$

which implies that,

$$\text{for } \omega \neq 0, \quad \widehat{\alpha}_s = \widehat{A}_s. \tag{14.25}$$

Similarly, the second equation of (14.24) can be re-written,

$$(\lambda + 2\mu)\left(k^2 - \frac{\omega^2}{V_P^2}\right)\widehat{\alpha}_p - \lambda k^2\widehat{\alpha}_p = 2\mu k^2\widehat{\alpha}_p - \frac{\lambda + 2\mu}{V_P^2}\omega^2\widehat{A}_p,$$

which gives,

$$\frac{\omega^2}{V_P^2}\left(\widehat{\alpha}_p - \widehat{A}_p\right) = 0,$$

and thus,

$$\text{for } \omega \neq 0, \quad \widehat{\alpha}_p = \widehat{A}_p. \tag{14.26}$$

Using (14.25) and (14.26) in the last two equations of (14.24) we obtain,

$$\begin{cases} i\dfrac{1}{V_S}(\xi_s + \xi_s^N)\widehat{\alpha}_s = 0, \\[2mm] i\dfrac{1}{V_P}(\xi_p + \xi_p^N)\widehat{\alpha}_p = 0, \end{cases}$$

which implies

$$\widehat{\alpha}_s = \widehat{\alpha}_p = 0,$$

and thus

$$\widehat{A}_s = \widehat{A}_p = 0,$$

because we know that $\xi_s + \xi_s^N \neq 0$ and $\xi_p + \xi_p^N \neq 0$.   ⬜

14.2.4   Accuracy Analysis

A simple way to quantify the precision of absorbing boundary conditions is to compute the reflection coefficients for plane waves as a function of the angle of incidence θ. More precisely, we consider an incident plane wave and we compute the reflected field. We call reflection coefficient the ratio of the amplitude of the reflected wave over the amplitude of the incident wave. It is then easy to show the following results:

THEOREM 14.2   *The reflection coefficients for the absorbing boundary conditions (14.12) are*

- *For an incident pressure plane wave,*

$$R_{pp} = \frac{\cos(\theta) - \cos_{app}(\theta)}{\cos(\theta) + \cos_{app}(\theta)},$$

$$R_{ps} = 0.$$

- *For an incident shear plane wave,*

$$R_{ss} = -\frac{\cos(\theta) - \cos_{app}(\theta)}{\cos(\theta) + \cos_{app}(\theta)},$$

$$R_{sp} = 0,$$

with $\cos_{app}(\theta) = 1 - \displaystyle\sum_{l=1}^{N} \frac{\beta_l \sin^2(\theta)}{1 - \alpha_l \sin^2(\theta)}.$

We omit the proof of this theorem as it is quite straightforward. Using the expression for the Padé coefficients (14.13) in R_{pp} and R_{ss}, it is easy to prove the following Lemma.

LEMMA 14.1 *For the absorbing boundary conditions (14.12) with α_l and β_l corresponding to the Padé approximation (14.13) we obtain*

- *For an incident pressure plane wave,*

$$R_{pp} = \frac{(\cos(\theta) - 1)^{2N+1}}{(\cos(\theta) + 1)^{2N+1}},$$

$$R_{ps} = 0.$$

- *For an incident shear plane wave,*

$$R_{ss} = -\frac{(\cos(\theta) - 1)^{2N+1}}{(\cos(\theta) + 1)^{2N+1}},$$

$$R_{sp} = 0.$$

This Lemma implies, in particular, that for an absorbing boundary condition that uses N rational fractions we get

$$R_{pp} = O(\theta^{4N+2}) \ R_{ps} = 0,$$

$$R_{ss} = O(\theta^{4N+2}), \ R_{sp} = 0.$$

This is a very nice and surprising result because it indicates that there is no wave conversion (from P to S or inversely) on the absorbing boundary.

Moreover for the reflected wave we obtain the same coefficient as for the scalar acoustic case. To our knowledge this property is not true for most of the existing local boundary conditions for elastodynamics [?, 86, 87, 92, 93]. We can also remark that in homogeneous, isotropic media we obtain exactly the same results if we first decompose the elastodynamic problem into two wave equations (using the compressional and shear potentials formulation) and then apply the higher order ABC's [?, ?, ?, 49] for the scalar wave equation to each one of the obtained equations. Although the results are the same, the approach presented in this section is more general and could be generalized to the anisotropic case, for which the above mentioned approach does not work as there is no compressional and shear potentials formulation.

14.3   Perfectly Matched Layers

14.3.1   Construction of the Perfectly Matched Layer for a General Evolution Problem

Following [50] we present in this section the basic principles of the PML model in the general case of a first order linear hyperbolic system. To simplify the presentation we restrict ourselfs to the two dimensional case. Consider a general evolution problem of the following form

$$(a) \ \partial_t v - A_x \partial_x v - A_y \partial_y v = 0 \ ,$$
$$(b) \ v(t = 0) = v_0 \ . \tag{14.27}$$

where v is an m-vector, A_x, A_y are $m \times m$ matrices with real coefficients, $(x, y) \in \mathbb{R}^2$ and assume that the initial condition v_0 is zero on the right half-space $(x > 0)$.

We would like to replace problem (14.27) by an equivalent one posed in the left half-space. The basic principle of the PML model, is *to couple the equation in the left half-space with an equation in the right half-space such that there is* **no reflection** *at the interface, and that the wave* **decreases exponentially** *inside the right half-space.* A formal construction of the split PML model consists in the following two steps: First we decompose the unknown $v(x, y, t)$ in,

$$v = v^x + v^y, \tag{14.28}$$

where v^x and v^y are solutions of,

$$\partial_t v^x - A_x \partial_x (v^x + v^y) = 0 \ , \ (i)$$
$$\partial_t v^y - A_y \partial_y (v^x + v^y) = 0 \ , \ (ii) \tag{14.29}$$

In this first step we have not changed anything yet, *i.e.*, if (v^x, v^y) is a solution of (14.29) then $v^x + v^y$ is solution of (14.27). Remark that in (14.29)-(i) we keep only the derivative which is normal to the interface, here the x-derivative, while in (14.29)-(ii) we keep only the derivative tangent to the interface, *i.e.*, the y-derivative. Note that for a PML in three dimensions the tangent derivatives would be the y, z-derivatives.

Secondly, we introduce a damping term $\zeta(x)v^x$ only in equation (14.29)-(i). $\zeta(x)$ is a damping function, which is zero on the left half-space and positive on the right half-space. Let us define, $u = u^x + u^y$, the solution of the following system

$$\partial_t u^x + \zeta(x)u^x - A_x\partial_x(u^x + u^y) = 0 \ ,$$
$$\partial_t u^y - A_y\partial_y(u^x + u^y) = 0 \ , \tag{14.30}$$
$$u^x = u^y = \frac{v_0}{2} \ ,$$

It is easy to see that u satisfies the same system of equations as v in the left half-space. Moreover we can prove that the plane wave solution of system (14.30) can be written in the form

$$u = v_0 \ \exp[-i(k_x x + k_y y - \omega t)] \ \exp[-\frac{k_x}{\omega}\int_0^x \zeta(s)ds] \ ,$$

and satisfies:

- $u \equiv v$ in the left half-space, $x \leq 0$, which means that we have no reflection at the interface: the layer model is perfectly matched.

- u is damped in the right half-space,

- the damping coefficient in the absorbing half-space is

$$\alpha_\zeta = \frac{\|u(x)\|}{\|v(x)\|} = \exp[-\frac{k_x}{\omega}\int_0^x \zeta(s)ds] \ . \tag{14.31}$$

REMARK 14.1 *Relation (14.31) implies that the waves decrease exponentially in the absorbing half-space. Moreover, we can see that the damping coefficient α_ζ depends on the direction of propagation of the wave (via the term k_x/ω), i.e., the damping is anisotropic. More precisely, α_ζ, decreases very fast for a wave propagating normally to the interface and decreases more and more slowly as the direction of propagation becomes parallel to the interface.*

REMARK 14.2 *In practice we do not solve system (14.30) everywhere, we instead keep the original system (14.27) in the physical domain (for $x < 0$) and solve system (14.30) only in the layer (for $x > 0$). The two domains are coupled at the interface $x = 0$ through the condition $v = u^x + u^y$.*

REMARK 14.3 *We can remark that the PML model consists in the simple substitution*

$$\partial_x \to \partial_{\tilde{x}} = \frac{i\omega}{i\omega + \zeta(x)}\partial_x \ ,$$

which implies the complex change of variables

$$\tilde{x}(x) = x - \frac{i}{\omega} \int_0^x \zeta(s)ds \ .$$

REMARK 14.4 *Although the plane wave decomposition indicates that the PML model is absorbing, this analysis is not complete because the solution is in general a superposition of plane waves, in which case one should consider the group velocity instead of the phase velocity. (cf. section 14.3.4).*

14.3.2   PML Model for Elastodynamics

In this section we present the PML model for the continuous elastodynamic problem. As we have seen in the previous section, we know how to construct a PML model for an evolution problem of the form (14.27). Thus, in order to apply the technique described previously to elastodynamics, we need to write the elastodynamic problem in the form of system (14.27). This can be easily done once we consider the mixed velocity-stress formulation,

$$\varrho\partial_t v - \mathbf{div}\sigma = 0 \ ,$$

$$A\partial_t \sigma - \varepsilon(v) = 0 \ , \tag{14.32}$$

together with initial conditions. We assume, as in the previous section, that the initial condition is supported in the left half-space. By identifying the stress tensor σ with the following vector (still denoted by σ)

$$\sigma = [\sigma_1, \ \sigma_2, \ \sigma_3]^T \ ,$$

$$\sigma_1 = \sigma_{xx} \ ; \ \ \sigma_2 = \sigma_{yy} \ ; \ \ \sigma_3 = \sigma_{xy} \ ,$$

we can write (14.32) in the following matrix form

$$\varrho\partial_t v = D^y \partial_y \sigma + D^x \partial_x \sigma,$$

$$A\partial_t \sigma = E^y \partial_y v + E^x \partial_x v,$$

with

$$D^y = \begin{bmatrix} 0 & 0 & 1 \\ 0 & 1 & 0 \end{bmatrix}, \qquad D^x = \begin{bmatrix} 1 & 0 & 0 \\ 0 & 0 & 1 \end{bmatrix},$$

$$E^y = \begin{bmatrix} 0 & 0 & 1 \\ 0 & 1/2 & 0 \end{bmatrix}^T, \quad E^x = \begin{bmatrix} 1 & 0 & 0 \\ 0 & 0 & 1/2 \end{bmatrix}^T \ .$$

We apply now the same technique as in the previous section and get the following system in the Perfectly Matched Layer $(x > 0)$

$$v = v^y + v^x \qquad \qquad ; \sigma = \sigma^y + \sigma^x \ ,$$

$$\varrho\partial_t v^y = D^y \partial_y \sigma \qquad \qquad ; A\partial_t \sigma^y = E^y \partial_y v \ ,$$

$$\varrho(\partial_t + \zeta(x))v^x = D^x \partial_x \sigma \ ; A(\partial_t + \zeta(x))\sigma^x = E^x \partial_x v \ .$$

In an homogeneous, isotropic elastic medium, the matrix A depends on the Lamé coefficients (λ, μ) of the medium. In this case the PML model can be written in the following form

$$
\begin{aligned}
&v = v^y + v^x &&; \sigma = \sigma^y + \sigma^x, \\
&\varrho\,(\partial_t + \zeta(x))v_x^x = \partial_x \sigma_{xx} &&; \varrho\,\partial_t v_x^y = \partial_y \sigma_{xy}, \\
&\varrho\,(\partial_t + \zeta(x))v_y^x = \partial_x \sigma_{xy} &&; \varrho\,\partial_t v_y^y = \partial_y \sigma_{yy}, \\
&(\partial_t + \zeta(x))\sigma_{xx}^x = (\lambda + 2\mu)\partial_x v_x &; \partial_t \sigma_{xx}^y = \lambda \partial_y v_y, \\
&(\partial_t + \zeta(x))\sigma_{yy}^x = \lambda \partial_x v_x &&; \partial_t \sigma_{yy}^y = (\lambda + 2\mu)\partial_y v_y, \\
&(\partial_t + \zeta(x))\sigma_{xy}^x = \mu \partial_x v_y &&; \partial_t \sigma_{xy}^y = \mu \partial_y v_x.
\end{aligned}
\tag{14.33}
$$

The corner case Up to now we considered the PML for the half-space problem. In practice we usually need to replace an unbounded domain by a finite one. To do so we introduce a rectangular domain that we surround by PML layers as illustrated in Figure 14.1. The system of equations for the

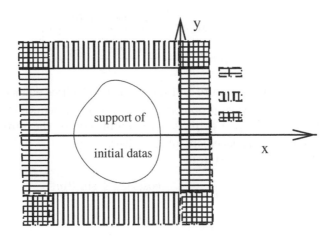

FIGURE 14.1: The different PML layers: in the corners both ζ^x and ζ^y are positive while $\zeta^y = 0$ (resp. $\zeta^x = 0$) inside the PML layers in the x (resp. y) direction.

PML in the upper right corner is,

$$
\begin{aligned}
&v = v^y + v^x &&; \sigma = \sigma^y + \sigma^x, \\
&\varrho(\partial_t + \zeta^y(y))v^y = D^y \partial_y \sigma \;\; &&; A(\partial_t + \zeta^y(y))\sigma^y = E^y \partial_y v, \\
&\varrho(\partial_t + \zeta^x(x))v^x = D^x \partial_x \sigma \;\; &&; A(\partial_t + \zeta^x(x))\sigma^x = E^x \partial_x v.
\end{aligned}
$$

where two damping coefficients are introduced ($\zeta^x(x)$, $\zeta^y(y)$) to distinguish damping in the x and y directions. When computing the solution inside the PML layers in the x direction we use the same system of equations with $\zeta^y = 0$, while for PML layers in the y direction we take $\zeta^x = 0$.

REMARK 14.5 *The PML model is very well adapted for computing the solution in rectangular domains and it is for this reason that it became so popular (namely it can be easily applied to any finite difference scheme). Although they become more complicated, PML models have been derived for different geometries as well (cf. [55] for a circular boundary and [138] for the more general case).*

14.3.3   Accuracy Analysis Using Plane Waves—Isotropic Medium

Infinite absorbing layer

In the case of a homogeneous, isotropic elastic medium, we recall that the displacement plane waves, U_j, $j = p,\ s$, corresponding to solutions of system (14.32) can be written in the following form

$$
\begin{aligned}
U_p &= a_p \vec{d_p} \exp[i\omega V_P(t - \frac{\cos(\theta)x + \sin(\theta)y}{V_P})]\ , \\
U_s &= a_s \vec{d_s} \exp[i\omega V_S(t - \frac{-\sin(\theta)x + \cos(\theta)y}{V_S})]\ ,
\end{aligned}
\tag{14.34}
$$

where V_P and V_S are the pressure and shear waves velocities defined by (9.50), θ gives the direction of wave propagation, $\vec{d_j}$, $j = p,\ s$, defines the direction of particle motion and a_j, $j = p,\ s$ is the amplitude of the wave. We can remark then that the displacement plane waves, $\tilde{U}_j$ (for $j = p,\ s$), solutions of system (14.33) can be written as

$$
\begin{aligned}
\tilde{U}_p &= a_p \vec{d_p} \exp[i\omega V_P(t - \frac{\cos(\theta)\tilde{x}_p + \sin(\theta)y}{V_P})]\ , \\
\tilde{U}_s &= a_s \vec{d_s} \exp[i\omega V_S(t - \frac{-\sin(\theta)\tilde{x}_s + \cos(\theta)y}{V_S})]\ ,
\end{aligned}
\tag{14.35}
$$

where we simply substituted $\tilde{x}_j$, $j = p,\ s$, for x in (14.34), with $\tilde{x}_j$, $j = p,\ s$, defined in the same way as in section 14.3.1 (ω replaced by ωV_j):

$$
\tilde{x}_j(x) = x - \frac{i}{\omega V_j} \int_0^x \zeta(s)ds, \quad j = p,\ s\ .
$$

Moreover it can be shown that $\tilde{U}_j, j = p, s$, satisfies:

- $\tilde{U}_j \equiv U_j$, for $j = p,\ s$, in the left half-space $x \leq 0$ (no reflection),

- $\tilde{U}_j$, $j = p,\ s$, is damped in the right half-space,

- the damping coefficient in the absorbing layer is

$$\frac{\|\tilde{U}_p(x)\|}{\|U_p(x)\|} = \exp[-\frac{\cos\theta}{V_P}\int_0^x \zeta(s)ds],$$

$$\frac{\|\tilde{U}_s(x)\|}{\|U_s(x)\|} = \exp[-\frac{\cos\theta}{V_S}\int_0^x \zeta(s)ds].$$

Finite absorbing layer

In practice, we take a finite absorbing layer by introducing a boundary at $x = \delta$, with a Dirichlet condition. This new boundary produces a reflection, but since the wave decreases exponentially in the layer, the reflection coefficient becomes quickly very small. This coefficient depends on the choice of $\zeta(x)$ and on the size δ of the layer. In order to study the properties of the model in this case, we recall some classical results for the elastodynamic problem.

Consider the elastodynamic problem with a homogeneous Dirichlet condition ($\vec{v} = 0$) on the boundary $x = 0$ as shown in Figure 14.2. Take for example the case of an incident plane wave P

$$U_p^{inc} = a_{inc}\vec{d_p}\exp[i\omega V_P(t - \frac{\vec{d_p}\cdot\vec{x}}{V_P})],\ \vec{d_p} = (\cos\theta, \sin\theta).$$

We know then (cf. [2], vol 1, §5.6, p. 173) that the incident wave is reflected

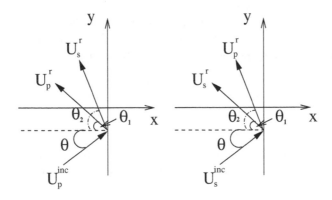

FIGURE 14.2: Schematic view of the reflection of an incident P wave (left) and of an incident S wave (right) in the case of a homogeneous Dirichlet boundary condition.

into a pressure wave U_p^r and a shear wave U_s^r given by

$$U_p^r = a_p^r \vec{d_p^r} \exp[i\omega V_P(t - \frac{\vec{d_p^r} \cdot \vec{x}}{V_P})], \quad \vec{d_p^r} = (-\cos\theta_1, \sin\theta_1) ,$$

$$U_s^r = a_s^r \vec{d_s^r} \exp[i\omega V_P(t - \frac{\vec{p_s} \cdot \vec{x}}{V_P})], \quad \vec{d_s^r} = (-\sin\theta_2, \cos\theta_2),$$

$$\vec{p_s} = (\cos\theta_2, \sin\theta_2) ,$$

and we have the reflection coefficients

$$R_{pp} = \frac{\|U_p^r\|}{\|U_p^{inc}\|} = \frac{\cos(\theta - \theta_2)}{\cos(\theta + \theta_2)},$$

$$R_{ps} = \frac{\|U_s^r\|}{\|U_p^{inc}\|} = \frac{\sin\theta_2}{\cos(\theta + \theta_2)},$$

with $\sin\theta_2 = V_S \sin\theta/V_P$. Similarly in the case of an incident plane wave S we have

$$R_{ss} = \frac{\|U_s^r\|}{\|U_s^{inc}\|} = \frac{\sin(\theta + \theta_2)}{\sin(\theta - \theta_2)} ,$$

$$R_{sp} = \frac{\|U_p^r\|}{\|U_s^{inc}\|} = \frac{\sin\theta_2}{\sin(\theta - \theta_2)} ,$$

with $\cos\theta_2 = V_P \cos\theta/V_S$. We can consider now the case of the finite layer. Given that the plane waves, U_j, $j = p, s$, corresponding to solutions of system (14.32) are given by equations (14.34), we can compute the reflection coefficients induced by the layer of width δ.

The case of a plane wave P

As we have shown previously there is no reflection at the interface $x = 0$ when the plane wave $\widetilde{U}_p$ penetrates the lossy medium. After traveling a length δ we can compute $\widetilde{U}_p$ using the formula (14.35), obtaining

$$\widetilde{U}_p = U_p \exp[-\frac{\cos\theta}{V_P} \int_0^\delta \zeta(s)ds] .$$

Then the plane wave $\widetilde{U}_p$, gives at the boundary $x = \delta$, two reflected waves $\widetilde{U}_p^r$ and $\widetilde{U}_s^r$, which are damped till penetrating again the elastic medium at the interface $x = 0$. It is easy to see that the reflection coefficient is given in this case by

$$R_{pp}^\delta = \frac{\|\widetilde{U}_p^r(x)\|}{\|U_p(x)\|} = R_{pp} \exp[-2\frac{\cos\theta}{V_P} \int_0^\delta \zeta(s)ds], \; x < 0 ,$$

$$R_{ps}^\delta = \frac{\|\widetilde{U}_s^r(x)\|}{\|U_p(x)\|} = R_{ps} \exp[-2\frac{\cos\theta}{V_P} \int_0^\delta \zeta(s)ds], \; x < 0 .$$

(14.36)

The case of a plane wave S

In the case of an incident plane wave S, we obtain, similarly,

$$
\begin{aligned}
R_{ss}^{\delta} &= \frac{\|\tilde{U}_s^r(x)\|}{\|U_s(x)\|} = R_{ss} \exp[-2\frac{\cos\theta}{V_S}\int_0^{\delta}\zeta(s)ds], \ x < 0 \ , \\
R_{sp}^{\delta} &= \frac{\|\tilde{U}_p^r(x)\|}{\|U_s(x)\|} = R_{sp} \exp[-2\frac{\cos\theta}{V_S}\int_0^{\delta}\zeta(s)ds], \ x < 0 \ .
\end{aligned}
\tag{14.37}
$$

REMARK 14.6 *We considered here the case of a Dirichlet boundary condition (v = 0) at the end of the layer. A similar analysis can be carried out for the Neumann boundary condition (σ · n = 0). To increase the effeciency of the method we could also use an absorbing boundary condition (ABC) at x = δ. In this case the results can be also described by (14.36) and (14.37) where the reflection coefficients R_{pp}, R_{ps}, R_{ss} and R_{sp} should be replaced by those corresponding to the ABC's used (cf. [59]).*

Relations (14.36) and (14.37) imply that the reflection can be made as weak as desired by choosing the damping factor $\zeta(x)$ large enough. However this is no longer true when we consider the discrete PML model. That is a consequence of the numerical dispersion, which introduces a reflection at the interface. For a second order scheme and for a damping factor of the following form

$$
\zeta(x) = \begin{cases} \zeta, & x > 0, \\ 0, & x \le 0, \end{cases}
$$

we can show that the reflection coefficient at the interface $x = 0$ is of the form (cf. [54, 56] for more details)

$$
R_1 = C_1 \ \zeta^2 \Delta x^2 + O(\Delta x^3),
\tag{14.38}
$$

Δx being the discretization step. Moreover, because of the finite length of the PML we get a reflection at $x = \delta$, which is of the same form as for the continuous problem. Without loss of generality let us look at the reflected P wave due to an incident P wave. In this case we have,

$$
R_2 = C_2 \exp^{[-2\frac{\cos(\theta)}{V_P}\zeta\delta]}(1 + O(\Delta x^2)).
\tag{14.39}
$$

Remark that R_1 is increasing with ζ while R_2 is decreasing. This means that we cannot chose an arbitrarily high value for the damping coefficient because the discrete waves do see a change in the medium. The appropriate way to deal with this problem is to take a variable ζ profile that is chosen so as to minimize the total reflection in the physical domain (cf. [54]). In practice, because the optimization problem can be quite expensive, it is sufficient to use a continuous, increasing convex function (most authors use a quadratic function).

14.3.4   Stability Analysis

Following [18] we consider first the general hyperbolic system (14.27). In order to study the stability of the PML model we consider the case of constant coefficients, *i.e.*, the matrices A_x and A_y do not depend on (x, y) and the damping coefficient is constant $\zeta(x) = \zeta$. Although this might not be realistic for the applications, it allows us to carry out the stability analysis (based on Fourier techniques) and it is the first step towards the more complicated variable coefficient case.

14.3.4.1   Stability and Well-Posedness for a General Hyperbolic System

Let us first briefly recall the definitions of hyperbolicity.

DEFINITION 14.1 *Set $\mathcal{A}(k) = k_x A_x + k_y A_y$. System (14.27) is hyperbolic if, $\forall k \in \mathbb{R}^2$, the eigenvalues of $\mathcal{A}(k)$ are real. It is strongly hyperbolic if in addition $\mathcal{A}(k)$ can be diagonalized and weakly hyperbolic if not. Finally, it is strictly hyperbolic if, $\forall k \in \mathbb{R}^2$, the eigenvalues of $\mathcal{A}(k)$ are real and distinct.*

We make the following assumption:

ASSUMPTION 14.1 *System (14.27) is strongly hyperbolic.*

We want to study the well-posedness and stability of the following Cauchy problem in $\mathbb{R}^2$,

$$\begin{cases} \partial_t u^x + \zeta u^x - A_x \partial_x u^x - A_x \partial_x u^y = 0, \\ \partial_t u^y - A_y \partial_y u^x - A_y \partial_y u^y = 0, \end{cases} \quad (14.40)$$

with $\zeta > 0$. Set $U = (u^x, u^y)$, and let us recall the definition of well-posedness.

DEFINITION 14.2 *The Cauchy problem (14.40) associated to the initial data U_0 is weakly (resp. strongly) well-posed if for any $U(\cdot, 0) = U_0$ given in the Sobolev space H^s, $s > 0$ (resp. $s = 0$), (14.40) admits a unique solution $U(t)$ that satisfies an estimate on the type*

$$\|U(\cdot, t)\|_{L^2} \le K e^{\alpha t} \|U_0\|_{H^s}. \quad (14.41)$$

We remark that the definition of well-posedness allows for the solution to be exponentialy increasing in time. This is of course not very useful when one is interested in absorbing boundaries, because using a PML that generates exponentialy increasing solutions does not seem a good idea. That is why we also introduce (cf. [18]) the following notion of stability.

DEFINITION 14.3 *The Cauchy problem (14.40) is weakly (resp. strongly) stable if it is weakly (resp. strongly) well-posed and if the solution $U(t)$ satisfies an estimate on the type*

$$\|U(\cdot, t)\|_{L^2} \le K(1 + t)^s \|U_0\|_{H^s} \quad (14.42)$$

with $s > 0$ (resp. $s = 0$).

In what follows we will say that a system is *stable* if it is at least weakly stable. We want to study the notions of well-posedness and stability using Fourier techniques. To do so, we seek plane wave solutions of system (14.40) of the form,

$$U(x,t) = e^{i(\omega t - k.x)} D, \quad k \in \mathbb{R}^2, \quad D \in \mathbb{R}^{2m}, \quad \omega \in \mathbb{C}. \tag{14.43}$$

Refering to the initial hyperbolic system (14.27)

$$\partial_t u - A_x \partial_x u - A_y \partial_y u = 0,$$

we know that it admits plane wave solutions of the form

$$u(x,t) = e^{i(\omega t - k.x)} d, \quad k \in \mathbb{R}^2, \quad d \in \mathbb{R}^m, \quad \omega \in \mathbb{C},$$

if and only if k and ω are related by the dispersion relation

$$F_1(\omega, k) = 0, \tag{14.44}$$

where

$$F_1(\omega, k) = \det(\omega I - k_x A_x - k_y A_y), \tag{14.45}$$

is a homogeneous polynomial in ω and k of degree m. Considering (14.44) as an equation in ω, we get from Assumption 14.1 (*i.e.*, the hyperbolicity of the system (14.27)) that the solutions of (14.44) are real,

$$\omega = \omega_j(k), \quad j = 1, ..., m, \quad \text{the eigenvalues of } \mathcal{A}(k).$$

Moreover, $\omega_j(k)$ are homogeneous functions of degree one. To simplify the presentation we also make the following assumption:

ASSUMPTION 14.2 *The initial system (14.27) admits:*

- N_e *non-zero eigenvalues of order one,* $\omega_j(k) \neq 0, \forall k \neq 0, \ j = 1, ..., N_e,$ *$(\omega_j(k) \neq \omega_i(k) \text{ for } i \neq j),$*

- *the zero eigenvalue of order* $\ell_0 = m - N_e, \ \omega_j = 0, \ j = N_e + 1, ..., m.$

The mode zero is a non-propagating mode, and we will call the other modes the *physical modes*. This leads to the following expression

$$F_1(\omega, k) = \omega^{\ell_0} \prod_{j=1}^{N_e} (\omega - \omega_j(k)). \tag{14.46}$$

As a consequence of Assumption 14.2, the physical modes $\omega(k) = \omega_j(k), j = 1, N_e$, are differentiable with respect to k, and we can define the following $(K = k/|k|)$

- The phase velocity

$$\mathcal{V}(K) = \frac{\omega(k)}{|k|} \equiv \omega(K). \qquad (14.47)$$

- The slowness vector

$$\mathcal{S}(K) = \frac{K}{\mathcal{V}(K)} = \frac{k}{\omega(k)}. \qquad (14.48)$$

- The group velocity

$$V_g(k) = V_g(K) = \nabla_k \omega(k) = -\left(\frac{\partial F_1}{\partial \omega}(\omega(k), k)\right)^{-1} \nabla_k F_1(\omega(k), k). \quad (14.49)$$

The group velocity is orthogonal to the slowness curves, defined as the set of points, in the plane of slowness vectors $S(K)$, that satisfy

$$F_1(1, \mathcal{S}(K)) = 0. \qquad (14.50)$$

We shall denote by $(V_g^x(k), V_g^y(k))$ the two components of $V_g(k)$ and by $(S_x(k), S_y(k))$ the ones of $\mathcal{S}(K)$ (see also section 9.4 for more details on group velocities and slowness vectors).

Going back to the Fourier analysis of the PML system (14.40), we see that (14.40) has solutions of the form (14.43) if and only if ω and k are related by the modified dispersion relation,

$$F_{pml}(\omega, k, \varsigma) \equiv F_1(\omega(\omega - i\varsigma), k_1\omega, k_2(\omega - i\varsigma)) = 0. \qquad (14.51)$$

(14.51) is a polynomial of degree $2m$ in ω and therefore the dispersion relation for the PML defines $2m$ modes, $\omega_j(k, \zeta)$, $j = 1, ..., 2m$. Note that since we have doubled the number of unknowns with the splitting, it is natural that the number of modes is also doubled.

We can now relate the definitions of well-posedness and stability to these modes (see [105] for more details).

Well-posedness. The system (14.40) is *strongly ill-posed* if there exists some exponentially growing modes, that is,

$$\text{Im}\, \omega(k, \zeta) \to -\infty, \quad \text{when } |k| \to +\infty, \qquad (14.52)$$

otherwise it is at least *weakly well-posed*.

Stability. The system (14.40) is *stable* in the sense of Definition 14.3 if and only if

$$\forall\, k \in \mathbb{R}^m, \text{ the solutions } w(k, \zeta) \text{ satisfy Imm } w(k, \zeta) \geq 0. \qquad (14.53)$$

Let us remark that by introducing $\omega = \omega_R + i\omega_I$ we can write the plane wave solution as,

$$u(x, t) = e^{-\omega_I t} e^{i(\omega_R t - k.x)} D.$$

We see now that the existence of solutions ω with negative imaginary parts would correspond to plane wave solutions with exponential growth in time. A *stable* system does not admit such solutions.

From (14.52) we see that the well-posedness and stability of the PML system (14.40) relies on a high frequency analysis of the modes (*i.e.*, the solutions of 14.51). This analysis was carried out in [18] and the following results were obtained.

THEOREM 14.1 *Under Assumptions 14.1 and 14.2 for system (14.27), the system (14.40) is well-posed.*

THEOREM 14.2 *Let us suppose that system (14.27) satisfies assumptions 14.1 and 14.2. A necessary condition for the stability of the PML model in the x direction (14.40) is that, for all physical modes of the unsplit system (14.27), we have*

$$\forall K = (K_x, K_y)\, /\, |K| = 1, \quad S_x(K) \cdot V_g^x \geq 0. \qquad (14.54)$$

Note that (14.54) is only a necessary but not sufficient condition. It is interesting to look at its geometric interpretation: (14.54) expresses the fact that, along the slowness curves, the slowness vector and the group velocity are oriented the same way with respect to the y axis. This result shows the

Stable   Not stable

FIGURE 14.3: Two different configurations. Left: the slowness vector $\mathcal{S}$ and the group velocity V_g are oriented in the same way with respect to the y axis. Right: $\mathcal{S}$ and V_g are not oriented in the same way with respect to the y axis.

importance of the group velocity in the stability analysis of PML models.

This was first pointed out by Trefethen in [142] (see also [143, 144]) for the stability analysis of finite difference schemes for linear hyperbolic systems and then by Higdon [91] for the well-posedness analysis of initial boundary value problems for linear hyperbolic systems. Concerning the stability analysis for PML models, we would like to mention [135] where the authors have related the instabilities observed with the PMLs for the linearized Euler equations, to the existence of waves for which the group velocity and the phase velocity travel in opposite directions.

14.3.4.2   Stability for Elastodynamics–Orthotropic Medium

To our knowledge there is no stability results for the PML in the case of general elastodynamics. In [18] the authors studied the stability in orthotropic media, *i.e.*, a particular class of anisotropic media. Without presenting the details of the calculations we will summarize in what follows the main results.

Let us first recall the definition of orthotropic media (cf. [11]): In an orthotropic medium whose principal axes coincide with (x, y), we have $c_{13} = c_{23} = 0$ so that the elasticity matrix becomes,

$$C = \begin{pmatrix} c_{11} & c_{12} & 0 \\ c_{12} & c_{22} & 0 \\ 0 & 0 & c_{33} \end{pmatrix}.$$

with (from the positivity of C),

$$c_{11} > 0, \quad c_{22} > 0, \quad c_{33} > 0, \quad c_{11}c_{22} - c_{12}^2 > 0. \tag{14.55}$$

For an isotropic medium, which is a particular orthotropic material, the coefficients can be expressed in terms of Lamé's coefficients $\lambda > 0$ and $\mu \geq 0$:

$$c_{11} = c_{22} = \lambda + 2\mu, \quad c_{12} = \lambda, \quad c_{33} = \mu. \tag{14.56}$$

The first result concerns the necessary stability condition of Theorem 14.2 which becomes:

THEOREM 14.3 *In an orthotropic elastic medium the necessary condition of stability (14.54) is equivalent to,*

$$(\mathcal{C}_1) \quad \{(c_{12} + c_{33})^2 - c_{11}(c_{22} - c_{33})\} \times \{(c_{12} + c_{33})^2 + c_{33}(c_{22} - c_{33})\} \leq 0.$$

Moreover, the following sufficient condition was derived (cf. [18]):

THEOREM 14.4 *The PML system in the x-direction for an orthotropic* *elastic medium is stable if one of the two following conditions is satisfied:*

$$(\mathcal{C}_{x_1})_1 \qquad (c_{12} + c_{33})^2 < (c_{11} - c_{33})(c_{22} - c_{33})$$

$$(\mathcal{C}_{x_1})_2 \begin{cases} \textit{(i)} \;\; (c_{11} - c_{33})(c_{22} - c_{33}) \leq (c_{12} + c_{33})^2 \\ \qquad\quad \leq \max\{-c_{33}(c_{22} - c_{33}), c_{11}(c_{22} - c_{33})\} \\[2mm] \textit{(ii)} \;\; (c_{11} - c_{33})(c_{11}c_{22} - c_{33}^2) < (c_{11} + c_{33})(c_{12} + c_{33})^2 \\[2mm] \textit{(iii)} \;\; (c_{12} + 2c_{33})^2 < c_{11}c_{22} \end{cases}$$

For the proof of these results we refer the reader to [18]. In the what follows we will illustrate their validity with some numerical simulations.

Numerical simulations in orthotropic elastic media   We consider here a 25m × 25m square (see Figure 14.4–left). The numerical method used to obtain the results in this section is based on a first-order original mixed formulation of the equations, described in [46]. We would like to point out that the results presented here are based on properties of the continuous problem and do not depend on the discretization used (similar results not shown here, were also obtained with the discretization method developed in [23, 24]). In

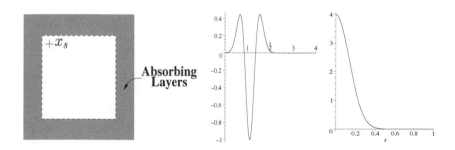

FIGURE 14.4: Computational domain (left) - Function $h(t)$ (center) - Function $g(r)$ (right).

all the simulations, the initial data are taken equal to zero and the source is introduced as a right hand side in equation (14.32),

$$f(x,t) = h(t)\, g(|x - x_S|)\, \vec{e}_x, \tag{14.57}$$

where $\vec{e}_x$ denotes the first vector of the canonical basis of $\mathbb{R}^2$ and h is the so-called second order Ricker signal with central frequency equal to $f_0 = 0.9$ Hz, namely (see also Figure 14.4 center),

$$h(t) = \left[2\pi^2(f_0 t - 1)^2 - 1\right] e^{-\pi^2(f_0 t - 1)^2}. \tag{14.58}$$

The function $g(r)$ is the Gaussian function defined by (see Figure 14.4 right):

$$g(r) = \frac{e^{-7(r/r_0)^2}}{r_0^2},$$

(14.59)

which is concentrated in a small disk of radius $r_0 = 0.5$m. In our experiments, the source point x_S is located close to the absorbing layer (2 meters away from each layer) (see Figure 14.4 left). In the following experiments the density is $\rho = 1$Kg.m^{-3} and give the elasticity coefficients in Pa.

The damping factor is chosen as follows:

$$\zeta(x) = \frac{3c}{2\delta^3} \log(1/R)x^2,$$

(14.60)

where $R = 10^{-3}$ is the theoretical reflection coefficient from the terminating reflection boundaries and $c = 4.5$m.s^{-1} is an upper bound of the wave velocities in all considered materials.

We represent, for each experiment:

(a) the slowness curves and the wavefronts of the material,

(b) the distribution in space of the norm of the displacement field (snapshots) at several times.

Simulation in medium (I).   In this example, the elasticity coefficients are given by:

$$c_{11} = 4, c_{22} = 20, c_{33} = 2 \text{ and } c_{12} = 3.8.$$

(14.61)

The slowness curves and wavefronts represented in Figure 14.5 illustrate the anisotropy of the medium. Note that in orthotropic media we refer to Quasi-Pressure and Quasi-Shear waves instead of pressure and shear waves. We remark that in medium (I) the sets enclosed by the slowness curves remain convex as in the isotropic case (in which we recall that they are circles). The snapshots of the corresponding numerical experiment are given in Figure 14.8. They show that the PMLs work pretty well. In particular, they are stable: the solution does not blow up, even after a long time.

Simulation in medium (II).   This time, the material is characterized by its elasticity coefficients:

$$c_{11} = c_{22} = 20, \quad c_{33} = 2, \quad c_{12} = 3.8.$$

(14.62)

In Figure 14.6, the medium appears to be much more anisotropic than the previous one. In particular, the set enclosed by the slowness curve of the QS wave is no longer convex, which gives rise to triplications of the wavefront. However, one can see in Figure 14.9 that the PML model still works very well and does not lead to any instability.

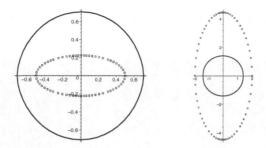

FIGURE 14.5: Slowness curves and wavefronts for medium (I).

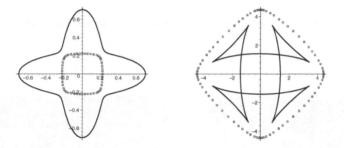

FIGURE 14.6: Slowness curves and wavefront for medium (II).

Simulation in medium (III). This medium is characterized by the following elasticity coefficients:

$$c_{11} = 4, \quad c_{22} = 20, \quad c_{33} = 2, \quad c_{12} = 7.5. \tag{14.63}$$

Once again, this is a medium which gives rise to triplications of the QS wavefront (see Figure 14.7). On the snapshots (see Figure 14.10), we can see two

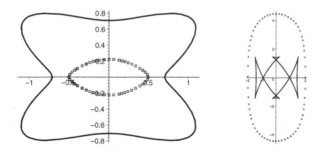

FIGURE 14.7: Slowness curves and wavefront for medium (III).

instabilities appearing very soon in the two PML layers. These instabilities clearly occur when the QS wave penetrates the absorbing layer.   These nu-

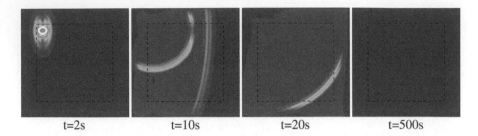

| t=2s | t=10s | t=20s | t=500s |

FIGURE 14.8: Some snapshots at different times for medium (I).

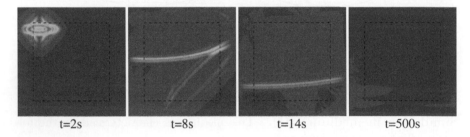

| t=2s | t=8s | t=14s | t=500s |

FIGURE 14.9: Some snapshots at different times for medium (II).

merical results illustrate that the PML model is stable for media (I) and (II) while it is clearly unstable for medium (III). These instabilities can be explained with Theorem 14.2 or equivalently Theorem 14.3. Indeed, we remark the following

- In media (I) and (II) for all K we have $S_x(K){\cdot}V_g^x \geq 0$ and $S_y(K){\cdot}V_g^y \geq 0$ (for both the QP and the QS waves) so that the necessary stability condition is satisfied for PML layers both in the x and y directions.

- In medium (III), for the QS waves there exist some vectors K (whose extremities describe the thickest line on figure 14.11) for which $V_g^x(K)S_x(K) < 0$. Also, there are some other vectors K (whose extremities describe the line of medium thickness on the figure) for which $V_g^y(K)S_y(K) < 0$ which means that the geometrical condition is not satisfied neither in x nor in y.

As we already mentioned the condition in Theorem 14.2 (or equivalently Theorem 14.3) is a necessary and not sufficient condition. It is easy to verify

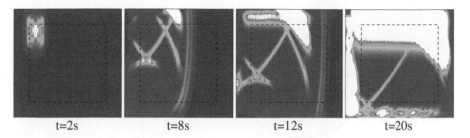

t=2s   t=8s   t=12s   t=20s

FIGURE 14.10: Some snapshots at different times for medium (III).

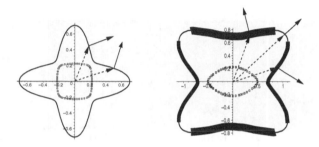

FIGURE 14.11: Slowness diagrams. Left:material (II), Right:material (III).

that media (I) and (II) also satisfy the sufficient condition given in Theorem 14.4. Let us now show an example of a material that satisfies the necessary condition but not the sufficient one.

Simulation in medium (IV). Our last medium is characterized by the following elasticity coefficients:

$$c_{11} = 10, \quad c_{22} = 20, \quad c_{33} = 6, \quad c_{12} = 2.5.$$

In Figure 14.12, we represent the slowness diagrams (left) and the wavefronts (right). It is clear that the high frequency necessary conditions are satisfied for both x and y layers since the slowness curves are convex. We consider the same computational domain as before, in which we now solve the PML model in the x direction with a constant damping coefficient equal to 2.6. The pulse is defined by (14.57)–(14.59) and the source point is located at the center of the computational domain.

The snapshots of the experiment (IV) are given in figure 14.13. At the beginning the x-PML absorbs very well the waves. But after a short while, an instability appears! In conclusion, given an orthotropic medium we know if the PML is going to be stable or not. We do not know however, how to derive stable PMLs for the media that do not satisfy the sufficient condition

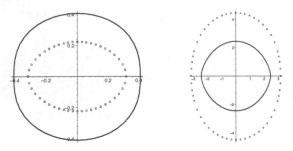

FIGURE 14.12: Slowness curves and wavefronts in medium (IV).

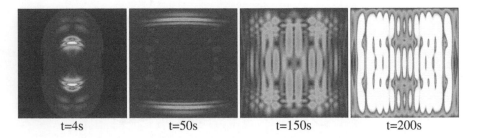

| t=4s | t=50s | t=150s | t=200s |

FIGURE 14.13: Experiment (IV): some snapshots at different times for medium (IV) with only x PML.

of Theorem 14.4. The good news is that for any isotropic medium the sufficient condition is satisfied and thus the PML model is stable. We show some numerical results in isotropic media in the following section.

14.3.5   Numerical Results

We present in this section some numerical results in order to illustrate the efficiency of the PML. For the discretization of the problem we use the mixed finite element method presented in section 11.3.1.

For our numerical experiments an explosive source located at point $S = (x^s, y^s)$ is used, that is

$$f(x,t) = F(t)\vec{g}(r).$$

The function $F(t)$ is defined by

$$F(t) = \begin{cases} -2\pi^2 f_0^2 (t - t_0) e^{-\pi^2 f_0^2 (t-t_0)^2} & \text{if } t \leq 2t_0 \\ 0 & \text{if } t > 2t_0 \end{cases}, \qquad (14.64)$$

where $t_0 = 1/f_0$, $f_0 = Vs/(hN_L)$ is the central frequency, and N_L is the number of points per S wave-length. The function $\vec{g}(r)$ is a radial function given by

$$\vec{g}(r) = \left(1 - \frac{r^2}{a^2}\right)^3 1_{B_a} \left(\frac{x - x^s}{r}, \frac{y - y^s}{r}\right) , \qquad (14.65)$$
$$r = \sqrt{(x - x^s)^2 + (y - y^s)^2}, \ a = 5h ,$$

where 1_{B_a} is one in B_a, the disk of center S and radius a, and zero elsewhere. In the absorbing layers we use the following model for the damping parameter $\zeta(x)$,

$$\zeta(x) = \zeta_0 \left(\frac{x}{\delta}\right)^N , \qquad (14.66)$$

where δ is the width of the layer and ζ_0 is a function of the theoretical reflection coefficient $(R = R_{pp}^\delta)$ (see relation (14.36) for $\theta = 0$):

$$\zeta_0 = \log\left(\frac{1}{R}\right) \frac{(N+1)V_P}{2\delta} . \qquad (14.67)$$

The degree of the polynomial (N) will be precised in each example.

Heterogeneous, isotropic elastic medium. We consider here a heterogeneous isotropic elastic medium that is characterized by the velocity model presented in Figure 14.14. For this medium we have $\dfrac{\max V_P}{\min V_P} = 2.1$ and $V_P = 1.6 V_S$. For the discretization V_P and V_S are piecewise constants (one value per element). The size of the grid is 200×200 with a discretization step $h = 0.15$m and $N_L = 10$. The source is located at point $(15\text{m}, 3\text{m})$.

We present, in Figure 14.15, the norm of the velocity scaled by a factor of 20 for three experiments corresponding to different absorbing layers. More precisely relations (14.66) and (14.67) are used with $N = 2$, $\delta = 5h - R = 0.01$, $\delta = 10h - R = 0.001$ and $\delta = 20h - R = 0.0001$. Let us recall that these values for R correspond to the theoretical (R_{pp}) reflection coefficients. The reflection coefficients obtained numerically are close to the theoretical ones. This is no longer true if the number of meshes inside the layer is too small: the smaller is R the larger layer width is required. This phenomenon is due to the numerical dispersion [54].

14.3.6   Reflection Coefficients

In this section we compute numerically the reflection coefficients obtained with the PML and we compare them to the ones obtained for the ABC's proposed by Higdon [93, 94]. To do so we consider a homogeneous isotropic elastic solid with $V_P = 5.710$m/s and $V_S = 2.93$m/s. To compute the reflection coefficients we first consider the solution of the elastodynamic problem at points M_i located near the upper boundary of domain D_1 (D_1 is a rectangular domain of size $200\text{m} \times 100\text{m}$ and a PML layer model is used on all boundaries of the domain). We denote by $(V_1)_i$ the velocity vector at point M_i. Then

The Vp velocity model

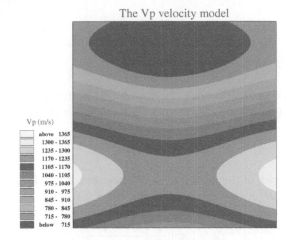

Vp (m/s)

above	1365
1300 -	1365
1235 -	1300
1170 -	1235
1105 -	1170
1040 -	1105
975 -	1040
910 -	975
845 -	910
780 -	845
715 -	780
below	715

FIGURE 14.14: The velocity model for the heterogeneous medium, $\dfrac{\max V_P}{\min V_P} =$ 2.1 and $V_P = 1.6 V_S$.

to obtain a reference solution at points M_i we move the upper boundary of the domain 50m above. That is, we solve the elastodynamic problem in the domain D_2 (a rectangle of size 200m $\times$ 150m). We also use the PML layer model on all boundaries of the domain D_2 and we denote by $(V_2)_i$ the velocity vector at point M_i. The geometry of the problem with the two computational domains is shown in Figure 14.16. In the case of a pure P-wave source we define the reflection coefficients by

$$
\begin{aligned}
(R_{pp})(\theta_i) &= \frac{\operatorname{div}(U_2)_i - \operatorname{div}(U_1)_i}{\operatorname{div}(U_2)_i} \;, \\
(R_{ps})(\theta_i) &= \frac{\operatorname{curl}(U_2)_i - \operatorname{curl}(U_1)_i}{\operatorname{curl}(U_2)_i} \;.
\end{aligned}
\tag{14.68}
$$

In the same way we define

$$
\begin{aligned}
(R_{ss})(\theta_i) &= \frac{\operatorname{curl}(U_2)_i - \operatorname{curl}(U_1)_i}{\operatorname{curl}(U_2)_i} \;, \\
(R_{sp})(\theta_i) &= \frac{\operatorname{div}(U_2)_i - \operatorname{div}(U_1)_i}{\operatorname{div}(U_2)_i} \;,
\end{aligned}
$$

in the case of a pure S wave source. For our examples we use the P-wave source function given by (14.64, 14.65) with $N_L = 16$. For this source function the curl of the velocity is smaller than the divergence by a factor of 10^{-10}: the reflection coefficient R_{ps} is neglected.

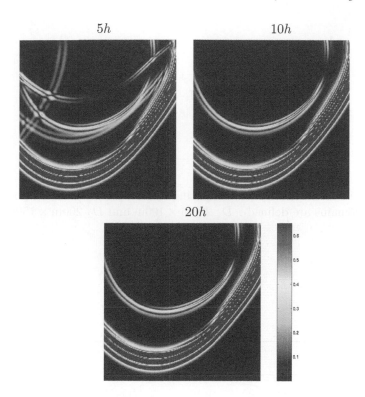

FIGURE 14.15: Snapshots of 2D elastic finite-elements simulations, in a heterogeneous medium, using the PML absorbing layer model with $\delta = 5h$, $\delta = 10h$ and $\delta = 20h$. The snapshots depict the norm of the velocity at $t = 14.2$ms scaled by a factor of 20.

REMARK 14.7 *Relation (14.68) is used to compute two reflection coefficients at each time step. The results presented in Figure 14.17 correspond to their mean value over $[0, T]$, for $T = 25sec$.*

The grid size is 400×200 for D_1, 400×300 for D_2 and $h = 0.5$m for both domains. The source is located at point $(200, 80)$ of the grid. In the PML layers, we take $N = 3$, $\delta = 5h - R = 10^{-4}$, $\delta = 10h - R = 10^{-6}$ and $\delta = 20h - R = 10^{-8}$. To examine the performance of the PML model, comparisons with the 1st, 2nd and 3rd order Higdon's absorbing boundary conditions are performed.

REMARK 14.8 *The reflection coefficients for the absorbing boundary conditions proposed by Higdon are the theoretical ones. We obtain them by ap-*

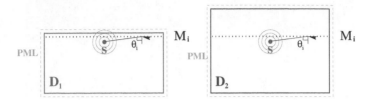

FIGURE 14.16: Geometry for computing reflection coefficients. Two computational domains are defined : D_1 200m × 100m and D_2 200m × 150m.

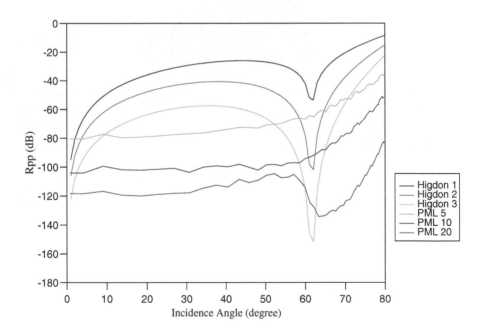

FIGURE 14.17: Comparison of R_{pp} for the Higdon ABC's and the PML.

plying an operator of the form

$$\prod_{i=1}^{m} \left(\beta_i \frac{\partial}{\partial t} + V_P \frac{\partial}{\partial n} \right) \,,$$

with $m = 1$; $\beta_1 = 1$, *or* $m = 2$; $\beta_1 = 1$, $\beta_2 = V_P/V_S (= 1.95)$, *or* $m = 3$; $\beta_1 = 1$, $\beta_2 = 1.5$, $\beta_3 = V_P/V_S (= 1.95)$.

As we can see in Figure 14.17, for almost all angles of incidence, the PML absorbing layer model gives better results than the absorbing boundary conditions of first and second order and this happens even for the $5h$ length absorbing layer. Compared to the third order absorbing condition the PML layers of $10h$ and $20h$ are substantially better except near the 60° angle. We want to point out that in this example we compare the reflection coefficients provided by the continuous Higdon model to the one given by the discrete PML model. It is known that the Higdon reflection coefficients increase when discretization is used.

References

[1] S. Abarbanel and D. Gottlieb. A mathematical analysis of the PML method. *J. Comput. Physics*, 134:357–263, 1997.

[2] J.D. Achenbach. *Wave Propagation in Elastic Solids*, volume I and II. Elsevier Science Publ. B.V., 1984.

[3] B. Alpert, L. Greengard, and T. Hagstrom. Rapid evaluation of nonreflecting boundary kernels for time-domain wave propagation. *SIAM J. Numer. Anal.*, 37(4):1138–1164 (electronic), 2000.

[4] M. Amara and J.M. Thomas. Equilibrium finite elements for the linear elastic problem. *Numer. Math.*, 33:367–383, 1979.

[5] D. N. Arnold, D. Boffi, and R. S. Falk. Approximation by quadrilateral finite elements. *Math. Comp.*, 71(239):909–922 (electronic), 2002.

[6] D. N. Arnold, F. Brezzi, B. Cockburn, and L. Donatella Marini. Unified analysis of discontinuous Galerkin methods for elliptic problems. *SIAM J. Numer. Anal.*, 39(5):1749–1779 (electronic), 2001/02.

[7] D.N. Arnold, F. Brezzi, and J. Douglas. PEERS: A new mixed finite element for plane elasticity. *Japan J. Appl. Math.*, 1:347–367, 1984.

[8] G.P. Astrakmantev. Methods of fictitious domains for a second order elliptic equation with natural boundary conditions. *U.S.S.R Comp. Math. and Math. Phys.*, 18:114–221, 1978.

[9] C. Atamian, R. Glowinski, J. Périaux, H. Stève, and G. Terrason. Control approach to fictitious domain in electro-magnetism. *Conférence sur l'approximation et les methodes numeriques pour la resolution des equations de Maxwell, Hotel Pullmann, Paris*, 1989.

[10] C. Atamian and P. Joly. An analysis of the method of fictitious domains for the exterior Helmholtz problem. *RAIRO Modèl. Math. Anal. Numér*, 27(3):251–288, 1993.

[11] B.A. Auld. *Acoustic Fields and Elastic Waves in Solids*, volume I and II. John Wiley & Sons, Inc., 1973.

[12] I. Babuska. The finite element method with lagrangian multipliers. *Numer. Math.*, 20:179–192, 1973.

[13] G.A. Baker and V.A. Dougalis. The effect of quadrature errors on finite element approximations for the second-order hyperbolic equations. *SINUM*, 13(4):577–598, 1976.

[14] A. Bamberger, G. Chavent, and P. Lailly. Étude de schémas numériques de l'élastodynamique linéaire. Technical Report 41, INRIA, October 1980.

[15] A. Bamberger, R. Glowinski, and Q.H. Tran. A domain decomposition method for the acoustic wave equation with discontinuous coefficients and grid change. *SIAM J. on Num. Anal.*, 34(2):603–639, April 1997.

[16] A. Bamberger, J.-C. Guillot, and P. Joly. Numerical diffraction by a uniform grid. *SIAM J. Numer. Anal.*, 25(4):753–783, 1988.

[17] E. Bécache, A. Ezziani, and P. Joly. A mixed finite element approach for viscoelastic wave propagation. *Computational Geosciences*, 8(3):255–299, 2004.

[18] E. Bécache, S. Fauqueux, and P. Joly. Stability of perfectly matched layers, group velocities and anisotropic waves. *J. Comput. Physics*, 188:399–433, 2003.

[19] E. Bécache and P. Joly. On the analysis of berengers perfectly matched layers for Maxwell equations. *M2AN*, 36(1):87–120, 2002.

[20] E. Bécache, P. Joly, and G. Scarella. A fictitious domain method for unilateral contact problems in non destructive testing. In *Computational Fluid and Solid Mechanics*, volume 1, pages 65–67. Elsevier, 2001.

[21] E. Bécache, P. Joly, and C. Tsogka. Eléments finis mixtes et condensation de masse en élastodynamique linéaire. (i) construction. *C.R. Acad. Sci. Paris*, t. 325, Série I:545–550, 1997.

[22] E. Bécache, P. Joly, and C. Tsogka. An analysis of new mixed finite elements for the approximation of wave propagation problems. *SIAM J. Numer. Anal.*, 37:1053–1084, 2000.

[23] E. Bécache, P. Joly, and C. Tsogka. Fictitious domains, mixed FE and PML for 2-D elastodynamics. *J. Comput. Acoustics*, 9(3):1175–1201, 2001.

[24] E. Bécache, P. Joly, and C. Tsogka. A new family of mixed finite elements for the linear elastodynamic problem. *SIAM J. Numer. Anal.*, 39:2109–2132, 2002.

[25] E. Bécache, J. Rodríguez, and C. Tsogka. On the convergence of the fictitious domain method for the wave equation. *Submitted to SINUM*, 2006.

[26] E. Bécache, J. Rodríguez, and C. Tsogka. On the convergence of the fictitious domain method for wave equation problems. Technical Report 5802, INRIA, January 2006.

[27] F. Ben Belgacem. The mortar finite element method with Lagrange multipliers. *Numer. Math.*, 84:173–197, 1999.

[28] F. Ben Belgacem, A. Buffa, and Y. Maday. The mortar finite element method for 3d Maxwell equations: First results. *SIAM J. Numer. Anal.*, 39(3):880–901, 2001.

[29] A. Bemberger, B. Engquist, L. Halpern, and P. Joly. Higher order paraxial approximations for the wave equation. *SIAM J. Appl. Math.*, 48(1):129–154, 1988.

[30] A. Bendali. *Approximation par éléments finis de surface de problèmes de diffraction des ondes électromagnétiques.* PhD thesis, *Université Paris VI*, 1984.

[31] J.P. Bérenger. A perfectly matched layer for the absorption of electromagnetic waves. *J. Comput. Physics*, 114:185–200, 1994.

[32] J.P. Bérenger. Three-dimensional perfectly matched layer for the absorption of electromagnetic waves. *J. Comput. Physics*, 127(2):363–379, 1996.

[33] J.P. Bérenger. Improved PML for the FDTD Solution of Wave-Structure Interaction Problems. *IEEE Trans. Antennas and Propagation*, 45(3):466–473, March 1997.

[34] C. Bernardi and Y. Maday. Raffinement de maillage en éléments finis par la méthode des joints. *C. R. Acad. Sci. Paris Sér. I Math.*, 320(3):373–377, 1995.

[35] C. Bernardi, Y. Maday, and A.T. Patera. A new nonconforming approach to domain decomposition: the mortar element method. In *Nonlinear partial differential equations and their applications. Collège de France Seminar, Vol. XI (Paris, 1989–1991)*, volume 299 of *Pitman Res. Notes Math. Ser.*, pages 13–51. Longman Sci. Tech., Harlow, 1994.

[36] S. Brenner and R.L. Scott. *The Mathematical Theory of Finite Element Methods*. Springer-Verlag, 1994.

[37] F. Brezzi. On the existence uniqueness and approximation of saddlepoint problems arising from Lagrangian multipliers. *RAIRO Ser. Rouge*, 8:129–151, 1974.

[38] F. Brezzi and M. Fortin. *Mixed and Hybrid Finite Element Methods*. Springer-Verlag, New York, 1991.

[39] B. Chalindar. *Conditions aux limites artificielles pour les équations de l'élastodynamique*. PhD thesis, Univ. Saint Etienne, 1987.

[40] Z. Chen. *Finite Element Methods And Their Applications*. Springer, 2005.

[41] M. W. Chevalier and R. J. Luebbers. FDTD local grid with material traverse. *IEEE Transactions on Antennas and Propagation*, 45(3):411–421, March 1997.

[42] M. J. S. Chin-Joe-Kong, W. A. Mulder, and M. Van Veldhuizen. Higher-order triangular and tetrahedral finite elements with mass lumping for solving the wave equation. *J. Engrg. Math.*, 35(4):405–426, 1999.

[43] P.G. Ciarlet. *The finite element method for elliptic problems*. North-Holland, 1982.

[44] P. Clément. Approximation by finite element functions using local regularization. *RAIRO Anal. Numer.*, 9:77–84, 1975.

[45] G. Cohen. *Higher-Order Numerical Methods for Transient Wave Equations*. Springer, 2002.

[46] G. Cohen and S. Fauqueux. Mixed finite elements with mass-lumping for the transient wave equation. *J. Comput. Acoustics*, 8:171–188, 2000.

[47] G. Cohen, P. Joly, J. E. Roberts, and N. Tordjman. Higher order triangular finite elements with mass lumping for the wave equation. *SIAM J. Numer. Anal.*, 38(6):2047–2078 (electronic), 2001.

[48] G. Cohen, P. Joly, and N. Tordjman. Higher Order Triangular Finite Elements with Mass Lumping for the Wave Equation. *Third International Conference on Mathematical and Numerical Aspects of Wave Propagation*, 1:270–279, 1995.

[49] F. Collino. High order absorbing boundary conditions for wave propagation models: straight line boundary and corner cases. In PA SIAM, Philadelphia, editor, *Second International Conference on Mathematical and Numerical Aspects of Wave Propagation (Newark, DE, 1993)*, pages 161–171, 1993.

[50] F. Collino. Perfectly matched absorbing layers for the paraxial equations. *J. Comput. Physics*, 131(1):164–180, 1996.

[51] F. Collino, T. Fouquet, and P. Joly. A conservative space-time mesh refinement method for the 1-D wave equation. II. Analysis. *Numer. Math.*, 95(2):223–251, 2003.

[52] F. Collino, P. Joly, and F. Millot. Fictitious domain method for unsteady problems: Application to electromagnetic scattering. Technical Report RR-2963, INRIA, 1996.

[53] F. Collino, P. Joly, and F. Millot. Fictitious domain method for unsteady problems: Application to electromagnetic scattering. *J. Comput. Physics*, 138(2):907–938, December 1997.

[54] F. Collino and P. Monk. Optimizing the perfectly matched layer. *Comput. Methods Appl. Mech. Enyry.*, 164:157–171, 1998.

[55] F. Collino and P. Monk. The Perfectly Matched Layer in curvilinear coordinates. *SIAM J. Scient. Comp.*, 164:157–171, 1998.

[56] F. Collino and C. Tsogka. Application of the PML absorbing layer model to the linear elastodynamic problem in anisotropic heteregeneous media. *Geophysics*, 66:294–305, 2001.

[57] R. Dautray and J.-L. Lions. *Analyse Mathématique et Calcul Numérique*. Masson, Paris, 1988.

[58] P.J. Davis and P. Rabinowitz. *Methods of Numerical Integration*. Academic Press, 1985.

[59] J. Diaz. *Approches analytiques et numériques de problèmes de transmission en propagation d'ondes en régime transitoire. Application au couplage fluide-structure et aux méthodes de couches parfaitement adaptées.* PhD thesis, Université Versailles Saint-Quentin, 2005.

[60] J. Diaz and P. Joly. Robust high order non-conforming finite element formulation for time domain fluid-structure interaction. *J. Comput. Acoustics*, 13(3):403–431, 2005.

[61] T. Ha Duong and P. Joly. On the stability analysis of boundary conditions for the wave equation by energy methods. part 1: The homogeneous case. Technical Report 1306, I.N.R.I.A., Domaine de de Voluceau Rocquencourt, B.P.105, 78153, Le Chesnay Cedex France, 1990.

[62] T. Dupont. l^2-estimates for galerkin methods for second order hyperbolic equations. *SIAM J. Numer. Anal.*, 10(5):880–889, 1973.

[63] M. Durufle. *Intégration numérique et éléments fins d'ordre élevé appliqués aux équations de Maxwell en régime harmonique.* PhD thesis, Université Paris IX Dauphine, 2006.

[64] G. Duvaut and J.L. Lions. *Inequalities in Mechanics and Physics.* Springer-Verlag New York, 1976.

[65] B. Engquist and A. Majda. Absorbing boundary conditions for the numerical simulation of waves. *Math. Comp.*, 31(139):629–651, July 1977.

[66] A. C. Eringen and E. S. Şuhubi. *Elastodynamics. Vol. I. Finite motions.* Academic Press [A subsidiary of Harcourt Brace Jovanovich, Publishers], New York-London, 1974.

[67] A.C. Eringen and E.S. Şuhubi. *Elastodynamics, Vol. II, Linear Theory.* Academic Press, 1975.

[68] A. Ern and J.L. Guermond. *Theory and Practice of Finite Elements.* Springer, 2004.

[69] Sandrine Fauqueux. *Eléments finis mixtes spectraux et couches absorbantes parfaitement adaptées pour la propagation d'ondes élastiques en régime transitoire.* PhD thesis, Université Paris IX Dauphine, 2003.

[70] S.A Finogenov and Y.A. Kuznetsov. Two stage fictitious components methods for solving the Dirichlet boundary value problem. *Sov. J. Num. Anal. Math. Modelling*, 3:301–323, 1988.

[71] B. Fornberg. *A practical guide to pseudospectral methods*, volume 1 of *Cambridge Monographs on Applied and Computational Mathematics.* Cambridge University Press, Cambridge, 1996.

[72] K. O. Friedrichs. On the boundary-value problems of the theory of elasticity and Korn's inequality. *Ann. of Math. (2)*, 48:441–471, 1947.

[73] S. Garcés. *Application des méthodes de domaines fictifs à la modélisation des structures rayonnantes tridimensionnelles Etude mathématique et numérique d'un modèle.* PhD thesis, *ENSAE (Toulouse)*, 1997.

[74] W. Gautschi. *Orthogonal polynomials: computation and approximation.* Numerical Mathematics and Scientific Computation. Oxford University Press, New York, 2004.

[75] T. Geveci. On the application of mixed finite element methods to the wave equations. M^2AN *Math. Model. Numer. Anal.*, 22(2):243–250, 1988.

[76] G. Geymonat and G. Gilardi. Contre-exemples à l'inégalité de Korn et au lemme de Lions dans des domaines irréguliers. In *Équations aux dérivées partielles et applications*, pages 541–548. Gauthier-Villars, Éd. Sci. Méd. Elsevier, Paris, 1998.

[77] V. Girault and R. Glowinski. Error analysis of a fictitious domain method applied to a Dirichlet problem. *Japan J. Ind. Appl. Math.*, 12(3):487–514, 1995.

[78] V. Girault and P.A. Raviart. *Finite element methods for Navier-Stokes equations*, volume 5 of *Springer Series in Computational Mathematics*. Springer-Verlag, Berlin, 1986. Theory and algorithms.

[79] D. Givoli. High-order local non-reflecting boundary conditions: a review. *Wave Motion*, 39(4):319–326, 2004.

[80] R. Glowinski, T.W. Pan, and J. Periaux. A fictitious domain method for Dirichlet problem and applications. *Comp. Meth. in Appl. Mech. and Eng.*, pages 283–303, 1994.

[81] R. Glowinski, T.W. Pan, and J. Periaux. A fictitious domain method for external incompressible viscous flow modeled by Navier-Stokes equations. *Comp. Meth. in Appl. Mech. and Eng.*, pages 283–303, 1994.

[82] P. Grisvard. Behavior of the solutions of an elliptic boundary value problem in a polygonal or polyhedral domain. In *Numerical solution of partial differential equations, III (Proc. Third Sympos. (SYNSPADE), Univ. Maryland, College Park, Md., 1975)*, pages 207–274. Academic Press, New York, 1976.

[83] P. Grob. *Méthodes num'eriques de couplage pour la vibroacoustique instationnaires : éléments finis spectraux d'ordre élevé et potentiels retardés*. PhD thesis, Université Paris IX Dauphine, 2006.

[84] M. Grote and J. Keller. Exact nonreflecting boundary conditions for the time dependent wave equation. *SIAM J. Appl. Math.*, 55:280–297, 1995.

[85] M. Grote and J. Keller. Nonreflecting boundary conditions for time dependent scattering. *J. Comput. Physics*, 127:52–81, 1996.

[86] T. Hagstrom. Radiation boundary conditions for the numerical simulation of waves. *Acta numerica*, 8:47–106, 1999.

[87] T. Hagstrom. New results on absorbing layers and radiation boundary conditions. In M. Ainsworth, P. Davies, D. Duncan, P. Martin, and B. Rynne, editors, *Topics in Computational Wave Propagation - Direct and Inverse Problems*, volume 31 of *Lecture Notes in Computational Science and Engineering*, pages 1–42. Springer, 2003.

[88] L. Halpern. *Étude de conditions aux limites absorbantes pour des schémas numériques relatifs à des équations hyperboliques linéaires.* PhD thesis, Paris IV, 1980.

[89] F. Hastings, J.B. Schneider, and S. L. Broschat. Application of the perfectly matched layer (PML) absorbing boundary condition to elastic wave propagation. *J. Acoust. Soc. Am.*, 100(5):3061– 3069, November 1996.

[90] S. Hesthaven and T. Warburton. Discontinuous galerkin methods for the time-domain Maxwell's equations: An introduction. *ACES Newsletter*, 19(1):10–29, 2004.

[91] R.L. Higdon. Initial-boundary value problems for linear hyperbolic systems. *Siam Review.*, 28:177–217, 1986.

[92] R.L. Higdon. Radiation boundary conditions for elastic wave propagation. *SIAM J. Numer. Anal.*, 27:831–870, 1990.

[93] R.L. Higdon. Absorbing boundary conditions for elastic waves. *Geophysics*, 56:231–241, 1991.

[94] R.L. Higdon. Absorbing boundary conditions for acoustic and elastic waves in stratified media. *J. Comput. Physics*, 101(2):386–418, 1992.

[95] Charles Hirsch. *Numerical Computation of Internal and External Flows, Volume 1, Fundamentals of Numerical Discretization.* John Wiley and Sons, 1989.

[96] L. Hörmander. *The analysis of linear partial differential operators. III*, volume 274 of *Grundlehren der Mathematischen Wissenschaften [Fundamental Principles of Mathematical Sciences]*. Springer-Verlag, Berlin, 1994. Pseudo-differential operators, corrected reprint of the 1985 original.

[97] M. Israeli and S.A. Orszag. Approximation of radiation boundary conditions. *J. Comput. Physics*, 41:115–135, 1981.

[98] Hans Johansen and Phillip Colella. A Cartesian grid embedded boundary method for Poisson's equation on irregular domains. *J. Comput. Physics*, 147(1):60–85, 1998.

[99] C. Johnson and B. Mercier. Some equilibrium finite element methods for two-dimensional elasticity problems. *Numer. Math.*, 30:103–116, 1978.

[100] P. Joly. Variational methods for time-dependent wave propagation problems. In *Topics in Computational Wave Propagation, Direct and Inverse Problems*, pages 201–264. LNCSE, 2003.

[101] P. Joly and L. Rhaouti. Domaines fictifs, éléments finis $h(div)$ et condition de neumann: le problème de la condition inf-sup. *C. R. Acad. Sci. Paris, Série I*, 328:1225–1230, 1999.

[102] P. Joly and J. Rodríguez. An error analysis of conservative space-time mesh refinement methods for the 1d wave equation. *SIAM J. Numer. Anal.*, 43:825–859, 2005.

[103] I. S. Kim and W. J. R. Hoefer. A local mesh refinement algorithm for the time-domain finite-difference method to solve Maxwell's equations. *IEEE Trans. Microwave Theory Tech.*, 38(6):812–815, June 1990.

[104] V. A. Kondratiev and O. A. Oleinik. On Korn's inequalities. *C. R. Acad. Sci. Paris Sér. I Math.*, 308(16):483–487, 1989.

[105] H-O. Kreiss and J. Lorenz. Initial-boundary value problems and the Navier-Stokes equations. In *Pure and Appl. Math.*, volume 136. Academic Press, Boston, USA, 1989.

[106] H.-O. Kreiss and N. Anders Petersson. A second order accurate embedded boundary method for the wave equation with Dirichlet data. *SIAM J. Sci. Comput.*, 27(4):1141–1167 (electronic), 2006.

[107] K. S. Kunz and L. Simpson. A technique for increasing the resolution of finite-difference solutions to the Maxwell equations. *IEEE Trans. Electromagn. Compat.*, EMC-23:419–422, Nov. 1981.

[108] R. J. Leveque. *Finite volume methods for hyperbolic problems*. Cambridge Texts in Applied Mathematics. Cambridge University Press, Cambridge, 2002.

[109] E. Lindman. Free space boundary conditions for time dependant wave equation. *J. Comput. Physics*, 18:66–78, 1975.

[110] J.L. Lions and E. Magenès. *Problèmes aux Limites non Homogènes et Applications*. Dunod, 1968.

[111] B. Lombard and J. Piraux. Numerical treatment of two-dimensional interfaces for acoustic and elastic waves. *J. Comput. Physics*, 195(1):90–116, 2004.

[112] Ch. G. Makridakis. On mixed finite element methods for linear elastodynamics. *Numer. Math.*, 61:235–260, 1992.

[113] G.I Marchuk, Y.A Kuznetsov, and A.M. Matsokin. Fictitious domain and domain decomposition methods. *Sov. J. Num. Anal. Math. Modelling*, 1:3–35, 1986.

[114] J. Miklowitz. *The theory of elastic waves and waveguides*, volume 22 of *North-Holland Series in Applied Mathematics and Mechanics*. North-Holland Publishing Co., Amsterdam, 1978.

[115] P. Monk. Sub-gridding FDTD schemes. *ACES Journal*, 11:37–46, 1996.

[116] P. Monk and G. R. Richter. A discontinuous Galerkin method for linear symmetric hyperbolic systems in inhomogeneous media. *J. Sci. Comput.*, 22/23:443–477, 2005.

[117] M.E. Morley. A family of mixed finite elements for linear elasticity. *Numer. Math.*, 55:633–666, 1989.

[118] W. A. Mulder. Higher-order mass-lumped finite elements for the wave equation. *J. Comput. Acoustics*, 9(2):671–680, 2001.

[119] J. Necas. *Les méthodes directes en théorie des équations elliptiques.* Masson, Paris, 1967.

[120] J.C. Nédélec. A new family of mixed finite elements in $\mathbb{R}^3$. *Numer. Math.*, 50:57–81, 1986.

[121] J.A. Nitsche. On Korn's second inequality. *RAIRO Anal. Numér.*, 15(3):237–248, 1981.

[122] C. Peng and M.N. Toksoz. An optimal absorbing boundary condition for elastic wave modeling. *Geophysics*, 60(1):296–301, January-February 1995.

[123] S. Pernet. *Etude de méthodes d'ordre élevé pour résoudre les équations de Maxwell dans le domaine temporel.* PhD thesis, Université Paris IX Dauphine, 2005.

[124] P.G. Petropoulos, L. Zhao, and A.C. Cangellaris. A reflectionless sponge layer absorbing boundary condition for the solution of Maxwell's equations with high-order staggered finite difference schemes. *J. Comput. Physics*, 139(1):184–208, 1998.

[125] D. T. Prescott and N. V. Shuley. A method for incorporating different sized cells into the finite-difference time-domain analysis technique. *IEEE Microwave Guided Wave Lett.*, 2:434–436, Nov. 1992.

[126] P.-A. Raviart and J. M. Thomas. A mixed finite element method for 2nd order elliptic problems. In *Mathematical aspects of finite element methods (Proc. Conf., Consiglio Naz. delle Ricerche (C.N.R.), Rome, 1975)*, pages 292–315. Lecture Notes in Math., Vol. 606. Springer, Berlin, 1977.

[127] R. D. Richtmyer and K. W. Morton. *Difference methods for initial-value problems.* Robert E. Krieger Publishing Co. Inc., Malabar, FL, second edition, 1994.

[128] J. Rodríguez. *Raffinement de Maillage Spatio-Temporel pour les Équations de l'Élastodynamique.* PhD thesis, Paris IX, 2004.

[129] A. Simone and S. Hestholm. Instability in applying absorbing boundary conditions to high-order seismic modeling algorithms. *Geophysics*, 63(3):1017–1023, May-June 1998.

[130] J. Sochacki, R. Kubichek, J. George, W.R. Fletcher, and S. Smithson. Absorbing boundary conditions and surface waves. *Geophysics*, 52(1):60–71, January 1987.

[131] R. Stenberg. On the construction of optimal mixed finite element methods for the linear elasticity Problem. *Numer. Math.*, 48:447–462, 1986.

[132] R. Stenberg. A family of mixed finite elements for the elasticity problem. *Numer. Math.*, 53:513–538, 1988.

[133] R. Stenberg. A Technique for analysing finite element methods for viscous incompressible flow. *Int. Jour. for Numer. Meth. in Fluids*, 11:935–948, 1990.

[134] A. H. Stroud. *Numerical quadrature and solution of ordinary differential equations.* Springer-Verlag, New York, 1974. A textbook for a beginning course in numerical analysis, Applied Mathematical Sciences, Vol. 10.

[135] C. K. W. Tam, L. Auriault, and F. Cambuli. Perfectly matched layer as an absorbing boundary condition for the linearized Euler equations in open and ducted domains. *J. Comput. Physics*, 144(1):213–234, 1998.

[136] M. E. Taylor. *Partial differential equations. I*, volume 115 of *Applied Mathematical Sciences*. Springer-Verlag, New York, 1996. Basic theory.

[137] F.L. Teixeira and W.C. Chew. Analytical derivation of a conformal perfectly matched absorber for electromagnetic waves. *Micro. Opt. Tech. Lett.*, 17:231–236, 1998.

[138] F.L. Teixeira and W.C. Chew. Unified analysis of perfectly matched layers using differential forms. *Microw. Optical Technol. Lett.*, 20(2):124–126, 1999.

[139] V. Thomée. *Galerkin finite elements methods for parabolic problems*, volume 1054. Springer-Verlag, Berlin, 1984.

[140] N. Tordjman. *Eléments finis d'ordre élevé avec condensation de masse pour l'équation des ondes.* PhD thesis, Univ. Paris IX, 1995.

[141] L. Trefethen and L. Halpern. Well posedness of one way equations and absorbing boundary conditions. *Math. Comp.*, 47:421–435, 1986.

[142] L.N. Trefethen. Group velocity in finite difference schemes. *SIAM Review*, 24(2), April 1982.

[143] L.N. Trefethen. Group velocity interpretation of the stability theory of Gustafsson, Kreiss and Sundström. *J. Comput. Physics*, 49(2), February 1983.

[144] L.N. Trefethen. Instability of difference models for hyperbolic initial boundary value problems. *Comm. Pure and Applied Math.*, 37:329–367, February 1984.

[145] C. Tsogka. *Modélisation mathématique et numérique de la propagation des ondes élastiques tridimensionnelles dans des milieux fissurés.* PhD thesis, Paris IX, 1999.

[146] E. Turkel and A. Yefet. Absorbing pml boundary layers for wave-like equations. *Appl. Numer. Math.*, 27(4):533–557, 1998.

[147] K. Yee. Numerical solution of inital boundary value problems involving Maxwell's equations in isotropic media. *IEEE Trans. Antennas and Propagation*, 14(3):302–307, 1966.

[148] C. Zhang and R.J. LeVeque. The immersed interface method for acoustic wave equations with discontinuous coefficients. *Wave Motion*, 25(3):237–263, 1997.

[149] L. Zhao and A.C. Cangellaris. GT-PML: generalized theory of perfectly matched layers and its application to the reflectionless truncation of finite-difference time-domain grids. *IEEE Trans. Microwave Theory Tech.*, 44:2555–2563, 1996.

[150] O. C. Zienkiewicz. *The finite element method in engineering science.* McGraw-Hill, London, 1971. The second, expanded and revised, edition of *The finite element method in structural and continuum mechanics.*

Part IV

Waves in Compressible Flows

Chapter 15

High-Order Accurate Space Discretization Methods for Computational Fluid Dynamics

John A. Ekaterinaris, School of Mechanical and Aerospace Engineering, University of Patras, Greece *and* Foundation for Research and Technology-Hellas, Institute of Applied and Computational Mathematics, 71110 Heraklion, Greece, ekaterin@iacm.forth.gr

15.1 Introduction

In recent years numerical methods have been widely used to effectively resolve complex flow features of aerodynamics flows with meshes that are reasonable for today's computers. High-order numerical methods were used mainly in direct numerical simulations and aeroacoustics. For many aeronautical applications, accurate computation of vortex-dominated flows is important because the vorticity in the flow field and the wake of swept wings at an incidence and rotor blades largely determines the distribution of loading. The main deficiency of widely available, second-order accurate methods for the accurate computation of these flows is the numerical diffusion of vorticity and acoustic disturbances to unacceptable levels. Application of high-order accurate, low diffusion numerical methods can significantly alleviate this deficiency of traditional second order methods. Furthermore, higher-order space discretizations have the potential to improve detached eddy simulation predictions of separated flows with significant unsteadiness. Recently developed high-order accurate finite-difference, finite-volume, and finite-element methods are reviewed. These methods can be used as an attractive alternative of traditional low-order central and upwind computational fluid dynamics methods for improved predictions of vortical and other complex, separated, unsteady flows. The main features of these methods are summarized, from a practical user's point of view, their applicability and relative strength is indicated, and examples from recent applications are presented to illustrate their performance on selected problems.

In the following chapters, applications of high-order methods are explored mainly in the field of aerodynamics, however use of high order methods developed for aerodynamics can benefit other disciplines, such as electromagnetics and aeroacoustics. High-order methods typically have at least third-order spatial accuracy. Traditionally, second-order accurate numerical methods are often preferred in practical aerodynamic calculations because of their simplicity and robustness. High-order accurate methods are often perceived as less robust, costly to run, and complicated to understand and code. As a result, there are very few working computational fluid dynamics (CFD) codes that use higher order accurate schemes for production numerical simulations of compressible flow. This is especially true for codes designed to compute steady flows. We will attempt to dispel this negative impression about high-order methods.

Before presenting the essential details of high-order methods, we point out that in many practical aerodynamic problems, the solution structures are so complicated and their time evolution is so long, that it is impossible to obtain an acceptable solution with today's computing speeds using high-grid density and low-order methods. These problems often involve regions of complicated but smooth flow structures, such as vortices interacting with each other or with shear layers, and regions that contain both shocks and complex smooth structures. Some examples of these flows are briefly discussed in the following paragraphs.

Vortical flow fields are especially challenging for the low-order numerical methods that are typically found in current Euler and Navier-Stokes flow solvers. The main cause for this deficiency is that these vortical flow features deform and dissipate prematurely due to excessive numerical diffusion in the solution algorithms. It is well recognized that low-order CFD algorithms require extremely fine grid resolution to accurately convect and preserve the strength of vortical flow fields. This fine grid resolution requirement leads to extremely large computational problems that cannot be solved on even the largest parallel computer architectures.

It is straightforward to demonstrate the power of high-order numerical methods by examining the numerical approximation of a smooth function on a regular grid. For a second-order accurate numerical scheme, the error in the functional approximation is proportional to h^2, where h the mesh size Δx. For an nth-order accurate numerical scheme, the error in the functional approximation is proportional to h^n. If we cut the grid spacing in half, the error in the second order scheme reduces by a factor of 4 and the error in the fourth order scheme reduces by a factor of 16. Provided that the computational requirements for the second and fourth order schemes are similar, the fourth order scheme is clearly more efficient. Efficiency improvements are even greater for higher order approximations. A demonstration of the superior performance of high order, low-dissipative, centered schemes compared to lower order more dissipative TVD schemes in preserving vorticity [204] is shown in Fig. 15.1. The conversion of a two-dimensional isentropic vortex is carried out for long time with the numerical solution of the inviscid compress-

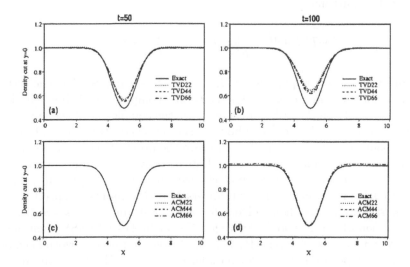

FIGURE 15.1: Isentropic vortex: Comparison of the various orders of TVD and ACM methods with the exact solution, illustrated by density profiles at the centerline y=0, at t=50 and t=100 for a 401 x 81 grid; (from Ref. [204]).

ible flow equations on a Cartesian mesh. At the absence of physical viscosity, the loss in vortex strength due to the numerical diffusion of the scheme is evident. For three-dimensional computations, grid refinement in all three directions which is necessary for better resolution of vortical structures that are isotropic becomes very expensive computationally. Therefore, application of high-order schemes is expected to significantly decrease computational cost.

To better demonstrate the potential of higher-order methods make the following assumptions: (1) The error in the solution is $\mathcal{O}(h^p)$ where h is the mesh length and p is the order of accuracy. (2) The number of intervals N_i in the grid, or elements for a finite element method, is related to the cell size by $N_i = \mathcal{O}(h^{-d})$, where d is the spatial dimension of the problem. (3) Higher order of accuracy is achieved by increasing the stencil, or the number of unknowns per element, $N_s = \mathcal{O}(p^d)$. Thus the total number of operations or unknowns, N, scales as $N = N_i N_s = \mathcal{O}((p/h)^d)$. (4) The operation count, W, required to solve the discrete problem scales as $W = \mathcal{O}(N^w)$, where w is the complexity of the discretization method. (5) The total time required for the numerical solution is $T = W/F$, where $1/F$ is the time for a single operation, which depends on the processor speed. These assumptions lead to the conclusion that the computational time required to achieve a specific error tolerance E scales as

$$T = \mathcal{O}\left((p/E^{1/p})^{wd}/F\right)$$

or taking the logarithm

$$\log T \approx wd\left(-\frac{1}{p}\log E + \log p \right) - \log F$$

For stringent accuracy requirements ($E \ll 1$), it is expected that the term $\log E$ dominates the $\log p$ terms. Therefore, the computational time will depend exponentially on the order of accuracy, p, the complexity of the method, w, and the grid resolution d. The above reasoning demonstrates that, from a practical point of view, it pays off to improve the order of accuracy provided that the operation count does increase dramatically. Furthermore, from the first equation it is evident that small changes of the ratio w/p can reduce computational time, which scales only inversely by the processor speed F.

The most widely used flow solvers for aerospace applications are based on the solution of the Reynolds Averaged Navier-Stokes (RANS) equations. These RANS methods have recently benefited from improved performance of workstations, supercomputers, and parallel processors. Hardware improvements combined with advances in areas such as grid generation and more computationally efficient solution algorithms made RANS a tool that often complements wind tunnel. RANS flow solvers have proven useful for airloads prediction, examination of detailed and gross flow features, as well as engineering design. The primary limitation of RANS is the inability to predict turbulent, separated flow. Turbulence models used in RANS model the entire spectrum of turbulence and they are unable to accurately predict phenomena dominated by turbulent eddies of massively separated flows. The deficiency of RANS to predict turbulent separated flow combined with the un-physical diffusion of vorticity by the numerical scheme in flow regions away from the wall, where the grid density is small, is responsible for inaccuracies of the computed flow field.

Attempts to overcome these deficiencies of RANS methods to accurately compute turbulent separated flow have led to the development of Large Eddy Simulation (LES) [66], [64], [123], [14], [65], [160], and more recently the Detached Eddy Simulation (DES) approach that was proposed by Spalart [170] and applied to complex separated flow computations [136], [172]. Both of these techniques, fully resolve the three-dimensional vortical structures and turbulent motions in detached flow regions. In addition, LES methods also resolve certain range of the small-scale turbulent flow structures in the attached or separated flow boundary layers. DES methods, on the other hand, model the attached flow regions with RANS techniques. As a result, the resolution requirements of DES are significantly lower than LES for the attached boundary layers, where the RANS turbulence model is used. In the separated flow field, however, the resolution requirements for DES are comparable to those for LES calculations. Like the RANS, LES and DES can also benefit from the application of high-order numerical algorithms to better resolve separated flow regions. Most of the high order numerical methods that we will discuss in these notes are therefore expected to have applications to RANS, LES and DES flow solvers.

In particular, the severe accuracy requirements that are encountered in typical LES of complex flows can be benefited from the application of high-order numerical methods. For LES of flows in simple domains, the resolution problem is solved with the use of the highly accurate spectral methods [26] and [111]. Spectral methods are not, however, easy to apply for complex domains and compressible flows with discontinuities. For flows with complex geometries that preclude the use of spectral methods, the use of high-order numerical methods is a necessity in order to minimize the overall computational cost. The continued increase of computing power and developments of parallel computers are expected to make possible DES or even LES of flows with increased complexity. For these applications, use of high-order accurate CFD methods is necessary.

While the use of high-order accurate numerical methods is relatively common for direct numerical simulations (DNS) and LES methods, high-order numerical methods have not been commonly used for DES. As a result, DES of complex flows often exhibit significant grid dependence. It is expected that use of high-order spatially accurate numerical schemes combined with improvements of DES models [133] will eliminate the grid dependencies in DES calculations. We hope that the resulting high-order RANS/DES hybrid models will yield reliable predictions of realistic, separated, unsteady flows in a computationally efficient manner.

15.2   Prior Work on High-Order Numerical Methods

The favorable effect of high order methods was early recognized even in several RANS numerical investigations of delta wing vortical flows [51], wind tip vortices [44], and helicopter rotors [173]. The sensitivity of the numerical solution on grid resolution [51], [173], and the order of accuracy of the numerical scheme [51], [44] on the resolution of vortices was demonstrated. Delta wing flows are dominated by the leading edge vortices and accurate capturing of the correct strength of the leading edge vortex strength is essential for the prediction of loads and the vortex breakdown location. Figure 15.2 shows the effects of increased grid resolution and high-order of accuracy for the computation of the leading edge vortices over a double delta wing [51]. It appears that the increase in order of accuracy can yield better resolution of the vortices even with lower grid resolution. Similar conclusions are reported for the numerical prediction of the wing-tip vortex [44]. In this study, increased grid

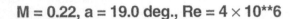

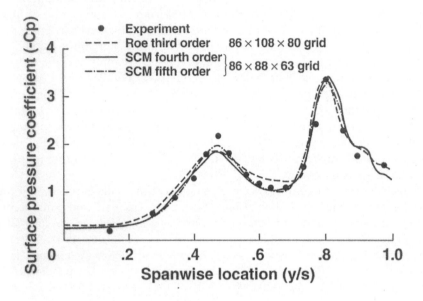

FIGURE 15.2: Computed pressure distributions with different grid densities and order of accuracy; (from Ref. [51]).

density and higher-order numerical schemes significantly improved the accuracy of the final results. The comparisons of the predicted peak velocity in the vortex core of Fig. 15.3 show that in addition to turbulence model the order of the scheme plays a very significant role.

Furthermore, computations of rotorcraft flows [173] and [174] demonstrated that the vortex is severely diffused by the first passage even in computations performed with 10 million grid points. These computations used curvilinear structured meshes and suffered from an inappropriate placement of the grid points and insufficient grid resolution for the tip vortex. This problem was addressed by the tetrahedral unstructured flow solver approach [174]. It was concluded, however, that adaptive tetrahedral mesh approaches could have only limited success for even the simplest hovering rotor cases because the adapted grid is anisotropic with computational cells of very large aspect ratio. The overset grid approach [6] appears to resolve some of the problems associated with grid topology but it still requires very high grid resolution (greater than 60 million grid points) for accurate capturing of the tip vortex. On the other hand, calculations performed with high-order accurate schemes

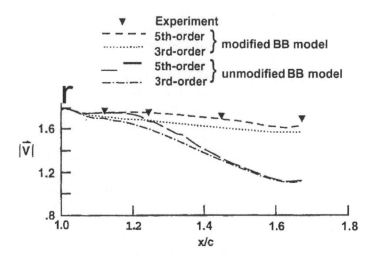

Fig. 3a   **Peak velocity magnitude at vortex core.**

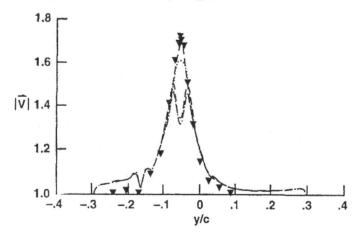

FIGURE 15.3: Effects of third- and fifth-order accurate differencing on the solution for a wake-case grid having 371,000 points (35 x 103 x 103); (from Ref. [44]).

in [145], [190] and [76] and more recently in [181] demonstrated that significant improvements in the accuracy of rotor performance predictions can be obtained with just a small increase in computational cost.

For the computation of flows with shocks, methods designed to regularize the numerical solution have been studied since the early attempts of von Neumann and Richtmyer who used finite-difference techniques combined with the so-called artificial viscosity or numerical dissipation. Use of numerical dissipation in the finite difference and finite volume context has found widespread application in solution methods for compressible aerodynamic flows. The main difficulty in the application of these methods in DNS, DES, and LES of compressible flows is the control of numerical dissipation necessary to capture discontinuities that occur in such flows. Too much numerical dissipation smears out important flow field features. Too little numerical dissipation yields unstable solutions.

In previous numerical investigations [118], [61], it was shown that even the reduced numerical dissipation of high-order, shock-capturing schemes can lead to significant damping of turbulence fluctuations and mask the effects of the subgrid-scale (SGS) models. For these cases, a local application of the shock-capturing scheme was found absolutely necessary in order to minimize numerical dissipation. In the study of [118], for example, this requirement was achieved by means of the application of an essentially non-oscillatory (ENO) scheme only in the shock-normal direction and over a few mesh points around the mean shock position. Unfortunately, in most cases, the shock position is unknown and one needs to introduce a sensor to detect possible discontinuities.

In recent years, efforts were made to alleviate the effects of numerical diffusion introduced by shock capturing schemes and upwind methods. It was shown [60], [155], [55], and [203] that for compressible flows without discontinuities, high-order centered finite-difference schemes are sufficiently stable and accurate for the computation of convecting vortical structures and aeroacoustic disturbances when they are combined with explicit, spectral-type filtering for numerical stability. In addition, centered schemes were found particularly useful for prediction of noise sources. These high-order centered schemes, which can be extended into finite volume context, and the spectral-type or characteristic based filters, which are necessary to stabilize centered schemes and suppress spurious modes, are summarized in the following sections.

For high speed flows with shocks, numerical solutions that are uniformly high order accurate up to the discontinuity can be obtained with ENO [164] and weighted ENO (WENO) [106], [15] schemes. These methods are presented in detail in Chapter 18. On the other hand, it was shown that explicit filters [55] could also be used with high-order centered schemes to obtain shock-capturing. Application of explicit filters [55] yielded improved computational efficiency of flows with discontinuities compared to ENO [164] or WENO methods [106] and [15]. Explicit filters can be easily implemented into existing codes because the filter step is essentially independent of the basic differencing scheme and is applied as post processing. In the same spirit, Yee et al. [204] showed that the dissipative part of a shock-capturing scheme could be applied after each time step to regularize the numerical solution and acts like a filter. Moreover, to meet the requirement of a local application of the numerical dissipation, the amplitude of the dissipation is evaluated with a sensor derived from the artificial compression method (ACM) of Harten [77]. The filters of Yee et al. [204] are referred to as characteristic-based filters and they are summarized in Chapter 18. The numerical test in Yee et al. [204] used total variation diminishing (TVD) schemes to construct the characteristic-based, non-linear filter. The possibility of using high-order non-linear filters based on essentially non-oscillatory (ENO) reconstruction has been demonstrated in [62].

In parallel with the finite-difference and finite-volume methods, which found widespread application in CFD, finite element methods [110], [13] and [91] were also used for convection-dominated problems. Application of the finite element method is far from trivial for non-linear convective problems, such as compressible flow with discontinuities. For such cases, the finite element numerical solution must capture the physically relevant discontinuities without introducing spurious oscillations. This issue is addressed successfully by the discontinuous Galerkin method [39]. The discontinuous Galerkin method (DG) assumes discontinuous approximate solutions. Subsequently, it treats discontinuities in a manner analogous to high-resolution finite-difference and finite-volume methods for nonlinear hyperbolic systems by incorporating suitably defined numerical fluxes and slope limiters into the finite element framework. Due to its local character, the DG finite element method is very suitable for local grid refinement and becomes highly parallelizable. This method is also presented and analyzed in Chapter 19.

Recently, another type of high-order accurate, conservative and computationally efficient scheme was introduced for the solution of conservation laws in unstructured grids. This scheme is known as the spectral volume (SV) method [193]. The concept of a "spectral volume" was introduced to achieve high-order accuracy in an efficient manner similar to spectral methods and at the same time retain the benefits of the finite volume formulations for problems with discontinuities. In the SV method, each spectral volume, which is the same as the traditional triangular or tetrahedral finite volume, is further subdivided into volumes called control volumes. Cell-averaged data from these control volumes are used to reconstruct a high-order approximation in the spectral volume, while Riemann solvers are used to compute the fluxes at the spectral volume boundaries. The main difference of the SV method and DG method is that for the SV method, the cell-averaged variables in the control volumes are updated independently. Similar to the DG method, the SV method uses total variation diminishing or total variation bounded limiters to eliminate/reduce spurious oscillations near discontinuities. A very desirable feature of the SV method is that the reconstruction is carried out analytically, and does not involve large stencils in contrast to the computationally intensive reconstruction in high-order finite volume methods.

The DG and SC methods are both suitable for high-order discretization of complex domains using unstructured meshes. In addition, they are fully conservative due to the use of Riemann fluxes across element boundaries. Another high-order conservative scheme for unstructured quadrilateral grids is the multidomain spectral method on a staggered grid recently developed by Kopriva and Kolias [112], [113], [111] [114]. The multidomain spectral method is similar to the spectral element method by Patera [139]. The spectral element method is high order accurate and more flexible compared to spectral methods for discretizations of complex domains. It is however not conservative. Although very high order of accuracy can be achieved with both the multidomain spectral and the spectral element method, these methods are difficult to extend to other cell types such as triangles or tetrahedral cells [86]. These spectral methods and other recent, less widely used, high order methods [2], [28] will not be presented here. Further information about these methods can be found in the original references. A detailed presentation of the theory and implementation of the spectral element method can also be found in the recent book by Karniadakis and Sherwin [110].

These notes are organized as follows: The governing equations in differential and integral form are presented in Chapter 16. In Chapter 17, high-order accurate discretization with finite-difference and finite-volume centered methods is presented. In Chapter 18, ENO and WENO reconstruction is explained and WENO and ENO high-order schemes are presented. The discontinuous Galerkin method is presented and analyzed in Chapter 19. At the end of each chapter selected examples from the application of the high-order methods are shown.

Chapter 16

Governing Equations

John A. Ekaterinaris, School of Mechanical and Aerospace Engineering, University of Patras, Greece *and* Foundation for Research and Technology-Hellas, Institute of Applied and Computational Mathematics, 71110 Heraklion, Greece, ekaterin@iacm.forth.gr

---

16.1 Governing Equations

In the following chapters we present high-order methods for high-order accurate in space discretizations of the compressible flow governing equations. These are the full Navier-Stokes (N-S) equations which govern non-linear fluid dynamics and aeroacoustics. For the majority of compressible flow simulations, these equations are cast in the strong conservation form [8]. Numerical solutions based on the Galerkin/least-square method, use different sets of variables [84]. This is not, however, true for finite difference and finite volume numerical methods traditionally used in aerodynamics. These methods use the conservative flow variables to ensure conservation and discontinuity capturing. The primitive variable formulation was used rarely in aerodynamics either with shock fitting schemes [31], or for the computation of subsonic compressible flows without discontinuities [147]. The transformation from conservative to primitive variables or other sets of variables is obtained by multiplying the conservative variables vector with the appropriate flux Jacobian. For example, primitive variables are obtained by multiplying with $M = \partial U / \partial V$, $U = [\rho, \rho u, \rho v, E_T]$, $V = [\rho, u, v, p]$ see [87] for more details.

16.2 Conservative Form of the N-S Equations

The conservative variable formulation of the N-S equations in divergence form is:

485

$$\frac{\partial \mathbf{U}}{\partial t} + \nabla \cdot \mathbf{F} = \frac{1}{Re} \nabla \cdot \mathbf{F_v} \qquad (16.1)$$

where $\mathbf{U}$ is the conservative variable vector, $\mathbf{F}$ is the inviscid flux vector, and $\mathbf{F_v}$ is the viscous flux vector. In Cartesian coordinates Eq. (16.1) is:

$$\frac{\partial U}{\partial t} + \frac{\partial F}{\partial x} + \frac{\partial G}{\partial y} + \frac{\partial H}{\partial z} = \frac{1}{Re}\left[\frac{\partial F_v}{\partial x} + \frac{\partial G_v}{\partial y} + \frac{\partial H_v}{\partial z}\right] \qquad (16.2)$$

with

$$U = \begin{bmatrix} \rho \\ \rho u \\ \rho v \\ \rho w \\ E_T \end{bmatrix}, \ F = \begin{bmatrix} \rho u \\ \rho u^2 + p \\ \rho uv \\ \rho uw \\ (E_T + p)u \end{bmatrix}, \ G = \begin{bmatrix} \rho v \\ \rho uv \\ \rho v^2 + p \\ \rho vw \\ (E_T + p)v \end{bmatrix}, \ H = \begin{bmatrix} \rho w \\ \rho uw \\ \rho vw \\ \rho w^2 + p \\ (E_T + p)w \end{bmatrix}$$

where E_T is the total energy $E_T = \rho E = \rho[e + (u^2 + v^2 + w^2)/2]$ related to pressure through the equation of state for a perfect gas $p = (\gamma - 1)\,[\rho E - \rho\,(u^2 + v^2 + w^2)/2]$, and the components of the viscous flux vector are given by:

and

$$F_v = [0, \tau_{xx}, \tau_{xy}, \tau_{xz}, u\tau_{xx} + u\tau_{xy} + w\tau_{xz} + q_x]^T$$
$$G_v = [0, \tau_{xy}, \tau_{yy}, \tau_{xz}, u\tau_{xy} + v\tau_{yy} + w\tau_{yz} + q_y]^T$$
$$H_v = [0, \tau_{xz}, \tau_{yz}, \tau_{zz}, u\tau_{xz} + v\tau_{yz} + w\tau_{zz} + q_z]^T$$

$$\tau_{xx} = \tfrac{2}{3}\mu\left(2\tfrac{\partial u}{\partial x} - \tfrac{\partial v}{\partial y} - \tfrac{\partial w}{\partial z}\right) \quad \tau_{xy} = \mu\left(\tfrac{\partial u}{\partial y} + \tfrac{\partial v}{\partial x}\right) \quad q_x = -k\,\tfrac{\partial T}{\partial x}$$
$$\tau_{yy} = \tfrac{2}{3}\mu\left(2\tfrac{\partial v}{\partial y} - \tfrac{\partial u}{\partial x} - \tfrac{\partial w}{\partial z}\right) \quad \tau_{xz} = \mu\left(\tfrac{\partial w}{\partial x} + \tfrac{\partial u}{\partial z}\right) \quad q_y = -k\,\tfrac{\partial T}{\partial y} \ .$$
$$\tau_{zz} = \tfrac{2}{3}\mu\left(2\tfrac{\partial w}{\partial x} - \tfrac{\partial u}{\partial x} - \tfrac{\partial v}{\partial y}\right) \quad \tau_{yz} = \mu\left(\tfrac{\partial v}{\partial z} + \tfrac{\partial w}{\partial y}\right) \quad q_z = -k\,\tfrac{\partial T}{\partial z}$$

Numerical solutions with the finite volume (FV) method are obtained with the integral form of the NS equations for a control volume Ω with boundary $\partial\Omega$

$$\frac{\partial}{\partial t}\int_\Omega U\,dx + \oint_{\partial\Omega}[F(\mathbf{U},\mathbf{n}) - F_v(U,\ \nabla U,\ \mathbf{n})]\,dS = 0 \qquad (16.3)$$

where

$$\mathbf{U} = [\rho, \rho\mathbf{V}, \rho E]^T, F(\mathbf{U},\ \mathbf{n}) = \mathbf{V}\cdot\mathbf{n}U\ , \ F_v(\mathbf{U},\ \nabla\mathbf{U},\ \mathbf{n}) = [0, \mathbf{t}, \mathbf{t}\cdot\mathbf{V} - \mathbf{q}\cdot\mathbf{n}]^T$$

with **t** and **q** representing the stress and heat flux vectors, $\mathbf{V} = (u, v, w)^T$ is the velocity vector, and **n** is the outward unit normal vector to the boundary $\partial\Omega$. For flows with discontinuities the integral form holds and any discrete scheme must obey both local and global conservation in order to capture correctly weak solutions [117]. Finite volume (FV) methods gained popularity in aerodynamics because they make possible numerical solutions in complex domains with unstructured or mixed type grids [187]. During the 1980s upwind mechanisms were introduced into FV algorithms leading to increased robustness of the FV method for applications in flow that include strong shocks and provided better resolution of viscous layers due to the decrease of numerical dissipation compared to FV methods that employed artificial dissipation [102], [103].

Finite-element (FE) methods uses the weak form of Eq. (16.1), gained popularity in aerodynamics with the development of the discontinuous Galerkin method [39] and the stabilized finite element methods [94], [92], [93]. The weak form is obtained by multiplying the strong form of Eq. (16.1) by a test function W^t and integrating over the domain. The weak form of Eq. (16.1) is

$$\frac{\partial}{\partial t} \int_\Omega W^t \mathbf{U} \, d\omega + \oint_{\partial\Omega} W^t [F(\mathbf{U}, \mathbf{n}) - F_v(U, \nabla U, \mathbf{n})] \cdot \mathbf{n} \, dS$$
$$- \int_\Omega \nabla W^t [F(\mathbf{U}, \mathbf{n}) - F_v(U, \nabla U, \mathbf{n})] \, d\Omega = 0.$$

The integral form, Eq. (2.3), is the weak form for unity weight function. The stabilized finite element methods augment the weak form of the governing equations with stabilization terms [92], [93], [162].

The real power of unstructured grid methods with FV or FE discretization is their ability to adapt not only to complex geometries but also to solution features without being constrained by considerations such as grid structure, orthogonality or topology. However, anisotropic grid adaptation for three-dimensional, high Reynolds number flows, which contain both discontinuities and smooth but complex flow features, such confluent boundary layer and wake roll-ups in high-lift systems, remains a challenge. Grid adaption for these flows, which are common in aerodynamic applications, combined with high-order methods offers the additional possibility of hp-type refinement [110] [167]. hp Refinement uses high grid density (h-refinement) at the neighborhood of discontinuities, while for the resolution of smooth but complex flow features and wave propagation increases the order of accuracy of the numerical solution (p-refinement) and uses a relatively coarse canonical mesh.

Finite difference methods on structured type of grids are used for the numerical solution of the governing equations because of their efficiency. The major difficulty with finite-difference methods is structured-type grid generation. The task of generating structured grids over complex configurations is still a serious challenge even with the multiblock structured grid approach or with the powerful approach of of overlapping or Chimera-type grids [23], where structured grids generated about different simple components are allowed to overlap. Numerical solutions in complex domains with finite-difference methods on body-fitted deformed meshes [171] are obtained by expressing, Eq. (16.1) (see Anderson et al., 1984 [8]) in terms of a generalized non-orthogonal curvilinear coordinate system (ξ, η, ζ) using general transformations $\xi = \xi\,(x, y, z)$, $\eta = \eta\,(x, y, z)$, $\zeta = \zeta\,(x, y, z)$ as follows:

$$\frac{\partial}{\partial t}\left(\frac{U}{J}\right) + \frac{\partial \widehat{F}}{\partial \xi} + \frac{\partial \widehat{G}}{\partial \eta} + \frac{\partial \widehat{H}}{\partial \zeta} = \frac{1}{Re}\left[\frac{\partial \widehat{F_v}}{\partial \xi} + \frac{\partial \widehat{G_v}}{\partial \eta} + \frac{\partial \widehat{H_v}}{\partial \zeta}\right] \qquad (16.4)$$

Here $U = [\rho, \rho u, \rho v, \rho w, \rho E]$ denotes the solution vector, and $J = \partial\,(\xi, \eta, \zeta)/\partial\,(x, y, z)$ is the Jacobian of the geometric transformation. The inviscid fluxes $\widehat{F}$, $\widehat{G}$ and $\widehat{H}$ are:

$$\widehat{F} = \begin{bmatrix} \rho \widehat{U} \\ \rho u \widehat{U} + \xi_x p \\ \rho v \widehat{U} + \xi_y p \\ \rho w \widehat{U} + \xi_z p \\ (\rho E + p)\widehat{U} - \widehat{\xi_t} p \end{bmatrix}$$

$$\widehat{G} = \begin{bmatrix} \rho \widehat{V} \\ \rho u \widehat{V} + \eta_x p \\ \rho v \widehat{V} + \eta_y p \\ \rho w \widehat{V} + \eta_z p \\ (\rho E + p)\widehat{V} - \widehat{\eta_t} p \end{bmatrix}$$

$$\widehat{H} = \begin{bmatrix} \rho \widehat{W} \\ \rho u \widehat{W} + \zeta_x p \\ \rho v \widehat{W} + \zeta_y p \\ \rho w \widehat{V} + \zeta_z p \\ (\rho E + p)\widehat{W} - \widehat{\zeta_t} p \end{bmatrix}$$

where

$$E = \frac{T}{(\gamma - 1)M_\infty^2} + \frac{1}{2}\left(u^2 + v^2 + w^2\right)$$

and $\widehat{U}$, $\widehat{V}$, and $\widehat{W}$ are the contravariant velocity components given by

$$\widehat{U} = \xi_t + \xi_x u + \xi_y v + \xi_z w$$
$$\widehat{V} = \eta_t + \eta_x u + \eta_y v + \eta_z w$$
$$\widehat{W} = \zeta_t + \zeta_x u + \zeta_y v + \zeta_z w.$$

In Eq. (16.4), $\xi_x = J^{-1} \, \partial\xi/\partial x$ are the metrics of the transformation with similar definitions for the other metric quantities. The viscous stress terms in transformed coordinates are:

$$\tau_{xx} = \frac{2}{3}\,\mu\,[2\,(\xi_x u_\xi + \eta_x u_\eta + \zeta_x u_\zeta) - (\xi_y v_\xi + \eta_y v_\eta + \zeta_y v_\zeta) - (\xi_z w_\xi + \eta_z w_\eta + \zeta_z w_\zeta)]$$

$$\tau_{yy} = \frac{2}{3}\,\mu\,[2\,(\xi_y v_\xi + \eta_y v_\eta + \zeta_y v_\zeta) - (\xi_x u_\xi + \eta_x u_\eta + \zeta_x u_\zeta) - (\xi_z w_\xi + \eta_z w_\eta + \zeta_z w_\zeta)]$$

$$\tau_{zz} = \frac{2}{3}\,\mu\,[2\,(\xi_z w_\xi + \eta_z w_\eta + \zeta_z w_\zeta) - (\xi_x u_\xi + \eta_x u_\eta + \zeta_x u_\zeta) - (\xi_y v_\xi + \eta_y v_\eta + \zeta_y v_\zeta)]$$

$$\tau_{xy} = \mu\,(\xi_y u_\xi + \eta_y u_\eta + \zeta_y u_\zeta + \xi_x v_\xi + \eta_x v_\eta + \zeta_x v_\zeta)$$

$$\tau_{xz} = \mu\,(\xi_z u_\xi + \eta_z u_\eta + \zeta_z u_\zeta + \xi_x w_\xi + \eta_x w_\eta + \zeta_x w_\zeta)$$

$$\tau_{yz} = \mu\,(\xi_z v_\xi + \eta_z v_\eta + \zeta_z v_\zeta + \xi_y w_\xi + \eta_y w_\eta + \zeta_y w_\zeta)$$

$$q_x = -k\,(\xi_x T_\xi + \eta_x T_\eta + \zeta_x T_\zeta)$$
$$q_y = -k\,(\xi_y T_\xi + \eta_y T_\eta + \zeta_y T_\zeta)$$
$$q_z = -k\,(\xi_z T_\xi + \eta_z T_\eta + \zeta_z T_\zeta).$$

For turbulent flow calculations, the molecular viscosity μ is replaced by the turbulent eddy viscosity $\mu + \mu_\tau$. The turbulent eddy viscosity μ_τ is obtained from turbulence models developed during the last decades. These turbulence models range from simple algebraic models to one- and two-equation turbulence models, or more sophisticated Reynolds stress models [197].

Chapter 17

High-Order Finite-Difference Schemes

John A. Ekaterinaris, School of Mechanical and Aerospace Engineering, University of Patras, Greece *and* Foundation for Research and Technology-Hellas, Institute of Applied and Computational Mathematics, 71110 Heraklion, Greece, ekaterin@iacm.forth.gr

---

17.1 High-Order Finite-Difference Schemes

In the past, efforts were made towards developing high-order finite-difference (FD) methods in the areas direct and large eddy simulations. For nonlinear problems, straightforward application of high-order accurate central difference schemes is not possible, because the spurious modes that develop from the unresolvable by the numcrical discretization high frequency modes lead to instabilities. Rai and Moin [146] found that high-order upwind schemes are more promising to simulate turbulent flows. However, early attempts to apply high-order finite differences were often frustrated because of lack of robustness of the proposed high-order (FD) schemes compared to spectral methods. For example, it was found [146] and references therein) that:

1. For DNS with energy conserving FD schemes the actual cnergy transfer toward large wave numbers in the simulation is too large;

2. Spectral methods have the advantage to signal their accuracy or inaccuracy through an inadequately resolved enstrophy dissipation spectrum whereas FD schemes do not and;

3. Spectral methods require roughly half as much resolution as FD schemes in each spatial direction to yield solutions of comparable accuracy.

In spite of these differences, some success was achieved for the computation of incompressible [146] and compressible [147] flows with high order, upwind FD schemes. The fifth-order accurate derivatives in [146] were computed with upwind-biased formulas based on the sign of the velocity as:

$$u_i = -\frac{1}{120} \left[+ 6u_{i+2} + 60u_{i+1} + 40u_i \right.$$
$$\left. - 120u_{i-1} + 30u_{i-2} - 4u_{i-3} \right], \qquad \text{for} \quad u_i > 0 \quad (17.1)$$

$$u_i = +\frac{1}{120} \left[+ 4u_{i+3} - 30u_{i+2} + 12u_{i+1} \right.$$
$$\left. - 40u_i - 60u_{i-1} + 6u_{i-2} \right], \qquad \text{for} \quad u_i < 0. \quad (17.2)$$

Upwinding alleviated some of the problems encountered with centered schemes and yielded some promising results for both incompressible [146] and compressible flow [147] direct numerical simulations.

Upwind biased, high-order accurate stencils (up to fifth-order) for the evaluation of the derivatives were also used successfully for the computation of complex incompressible flows [157], [53] and wave propagation [52]. Upwind biased schemes, however, based only on formal accuracy (truncation error) inherently introduce some form of artificial smoothing that makes them inappropriate for long time integration and direct simulation or large eddy simulation or turbulence.

The first systematic attempt to develop high-order accurate finite difference schemes appropriate for problems with a wide range of scales was presented by Lele [119]. Compared to the traditional finite-difference approximations the compact schemes presented by Lele [119] provided a better representation of the short length scales. As a result, compact high-order schemes are closer to spectral methods and at the same time maintain the freedom to retain accuracy in complex stretched meshes. Emphasis in the development of compact schemes was given on the resolution characteristics of the difference approximations rather than formal accuracy (i.e., truncation error). The notion of resolution was quantified by Lele [119] using a Fourier analysis of the differencing scheme [188], [189]. This analysis compares the resolving power of different schemes based on a more general notion of intervals per wavelength of Kreiss and Oliger [116]. Using these ideas, Lele [119] analyzes the resolution characteristics of the schemes based on the accuracy with which the difference approximation represents the exact result over the full range of length scales that can be realized for a given mesh.

17.2   Explicit Centered High-Order FD Schemes

Application of Taylor series expansion yields centered, explicit and compact finite difference formulas for space differentiation on equally spaced meshes. Central-difference schemes gained popularity in the simulation of wave propagation phenomena because in contrast to upwind methods, which in addition to dispersion naturally introduce artificial dissipation, the dominant error in centered discretizations is only depressive. The stencils for the fourth-, sixth-, and eight-order accurate symmetric, explicit, centered schemes are given by:

4th order (E4)

$$\frac{\partial u}{\partial x}\bigg|_i = \frac{1}{12\Delta x}\left(u_{i+2} + 8u_{i+1} - 8u_{i-1} - u_{i-2}\right) + \mathcal{O}\left(\Delta x^4\right) \qquad (17.3)$$

6th order (E6)

$$\frac{\partial u}{\partial x}\big|_i = (u_{i+3} - 9u_{i+2} + 45u_{i+1}$$
$$- 45u_{i-1} + 9u_{i-2} - u_{i-3})/60\Delta x + \mathcal{O}\left(\Delta x^6\right). \qquad (17.4)$$

8th order (E8)

$$\frac{\partial u}{\partial x}\big|_i = (- u_{i+4} + \frac{32}{3}\,u_{i+3} - 56u_{i+2} + 224u_{i+1}$$
$$+ u_{i-4} - \frac{32}{3}\,u_{i-3} + 56u_{i-2} - 224u_{i-1})/280\Delta x$$
$$+ \mathcal{O}\left(\Delta x^8\right) \qquad (17.5)$$

Explicit high order finite difference formulas of Eqs. (17.3)–(17.4) are often used to discretize the second derivatives of the viscous terms by taking the first derivative twice. In order to reduce the stencil width the inner derivative is evaluated at half points.

17.3   Centered Compact Schemes

Of particular interest in recent applications has been the class of centered schemes that require small stencil support. These "compact" schemes can also be derived from Taylor series expansions and compute simultaneously the derivatives along an entire line in a coupled fashion. The main advantage of compact schemes is simplicity in boundary condition treatment and smaller truncation error compared to the their noncompact counterparts (see, for example, Eqs. (17.3)–(17.5)) of equivalent order. Compact schemes, however, are more intensive computationally compared to explicit schemes because they require matrix inversion.

A seven-point wide stencil, finite-difference discretization of a spatial derivative f' of a scalar pointwise discrete quantity, f, in the governing equations, such as metric terms or flow variables, is obtained in the computational domain on an equally spaced mesh by:

$$Bf'_{j+2} + Af'_{j-1} + f'_j + Af'_{j+1} + Bf'_{j+2} =$$
$$a\,\frac{f_{j+1} - f_{j-1}}{2h} + b\,\frac{f_{j+2} - f_{j-2}}{4h} + c\,\frac{f_{j+3} - f_{j-3}}{6h} \tag{17.6}$$

where A, B, a, b and c determine the spatial accuracy of the discretization. Different values of the coefficients in the formula of Eq. (17.6) yield schemes of different accuracy ranging from the fourth-order explicit method (E4) to the compact tenth-order accurate scheme (C10). The values of the coefficients in Eq. (17.6) for schemes of different order of accuracy are shown in Table I.

TABLE 17.I

Schemes with five-point stencil of Eq. (17.6)

Scheme	A	B	a	b	c
E4	0	0	4/3	-1/3	0
C4	1/4	0	3/2	0	0
C6	1/3	0	14/9	1/9	0
C8/5	4/9	1/36	40/27	25/54	0
C8/3	3/8	0	75/48	1/5	-1/80
C10	1/2	1/20	17/12	101/150	1/100

In Table 17.I, C8/3 refers to the eight-order compact scheme that requires tridiagonal matrix inversion and C8/5 refers to the eight-order compact scheme that requires pentadiagonal matrix inversion.

Compact approximations for the second derivative are obtained from the following general form

$$Bf''_{j-2} + Af''_{j-1} + f''_j + Af''_{j+1} + Bf''_{j+2} =$$

$$= \frac{a}{h^2} \left(f_{j+1} - 2f_j + f_{j-1} \right)$$

$$+ \frac{b}{4h^2} \left(f_{j+2} - 2f_j + f_{j-2} \right)$$

$$+ \frac{c}{9h^2} \left(f_{j+3} - 2f_j + f_{j+3} \right).$$

$$(17.7)$$

The main advantage of compact schemes over their equivalent order explicit schemes is better resolution in wave-space as it will be discussed later. Explicit schemes may be used for the evaluation of the viscous terms because application of the compact schemes is more expensive computationally. Furthermore, there are no very significant improvements in wave-space resolution with the use of compact schemes [119].

The wave space resolution of various explicit and compact schemes is obtained using Fourier analysis for Eqs. (17.6) or (17.7). Considering that the exact result is a sinusoidal function $f_j = e^{ikh}$, where $h = \Delta x_j$ is a uniform grid spacing and k is the wave number, the exact value of the derivative f' is $f'_j = ik f_j$. On the other hand, the derivative computed with finite difference formulas is given by $\widehat{f'_j} = i\widehat{k} f_j$, where $\widehat{k}$ the modified wave number, which depends on the form of the finite-difference formulas used for the evaluation of the first-order derivative. The difference between the true wave number k and the modified wave number $\widehat{k}$ is a measure of the scheme's resolving ability. The modified wave number of various finite difference schemes can be obtained using standard shift operators $f'_{j\pm n} = f'_j e^{\pm ik}$. For example, the modified wave number of the fourth-order accurate explicit scheme of Eq. (17.6) is given by $\widehat{k} = i(8\sin k - \sin 2k)/6$, with analogous expression for the other methods. A comparison of the modified wave numbers of the first derivative for several central compact and non-compact schemes is shown in Fig. 17.1. Using the scaled wave number $\omega = 2\pi kh/\lambda$, where λ is the wavelength, it is found that the number of intervals or grid points per wavelength is $2\pi k/\omega$. Therefore, the lower the scheme's resolving ability the higher is the number of points per wavelength required to resolve accurately certain predetermined portion of the range $[0, 2\pi]$.

Stable and accurate formulas must be used for the boundary points for the evaluation of the derivatives with explicit and compact FD schemes [27]. These boundary closures are given in the next chapter. In addition, the numerical solutions of nonlinear hyperbolic equations with central-difference methods develop spurious modes arising from unresolvable scales and inaccuracies in the application of boundary conditions. Spectral-type [60] or characteristic-based [204] filters that can be used to stabilize numerical solutions performed with central-difference methods are also presented in the following sections.

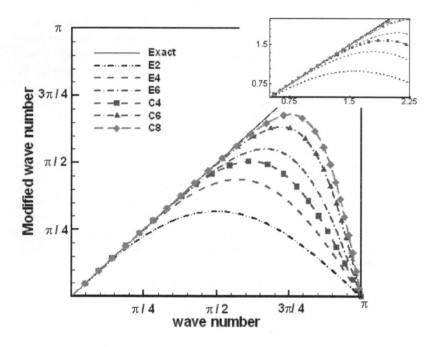

FIGURE 17.1: Wave space resolution of explicit and compact centered schemes for the first order derivative.

17.4   Boundary Closures of High-Order Schemes

The primary difficulty in using higher order schemes is identification of stable boundary schemes that preserve their formal accuracy. Boundary closures for various explicit and compact high-order centered schemes were presented by Carpenter et al. [27]. The stability characteristics of compact fourth- and sixth-order spatial operators with boundary closures were assessed [27] with the theory of Kreiss [115] and Gustafsson, Kreiss and Sundstrom (GKS) [72] for the semidiscrete initial value problem.

Numerical solutions of hyperbolic systems preserve their formal spatial accuracy when an Nth-order inner scheme is closed with at least an $(N-1)$th-order boundary scheme. Furthermore, determination of the numerical stability of a fully discrete approximation (including boundary schemes) for a linear hyperbolic partial differential equation is a difficult task. Fourier techniques are not straight forward to apply and do not provide sufficient conditions for numerical stability. Gustafsson et al. [72] developed stability analysis techniques based on normal modal analysis. The theory of [72], generally referred to as GKS stability theory, established conditions that the inner and boundary schemes must satisfy to ensure stability. The GKS theory was used by Carpenter et al. [27] to assess the stability of boundary schemes proposed for the fourth- and sixth-order compact schemes. The boundary closure schemes of Carpenter et al. [27] are summarized next.

Consider the finite-difference representation of the continuous derivative $U_x = U'_j$ on an equally spaced mesh. The discrete form of the first derivative U'_j involves functional values U_j, $j = 1, ..., N$ at discrete points. According to [27] for an explicit uniformly fourth-order accurate in space scheme the spatial discretization is obtained as

$$U'_1 = \frac{1}{12\Delta x} \left(-25U_1 + 48U_2 - 36U_3 + 16U_4 - 3U_5 \right)$$

$$U'_2 = \frac{1}{12\Delta x} \left(-3U_1 - 10U_2 + 18U_3 - 6U_4 + U_4 \right)$$

$$U'_j = \frac{1}{12\Delta x} \left(U_{j-2} - 8U_{j-1} + 8U_{j+1} - U_{j+2} \right) \quad j = 3, ..., N-2 \quad (17.8)$$

$$U'_{N-1} = \frac{1}{12\Delta x} \left(-U_{N-4} + 6U_{N-3} - 18U_{N-2} + 10U_{N-1} + 3U_N \right)$$

$$U'_N = \frac{1}{12\Delta x} \left(3U_{N-4} - 16U_{N-3} + 36U_{N-2} - 48U_{N-1} + 25U_N \right).$$

The derivative computed with the fourth-order, explicit, centered scheme of Eq. (17.8) is written in matrix form as

$$\mathbf{U}' = [M]\,\mathbf{U} \tag{17.9}$$

where

$$\mathbf{U}' = \begin{bmatrix} U_1' \\ U_2' \\ \vdots \\ U_j' \\ \vdots \\ U_{N-1}' \\ U_N' \end{bmatrix}, \quad M = \frac{1}{12\Delta x} \begin{bmatrix} -25 & 48 & -36 & 16 & -3 \\ -3 & -10 & 18 & -6 & 1 \\ 1 & -8 & 0 & 8 & -1 & \ddots \\ \ddots & 1 & -8 & 0 & 8 & 1 \\ \ddots & \ddots & 1 & -8 & 0 & 8 & -1 \\ & & -1 & 6 & -18 & 10 & 3 \\ & & 3 & -16 & 36 & -48 & 25 \end{bmatrix},$$

$$U_j = \begin{bmatrix} U_1 \\ U_2 \\ \vdots \\ U_j \\ \vdots \\ U_{N-1} \\ U_N \end{bmatrix}.$$

The fourth order, compact FD scheme for the approximation of the first derivative, U_j', that has narrower stencil than the explicit fourth order scheme requires closures only at $j = 1$ and $j = N$. The fourth order compact scheme with boundary closures is:

$$U_1' + 3U_2' = \frac{1}{6\Delta x}\left(-17U_1 + 9U_2 + 9U_3 - U_4\right)$$

$$U_{j-1}' + 4U_j' + U_{j+1}' = \frac{1}{\Delta x}\left(-3U_{j-1} + 3U_{j+1}\right) \tag{17.10}$$

$$3U_{N-1}' + U_N' = \frac{1}{6\Delta x}\left(U_{N-3} - 9U_{N-2} - 9U_{N-1} + 17U_N\right)$$

or in matrix form

$$[M']\,\mathbf{U}' = [M]\,\mathbf{U} \tag{17.11}$$

where

$$[M'] = \begin{bmatrix} 6 & 18 & & 0 \\ 1 & 4 & 1 & 0 \\ \ddots & \ddots & \ddots \\ 0 & 1 & 4 & 1 \\ 0 & & 18 & 6 \end{bmatrix}, \quad M = \begin{bmatrix} -17 & 9 & 9 & -1 & 0 \\ -3 & 0 & 3 & & 0 \\ \ddots & \ddots & \ddots & \ddots \\ 0 & & -3 & 0 & 3 \\ & & 1 & -9 & -9 & 17 \end{bmatrix}$$

$$\mathbf{U}' = \begin{bmatrix} U_1' \\ U_2' \\ \vdots \\ U_{N-1}' \\ U_N' \end{bmatrix}, \quad \mathbf{U} = \begin{bmatrix} U_1 \\ U_2 \\ \vdots \\ U_{N-1} \\ U_N \end{bmatrix}.$$

The sixth-order compact scheme has a five point-wide stencil and utilizes information from all five points explicitly and three points implicitly (tridiagonal system). Boundary closures must be provided at two points at each end of the domain, e.g., at j=1,2 and at j=N-1, N. To ensure for a GKS stable schemes with sixth-order formulas we require fifth order at the boundaries j=1,2 and j=N-1, N and sixth order in the interior. This is written in shorthand notation as (5, 5-6-5, 5). Formally sixth-order accurate GKS stable schemes are difficult to construct. Therefore, the (3, 5-6-5, 3) and the (4, 5-6-5, 4) schemes may be also used for full discretization with the sixth-order compact scheme. The third order closure at j=1 is:

$$U_1' + 2U_2' = \frac{1}{2\Delta x}\left(-5U_1 + 4U_2 + U_3\right).$$

The fourth-order closure at j=1 is

$$U_1' + 3U_2' = \frac{1}{6\Delta x}\left(-17U_1 + 9U_2 + 9U_3 - U_4\right)$$

and the fifth-order closure at j=2 is accomplished by

$$U_1' + 6U_2' + 3U_3' = \frac{1}{3\Delta x}\left(-10U_1 - 9U_2 + 18U_3 + U_4\right).$$

The matrices M' and M for the sixth order scheme are:

$$M'_{356} = \begin{bmatrix} 2 & 4 & & & & \\ 3 & 18 & 9 & & & \\ & 12 & 36 & 12 & & \\ & & \ddots & \ddots & \ddots & \\ & & 12 & 36 & 12 & \\ & & & 9 & 18 & 3 \\ & & & & 4 & 2 \end{bmatrix}, \quad M'_{456} = \begin{bmatrix} 6 & 18 & & & & \\ 3 & 18 & 9 & & & \\ & 12 & 36 & 12 & & \\ & & \ddots & \ddots & \ddots & \\ & & 12 & 36 & 12 & \\ & & & 9 & 18 & 3 \\ & & & & 18 & 6 \end{bmatrix}$$

$$M_{356} = \begin{bmatrix} -5 & 4 & 1 & & & & \\ -10 & -9 & 18 & 1 & & & \\ -1 & -28 & 28 & 1 & & & \\ & \ddots & \ddots & \ddots & \ddots & & \\ & & -1 & -28 & 28 & 1 & \\ & & & -1 & -18 & 9 & 10 \\ & & & & -1 & -4 & 5 \end{bmatrix}, \quad M_{456} = \begin{bmatrix} -17 & 9 & 9 & 1 & & & \\ -10 & -9 & 18 & 1 & & & \\ -1 & -28 & 28 & 1 & & & \\ & \ddots & \ddots & \ddots & \ddots & & \\ & & -1 & -28 & 28 & 1 & \\ & & & -1 & -18 & 9 & 10 \\ & & & & -1 & -9 & -9 & 17 \end{bmatrix}$$

The two key issues encountered with higher-order finite-difference schemes used in CFD are boundary treatments and grid uniformity. The boundary treatment was carefully addressed by Carpenter et al. [27]. It was found that the effect of boundary closures to the overall resolution is indeed small [4] even for highly accurate DNS. The grid non-uniformity issue is, however, more important and its effect on the overall accuracy of higher-order finite-difference schemes on nonuniform grids was recently assessed in [33].

The use of nonuniform grids in turbulent flow simulations is inevitable. The typical ratio of the maximum to the minimum grid spacing is about 100. The behavior of the second- and fourth-order explicit centered schemes and the fourth- and sixth-order compact schemes was assessed in [33] for smooth stretched grids. It was found that grid quality has stronger effects on the higher-order compact schemes than on the explicit schemes. Furthermore, an accuracy deterioration of higher-order compact schemes with low grid density was observed for nonuniform meshes.

17.5 Compact Schemes for the Simultaneous Evaluation of the First and Second Derivative

A more general version of the standard compact schemes presented by Lele [119] was developed by Mahesh [132]. These schemes are symmetric and Hermitian and differ from standard compact schemes in that the first and second derivatives are evaluated simultaneously. In addition, for the same stencil width the schemes proposed by Mahesh are two orders higher in accuracy, they have significantly better spectral representation, and the computational cost for the evaluation of both derivatives is shown to be essentially the same as standard compact schemes. As a result, the proposed schemes appear to be attractive alternative to standard compact schemes for the Navier-Stokes equations that require second-order derivative evaluation in the viscous terms. The schemes that compute simultaneously the first f' and the second derivative f'' of a function f given at uniform mesh, $\Delta x = h$, are defined by

$$a_1 f'_{j-1} + a_0 f'_j + a_2 f'_{j+1} + h \left(b_1 f''_{j-1} + b_0 f''_j + b_2 f''_{j+1} \right) =$$
$$= \frac{1}{h} \left(c_1 f_{j-2} + c_2 f_{j-1} + c_0 f_j + c_3 f_{j+1} + c_4 f_{j+2} \right). \qquad (17.12)$$

For symmetric (centered) schemes, $a_1 = a_2$, $b_1 = -b_2$, $c_1 = -c_4$ and $c_2 = -c_3$ and considering $a_0 = 1$ obtain from Eq. (18.11)

$$a_1 f'_{j_1} + f'_i + a_1 f'_{j+1} + h \left(-b_2 f''_{j-1} + b_0 f''_i + b_2 f''_{j+1} \right) =$$
$$= \frac{1}{h} \left[c_0 f_j + c_3 \left(f_{j+1} - f_{j-1} \right) + c_4 \left(f_{j+2} - f_{j-2} \right) \right].$$

$$(17.13)$$

Similarly, for $b_0 = 1$ and requiring for symmetry $\tilde{b}_1 = \tilde{b}_2$, $\tilde{c}_1 = \tilde{c}_4$, $\tilde{c}_2 = \tilde{c}_3$, and $\tilde{a}_1 = -\tilde{a}_2$, where the tilde denotes that $\tilde{a}_1$,...etc (obtained for $b_0 = 1$) and different from a_1 etc (obtained for $a_0 = 1$) have

$$\tilde{a}_0 f'_j + \tilde{a}_2 (f'_{j+1} - f'_{j-1}) + h \left(\tilde{b}_1 f''_{j-1} + f''_j + b_1 f''_{j+1} \right) =$$
$$= \frac{1}{h} \left[\tilde{c}_1 (f_{j-2} + f_{j+2}) + \tilde{c}_2 (f_{j-1} + f_{j+1}) + \tilde{c}_0 f_j \right].$$

$$(17.14)$$

Expanding both sides of Eqs. (18.12) and (18.13) in a Taylor series and collecting terms of the same order obtain a sixth and an eight order that are given next

$6^{th} - Order Scheme$

$$7 f'_{j-1} + 16 f'_j + 7 f'_{j+1} + h \left(f''_{j-1} - f''_{j+1} \right)$$
$$= \frac{15}{h} \left(f_{j+1} - f_{j-1} \right)$$
$$-9 f_{j-1} - 9 f_{j+1} - h \left(f''_{j-1} - 8 f''_j + f_{j+1} \right)$$
$$= \frac{24}{h} \left(f_{j-1} - 2 f_j + f_{j+1} \right).$$

$$(17.15)$$

Comparing the new scheme of Eq. (18.14) with the standard compact sixth-order compact scheme

$$f'_{j-1} + 3 f'_j + f'_{j+1} = \frac{7}{3h} \left(f_{j+1} - f_{j-1} \right) + \frac{1}{12h} \left(f_{j+2} - f_{j-2} \right)$$
$$2 f''_{j-1} + 11 f''_j + 2 f''_{j+1} = \frac{12}{h^2} \left(f_{j-1} - 2 f_j + f_{j+1} \right)$$
$$+ \frac{3}{4h^2} \left(f_{j-2} - 2 f j + f_{j+2} \right).$$

$$(17.16)$$

It is evident that the scheme, which has been developed by Mahesh [132] and computes the first and second derivative in a coupled fashion uses a more narrow stencil compared to the standard compact scheme of the previous chapter.

$6^{th} - Order Scheme : Periodic$

The matrix form of the sixth-order accurate coupled, compact scheme on a periodic domain is given by

$$
\begin{bmatrix}
16 & 0 & 7 & -h & & & & 0 & 7 & h \\
0 & 8h & 9 & -h & 0 & & & & -9 & -h \\
\ddots & \ddots & \ddots & \ddots & \ddots & \ddots & & & & \\
& & 0 & 7 & h & 16 & 0 & 7 & -h & \\
& & -9 & -h & 0 & 8h & 9 & -h & 0 & \\
& & & \ddots & \ddots & \ddots & \ddots & \ddots & \ddots & \ddots \\
7 & -h & & & & 0 & 7 & h & 16 & 0 \\
9 & -h & 0 & & & & -9 & -h & 0 & 8h
\end{bmatrix}
\begin{bmatrix}
f_1' \\
f_1'' \\
\vdots \\
f_i' \\
f_i'' \\
\vdots \\
f_N' \\
f_N''
\end{bmatrix}
\begin{bmatrix}
15(f_2 - f_N) \\
24(f_N - 2f_1 + f_2) \\
\vdots \\
15(f_{i+1} - f_{i-1}) \\
24(f_{i-1} - 2f_i + f_{i+1}) \\
\vdots \\
15(f_1 - f_{N-1}) \\
24(f_{N-1} - 2f_N + f_1)
\end{bmatrix} .
$$

$$(17.17)$$

$6^{th} - Order Scheme : (3,3) Boundary\ Closure$

For non-periodic domains, the sixth-order interior scheme is used at the nodes $j = 2$ to $N - 1$, and third-order boundary closures are used at $j = 1$ and N. The resulting scheme is given by

$$
\begin{bmatrix}
1 & 0 & 2 & -h/2 & & & & \\
0 & h & -6 & 5h & 0 & & & \\
7 & h & 16 & 0 & 7 & -h & & \\
0 & 7 & h & 16 & 0 & 7 & -h & \\
& -9 & -h & 0 & 8h & 9 & -h & 0 \\
& & \ddots & \ddots & \ddots & \ddots & \ddots & \ddots & \ddots \\
& & & -9 & -h & 0 & 8h & 9 & -h \\
& & & & 0 & 2 & h/2 & 1 & 0 \\
& & & & & -6 & -5h & 0 & -h
\end{bmatrix}
\begin{bmatrix}
f_1' \\
f_1'' \\
\vdots \\
f_i' \\
f_i'' \\
\vdots \\
f_N' \\
f_N''
\end{bmatrix}
= \frac{1}{h}
\begin{bmatrix}
3(f_2 - f_1) \\
9f_1 - 12f_2 + 3f_3 \\
\vdots \\
15(f_{i+1} - f_{i-1}) \\
24(f_{i-1} - 2f_i + f_{i+1}) \\
\vdots \\
3(f_N - f_{N-1}) \\
-9f_N + 12f_{N-1} - 3f_{N-2}
\end{bmatrix} .
$$

$$(17.18)$$

The eight-order coupled compact scheme is:

$$51f'_{j+1} + 108f'_j + 51f'_{j-1} + 9h\,(f''_{j+1} \qquad + f''_{j-1}) =$$
$$\frac{107}{h}\,(f_{j+1} - f_{j-1}) - \frac{1}{h}\,(f_{j+2} - f_{j-2}) \qquad (17.19)$$
$$138f'_{j+1} - 138f_{j-1} - h\,(18f''_{j+1} - 108f''_j + 18f''_{j-1}) =$$
$$\frac{352}{h}\,(f_{j+1} - f_{j-1}) + \frac{1}{h}\,(f_{j+2} - f_{j-2}) - \frac{702}{h}f_j$$

$8^{th} - Order\,Scheme : Periodic$

The matrix form of the eight-order coupled compact scheme on a periodic domain is given by

$$
\begin{bmatrix}
108 & 0 & 51 & -9h & & & & 0 & 51 & 9h \\
0 & 108h & 138 & -18h & 0 & & & & -138 & -18h \\
\ddots & \ddots & \ddots & \ddots & \ddots & \ddots & & & & \\
& 0 & 51 & 9h & 108 & 0 & 51 & -9h & & \\
& -138 & -18h & 0 & 108h & 138 & -18h & 0 & & \\
& & \ddots & \ddots & \ddots & \ddots & \ddots & \ddots & \ddots & \\
51 & -9h & & & & 0 & 51 & 9h & 108 & 0 \\
138 & -18h & 0 & & & & -138 & -18h & 0 & 108h
\end{bmatrix}
\begin{bmatrix}
f'_1 \\ f''_1 \\ \vdots \\ f'_i \\ f''_i \\ \vdots \\ f'_N \\ f''_N
\end{bmatrix}
$$

$$
= \frac{1}{h}
\begin{bmatrix}
107(f_2 - f_N) - (f_3 - f_{N-1}) \\
-(f_3 + f_{N-1}) + 352(f_2 + f_N) - 702f_1 \\
\vdots \\
107(f_{i+1} - f_{i-1}) - (f_{i+2} - f_{i-2}) \\
-(f_{i+2} + f_{i-2}) + 352(f_{i+1} + f_{i-1}) - 702f_i \\
\vdots \\
107(f_1 - f_{N-1}) - (f_2 - f_{N-2}) \\
-(f_2 + f_{N-2}) + 352(f_1 + f_{N-1}) - 702f_N
\end{bmatrix}
\cdot \qquad (17.20)
$$

$8^{th} - Order\,Scheme : (3,3)\,Boundary\,Closure$

For non-periodic domains the eighth-order interior scheme is used at the nodes $j = 3$ to $N - 2$. The sixth-order interior scheme is used at $j = 2$ and $N - 1$, and third-order boundary closures are used at $j = 1$ and N. The resulting scheme is given by

$$
\begin{bmatrix}
1 & 0 & 2 & -h/2 & & & & \\
0 & h & -6 & 5h & 0 & & & \\
7 & h & 16 & 0 & 7 & -h & & \\
9 & -h & 0 & 8h & 9 & -h & 0 & \\
& \ddots & \ddots & \ddots & \ddots & \ddots & \ddots & \ddots \\
& & 0 & 51 & 9h & 108 & 0 & 51 & -9h \\
& & & -138 & -18h & 0 & 108h & 138 & -18h & 0 \\
& & & & \ddots & \ddots & \ddots & \ddots & \ddots & \ddots & \ddots \\
& & & & & 0 & 7 & h & 16 & 0 & 7 & -h \\
& & & & & & -9 & -h & 0 & 8h & 9 & -h \\
& & & & & & & 0 & 2 & h/2 & 1 & 0 \\
& & & & & & & & -6 & -5h & 0 & -h
\end{bmatrix}
\begin{bmatrix}
f_1' \\
f_1'' \\
f_2' \\
f_2'' \\
\vdots \\
f_i' \\
f_i'' \\
\vdots \\
f_{N-1}' \\
f_{N-1}'' \\
f_N' \\
f_N''
\end{bmatrix}
$$

$$
= \frac{1}{h}
\begin{bmatrix}
3(f_2 - f_1) \\
9f_1 - 12f_2 + 3f_3 \\
15(f_3 - f_1) \\
24(f_1 - 2f_2 + f_3) \\
\\
107(f_{i+1} - f_{i-1}) - (f_{i+2} - f_{i-2}) \\
-(f_{i+2} + f_{i-2}) + 352(f_{i+1} + f_{i-1}) - 702f_i \\
\\
15(f_N - f_{N-2}) \\
24(f_{N-2} - 2f_{N-1} + f_N) \\
3(f_N - f_{N-1}) \\
-9f_N + 12f_{N-1} - 3f_{N-2}
\end{bmatrix}. \tag{17.21}
$$

17.6   Modified High-Order Finite-Difference Schemes

The idea of modifying or optimizing a finite difference schemes by calculating values of the coefficients that introduce upwinding [207] or minimize a particular type of error instead of the truncation error [175] has been used successfully in the design of new schemes with desired properties. Some form of upwinding is often needed for the computation of flows with discontinuities and nonlinearities. Modifications of standard centered, explicit and compact schemes were carried out by Zhong [207]. For the modified schemes, the formal order of the scheme for certain stencil size was sacrificed and high-order upwinding with low dissipation was introduced.

Other optimized schemes have also been developed [128] in the field of computational aeroacoustics [175], [208], [209]. The rationale for optimizing numerical schemes for short waves is that for long waves, even lower-order schemes can do well. The short waves, however, require high resolution in order to obtain accurate representation of the broadband acoustic waves. The optimized finite-difference scheme of Tam and Webb [175], for example, referred to as the dispersion relation preserving (DPR) scheme, uses central differences to approximate the first derivative. The approximation is therefore, nondissipative in nature. The maximum formal order of accuracy of the centered scheme [175] for certain stencil size is again sacrificed in order to optimize resolution of the high wave numbers. Although nondissipative schemes are ideal for aeroacoustics, numerical dissipation is often required to damp nonphysical waves generated by boundary and/or initial conditions. In many practical applications, therefore, high order dissipative terms were added to the centered scheme of Ref. [175]. To remedy this problem optimized DPR schemes were developed by Zhuang and Chen [208] and Lockard et al. [128]. In the following sections, the upwind high order schemes of Zhong [207] and the DPR scheme of Tam and Webb [175] are presented.

17.6.1 Upwind High-Order Schemes

A family of finite-difference high-order upwind compact and explicit schemes for the discretization of convective terms was derived in [207]. The general compact and explicit finite-difference approximation of $\partial u / \partial x = u'$ is

$$\sum_{k=-M+M_0+1}^{M_0} b_{i+k} \, u'_{i+k} = \frac{1}{h} \sum_{k=-N+N_0+1}^{N_0} a_{i+k} \, u_{i+k} \qquad (17.22)$$

where h is the uniform grid spacing, u'_{i+k} is the numerical approximation of the first derivative at the $(i+k)$th grid point, and N_0, M_0 are biases with respect to the base point i. The family of upwind compact and explicit schemes with central stencils $N = 2N_0 + 1$, $M = 2M_0 + 1$, was also considered in [207]. The coefficients a_{i+k} and b_{i+k} of these upwind schemes were determined such that the order of the schemes is one order lower than the maximum achievable order for the standard central stencil. As a result, the orders of the upwind schemes are always odd integers $p = 2\,(N_0 + M_0) - 1$, and there is a free parameter θ in the coefficients a_{i+k} and b_{i+k}. The value of θ is set to be the coefficient of the leading truncation term, which is an even order derivative.

$$\sum_{k=-M_0}^{M_0} b_{i+k} \, u'_{i+k} = \frac{1}{h} \sum_{k=-N_0}^{N_0} a_{i+k} \, u_{i+k} - \frac{\theta}{(p+1)!} \, h^p \left(\frac{\partial U^{p+1}}{\partial x^{p+1}} \right)_i + ... \quad (17.23)$$

Schemes based on Eq. (18.22) are $(p+1)$th order accurate for $\vartheta = 0$ and p-th order accurate for $\partial \neq 0$. The choice of ϑ affects the magnitude of numerical dissipation and the stability of the scheme. The third-, fifth- and seventh-order schemes derived in [207] are

Third-Order Explicit Scheme

$$u' = \frac{1}{h\,b_i} \sum_{k=-2}^{2} a_{i+k}\, u_{i+k} - \frac{\theta}{4!\,b_i}\, h^3 \left(\frac{\partial U^4}{\partial^4 x} \right)_i + \dots \qquad (17.24)$$

with

$$a_{i+2} = \mp 5 + \frac{5}{2}\theta$$
$$a_{i\pm1} = \pm 40 - 10\theta$$
$$a_i = 15\theta$$
$$b_i = 60$$

the recommended value of θ in [207] is $\theta = 1/4$.

Fifth-Order Upwind Compact Scheme

$$b_{i-1}\, u'_{i-1} + b_i\, u_i + b_{i+1}\, u'_{i+1} =$$
$$\frac{1}{h} \sum_{k=-2}^{2} a_{i+k}\, u_{i+k} - \frac{\theta}{6!}\, h^5 \left(\frac{\partial U^6}{\partial^6 x} \right) \qquad (17.25)$$

with

$$a_{i\pm2} = \pm\frac{5}{3} + \frac{5}{6}\theta$$
$$a_{i\pm1} = \pm\frac{140}{3} + \frac{20}{3}\theta \quad b_{i\pm1} = 20 \pm \theta.$$
$$a_i = -15\theta \quad b_i = 60$$

Fourier analysis shows that dissipation errors increase as θ increases. For θ satisfying $-1 \leq \theta \leq 0$ the dissipation errors of the upwind compact schemes are smaller or comparable to the corresponding phase errors.

The boundary closures for the fifth-order upwind compact scheme are

$$60u_1' + 180u_2' = \frac{1}{h}(-170u_1 + 90u_2 + 90u_3 - 10u_4)$$

$$15u_1' + 60u_2' + 15u_3' = \frac{45}{h}(u_3 - u_1)$$

$$15u_N' + 60u_{N-1}' + 15u_{N-2}' = \frac{45}{h}(u_N - u_{N-2})$$

$$60u_N' + 180u_{N-1}' = -\frac{1}{h}(-170u_N + 90u_{N-1} + 90u_{N-2} - 10u_{N-3}).$$

Fifth-Order Upwind Explicit Scheme

$$u' = \frac{1}{h\,b_i}\sum_{k=-3}^{3} a_{i+k}\,u_{i+k} - \frac{\theta}{6!\,b_i}\,h^5\left(\frac{\partial u^6}{\partial^6 x}\right) + \dots \tag{17.26}$$

with

$$a_{i+3} = \pm 1 + \frac{\theta}{12} \qquad a_{i\pm 2} = \mp 9 - \frac{\theta}{2}$$

$$a_{i\pm 1} = \pm 45 + \frac{5\theta}{4} \qquad a_i = -\frac{5\theta}{3}\ ,\qquad b_i = 60.$$

The recommended θ value $\theta = -6$ yields

$$u' = \frac{1}{60h}\left(-\frac{3}{2}u_{i-3} + 12u_{i-2} - \frac{105}{2}u_{i-1} + 10u_i + \frac{75}{2}u_{i+1} - 6u_{i+2} + \frac{1}{2}u_{i+3}\right). \tag{17.27}$$

Seventh-Order Explicit Scheme

$$u_i' = \frac{1}{h\,b_i}\sum_{k=-4}^{4} a_{i+k}\,u_{i+k} - \frac{\theta}{8!\,b_i}\,h^7\left(\frac{\partial u^8}{\partial^8 x}\right) \tag{17.28}$$

with

$$a_{i\pm 4} = \mp\frac{3}{14} + \frac{\theta}{672}\ ,\quad a_{i\pm 3} = \pm\frac{16}{7} - \frac{\theta}{84}$$

$$a_{i\pm 2} = \mp 12 + \frac{\theta}{24}\ ,\quad a_{i\pm 1} = \pm 48 - \frac{\theta}{12}$$

$$a_i = \frac{5\theta}{48}\ ,\quad b_i = 60$$

and the recommended value for θ is $\theta = 36$.

17.7   Dispersion-Relation-Preserving (DPR) Scheme

Numerical solutions of the linearized Euler equations with high-order finite-difference schemes are used to assure that the computed results have the same number of wave modes (acoustic, vorticity and entropy), the same propagation characteristics (isotropic, nondissipative and nondispersive), and the same wave speeds as nearly as possible with those of the exact solution of equations, which govern acoustic disturbance propagation. These requirements are fulfilled by the numerical solution if the discrete equations have the same dispersion relation with the continuous equation. Finite difference schemes which yield the same dispersion relations as the original partial differential equations are referred to as dispersion-relation-preserving (DPR) schemes. A way to construct time marching DPR schemes was proposed in [175] by optimizing the finite difference approximations of the space and time derivatives in the wave number and frequency space. The new optimized, high order, finite difference scheme [175] designed with these criteria meets the usual conditions of consistency, stability, and hence convergence. In addition, it supports, in the case of small amplitude waves, wave solutions which have the same characteristics as those of the linearized Euler equations as nearly as possible.

17.7.1   Optimized Spatial Discretization

Consider the approximation of the first derivative $\partial f / \partial x$ at the j-th node on a uniform grid using N values of f to the left. The finite difference approximation is

$$\left(\frac{\partial f}{\partial x}\right)_i \simeq \frac{1}{\Delta x} \sum_{j=-N}^{M} a_j \, f_{i+j} \tag{17.29}$$

The usual way to determine the coefficients a_j in Eq. (18.28) is to expand in Taylor series and equate coefficients of the same powers in Δx.

Instead of the straight forward approach that minimizes the truncation error Tam and Web [175] proposed to determine the coefficients by requiring the Fourier transform of the finite difference scheme on the right of Eq. (18.28) to be a close approximation of the partial derivative.

The finite difference representation of Eq. (18.28) can be written as:

$$\frac{\partial f}{\partial x} \simeq \frac{1}{\Delta x} \sum_{j=-N}^{M} a_j \, f \, (x + j\Delta x) \tag{17.30}$$

considering the Fourier transform of the left and right of Eq. (18.29) obtain:

$$ik\widetilde{f} \simeq \left(\frac{1}{\Delta x} \sum_{j=-N}^{M} a_j \, e^{ikj\Delta x} \right) \widetilde{f} \tag{17.31}$$

Clearly the quantity

$$\widehat{k} = -\frac{i}{\Delta x} \sum_{j=-N}^{M} a_j \, e^{ijk\Delta x} \tag{17.32}$$

is effectively the wavenumber of the Fourier transform for the finite-difference schemes of Eqs. (18.28) and (18.29).

The Fourier transform of the finite difference scheme is a good approximation of that of the partial derivative over the range of wavenumbers $|k\Delta x| \leq \pi/2$ when the coefficients a_j are chosen so that they minimize the integral of the error. This integral is defined by

$$E = \int_{\pi/2}^{\pi/2} \left| k\Delta x - \widehat{k}\Delta x \right|^2 d\,(k\Delta x) = \int_{-\pi/2}^{\pi/2} \left| iK - \sum_{j=-N}^{M} a_j \, e^{ijK} \right| dK \tag{17.33}$$

The condition of the minimum is

$$\frac{\partial E}{\partial a_j} = 0 \quad j = -N, ..., M \tag{17.34}$$

The solution of the algebraic system given by Eqs. (18.33) determines the values of the coefficients a_j that give a good approximation of the derivative for $|k\Delta x| \leq \pi/2$.

In Ref. [175], the condition of Eq. (18.29) was imposed for $n = M = 3$ (fourth-order accuracy) and a_1 was left as free parameter. Minimization of Eq. (18.32) with a_1 yields the following values for the coefficients

$$
\begin{aligned}
a_0 &= 0 \\
a_1 &= -a_{-1} = 0.79926643 \\
a_2 &= -a_{-2} = -0.18941314 \\
a_3 &= -a_{-3} = 0.02651995
\end{aligned}
\tag{17.35}
$$

The formal accuracy of the scheme with the coefficients from Eq. (18.32) was sacrificed, since a fourth order accurate scheme was obtained for a seven-point-wide stencil, but the resolution in wavespace was improved.

17.8   Spectral-Type Filters

High-order accurate centered schemes are non-dissipative and they are particularly suitable for the convection of small-scale disturbances governed by the linearized Euler equations. Non-dissipative, central-difference discretizations for nonlinear problems, however, produce high-frequency spurious modes that originate from mesh nonuniformities, inaccuracies of the boundary conditions, and mainly from nonlinear interactions. In order to prevent numerical instabilities due to growth of high-frequency modes while retaining the high-order accuracy of the compact or non-compact central discretizations, filtering of the computed solution is required. Filtering of the solution with explicit-type filters was proposed by Lele [119]. More recently, high-order compact filters were introduced by Gaitonde and Visbal [60]. These compact filters are applied on the components of the computed solution vector. Denoting by ϕ, the computed value, the filtered value $\widehat{\phi}$ is obtained by solving the system

$$a_f\widehat{\phi}_{j-1} + \widehat{\phi}_j + a_f\widehat{\phi}_{j+1} = \sum_{n=0}^{N} \frac{a_n}{2}\left(\phi_{j+n} + \phi_{j-n}\right). \qquad (17.36)$$

The compact filter [60] of Eq. (18.35) provides a $2N$th-order accurate formula on a $2N + 1$ point stencil. Application of the compact filter makes possible high-resolution, low-diffusion numerical solutions of flows without discontinuities. The coefficients a_n of the filter are functions of the filtering parameter a_f. These coefficients for different order filters are given in Table 17.II.

TABLE 17.II

Coefficients for the spectral-type filter of Eq. (18.35)

	$F2$	$F4$	$F6$	$F8$	$F10$
a_0	$\frac{1}{2} + a_f$	$\frac{5}{8} + \frac{3a_f}{4}$	$\frac{11}{16} + \frac{5a_f}{8}$	$\frac{93}{128} + \frac{70a_f}{128}$	$\frac{193-126a_f}{256}$
a_1	$\frac{1}{2} + a_f$	$\frac{1}{2} + a_f$	$\frac{15}{32} + \frac{17a_f}{16}$	$\frac{7}{16} + \frac{18a_f}{16}$	$\frac{105-302a_f}{256}$
a_2	0	$-\frac{1}{8} + \frac{a_f}{4}$	$\frac{-3}{16} + \frac{3a_f}{8}$	$\frac{-7}{32} + \frac{14a_f}{32}$	$\frac{-15+30a_f}{64}$
a_3	0	0	$\frac{1}{32} - \frac{a_f}{16}$	$\frac{1}{16} - \frac{a_f}{8}$	$\frac{45-90a_f}{512}$
a_4	0	0	0	$\frac{-1}{128} + \frac{a_f}{64}$	$\frac{-5+10a_f}{256}$
a_5	0	0	0	0	$\frac{1-2a_f}{512}$
Order of Accuracy	2nd	4th	6th	8th	10th

The filtering ability of the filters given by Eq. (18.35) is determined by the transfer function in wave space. Therefore, the compact filters of Eq. (18.35)

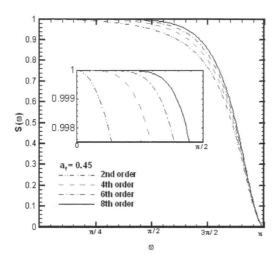

FIGURE 17.2: Transfer function for filters of different order for constant value of the filtering parameter $A_f = 0.45$

are referred to from now on as spectral-type filters. The transfer function for the spectral-type filter of Eq. (18.35) is given by

$$S\left(\omega\right) = \frac{\sum_{n=0}^{N} a_n \cos\left(n\omega\right)}{1 + 2a_f \cos\left(\omega\right)} \tag{17.37}$$

The parameter, a_f, which is in the range $0.5 < a_f < 0.5$ determines the filtering properties. High values of the parameter a_f yield less dissipative filters.

At the boundaries, the order of the filter must be dropped in order to reduce the stencil size. It was demonstrated [60] that the filtering ability of the low order filters could approximate the filtering performance of high-order filters by varying the value of the filtering parameter a_f. The variation of the spectral function of Eq. (18.36) for constant value of the filtering parameter $a_f = 0.45$ for the second up to eighth order filters is shown in Fig. 17.2. The spectral function of Eq. (18.36) for the second-, and eight-order filter is plotted in Fig. 17.3 for different values of the filtering parameter $a_f = 0.45 - 0.49$. It is evident that as the value of the filtering parameter a_f becomes larger low-pass filtering is obtained even with the second-order filter.

Centered schemes with spectral-type filters are not appropriate for computations of flows with shock waves and other discontinuities. High-order accurate computation of transonic or supersonic flows with strong discontinuities can be performed using high order explicit or compact centered schemes and

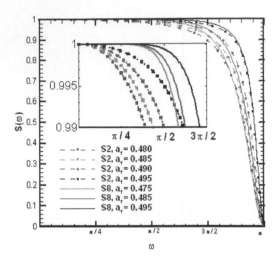

FIGURE 17.3: Transfer function of the second and eighth order filter for different values of the filtering parameter A_f.

the application of characteristic-based filters [204]. These filters are described in the following chapter.

17.9   Characteristic-Based Filters

Characteristic filters of Yee et al. [204] can be applied instead of spectral-type filters with implicit and explicit methods for time discretization. Characteristic-based filters remove spurious oscillations and in addition can be used for shock capturing. They can be applied at every stage of an RK method or after each Newton-type subiteration of an implicit-type time integration scheme for flows with strong shock interactions. For computational efficiency, however, the filter is often applied at the end of the full RK step or at the final update of an implicit time scheme.

Let L_f be the filter operator defined as

$$L_f \left(F^*, G^* \right)_{i,j} = \frac{1}{\Delta x} \left[\widetilde{F}_{i+1/2,j}^* - \widetilde{F}_{i-1/2,j}^* \right] + \frac{1}{\Delta y} \left[\widetilde{G}_{i,j+1/2}^* - \widetilde{G}_{i,j+1/2}^* \right] \quad (17.38)$$

where $\widetilde{F}_{i+1/2,j}^*$ and $\widetilde{G}_{i,j+1/2}^*$ are the dissipative numerical fluxes of the filter operator to be discussed below. Then the new time level $n + 1$ is defined as

$$U^{n+1} = \widehat{U}^{n+1} + \Delta t \; \mathcal{L}_f \; (F^*, G^*)_{i,j} \qquad (17.39)$$

where the filter numerical fluxes $\widetilde{F}_{i+1/2,j}$ and $G^*_{i,j+1/2}$ are evaluated at $\widehat{U}^{n+1}$. The simplest form for L_f is the linear filter proposed by Gustafsson and Olsson [73] where a switch similar to that of Harten [77] was used.

According to Yee et al. [204] the filter numerical flux may be written in a form which is similar to TVD schemes of [78], [80], [200] and [201], as follows

$$\widetilde{F}_{i+1/2,j} = \frac{1}{2} \left[F_{i+1,j} + F_{i,j} + \mathcal{D}_{i,j} \right] \qquad (17.40)$$

where $\frac{1}{2} \left[F_{i+1,j} + F_{i,j} \right]$ is the central difference portion of the numerical flux. This flux is evaluated by a high-order centered approximation in the schemes proposed by Yee et al. [204], and the term $\mathcal{D}_{i,j}$ is the nonlinear dissipation. For characteristic-based methods, e.g., methods where the dissipative term is evaluated in characteristic variable space $\mathcal{D}_{i,j} = R_{i+1/2,j} \; \Phi_{i+1/2,j}$ where $R_{i+1/2,j}$ is the right eigenvector of the flux Jacobian matrix $\partial F / \partial U$. In order to introduce some upwinding the elements of $R_{i+1/2}$ are computed at Roe's approximate average state.

The artificial compression method (ACM) of Harten's [77] was generalized in [204] to achieve a low-dissipative high-order shock-capturing scheme by nearly maintaining the accuracy to high-order. The ACM filter numerical flux $\widetilde{F}_{i+1/2,j}$ has the form:

$$\widetilde{F}^*_{i+1/2,j} = \frac{1}{2} \; R_{i+1/2} \; \Phi^*_{i+1/2}, \qquad (17.41)$$

where $R_{i+1/2}$ is the right eigenvector matrix of the flux Jacobian $A = \partial F / \partial U$ at Roe's approximate average state and the elements of the matrix $\Phi^*_{i+1/2}$ denoted by $\phi^{*l}_{i+1/2}$ are:

$$\phi^{*l}_{i+1/2} = \kappa \; \vartheta^l_{i+1/2} \; \phi^l_{i+1/2}. \qquad (17.42)$$

The function $\kappa \; \vartheta^l_{i+1/2}$ in Eq. (17.42) is the key mechanism for achieving high accuracy of the fine scale flow structures as well as capturing of shock waves in a stable manner. The elements of $\Phi^*_{i+1/2}$ can be identified as the nonlinear dissipation portion of a TVD, ENO, or WENO scheme with the exception that they are premultiplied by $\kappa \; \vartheta^*_{i+1/2}$. Yee et al. [204] defined $\Phi^*_{i+1/2}$ using a TVD scheme. Garnier et al. [62], on the other hand, defined $\Phi^*_{i+1/2}$ using the dissipative part of ENO or WENO schemes.

The main disadvantage with characteristic-based filters is that the parameter κ is problem depended. Numerous simulations and tests were carried out by Yee et al. [204] for different flows. It was found, however, that different examples require a different value of κ. The suggested range of κ in [204] was $0.03 \le \kappa \le 2$ where larger values of κ are used for flows with discontinuities and smaller values are required for smooth flows including complex features, such as vortex convection or vortex pairing. It will be shown later that in order to remedy this problem, Sjogreen and Yee [169] used a regularity estimate obtained from the wavelet coefficients of the solution to obtain a better estimation for the value for the filter sensor.

The function $\theta^l_{i+1/2}$ is the Harten switch. This switch for a general $2m + 1$ points scheme is given by

$$\theta^l_{i+1/2} = \max (\widehat{\theta}_{i-m+1}, ..., \widehat{\theta}_{i+m}), \tag{17.43}$$

$$\widehat{\theta}^l_i = \left| \frac{|e^l_{i+1/2}| - |e^l_{i-1/2}|}{|e^l_{i+1/2}| - |e^l_{i-1/2}|} \right|^p. \tag{17.44}$$

In Eq. (17.44), p is a second parameter that determines the performance of the filter and can be varied to better capture the particular physics instead of varying κ. The higher the parameter p the less is the amount of numerical dissipation added to the numerical solution. For $p \geq 1$ the order of accuracy of the dissipation term is essentially increased for all numerical examples in [204] and [169] a constant value $p = 1$ was used. Furthermore, in order to keep the stencil of the scheme compact Harten's switch was computed as

$$\theta^l_{i+1/2} = \max (\theta^l_i, \theta^l_{i+1}). \tag{17.45}$$

In Eq. (17.44), $\rho^l_{i+1/2}$ are elements of $R^{-1}_{i+1/2} (U_{i+1} - U_i)$ where $R^{-1}_{i+1/2}$ is the left eigenvector of the flux Jacobian $\partial F/\partial U$ that transforms back to the conservative variable space the filter operator.

The elements $\phi^l_{i+1/2}$ in Eq. (17.40) are evaluated based on TVD schemes [200], [202] and [201]. Choosing Harten-Yee upwind TVD for example obtain

$$\phi^l_{i+1/2} = \frac{1}{2} \ \psi \ (c^l_{i+1/2}) \ (g^l_{i+1}) - \psi \ (c^l_{i+1/2} + \gamma^l_{i+1/2}) \ e^l_{i+1/2} \tag{17.46}$$

where

$$\gamma^l_{i+1/2} = \frac{1}{2} \ \psi \ (c^l_{i+1/2}) = \begin{cases} (g^l_{i+1} - g^l_i)/e^l_{i+1/2} & e^l_{i+1/2} \neq 0 \\ 0 & e^l_{i+1/2} = 0 \end{cases} \tag{17.47}$$

$$\psi \ (c^l_{i+1/2}) = \begin{cases} |c^l_{i+1/2}| & |c^l_{i+1/2}| \geq \delta_1 \\ \frac{c^l_{i+1/2} + \delta^2_1}{2\delta_1} & |c^l_{i+1/2}| < \delta_1. \end{cases} \tag{17.48}$$

In Eq. (17.46), $c^l_{i+1/2}$ are the characteristic speeds of the flux Jacobian matrix $\partial F/\partial U$ evaluated at the Roe's average state [156], ψ is an entropy correction [79] where $0 < \delta < 1$, and g^l_i is a limiter function.

Examples of commonly used limiter functions are:

(a) $g^l_j = \text{minmod } (e^l_{j-1/2}, e^l_{j+1/2})$,

(b) $g^l_j = (e^l_{j+1/2} \, e^l_{j-1/2} + |e^l_{j+1/2} \, e^l_{j-1/2}|)/(e^l_{j+1/2} + e^l_{j-1/2})$,

(c) $g^l_j = \{e^l_{j-1/2} \, [(e^l_{j+1/2})^2 + \delta_2] + e^l_{j+1/2} \, [(e^l_{j-1/2})^2 + \delta_2]\}/$
$\qquad [(e^l_{j+1/2})^2 + (e^l_{j-1/2})^2 + 2\delta_2]$,

(d) $g^l_j = \text{minmod } (e^l_{j-1/2}, \, 2e^l_{j+1/2}, \, \dfrac{1}{2} \, (e^l_{j+1/2} + e^l_{j-1/2}))$,

(e) $g^l_j = S \cdot \max \, [0, \, \min \, (2|e^l_{j+1/2}|, \, S \cdot e^l_{j-1/2}), \, \min \, (|e^l_{j+1/2}|, \, 2S \cdot e^l_{j-1/2})]$;
$\qquad S = \text{sgn } (e^l_{j+1/2})$.

Here δ_2 is a small dimensionless parameter to prevent division by zero and sgn $(e^l_{j+1/2}) = \text{sign } (e^l_{j+1/2})$. In practical calculations $10^{-7} \le \delta_2 \le 10^{-5}$ is a commonly used range. For $e^l_{j+1/2} + e^l_{j-1/2} = 0$, g^l_j is set to zero in (b). The minmod function of a list of arguments is equal to the smallest number in absolute value if the list of arguments is of the same sign, or is equal to zero if any arguments are of opposite sign.

To facilitate computer implementation the entropy correction ψ (c) in Eq. (17.48) is evaluated as

$$\psi \, (c) = \sqrt{(c^2 + \delta^2)}, \quad \delta = 1/16 \qquad (17.49)$$

and $\gamma^l_{i+1/2}$ in Eq. (17.47) is computed by

$$\gamma^l_{i+1/2} = \frac{\psi \, (c^l_{i+1/2}) \, (\vartheta^l_{i+1/2} - g^l_i) \, e^l_{i+1/2}}{2 \, (c^l_{i+1/2})^2 + \varepsilon}, \quad \varepsilon = 10^{-7} \qquad (17.50)$$

In addition, the switch in Eq. (17.44) is modified in order to avoid division by zero and is evaluated by the following formula

$$\theta^l_i = \frac{| \, |e^l_{i+1/2}| - |e^l_{i-1/2}| \, |}{|e^l_{i+1/2}| + |e^l_{i-1/2}| + \varepsilon}. \qquad (17.51)$$

17.9.1   Other Filter Formulations

The dissipative filter numerical flux can be expressed through an upwind MUSCL-type scheme and an approximate Riemann solver as follows

$$\widetilde{F}^*_{i+1/2} = \frac{1}{2}\, R_{i+1/2}\, \Phi^o_{i+1/2} \qquad (17.52)$$

In this case, the elements of $\Phi^o_{i+1/2}$ are given by

$$(\phi^o)^l_{i+1/2} = -\kappa\, (\theta^o)^l_{i+1/2}\, \psi\, (c^l_{i+1/2})\, e^l_{i+1/2} \qquad (17.53)$$

where again $e^l_{i+1/2}$ are the element of $R^{-1}_{i+1/2}\, (U^R_{i+1/2} - U^L_{i+1/2})$, $c^l_{i+1/2}$ are the eigenvalues of $A = \partial F/\partial U$ or the characteristic speeds, and $R_{i+1/2}$ is the right eigenvector matrix of A.

The same switch of Eq. (17.44) or (17.51) is used but evaluated a symmetric average between $U^R_{i+1/2}, U^L_{i+1/2}$. Where U^R and U^L are the upwind-biased interpolation of the neighboring $U_{i,j}$ values with slope limiters imposed. The slope limiters can be imposed on conservative characteristic, or primitive variables.

The function Φ of the filter numerical flux with the Lax-Friedrichs numerical flux becomes

$$\Phi_{i+1/2} = -c_{i+1/2}\, (U^R_{i+1/2} - U^L_{i+1/2}) \qquad (17.54)$$

where

$$c_{i+1/2} = \frac{1}{2}\, (|u_{i+1/2}| + a_{i+1/2})$$

and $a = \sqrt{\gamma p/\rho}$

17.9.2   ENO and WENO ACM Filters

A recent improvement of ACM filters is application of ENO and WENO procedure in the evaluation of the dissipative fluxes. The dissipative numerical fluxes for TVD-MUSCL schemes (see Eq. (17.41)) are

$$\widetilde{F}^{*M}_{i+1/2} = \frac{1}{2}\, R_{i+1/2}\, \Phi^M_{i+1/2} \qquad (17.55)$$

where the elements $\phi^l_{i+1/2}$ of $\Phi^{*M}_{i+1/2}$ are given by

$$\phi^l_{i+1/2} = \kappa\, \vartheta^l_{i+1/2}\, |c^l_{i+1/2}|\, e^l_{i+1/2} \qquad (17.56)$$

where

$$e^l_{i+1/2} = R^{-1}_{i+1/2}\, (U^R_{i+1/2} - U^L_{i+1/2})$$

and $c^l_{i+1/2}$ are the eigenvalues of the flux Jacobian $U^R_{i+1/2}$ and $U^L_{i+1/2}$ are the upwind-biased interpolation of the neighboring U_i values with the slope limiters imposed. The MUSCL approach can be extended to r-th order accurate ENO schemes [204] as follows. The dissipative numerical flux is written as

$$\widetilde{F}^{*\,ENO}_{i+1/2} = R_{i+1/2}\; \Phi^{*\,ENO}_{i+1/2} \tag{17.57}$$

where the element $\phi^l_{i+1/2}$ of $\Phi^{*\,ENO}_{i+1/2}$ is obtained from the dissipative part of the ENO scheme, which results (see next chapter) by subtracting an mth-order accurate, centered scheme from an rth-order accurate ENO approximation as

$$\phi^l_{i+1/2} = \theta^l_{i+1/2}\; \Big(\sum_{p=0}^{r-1} c^r_{k,p}\; R^{-1}_{i+1/2}\; F_{i-r+1+k+p} - \tag{17.58}$$

$$\sum_{p=0}^{m-1} c^m_{\frac{m}{2},p}\; R^{-1}\; F_{i-m+1+\frac{m}{2}+p}\Big)$$

where $c^r_{k,p}$ are the reconstruction coefficients of the ENO reconstruction, and k is the stencil index selected among the r candidate stencils S_k that are defined as

$$S(k) = \{I_{i-r+k+1}, ..., I_i, ..., I_{i+k}\} \tag{17.59}$$
$$= (x_{i-r+k+1}, ..., x_i, ..., x_{i+k}), \quad k = 0, ..., r-1.$$

The mth-order accurate centered scheme is a subclass of ENO stencils with $k = m/2$ and m even. All choices $m \geq 2l$, $l = 1, 2, ...$ are valid for the construction of error dissipative terms. However, in order to keep the accuracy of the base scheme the same with the order of the dissipative terms $m = q$ since larger values of m do not improve the formal accuracy and increase the computational cost.

An increased order of accuracy can be achieved by exploiting the WENO idea in the construction of the $\phi^l_{i+1/2}$ dissipative terms as follows:

$$\phi^l_{i+1/2} = \sum_{k=0}^{r-1} \omega_k\; \Big[\theta^l_{i+1/2}\; \Big(\sum_{p=0}^{r-1} c^r_{k,p}\; R^{-1}_{i+1/2}\; F_{i-r+1+k+p} - \tag{17.60}$$

$$\sum_{p=0}^{m-1} c^m_{m/2,p}\; R^{-1}_{i+1/2}\; F_{i-m+1+m/2+p}\Big)\Big].$$

The WENO approach (see next chapter) achieves $(2r-1)$th-order of accuracy by performing linear, convex combination with weights ω_k of the r possible r^{th} order ENO stencils.

17.10   Wavelet Estimation of the Sensor

Sjogreen and Yee [168] proposed further improvements for better estimation of the filter numerical flux introducing regularity estimates from the wavelet coefficients of the solution as follows.

Consider the l-th element of the filter numerical flux function $\widetilde{f}^l_{i+1/2}$ (see Eq. (17.42)) as the product of the sensor $w^l_{i+1/2}$ and a nonlinear dissipation function $\phi^l_{i+1/2}$

$$\widetilde{f}^l_{i+1/2} = w^l_{i+1/2}\, \phi^l_{i+1/2}. \tag{17.61}$$

The sensor $w^l_{i+1/2}$ in Eq. (17.44) is modified by changing the value of κ while the numerical dissipation portion $\phi_{i+1/2}$ of Eq. (17.46) of Harten and Yee TVD scheme remains the same.

17.10.1   Wavelet Analysis

Wavelet analysis for a given function f yields an estimate of the local Lipschitz exponent α, defined as the largest α satisfying

$$\sup_{h \neq 0} \frac{|f(x+h) - f(x)|}{h^\alpha} \leq c \tag{17.62}$$

The exponent α is a measure of regularity for the function f where small α indicates poor regularity.

A wavelet function ψ with compact support can be used to estimate the exponent α from the wavelet coefficients,

$$w_{m,j} = \langle f, \psi_{m,j} \rangle = \int f(x)\, \psi_{m,j}(x)\, dx. \tag{17.63}$$

Using

$$\psi_{m,j} = 2^m\, \psi \left(\frac{x-j}{2^m} \right) \tag{17.64}$$

as wavelet function, located at the point j in a space of scale m, it can be shown that the coefficients $max|w_{m,j}|$ decay as $2^{m\alpha}$ in a neighborhood j_o as the scale is refined, where α is the Lipschitz exponent at j_o. Evaluating $w_{m,j}$ on the smallest scale m_o determined by the grid size and few coarser scales $m_o + 1$, $m_o + 2$ the exponent α at j_o is estimated by least square fit to the line by

$$\max_{\substack{j \text{ near } j_o}} \log_2 |w_{m,j}| = m\, \alpha_{j_o} + c. \tag{17.65}$$

The sensor $w^l_{i+1/2}$ is then modified as

$$w_{i+1/2} = \max\left(\tau(\alpha_j),\ \tau(\alpha_{j=1})\right). \qquad (17.66)$$

$$\tau(\alpha) = \begin{cases} 1 & \alpha \leq 0.5 \\ 0 & \alpha > 0.5 \end{cases}$$

For more details see [168].

Chapter 18

ENO and WENO Schemes

John A. Ekaterinaris, School of Mechanical and Aerospace Engineering, University of Patras, Greece *and* Foundation for Research and Technology-Hellas, Institute of Applied and Computational Mathematics, 71110 Heraklion, Greece, ekaterin@iacm.forth.gr

---

18.1 ENO and WENO Schemes

ENO and WENO schemes are high-order accurate finite-difference or finite-volume numerical methods designed for the solution of hyperbolic problems with piecewise smooth solutions containing discontinuities. The key idea of these methods is the design of the locally smoothest stencil that avoids crossing discontinuities as much as possible.

The ENO idea, proposed by Harten and Osher [81] and Harten et al. [82], is the first successful attempt to obtain no-mesh-size dependent (self-similar), uniformly high order accurate, yet essentially non-oscillatory interpolation for piecewise smooth functions using an adaptive local stencil that satisfies certain measures of local smoothness. ENO offered significant improvements over earlier approaches, which attempt to eliminate or reduce spurious oscillations generated at discontinuities by the fixed stencil, second, or higher order accurate methods, which used artificial viscosity or limiters.

ENO and WENO schemes were proven suitable for CFD of compressible turbulence, aeroacoustics, and other applications where the solution contains both smooth but complex features and discontinuities. Several of ENO and WENO applications shown at the end of this chapter demonstrate the enhanced resolution of these schemes and their potential to replace traditional methods in large scale computations. The presentation of ENO and WENO schemes starts with a review of interpolation and approximation theory used with these schemes.

18.2   High-Order Reconstruction

A basic approximation problem encountered in the numerical solution of hyperbolic conservation laws is to obtain high-order (second-order accurate or higher) reconstruction (finite-volume FV approximation) or *conservative* approximation of the derivatives (FD approximation) from cell averaged or point values of the state variables, respectively. Numerical algorithms for conservation laws that have been extensively investigated in the past decades laid the foundation for the development of modern schemes, [138], [184], [185] including TVD schemes [78] and the piecewise parabolic method [41] that reduce or eliminate spurious numerical oscillations at discontinuities. In the following sections, the basic ideas of ENO approximation and reconstruction, and the conservative approximation of the derivatives with an adaptive, smooth stencil that uses the smoothest possible data for the reconstruction are explained. Based on this idea, ENO and WENO schemes have a distinct advantage over TVD schemes that near every extrema, even smooth ones, degrade to first order accuracy to suppress any spurious numerical oscillations. ENO schemes, on the other hand, are based on high order accurate conservative approximations of derivatives (FD) or polynomial reconstructions from the average state (FV), are high-order accurate, and essentially nonoscillatory up to the discontinuity; i.e., the numerical oscillations, if any, decay with the order of the truncation error.

The one-dimensional high-order approximation problem of a function from given cell average values is described first. Approximation is obtained with polynomials. This polynomial approximation procedure is the basis of finite volume ENO schemes for arbitrary spacing in one-dimension. The conservative approximation of the derivative, which is also based on the one-dimensional polynomial approximation is the basis of the finite-difference ENO schemes is presented next. In the subsequent sections, the ENO and WENO reconstruction procedures are presented. The implementation of ENO and WENO schemes is given and recent improvements of these schemes are discussed.

18.3   Approximation in One Dimension

In this chapter the basic information about polynomial interpolation and approximation is reviewed. This background is fundamental for the understanding ENO interpolation and other numerical methods. The presentation is

$$u_{j+1/2} = \sum_{\ell=0}^{k-1} c_{r\ell}\,\overline{u}_{j-r+\ell}, \qquad r+s = k-1$$

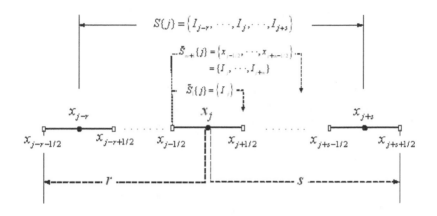

FIGURE 18.1: ENO $r + s = k - 1$ wide stencil for the kth order accurate approximation of $u_{j+1/2}$ from average cell values $\overline{u}_j$.

given in one space dimension. The formulation of the basic approximation problem is as follows:

Given a mesh in the interval between a and b

$$a = x_{1/2} < x_{3/2} < ...x_{i+1/2}... < x_{N-1/2} < x_{N+1/2} = b$$

with subintervals or cells $I_i \equiv [x_{i-1/2}, x_{i+1/2}]$, cell centers $x_i \equiv \frac{1}{2}(x_{i-1/2} + x_{i+1/2})$ and cell size $\Delta x_i \equiv x_{i+1/2} - x_{i-1/2}$, $i = 1, 2, ..., N$, (see Fig. 18.1), consider the following reconstruction problem. Given the cell averages of a function $u(x)$, which represents the solution at the cell centers x_i

$$\overline{u}_i \equiv \frac{1}{\Delta x_i} \int_{x_{i-1/2}}^{x_{i+1/2}} u\left(\xi\right) d\xi, \qquad i = 1, 2, ..., N \tag{18.1}$$

find a polynomial $p_i(x)$, of degree $k - 1$ at most, such that $p_i(x)$ is the k-th order accurate approximation to the function $u(x)$ inside each cell I_i, e.g., a polynomial satisfying

$$p_i(x) = u(x) + \mathcal{O}(\Delta x^k), \quad x \in I_i, \quad i = 1, 2, ..., N \tag{18.2}$$

The approximations of the solution $u(x)$ with the polynomial $p_i(x)$ at the I_i cell boundaries, $i - 1/2$ and $i + 1/2$, are also k-th order accurate, e.g.,

$$u_{i-1/2} = p_i(x_{i-1/2}) = u(x_{i-1/2}) + \mathcal{O}(\Delta x^k)$$
$$u_{i+1/2} = p_i(x_{i+1/2}) = u(x_{i+1/2}) + \mathcal{O}(\Delta x^k) \qquad i = 1, 2, ..., N. \quad (18.3)$$

The procedure to solve this problem is: Given the location I_i and the order of accuracy k, choose a stencil $S(i) = \{I_{i-r}, ..., I_{i+s}\}$ including I_i itself, r cells to the left of I_i, and s cells to the right, with $r + s + 1 = k$, (see Fig. 18.1), and use the average values $\bar{u}_i$ in $S(i)$ to obtain the polynomial $p_i(x)$ that yields the k-th order accurate approximation of the solution $u(x)$ as given in Eq. (18.2).

For a smooth function $u(x)$ in the region covered by the stencil $S(i)$, there is a unique polynomial of degree $k - 1 = r + s$ at most whose cell average in each cell of $S(i)$ agrees with the average of $u(x)$, e.g.,

$$\frac{1}{\Delta x_i} \int_{x_{j-1/2}}^{x_{j+1/2}} p\,(\xi)\,d\xi = \bar{u}_i \qquad j = i - r, ..., i + s. \quad (18.4)$$

The polynomial $p(x)$ is the k-th order approximation of the solution we are looking for. Furthermore, since the mappings from the given cell average values $\bar{u}_j$ in the stencil $S(i)$ to the values $u_{i-1/2}$ and $u_{i+1/2}$ at the cell boundary are linear, there exist constants c_{rj}, which depend on the left shift r, on the order of accuracy k and on the cell size Δx_i that yield k-th order approximations $u_{i-1/2}$ and $u_{i+1/2}$. For example,

$$u_{i+1/2} = \sum_{j=0}^{k-1} c_{rj}\,\bar{u}_{i-r+j} = u\,(x_{i+1/2}) + \mathcal{O}(\Delta x^k). \quad (18.5)$$

The constants c_{rj} are obtained using the primitive function $U(x)$ of $u(x)$ defined as:

$$U(x) \equiv \int_{-\infty}^{x} u\,(\xi)\,d\xi. \quad (18.6)$$

The primitive function can be expressed by cell average values $\bar{u}_i$ of $u(x)$ using the definition of Eq. (18.1) to obtain

$$U(x_{i+1/2}) = \sum_{j=-\infty}^{i} \int_{x_{j-1/2}}^{x_{i+1/2}} u\,(\xi)\,d\xi = \sum_{j=-\infty}^{i} \bar{u}_j\,\Delta x_j. \quad (18.7)$$

This equation shows that the cell averages $\bar{u}_i$ yield the exact values of the primitive function $U(x)$ at the cell boundaries.

Considering the unique polynomial $P(x)$ of degree k which interpolates the primitive function at the $k + 1$ cell boundaries $x_{i-r-1/2}, ..., x_{i+s+1/2}$ and denoting its derivative $P'(x)$ by $p(x)$ it is easy to verify that $p(x)$ is the polynomial we are looking for and satisfies Eq. (18.4) as follows:

$$\frac{1}{\Delta x_j} \int_{x_{j-1/2}}^{x_{j+1/2}} p(\xi) \, d\xi = \frac{1}{\Delta x_j} \int_{x_{j-1/2}}^{x_{j+1/2}} P'(\xi) \, d\xi \tag{18.8}$$

$$= \frac{1}{\Delta x_j} \int_{x_{j-1/2}}^{x_{j+1/2}} u(\xi) \, d\xi = \bar{u}_j \,, \quad j = i - r, ..., i + s$$

in addition, have

$$p(x) = P'(x) = U'(x) + \mathcal{O}(\Delta x^k) \,, \quad \forall x \in I_i.$$

Since the exact value of the primitive function $U(x)$ of $u(x)$ at the cell boundaries are obtained from the cell averages (see Eq. 18.7) one can use Lagrange interpolation polynomials to obtain the constants c_{rj} in Eq. (18.5) as follows:

$$P(x) = \sum_{m=0}^{k} U(x_{i-r+m-1/2}) \prod_{\substack{l=0 \\ l \neq m}}^{k} \frac{x - x_{i+r+l-1/2}}{x_{i-r+m-1/2} - x_{i-r+l-1/2}}. \tag{18.9}$$

Subtracting $U(x_{i-r-1/2})$ form both sides of Eq. (18.9) and after some algebra (see article by Shu in [17] pages 439-582) obtain the values

$$c_{rj} = \Delta x_{i-r+j} \sum_{m=j+1}^{k} \frac{\sum_{\substack{l=0 \\ l \neq m}}^{k} \prod_{\substack{q=0 \\ q \neq m,l}}^{k} (x_{i+1/2} - x_{i-r+q-1/2})}{\prod_{\substack{l=0 \\ l \neq m}}^{k} (x_{i-r+m-1/2} - x_{i-r+l-1/2})} \tag{18.10}$$

that simplifies for uniform grid to

$$c_{rj} = \sum_{m=j+1}^{k} \frac{\sum_{\substack{l=0 \\ l \neq m}}^{k} \prod_{\substack{q=0 \\ q \neq m,l}}^{k} (r - q + 1)}{\prod_{\substack{l=0 \\ l \neq m}}^{k} (l - m)}. \tag{18.11}$$

The values of the constants c_{rj} for uniform grid and order of accuracy between $k = 1$ and $k = 6$ are given in Table 18.I.

TABLE 18.I

k	r	$j=0$	$j=1$	$j=2$	$j=3$	$j=4$	$j=5$
1	-1	1					
	0	1					
2	-1	3/2	-1/2				
	0	1/2	1/2				
	1	-1/2	3/2				
3	-1	11/6	-7/6	1/3			
	0	1/3	5/6	-1/6			
	1	-1/6	5/6	1/3			
	2	1/3	-7/6	11/6			
4	-1	25/12	-23/12	13/12	-1/4		
	0	1/4	13/12	-5/12	1/12		
	1	-1/12	7/12	7/12	-1/12		
	2	1/12	-5/12	13/12	1/4		
	3	-1/4	13/12	-23/12	25/12		
5	-1	137/60	-163/60	137/60	-21/20	1/5	
	0	1/5	77/60	-43/60	17/60	-1/20	
	1	-1/20	9/20	47/60	-13/60	1/30	
	2	1/30	-13/60	47/60	9/20	-1/20	
	3	-1/20	-17/60	-43/60	77/60	1/5	
	4	1/5	-21/20	137/60	-163/60	137/60	
6	-1	49/20	-71/20	79/20	-163/60	31/30	-1/6
	0	1/6	29/20	-21/20	37/60	-13/60	1/30
	1	-1/30	11/30	19/20	-23/60	7/60	-1/60
	2	1/60	-2/15	37/60	37/60	-2/15	1/60
	3	-1/60	7/60	-23/60	19/20	11/30	-1/30
	4	1/30	-13/60	37/60	-21/20	29/20	1/6
	5	-1/6	31/30	-163/60	79/20	-71/20	49/20

From Table 18.I and Eq. (18.5) obtain, for example,

$$u_{i+1/2} = -\frac{1}{12}\,\overline{u}_{i-1} + \frac{7}{12}\,\overline{u}_i + \frac{7}{12}\,\overline{u}_{i+1} - \frac{1}{12}\,\overline{u}_{i+2} + \mathcal{O}(\Delta x^4)$$

$$u_{i+1/2} = \frac{49}{20}\,\overline{u}_{i+1} - \frac{71}{20}\,\overline{u}_{i+2} + \frac{79}{20}\,\overline{u}_{i+3} - \frac{163}{60}\,\overline{u}_{i+4} + \frac{31}{30}\,\overline{u}_{i+5} - \frac{1}{6}\,\overline{u}_{i+6} + \mathcal{O}(\Delta x^6).$$

$$\hat{u}_{j+1/2} = \sum_{\ell=0}^{k-1} c_{r\ell} u_{j-r+\ell}, \qquad r+s = k-1$$

FIGURE 18.2: ENO $r + s = k - 1$ wide stencil for the kth order accurate approximation of the numerical flux $\hat{u}_{j+1/2}$ from nodal values u_j.

18.4 One-Dimensional Conservative Approximation of the Derivative

For finite-difference schemes, the basic problem of high-order accurate discretization is the conservative approximation of the derivative of a function $u(x)$

$$u_i \equiv u(x_i), \qquad i = 1, 2, ...N \qquad (18.12)$$

from given point values at the nodes x_i of the mesh.

In the ENO and WENO framework, this is accomplished through the use of a numerical flux function $\hat{u}_{i+1/2}$ at the cell centers or half nodes $x_{i+1/2}$

$$\hat{u}_{i+1/2} \equiv \hat{u}(u_{i-r}, ..., u_{i+s}), \qquad i = 0, 1..., N. \qquad (18.13)$$

This numerical flux depends on r point values on the left and s point values on the right (see Fig. 18.2) such that the flux difference approximates the derivative $u'(x) = du(x)/dx$ to k-th order accuracy

$$\frac{1}{\Delta x_i} \left(\widehat{u}_{i+1/2} - \widehat{u}_{i-1/2} \right) = u'(x_i) + \mathcal{O}\left(\Delta x^k\right), \quad i = 0, 1, ..., N. \quad (18.14)$$

Assuming that the grid is uniform $\Delta x_i = \Delta x$, this problem can be solved with the same technique used to obtain k-th order accurate reconstruction, which was presented in a previous chapter. The assumption of uniform grid spacing for the conservative evaluation of the derivative in the finite-difference formulation is essential. Note, however, that the reconstruction for the finite-volume formulation of the previous chapter is valid for nonuniform grid spacing. Assuming that a function $h(x)$ exists, which depends on the uniform grid spacing Δx, such that

$$u_i \equiv \overline{u}_i = \frac{1}{\Delta x} \int_{x-\frac{\Delta x}{2}}^{x+\frac{\Delta x}{2}} h\left(\xi\right) d\xi \quad (18.15)$$

then

$$u'(x) = \frac{1}{\Delta x} \left[h\left(x + \frac{\Delta x}{2}\right) - h\left(x - \frac{\Delta x}{2}\right) \right] \quad (18.16)$$

therefore, the numerical flux function we are looking for must satisfy

$$\widehat{u}_{i+1/2} = h\left(x_{i+1/2}\right) + \mathcal{O}\left(\Delta x^k\right). \quad (18.17)$$

It is not straightforward to find $h(x)$ because Eq. (18.15) defines the unknown function $h(x)$ implicitly. However, observing that the known function $u(x)$ is the cell average of $h(x)$, (see Eq. (18.15)), one can use the same reconstruction procedure of Section 18.2, to approximate $h(x)$.

Considering the primitive $H(x)$ of $h(x)$, where $H(x)$ satisfies

$$H(x) = \int_{-\infty}^{x} h\left(\xi\right) d\xi \quad (18.18)$$

then Eq. (18.15) implies

$$H(x_{i+1/2}) = \sum_{j=-\infty}^{i} \int_{x_{j-1/2}}^{x_{j+1/2}} h\left(\xi\right) d\xi = \Delta x \sum_{j=-\infty}^{i} u_i \quad (18.19)$$

Summarizing: The conservative approximation of the derivative in the finite-difference context becomes equivalent to the reconstruction problem in the finite-volume formulation when the given point values u_i are identified as cell averages of an unknown function $h(x)$ that satisfies Eq. (18.15). The primitive function, $H(x)$, of $h(x)$ is then exactly known at the cell interfaces (see Eq. (18.19)), or half points $x_{i+1/2}$ from the point values u_i at the nodes. Using the same approximation procedure described in Section 18.2, obtain the k-th order approximation to $h(x_{i+1/2})$, which according to Eq. (18.17) is the numerical flux $\widehat{u}_{i+1/2}$ we are looking for.

For a stencil $S(i)$ around the point i (see Fig. 18.2), r points to the left and s points to the right, $(x_{i-r}, ..., x_i, ..., x_{i+s})$, where $r + s = k + 1$ the numerical flux is expressed as:

$$\widehat{u}_{i+1/2} = \sum_{j=0}^{k-1} c_{rj} \, u_{i-r+j} \qquad (18.20)$$

where the values of the constants c_{rj} are given in Table 18.I.

For a globally smooth function $u(x)$, the best approximation is obtained for even k by centered approximation $r = s - 1$. For example, the fourth-order accurate centered flux approximation from Eq. (18.20) and $k = 4$ is obtained by:

$$\widehat{u}_{i+1/2} = \frac{1}{12} \left[u_{i-1} + 7u_i + 7u_{i+1} - u_{i+2} \right] + \mathcal{O} \left(\Delta x^4 \right)$$

For k odd, the best approximation is obtained by one point upwind-biased stencils $r = s$ or $r = s - 2$. For example, 3rd order upwind fluxes for $k = 3$ are

$$\widehat{u}_{i+1/2} = \frac{1}{6} \left[-u_{i-1} + 5u_i + 2u_{i+1} \right] + \mathcal{O} \left(\Delta x^3 \right)$$

$$\widehat{u}_{i-1/2} = \frac{1}{6} \left[-u_{i-2} + 5u_{i-1} + 2u_i \right] + \mathcal{O} \left(\Delta x^3 \right)$$

which yield a third order accurate conservative approximation of the derivative

$$\frac{1}{\Delta x} \left[\widehat{u}_{i+1/2} - \widehat{u}_{i-1/2} \right] = \frac{1}{6} \left[u_{i-2} - 6u_{i-1} + 3u_i + 2u_{i+1} \right] = u'(x_i) + \mathcal{O} \left(\Delta x^3 \right)$$

The selection of the most suitable stencils among different possible stencils of Sections 18.2 and 18.3 for fixed k is accomplished through the ENO or WENO reconstruction procedures that are described next.

18.5   ENO Reconstruction

Approximation of discontinuous solutions that occur in hyperbolic conservation laws is accomplished with piecewise smooth functions. These functions $u(x)$ posses derivatives at all points except at discontinuities where the function and its derivatives are assumed to have finite left and right limits. For such piecewise smooth functions, the order of accuracy is determined by the local truncation error in smooth regions of the definition of the function. A fixed stencil high-order approximation of a piecewise smooth function is not adequate near discontinuities, because stencils that contain discontinuous cells cause oscillations (Gibbs phenomenon) in the numerical solution.

The basic idea of the ENO approximation is to avoid including discontinuous cells in the stencil as much as possible. This is accomplished by the "adaptive stencil" where the left shift changes with the location x_i. In the ENO approximation, this is achieved by using the Newton divided differences of the interpolation polynomial as a smoothness indicator of the stencil.

The $j-th$ degree divided differences $F\left[x_{i-1/2}, ..., x_{i+j-1/2}\right]$ of the primitive function $F(x)$, of $f(x)$ (see Eq. 18.18), that is defined at the cell faces or half points $x_{i-1/2}, ..., x_{i+j-1/2}$ are given by the recursive formula

$$F\left[x_{i-1/2}\right] = F\left(x_{i-1/2}\right)$$

$$F\left[x_{i-1/2}, ..., x_{i+j-1/2}\right] \equiv$$

$$\equiv \frac{F\left[x_{i+1/2}, ..., x_{i+j-1/2}\right] - F\left[x_{i-1/2}, ..., x_{i+j-3/2}\right]}{x_{i+j-1/2} - x_{i-1/2}}$$

$$(18.21)$$

A similar formula defines the divided differences of the cell averages $\overline{f}_i \equiv \overline{f}\left[x_i\right]$ of the function $f(x)$ defined at the cell centers (FV), or nodes $x_i, ..., x_{i+j}$ (FD) approach. The divided differences for $\overline{f}_i$ are:

$$\overline{f}\left[x_i, ..., x_{i+j}\right] = \frac{\overline{f}\left[x_{i+1}, ..., x_{i+j}\right] - \overline{f}\left[x_i, ..., x_{i+j-1}\right]}{x_{i+j} - x_i}$$

The function $f(x)$ and its primitive $F(x) = \int_{-\infty}^{x} f\left(\xi\right) d\xi$ are related by

$$F\left(x_{i+1/2}\right) = \sum_{j=-\infty}^{i} \int_{x_{j-1/2}}^{j+1/2} f\left(\xi\right) d\xi = \sum_{j=-\infty}^{i} \overline{f}_j \Delta x_j \qquad (18.22)$$

therefore

$$F\left[x_{i-1/2}, x_{i+1/2}\right] = \frac{F\left(x_{i+1/2}\right) - F\left(x_{i-1/2}\right)}{x_{i+1/2} - x_{i-1/2}} = \overline{f}_i \qquad (18.23)$$

As a result Eq. (18.21) can be expressed in terms of $\overline{f}$, since the first degree divided differences of $F(x)$ are the zeroth degree divided differences of $\overline{f}$, and the computation of the primitive function F can be completely avoided.

Using the above definitions the k-th degree interpolation polynomial $P(x)$ that interpolates the primitive function $F(x)$ at $k+1$ points is expressed with divided differences by

$$P\left(x\right) = \sum_{j=0}^{k} F\left[x_{i-r-1/2}, ..., x_{i-r+j-1/2}\right] \prod_{m=0}^{j-1} \left(x - x_{i-r+m-1/2}\right) \qquad (18.24)$$

and the polynomial $p(x) = P'(x)$ is expressed as

$$p(x) = \sum_{j=0}^{k} F[x_{i-r-1/2}, ..., x_{i-r+j-1/2}] \sum_{m=0}^{j-1} \prod_{\substack{l=0 \\ l \neq m}}^{j-1} (x - x_{i+r+l-1/2}) \quad (18.25)$$

where again $p(x)$ can be expressed by divided differences of $\overline{f}$.

The ENO selection process of the smoothest stencil of $k+1$ consecutive points that include $x_{i-1/2}$ and $x_{i+1/2}$ is based on Eq. (18.24) and is performed with the following steps:

1. Start with the two point stencil $\widetilde{S}_2(i) = \{x_{i-1/2}, x_{i+1/2}\}$ (see Figs. 18.1 and 18.2) of the primitive function U of u, which has a corresponding single cell stencil $S(i) = \{I_i\}$ in terms of $\overline{u}$ (see Eq. (18.23) and Figs. 18.1 and 18.2).

2. Obtain the linear (first degree) interpolation polynomial P^1 on the stencil $\widetilde{S}_2(i)$ using Newton forms as

$$P^1(x) = U[x_{i-1/2}] + U[x_{i-1/2}, x_{i+1/2}](x - x_{i-1/2}). \quad (18.26)$$

3. Obtain higher order interpolation polynomials P_R^2 and P_S^2 by expanding the stencil to the left including $x_{i-3/2}$ or to the right including $x_{i+3/2}$, respectively.

$$P_R^2 = P^1(x) + U[x_{i-3/2}, x_{i-1/2}, x_{i+1/2}](x - x_{i-1/2})(x - x_{i+1/2})$$
$$P_S^2 = P^1(x) + U[x_{i-1/2}, x_{i+1/2}, x_{i+3/2}](x - x_{i-1/2})(x - x_{i+1/2})$$
$$(18.27)$$

The deviations from the linear approximation of the quadratic approximations P_R^2 and P_S^2 depend on the divided differences $U[x_{i-3/2}, x_{i-1/2}, x_{i+1/2}]$ and $U[x_{i-1/2}, x_{i+1/2}, x_{i+3/2}]$, respectively.

However, a divided difference is a measure of smoothness of the function in the stencil because for a smooth function $U(x)$ have $U[x_{i-1/2}, ..., x_{i+j-1/2}] = \frac{V^{(j)}(\xi)}{j!}$ for some $x_{i-1/2} < \xi < x_{i+j-1/2}$, while if $U(x)$ is discontinuous inside the stencil $U[x_{i-1/2}, ..., x_{i+j-1/2}] = \mathcal{O}\left(\frac{1}{\Delta x^j}\right)$.

Therefore, if

$$|U[x_{i-3/2}, x_{i-1/2}, x_{i+1/2}]| < |U[x_{i-1/2}, x_{i+1/2}, x_{i+3/2}]| \quad (18.28)$$

select the stencil

$$\widetilde{S}_3(i) = \{x_{i-3/2}, x_{i-1/2}, x_{i+1/2}\} \quad or \quad S_2(i) = \{I_{i-1}, I_i\} \qquad (18.29)$$

otherwise, select the stencil

$$\widetilde{S}_3(i) = \{x_{i-1/2}, x_{i+1/2}, x_{i+3/2}\} \quad or \quad S_2(i) = \{I_i, I_{i+1}\}. \qquad (18.30)$$

4. Continue this process until the stencil $\widetilde{S}_k(i)$ with the desired number of points is reached.

Note again that all divided differences of the primitive are computed in terms of averages and for uniform mesh the divided differences are replaced by undivided differences. The finite-volume and finite-difference ENO algorithms based on the approximation and ENO reconstruction procedures of the previous sections are summarized next.

18.6   1D Finite-Volume ENO Scheme

From the cell average values $\{\overline{u}_i\}$ of the function $u(x)$, (see Fig. 18.1) obtain a piecewise polynomial reconstruction of $u(x)$ of degree $k-1$ at most as follows:

1. Using $\{\overline{u}_i\}$ compute the divided (or undivided differences for uniform mesh) of the primitive function $U(x)$ for degrees 1 to k with Eq. (18.21) and (18.23).

2. Start with the two point stencil for the primitive $\widetilde{S}_2(i) = \{x_{i-1/2}, x_{i+1/2}\}$ which is equivalent to the one point stencil $S_1(i) = \{I_i\}$ for the cell average.

3. Add one of the two neighboring points to the stencil $\widetilde{S}_l(i)$, $l = 2, ..., k$ following the ENO procedure of Eqs. (18.28) - (18.30).

4. Use the Lagrange form, Eq. (18.9), or the Newton divided differences form Eq. (18.25) to obtain the polynomial $p_i(x)$ of degree $k-1$ the most that satisfies the accuracy requirement.

   In practice, however, once the stencil is known, it is more convenient to find the approximation at the cell boundaries using Eq. (18.5)

$$u_{i+1/2} = \sum_{j=0}^{k-1} c_{rj}\, \overline{u}_{i-r+j}$$

where the values of the constant c_{rj} are given by Eqs. (18.10) and (18.11), for nonuniform and uniform mesh, respectively.

18.7   1D Finite-Difference ENO Scheme

The finite-difference ENO reconstruction is valid only for fixed mesh size and includes the following steps:

1. Compute the numerical flux $\widehat{u}_{i+1/2}$ using the given point values $\{u_j\}$ (see Fig. 18.2) and all k points fixed stencils, where $r + s = k - 1$, by

$$\widehat{u}_{i+1/2} = \sum_{j=0}^{k-1} c_{rj}\, u_{i-r+j}.$$

   Note that in the finite-difference ENO approximation the given point values $\{u_i\}$ are identified as cell averages of another function $h(x)$, which has a primitive $H(x)$ exactly known at the cell interfaces $x_{i+1/2}$, and the k-th order approximation $h(x_{i+1/2})$ is the numerical flux $\widehat{u}_{i+1/2}$.

2. Perform steps 1 through 4 of the finite-volume ENO reconstruction treating $\{u_i\}$ as cell averages to select the smoothest stencil.

3. Obtain a conservative k-th order accurate approximation of the derivative as

$$\frac{1}{\Delta x}\left(\widehat{u}_{i+1/2} - \widehat{u}_{i+1/2}\right) = u'(x_i) + \mathcal{O}(\Delta x^k).$$

18.8   WENO Approximation

In the previous sections, it was shown that the ENO reconstruction is uniformly high order accurate right up to the discontinuity, and achieves this by adaptively choosing the smoothest stencil using the absolute values of divided differences. However, ENO reconstruction in practice may face the following problems:

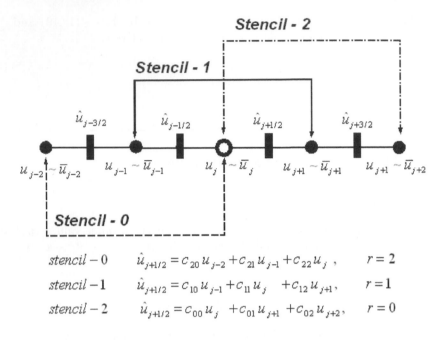

$$\text{stencil} - 0 \qquad \hat{u}_{j+1/2} = c_{20} u_{j-2} + c_{21} u_{j-1} + c_{22} u_j \,, \qquad r = 2$$

$$\text{stencil} - 1 \qquad \hat{u}_{j+1/2} = c_{10} u_{j-1} + c_{11} u_j + c_{12} u_{j+1}, \qquad r = 1$$

$$\text{stencil} - 2 \qquad \hat{u}_{j+1/2} = c_{00} u_j + c_{01} u_{j+1} + c_{02} u_{j+2}, \qquad r = 0$$

FIGURE 18.3:   Candidate stencils for (2k-1)-th order accurate WENO; WENO5 for $k = 3$. The nodal values of u_j for $j - r < j < j - r + k - 1$ are identified as cell averages for the computation of the WENO reconstruction of the numerical fluxes $\widehat{u}_{j+1/2}$ at half point nodes.

1. For finite-volume (FV) reconstructions, round-off error perturbations may change the stencil because they can result in sign change of divided differences even in smooth regions [158], [166]. The stencil free adaption results into a non-smooth numerical flux in finite-difference (FD) ENO.

2. The reconstruction in (FV) or the numerical flux evaluation in (FD) ENO of k-th order accuracy uses only one of the k candidate stencils that cover $2k - 1$ cells. However, if all candidate stencils were used, then $(2k - 1)$th order accuracy could be achieved.

It was proposed in [166] and [56] to remedy the free adaption problem using a biasing strategy for the stencil selection process. However, the most recent improvement of ENO is the WENO (weighted ENO) approximation [126], [106]. The basic idea of WENO is to use a convex combination of all candidate stencils to form the reconstruction. The WENO approximation is explained more precisely in this chapter.

Suppose that the k candidate stencils $S_r(i) = \{x_{i-r}, ..., x_{i-r+k-1}\}$ (see Fig. 18.3 for $k = 3$) that produce k different reconstructions for the value $u_{i+1/2}$ at the cell interface or the numerical flux $\widehat{u}_{i+1/2}$ are available. In both cases the reconstruction is obtained by

$$u^{(r)}_{i+1/2} = \sum_{j=0}^{k-1} c_{rj} \, \overline{u}_{i-r+j} \qquad r = 0, ..., k-1 \qquad (18.31)$$

where the superscript r denotes the shift to the left.

Reconstruction with WENO considers a convex combination of all $u^{(r)}_{i+1/2}$ as a new approximation at the cell boundary as follows:

$$u_{i+1/2} = \sum_{r=0}^{k-1} \omega_r \, u^{(r)}_{i+1/2} \qquad (18.32)$$

$$\sum_{r=0}^{k-1} \omega_r = 1 \,, \quad \omega_r \geq 0 \qquad (18.33)$$

When the function $u(x)$ includes a discontinuity in one or more of the stencils $S_r(i)$ then the corresponding weight ω_r must be essentially zero in order to follow as closely as possible the successful ENO idea. Furthermore, the weights should be smooth functions of the cell averages (FV) or the point values (FD) approximation. All these considerations [106] lead to the following forms of the weights:

$$\omega_r = \frac{\alpha_r}{\sum_{S=0}^{k-1} \alpha_S} \,, \qquad r - 0, ..., k-1 \qquad (18.34)$$

$$\alpha_r = \frac{d_r}{(\varepsilon + \beta_r)^2} \qquad (18.35)$$

where $\varepsilon > 0$ taken as $\varepsilon = 10^{-6}$ to avoid division by zero and β_r are the smoothness indicators of the stencil $S_r(i)$. The smoothness indicators are obtained by minimizing the total variation of the $(k-1)$-th degree reconstruction polynomial $p(x)$ constructed on each stencil $S_r(i)$, which if evaluated at $x_{i+1/2}$ yields the k-th order approximation of $u(x_{i+1/2})$.

Then the smoothness indicators β_r are defined by

$$\beta_r = \sum_{m=1}^{k-1} \int_{x_{i-1/2}}^{x_{i+1/2}} \Delta x^{2m-1} \left(\frac{\partial^m p_r(x)}{\partial^m x} \right)^2 dx \qquad (18.36)$$

The smoothness indicators for $k = 2$ and 3 obtained from Eq. (18.36) are:

Smoothness indicators for $k = 2$, 3-order WENO reconstruction

$$\beta_0 = (\overline{u}_{i+1} - \overline{u}_i)^2$$
$$\beta_1 = (\overline{u}_i - \overline{u}_{i-1})^2 \qquad (18.37)$$

Smoothness indicators for $k = 3$, 5-th order WENO reconstruction

$$\beta_0 = \frac{13}{12}\,(\overline{u}_i - 2\overline{u}_{i+1} + \overline{u}_{i+2})^2 + \frac{1}{4}\,(3\overline{u}_i - 4\overline{u}_{i+1} + \overline{u}_{i+2})^2$$

$$\beta_1 = \frac{13}{12}\,(\overline{u}_{i-1} - 2\overline{u}_i + \overline{u}_{i+1})^2 + \frac{1}{4}\,(\overline{u}_{i-1} - 4\overline{u}_{i+1})^2 \tag{18.38}$$

$$\beta_3 = \frac{13}{12}\,(\overline{u}_{i-2} - 2\overline{u}_i + u_i)^2 + \frac{1}{4}\,(\overline{u}_{i-2} - 4\overline{u}_{i-1} + 3\overline{u}_i)^2.$$

Note that for $k = 2$ obtain $(2k-1) = 3^{rd}$ order accuracy and for $k = 3$ obtain $(2k-1) = 5^{th}$ order accuracy. The WENO idea presented for FV reconstruction carries over to the finite-difference context once the cell averages $\overline{u}_i$ are replaced by nodal values u_i and the approximation at the cell boundaries of Eq. (18.32) are replaced by the numerical flux function $\widehat{u}_{i+1/2}$ at half point nodes.

Implementation of WENO schemes is more convenient in practice with the formulation of Jiang and Wu [107]. For example, for $k = 3$, 5th-order scheme, the numerical flux $\widehat{u}_{i+1/2}$ is taken as the weighted average of the numerical fluxes in the three substencils S_0, S_1, and S_2 of Fig. 18.3. The third-order accurate approximations $\widehat{u}^s_{i+1/2}$, $s = 0, 1, 2$ are:

$$\widehat{u}^0_{i+1/2} = \frac{1}{3}\,\overline{u}_{i-2} - \frac{7}{6}\,\overline{u}_{i-1} + \frac{11}{6}\,\overline{u}_i$$

$$\widehat{u}^1_{i+1/2} = -\frac{1}{6}\,\overline{u}_{i-1} + \frac{5}{6}\,\overline{u}_i + \frac{1}{3}\,\overline{u}_{i+1} \tag{18.39}$$

$$\widehat{u}^2_{i+1/2} = \frac{1}{3}\,\overline{u}_i + \frac{5}{6}\,\overline{u}_{i+1} - \frac{1}{6}\,\overline{u}_{i+2}$$

where $\overline{u}_i \equiv u_i$ denotes the point value at the nodes i for the FD WENO formulation.

The 5th-order accurate WENO approximation of the numerical flux is

$$\widehat{u}_{i+1/2} = \omega_0\,\widehat{u}^0_{i+1/2} + \omega_1\,\widehat{u}^1_{i+1/2} + \omega_2\,\widehat{u}^2_{i+1/2} \tag{18.40}$$

where the $(2k-1) = 5^{th}$ order approximation to $\widehat{u}_{i+1/2}$ is based on the five point stencil $i - 2 \le k \le i + 2$.
For $\omega_0 = 1/10$, $\omega_1 = 6/10$, $\omega_2 = 3/10$, which are referred to as optimal weights in Balsara and Shu [15], the approximation of Eq. (18.40) becomes

$$\widehat{u}_{i+1/2} = \frac{1}{30}\,\overline{u}_{i-2} - \frac{13}{60}\,\overline{u}_{i-1} + \frac{47}{60}\,\overline{u}_i + \frac{9}{20}\,\overline{u}_{i+1} - \frac{1}{20}\,\overline{u}_{i+2} \tag{18.41}$$

or

$$\widehat{u}_{i+1/2} = -\frac{1}{12}\,\overline{u}_{i-1} + \frac{7}{12}\,\overline{u}_i + \frac{7}{12}\,\overline{u}_{i+1} - \frac{1}{12}\,\overline{u}_{i+1}$$

$$- \frac{1}{30}\,(-\overline{u}_{i-2} + 4\overline{u}_{i-1} - 6\overline{u}_i + 4\overline{u}_{i+1} - \overline{u}_{i+2}). \tag{18.42}$$

The last form of Eq. (18.42) shows that the WENO approximation of the numerical flux $\widehat{u}_{i+1/2}$ is a sum of a centered flux

$$\widehat{u}^c_{i+1/2} = \frac{1}{12} \left(-\bar{u}_{i-1} + 7\bar{u}_i + 7\bar{u}_{i+1} - \bar{u}_{i+2} \right)$$

plus a dissipative portion of the WENO scheme.

Replacing w_1 by $w_1 = 1 - w_0 - w_2$ and using Eqs. (18.39) in Eq. (18.40) obtain

$$\begin{aligned}
\widehat{u}_{i+1/2} &= \frac{1}{12} \left(-\bar{u}_{i-1} + 7\bar{u}_i + 7\bar{u}_{i+1} - \bar{u}_{i+2} \right) \\
&+ \frac{1}{3} \left(\bar{u}_{i-2} - 3\bar{u}_{i-1} + 3\bar{u}_i - \bar{u}_{i+1} \right) w_0 \\
&+ \frac{1}{6} \left(\bar{u}_{i-1} - 3\bar{u}_i + 3\bar{u}_{i+1} - \bar{u}_{i+2} \right) \left(w_2 - \frac{1}{2} \right) \\
&= \frac{1}{12} \left(-\bar{u}_{i-1} + 7\bar{u}_i + 7\bar{u}_{i+1} - \bar{u}_{i+2} \right) \qquad (18.43) \\
&- \phi_w \left(\Delta \bar{u}_{i-3/2}, \ \Delta \bar{u}_{i-1/2}, \ \Delta \bar{u}_{i+1/2}, \ \Delta \bar{u}_{i+3/2} \right) \\
&= \frac{1}{12} \left(-\bar{u}_{i-1} + 7\bar{u}_i + 7\bar{u}_{i+1} - \bar{u}_{i+2} \right) \\
&- \frac{1}{3} w_0 \left(D_0 - 2D_1 + D_2 \right) + \frac{1}{6} \left(w_2 - \frac{1}{2} \right) \left(D_1 - 2D_2 + D_3 \right)
\end{aligned}$$

where $\Delta \bar{u}_{i-3/2} = D_0 = \bar{u}_{i-2} - \bar{u}_{i-1}$, $\Delta \bar{u}_{i-1/2} = D_1 = \bar{u}_{i-1} - \bar{u}_i$, $\Delta \bar{u}_{i+1/2} = D_2 = \bar{u}_i - \bar{u}_{i+1}$, $\Delta \bar{u}_{i+3/2} D_3 = \bar{u}_{i+1} - \bar{u}_{i+2}$.
The weights are defined as

$$w_0 = \frac{\alpha_0}{\alpha_0 + \alpha_1 + \alpha_2} \ , \quad w_2 = \frac{\alpha_2}{\alpha_0 + \alpha_1 + \alpha_2} \qquad (18.44)$$

$$\alpha_r = \frac{d_r}{(SI_r^k + \varepsilon)}, r = 0, 1, 2.$$

Summarizing:

For $k = 3$ the centered stencil of the $(2k - 1)$, 5th order scheme is:

$$\widehat{u}^c_{i+1/2} = \frac{1}{12} \left(-\bar{u}_{i-1} + 7\bar{u}_i + 7\bar{u}_{i+1} - \bar{u}_{i+2} \right) + \mathcal{O} \left(\Delta x^4 \right) \qquad (18.45)$$

and the optimal weights are $d_0 = 1/10$, $d_1 = 6/10$, $d_2 = 3/10$. For the smooth regions the numerical flux is evaluated using optimal weights by

$$\widehat{u}_{i+1/2} = \frac{1}{30} \bar{u}_{i-2} - \frac{13}{60} \bar{u}_{i-1} + \frac{47}{60} \bar{u}_i + \frac{9}{20} \bar{u}_{i+1} - \frac{1}{20} \bar{u}_{i+2} + \mathcal{O} \left(\Delta x^5 \right). \qquad (18.46)$$

The dissipative WENO part is:

$$\phi_\omega^3 = -\frac{1}{3}\,\omega_0\,(D_0 - 2D_1 + D_2) + \frac{1}{6}\left(\omega_2 - \frac{1}{2}\right)(D_1 - 2D_2 + D_3). \quad (18.47)$$

The smoothness indicators $\beta_r^3 = SI_r^3,\ r = 0, 1, 2$ are given by:

$$
\begin{aligned}
SI_0^3 &= D_0\,(4D_0 - 11D_1) + 10D_1^2 \\
SI_1^3 &= D_1\,(4D_1 - 5D_2) + 4D_2^2 \\
SI_2^3 &= D_2\,(10D_2 - 11D_3) + 4D_3^2
\end{aligned}
\quad (18.48)
$$

Using this notation the high-order WENO reconstructions $\widehat{u}_{i+1/2} = \sum_{r=i-k+2}^{i+k-1} c_r^k\,\overline{u}_r$ of Balsara and Shu [15] for $k = 4$ and $k = 5$ with weights

$$\omega_r = \frac{\alpha_r}{\alpha_0 + \alpha_1 + \ldots + \alpha_{k-1}}\ ,\quad \alpha_r = \frac{d_r^k}{(\varepsilon + SI_r^k)^2}$$

are given next.

For $k = 4$, the centered stencil of the $(2k - 1)$, 7th-order scheme is

$$\widehat{u}_{i+1/2}^c = \frac{1}{60}\,(\overline{u}_{i-2} - 8\overline{u}_{i-1} + 37\overline{u}_i + 37\overline{u}_{i+1} - 8\overline{u}_{i+2} + \overline{u}_{i+3}) + \mathcal{O}\left(\Delta x^6\right)$$

the optimal weights are:

$$d_0^4 = 1/35\ ,\ d_1^4 = 12/35\ ,\ d_2^4 = 18/35\ ,\ d_3^4 = 4/35.$$

For smooth regions the numerical flux $\widehat{u}_{i+1/2}$ is evaluated using optimal weights by

$$
\begin{aligned}
\widehat{u}_{i+1/2} = {}&-\frac{1}{60}\,u_{i-3} + \frac{5}{84}\,u_{i-2} - \frac{101}{420}\,u_{i-1} + \frac{319}{420}\,u_i \\
&+ \frac{107}{210}\,u_{i+1} - \frac{19}{210}\,u_{i+2} + \frac{1}{105}\,u_{i+3} + \mathcal{O}\left(\Delta x^7\right). \quad (18.49)
\end{aligned}
$$

The dissipative WENO part is:

$$
\begin{aligned}
\phi_\omega^4 = {}&-\frac{1}{12}\,[\,\omega_0(3D_0 - 10D_1 + 12D_2 - 6D_3 + D_4) \\
&+\left(\omega_1 - \frac{1}{5}\right)(-D_1 + 3D_2 - 3D_3 + D_4) \\
&+\left(\omega_3 - \frac{1}{5}\right)(-D_2 + 3D_3 - 3D_4 + D_5)\,]
\end{aligned}
\quad (18.50)
$$

with analogous as before definitions for D_r, $r = 0,...5$ as before, and the smoothness measures given by:

$$
\begin{aligned}
SI_0^4 &= D_0 \left(547D_0 - 2788D_1 + 1854D_2\right) \\
&\quad + D_1 \left(3708D_1 - 5188D_2\right) \\
&\quad + D_2 2107D_2 \\
SI_1^4 &= D_1 \left(267D_1 - 1108D_2 + 494D_3\right) \\
&\quad + D_2 \left(1468D_2 - 1428D_3\right) \\
&\quad + D_3 547D_3 \\
SI_2^4 &= D_2 \left(547D_2 - 1428D_3 + 494D_4\right) \\
&\quad + D_3 \left(1468D_3 - 1108D_4\right) \\
&\quad + D_4 267D_4 \\
SI_3^4 &= D_3 \left(2107D_3 - 5188D_4 + 1854D_5\right) \\
&\quad + D_4 \left(3708D_4 - 2788D_5\right) \\
&\quad + D_5 547D_5.
\end{aligned}
\tag{18.51}
$$

For $k = 5$, the centered stencil for the $(2k - 1)$, 9th-order scheme is

$$
\begin{aligned}
\widehat{u}_{i+1/2}^c = \frac{1}{840} \left(-3\bar{u}_{i-3} + 29\bar{u}_{i-2} - 139\bar{u}_{i-1} + 533\bar{u}_i \right. \\
\left. + 533\bar{u}_{i+1} - 139\bar{u}_{i+1} + 29\bar{u}_{i+2} - 3\bar{u}_{i+3}\right) + \mathcal{O}\left(\Delta x^8\right).
\end{aligned}
\tag{18.52}
$$

The optimal weights are:

$$
d_0^5 = \frac{1}{126} \,,\ d_1^5 = \frac{10}{63} \,,\ d_2^5 = \frac{10}{21} \,,\ d_3^5 = \frac{20}{63} \,,\ d_4^5 = \frac{5}{126}.
$$

For smooth regions $\widehat{u}_{i+1/2}$ is evaluated using optimal weights by

$$
\begin{aligned}
\widehat{u}_{i+1/2} &= (1/630)\, \bar{u}_{i-4} - (41/2520)\, \bar{u}_{i-3} + (199/2520)\, \bar{u}_{i-2} \\
&\quad - (641/2520)\, \bar{u}_{i-1} + (1879/2520)\, \bar{u}_i + (275/540)\, \bar{u}_{i+1} \\
&\quad - (61/504)\, \bar{u}_{i+2} + (11/504)\, \bar{u}_{i+3} - (1/504)\, \bar{u}_{i+4} + \mathcal{O}\left(\Delta x^9\right).
\end{aligned}
\tag{18.53}
$$

The dissipative WENO part is

$$
\begin{aligned}
\phi_N^5 = - &\left[\frac{1}{20} (-4D_0 + 17D_1 - 28D_2 + 22D_3 - 8D_4 + D_5)\, \omega_0 \right. \\
&+ \frac{1}{20} \left(D_1 - 4D_2 + 6D_3 - 4D_4 + D_5\right)\left(\omega_1 - \frac{1}{14}\right) \\
&+ \frac{1}{30} \left(D_2 - 4D_3 + 6D_4 - 4D_5 + D_6\right)\left(\omega_3 - \frac{3}{7}\right) \\
&\left. + \frac{1}{60} \left(2D_2 - 11D_3 + 24D_4 - 26D_5 + 14D_6 - 3D_7\right)\left(\omega_4 - \frac{1}{14}\right) \right]
\end{aligned}
\tag{18.54}
$$

and the smoothness indicators are given by:

$$
\begin{aligned}
SI_0^5 &= D_0 \left(22658D_0 - 163185D_1 + 201678D_2 - 86329D_3\right) \\
&\quad + D_1 \left(297120D_1 - 745293D_2 + 325158D_3\right) \\
&\quad + D_2 \left(478980D_2 - 433665D_3\right) \\
&\quad + D_3 107918D_3 \\
SI_1^5 &= D_1 \left(6908D_1 - 47055D_2 + 52158D_3 - 18079D_4\right) \\
&\quad + D_2 \left(84600D_2 - 196563D_3 + 70218D_4\right) \\
&\quad + D_3 \left(125130D_3 - 94935D_4\right) \\
&\quad + D_4 22658D_4 \\
SI_2^5 &= D_2 \left(6908D_2 - 37185D_3 + 30738D_4 - 8209D_5\right) \\
&\quad + D_3 \left(60870D_3 - 109413D_4 + 30738D_5\right) \\
&\quad + D_4 \left(60870D_4 - 37185D_5\right) \\
&\quad + D_5 6908D_5 \\
SI_3^5 &= D_3 \left(22658D_3 - 94935D_4 + 70218D_5 - 18079D_6\right) \\
&\quad + D_4 \left(125130D_4 - 196563D_5 + 52158D_6\right) \\
&\quad + D_5 \left(84600D_5 - 47055D_6\right) \\
&\quad + D_6 6908D_6 \\
SI_4^5 &= D_4 \left(107918D_4 - 433665D_5 + 325158D_6 - 86329D_7\right) \\
&\quad + D_5 \left(478980D_5 - 745293D_6 + 201678D_7\right) \\
&\quad + D_6 \left(297120D_6 - 163185D_7\right) \\
&\quad + D_7 22658D_7.
\end{aligned}
\tag{18.55}
$$

18.9   Application of ENO and WENO in One Dimension

Consider the one-dimensional hyperbolic conservation law

$$
u_t(x,t) + f_x\left(u\left(x,t\right)\right) = 0 \tag{18.56}
$$

with suitable initial and boundary conditions. The finite-volume and finite-difference schemes for the numerical solution of Eq. (18.56) are applied next.

18.9.1 Finite-Volume Formulation

Cell averaged based finite volume schemes solve the integrated version of Eq. (18.56) over the interval $\{I\}$ by

$$\frac{d\bar{u}}{dt} = -\frac{1}{\Delta x_i} \left[f(u(x_{i+1/2}, t)) - f(u(x_{i+1/2}, t)) \right] \tag{18.57}$$

where $\bar{u}$ is the cell average $\bar{u} = \frac{1}{\Delta x_i} \int_{x_{i-1/2}}^{x_{i+1/2}} u\,(\xi, t)\, d\xi$.

Using a numerical flux $\widehat{f}_{i+1/2} = h\,(u_{i+1/2}^-, u_{i+1/2}^+)$ with $u_{i+1/2}^{\pm}$ obtained by ENO or WENO reconstruction Eq. (18.57) is approximated by the following conservative scheme

$$\frac{d\bar{u}}{dt} = -\frac{1}{\Delta x} (\widehat{f}_{i+1/2} - \widehat{f}_{i-1/2}). \tag{18.58}$$

Examples of motone fluxes for the two argument function h are:

1. The Gudunov flux

$$\begin{aligned} h(a,b) = min_{a<u<b} f(u) && if\ a \le b \\ max_{b \le u \le a} f(u) && if\ a > b. \end{aligned} \tag{18.59}$$

2. The Engquist-Osher flux

$$h(a,b) = \int_0^a max\,(f'(u), 0)\, du + \int_0^b min\,(f'(u), 0)\, du + f(0). \tag{18.60}$$

3. The Lax-Friedrichs (LF) flux

$$h(a,b) = \frac{1}{2} \left[f\,(a) + f\,(b) - m\,(b - a) \right] \tag{18.61}$$

where the constant m is $m = max_u |f'(u)|$.

Among these fluxes the Godunov flux is the less diffusive and the LF flux the most diffusive. For low order method (order of reconstruction 2) the selection of the flux is critical. This difference becomes very small for higher order reconstruction. Therefore, for high order calculations it is preferable to use the computationally efficient LF flux.

18.9.2 Finite-Difference Formulation

The ENO and WENO finite-difference formulations are valid for uniform grid spacing and solve Eq. (18.56) with direct conservative approximation of the space derivative

$$\frac{du}{dt} = \frac{1}{\Delta x} \left(\widehat{f}_{i+1/2} - \widehat{f}_{i-1/2} \right) \tag{18.62}$$

where the numerical flux $\widehat{f}_{i+1/2}$ in Eq. (18.59) is obtained by ENO or WENO reconstruction.

18.10   ENO and WENO for Characteristic Variables

The component-wise application of the finite-volume ENO and WENO schemes is straightforward. Furthermore, the characteristic variables can be used for ENO and WENO approximation in the following steps:

1. Compute the divided differences (or undivided differences for equal grid spacing) for all cell averages $\overline{u}_i$

2. At the cell interfaces $x_{i+1/2}$

    2.1 Compute an average state $u_{i+1/2}$ using simple mean $u_{i+1/2} = (u_i + u_{i+1})/2$ or a Roe average

    2.2 Compute the right, R, and left, R^{-1}, eigenvectors and eigenvalues of the flux Jacobian $f'(u) = \partial f / \partial u$
    $$R = R(u_{i+1/2}), \ R^{-1} = R^{-1}(u_{i+1/2}), \ \Lambda = \Lambda(u_{i+1/2})$$

    2.3 Transform the differences computed in step 2.1 to the local characteristic field
    $$\overline{w}_i = R^{-1}\overline{u}_i$$

    2.4 Perform scalar ENO or WENO reconstruction for each component of the characteristic variables vector $\mathbf{w}$ to obtain the reconstruction $\mathbf{w}^{\pm}_{i+1/2}$

    2.5 Transform back into physical space $u^{\pm}_{i+1/2} = R \, \mathbf{w}^{\pm}_{i+1/2}$ at the cell faces

3. Apply on exact or approximate Riemann solver to compute the flux $\widehat{f}_{i+1/2}$

The finite-difference ENO or WENO for uniform meshes follows the same steps considering the point values u_i as cell averages. Using then u_i compute the undivided differences for the fluxes $f(u_i)$.
Repeat steps 2.1-2.3 and replace 2.4 and 2.5 by

    2.4 Perform scalar ENO or WENO to each component of the characteristic field w_i to obtain the numerical fluxes $\widehat{w}_{i+1/2}$ at half points

2.5 Transform back into physical space $\widehat{f}$ to obtain

$$\widehat{f}_{i+1/2} = R\,\widehat{w}_{i+1/2}$$

and using $\widehat{f}_{i+1/2}$ compute the conservative approximation of the flux derivative.

18.11 Multidimensional ENO and WENO Reconstruction

The multidimensional ENO and WENO schemes are based on the preliminaries of the previous sections. For fully unstructured meshes, however, the identification of suitable stencils is not straightforward. Presentation of multidimensional algorithms starts from the ENO in Cartesian meshes.

18.11.1 Finite-Volume Reconstruction for Cartesian Mesh

Multidimensional reconstruction and approximation without loss of generality is considered in two dimensions. The ideas described carry over to three dimensions as well. Initially, consider Cartesian grid covered by cells I_{ij} with faces $x_{i+1/2}$, $y_{j+1/2}$ denoted as $x_{i+1/2}\,y_{j+1/2}$ and cell centers $x_{i,j}$, $y_{i,j}$ denoted by x_i, y_j as shown in Fig. 18.4. This Cartesian grid is defined as:

$$I_{ij} = [x_{i-1/2}, x_{i+1/2}] \times [y_{j-1/2}, y_{j+1/2}]$$

$$x_i = (x_{i-1/2} + x_{i+1/2})/2 \qquad\qquad y_j = (y_{j-1/2} + y_{j+1/2})/2$$

The problem of two dimensional reconstruction for the rectangular cells of the Cartesian mesh defined above is as follows:

Given the cell averages

$$\overline{u}_{ij} = \frac{1}{\Delta x_i \Delta y_j} \int_{x_{i-1/2}}^{x_{i+1/2}} \int_{y_{j-1/2}}^{y_{j+1/2}} u(x,y)\,dx\,dy \tag{18.63}$$

find a polynomial $p_{ij}(x,y)$ of degree $k-1$ for each cell I_{ij} which is the k-th order accurate approximation of $u(x,y)$ in the cells I_{ij}, e.g.,

$$p_{ij}(x,y) = u(x,y) + \mathcal{O}\left(\Delta^k\right)$$
$$(x,y) \in I_{ij}\,, \qquad i = 1, ..., N_x, \qquad j = 1, ..., N_y \tag{18.64}$$

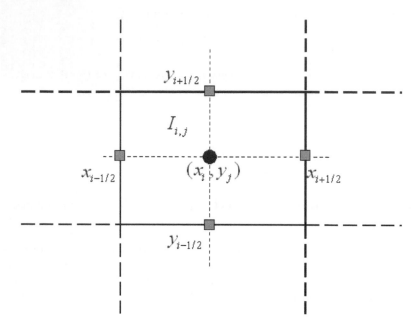

FIGURE 18.4: Cell definition for ENO reconstruction in two dimensional Cartesian mesh.

where $\Delta = max\ (\Delta x, \Delta y)$

The polynomial $p_{ij}(x, y)$ gives the following approximations of the function at the cell boundaries

$$u_{i+1/2,y}^{\pm} = p_{ij}\ (x_{i\mp 1/2}, y) \qquad \begin{matrix} i = 1, ..., N_x \\ y_{j-1/2} \leq y \leq y_{j+1/2} \end{matrix} \qquad (18.65)$$

$$u_{x,j+1/2}^{\pm} = p_{ij}\ (x, y_{j\mp 1/2}) \qquad \begin{matrix} j = 1, ..., N_y \\ x_{i-1/2} \leq x \leq x_{i+1/2}. \end{matrix} \qquad (18.66)$$

The approximations at the cell boundaries are k-th order accurate satisfying

$$u_{i+1/2,y}^{\pm} = u\ (x_{i+1/2}, y) + \mathcal{O}\ (\Delta^k) \qquad \begin{matrix} i = 0, 1, ..., N_x \\ y_{j-1/2} \leq y \leq y_{j+1/2} \end{matrix}$$

$$u_{x,j+1/2}^{\pm} = u\ (x, y_{j+1/2}) + \mathcal{O}\ (\Delta^k) \qquad \begin{matrix} j = 0, 1, ..., N_y \\ x_{i-1/2} \leq x \leq x_{i+1/2}. \end{matrix} \qquad (18.67)$$

The two dimensional ENO reconstruction problem is as follows:
Given the cell I_{ij} and the order of accuracy k, choose a stencil $S(i, j)$ based on $k(k+1)/2$ neighboring cells, and find a polynomial $p(x, y)$ of degree $k - 1$

at most whose cell average in each of the cells in the stencil $S(i, j)$ agrees with $u(x, y)$

$$\bar{u}_{ij} = \frac{1}{\Delta x_i \Delta y_j} \int_{y_{j-1/2}}^{y_{j+1/2}} \int_{x_{i-1/2}}^{x_{i+1/2}} p(\xi, \eta) \, d\xi d\eta \qquad I_{ij} \in S(i, j) \qquad (18.68)$$

The multidimensional ENO reconstruction is more complex than the one dimensional case because there are many more candidate stencils $S(i, j)$ and has the following difficulties:

- Some candidate stencils cannot be used to obtain the polynomial $p(x, y)$ that satisfies Eq. (18.68).

- Some of the polynomials do not satisfy the accuracy condition of Eq. (18.64).

These difficulties are more profound for unstructured meshes [1]. For rectangular meshes [29] the tensor products of one dimensional polynomials are used and the polynomial $p(x, y)$ is written as

$$p(x, y) = \sum_{m=0}^{k-1} \sum_{n=0}^{k-1} a_{mn} \, x^n y^m. \qquad (18.69)$$

Furthermore, by restricting the search in the following tensor product stencils (see Fig. 18.5)

$$S_{rs}(i, j) = \{I_{mn} : i - r \le m \le i + k - 1 - r, \; j - s \le n \le j + k - 1 - s\}$$

the reconstruction proceeds as in one dimension as follows:
Introduce the two dimensional primitive function

$$U(x, y) = \int_{-\infty}^{y} \int_{-\infty}^{x} u(\xi, \eta) d\xi d\eta \qquad (18.70)$$

that satisfies

$$U(x_{i+1/2}, y_{j+1/2}) = \int_{-\infty}^{y_{j+1/2}} \int_{-\infty}^{x_{i+1/2}} u(\xi, \eta) d\xi d\eta$$

$$= \sum_{m=-\infty}^{j} \sum_{n=-\infty}^{i} \bar{u}_{nm} \Delta x_n \Delta y_m \qquad (18.71)$$

e.g., as in the one-dimensional case once the cell averages $\bar{u}_{nm}$ are known the primitive function is exactly known at the cell face point $x_{i+1/2}, y_{j+1/2}$.

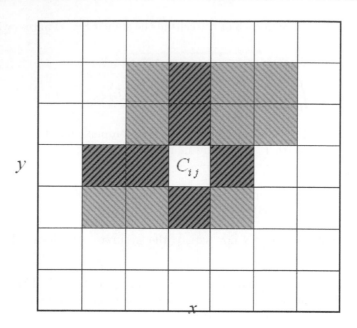

FIGURE 18.5: Possible reconstruction stencils in a Cartesian mesh.

Considering the tensor product stencil $\widetilde{S}_{rs}(i,j)$

$$\widetilde{S}_{rs}(i,j) = (x_{n+1/2}, y_{m+1/2}), \quad j - s - 1 \le m \le j + k - 1 - s$$
$$i - r - 1 \le n \le i + k - 1 - r$$

that interpolates U at every point in $\widetilde{S}_{rs}(i,j)$ with the tensor product polynomial $P(x,y)$ where the derivative of $P(x,y)$ satisfies

$$p(x,y) = \frac{\partial^2 P(x,y)}{\partial x\,\partial y}. \tag{18.72}$$

This is the $(k-1)$-th degree polynomial $p(x,y)$ which approximates $u(x,y)$ to k-th order

$$u(x,y) = p(x,y) + \mathcal{O}\left(\Delta^k\right)$$

as required in Eq. (18.64) and satisfies

$$\frac{1}{\Delta x_n \Delta y_m} \int_{y_{m-1/2}}^{y_{m+1/2}} \int_{x_{n-1/2}}^{x_{n+1/2}} p\left(\xi, \eta\right)\, d\xi d\eta = \bar{u}_{nm} \tag{18.73}$$

$$i - r \le n \le i + k - 1 - r, \quad j - s \le m \le j + k - 1 - s.$$

The computational cost for two-dimensional reconstruction is high because for each point the reconstruction cost is double. Casper and Atkins [29] give details about possible reconstruction stencils (see Fig. 18.5).

18.11.2   Finite-Volume ENO for the Euler Equations

The finite volume formulation of the two dimensional Euler equations is

$$\frac{d\bar{u}_{ij}}{dt} = -\frac{1}{A_{ij}} \left[(\widehat{f}_{i+1/2,j} - \widehat{f}_{i-1/2,j}) + (\widehat{g}_{i,j+1/2} - \widehat{g}_{i,j-1/2}) \right] \qquad (18.74)$$

where A_{ij} is cell area and $\bar{u}_{ij}$ is defined in Eq. (18.63) and $\widehat{f}_{i+1/2,j}$ etc. given by

$$\widehat{f}_{i+1/2,j} = \int_{y_{j-1/2}}^{y_{j+1/2}} f\left(u\left(x_{i+1/2}, y\right)\right) dy. \qquad (18.75)$$

The integral in Eq. (18.75) is approximated by Gaussian quadrature as

$$\int_{y_{0-1/2}}^{y_{j+1/2}} f\left(u\left(x_{i+1/2}, y\right)\right) dy \simeq \frac{\Delta_j y}{2} \sum_{k=1}^{K} w_k\, f\left(u\left(x_{i+1/2}, y_k\right)\right) \qquad (18.76)$$

where w_k are the weights of the quadrature formula, $2K \geq k$ with k denoting the ENO approximation order of accuracy (see Eq. (18.64)) and K is related to order of the quadrature, e.g., the quadrature that integrates exactly a polynomial of degree less than or equal to $2K - 1$.

The characteristic-wise implementation for the fourth order accurate ENO or WENO scheme for a Cartesian-type quadrilateral mesh shown in Fig. 18.6 is

$$\widehat{f}_{i+1/2,j} = \frac{1}{2} \sum_{p=1,2} \mathbf{h}\left(\mathbf{u}^-\left(G_p\right), \mathbf{u}^+\left(G_p\right)\right)$$

where $\mathbf{h}$ is a high order pointwise approximation of the flux $\mathbf{f}$ and G_p denotes Gaussian points at $(x_{i+1/2}, y + \alpha_p \Delta y), \alpha_1 = \alpha_2 = -\frac{\sqrt{3}}{6}$ (see Fig. 18.6). Evaluating the numerical flux function $\mathbf{h}$ with the Roe's approximate Riemann solver, for example, obtain

$$h\left(G_p\right) = \frac{1}{2} \left[\mathbf{f}\left(\bar{u}^-\left(G_\alpha\right)\right) + f\left(\bar{u}^+\left(G_p\right)\right)\right]$$

$$= -\frac{1}{2} R\left|\Lambda\right| R^{-1} \left[\mathbf{u}^-\left(G_p\right) - \mathbf{u}^+\left(G_p\right)\right]$$

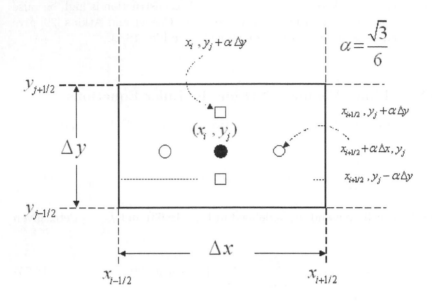

FIGURE 18.6: Computational cell and quadrature points for FV ENO reconstruction.

where R and R^{-1} are the right and left eigenvector matrices and Λ is the diagonal eigenvalue matrix. At the Gaussian points G_p these matrices are evaluated at the Roe's average state.

Using the cell average values of Eq. (18.63), the averages $\bar{u}_{i,j}^{(x)}$ are evaluated by performing one dimensional ENO or WENO reconstruction in the y-direction. The left and right states $\bar{u}^{\pm}$ are determined after a second one-dimensional characteristic-wise reconstruction of $\bar{\mathbf{u}}^{(x)}$ in the x direction as follows:

1. The reconstructed average $\bar{\mathbf{u}}^{(x)}$ is first projected on the characteristic variables $\bar{\mathbf{w}}^{(x)}$.

2. The left and right states $\mathbf{w}^{\pm}$ are determined by reconstruction at the Gaussian quadrature points.

3. The conservative values $\mathbf{u}^{\pm}(G_p)$ are obtained by projecting back to the conservative variables space along the right eigen directions.

18.11.3   Two-Dimensional Reconstruction for Triangles

The reconstruction problem for triangular meshes is:

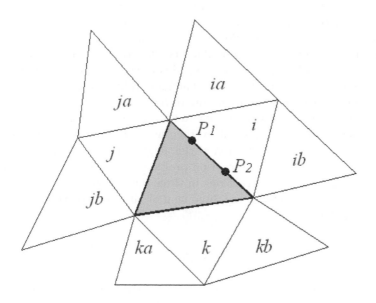

FIGURE 18.7: Stencil definition for reconstruction in a triangular mesh.

Given the cell averages $\bar{u}_i$ of $u(x, y)$ on a triangulation T_i

$$\bar{u}_i = \frac{1}{|T_i|} \int_{T_i} u\ (\xi, \eta)\ d\xi d\eta \tag{18.77}$$

where $|T_i|$ is the area of the triangle T_i find a polynomial $p_i(x, y)$ of degree $k-1$ at most for each triangle T_i such that the k-th order accurate approximation of $u(x, y)$ inside the triangles T_i is

$$p_i(x, y) = u(x, y) + \mathcal{O}\ (\Delta^k)\ , \quad (x, y) \in T_i \quad i = 1, ..., N \tag{18.78}$$

where Δ denotes a typical length of the triangle for example the longest edge. For finite volume schemes, the approximation of $u(x, y)$ at the triangle boundaries (see points P_1, P_2 in Fig. 18.7) is needed to apply quadratures as in Eq. (18.76) and Eq. (18.78) yields the approximation of $u(x, y)$ at these points.

Similar to the finite volume for Cartesian mesh, given the triangle T_i and the order of accuracy k, again choose a stencil based on $m = k(k+1)/2$ neighboring triangles, which form the stencil $S(i)$. For $S(i)$, find a polynomial $p(x, y)$ of degree $k-1$ the most whose cell average in each of the triangles in the stencil $S(i)$ agrees with the average of $u(x, y)$ given by Eq. (18.77). This condition yields an $m \times m$ linear system, and if this system has a unique solution then the stencil $S(i)$ is admissible. For second order linear reconstruction, $k = 1$

the stencil formed by T_i plus two immediate neighbors is admissible for most triangulations. For third order reconstruction ($k = 3$) a quadratic polynomial $k - 1 = 2$ is needed and $m = 6$. Some of the stencils consisting of T_i and five of its neighbors may not be admissible. For fourth order reconstruction $k = 4$, (cubic polynomial), $m = 10$, the stencil consists of T_i plus nine of the immediate neighbors shown in Fig. 18.7. The most robust way for third and fourth order reconstruction is the least square reconstruction procedure suggested by Barth and Frederickson [16] where the polynomial p is determined by requiring that p has the same cell average as u on T_0 and also has the same cell average as u on the collection of the neighboring triangles but only in a least-square sense. The reconstruction problem becomes extremely time consuming for high order reconstructions in three dimensions.

Friedrich [58] proposed to use a weighted sum of all reconstruction polynomials $p_1, ..., p_m$ as follows:

$$p = \sum_{i=1}^{m} \omega_i \, p_i \qquad (18.79)$$

where the weights of p_i are chosen such that ω_i is low if the oscillation of p_i is high. As computational cells in Friedrich formulation [58] were considered the dual cells resulting from the lines joining the barycenters with the midpoints of the edges (see Fig. 18.8). The cell averages are defined as

$$\overline{u}_k = \frac{1}{|\Omega_k|} \int_{\Omega_k} u(\mathbf{x}) \, d\mathbf{x} \qquad (18.80)$$

where Ω_k are polygonally bounded by a finite number of line segments.

The expression for the polynomial of the cell Ω_i is

$$p(\mathbf{x}) = \sum_{|\alpha| \leq n} c_\alpha \, (\overline{x} - b)^\alpha \qquad (18.81)$$

where $\alpha = (\alpha_1, \alpha_2)$, $\mathbf{x} = (x, y)$, $|\alpha| = \alpha_1 + \alpha_2$, and $\mathbf{x}^\alpha = \mathbf{x}(\alpha_1, \alpha_2) = x^{\alpha_1} y^{\alpha_2}$. In Eq. (18.81), $c_\alpha \, |\alpha| \leq n$ are the unknown coefficients and the stencils are defined as

$$S = \{\Omega_1, \Omega_2, ..., \Omega_{N(n)}\} \qquad (18.82)$$

The interpolation conditions

$$\langle p\,(\mathbf{x})\rangle_{\Omega_1} = \frac{1}{|\Omega_1|} \int_{\Omega_1} p(\mathbf{x}) \, d\mathbf{x} = \overline{u}_1$$

$$\langle p\,(\mathbf{x})\rangle_{\Omega_2} = \overline{u}_2$$

$$\langle p\,(\mathbf{x})\rangle_{\Omega_{N(n)}} = \overline{u}_{N(n)}$$

yield a linear system of $N(n)$ equations for $N(n)$ unknowns. The stencil S is admissible if this system has a unique solution.

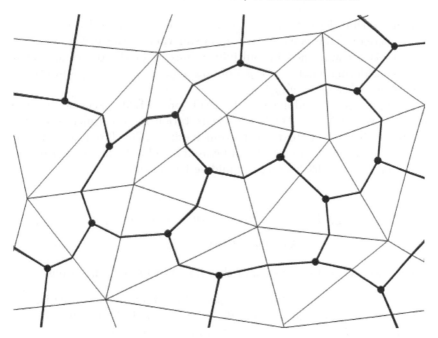

FIGURE 18.8: Dual mesh definition for finite volume WENO.

Further details on the full reconstruction algorithm the selection of admissible stencils and the choice of smoothness indicators and the WENO reconstruction weights ω_i in Eq. (18.78) can be found in [58].

18.11.4 Multidimensional Finite-Difference ENO

For finite-difference methods for hyperbolic conservation laws, the problem is to obtain high order, conservative approximation of the derivative from point values [164], [165]. Similar to the one-dimensional case, in multidimensions the uniform mesh assumption is essential. For finite-difference methods in two dimensions the statement of the problem is:
Given the point u_{ij} values of $u(x,y)$ on a uniform mesh

$$u_{ij} \equiv u\,(x_i, y_j) \quad \begin{matrix} i = 1, 2, ..., N_x \\ j = 1, 2, ..., N_y \end{matrix} \tag{18.83}$$

find the numerical flux functions

$$\begin{aligned} \widehat{u}_{i+1/2,j} &= \widehat{u}\,(u_{i-r,j},..., u_{i+k-1-r,j}) \quad i = 0, 1, ..., N_x \\ u_{i,j+1/2} &= \widehat{u}\,(u_{i,j-s},..., u_{i,j+k-1-s}) \quad j = 0, 1, ..., N_y \end{aligned} \tag{18.84}$$

and obtain a k-th order conservative approximation of the derivative by Eq. (18.17). Therefore, the conservative approximation of the derivative from point values is accomplished in multidimensions as in the one-dimensional case, e.g., considering $w(x) = u(x, y_j)$ obtain $u_x(x_i, y_j) = w'(x_i)$ and $v(y) = u(x_i, y)$ obtain $v_y(x_i, y_j) = v'(y_j)$

An example of multidimensional ENO or WENO scheme is the characteristic-wise ENO or WENO implementation for the Euler equations in the finite-difference context on uniform meshes. The Lax-Friedrich's or Roe's approximate Riemann solver can be used to split the fluxes. For example, using the Lax-Friedrich's flux as building block obtain

$$f_{LF}(x, y) = \frac{1}{2} [f(x) + f(y) - \lambda(y - x)]$$
$$\lambda = max \, |f'(u)|$$

Perform the following steps in each of the (i, j, k) or (x, y, z) directions

1. Project the positive and negative part of the flux in the i direction, to the characteristic field in the i direction by multiplying with the left eigenvectors l_i^m ($m = 1 - 4$ for 2D or m $= 1 - 5$ for 3D) to obtain f_{LF}^W as follows:

$$(f_{LF}^W)_{i+1/2}^m = \sum_{n=1}^{m} (f_{LF})_{i+1/2}^n \, l_{i+1/2}^n$$

2. Perform k-th order ENO or WENO reconstruction to obtain the numerical fluxes $\widehat{f}_{LF}^W$ in the characteristic field.

3. Project the numerical $(\widehat{f}_{LF}^W)$ fluxes to the physical space by multiplying with the right eigenvectors r_i^m to obtain the numerical fluxes $\widehat{f}$ in the conservative variable space

$$(\widehat{f})_{i+1/2}^m = \sum_{n=1}^{m} r_{i+1/2}^n \, (\widehat{f}_{LF}^W)_{i+1/2}^n$$

4. Compute the k-th order accurate conservative approximation of the derivative f_x as

$$f_x = \frac{\widehat{f}_{i+1/2} - \widehat{f}_{i-1/2}}{\Delta x} + \mathcal{O}(\Delta x^k)$$

Repeat steps 1 to 4 for the other directions to obtain the flux derivatives g_y and h_z.

18.12   Optimization of WENO Schemes

WENO schemes have been successfully applied to problems with shocks and complex smooth flow features [163]. Direct application of WENO schemes to wave propagation problems, such as computational aeroacoustics (CAA) and computational electromagnetics (CEM), where resolution of short waves is important is not optimal because WENO schemes are designed for high resolution of discontinuities and to achieve the formal order of accuracy of the reconstruction. Efficient and accurate resolution of short waves is achieved in CAA with the optimized schemes where the coefficients of the scheme are altered to minimize a particular type of error instead of the truncation error. These optimized schemes [59], [119], [127], [195], [208], [209] have been used successfully for better resolution of short waves in broadband acoustic wave propagation.

Wang and Chen [192], using ideas of CAA optimized schemes proposed modification of the WENO smoothness measures and developed an optimized WENO (OWENO) scheme following the practice of the DPR schemes [175] to achieve high resolution for short waves. The OWENO schemes of Wang and Chen [192] optimizes all candidate stencils and finds the best weights to combine them. The approach of Wang and Chen [192] was a significant improvement over previous attempts to optimize WENO schemes [125], where only the weights of the WENO schemes were optimized.

The development of the OWENO scheme is based on the DRP idea [175]. The OWENO scheme instead of achieving the maximum order of accuracy k, compromises the accuracy requirement by setting $p_{opt} < k$ in the conservative approximation of the derivative

$$\frac{\partial u_i}{\partial x} = \frac{(\widehat{u}^r_{i+1/2} - \widehat{u}^r_{i-1/2})}{\Delta x} + \mathcal{O}(\Delta x^{p_{opt}}), \quad r + s + 1 = k \qquad (18.85)$$

and minimizes the difference between the numerical wavenumber and the actual wavenumber. The numerical wavenumber for the scalar wave equation $u + \alpha u_x = 0$ where the spatial derivative is evaluated as in Eq. (18.85) becomes

$$\overline{a}^r = \frac{-i}{\Delta x} \sum_{m=-r}^{s} c_{r,j+r} \exp(im\,\alpha\Delta x)\,[1 - \exp(-i\alpha\Delta x)] = \alpha + \mathcal{O}\,(\alpha\Delta x)^{p_{opt}}$$

$$(18.86)$$

The optimized coefficients c_{rj} in the approximation of the numerical flux $\widehat{u}_{i+1/2} = \sum_{j=0}^{k-1} c_{rj}\,u_{i-r+j}$ are obtained by minimizing the L_2 norm of the difference between the numerical wavenumber of Eq. (18.86) and the actual wavenumber for a particular range $[-\alpha\Delta x\,, \alpha_0\Delta x]$. These coefficients c_{rj} are obtained by minimizing the integral

$$E_r = \int_{-\alpha_0\Delta x}^{\alpha_0\Delta x} \{\lambda \left[R_e\left(\overline{\alpha}^r \Delta x\right) - \alpha\Delta x\right]^2 + (1-\lambda)\left[Im\left(\overline{\alpha}^r \Delta x\right)\right]^2 d\left(\alpha\Delta x\right) \quad (18.87)$$

The OWENO scheme is then obtained in two steps:

1. For p_{opt}-th order of accuracy with Eq. (18.85) the following p_{opt} linear equations must be satisfied

$$\sum_{j=0}^{k-1} b_{mj}\, c_{rj} = z_m \qquad m = 1, ..., p_{\text{opt}} \qquad (18.88)$$

where b_{mj} and z_m are constants the rest $k - p_{\text{opt}}$ are obtained from the minimization of E_r in Eq. (18.87)

$$\frac{\partial E_r}{\partial c_{rj}} = 0 \qquad j = p_{\text{opt}}, ..., k-1 \qquad (18.89)$$

2. Perform a convex combination of the k candidate p_{opt}-th order accurate stencils to obtain OWENO as

$$\widehat{u}_{i+1/2}^{OWENO} = \sum_{r=o}^{k-1} h_r\, \widehat{u}_{i+1/2}^r$$

so that $\sum_{r=0}^{k-1} h_r = 1$, $\quad h_r \geq 0$, $\quad p_2 \leq k+1$

$$\frac{1}{\Delta x}\left(\widehat{u}_{i+1/2}^{OWENO} - \widehat{u}_{i+1/2}^{OWENO}\right) = \left(\frac{\partial u}{\partial x}\right)_i + \mathcal{O}\left(\Delta x^{p_{\text{opt}}+p_2}\right) \qquad (18.90)$$

The $k - 1 - p_2$ free weight parameters of Eq. (18.90) are $p_2 \leq k - 1$ then determined by minimizing again the integral of Eq. (18.87) but with $\overline{\alpha}^r$ replaced by $\overline{\alpha} = \sum_{r=0}^{k-1} h_r\, \overline{\alpha}^r$

The coefficients of several high-order accurate OWENO schemes are shown in Tables 18.II–18.IV.

TABLE 18.II
$p_{\text{opt}} = 2$, $p_2 = 1$ (3rd Order OWENO)

	j	$r=0$	$r=1$	$r=2$	$r=3$
c_{rj}	0	0.28418590	-0.10076912	0.10019444	-0.27941025
	1	1.0318226	0.60076912	-0.41620299	0.98165507
	2	-0.41620299	0.60076912	1.0318226	-1.6250794
	3	0.10019444	-0.10076912	0.28418590	1.9228346
h_r		0.14150117	0.48616615	0.33383476	0.038497919

TABLE 18.III

$p_{opt} = 3$, $p_2 = 2$ (5th Order OWENO)

	j	$r = 0$	$r = 1$	$r = 2$	$r = 3$
c_{rj}	0	0.25866239	-0.083333333	0.074670939	-0.18445575
	1	1.0573462	0.58333333	-0.39067948	0.88670057
	2	-0.39067948	0.58333333	1.0573462	-1.7200339
	3	0.074670939	-0.083333333	0.25866239	2.0177891
h_r		0.14196688	0.51976365	0.31535440	0.022915068

TABLE 18.IV

$p_{opt} = 4$, $p_2 = 3$ (7th Order OWENO)

	j	$r = 0$	$r = 1$	$r = 2$	$r = 3$
c_{rj}	0	1/4	-1/12	1/12	-1/4
	1	13/12	7/12	-5/12	13/12
	2	-5/12	7/12	13/12	-23/12
	3	1/12	-1/12	1/4	25/12
h_r		4/35	18/35	12/35	1/35

The new smoothness indicators of the k=4, OWENO scheme that satisfy

$$\beta_r = \sum_{\substack{m=2 \ k>2}}^{k-1} \left[\int_{x_{i-1/2}}^{x_{i+1/2}} \Delta x^{m-1} \frac{\partial^m p_r(x)}{\partial^m x \, dx} \right]^2 \ , \quad r = 0, ..., k-1 \qquad (18.91)$$

instead of Eq. (18.36) are given by

$$\beta_0 = (2u_i + 5u_{i+1} + 4u_{i+2} - u_{i+3})^2 - (u_i + 3u_{i+1} - 3u_{i+2} + u_{i+3})^2$$
$$\beta_1 = (u_{i-1} - 2u_i + u_{i+1})^2 + (-u_{i-1} + 3u_i - 3u_{i+1} + u_{i+2})^2 \qquad (18.92)$$
$$\beta_2 = (u_{i-1} - 2u_i + u_{i+1})^2 + (-u_{i-2} + 3u_{i-1} - 3u_i + u_{i+1})^2$$
$$\beta_3 = (-u_{i-3} + 4u_{i-2} - 5u_{i-1} + 2u_i)^2 + (-u_{i-3} + 3u_{i-2} - 3u_{i-1} + u_i)^2$$

Taylor expansion of $\beta_0, \beta_1, \beta_2, \beta_3$ yields

$$\beta_r = (u_i'' \Delta x^2) + (u_i''' \Delta x^3)^2 + \partial(\Delta x^6), \quad r = 0, ..., 3 \qquad (18.93)$$

therefore in the case $u_i'' = u_i''' = 0$ the OWENO scheme for $k = 4$ does not have the formal accuracy of standard WENO scheme $(2k-1) = 7$ but achieves better resolution of high wave numbers.

18.13   Compact WENO Approximation

It was pointed out in the previous chapter that two advantages of compact schemes compared to non-compact (explicit) counterpoints are: (i) they yield better accuracy and wavespace resolution for a smaller number of points in the stencil, and (ii) they offer an advantage in the implementation of boundary conditions because fewer boundary points must be handled. These advantages are achieved at a moderate increase in computational cost resulting from the matrix inversion. Pirozzoli [142], exploited the advantages of compact schemes and developed a narrow stencil fifth order accurate WENO compact scheme referred to from now on as CWENO5 scheme.

Starting with the same formulation as in the beginning of this chapter (see Eq. (18.12) through (18.17)) considered the following compact representation for the reconstruction of the numerical flux $\widehat{u}_{i+1/2}$

$$\sum_{l=-L_1}^{L_2} A_l \, \widehat{u}_{i+1/2+l} = \sum_{m=M_1}^{M_2} a_m \, \overline{u}_{i+m} \tag{18.94}$$

The Taylor series expansion of u up to order K around $x_{i+1/2}$ is

$$u(x) = \sum_{n=0}^{K-1} \frac{\partial^n u}{\partial x^n}\big|_{i+1/2} \, \frac{(x - x_{i+1/2})^n}{n!} + \mathcal{O}\left(h^K\right) \tag{18.95}$$

Recalling that the points values at the nodes, u_i which are considered as cell averages for the finite difference ENO reconstruction, $u_i \equiv \overline{u}_i$ and satisfy

$$\overline{u}_{i+m} = \frac{U_{i+m+1/2} - U_{i+m-1/2}}{\Delta x} \tag{18.96}$$

Considering the primitive function, $U(x)$, of $u(x)$ that satisfies

$$U(x) = \sum_{n=0}^{K-1} \frac{\partial^n u}{\partial x^n}\big|_{i+1/2} \, \frac{(x - x_{i+1/2})^{n+1}}{(n+1)!} + \mathcal{O}\left(\Delta x^{K+1}\right) \tag{18.97}$$

we obtain

$$\overline{u}_{i+m} = \sum_{n=0}^{K-1} \frac{\partial^n u}{\partial x^n}\big|_{i+1/2} \, \frac{1}{(n+1)!} \, [m^{n+1} - (m-1)^{n+1}]\Delta x^n + \mathcal{O}\left(\Delta x^{K+1}\right) \tag{18.98}$$

Inserting the Taylor series expansions for $\widehat{u}_{i+1/2}$ and $\overline{u}_{i+m}$ from Eq. (18.95) and Eq. (18.98), respectively, in Eq. (18.94) and matching the coefficients of like powers of h requiring k-th order accurate approximation (with $k \leq K$) obtain

$$(n+1) \sum_{l=-L_1}^{L_2} A_l \, l^n - \sum_{m=-M_1}^{M_2} a_m \, [m^{n+1} - (m-1)^{n+1}] = 0 \quad n = 0, ..., k-1$$

(18.99)

Solving the system of Eq. (18.99) for L_1+L_2+1, A_l unknowns and M_1+M_2+1, a_m unknowns obtain the coefficients of the compact upwind scheme. The only fifth order accurate compact scheme ($k = 5$) is obtained for $L_1 = L_2 = M_1 = M_2 = 1$. This is a compact scheme for the evaluation of the numerical flux that involves tridiagonal matrix inversion as follows

$$3\widehat{u}_{i-1/2} + 6\widehat{u}_{i+1/2} + \widehat{u}_{i+3/2} = \frac{1}{3} u_{i-1} + \frac{19}{3} u_i + \frac{10}{3} u_{i+1}$$

(18.100)

The numerical fluxes $u_{i+1/2}$ are related to the point values $u_i \equiv \overline{u}_i$ through the solution of the system of Eq. (18.100) yield a fifth order accurate, conservative evaluation of the derivative as

$$\frac{1}{\Delta x} \left(\widehat{u}_{i+1/2} - \widehat{u}_{i-1/2}\right) = u'(x_i) + \mathcal{O}\left(\Delta x^k\right)$$

(18.101)

Furthermore, the following explicit WENO approximations were obtained
5-th order upwind-biased approximation

$$\widehat{u}_{i+1/2} = \frac{1}{60} \left(2u_{i-2} - 13u_{i-1} + 47u_i + 27u_{i+1} - 3u_{i-2}\right)$$

(18.102)

7-th order upwind-biased approximation

$$\widehat{u}_{i+1/2} = \frac{1}{420} \left(-3u_{i-3} + 25u_{i-2} - 101u_{i-1} \right.$$
$$\left. + 319u_i + 214u_{i+1} - 38u_{i+2} + 4u_{i+3}\right)$$

(18.103)

The approximations of Eqs. (18.102) and (18.103) can be used as fifth- or seventh-order WENO schemes in smooth flow regions.

The resolution properties of the WENO approximations of Eqs. (18.100), (18.102), and (18.103) were demonstrated in [142]. The linear advection equation $u_x + au_x = 0$, $a > 0$ was considered for the evaluation of the resolving ability. The dispersion properties of the new schemes are compared with the classical, symmetric compact schemes in Fig. 18.9.

The upwind-biased approximations of Eqs. (18.97), (18.102), and (18.103) introduce in addition dissipation errors. The dissipation error is shown in Fig. 18.10. It can be seen that the compact WENO scheme of Eq. (18.100) has very small dissipation error, smaller than the fifth-order and seventh-order explicit schemes of Eqs. (18.102) and (18.103), for the range $0 < \widehat{k} < \pi/2$.

For non-periodic domains, the following fourth-order accurate boundary closures were proposed for the implementation of Eq. (18.100).

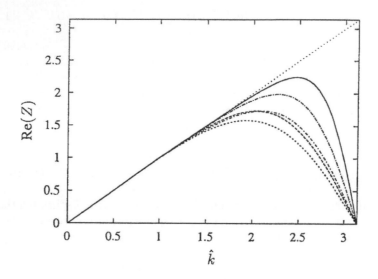

FIGURE 18.9: Dispersion of compact WENO schemes.

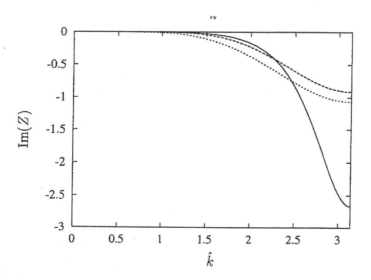

FIGURE 18.10: Dissipation of compact WENO schemes.

$$\widehat{u}_{1/2} = \frac{1}{12} \left(3u_0 + 13u_1 - 5u_2 + u_3 \right)$$

$$\widehat{u}_{N+1/2} = \frac{1}{12} \left(25u_N - 23u_{N-1} + 13u_{N-2} - 3u_{N-3} \right) \qquad (18.104)$$

A Roe-type, characteristic-wise, finite-difference, implementation of Piroz-zoli's compact CWENO5 scheme was recently presented by Ren et al. [153]. The formulation of Ren et al. [153] closely follow the hybrid compact-ENO ideas introduced by Adams and Shariff (1996) [3]. The implementation proposed in [153] for the scalar hyperbolic conservation law

$$\frac{\partial u}{\partial t} + \frac{\partial f(u)}{\partial x} = 0 \qquad (18.105)$$

is as follows
The numerical flux function $\widehat{f}_{i+1/2}$ for the evaluation of the conservative approximation of the derivative f_x is obtained by

$$3\widehat{f}_{i-1/2} + 6\widehat{f}_{i+1/2} + \widehat{f}_{i+3/2} =$$

$$= \frac{1}{3} \left(f_{i-1} + 19f_i + 10f_{i+1} \right) \quad \text{if } \widetilde{a}_{i+1/2} \geq 0 \qquad (18.106)$$

$$\widehat{f}_{i-1/2} + 6\widehat{f}_{i+1/2} + 3\widehat{f}_{i+3/2} =$$

$$= \frac{1}{3} \left(10f_i + 19f_{i+1} + f_{i+2} \right) \quad \text{if } \widetilde{a}_{i+1/2} < 0 \qquad (18.107)$$

where $\widetilde{a}_{i+1/2}$ is the numerical wave speed defined by

$$\widetilde{a}_{i+1/2} = \begin{cases} \frac{\widehat{f}_{i+1} - \widehat{f}_i}{u_{i+1} - u_i} & if u_{i+1} - u_i \neq 0 \\ \left(\frac{\partial f}{\partial u} \right)_i & \text{otherwise} \end{cases} \qquad (18.108)$$

Denoting $S_{i+1/2} = sign \left(\widetilde{a}_{i+1/2} \right)$ Eqs. (18.106) can be combined as

$$A_{i+1/2} \, \widehat{f}_{i-1/2} + \widehat{f}_{i+1/2} + B_{i+1/2} \, \widehat{f}_{i+3/2} = b_{i+1/2} \qquad (18.109)$$

where

$$A_{i+1/2} = \frac{1}{3} + \frac{S_{i+1/2}}{6} \ , \qquad B_{i+1/2} = \frac{1}{3} + \frac{S_{i+1/2}}{6}$$

$$b_{i+1/2} = \frac{1 + S_{i+1/2}}{2} \left(\frac{1}{18} f_{i-1} + \frac{19}{18} f_i + \frac{5}{9} f_{i+1} \right)$$

$$+ \frac{1 - S_{i+1/2}}{2} \left(\frac{5}{9} f_i + \frac{19}{18} f_{i+1} + \frac{1}{18} f_{i+2} \right)$$

The compact WENO scheme of Eq. (18.109) gives very satisfactory results for smooth flow regions. However, at the discontinuities of the solution the Gibbs phenomenon will occur that contaminates the solution and eventually leads to nonlinear instability.

For these regions, a hybrid method is proposed in [153] where the compact scheme of Eq. (18.109) is coupled with the WENO procedure. The hybrid scheme is the weighted average of the compact scheme in Eq. (18.109) and WENO of Section 18.7.

This hybrid scheme [153] has the following form

$$\sigma_{i+1/2} \, A_{i+1/2} \, \widehat{f}_{i-1/2} + \widehat{f}_{i+1/2} + \sigma_{i-1/2} \, B_{i+1/2} \, \widehat{f}_{i+3/2} = \widehat{c}_{i+1/2} \qquad (18.110)$$

where σ is the weight and

$$c_{i+1/2} = \sigma_{i+1/2} \, b_{i+1/2} + (1 - \sigma_{i+1/2}) \, f_{i+1/2}^{WENO}$$

For $\sigma_{i+1/2} = 1$ Eq. (18.110) reduces to the compact scheme of Eq. (18.109) and for $\sigma_{i+1/2} = 0$ becomes a WENO scheme. It is necessary, therefore, that the weight be directly related to the smoothness of the numerical solution. This is achieved through the definition of a smoothness indicator $r_{i+1/2}$ are given by

$$r_{i+1/2} = min \, (r_j, r_{j+1}) \qquad (18.111)$$

$$r_j = \frac{2 \, |\Delta f_{i+1/2} \, \Delta f_{i-1/2}| + \varepsilon}{(\Delta f_{i+1/2})^2 + (\Delta f_{i-1/2})^2 + \varepsilon}$$

$$\Delta f_{i+1/2} = f_{i+1} - f_i$$

$$\varepsilon = \frac{0.9 r_c}{1 - 0.9 r_c} \, \xi^2 \, , \, \xi > 0 \quad r_c \simeq 1$$

Note that Pirozzoli simply defined the smoothness indicator as $r_{i+1/2} = |f_{j+1} - f_j|$ and determined $\sigma_{i+1/2}$ by

$$\sigma_{j+1/2} = \begin{cases} 1 \; if \; \; r_{j-1/2} \le \tilde{r}_c \; \; and \; r_{j+1/2} \le \tilde{r}_c \; \; and \; r_{j+3/2} \le \tilde{r}_c \\ 0 \qquad\qquad\qquad\qquad otherwise \end{cases} \qquad (18.112)$$

where r_c is a problem dependent threshold. Using the definition of Ren et al. [153] for $r_{i+1/2}$ the weight $\sigma_{i+1/2}$ is computed by

$$\sigma_{i+1/2} = min \, \left(1, \, \frac{r_{i+1/2}}{r_c} \right) \qquad (18.113)$$

18.14   Application of the Hybrid Compact-WENO Scheme for the Euler Equations

A characteristic-wise approach is used in [153] with the hybrid, compact WENO scheme for the numerical solution of the Euler equations. The numerical flux evaluation is performed in the following steps:

1. Compute an average state $u_{i+1/2}$ by the simple mean $u_{i+1/2} = (u_i + u_{i+1})/2$ or the Roe average

2. Compute the eigenvalues $\lambda_{i+1/2}$ $i + 1/2 = 1, 2, 3, 4$ and the left eigenvectors $\mathbf{l}_{j+1/2}^{(m)}$ at the average state.

3. Perform local characteristic decomposition

$$\mathbf{w}_n^{(m)} = \mathbf{l}_{i+1/2} \, F_n \qquad \begin{matrix} m = 1, 2, 3, 4 \\ n = i - 1, ..., i + 2. \end{matrix}$$

4. Define

$$s_{i+1/2}^{(m)} = sign \, (\lambda_{i+1/2}^{(m)})$$
$$r_{i+1/2}^{(m)} = min \, (r_j^{(i)}, \, r_{i+1}^{(i)})$$
$$r_i^{(m)} = \frac{|2\Delta w_{i+1/2}^{(m)} \, \Delta w_{i-1/2}^{(m)}| + \varepsilon}{(\Delta w_{i+1/2}^{(m)})^2 + (\Delta w_{i-1/2}^{(m)})^2 + \varepsilon}$$
$$\sigma_{i+1/2}^{(m)} = min \, \left(1, \, \frac{r_{i+1/2}^{(m)}}{r_c} \right).$$

5. Apply the hybrid, compact-WENO scheme for the local characteristic variables as

$$\sigma_{i+1/2}^{(m)} \quad A_{i+1/2}^{(m)} \, w_{i-1/2}^{(m)} + w_{i+1/2}^{(m)} + \sigma_{i+1/2}^{(m)} \, B_{i+1/2}^{(m)} \, w_{i+3/2}^{(m)} = c_{i+1/2}^{(m)}$$

$$A_{i+1/2}^{(m)} = \frac{2 + s_{i+1/2}^{(m)}}{6} \quad , \quad B_{i+1/2}^{(m)} = \frac{2 - s_{i+1/2}^{(m)}}{6}$$

$$c_{i+1/2}^{(m)} = \sigma_{i+1/2}^{(m)} \, b_{i+1/2} + (1 - \sigma_{i+1/2}^{(m)}) \, w_{i+1/2}^{(m)WENO} \qquad (18.114)$$

$$b_{i+1/2}^{(m)} = \frac{1}{18} \left(\frac{1 + s_{i+1/2}^{(m)}}{2} \right) \, (w_{i-1}^{(m)} + 19w_i^{(m)} + 10w_{i+1}^{(m)}) +$$

$$= + \frac{1}{18} \left(\frac{1 - s_{i+1/2}^{(m)}}{2} \right) \, (10w_i^{(m)} + 19w_i^{(m)} + w_{i+2}^{(m)})$$

where $w^{(m)WENO}$ is computed with the WENO scheme of Section 18.5.

6. Project Eq. (18.114) back to conservative variables and solve the following block-tridiagonal system of equations to obtain the numerical flux in the conservative variable space

$$[A]_{i_{1}/2} \, \mathbf{F}_{i-1/2} + L_{i+1/2} \, \mathbf{F}_{i+1/2} + [B]_{i+1/2} \, \mathbf{F}_{i+3/2} = \mathbf{c}_{i+1/2} \qquad (18.115)$$

$$[\mathbf{A}]_{i+1/2} = \begin{bmatrix} \sigma_{i+1/2}^{(1)} \, A_{i+1/2}^{(1)} \, \mathbf{l}_{i+1/2}^{(1)} \\ \\ \sigma_{i+1/2}^{(4)} \, A_{i+1/2}^{(4)} \, \mathbf{l}_{i+1/2}^{(4)} \end{bmatrix} \quad , \quad [\mathbf{B}]_{i+1/2} = \begin{bmatrix} \sigma_{i+1/2}^{(1)} \, B_{i+1/2}^{(1)} \, \bar{l}_{i+1/2}^{(1)} \\ \\ \sigma_{i+1/2}^{(4)} \, B_{i+1/2}^{(4)} \, \bar{l}_{i+1/2}^{(4)} \end{bmatrix}$$

$$L_{i+1/2} = \begin{bmatrix} \mathbf{l}_{i+1/2}^{(1)} \\ \mathbf{l}_{i+1/2}^{(2)} \\ \mathbf{l}_{i+1/2}^{(3)} \\ \mathbf{l}_{i+1/2}^{(4)} \end{bmatrix} \qquad \mathbf{c}_{i+1/2} = \begin{bmatrix} c_{i+1/2}^{(1)} \\ c_{i+1/2}^{(2)} \\ c_{i+1/2}^{(3)} \\ c_{i+1/2}^{(4)} \end{bmatrix} .$$

18.15   Applications of ENO and WENO

It is well known in CFD that the solution error for a given problem strongly depends on the smoothness of the computational grid. Uniform or smoothly varying grids always yield smaller solution errors than non-uniform or non-smooth grids. For example, many second-order structured or unstructured grid CFD algorithms often degrade into first-order for non-smooth grids. Analogous behavior should be expected for higher order methods. The dependence on grid smoothness of the numerical solutions obtained with ENO was investigated by Casper and Atkins [30].

The two basic formulations, finite volume (FV) and finite difference (FD), of ENO were considered by Casper and Atkins (1994) [30]. The FD and FV fourth-order accurate algorithms were compared for accuracy, sensitivity to grid irregularities, wave resolution, and computational efficiency. It was found [30] (see Fig. 18.11) that for fourth order design accuracy the two-dimensional FD ENO numerical solution is approximately two-times more efficient than the FV ENO numerical solution. The CPU time for three dimensional implementation (see Fig. 18.12) shows more dramatic increase for the FV formulation. For example, the CPU time required for FV ENO with fourth-order formal accuracy is approximately three times higher than the FD ENO with equivalent formal accuracy. It was shown, however, in [30] that the formal accuracy of the FD ENO can only be achieved with smooth grids. The finite volume implementation was found [30] less sensitive to derivative discontinuities, whether in the computational mesh or the solution. Therefore, for applications where the computational domain is known to be sufficiently smooth and can be suitably structured the FD ENO algorithm must be the method of choice. Taking, however, into account that the formal accuracy of the FD ENO can significantly degrade for non-smooth meshes [30] (see Fig. 18.13), for problems with complex geometries it may pay to use the more expensive FV algorithm.

The resolving ability and the performance for long time integration of the WENO and other numerical schemes presented in the previous sections was evaluated first for simple linear problems by Ekaterinaris [54]. The performance of centered and WENO schemes for aeroacoustics using the full Euler equations was also considered. In addition, numerical solutions for problems with strong shocks are obtained on curvilinear coordinates with centered schemes and characteristic based filters and compared with solutions obtained with WENO schemes. The computed solutions are compared with available exact solutions.

Long time integration is important in many practical applications, such as aeroacoustics, LES and helicopter rotor aerodynamics where the tip vortex and the rotor wake need to propagate for long distances. Therefore the ability of symmetric, centered compact and non-compact schemes as well as several WENO high-order accurate stencils was also evaluated in [54], for wave convection. Sufficiently accurate convection of simple Gaussian pulses, $u(x, t = 0) = e^{-ax^2}$, with twelve points per waveform (not shown here) was achieved even with explicit fourth-order accurate in space methods.

Further accuracy test are shown for linear wave convection by the one-dimensional wave equation $u_t + cu_x = 0$. Numerical solutions with unit wave

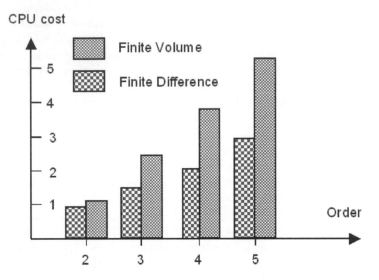

FIGURE 18.11: Comparison of the required CPU time for FV and FD ENO in two dimensions.

speed, $c = 1$, are obtained for long propagation times. Time marching is performed with the third-order accurate Runge-Kutta method. A time step of $\Delta t = 0.1$, which is below the stability limit of the method, is used for all tests in order to keep time integration errors at low level.

The first test is propagation of a high frequency sinusoidal wave $u_0(x) = \sin(\pi x/6)$. Convection of the sinusoidal wave is obtained with $\Delta x = 0.1$ (twelve points per wavelength) and periodic boundary conditions. The mean square error obtained from explicit space discretizations with the symmetric fourth-, and sixth-order accurate schemes and with WENO schemes of fifth- and seventh-order accuracy Eqs. (18.46) and (18.49) is shown in Table 18.V. It appears that only the schemes with formal accuracy more than five are capable to obtain sufficiently accurate solution for long time integration.

TABLE 18.V

L_2 **error at** $T = 200$ **for the convection of** $\sin(\pi x/6)$ **with explicit schemes**

Explicit schemes	4th order centered	6th order centered	5th order WENO	7th order WENO
L_2 error	0.2E00	0.11E-2	0.12E-1	0.14E-3

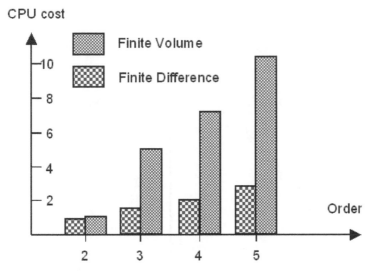

FIGURE 18.12: Comparison of the required CPU time for FV and FD ENO in three dimensions.

Further evaluation of high-order accurate symmetric explicit and compact schemes and WENO stencils to perform linear wave convection is carried out using the following modulated wave $u_0(x) = \cos(a|x|) \, e^{-b|x|}$ with $a = 3/4$ and $b = 1/10$ as initial condition. For explicit schemes, time integration is performed until $T = 200$. For compact schemes, which have increased resolving ability, time integration is performed until the final time $T = 500$. A comparison of the solutions computed using explicit high-order symmetric schemes and WENO stencils with the exact result is shown in Fig. 18.14. It can be seen that at least sixth order of accuracy is needed for long time propagation. The mean square error of the solutions computed with different schemes is shown in Tables 18.VI and 18.VII. It appears that the 9th order accurate WENO stencil provides uniformly high order of accuracy for smooth initial data.

TABLE 18.VI

L_2 **error at** $T = 200$ **for the convection of** $\cos(a|x|) \, e^{-b|x|}$ **with explicit schemes**

Scheme	6th order centered	5th order WENO	7th order WENO	9th order WENO
L_2 error	0.4E-3	0.1E-2	0.7E-4	0.7E-5

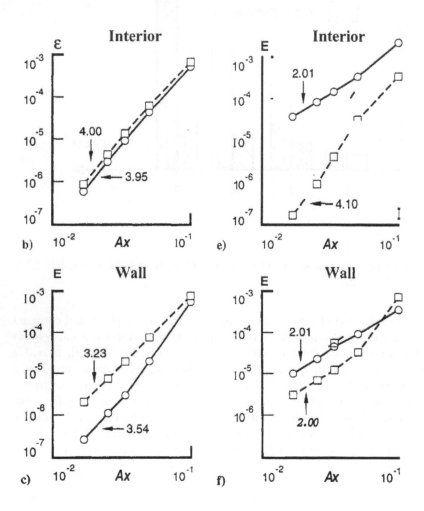

FIGURE 18.13: L1 entropy errors, steady two-dimensional channel flow; (rom Ref. [30]).

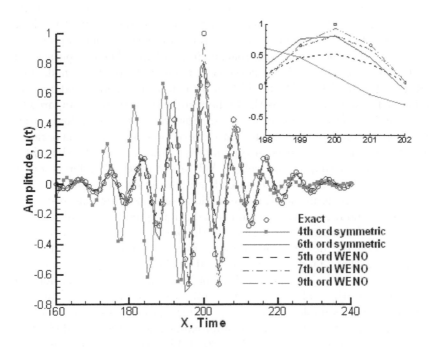

FIGURE 18.14: Comparison of the exact solution $u(x,0) = cos(a|x|exp(-b|x|))$ at $T = 200$ with results computed with explicit schemes.

TABLE 18.VII

L_2 error at $T = 500$ for the convection of $\cos(a|x|)\ e^{-b|x|}$ with compact schemes

Compact schemes	5th order WENO	6th order centered	8th order centered	10th order centered
L_2 error	0.19E-3	0.57E-4	0.92E-5	0.53E-5

The mean square error for long time integration, $T = 500$, of high order compact schemes (see Table 18.VII) also remains at low levels. It can be seen that compact schemes with formal order of accuracy more that four perform adequately. The high-order accurate compact schemes appear to be particularly suitable for linear aeroacoustic problems. The computational time of the explicit schemes is proportional to the width of the stencil. The computational cost of the 8th and 10th order compact schemes that require pentadiagonal matrix inversion is the highest and almost double compared to the time required by the explicit schemes. However, use of very high order centered methods may be required for wave convection over long time periods.

Symmetric schemes with spectral-type filtering and characteristic-based filters of the previous chapter, as well as WENO schemes of different order of accuracy are used to compute spread and reflection of a pressure disturbance. The full nonlinear Euler equations are used for this test. At the far field boundaries of the domain a radiation boundary condition was used. On the solid surface the normal to the wall velocity component was set to zero while the density and pressure were extrapolated from the interior assuming that $\partial\rho/\partial n = \partial p/\partial n = 0$ or $\partial\rho/\partial y = \partial p/\partial y = 0$. It was found that it is required to use high-order accurate approximations of the derivatives at the wall in order to retain the accuracy of the numerical solution. For example, the pressure is extrapolated using the following one-seeded, fourth-order accurate approximation of the first derivative $(dp/dy)_1 = (-25p_1 + 48p_2 - 36p_3 + 16p_4 - 3p_5)/12$. The computed results are compared with the exact solution which gives the time variation of an initial pressure disturbance $p\ (x,y) = \exp\left\{-\ln 2\ [x^2 - (y - y_0)^2]\right\}$. The initial disturbance is located at $y = y_0$, and as it spreads, reflects from a solid wall at $y = 0$.

A comparison of the solution computed on an artificially distorted mesh by WENO schemes with $r = 3$, and 5 is shown in Fig. 18.15. It can be seen that the solution computed with the $(r = 5)$ 9th order accurate WENO scheme on a the baseline, 100x50 point grid, which provides twelve points per wave, is almost indistinguishable from the solution computed with the $(r = 3)$ 5th order accurate WENO scheme on a 200x100 point grid, refined in both directions. It is important to note that the full WENO scheme with the appropriate smoothness measures must be used for the propagation of the pressure disturbance with the nonlinear Euler equations. Numerical solutions

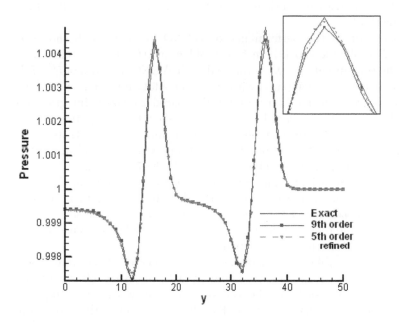

FIGURE 18.15: Comparison of WENO scheme computations with the exact solution.

of the linearized Euler equations that describe the propagation of the acoustic-type pressure disturbance may be possible only with the optimal WENO stencils of Eqs. (18.46) and (18.49).

Several detailed comparisons demonstrated that both high-order centered schemes and the finite-difference WENO of 7th order or higher are appropriate for aeroacoustic computations of subsonic flows. For flows with shocks, however, WENO schemes appear to be more appropriate for aeroacoustics, because computation of these flows require high values of the ACM filter parameter ($\kappa > 0.5$) in order to prevent numerical oscillations. Numerical solutions obtained for the same problem on artificially distorted meshes have demonstrated that the accuracy of the solution does not deteriorate when the definition of the metrics is consistent and the metrics are computed with a high order method.

A demonstration of the performance of WENO for flows with shocks is given next. The oblique shock reflection problem at $M_\infty = 2.9$ is chosen as test case. The pressure at $y = 0.5$ computed with WENO schemes of 5th, 7th, and 9th order accuracy is compared with the exact solution in Fig. 18.16. The computations were performed on a uniformly spaced 200x50 point grid in a domain $-2.0 \leq x \leq 2.0$, $0 \leq y \leq 1.0$. At the left inflow boundary, free stream was specified. At the right outflow boundary all quantities were extrapolated. On the solid wall at $y = 0$, slip boundary condition was specified, and at the top the flow quantities were specified as: $\rho = 1.69997$, $u = 2.61934$, $v = -0.506$, $p = 1.528$. Sufficient number of ghost points, depending on the order of the scheme, were used at the edges of the domain in order to retain the formal order of the scheme. For example, computations with the seven-point wide WENO5 scheme required three ghost points while solutions with the 6th order accurate ACM method require two ghost points.

At steady state, an oblique incident shock and a reflected shock were generated. The comparison of the computed pressure with the exact solution of Fig. 18.16 shows that as the order of the WENO scheme increases the computed solution approaches the exact solution. For all computations, the shocks are captured within two cells. The pressure field obtained from the numerical solutions with the 4th order accurate compact centered scheme with $\kappa = 0.7$ and the 5th order accurate WENO scheme on an artificially distorted mesh is shown in Fig. 18.17. Both solutions were computed with the explicit time marching scheme. It can be seen that both methods can capture the oblique strong shock without oscillations. Furthermore, the artificially distorted mesh does not cause oscillations.

A comparison of the computed pressure at $y = 0.5$ from the solution obtained with the 5th order accurate WENO scheme and the solutions obtained with different values of the ACM parameter is shown in Fig. 18.18. It appears that the computed solution is sensitive to the selection of the ACM parameter. Furthermore, the choice of the upwind TVD limiter affects the solution.

The numerical solution for the same problem was computed with the implicit time marching schemes. The convergence rates of the numerical solu-

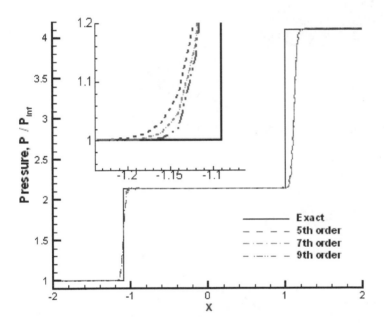

FIGURE 18.16: Comparison of the computed pressure at y = 0.5 with the exact solution.

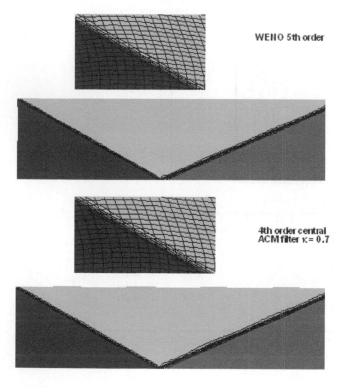

FIGURE 18.17: Pressure field computed with WENO 5th order and with the 4th order centered scheme with characteristic-based filter (ACM parameter$\kappa = 0.7$) on an artificially distorted mesh.

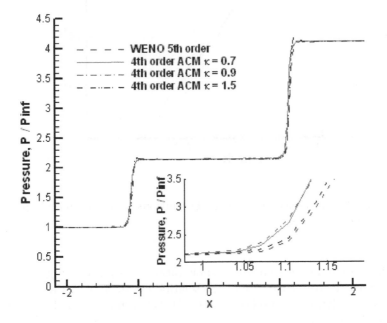

FIGURE 18.18: Effect of ACM parameter on the computed pressure at y = 0.4.

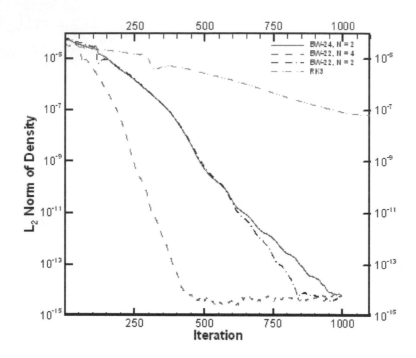

FIGURE 18.19: Convergence rate of WENO solutions computed with implicit and explicit time marching.

tions obtained for different number of subiterations is shown in Fig. 18.19. For reference, the convergence rate of the solution obtained with the explicit third order Runge-Kutta method of Ref. [68] is shown in the same figure. At convergence all solutions were the same and computed pressure and density obtained from implicit or explicit time marching were almost identical.

Computations of supersonic flows over a cylinder at various Mach numbers are shown next. These solutions are obtained using 5th and 7th order accurate WENO schemes. An algebraically generated 181 x 51 point grid was used for this computation of supersonic flows over the cylinder. Similarly to the shock reflection case sufficient number of ghost points depending on the order of the scheme was used at the edges of the domain. At the inflow free stream supersonic flow was specified. At the outflow all quantities were extrapolated from the interior. On the cylinder solid surface the normal to the surface velocity component was set to zero and all the other quantities were extrapolated from the interior using high order extrapolation and assuming that the normal derivative is zero.

The computed pressure and entropy fields at $M_\infty = 5.0$ are shown in Fig.

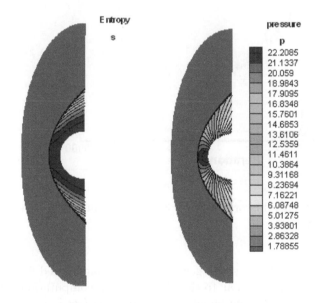

FIGURE 18.20: Computed entropy and pressure fields with the 9th order accurate WENO scheme at $M = 5$.

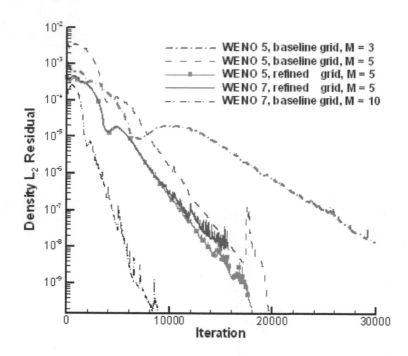

FIGURE 18.21: Convergence history at $M = 3$, $M = 5$, and $M = 10$ with the baseline and refined grids.

18.20. The resolution of the strong shock generated by the high-speed flow is captured without oscillations. Similar to the oblique shock computations of the previous chapter the shock for the supersonic cylinder flow (see Fig. 18.20) is captured within two cells. The convergence rate obtained at different Mach numbers, order of accuracy, and for baseline (91 x 51) and refined (181 x 101) grids is shown in Fig. 18.21. All computations were obtained at the same time step and the third order TVD Runge-Kutta method of Ref. [68]. For all cases, the convergence was satisfactory and the solution practically remained unchanged when the residuals drop four orders of magnitude. A comparison of the computed pressure distributions for the grid line on the symmetry axis that passes through the stagnation point is shown in Fig. 18.22. It can be seen that the shock is captured within two computational cells and the solution is free from oscillations.

Inviscid flows solutions over a NACA-0015 airfoil are computed using the 5th order accurate WENO scheme. The computed pressure fields at transonic and supersonic speed are shown in Figs. 18.26 and 18.27. The solutions were computed with the explicit time marching scheme on a 261 x 51 point, C-type

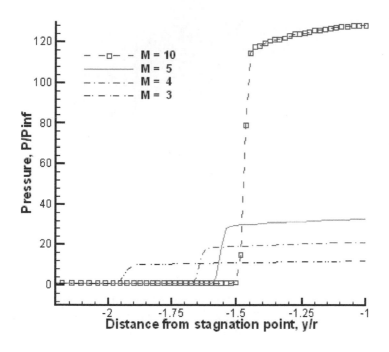

FIGURE 18.22: Computed pressure on the symmetry line at $M = 3$, $M = 5$, and $M = 10$.

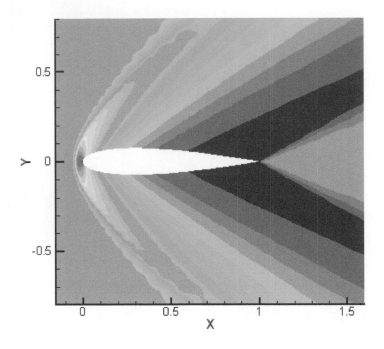

FIGURE 18.23: Computed pressure field over a NACA-0015 airfoil at $M = 2$, and $\alpha = 2$ deg.

grid. The airfoil grid included three ghost points at the edges of the domain in order to use the WENO5 scheme for the entire domain without dropping the stencil accuracy at the airfoil surface and the wake. The supersonic and transonic flow computations were obtained on the same grid. For both flow speeds, a smooth solution is obtained on the highly stretched, high aspect ratio C-type grid.

The computed pressure field at $M_\infty = 2.0$ is shown in Fig. 18.23. The computed pressure field shows adequate resolution of the leading edge bow shock and the two shocks at the trailing edge despite of the coarseness of the grid. The same type of inflow/outflow and solid wall boundary conditions as the supersonic cylinder flow was used. At the wake of the C-type grid averaging was used. For the transonic flow computation at $M_\infty = 0.8$, the shocks of the upper and lower surface shown by the pressure contours of Fig. 18.24 are well resolved. For this computation the inflow and outflow boundary conditions were specified using one-dimensional Riemann invariants. A comparison of the computed surface pressure coefficient for transonic flow over the NACA-0012 airfoil with the experimental data is shown in Fig. 18.25.

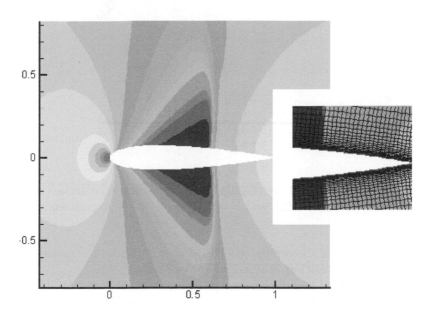

FIGURE 18.24: Computed pressure field over a NACA-0015 airfoil at $M = 0.8$, and $\alpha = 0.1$ deg.

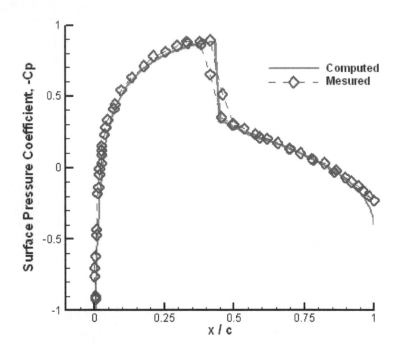

FIGURE 18.25: Computed of the computed surface pressure coefficient over a NACA-0015 airfoil at $M = 0.8$, and $\alpha = 0.1$ deg. with the experiment.

The overall agreement of the computed inviscid solution with the experiment is satisfactory and the shock is resolved within two cells.

Numerical solutions of two-dimensional viscous transonic flow over airfoils were obtained using a second-order accurate FV ENO or WENO scheme by Yang et al. [199]. The LU-SGS scheme of Section 18.3 was used for time integration to avoid stability limitations from the highly clustered viscous meshes. The computations of Yang et al. [199] showed sharp capturing of discontinuities without oscillations. It was found that after the residuals of the ENO2 scheme have decayed for three orders of magnitude the convergence levelled off. In contrast, as expected (see remarks in Section 18.4) a monotone convergence was achieved with the WENO2 scheme. The computed surface pressure distributions from the ENO and WENO algorithms were almost indistinguishable. Very good agreement with the measured pressure coefficient distribution was found in [199] for three-dimensional flow over the ONERA M6 computed with the WENO2 scheme. Comparisons of the surface pressure distribution are shown in Fig. 18.26.

The resolution and efficiency of high-order accurate WENO schemes for

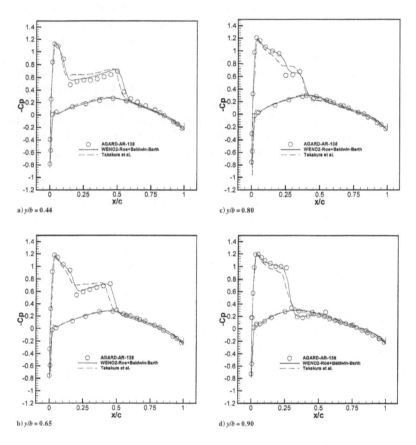

FIGURE 18.26: Steady pressure distributions for ONERA M6 wing at, $M = 0.8395, \alpha = 3.06$ deg., and $Re = 2.6 \times 10^6$; (from Ref. [199]).

the computation of flows containing both discontinuities and complex flow features was recently demonstrated by Shi et al. (2003) [163]. The first representative numerical example was the double Mach reflection, a problem that includes strong shock waves and very complex flow features. This problem was initially proposed by Woodward and Colella (1984) [198] and has been used extensively in the literature as a test for high resolution schemes. A Mach 10, right moving shock at an angle of 60° degrees is reflected from a wall (for more details see [198]). The flow is computed as inviscid and the results are displayed in Fig. 18.27 at $t = 0.2$ in the domain $[0, 3] \times [0, 1]$. Three different uniform meshes $h = 1/240$, $h = 1/480$, and $h = 1/960$ were used in [163]. The computed solutions are compared in Fig. 18.27. It is clear that *WENO9* with $h = 1/240$ produces qualitatively the same resolution as *WENO5* with $h = 1/480$. The same is true for *WENO5* with $h = 1/960$ and *WENO9* with $h = 1/480$. It is evident that the resolution increases consistently with the order of accuracy and mesh refinement. Furthermore, the ability to obtain the same resolution with half the mesh by increasing the order of the method (compare *WENO5* with $h = 1/960$ and *WENO9* with $h = 1/480$) is clearly demonstrated.

The second problem considered in [163] is the Reyleigh-Taylor (RT) instability. This instability happens on an interface between fluids with different densities due to the motion of the heavy fluid. The numerical solution of the RT problem demonstrates the ability of WENO schemes to consistently capture smooth, complex flow features with increased resolution either by grid refinement or the increase of the order of the method. Inviscid flow in a computational domain $[0, 1/4] \times [0, 1]$ was computed in [163] with the heavy fluid with density $\rho = 2$ below, the interface at $y = 1/2$, and the light fluid with $\rho = 1$ above the interface. The solutions of Fig. 18.28 were computed on uniform meshes with $h = 1/240$, $1/480$, $1/960$, and $1/1920$. Again the *WENO9* scheme yields the same resolution with the *WENO5* scheme at double grid resolution. In both cases of Figs. 18.27 and 18.28, the comparable resolution obtained with the *WENO9* scheme using half the number of grid points in each direction than *WENO5* implies significant saving in computational time since the *WENO9* scheme needs only approximately 30% more CPU time than the *WENO5* scheme.

The dynamics of shock vortex interactions and the coupling of counter-rotating compressible vortices interacting with a planar shock wave were investigated by Grasso and Pirozzoli in [70], [71], and [141], respectively. The numerical solutions of the inviscid, compressible flow equations was obtained using a finite volume WENO scheme for Cartesian type meshes. The simulations for the single vortex-shock wave interaction [70] initiated with a homoentropic Taylor vortex. Results from the simulations of [70] are shown in Fig. 18.37 and 18.38. Fig. 18.29 shows the evolution of shock-vortex interaction with the computed Schlieren plot for a shock at $M_s = 1.2$ and vortex Mach number $M_v = 0.8$. Fig. 18.30 shows the computed pressure contour plots at the same times with Fig. 18.29.

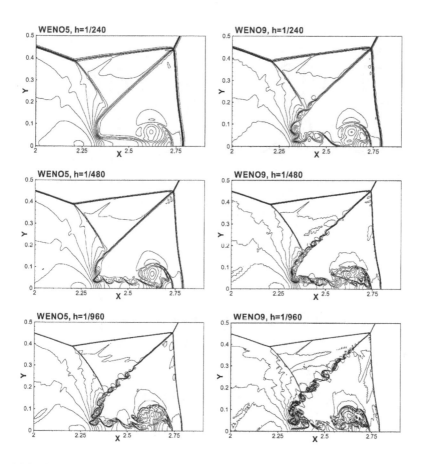

FIGURE 18.27: Double Mach reflection problem. Blown-up region around the double Mach stems. Density; 30 equally spaced contour lines from $p = 1.5$ to $p = 22.9704$. Left from top to bottom: fifth order WENO results with $h = 1/240, 1/480, 1/960$; right from top to bottom: ninth order WENO results with $h = 1/240, 1/480, 1/960$; (from Ref. [163]).

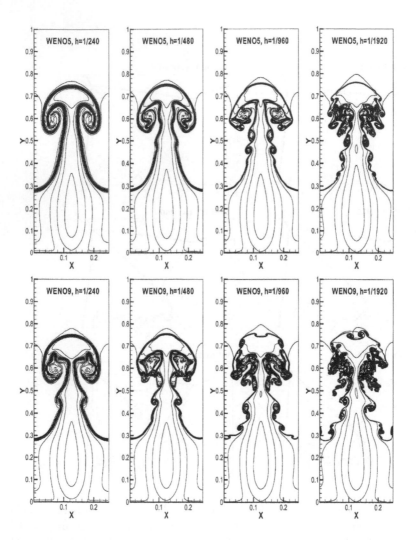

FIGURE 18.28: Rayleigh-Taylor Instability. Density; 15 equally spaced contour lines from $\rho = 0.952269$ to $\rho = 2.14589$. Top from left to right: fifth order WENO results with $h = 1/240, 1/480, 1/960, 1/1920$; bottom left to right: ninth order WENO results with $h = 1/240, 1/480, 1/960, 1/1920$; (from Ref. [163]).

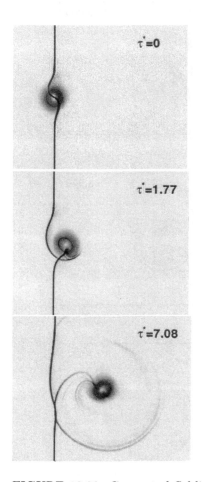

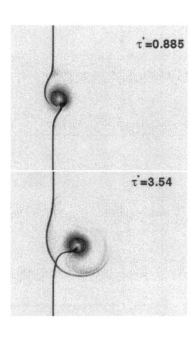

FIGURE 18.29: Computed Schlieren for shock-vortex interaction (from Ref. [70]).

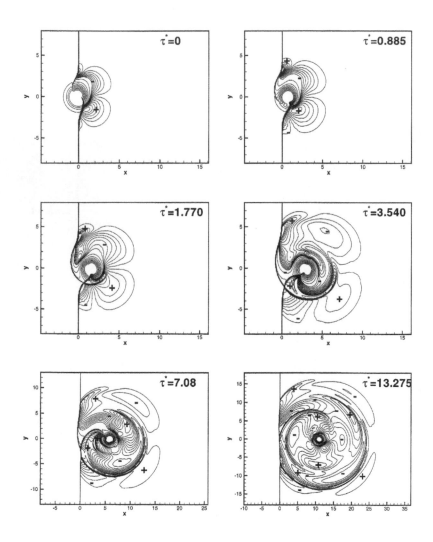

FIGURE 18.30: Computed pressure field for shock vortex interaction (from Ref. [70]).

The dynamics of the interaction of shock-wave at $M_s = 5.0$ colliding with a counter-rotating vortex pair at $M_v = 0.26$ from [71] are shown in Fig. 18.31. The numerical Schlieren for shock wave-colliding vortex pair interaction at $M_v = 0.9$, $M_s = 1.2$ [141] is shown in Fig. 18.32. The acoustic pressure field from [141] at $M_v = 0.1$, $M_s = 1.2$ is shown in Fig. 18.41. It can be seen that in all cases the dynamics of the interaction and the flow filed structure are very complex. The computed solutions in [70] and [71] were found in very good agreement with measurement and analytic solutions from acoustic analogies. These results show that high order WENO schemes are appropriate for the computation of noise. Further investigation of three dimensional shock ring-vortex interaction was carried out by Pirozzoli (2004) [143] using the compact WENO scheme he developed in [142]. The computed complex interaction of the impact of a vortex ring $M_v = 0.25$ on a planar shock $M_s = 1.5$ is shown in Fig. 18.42.

Adams [5] used the hybrid compact-ENO finite difference scheme of [3] to perform direct simulation of the flow over a compression ramp. The scheme used for this DNS is 5th order accurate in smooth flow regions and around the discontinuities becomes the 4th order ENO scheme. The viscous terms in [3] are discretized with the 6th order accurate compact finite-difference scheme of Lele [119]. Time marching of the DNS in [5] is performed with a RK3 method. Computed Schlieren from the DNS in Fig. 18.34 are in good agreement with similar measurement shown for comparison in Fig. 18.35. The computed density field is shown in Fig. 18.36. The good agreement of the DNS of [5] with the experiments demonstrates that high order accurate ENO discretization yields the necessary resolution and keeps the numerical diffusion at a low level. These are key ingredients required for accurate capturing of turbulent fluctuations in compressible turbulence DNS and LES. Additional computations with ENO and WENO schemes can be found in [144], [46], [74], [121], [122], [32].

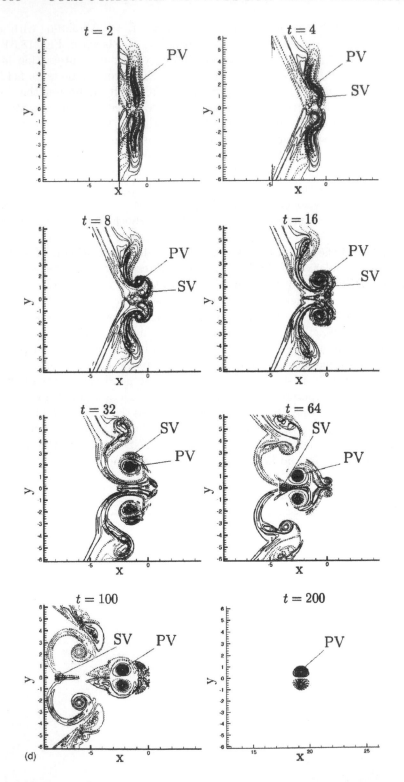

FIGURE 18.31: Computed pressure field for shock counter-rotating vortex
pair interaction (from Ref. [71])

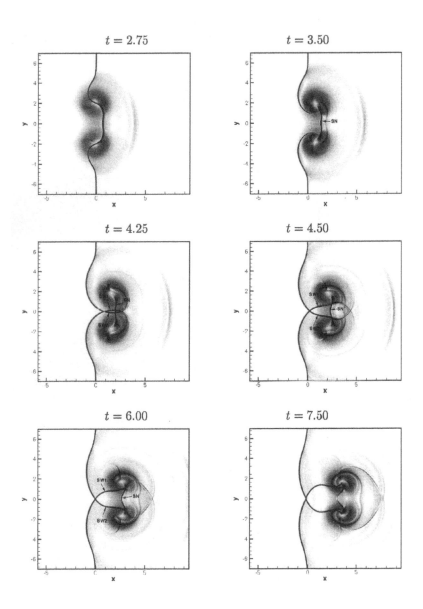

FIGURE 18.32: Shock wave-colliding vortex pair interaction: numerical Schlieren at $M_v = 0.9$, $M_s = 1.2$, $d = 4r_v$, Type-V interaction; (SW), shock wave; (SN), diffracted shock (from Ref. [71]).

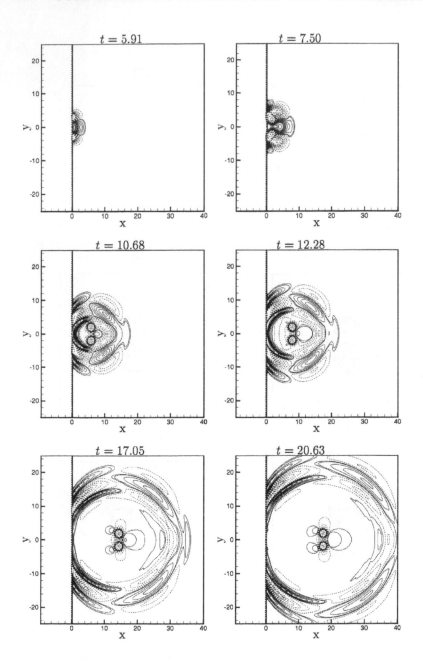

FIGURE 18.33: Shock wave-colliding vortex pair interaction: acoustic pressure field at $M_v = 0.1$, $M_s = 1.2$, $d = 4r_v$, Type-I interaction, $-0.1 < p' < 0.5$, 72 contour levels. Dashed lines stand for $p' < 0$ while solid lines stand for $p' > 0$ (from Ref. [71]).

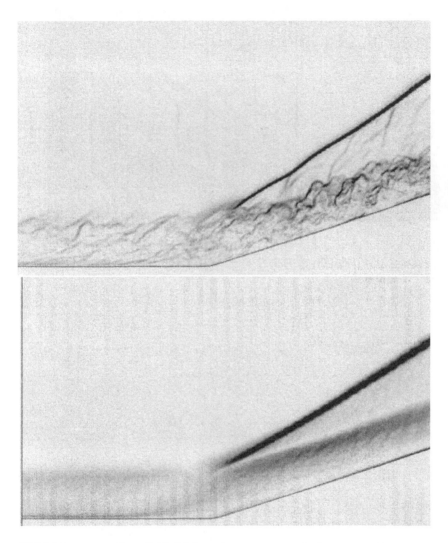

FIGURE 18.34: Flow-field Schlieren imitation with density gradient contours, (a) instantaneous x-average, (b) x-average and time average using 100 samples (from Ref. [5]).

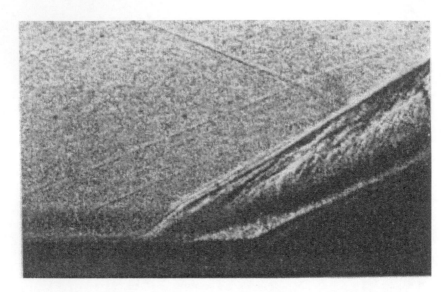

FIGURE 18.35: Flow-field experimental Schlieren visualization of a 25 deg. compression ramp at $M = 2.9, Re_\theta = 9600$ provided by A. Zheltovodov, ITAM, Novosibirsk.

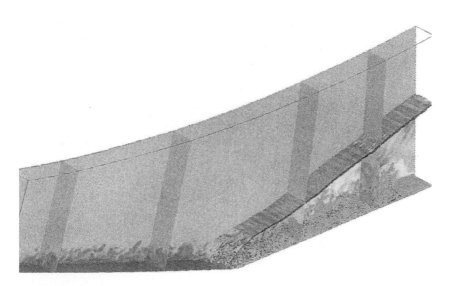

FIGURE 18.36: Density at the wall, in the plane $x = 2.9$ and four cross-flow planes; shock-surface with $\partial_{x_j} u_j = -0.4$ (from Ref. [5]).

Chapter 19

The Discontinuous Galerkin (DG) Method

John A. Ekaterinaris, School of Mechanical and Aerospace Engineering, University of Patras, Greece *and* Foundation for Research and Technology-Hellas, Institute of Applied and Computational Mathematics, 71110 Heraklion, Greece, ekaterin@iacm.forth.gr

19.1   The Discontinuous Galerkin (DG) Method

Generation of structured three-dimensional meshes even in domains with moderate complexity, such as a wing body junction, is not trivial. On the other hand, the performance of most high-resolution accurate methods depends on the smoothness of the grid. The difficulty in generating smooth structured grids for complex geometries has promoted the development of finite-volume algorithms for unstructured grids [83], [108], [131], [186], [191]. These unstructured grid methods are second order accurate. Higher-order finite-volume schemes were pioneered by Barth and Frederickson [16] with the k-exact finite volume scheme that can be used for arbitrary high order reconstruction in triangular or tetrahedral meshes. The implementation of ENO for unstructured grids was developed by Abgrall [1], while WENO schemes for triangular meshes were developed by Friedrich [58] and Hu and Shu [90]. Theoretically, the finite volume approaches of [16] and [90], [1] can be used to obtain arbitrarily high-order accurate finite volume schemes with high-order polynomial data reconstructions. However, in practice higher than linear reconstructions are not used in three dimensions because of the difficulty to construct nonsingular stencils and the large memory required to store the reconstruction coefficients. This was clearly shown by Delanaye and Liu [45] who found that for the third-order (quadratic reconstruction) FV scheme in three dimensions, the average size of the reconstruction stencils is about 50-70. The size of reconstruction stencils increases nonlinearly with the order of accuracy and it was estimated that for the fourth-order FV scheme the stencil size would be approximately 120.

The high-order accurate conservative scheme called the discontinuous Galerkin (DG) method developed by Cockburn et al. in a series of papers

([34], [35], [36], [40]) shows promise for high resolution simulations in fully unstructured meshes and shows promise to overcome the problems encountered with the unstructured finite volume methods. The DG method assumes a high-order expansion for the distribution of the state variables for each element and solves for the coefficients of the expansion polynomials. The resulting state variables are usually not continuous across the element boundaries. Therefore, the fluxes through the element boundaries are computed using an approximate Riemann solver and the residual is minimized with a Galerkin approach. It is the use of Riemann fluxes across element boundaries makes the DG method fully conservative.

19.2   Space Discretization

The basic idea of the DG method is described in this chapter. Further information and more details can be found in the original references [36], [151], [38] and the reviews [39], [40]. For each time $t \in [0,T]$ the approximate solution, u_h, of the governing equations in conservation law form, $\partial_t \mathbf{u} + \text{div} \mathbf{F}(\mathbf{u}) = 0$, is sought in the finite element space of discontinuous functions V_h

$$\mathbf{V}_h = \{\phi_h \in \mathbf{L}^\infty(\Omega) : \phi_h|_\mathbf{K} \in P^k(\mathbf{K}), \; \forall \mathbf{K} \in T_h\} \qquad (19.1)$$

where T_h is a discretization of the domain Ω using triangular or quadrilateral elements and $P^k(\mathbf{K})$ is the local space that contains the collection of polynomials up to degree k.

The development of the DG method starts from weak formulation of the governing equations

$$\frac{d}{dt} \int_K u\,(x,t)\,\phi(x)\,d\Omega = \int_K F\,(u\,(x,t)) \cdot \nabla(\phi\,(x))\,d\Omega - \\ \sum_{e \in \partial K} \oint_e F\,(u\,(x,t)) \cdot \mathbf{n}_{e,K}\,\phi\,(x)\,d\,S \qquad (19.2)$$

where $\phi\,(x)$ is any sufficiently smooth test function and $\mathbf{n}_{e,K}$ denotes the outward, unit normal to the face or edge e of an element.

The stiffness matrix integral at the left hand side of Eq. (19.2) is evaluated numerically using Gauss-Radau integration rules. The integrals on the right hand side of Eq. (19.2) are evaluated using appropriate quadrature rules as follows:

$$\int_e F\left(u\left(x,t\right)\right)\cdot\mathbf{n}_{e,K}\,\phi\left(x\right)dS \approx \sum_{l=1}^{L}\psi_l\,F\left(u\left(x_{el},t\right)\right)\cdot\mathbf{n}_{e,K}\,\phi\left(x_{el}\right)\left|e\right|\quad(19.3)$$

$$\int_K F\left(u\left(x,t\right)\right)\cdot\nabla\left(\phi\left(x\right)\right)d\Omega \approx \sum_{j=1}^{J}w_j\,F\left(u\left(x_{K\,j},t\right)\right)\cdot\nabla\left(\phi\left(x_{K\,j}\right)\right)\left|K\right|.\quad(19.4)$$

For example, for a third-order polynomial basis, a Gaussian quadrature rule that integrates exactly at least sixth-order polynomial is used for the line integrals of Eq. (19.3). In general, a kth order DG method ($k-1$ order polynomial reconstruction) requires a $2k$-th order quadrature formula for the surface (line) integrals of Eq. (19.3), and a $(2k-1)$th order quadrature formula for the volume integrals of Eq. (19.4).

The data are assumed discontinuous across the interfaces of the continuous domain and at each interface two values are available. Therefore, the flux $F\left(u\left(x,t\right)\right)\cdot\mathbf{n}_{e,K}\,\phi(x)$ is replaced by a suitable numerical flux $\widetilde{F}_{e,K}(x,t)$ for the approximate solution u_h and the test function $\psi_h \in V(K)$. Using $\widetilde{F}_{e,K}(x,t)$ in Eqs. (19.3) and (19.4) the approximate solution u_h is given by

$$\frac{d}{dt}\int_K u_h\left(x,t\right)\phi_h(x)\,dx =$$

$$= \sum_{j=1}^{J}w_j\,F\left(u_h\left(x_{K\,j},t\right)\right)\cdot\nabla(\phi_h\left(x_{K\,j}\right))\left|K\right|$$

$$- \sum_{e\in\partial K}\sum_{l=1}^{L}\psi_l\,\widetilde{F}_{e,K}(u\left(x_{el},t\right))\cdot\mathbf{n}_{e,K}\,\phi\left(x_{el}\right)\left|e\right|,$$

$$\forall\phi_h \in V\left(K\right),\ \forall K \in T_h \quad(19.5)$$

where time advancement of Eq. (19.5) is performed with the third order accurate Runge-Kutta method.

The major difference of the DG formulation with a standard node-based Galerkin finite element method is that the expansion in each element is local without any continuity across the element boundaries. The value of the numerical flux $\widetilde{F}_{e,K}(x,t)$ at the edge of the boundary of the element K depends on two values of the approximate solution, one from the interior (right) of the element K, $u^R = u_h(x^{int(K)},t) = u_h^-$, and the other from the exterior (left) of the element K, $u^L = u_h(x^{ext(K)},t) = u_h^+$. Any consistent, conservative exact or approximate Riemann solver can be used to obtain the numerical flux $\widetilde{F}_{e,K}\left(u\left(x^{int(K)},t\right),\ u\left(x^{ext(K)},t\right)\right)$ or $\widetilde{F}_{e,K}\left(u^-,u^+\right)$ as follows

$$\widetilde{F}_{e,K}(u^-,\ u^+) = \frac{1}{2}\left[F\left(u^-\right)\cdot\mathbf{n}_{e,K} + F\left(u^+\right)\cdot\mathbf{n}_{e,K} - F^*(u^-,\ u^+)\right]\quad(19.6)$$

where $F^*(u^-, u^+)$ is the dissipative part of the numerical flux. The computationally efficient local Lax-Friedrichs flux is used in many applications. The flux F is split as $F = F^+ + F^-$ where $F^\pm = F + \alpha q$ where $\alpha = \lambda$ and λ is the maximum eigenvalue of the flux Jacobian. For the linearized Euler equations the eigenvalues are constant and the derivatives are continuous for the nonlinear case, however, in order to obtain continuous higher derivatives $\alpha = \sqrt{\varepsilon^2 + \lambda^2}$ with $\varepsilon = 0.05$. The DG method can be applied for discretizations on triangular, quadrilateral or mixed-type of elements. Expansion bases of different order for triangular and quadrilateral elements are shown next.

19.3   Triangular Element Bases

For the general nonlinear case, the order of accuracy of the DG method (see Ref. [39] and references therein) is at least $k + 1/2$ if polynomials of degree at most k are used as basis functions. Furthermore, it was shown (see Ref. [39]) that for linear problems that for canonical semi-uniform triangular grids the order of accuracy is $(k + 1)$. For simplicity in the rest of this chapter, the method is called $(k+1)^{th}$ order accurate if the basis functions are polynomials of degree at most k.

The approximate solution within each element is expanded in a series of local bases functions (polynomials) as follows

$$u_h(x, y, t) = \sum_{j=1}^{d} c_j(t) \, P_j^k(x, y) \tag{19.7}$$

where $c_j(t), j = 1, 2, ..., d$ are expansion coefficients or degrees of freedom for each element, to be evolved in time, and $P_j^k(x, y)$ are polynomial bases of degree k the most. The range of d number of polynomials (also number of nodes on a triangle) in Eq. (19.7) in two dimensions is related to the degree of the polynomial by $d = (k+1)(k+2)/2$. For example, cubic reconstruction $k = 3$ requires $d = 10$ e.g., 10 nodes (see Fig. 19.1).

The value of d for bases in two and three dimensions, e.g., triangles or tetrahedra is given by

$$d = \binom{n+k}{k} = \frac{(n+k)!}{n! \, k!} \tag{19.8}$$

where k is the order of the polynomial $P_j^k = 1, ..., d$ and ($n = 2$ in the rest of this chapter) is the spatial dimension of the problem. It can be seen that in three dimensions the third order basis includes 20 polynomials and the

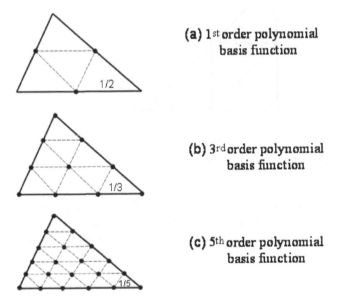

(a) 1ˢᵗ order polynomial basis function

(b) 3ʳᵈ order polynomial basis function

(c) 5ᵗʰ order polynomial basis function

FIGURE 19.1: Nodal points for polynomial bases for P^1, P^2, and P^3 bases on triangular elements.

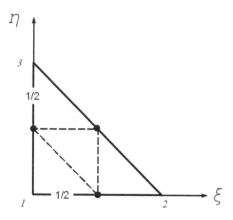

FIGURE 19.2: Nodal points on the P^1 reference element.

fifth order basis 56 polynomials. It appears therefore, that for realistic, three-dimensional problems, it could be very intensive computationally to achieve accuracy higher than fourth order. Simple nodal bases of first-, third-, and fifth-order polynomials for triangular elements are given next.

19.3.1   First-Order Polynomials

The following first order basis $(d = 3)$, P_j^1 $j = 1, 2, 3$, is used to achieve second order accuracy.

$$P_1^1 = 1 - 2y, \ P_2^1 = 2x + 2y - 1, \ P_3^1 = 1 - 2x \qquad (19.9)$$

Each of these polynomials (see Fig. 19.2) takes unit value at one node, located in the middle of an edge, and zero value at the other nodes located at the middle of the other edges. The polynomials P_1^1, P_2^1, P_3^1 of Eq. (19.9) are orthogonal $(\int_{el} P_i P_j = 0, \ i \neq j)$ and the mass matrix resulting from the integration at the left hand side of Eq. (19.2) is diagonal. For the first- as well as the higher-order bases all calculations of Eq. (19.5), are carried out on the reference element. The numerical flux computation along the edges of the element, which depends on the neighboring element, may be carried out at the physical space or at the transformed space.

19.3.2 Third-Order Polynomials

The following third-order basis ($d = 10$), P_j^3 $j = 1, ..., 10$, can be used to achieve fourth-order accuracy.

$$
\begin{aligned}
P_1^3 &= 9x \ (x^2 - x - 2/9)/2 \\
P_2^3 &= 9y \ (y^2 - y - 2/9)/2 \\
P_3^3 &= 9 \ (1 - x - y) \ (2/3 - x - y) \ (1/3 - x - y)/2 \\
P_4^3 &= 9xy \ (3x - 1)/2 \\
P_5^3 &= 27xy \ (y - 1/3)/2 \\
P_6^3 &= 27y \ (1 - x - y) \ (y - 1/3)/2 \\
P_7^3 &= 27y \ (1 - x - y) \ (2/3 - x - y)/2 \\
P_8^3 &= 27x \ (1 - x - y) \ (2/3 - x - y)/2 \\
P_9^3 &= 27x \ (x - 1/3) \ (1 - x - y)/2 \\
P_{10}^3 &= 27xy \ (1 - x - y)
\end{aligned}
$$

$$(19.10)$$

At the reference element, each of these polynomials (see Fig. 19.2) takes unit value at one of the 1/3-distance nodes and zero value at all other nodes. The third-order polynomial basis, P_j^3, $j = 1, ..., 10$, of Eq. (19.10) are nonorthogonal and the mass matrix of Eq. (19.5) is computed using high-order accurate Gauss-Radau integration. In addition, orthogonal, Jacobi polynomial, modal bases [110] may be defined for the reference triangle as follows

P^1 Jacobi basis

$$1, \ 1 + 2\xi + \eta, \ -1 + 3\eta.$$

P^2 Jacobi basis

$$1, \ 1 + 2\xi + \eta, \ -1 + 3\eta$$

$$1 - 2\eta + \eta^2 + 6\xi\eta - 6\eta + 6\eta^2, \ 1 + 5\eta^2 + 10\xi\eta - 6\eta - 2\xi, \ 1 - 8\eta + 10\eta^2.$$

P^3 Jacobi basis

$$1, \ 1 + 2\xi + \eta, \ -1 + 3\eta$$

$$1 - 2\eta + \eta^2 + 6\xi\eta - 6\eta + 6\eta^2, \ 1 + 5\eta^2 + 10\xi\eta - 6\eta - 2\xi, \ 1 - 8\eta + 10\eta^2$$

$$-1 + \eta^3 - 24\xi\eta + 30\xi^2\eta + 12\xi\eta^2 + 20\xi^3 + 12\xi + 3\eta - 3\eta^2 - 30\xi^2,$$

$$-1 + 7\eta^3 + 42\xi\eta^2 - 15\eta^2 - 48\xi\eta + 9\eta + 42\xi^2\eta - 6\xi^2 + 6\xi,$$

$$-1 + 21\eta^3 + 42\xi\eta^2 - 33\eta^2 - 24\xi\eta + 13\eta + 2\xi$$

$$-1 + 15\eta - 45\eta^2 + 35\eta^3.$$

Clearly these bases are hierarchical and offer an advantage for $p-$ type multigrid and p refinement.

19.3.3   Fifth-Order Polynomials

Sixth-order accurate solutions in two dimensions are obtained with a polynomial basis that contains $6 \times 7/2$ or $7!/(2!\ 5!) = 21$ polynomials. These polynomials (or higher order polynomial bases [91], [110]) may be systematically defined through the use of triangular coordinates $x, y, t = 1 - x - y$. The general form of Lagrange interpolation in triangular coordinates is given by

$$P_L(x, y, t) = L_i\ (x)\ L_j\ (y)\ L_k\ (t) \qquad (19.11)$$

where $L_m\ (x)$ are one-dimensional $(m - 1)$ order Lagrange polynomials [110] and $L = L\ (i, j, k)$ (see Fig. 19.3) denotes the nodal index. For example, in Fig. 19.4 at the nodal index $L\ (2, 3, 3)$ the base polynomial is defined as

$$P_{(2,3,3)}\ (x, y, t) = 5x\ \frac{y\ (y - 1/5)}{\left(\frac{2}{25}\right)}\ \frac{t\ (t - 1/5)}{\left(\frac{2}{25}\right)}. \qquad (19.12)$$

This fifth order polynomial Lagrange basis is also nonorthogonal and the mass matrix is computed using Gauss Radau integration. Note that the P_j^1 and P_j^3 bases can be constructed using procedure followed for the construction of the P_j^5 basis. The Lagrange bases for triangular elements offer an implementation advantage compared to hierarchical bases, which are more appropriate for multigrid [57], because for the selected bases the nodal values are known. For nodal Lagrange bases it is more efficient to perform the numerical flux computation on the physical space.

19.4   Quadrilateral Element Bases

Space discretization can also be obtained using quadrilateral isoparametric elements (see Fig. 19.5) of arbitrary shape. The reference element is a square (see Fig. 19.5) with vertices at the points $(-1, -1)$, $(1, -1)$, $(1, 1)$, $(-1, 1)$. In Fig. 19.5, the physical space arbitrary shape quadrilateral and the nodal points of the third order accurate polynomial basis on the reference element are also shown. The coordinates x_j^k, $j = 1, 2$, $k = 1, 2, 3, 4$ ($x_1^k = x^k$, $x_2^k = y^k$)

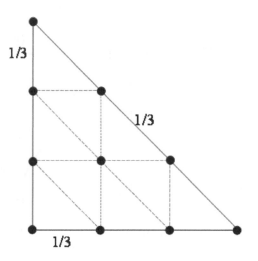

FIGURE 19.3: Nodal points on the P^3 reference element.

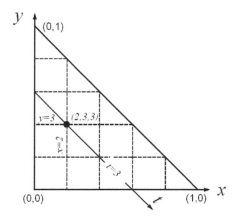

FIGURE 19.4: Nodal points on the P^5 reference element.

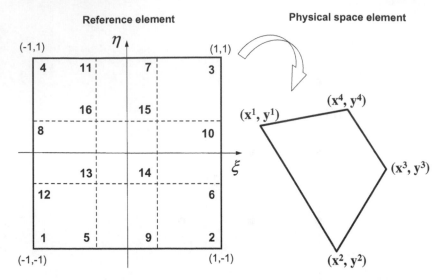

FIGURE 19.5: Reference element with nodal points for third-order polynomial basis and isoparametric mapping for the arbitrary shape quadrilateral element in physical space.

of the arbitrary shape quadrilateral elements in the physical space are related to the square reference element coordinates $\widehat{x}_j$ ($\widehat{x}_1 = \widehat{x}$, $\widehat{x}_2 = \widehat{y}$) through the map

$$x_j = \sum_{k=1}^{4} \widehat{f}_k (\widehat{x}_1, \widehat{x}_2) \, x_j^k \tag{19.13}$$

where $\widehat{f}_k (\widehat{x}_1, \widehat{x}_2)$ are first order Lagrange polynomials $\widehat{f}_k (\widehat{x}_1, \widehat{x}_2) = (1 + x_1^k \, \widehat{x}_1) (1 + x_2^k \, \widehat{x}_2)/4$. The basis polynomials $P_i^k (x, y) \in Q^k$ for finite element discretization with quadrilateral elements are tensor products of appropriate order one-dimensional Lagrange polynomials. For example, the basis with third order accurate polynomials is $P_i^3 (x, y) = L_j (x) L_k (y)$, $i = 1, ..., 16$, j, $k = 1, 2, 3, 4$.

19.5 Implementation of a Quadrature-Free DG Method for Systems of Linear Hyperbolic Equations

Large portion of computational resources during the implementation of the DG method is devoted for the evaluation of integrals, such as the mass and stiffness matrices using quadrature rules. For linear problems, the computational cost of the DG method can be significantly reduced using the quadrature-free DG method.

For simplicity, and without loss of generality, the quadrature-free DG method is presented for the following two-dimensional system of linear hyperbolic equations

$$\frac{\partial u_p}{\partial t} + A_{pq} \frac{\partial u_q}{\partial x} + B_{pq} \frac{\partial u_q}{\partial y} = 0 \tag{19.14}$$

The numerical solution in each triangular element $T^{(m)}$ is approximated as

$$(u_h^{(m)})_p\,(\xi,\eta,t) = \widehat{u}_{pl}^{(m)}\,(t)\,\Phi_l\,(\xi,\eta), \qquad \Phi_l(\xi,\eta) \in V \tag{19.15}$$

where ξ and η are the coordinates of the reference element (see Fig. 19.6) defined by

$$x = x_1 + (x_2 - x_1)\,\xi + (x_3 - x_1)\,\eta$$
$$y = y_1 + (y_2 - y_1)\,\xi + (y_3 - y_1)\,\eta$$
$$\xi = \frac{1}{|J|}\,[(x_3 y_1 - x_1 y_3) + x\,(y_3 - y_1) + y\,(x_1 - x_3)]$$
$$\eta = \frac{1}{|J|}\,[(x_1 y_2 - x_2 y_1) + x\,(y_1 - y_2) + y\,(x_2 - x_1)]$$
$$|J| = (x_2 - x_1)\,(y_3 - y_1) - (x_3 - x_1)\,(y_2 - y_1).$$

After multiplication with a test function $\Phi_k \in V$ and integration by parts obtain

$$\int_{T^{(m)}} \Phi_k\,\frac{\partial u_p}{\partial t}\,dV + \int_{\partial T^{(m)}} \Phi_k\,F_p^h\,dS -$$
$$- \int_{T^{(m)}} \left(\frac{\partial \Phi_k}{\partial x}\,A_{pq}\,u_q + \frac{\partial \Phi_k}{\partial y}\,B_{pq}\,u_q \right)\,dV = 0 \tag{19.16}$$

where F_p^h is the numerical flux.

The global system solution vector u_p (see Fig. 19.6) is transformed to the edge-aligned coordinate system through the transformation

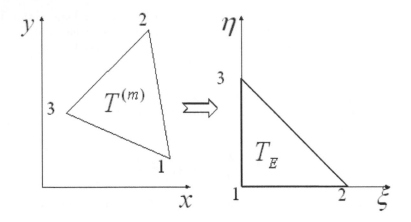

FIGURE 19.6: Physical and transformed space elements.

$$u_p = T_{pq} \, u_q^T \quad , \quad T_{pq} = \begin{pmatrix} 1 & 0 & 0 & 0 \\ 0 & n_x & -n_y & 0 \\ 0 & n_y & n_x & 0 \\ 0 & 0 & 0 & 1 \end{pmatrix} \qquad (19.17)$$

where $\mathbf{n} = (n_x, n_y)^T$ is the edge normal unit vector.

Using the exact Godunov's flux [67] or any other approximate flux to evaluate the numerical flux F_p^h between two triangles $T^{(m)}$ and $T^{(n)}$, the numerical flux for the global x-y system is

$$F_p^h = \frac{1}{2} \, T_{pq} \, (A_{qr}^T + |A_{qr}^T|) \, (T_{rs})^{-1} \widehat{u}_{sl}^{(m)} \, \Phi_l^{(m)}$$
$$+ \frac{1}{2} \, T_{pq} \, (A_{qr}^T - |A_{qr}^T|) \, (T_{rs})^{-1} \widehat{u}_{sl}^{(n)} \, \Phi_l^{(n)}. \qquad (19.18)$$

With the notation introduced, the physical x-y space DG method becomes

$$\frac{\partial}{\partial t}\,\widehat{u}_{pl}^{(m)} \int_{T^{(m)}} \Phi_k\,\Phi_l\,dV +$$

$$\sum_{j=1}^{3} T_{pq}^{j}\,\frac{1}{2}\,(A_{qr}^{T} + |A_{qr}^{T}|)\,(T_{rs}^{j})^{-1}\,u_{sl}^{(m)} \int_{(\partial T^{(m)})_j} \Phi_k^{(m)}\,\Phi_l^{(m)}\,dS +$$

$$\sum_{j=1}^{3} T_{pq}^{j}\,\frac{1}{2}\,(A_{qr}^{T} - |A_{qr}^{T}|)\,(T_{rs}^{j})^{-1}\,u_{sl}^{(n)} \int_{\partial T^{(m)})_j} \Phi_k^{(m)}\,\Phi_l^{(m)}\,dS -$$

$$A_{pq}\,\widehat{u}_{ql}^{(m)} \int_{T^{(m)}} \frac{\partial \Phi_k}{\partial x}\,\Phi_l\,dV - B_{pq}\,\widehat{u}_{ql}^{(m)} \int_{T^{(m)}} \frac{\partial \Phi_k}{\partial n}\,\Phi_l\,dV = 0.$$

$$(19.19)$$

For integration in the reference space, the semi-discrete DG formulation becomes

$$\frac{\partial}{\partial t}\,\widehat{u}_{pl}^{(m)}|J| \int_{T_E} \Phi_k\Phi_l\,d\xi d\eta +$$

$$\sum_{j=1}^{3} T_{pq}^{j}\,\frac{1}{2}\,(A_{qr}^{T} + |A_{qr}^{T}|)\,(T_{rs}^{j})^{-1}\,\widehat{u}_{sl}^{(m)}|S_j| \int_0^1 \Phi_k^{(m)}(x_j)\,\Phi_l^{(m)}(x_j)\,dx_j +$$

$$\sum_{j=1}^{3} T_{pq}^{j}\,\frac{1}{2}\,(A_{qr}^{T} - |A_{qr}^{T}|)\,(T_{rs}^{j})^{-1}\,u_{sl}^{n}|S_j| \int_0^1 \Phi_k^{(m)}(x_j)\,\Phi^{(n)}(x_j)\,dx_j -$$

$$A_{pq}^{*}\,\widehat{u}_{ql}|J| \int_{T_E} \frac{\partial \Phi_k}{\partial \xi}\,\Phi_l\,d\xi d\eta - B_{pq}^{*}\,\widehat{u}_{ql}|J| \int_{T_E} \int_0^0 \frac{\partial \Phi_k}{\partial \eta}\,\Phi_l\,d\xi d\eta = 0.$$

$$(19.20)$$

$$A_{pq}^{*} = A_{pq}\xi_x + B_{pq}\xi_y \quad , \qquad B_{pq}^{*} = A_{pq}\eta_x + B_{pq}\eta_y$$

where $0 < x_j < 1$ is the parametrization of the j-th edge of the reference element and $|S_j|$ is the length of the edge in physical space.

The following integrals may be precalculated using appropriate order quadrature rules

$$M_{kl} = \int_{T_E} \Phi_k\Phi l\,d\xi d\eta$$

$$F_{kl}^{j,0} = \int_0^1 \Phi_k^{(m)}(x_j)\,dx_j \quad , \quad F_{kl}^{j,1} = \int_0^1 \Phi_k^{(m)}(x_j)\,\Phi_l^n(x_j)\,dx_j \quad (19.21)$$

$$K_{kl}^{\xi} = \int_{T_0} \frac{\partial \Phi_k}{\partial \xi}\,\Phi_l\,d\xi d\eta \quad , \quad K_{kl}^{\xi} = \int_{T_E} \int_0^0 \frac{\partial \Phi_k}{\partial \eta}\,\Phi_l\,d\xi d\eta.$$

Then time integration of the semi-discrete DG formulation with RK3 or RK4 yields the quadrature-free Runge-Kutta DG approach of Atkins and Shu (1998) [12].

19.6   Arbitrary High Order DG Schemes

The ADER (Arbitrary high order scheme which utilizes the hyperbolic Riemann problem for the advection, of the higher order DERivatives) can be applied for DG numerical solutions of linear hyperbolic systems to obtain a quadrature-free, explicit single step method of arbitrary order of accuracy in both space and time.

Note that for the nonlinear case, the ADER-DG scheme requires Gaussian quadratures of suitable order of accuracy. The ADER approach of Titarev and Toro [179] is based on the solution of the generalized Riemann problems (GRPs) at the cell boundaries and application of the Lax-Wendroff procedure for highly accurate time integration of the numerical flux. It was shown numerically by Dumbser et al. [49] that the ADER-DG scheme is $3N + 3$ order accurate for N-th order basis functions.

19.6.1   ADER-DG-Discretization

The main ingredients of the method are:

- Taylor expansion in time of the solution

- Use of the Lax-Wendroff procedure to replace time derivatives by space derivatives

- Solution of generalized Riemann problems (GRP) to approximate space derivatives

Consider the original linear PDE for the reference element

$$\frac{\partial u_p}{\partial t} + (A_{pq}\xi_x + B_{pq}\xi_y)\,\frac{\partial u_q}{\partial \xi} + (A_{pq}\eta_x + B_{pq}\eta_y)\,\frac{\partial u_q}{\partial \eta} = 0 \qquad (19.22)$$

or in compact form using the abbreviation $A_{pq}^* = A_{pq}\xi_x + B_{pq}\xi_y$

$$\frac{\partial u_p}{\partial t} + A_{pq}^*\,\frac{\partial u_q}{\partial \xi} + B_{pq}^*\,\frac{\partial u_q}{\partial \eta} = 0. \qquad (19.23)$$

The k-th time derivative in the $\xi - \eta$ reference space is given by

$$\frac{\partial^k u_p}{\partial t^k} = (-1)^k \, (A_{pq}^* \partial_\xi + B_{pq}^* \partial_\eta) \; u_q. \tag{19.24}$$

The expansion of the solution in a Taylor series in time up to order N is

$$u_p(x,t) = \sum_{k=0}^{N} \frac{t^k}{k!} \frac{\partial^k}{\partial t^k} \, u_p(x,0) = \sum_{k=0}^{N} \frac{t^k}{k!} \, (-1)^k (A_{pq}^* \partial_\xi + B_{pq}^* \partial \eta)^k \, u_q(x,0). \tag{19.25}$$

The discontinuous Galerkin approximation is

$$u_p(x,t) = \sum_{k=0}^{N} \frac{t^k}{k!} \, (-1)^k (A_{pq}^* \partial_\xi + B_{pq}^* \partial_\eta)^k \, \Phi_l(\xi) \, \widehat{u}_{pl}(0). \tag{19.26}$$

Projection of the DG approximation onto each basis function yields the following approximation of the evolution of the degrees of freedom during one time step from level n to level $n+1$.

$$\widehat{u}_{pl}(t) = \frac{\left\langle \Phi_n, \sum_{k=0}^{N} \frac{t^k}{k!} \, (-1)^k (A_{pq}^* \partial_\xi + B_{pq}^* \partial_n \Phi_m^{(\xi)}) \right\rangle}{\langle \Phi_n, \Phi_l \rangle} \; \dot{u}_{qm}(0) \tag{19.27}$$

analytic integration in time yields

$$\int_0^{\Delta t} \widehat{u}_{pl}(t) \, dt = \underbrace{\frac{\left\langle \Phi_n, \sum_{k=0}^{N} \frac{\Delta t^{(k+1)}}{(k+1)!} \, (-1)^k (A_{pq}^* \partial_\xi + B_{pq}^* \partial_n)^k \, \Phi_m(\xi) \right\rangle}{\langle \Phi_n, \Phi_l \rangle}}_{I_{plqm}(\Delta t)} \; \widehat{u}_{qm}(0)$$

$$= I_{plqm}(\Delta t) \, \widehat{u}_{qm}(0). \tag{19.28}$$

The ADER-DG scheme for linear hyperbolic systems becomes

$$\left[(\widehat{u}_{pl}^{(m)})^{n+1} - (\widehat{u}_{pl}^{(m)})^n \right] |J| \, M_{kl}$$

$$+ \frac{1}{2} \sum_{j=1}^{3} T_{pq}^j \, (A_{qr}^T + |A_{qr}^T|) \, (T_{rs}^j)^{-1} |S_j| \, |F_{kl}^j \cdot I_{slmn}(\Delta t) \, (\widehat{u}_{mn}^{(m)})^n \tag{19.29}$$

$$+ \frac{1}{2} \sum_{j=1}^{3} T_{pq}^j \, (A_{qr}^T - |A_{qr}^T|) \, (T_{rs}^j)^{-1} |S_j| \, F_{kl}^j \cdot I_{slmn}(\Delta t) \, (\widehat{u}_{mn}^{(k)})^n$$

$$- A_{pq}^* \, |J| \, K_{kl}^\xi \cdot I_{qlmn}(\Delta t) \, (\widehat{u}_{mn}^{(m)})^n - B_{pq}^* \, |J| \, K_{pq}^\eta \cdot I_{qlmn}(\Delta t) \, (\widehat{u}_{mn}^{(m)})^n = 0.$$

19.6.2 ADER-DG Schemes for Nonlinear Hyperbolic Systems

Consider the nonlinear hyperbolic system of the form

$$\frac{\partial u}{\partial t} + \nabla \cdot \mathbf{F}(u) = 0 \quad, \qquad \mathbf{F} = [\mathbf{f}, \mathbf{g}]. \tag{19.30}$$

The weak form of the nonlinear problem is

$$|J| \int_{T_E} (u_h)_t \, \Phi_k d\xi d\eta + \int_{\partial T^{(m)}} F^h \cdot \mathbf{n} \Phi_k \, dS - |J| \int_{T_E} F(u_h) \nabla \Phi_k \, d\xi d\eta = 0 \tag{19.31}$$

where $F^h \cdot \mathbf{n}$ denotes the numerical flux over the element boundary. Using N_S^{GP} Gaussian quadrature points for the boundary integral and N_V^{GP} Gaussian quadrature points for the volume integral obtain

$$\int_{\partial T^{(m)}} F^h \cdot \mathbf{n} \, \Phi_k dS \approx \sum_{j=1}^{3} |S_j| \sum_{n=1}^{N_S^{GP}} \omega_n F^h \left(u^{(m)}(x_n), u^{(k)}(x_n) \right) \mathbf{n}_j \Phi_k(x_n) \tag{19.32}$$

$$\int_{T_E} F(u_h) \nabla \Phi_k \, d\xi d\eta \approx \sum_{n=1}^{N_V^{GP}} \psi_n \, F \left(u_h(\xi_n, \eta_n) \right) \nabla \left(\Phi_k \left(\xi_n, \eta_n \right) \right). \tag{19.33}$$

For the nonlinear case, the evaluation of higher order time derivatives is more complicated, for example:

$$
\begin{aligned}
u_t &= -A(u) \, u_x - B(u) \, u_y \\
u_{tt} &= -A(u) \, u_{xt} - [A'(u)u_t] \, u_x - [B'(u) \, u_t] \, u_y - B(u) \, u_{yt} \\
u_{tx} &= -A(u) \, u_{xx} - B(u) \, u_{yx} - [A'(u)u_x] \, u_x - [B'(u) \, u_x] \, u_y \quad (19.34) \\
u_{ty} &= -A(u) \, u_{xy} - B(u) \, u_{yy} - [A'(u)u_y] \, u_x - [B'(u) \, u_y] \, u_y.
\end{aligned}
$$

The last two equations can be used to express u_{tt} in terms of pure space derivatives.

The essential part of the nonlinear ADER-DG scheme is the solution of the generalized Riemann problem (GRP) posed by the DG polynomials $u_h^{(m)}(x, y)$. The solution of the GRP is determined by the solution u_h^{RP} of a conventional (piecewise constant data) nonlinear Riemann problem of the boundary extrapolated values $u_h^{(m)}(x_n)$ and $u_h^{(k)}(x_n)$ on the left and right hand side of the quadrature point x_n.

Consider again a hyperbolic PDE with appropriate initial and boundary conditions.

PDE $u_t + \nabla \cdot \mathbf{F}(u) = 0$

IC $u_h\,(x, y, 0) = \begin{cases} u_h^{(m)}(x_n) & if\ (x, y) \in T^{(m)} \\ u_h^{(k)}(x_n) & if\ (x, y) \in T^{(k)} \end{cases}$

Furthermore, it is determined by the set of solutions $u_h^{RP_0(q,r)}(x_n)$ of a sequence of linearized conventional Riemann problems $RP_{q,r}$ for all space derivatives of $u^{(q,r)}$
$1 \leq p \leq N$ and $q + r = p$

PDE $u_t^{(q,r)} + A\left(u_h^{RP}(x_n)\right)\,u_x^{(q,r)} + B\left(u_h^{RP}(x_n)\right)\,u_y^{(q,r)} = 0$

IC $u_h^{(q,r)}\,(x, y, 0) = \begin{cases} \frac{\partial^p}{\partial x^q\,\partial y^r}\,u_h^{(m)}(x_n) & if\ (x, y) \in T^{(m)} \\ \frac{\partial^p}{\partial x^q\,\partial y^r}\,u_h^{(m)}(x_n) & if\ (x, y) \in T^{(k)} \end{cases}$

As before, the solution of the generalized Riemann problem is finally given by a Taylor series where time derivatives have been replaced by space derivatives as

$$u_h^{GRP}(x_n, \tau) = u^{RP_0}(x_n) + \sum_{k=1}^{N} \frac{\tau^k}{k!}\,G_k\left(u_h^{RP_{(q,r)}}(x_n)\right) \tag{19.35}$$

$$\forall\ \ 0 \leq p \leq k\ \ ,\ \ \ q + r = p$$

where G_k denotes the nonlinear function that expresses the k-th time derivative in terms of space derivatives.
    The ADER solution is

$$u_h^{ADER}\,(x_n, y_n, \tau) = \sum_{k=0}^{N} \frac{\tau^k}{k!}\,G_k\left(\frac{\partial^p u_h^{(m)}\,(x_n, y_n, 0)}{\partial x^p\,\partial y^r}\right). \tag{19.36}$$

$$\forall\ \ 0 \leq p \leq k\ \ ,\ \ \ q + r = p$$

The line and volume integrals in the interval $[0, \Delta t]$ are obtained by

$$\int_0^{\Delta t} \left(\int_{\partial T^{(m)}} \Phi_k\,F^h \mathbf{n}\,dS\right)\,d\tau \approx$$

$$\sum_{l=1}^{N} \alpha_l \sum_{j=1}^{3} |S_j| \sum_{n=1}^{N_S^{GP}} \omega_n\,\Phi_k\,(x_n)\,F\left(u_h^{GRP_N}(x_n, \tau_l)\right)\,\mathbf{n}_j. \tag{19.37}$$

In this quadrature, the flux function F is the physical flux which is evaluated by using the solution of the GRP_N on time τ_l. Any approximate or exact Riemann solver might be used to deliver the solution $u_h^{RP_0}(x_n)$ at the interface.

$$\int_0^{\Delta t} \left(\int_{T_E} F\left(u_h\right) \nabla \Phi_k \, d\xi d\eta \right) d\tau \approx$$

$$\sum_{l=1}^{N} \alpha_l \sum_{n=1}^{N_V^{GP}} \omega_n \, F\left(u_h^{ADER}(\xi_n, \eta_n, \tau_l)\right) \, \nabla \Phi_k\left(\xi_n, \eta_n\right) \quad (19.38)$$

19.7   Analysis of the DG Method for Wave Propagation

In this chapter the analysis of Hu et al. [88] for the wave propagation properties of the semi-discrete DG method for conservation laws with linear flux is summarized. Consider the liner conservation law

$$\frac{\partial \mathbf{u}}{\partial t} + \nabla \cdot \mathbf{F}\left(\mathbf{u}\right) = 0 \qquad (19.39)$$

$$\mathbf{F}\left(\mathbf{u}\right) = \left(A_1 \mathbf{u}, ..., A_d \mathbf{u}\right), \quad \mathbf{u} = \left(u_1, ..., u_m\right)$$

where d is the number of space dimensions and A_i are constant $m \times m$ real matrices.

For the linear problem the numerical fluxes are:

$$\mathbf{F}_{nun}\left(\mathbf{u_1}, \mathbf{u_2}, \mathbf{n}\right) = \tilde{A}_{av}^{+} \mathbf{u_1} + \tilde{A}_{av}^{-} \mathbf{u_2} \qquad (19.40)$$

$$\tilde{A} = \sum_{k=1}^{d} A_k n_k \;, \quad \tilde{A}_{av}^{+} = \frac{\tilde{A} + \alpha |\tilde{A}|}{2} \;, \quad \tilde{A}_{av}^{-} = \frac{\tilde{A} - \alpha |\tilde{A}|}{2}$$

where α is a real positive number and $\alpha = 0$ yields a centered flux while $\alpha = 1$ yields the Roe flux.

The approximate solution $\mathbf{u_h}$ is written in the usual way as an expansion of the local basis set as

$$\mathbf{u_h}\left(\mathbf{x}, t\right) = \sum_{l=0}^{N-1} \mathbf{c_l}\left(t\right)\phi_l\left(\mathbf{x}\right). \qquad (19.41)$$

The dispersion relation of the scheme is obtained for a solution of the form

$$\mathbf{u}\left(\mathbf{x}, t\right) = \hat{\mathbf{u}} \, e^{i\,(\mathbf{k}\cdot\mathbf{x} - \omega t)} \qquad (19.42)$$

which represents a sinusoidal wave train with wavenumber $\mathbf{k}$ and frequency ω.

The expansion coefficients c_l are calculated by projecting the sinusoidal solution onto the local basis

$$c_l(t) = \hat{u} \frac{\int_{\Omega_e} \phi_l(\xi) e^{i(k\xi - \omega t)}}{\int_{\Omega_e} (\phi_l(\xi))^2 d\xi} d\xi, \qquad l = 0, ..., N-1. \qquad (19.43)$$

For structured meshes $c_l(t) = \hat{c}_l e^{i(\mathbf{k} \cdot \mathbf{x} - \omega t)}$ with

$$\hat{c}_l = \hat{u} \frac{\int_{\Omega_e} \phi_l(\xi) e^{ik\xi} d\xi}{\int_{\Omega_e} (\phi_l(\xi))^2 d\xi}. \qquad (19.44)$$

19.7.1 One-Dimensional Analysis

Consider the one-dimensional scalar advection equation

$$\frac{\partial u}{\partial t} + c \frac{\partial u}{\partial x} = 0 \qquad -\infty < x < \infty \qquad (19.45)$$
$$u(x, 0) = e^{ikx}$$

with exact dispersion relation $\omega = \alpha k$.

For an upwind numerical flux ($\alpha = 1$) the DG discretization is

$$\frac{\Delta x_n}{2} \mathbf{Q} \frac{\partial \mathbf{c}^n}{\partial t} + c \, \mathbf{N_1} \, \mathbf{c}^{n-1} + c \, \mathbf{N_0} \, \mathbf{c}^n = 0 \qquad (19.46)$$

where $\mathbf{c}^n = (c_0^n, c_1^n, ..., c_{N-1}^n)^T$ are the expansion coefficients in the element e_n and $\mathbf{Q}, \mathbf{N_1}$ and $\mathbf{N_0}$ are matrices involving base polynomials [88].

For uniform mesh $\Delta x_n = \delta$ and solution of the form

$$\mathbf{c}^n(t) = \hat{\mathbf{c}} \, e^{i(kn\delta - \omega t)} \qquad (19.47)$$

obtain the following algebraic system for the expansion coefficients $\widehat{\mathbf{c}^n} = (c_0^n, c_1^n, ..., c_{N-1}^n)^T$

$$\left(-\frac{i\omega\delta}{2} \mathbf{Q} + c \, e^{-ik\delta} \, \mathbf{N_1} + c \, \mathbf{N_0}\right) \hat{\mathbf{c}} = 0 \qquad (19.48)$$

the matrices $\mathbf{Q}, \mathbf{N_1}$ and $\mathbf{N_0}$ that involve integrals of the polynomial basis [88]. For sinusoidal solutions $\mathbf{c}^n(t) = \hat{\mathbf{c}} \, e^{i(kn\delta - \omega t)}$ on a uniform mesh $\Delta x_n = \delta$ obtain the following algebraic system for $\hat{\mathbf{c}}$

$$\left(-\frac{i\omega\delta}{2} \mathbf{Q} + c \, e^{ik\delta} \, \mathbf{N_1} + c \, \mathbf{N_0}\right) \hat{\mathbf{c}} = 0 \qquad (19.49)$$

and $\hat{\mathbf{c}}$ is an eigenvector corresponding to the eigenvalue ω of the matrix

$$\mathbf{M} = \frac{2}{i\delta} \mathbf{Q}^{-1} (c \, e^{-ik\delta} \, \mathbf{N_1} + c \, \mathbf{N_0}) \qquad (19.50)$$

The matrix $\mathbf{M}$ has N eigenvalues ω_m, $m = 0$, ..., $N-1$ and N eigenvectors $\widehat{\mathbf{c}_m}$, $l = 0$, ..., $N-1$ and the sinusoidal solution (see for example LeVeque [120]) is written as

$$\widehat{\mathbf{c}^n}(t) = \sum_{m=0}^{N-1} \lambda_m \, \widehat{\mathbf{c}_l} \, e^{i \, (kn\delta - \omega_m t)} \tag{19.51}$$

where λ_m are suitable scaling coefficients so that this eigenvector expansion satisfies the initial conditions.

The eigenvector expansion of the numerical solution is a superposition of N waves travelling at different phase speeds. The physical modes are the waves with frequencies that approximate the exact dispersion relation for a range of wave numbers while the others are the parasitic modes resulting from the DG discretization and the particular choice of the numerical flux.

The exact dispersion relation is $\Omega = K$ where $K = k\delta$ and $\Omega = \frac{\omega\delta}{c}$ are the nondimensional wave number and frequency, respectively. The numerical dispersion relation is determined by

$$\det\left(-i\Omega\mathbf{Q} + 2e^{-iK}\,\mathbf{N}_1 + 2\mathbf{N}_0\right) = 0 \tag{19.52}$$

and yields complex values of $\Omega = \Omega_r + i\Omega_i$ with negative imaginary part that represents the numerical damping inherent in the DG discretization process. Based on this analysis Hu et al. [88] arrived to the following important conclusions:

- Increase of the order of the scheme significantly reduces the dissipation error.

- The sixth-order scheme is optimal for scalar advection in the sense that it is the minimal order for which the dispersion and dissipation errors are less than 0.5% for K up to approximately N.

- The dissipation error imposes a relatively more stringent condition on the accuracy of the scheme than does the dispersion error.

The analysis of Hu et al. [88] for the two dimensional case is summarized next.

19.7.2   Two-Dimensional Advection (Wave Equation)

Consider the two-dimensional wave equation

$$\frac{\partial^2 \phi}{\partial t^2} - c^2 \nabla^2 \phi = 0 \tag{19.53}$$

which can be written as a system of first order equations

$$\mathbf{u} = \begin{bmatrix} u_1 \\ u_2 \end{bmatrix} \begin{bmatrix} \frac{\partial \phi}{\partial t} & -c \frac{\partial \phi}{\partial x} \\ 0 & -c \frac{\partial \phi}{\partial y} \end{bmatrix} \tag{19.54}$$

$$\frac{\partial \mathbf{u}}{\partial t} + \nabla \cdot \mathbf{F}(\mathbf{u}) = 0 \quad \mathbf{F}(\mathbf{u}) = \begin{bmatrix} A_1 \mathbf{u} \\ A_2 \mathbf{u} \end{bmatrix}, \quad A_1 = c \begin{bmatrix} 1 & 0 \\ 0 & -1 \end{bmatrix}, \quad A_2 = c \begin{bmatrix} 0 & 1 \\ 1 & 0 \end{bmatrix}.$$

For the DG discretization consider rectangular elements $E^{nm} = [x_n, x_{n+1}] \times [y_m, y_{m+1}]$ and the Roe flux. Then the DG discretization in matrix form is as follows

$$\mathbf{Q} \frac{\partial \mathbf{c}^m}{\partial t} + \frac{2c}{\Delta x_n} \left[\mathbf{N}_0 \, \mathbf{c}^{n,m} + \mathbf{N}_{-1} \, \mathbf{c}^{n-1,\, m} + \mathbf{N}_{+1} \, \mathbf{c}^{n+1,\, m} \right] \tag{19.55}$$

$$+ \frac{2c}{\Delta y_m} \left[\mathbf{M}_0 \, \mathbf{c}^{n,m} + \mathbf{M}_{-1} \, \mathbf{c}^{n-1,\, m} + \mathbf{M}_{+1} \, \mathbf{c}^{n,\, m+1} \right] = 0.$$

For sinusoidal plane wave solutions with wave number k and propagation angle ϑ the expansion coefficients $\mathbf{c}^{n,m}$ have the form

$$\mathbf{c}^{n,m}(t) = \widehat{\mathbf{c}} \, e^{i \, [k(\cos \vartheta x_n + \sin \vartheta y_m) - \omega t]} \tag{19.56}$$

where $\mathbf{k} = (k \cos \vartheta, k \sin \vartheta)$ and the vector $\widehat{\mathbf{c}}$ is complex and independent of n, m and t.

For these plane wave solutions the DG discretization is

$$\left[-i\omega \mathbf{Q} + \frac{2c}{\delta x} \left(\mathbf{N}_0 + e^{-ik \cos \vartheta \delta x} \mathbf{N}_{-1} + e^{ik \cos \vartheta \delta x} \mathbf{N}_{+1} \right) \right. \tag{19.57}$$

$$\left. + \frac{2c}{\delta y} \left(\mathbf{M}_0 + e^{-ik \sin \vartheta \delta y} \mathbf{M}_{-1} + e^{ik \sin \vartheta \delta y} \mathbf{M}_{+1} \right) \widehat{\mathbf{c}} \right] = 0.$$

Similar to the one-dimensional case, the exact dispersion relation is $\Omega = \pm K$ where $\Omega = \frac{\omega \delta x}{c}$, $K = k \delta x$, $\frac{\delta y}{\delta x} = \gamma$ while the numerical dispersion relation is given by

$$\det \left(-i\Omega \mathbf{Q} + 2 \left[\mathbf{N}_0 + e^{-iK \cos \vartheta} \mathbf{N}_{-1} + e^{iK \cos \vartheta} \mathbf{N}_{+1} \right] \right. \tag{19.58}$$

$$\left. + 2\gamma \left[\mathbf{M}_0 + e^{-iK \sin \gamma \vartheta} \mathbf{M}_{-1} + e^{iK \sin \gamma \vartheta} \mathbf{M}_{+1} \right] \right) = 0$$

where the value of Ω is function of the wavenumber K and the angle ϑ. An important conclusion of this analysis is that the wave propagation is anisotropic and the dependence on wave propagation angle is stronger for the higher wave numbers.

19.8   Dissipative and Dispersive Behavior of High-Order DG Discretizations

In the previous chapter, the dissipative and dispersive properties of the DG discretizations were analyzed and demonstrated for linear aeroacoustic problems. A systematic analysis of the DG method for linear wave propagation with very short wave lengths was recently presented by Ainsworth [7]. Ainsworth's analysis targets electromagnetic wave propagation. However, efficient and accurate resolution of electromagnetic waves without excessive numerical dissipation or dispersion is important in the context of other high frequency applications such as magneto gas dynamics MGD.

The most promising approach for wave propagation is obtained with higher order schemes such as spectral element methods (Gottlieb and Hesthaven, [69], Dyson, [50]). Higher order standard Galerkin finite element methods were also used in the past by Astley et al. [11], Ihlenburg [99], Thomson and Pinsky [178]. More recently, high order DG finite element methods (Bey and Oden, [24], Biswas et al., [25], Cockburn et al., [39], Warburton et al, [194], Hu and Atkins, [89]) were applied and analyzed for wave propagation.

It was found that for the small wave-number limit $hk \to 0$ the discontinuous Galerkin method gives a higher order of accuracy than the standard Galerkin methods. Hu and Atkins [89], for example, concluded that the dispersion relation of the scalar advection equation for an $N - th$ order DG method is accurate to order $2N + 3$ in hk for the dispersion error and $2N + 2$ for the dissipation error. The more systematic study of Ainsworth [7] for the dissipative and dispersive behavior of the DG method that gives sharp error estimates is outlined next.

19.8.1   Dispersive Properties of DG for the Advection Equation

Consider the linear advection equation

$$u_t + \mathbf{a} \cdot \nabla u = 0 \tag{19.59}$$

with nontrivial solutions of the form

$$u\left(\mathbf{x}, t\right) = ce^{i\ (\mathbf{k} \cdot \mathbf{x} - \omega t)} \tag{19.60}$$

provided that ω and $\mathbf{k}$ satisfy the dispersion relation

$$\omega = \mathbf{a} \cdot \mathbf{k}.$$

Consider an N-th order DG approximate solution u^{DG} for the tensor product polynomial space P_N.

Continuity between the neighboring elements k and k' is enforced in a weak sense through the use of a numerical flux function $\tilde{\sigma}_\gamma$ on the interface

$$\tilde{\sigma}_\gamma\,(\mathbf{n}_k, u^{DG}) = \Lambda_\gamma^+\,(\mathbf{n}_k)\,u_k^{DG} + \Lambda_\gamma^-\,(\mathbf{n}_k)\,u_{k'}^{DG} \qquad (19.61)$$

$$\Lambda_\gamma^\pm\,(\mathbf{n}) = \frac{1}{2}\,(\mathbf{n}\cdot\alpha \pm \gamma\,|\mathbf{n}\cdot\alpha|)$$

where γ is the upwind parameter and $\Lambda_\gamma^\pm$ satisfies

$$\Lambda_\gamma^+\,(\mathbf{n}) + \Lambda_\gamma^-\,(\mathbf{n}) = \mathbf{n}\cdot\mathbf{a} \qquad (19.62)$$

$$\Lambda_\gamma^\pm\,(-\mathbf{n}) = -\Lambda_\gamma^\pm\,(\mathbf{n}).$$

The first property guarantees that the true flux satisfies

$$\sigma\,(\mathbf{n}, u) = \mathbf{n}\cdot\alpha u = \tilde{\sigma}_\gamma\,(\mathbf{n}, u). \qquad (19.63)$$

The second property implies flux balance from element k to k'

$$\tilde{\sigma}_\gamma\,(\mathbf{n_k}, u^{DG}) = -\tilde{\sigma}_\gamma\,(\mathbf{n}_{k'}, u^{DG}). \qquad (19.64)$$

Replacing the true flux by the numerical flux, the DG solution for every element K satisfies

$$\int_K v\,(u_{K,t}^{DG} + \mathbf{a}\cdot\nabla u_K^{DG}) + \int_{\partial K} v\,\Lambda_\gamma^-\,(\mathbf{n_K})\,(u_{K'}^{DG} - u_K^{DG}) = 0 \qquad \forall\, v \in P_N.$$
$$(19.65)$$

Ainsworth obtained an expression for the error proving that if ω and k satisfy the dispersion relation $\omega = \mathbf{a}\cdot\mathbf{k}$ then the DG solution u^{DG} satisfies

$$u^{DG}\,(\mathbf{x} + h\mathbf{m}, t + \tau) = e^{i\,(\mathbf{k}\cdot h\mathbf{m} - \omega\tau)}u^{DG}\,(\mathbf{x}, t) \qquad \forall\,\mathbf{x} \in \mathbf{R}^\mathbf{d} \qquad (19.66)$$
$$t \in \mathbf{R}$$

and each component of the discrete wave-vector $\widetilde{k}_l$ may take values corresponding to a

physical mode $e^{ih\widetilde{k}_l} \approx e^{ihk_l}$

or a spurious mode $e^{ih\widetilde{k}_l} \approx (-1)^{N+1}\frac{(1+\gamma)}{(1-\gamma)}\frac{H_N^*}{H_N}\,e^{ihk_l},\ \gamma \neq 1$

where $H_N =_1 F_1\,(-N;\ -2N-1;\ ihk_l)$,

$_1F_1$ denotes the hypergeometric function defined by

$$_1F_1\,(a,b,z) = 1 + \frac{a}{b}\,z + \frac{a}{b}\,\frac{a+1}{b+1}\,\frac{z^2}{2!} + \frac{a}{b}\,\frac{a+1}{b+1}\,\frac{a+2}{b+2}\,\frac{z^3}{3!} + ... \qquad (19.67)$$

in both cases the relative error ρ_N is

$$\rho_N = \frac{(1-\gamma)\,H_N e^{ihk_l}\varepsilon_N + (-1)^{N+1}\,(1+\gamma)\,H_N^* e^{-ihk_l}\varepsilon_N^*}{(1-\gamma)\,H_N e^{ihk_l} + (-1)^N\,(1+\gamma)\,H^* e^{-ihk_l}} + 0\,(|\varepsilon_N|^2)$$
$$(19.68)$$

where ε_N is the relative error in the N-Padé approximation of e^{ihk_l}.
  For $hk \ll 1$ obtain

$$\gamma \neq 0 \quad \rho_N \approx \frac{1}{2}\,(hk)^{2N+2}\left[\frac{N!}{(2N+1)!}\right]\,Q_N\,\left(\gamma^{(-1)^N}\right)$$

$$\gamma = 0 \quad \rho_N \approx \frac{1}{2}\left[\frac{N!}{(2N+1)!}\right]^2\begin{cases}-(hk)^{2N+3}\,\frac{N+1}{2N+3} & N \text{ even} \\ (hk)^{2N+1}\,\frac{2N+1}{N+1} & N \text{ odd}\end{cases} \qquad (19.69)$$

dispersion error

$$R\,(h\widetilde{\mathbf{k}}) - R\,(hk) \approx \frac{(hk)^{2N+3}}{2}\left[\frac{N!}{(2N+1)!}\right]^2\left[\frac{N+1}{2N+1}\,\gamma^{2(-1)^N} - \frac{N+1}{2N+3}\right]$$

dissipation error

$$J\,(h\widetilde{\mathbf{k}}) \approx \frac{(hk)^{2N+2}}{2}\left[\frac{N!}{(2N+1)!}\right]\,\gamma^{(-1)^N}.$$

Ainsworth estimates are very sharp and agree with the a posteriori error estimates of Hu and Atkins [89].

19.8.2   Relative Error for Fixed Mesh-Size and Large Wave Number kh

For increased order N of the method the analysis summarized in the previous chapter obtains the following important estimates:
As the order N is increased, the behavior of the error passes through three different phases depending on the size N relative to hk:

1. Pre-asymptotic regime $2N + 1 < hk - \mathcal{O}\,(hk)^{1/3}$.

    The resolving ability of the method is inadequate and the relative error tends to oscillate without decay.

2. The transition zone where $hk - \mathcal{O}\,(hk)^{1/3} < 2N + 1 < hk + \mathcal{O}\,(hk)^{1/3}$.

   The relative error is of order unity and decreases at an algebraic rate $N^{-1/3}$.

3. Asymptotic regime where N is large compared to hk $2N + 1 > hk + \mathcal{O}\,(hk)^{1/3}$.

   The relative error reduces at a super exponential rate.

19.8.3 Exponential Convergence on the Envelope $2N + 1 \approx hk$

Reduction of the mesh-size to the limit $hk << 1$ is not practical for accurate capturing of high frequencies. It was shown, however, that increasing the order N on a fixed mesh is more effective than reducing the mesh size. The conclusions of Ainsworth's analysis [7] are:

1. It is inefficient to increase the order N much beyond the threshold $2N + 1 > hk + \mathcal{O}\,(hk)^{1/2}$.

2. A more practical approach is to work on the envelope of the region where the super-exponential convergence sets in.

3. To resolve problems where $hk >> 1$, the order must be chosen so that $2N + 1 \approx \kappa hk$ where $\kappa > 1$ is a constant.

4. In addition, Ainsworth [7] showed that the relative error ρ_N decays at an exponential rate as $N \to \infty$, i.e.,

$$\rho_N \approx -e^{-\beta(N+1/2)}\left(1 - \sqrt{1 - \frac{1}{\kappa^2}}\right)\left(\sqrt{\kappa^2 - 1} - 1\right) \qquad (19.70)$$

$$\beta = \ln\frac{1 + \sqrt{1 - 1/\kappa^2}}{1 - \sqrt{1 - 1/\kappa^2}} - 2\sqrt{1 - 1/\kappa^2}. \qquad (19.71)$$

19.9 Limiting of DG Expansions

Limiting operators $\Lambda\Pi_h$ on piecewise linear DG expansions u_h are constructed in such a way that they satisfy the following properties:

1. Accuracy: if u_h is linear, then $\Lambda\Pi_h u_h = u_h$.

2. Conservation of mass: for every element K have

$$\int_k \Lambda\Pi_k u_h \; dV = \int_k u_h \; dV. \qquad (19.72)$$

3. Slope limiting: The gradient of $\Lambda\Pi_h u_h$ is not bigger than that of u_h for each element K.

Theoretical analysis of the slope limiting operators can be found in Cockburn and Shu [34] and Cockburn et al. [36].

19.9.1   Rectangular Elements

The P^1 expansion of the approximate solution $u_h\,(x,y,t)$ inside rectangular elements $[x_{i-\frac{1}{2}}, x_{i+\frac{1}{2}}] \times [y_{i-\frac{1}{2}}, y_{i+\frac{1}{2}}]$ is

$$u_h(x,y,t) = \overline{u}\,(t) + u_x\,(t)\,\phi_i\,(x) + u_y\,(t)\,\psi_j\,(y) \qquad (19.73)$$

where $\overline{u}(t)$, $u_x(t)$, $u_y(t)$ are the expansion coefficients or the degrees of freedom to be evolved in time and

$$\phi_i\,(x) = \frac{x - x_i}{(\Delta x_i/2)}, \qquad \psi_j\,(y) = \frac{y - y_i}{(\Delta y_j/2)} \qquad (19.74)$$

For the scalar equation, limiting for these quadrilateral element expansion is performed on u_x and u_y using the difference of the means.
For example, u_x is replaced by

$$\overline{m}\,(u_x\,,\,\overline{u}_{i+1,j} - \overline{u}_{i,j}\,,\,\overline{u}_{i,j} - \overline{u}_{i-1,j}) \qquad (19.75)$$

where $\overline{m}$ is the total variation bounded (TVB) corrected minmod function defined as

$$\overline{m}\,(\alpha_1\,,\alpha_2\,,...,\alpha_m) = \begin{cases} \alpha_1 & \text{if } |\alpha_1| \leq M\,(\Delta x)^2 \\ m\,(\alpha_1\,,\alpha_2\,,...,\alpha_n) & \text{otherwise} \end{cases} \qquad (19.76)$$

and m is the total variation diminishing (TVD) minmod function defined as

$$m\,(\alpha_1\,,\alpha_2\,,...\alpha_n...,\alpha_N) = \begin{cases} s\;min_{1\leq n\leq N}|\alpha_n| \\ \qquad\qquad \text{if} \quad s = sign\,(\alpha_1) = ... = sign\,(\alpha_N) \\ \qquad 0 \qquad\qquad\qquad \text{otherwise.} \end{cases}$$
$$(19.77)$$

The TVB correction is introduced in order to avoid unnecessary limiting near smooth extrema where the expansion coefficients (or degrees of freedom)

u_x, u_y are on the order of $\mathcal{O}\left(\Delta x^2\right)$, $\mathcal{O}\left(\Delta y^2\right)$, respectively. The numerical results are not usually sensitive to the choice of the constant M. The suggested value for this constant [38] is $M = 50$

Similarly, u_y is replaced (limited) by

$$\overline{m}\left(u_y, \overline{u}_{i,j+1} - \overline{u}_{i,j}, \overline{u}_i, -\overline{u}_{i-1,j}\right). \tag{19.78}$$

For systems, such as the Euler equations, limiting is performed in the local characteristic variables as follows:

- The left and right eigenvector matrices R^{-1} and R, which diagonalize the Jacobian $A = \frac{\partial f\,(u)}{\partial u}$, are evaluated at the average state $\overline{u}_{i,j}$ in the element i, j in the x and y directions as follows

$$R^{-1}AR = \Lambda \tag{19.79}$$

  where Λ is the diagonal matrix that contains the eigenvalues of the flux Jacobian. The columns of R are the right eigenvectors of A and the rows of R^{-1} are the left eigenvectors.

- All quantities needed for limiting are transformed to the characteristic field by left multiplying by R^{-1}. For the system version of the limiter given by Eq. (19.75) for example, the vectors $u_{x_{i,j}}$, $\left(\overline{u}_{i+1,j} - \overline{u}_{i,j}\right)$, $\left(\overline{u}_{i,j} - \overline{u}_{i-1,j}\right)$ are transformed to the characteristic field.

- The limiter of Eq. (19.75) or Eq. (19.78) is applied on each component of the characteristic variables vector.

- The limited (replaced) values are transformed back to the original conservative variables by multiplying by R.

19.9.2   Triangular Elements

For the P^1 case, the following expansion inside the triangle K is used for the approximate solutions $u_h\,(x, y, t)$

$$u_h\,(x, y, t) = \sum_{i=1}^{3} u_i\,(t)\,\phi_i\,(x, y) \tag{19.80}$$

where the degrees of freedom or expansion coefficients $u_i(t)$ are the values of the numerical solution at the midpoints of the edges. The basis function is a linear function that takes unit value at the midpoints m_i of the $i - th$ edge and zero value at the midpoints of the other two edges.

The slope limiting operator for triangular elements is constructed as follows. Consider the triangle K_0 of Fig. 19.7 where limiting is performed and

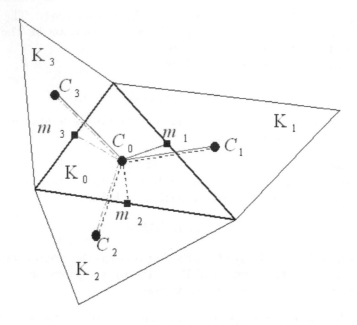

FIGURE 19.7: Geometric definitions for the limiter on an arbitrary triangular mesh.

the neighboring triangles K_1, K_2 and K_3. Suitable choice of the triads of the vectors c_0m_1, c_0m_2, c_0m_3, c_0c_1, c_0c_2, c_0c_3 are shown in Fig. 19.7. In this figure the vector joining the barycenter c_0 with the middle of an edge is between the vectors joining the barycenter c_0 with the barycenters of the neighboring triangles. Based on the schematic of Fig. 19.7 perform the following geometrical decomposition:

$$(\overrightarrow{c_0m_1}) = \alpha_1 \left(\overrightarrow{c_0c_1}\right) + \alpha_2 \left(\overrightarrow{c_0c_2}\right) \tag{19.81}$$

where α_1 and α_2 are positive coefficients that depend on the geometry.

Furthermore, for any linear function have the expansion

$$u_h\left(m_1\right) - u_h\left(c_0\right) = \alpha_1 \left[u_h\left(c_1\right) - u_h\left(c_0\right)\right] + \alpha_2 \left[u_h\left(c_2\right) - u_h\left(c_0\right)\right] \tag{19.82}$$

and the averages over the triangles are

$$\overline{u}^{K_i} = \frac{1}{|K_i|} \int_{K_i} u_h \, dV = u_h\left(c_i\right) , \qquad i = 0,1,2,3. \tag{19.83}$$

Therefore

$$\overline{u_h}^{K_0} (m_1) \equiv u_h (m_1) - \overline{u}^{K_0}$$
$$\equiv \alpha_1 (\overline{u}^{K_1} - \overline{u}^{K_0}) + \alpha_2 (\overline{u}^{K_2} - \overline{u}^{K_0})$$
$$\equiv \Delta\overline{u}^{K_0} (m_1). \tag{19.84}$$

For a piecewise linear function u_h in K_0 and the three midpoints m_1, m_2, m_3 obtain

$$u_h^{K_0} (x,y) = \sum_{i=1}^{3} u_h (m_i) \, \phi_i (x,y)$$

$$= \overline{u}^{K_0} + \sum_{i=1}^{3} \overline{u}_h^{K_0} (m_i) \, \phi_i (x,y). \tag{19.85}$$

The limited value $\Lambda\Pi_h u_h^{K_0} (x,y)$ is obtained by first evaluating the quantities

$$\Delta_i = \overline{m} \left(\overline{u}_h^{K_0} (m_i) \,,\, \omega \, \Delta\overline{u}^{K0}(m_i) \right) \tag{19.86}$$

where $\overline{m}$ is the total variation bounded (TVB) modified minmod function of Eq. (19.76) and $\omega > 1$ is a weight taken $\omega = 1.5$.
Then

$$\text{if} \quad \sum_{i=1}^{3} \Delta_i = 0 \quad \Lambda\Pi_h u_h^{K_0} = \overline{u}^{K_0} + \sum_{i=1}^{3} \Delta_i \phi_i (x,y)$$

$$\text{if} \quad \sum_{i=1}^{3} \Delta_i \neq 0 \quad \text{compute}$$

$$\text{pos} = \sum_{i=1}^{3} \max (0, \Delta_i) \,,\, \text{neg} = \sum_{i=1}^{3} \max (0, -\Delta_i)$$

$$\theta^+ = \min \left(1 - \frac{\text{neg}}{\text{pos}} \right) \,,\, \theta^- = \min \left(1, \frac{\text{pos}}{\text{neg}} \right)$$

and obtain the limited value by

$$\Lambda\Pi_h u_h (x,y) = \overline{u}^{K_0} + \sum_{i=1}^{3} \widehat{\Delta}_i \, \phi_i (x,y) \tag{19.87}$$

where

$$\widehat{\Delta}_i = \theta^+ max (0, \Delta_i) - \theta^- max (0, -\Delta_i). \tag{19.88}$$

For systems of equations on triangular meshes, limiting is performed on the local characteristic variables. For triangular meshes however the following flux Jacobian

$$\frac{\partial}{\partial u} f\left(\overline{u}^{K_0}\right) \frac{\overrightarrow{m_i c_0}}{|m_i c_0|} \tag{19.89}$$

e.g., the flux Jacobian along the direction of the unit vector, $\mathbf{k} = \frac{\overrightarrow{m_i c_0}}{|m_i c_0|}$, must be diagonalized to evaluate the left and right eigenvector matrices R^{-1} and R.

The matrices R and R^{-1} of $\mathbf{A} \cdot \vec{k}$ are:

$$R = \begin{bmatrix} \hat{k}_x & 0 & \frac{\rho}{2c} & \frac{\rho}{2c} \\ u\hat{k}_x & \rho\hat{k}_y & \frac{\rho}{2c}(u + c\hat{k}_x) & \frac{\rho}{2c}(u - c\hat{k}_x) \\ v\hat{k}_x & -\rho\hat{k}_y & \frac{\rho}{2c}(v + c\hat{k}_x) & \frac{\rho}{2c}(v - c\hat{k}_y) \\ \hat{k}_x \frac{\vec{v}^2}{2} & \rho(\vec{v} \times \vec{1}_k) & \frac{\rho}{2c}(H + c\vec{v} \cdot \vec{1}_k) & \frac{\rho}{2c}(H - c\vec{v} \cdot \vec{1}_k) \end{bmatrix} \tag{19.90}$$

and for R^{-1} we have

$$R^{-1} = \begin{bmatrix} (1 - \frac{\gamma-1}{2}M^2)\hat{k}_x & \frac{(\gamma-1)u\hat{k}_x}{c^2} & \frac{(\gamma-1)v\hat{k}_x}{c^2} & -\frac{(\gamma-1)\hat{k}_x}{c^2} \\ \frac{\vec{v}\times\vec{1}_k}{\rho} & \frac{\hat{k}_y}{\rho} & -\frac{\hat{k}_x}{\rho} & 0 \\ \frac{c}{\rho}(\frac{(\gamma-1)M^2}{2} - \frac{\vec{v}\cdot\vec{1}_k}{c}) & \frac{1}{\rho}(\hat{k}_x - \frac{(\gamma-1)u}{c}) & \frac{1}{\rho}(\hat{k}_y - \frac{(\gamma-1)v}{c}) & \frac{(\gamma-1)}{\rho c} \\ \frac{c}{\rho}(\frac{(\gamma-1)M^2}{2} + \frac{\vec{v}\cdot\vec{1}_k}{c}) & -\frac{1}{\rho}(\hat{k}_x + \frac{(\gamma-1)u}{c}) & \frac{1}{\rho}(\hat{k}_y + \frac{(\gamma-1)v}{c}) & \frac{(\gamma-1)}{\rho c} \end{bmatrix} \tag{19.91}$$

where $\vec{1}_k = (\hat{k}_x, \hat{k}_y)$ be the unit vector along direction $\vec{k}$, $H = \frac{\rho e + p}{\rho}$, $\vec{v} \cdot \vec{1}_k = u\hat{k}_x + v\hat{k}_y$ the inner product between the vector $\vec{1}_k$ and the velocity $\vec{v} = (u, v)$, $\rho(\vec{v} \times \vec{1}_k) = \rho(u\hat{k}_y - v\hat{k}_x)$, $\vec{v}^2 = u^2 + v^2$.

19.10   Component-Wise Limiters

The first way to apply a limiter to each characteristic variable was presented in the previous sections for both quadrilateral and triangular elements. The other way is to apply a limiter to each of the conservative variables. The characteristic-wise application of limiters for one-dimensional linear hyperbolic systems has the nice property of naturally degenerating to the scalar case. In multiple dimensions, the characteristic variables must be defined in a particular direction. For unstructured meshes, there is no coordinate direction to define a characteristic variable and these variables are defined in the face normal direction. The design of characteristic-based limiters in multiple directions is difficult and time consuming.

In this chapter, the component-wise approach for the limiter of the DG method is shown. This approach is expected to be more efficient than the characteristic approach of the previous sections. The component-wise approach is based on the following numerical monotonicity criterion for each element.

$$\bar{u}_i^{\min} < u_i\left(\mathbf{r}_r\right) < \bar{u}_i^{\max} \tag{19.92}$$

where $\bar{u}_i^{\min}$ and $\bar{u}_i^{\max}$ are the minimum and maximum cell-averaged solutions among all its neighboring elements sharing a face with the triangle T_i and $u_i\left(\mathbf{r}_s\right)$ is the solution at any of the quadrature points. Violation of Eq. (19.92) for any quadrature point indicates that the element is close to a discontinuity and the solution in the element is forced locally linear, i.e.,

$$u_i\left(\mathbf{r}\right) = \bar{u}_i + \nabla u_i\left(\mathbf{r} - \mathbf{r}_i\right) \quad \forall\, \mathbf{r} \in T_i \tag{19.93}$$

where $\mathbf{r}_i$ is the position vector of the centroid of T_i. The magnitude of the solution gradient is maximized subject to the monotonicity condition given in Eq. (19.92). The original DG polynomial basis is used to compute on initial guess for the gradient as

$$\nabla u_i = \left(\frac{\partial u_i}{\partial x}, \frac{\partial u_i}{\partial y}\right)\Bigg|_{r_2}. \tag{19.94}$$

Using Eq. (19.93) and the gradient computed by Eq. (19.94), Eq. (19.92) may not still be satisfied. Therefore, $u_i\left(\mathbf{r}\right)$ is limited by multiplying by a scalar limiter $\phi \in [0, 1]$ so that the solution vector obtained from

$$u_i\left(\mathbf{r}\right) = \bar{u}_i + \phi \nabla u_i\left(\mathbf{r} - \mathbf{r}_i\right) \tag{19.95}$$

satisfies Eq. (19.92). The scalar limiter ϕ in Eq. (19.95) is obtained by examining the numerical solutions at all quadrature points as follows: Denoting by

$$\Delta u_r = p_i\left(\mathbf{r}\right) - \bar{u}_i \tag{19.96}$$

$$\phi = \begin{cases} \min\left(1, \dfrac{\Delta u_r}{\bar{u}_i^{\max} - \bar{u}_i}\right) & \text{if } \Delta u_r > 0 \\ 1 \\ \min\left(1, \dfrac{\Delta u_r}{\bar{u}_i^{\min} - \bar{u}_i}\right) & \begin{array}{l}\text{if } \Delta u_r < 0 \\ \text{otherwise.}\end{array} \end{cases}$$

19.11 DG Stabilization Operator

High-order accurate DG finite element solutions (with polynomial bases of degree one or higher) do not guarantee monotonicity around discontinuities and sharp gradients. The slope limiter of Cockburn et al. (1990) [36] of the previous chapter guaranties monotone solutions for multidimensional scalar conservation laws and its extension to the Euler equations. This slope limiter results in a robust numerical discretization and has become quite popular. Despite its robustness the slope limiter of Cockburn et al. [36] has the following serious disadvantages. It may result in an unnecessary reduction in accuracy in smooth parts of the flow field and slows down or prevents convergence to steady state. Furthermore, its implementation for multidimensional cases and high order discretizations is very intensive computationally.

Recently, van der Vegt and van der Ven [183] suggested that a better alternative for stabilization of the DG method is addition of artificial dissipation. This approach was also followed in the past by Cockburn and Grenaud [37], [140], and Jaffre et al. [101]. The stabilization operators make optimal use of the information contained in a DG discretization and preserve the compactness of the DG method because the use the jump in the polynomial representation only at the element faces.

The stabilization operator $\mathcal{D} \in \mathbf{R}^{4 \times 4}$ of van der Vegt and van der Ven [183] is for quadrilateral elements and is defined as

$$\mathcal{D}_{lm}\left(u_h^{K_j^n}, u_h^{*\,K_j^n}\right) = \int_{K_j^n} \frac{\partial \psi_l}{\partial x_k} \, \mathbf{D}_{kp}\left(u_h^{K_j^n}, u_h^{*\,K_j^n}\right) \frac{\partial \psi_m}{\partial x_p} \, dK \qquad (19.97)$$

where $u_h^{K_j^n}$ is the expansion in the element K_j^n.

$$u_h(x,t) = \sum_{k=1}^{K} \widehat{u}_k\left(K_j^n\right) \psi_k\left(x\right)$$

and $u_h^{*\,K_j^n}$ denotes the solution in the elements which connect to the element K_j^n.

The definition of the artificial viscosity matrix is more straightforward when the stabilization is introduced independently in all computational coordinate directions by introducing the artificial viscosity matrix $\widetilde{\mathcal{D}} \in \mathbf{R}^{4 \times 4}$ into the computational space of the reference elements using the relation

$$\mathcal{D}\left(u_h^{K_j^n}, u_h^{*\,K_j^n}\right) = R^T \, \widetilde{\mathcal{D}}\left(u_h^{K_j^n}, u_h^{*\,K_j^n}\right) R \qquad (19.98)$$

where the matrix R is defined as

$$R = 2H^{-1}\nabla G_K \qquad (19.99)$$

and

$$H = diag\,(h_1, h_2, h_3, h_4)$$

with h_i denoting the leading terms of the expansion of the mapping G_K given by

$$G^K : \widehat{K} \to K^n : \xi \to (\overline{x}) = \left(\frac{1}{2}\,(1 - \xi_4)\,F_K^n\,(\overline{\xi}) + \frac{1}{2}\,(1 + \xi_4)\,F_K^n\,(\xi) \right)$$

(19.100)

where $\widehat{K}$ is the reference element related through the mapping F_K^n.

$$F_K^n : \widehat{K} \to K^n : \overline{\xi} \to \overline{x} = \sum_{i=1}^{8} x_i\,(K^n)\,\mathcal{X}\,(\overline{\xi})$$

(19.101)

with $x_i(K^n)$ denoting the spatial coordinates of the vertices of a quadrilateral or hexahedron K^n and $\mathcal{X}_i(\overline{\xi})$ are the finite element shape functions for quadrilaterals or hexahedra.

The integral in the stabilization operator of Eq. (19.97) is evaluated as

$$
\begin{aligned}
\mathcal{D}_{nm}\left(u_h^{K_j^n}, u_h^{*K_j^n}\right) &= \int_{K_j^n} \frac{\partial \psi_n}{\partial x_k}\,R_{pk}\,\widetilde{\mathcal{D}}_{pq}\left(u_h^{K_j^n}, u_h^{*K_j^n}\right) R_{ql}\,\frac{\partial \psi_m}{\partial x_l}\,dK \\
&= 4 \int_{\widehat{K}} (H^{-1})_{pn}\,\widetilde{\mathcal{D}}\left(u_h^{K_j^n}, u_h^{*K_j^n}\right) (H_{qm}^{-1})\,|J_{G_K}|\,d\widehat{K} \\
&= \frac{4|K_j^n|}{h_n^2}\,\delta_{nm}\,\widetilde{\mathcal{D}}_{nm}\left(u_h^{K_j^n}, u_h^{*K_j^n}\right)
\end{aligned}
$$

(19.102)

where the relations $(\nabla G_K)_{ij} = \partial x_j / \partial \xi_i$ and $\partial \psi_n / \partial \xi_p = \delta_{np}$ were used and $\widetilde{\mathcal{D}}$ is assumed constant in each element.

The stabilization operator must be applied in areas of discontinuities and regions where the residual is large due to insufficient grid resolution. This information is, however, available in the DG discretizations, and is coupled to the jump in the solution across element faces and the element residual. In the regions of smooth solution both the jumps and the residual are on the order of the truncation error. Two artificial viscosity approaches were tested:

19.11.1   Subsonic and Transonic Flow Model

This model works well in subsonic flows and transonic flows with weak shocks. In this artificial viscosity model only the jump in pressure across the elements influences the stabilization matrix that is defined as

$$\widetilde{D}_{qq}\left(u_h^{K_j^n}, u_h^{*K_j^n}\right) = \frac{c'\lambda h_q}{|\Omega_j^n|} \sum_{m=1}^{6} \frac{|p^+(x_m) - p^-(x_m)|\,|S_m|}{p^+(x_m) + p^-(x_m)} , \quad q = 1, 2, 3$$

(19.103)

with $p^\pm(x_m) = \gamma^\pm(p(x_m))$ denoting the pressure at the edge midpoints or the centers of the faces and is γ the trace operator. The scaling factor $\lambda = |\mathbf{n}_k \cdot \mathbf{v}|\,\alpha$ is the maximum of the eigenvalues of the Jacobian $\partial f/\partial u$ at the midpoints x_m, $\mathbf{v}$ is the fluid velocity, and α is the local speed of sound $\alpha = \sqrt{\gamma p/\rho}$. The constant c' is of order one and Ω_j^n is the element area or volume.

19.11.2 Supersonic Flow Model

The artificial viscosity model for problems with strong discontinuities such as supersonic flow uses both jumps at element interfaces and the element residual to apply artificial smoothing. This model was proposed and analyzed by Jaffre et al. [101] and defines $\widetilde{D}$ as

$$\widetilde{D}_{qq}\left(u_h^{K_j^n}, u_h^{*K_j^n}\right) = \max\left(c_2\,h_K^{2-\beta}\,R_q\,(u_h^{K_j^n}, u_h^{*K_j^n}),\ c_1 h_K^{3/2}\right), \quad q = 1, 2, 3$$

(19.104)

with

$$R\left(u_h^{K_j^n}, u_h^{*K_j^n}\right) = \left|\sum_{m=0}^{3} \frac{\partial F(u_n)}{\partial u_{n,i}} \frac{\partial u_{h,i}(G_m(0))}{\partial x_m}\right|$$

$$+ c_0\left|u_h^r(x_i) - u_h^-(x_i)\right|/h_K$$

(19.105)

$$+ \sum_{m=1}^{6} \frac{1}{h_K}\left|\mathbf{n}_K\,\widetilde{F}\left(u_h^+(x_m)\right) - \mathbf{n}_K\,\widetilde{F}\left(u_h^+(x_m)\right)\right|$$

with $h_K = \sqrt{h_1^2 + h_2^2 + h_3^2 + h_4^2}$, $c_0 = 1.2$, $c_1 = 0.1$, $c_2 = 1$, $\beta = 0.1$.

19.12   DG Space Discretization of the NS Equations

The compressible Navier-Stokes (NS) equations can be written in compact vector form as follows:

$$\frac{\partial \mathbf{u}}{\partial t} + \nabla \cdot \mathbf{f_i}(\mathbf{u}) + \nabla \mathbf{f_v}(\mathbf{u}, \nabla \mathbf{u}) = 0$$

(19.106)

where $\mathbf{u}$ is the vector of the conservative variables and $\mathbf{f_i}$, $\mathbf{f_v}$ denote the inviscid and viscous flux functions. The viscous flux $\mathbf{f_v}$ is a linear function of the velocity gradient ∇u. Therefore Eq. (19.106) can be also written as

$$\frac{\partial \mathbf{u}}{\partial t} + \nabla \cdot \mathbf{f_i}\,(\mathbf{u}) + \nabla \cdot [\mathcal{A}(\mathbf{u}),\,\nabla \mathbf{u}] = 0. \qquad (19.107)$$

The discretization of the viscous, diffusive part of the NS equations with the DG method is less well known and different than the method described previously for the convective, inviscid part. A simple way to extend the scheme of Eq. (19.5), which was developed for convective problems of the form $u_t + u_x = 0$, for the diffusion equation $u_t + u_{xx} = 0$ is to simply replace u by u_x and find $u \in V_h$ such that

$$\frac{d}{dt} \int_K u\,(x,t)\,\phi\,(x)\,dx = \int_K u_x\,\phi_x\,dx \qquad (19.108)$$
$$-(\widehat{u}_x)_{j+1/2}\,\phi^-_{j+1/2} + (\widehat{u}_x)_{j-1/2}v^+_{j-1/2} = 0$$

where for the lack of an upwind mechanism for the diffusive term the numerical flux is the centered flux $(\widehat{u}_x)_{j+1/2} = 1/2\,[(u_x)^L_{j+1/2} + (u_x)^R_{j+1/2}]$.

Unfortunately, this simple but naive formulation of Eq. (19.108) leads to numerically stable but inconsistent solutions [40], [206]. The numerical solutions seem to converge with mesh refinement but have $\mathcal{O}(1)$ errors with exact solutions of the heat equation [206]. This is a pitfall of the DG method applied to diffusion equations and it is very undesirable that the scheme of Eq. (19.108) produces stable but completely incorrect solutions.

A formulation of the DG method that is convergent and consistent was used by Bassi and Rebay [18] for the compressible Navier-Stokes equations. A second successful method that avoids the inconsistencies of the simple formulation of Eq. (19.108) was presented by Baumann and Oden [21] who added extra penalty terms to the inner boundaries.

The consistent formulation of Bassi and Rebay [18] for the spatial discretization of the viscous term in the NS equations was constructed by resorting to a mixed finite element formulation. The second-order derivatives of the conservative variables required for the viscous terms were obtained by using the gradient of the conservative variables, $\nabla \mathbf{u} = \mathbf{S}$, as auxiliary unknowns of the NS equations. The NS equations were therefore reformulated as the following coupled system to the unknowns $\mathbf{S}$ and $\mathbf{u}$.

$$\mathbf{S} - \nabla \mathbf{u} = 0 \qquad (19.109)$$
$$\partial_t \mathbf{u} + \nabla \cdot \mathbf{f}_i\,(\mathbf{u}) + \nabla \cdot \mathbf{f}_v\,(\mathbf{u}, \mathbf{S}) = 0$$

The weak formulation of the first equation of the system of Eq. (19.109) is

$$\int_K S_h\,\phi\,d\Omega - \oint_e \mathbf{u}_h\,\mathbf{n}\,\phi\,d\Gamma + \int_K u_h\,\nabla\phi\,d\Omega = 0 \qquad (19.110)$$

where the term $\mathbf{u}_h\,\mathbf{n}$ in the second (contour) integral of Eq. (19.110) is replaced by a numerical flux $H_s\,(\mathbf{u}^-,\mathbf{u}^+,\mathbf{n})$. This numerical flux is the average between the two interface states

$$H_s\,(\mathbf{u}^-,\mathbf{u}^+,\mathbf{n}) = \frac{1}{2}\,(\mathbf{u}^- + \mathbf{u}^+)\,\mathbf{n}. \qquad (19.111)$$

The computed auxiliary variables $\mathbf{S}_h$ are used to form the second equation of the system in Eq. (19.109) as follows:

$$\frac{d}{dt}\int_K \mathbf{u}_h\,\phi\,dx + \oint_e \mathbf{f}_i\,(\mathbf{u}_h)\cdot\mathbf{n}\,\phi\,d\Gamma - \int_K \mathbf{f}_i\,(\mathbf{u}_h)\,\nabla\,\phi\,dx$$

$$\qquad (19.112)$$

$$+ \oint_e \mathbf{f}_v\,(\mathbf{u}_h,\mathbf{S}_h)\cdot\mathbf{n}\,d\Gamma - \int_K \mathbf{f}_v\,(\mathbf{u}_h,\mathbf{S}_h)\,\nabla\phi\,dx.$$

In Eq. (19.112), the term $\mathbf{f}_v\,(\mathbf{u}_h,\mathbf{S}_h)\cdot\mathbf{n}$ is replaced with the following centered numerical flux

$$\mathbf{h}_v\,(\mathbf{u}^-,\mathbf{S}^-,\mathbf{u}^+,\mathbf{S}^+,\mathbf{n}) = \frac{1}{2}\,[\mathbf{f}_v\,(\mathbf{u}^-,\mathbf{S}^-) + \mathbf{f}_v\,(\mathbf{u}^+,\mathbf{S}^+)]\cdot\mathbf{n}.$$

The disadvantage of the above approach for the DG viscous term discretization is (see Eqs. (19.111) and (19.112)) that it extends the compact DG stencil beyond the neighboring cells. A more appropriate treatment of higher order derivatives, which does not enlarge the compact DG stencil, a detailed presentation and an analysis of the DG discretization of the viscous part is given in the next chapter.

19.12.1   DG Discretization of Elliptic Problems

First, the basic ideas are outlined using a simple purely elliptic problem the Helmholtz equation with Dirichlet and Neumann boundary conditions and then extension to the full NS equations is carried out. Consider the Helmholtz equation with Dirichlet and Neumann boundary conditions

$$\begin{aligned}
\nabla^2 u - \alpha u &= S \quad \text{in } \Omega \\
u &= u^b \quad \text{on } \Gamma^o \in \partial\Omega \\
\partial_n u &= z^b \quad \text{on } \Gamma^n \in \partial\Omega
\end{aligned} \qquad (19.113)$$

In order to use the same techniques already developed for the discretization of the convective terms, the Helmholtz problem of Eq. (19.113) is reformulated as a first order system

$$\mathbf{z} = \nabla u \qquad u = u^b \qquad \text{on } \Gamma^o$$
$$\nabla \cdot \mathbf{z} - \alpha u = S \qquad \mathbf{z} \cdot \mathbf{n} = z^b \qquad \text{on } \Gamma^n \qquad (19.114)$$

the weak form of this system is

$$\int_\Omega \mathbf{g} \cdot \mathbf{z} \, d\Omega = - \int_\Omega u \nabla \cdot \mathbf{g} \, d\Omega + \oint_\Gamma u^* \mathbf{g} \cdot \mathbf{n} \, dS \quad \forall \, \mathbf{g}$$
$$- \int_\Omega \nabla v \cdot \mathbf{z} \, d\Omega + \oint v \mathbf{z}^* \cdot \mathbf{n} \, dS - \alpha \int_\Omega v \, u \, d\Omega$$
$$= \int_\Omega v \, s \, d\Omega \quad \forall \, v \qquad (19.115)$$

where $\mathbf{g}$ and v are arbitrary test functions for the first and second equations, respectively. The boundary conditions for the first order system are:

$$u^* = u^b \, , \, \mathbf{z}^* = \mathbf{z} \qquad \text{on } \Gamma^o \qquad (19.116)$$
$$u^* = u \, , \, \mathbf{z}^* \cdot \mathbf{n} = z^b \quad \text{on } \Gamma^n.$$

Consider the finite element spaces

$$V_h = v|_e \in P^k(e) \; \forall \, e \in T_n \qquad (19.117)$$
$$\mathbf{G_h} = \mathbf{g}|_e \in P^k(e) \; \forall \, e \in T_n$$

where $P^k(e)$ denotes the space of polynomial functions of degree at most K in the element e.

The DG formulation of Eq. (19.115) for $u_h \in V_h$ and $\mathbf{z_h} \in \mathbf{G_h}$ is

$$\sum_e \int_{\Omega_e} \mathbf{g_h} \cdot \mathbf{z_h} \, d\Omega = - \sum_e \int_{\Omega_e} u_h \nabla \cdot \mathbf{g_h} \, d\Omega + \sum_e \oint_{\partial\Omega_e} \widehat{u_h \mathbf{g_h}} \cdot \mathbf{n} \, dS$$
$$- \sum_e \int_{\Omega_e} \nabla v_h \cdot \mathbf{z_h} \, d\Omega + \sum_e \oint_{\partial\Omega_e} v_h \widehat{\mathbf{z_h}} \cdot \mathbf{n} \, dS + \sum_e \int_{\Omega_e} v_h u_h \, d\Omega$$
$$= \sum_e \int_{\Omega_e} v_h s \, d\Omega$$
$$(19.118)$$

where as in the convective inviscid flux case the numerical flux functions $\widehat{u_h}$ and $\widehat{z_h}$ have been introduced to unquietly define the contour integrals

$$\sum_e \int_{\partial\Omega_e} \hat{u}_h \mathbf{g_h} \cdot \mathbf{n} \, dS = \sum_{i \in T} \int_{\Sigma_i} \hat{u}[(\mathbf{g_h} \cdot \mathbf{n})^- + (\mathbf{g_h} \cdot \mathbf{n})^+] \, dS$$

$$\sum_e \int_{\partial\Omega_e} v_h \widehat{\mathbf{z_h}} \cdot \mathbf{n} \, dS = \sum_{i \in T} \int_{\Sigma_i} \widehat{\mathbf{z_h}}[(v_h \mathbf{n})^- + (v_h \mathbf{n})^+] \, dS \qquad (19.119)$$

where $(\cdot)^+$ and $(\cdot)^-$ denote values on Σ_i of any quantity evaluated for two elements e^+ and e^- which share the edge i. Introducing in addition the jump operator $\mathcal{J}$ defined as

$$\mathcal{J}x = x^- \mathbf{n}^- + x^+ \mathbf{n}^+ \qquad (19.120)$$
$$\mathcal{J}\mathbf{x} = \mathbf{x}^- \mathbf{n}^- + \mathbf{x}^+ \mathbf{n}^+$$

the DG formulation of Eq. (19.118) with Eqs (19.119) and (19.120) is written as

$$\int_{\Omega_h} \mathbf{g_h} \cdot \mathbf{z_h} \, d\Omega = -\int_{\Omega_h} u_h \nabla \cdot \mathbf{g_h} \, d\Omega + \int_{\Sigma_h} \widehat{u_h} \mathcal{J} \cdot \mathbf{g_h} \, dS$$

$$+ \int_{\Gamma_h^o}^b u^b \mathbf{g_h} \cdot \mathbf{n} \, dS + \int_{\Gamma_h^n} u_h \mathbf{g_h} \cdot \mathbf{n} \, dS$$

$$- \int_{\Omega_h} \nabla v_h \cdot \mathbf{z_h} \, d\Omega + \int_{\Sigma_h} \widehat{\mathbf{z_h}} \cdot \mathcal{J} v_h \, dS + \int_{\Gamma_h^o} \mathbf{z} \cdot \mathbf{n} v_h \, dS \qquad (19.121)$$

$$+ \int_{\Gamma_h^n} z^b v_h \, dS - \alpha \int_{\Omega_h} v_h u_h \, d\Omega$$

$$= \int_{\Omega_h} v_h s \, d\Omega.$$

19.12.2   Approximation of the Numerical Fluxes $\widehat{\mathbf{z_h}}$ and $\widehat{u_h}$

The approximation of the numerical flux is carried out for each equation separately. First, the approximation for the first of the equations, containing the grad operator $\mathbf{z} = \nabla u$, is obtained. Next, approximation of the second equation is described.

Approximation of the $\widehat{u_h}$ flux

For the first equation of the system Eq. (19.121), consider a "centered" numerical flux $\widehat{u_h} = u_h$ where $\{\cdot\}$ denotes the average operator for internal edges and the identity operator at the boundary edges, i.e.,

$$\{(\cdot)\} = \begin{cases} \frac{1}{2}\left[(\cdot)^- + (\cdot)^+\right] & on \ \Sigma_n \\ (\cdot) & on \ \Gamma_n \end{cases}$$

using this numerical flux obtain

$$\int_{\Omega_h} \mathbf{g_h} \cdot \mathbf{z_h} \ d\Omega - \int_{\Omega_h} \mathbf{g_h} \cdot \nabla u_h \ d\Omega + \int_{\Sigma_h^o} \{\mathbf{g_h}\} \cdot \mathcal{J}^0 u_n \ dS = 0 \qquad (19.122)$$

where $\Sigma_h^0 = \Sigma_h \bigcup \Gamma_h^0$ and $\mathcal{J}^0$ is an expanded jump operator defined by

$$\mathcal{J}^0 x = \begin{cases} (x - x^b)\mathbf{n} & on \ \Gamma_h^0 \\ \mathcal{J}x = x^-\mathbf{n} + x^+\mathbf{n} & on \ \Sigma_h. \end{cases}$$

Approximation of the $\widehat{\mathbf{z_h}}$ flux

The vector numerical flux $\widehat{\mathbf{z_h}}$ is evaluated as an average (centered) flux $\widehat{\mathbf{z_h}} = \{\mathbf{z_h}\}$.

The DG discretization for the second equation is

$$-\int_{\Omega_h} \nabla v_h \cdot \mathbf{z_h} \ d\Omega + \int_{\Sigma_h^0} \{\mathbf{z}\} \mathcal{J}^0 v_h \ dS - \alpha \int_{\Omega_h} v_h u_h \ d\Omega$$

$$= \int_{\Omega_h} v_h s \ d\Omega - \int_{\Gamma_h^n} z^b v_h \ dS. \qquad (19.123)$$

Based on the above formulation, an alternative scheme for "compact" DG approximation of elliptic problems, was proposed by Bassi et al. [20]. This alternative scheme overcomes limitations of the method described in the beginning of this chapter, which are associated with not optimal accuracy for approximations with odd order polynomials and the additional computational cost occurring form the introduction of the auxiliary variable $\widehat{\mathbf{z_h}}$. Note that $\widehat{\mathbf{z_h}}$ is obtained in terms of u_h at the cost of a block diagonal mass matrix inversion from the first equation. However, the formulation of the second equation, which is obtained in this manner, and contains the primal variable alone involves an "enlarged stencil". This enlarged stencil occurs because the primal unknown u_h for any internal element e is coupled not only with unknowns of the neighboring elements but also with unknowns associated to the neighbors of the neighbors (because the second derivative is evaluated as the first derivative of a first derivative). The enlarged stencil implies additional computational cost and loss of the parallel implementation advantage of the DG method. Bassi et al. [20] show that only the jump contribution to the auxiliary variable $\widehat{\mathbf{z_h}}$ is responsible for the non-compact support of the scheme described and they propose modifications that one can use to arrive at a scheme with compact support for elliptic problems and for DG discretizations of the viscous terms in the NS equations.

19.12.3   A DG Scheme with Compact Support

The second equation of the system in Eq. (19.121) can be written as

$$
-\int_{\Omega_h} \nabla v_h \cdot \nabla u_n \, d\Omega + \int_{\Sigma_h^0} \left[\{\nabla v_h\} \cdot \mathcal{J}^0 u_h + \{\nabla u_n\} \mathcal{J}^0 v_n \right] \, dS
$$

$$
+ \int_{\Sigma_h^0} \left\{ \mathbf{R_h} \left(\mathcal{J}^0 u_h \right) \right\} \cdot \mathcal{J}^0 v_h \, dS - \alpha \int_{\Omega_h} v_h u_n \, d\Omega =
$$

$$
= \int_{\Omega_n} v_n s \, d\Omega - \int_{\Gamma_h^n} z^b v_h \, dS
$$

$$
\tag{19.124}
$$

where $\{\cdot\}$ and $\mathcal{J}^0$ denote the average and the jump operator defined previously, and $\mathbf{R}_h$ is the so-defined lift operator.

$$
\int_{\Omega_h} \mathbf{g}_h \cdot \mathbf{R_h} \left(\mathcal{J}^0 u_h \right) \, d\Omega = - \int_{\Sigma_h^0} \{\mathbf{g}_n\} \cdot \mathcal{J}^0 u_n \, dS. \tag{19.125}
$$

This lift operator can be used to express both $\mathbf{z_h}$ and $\widehat{\mathbf{z_h}}$ in finite form as

$$
\mathbf{z_h} = \nabla u_n + \mathbf{R}_h \left(\mathcal{J}^0 u_n \right) , \quad \widehat{\mathbf{z_h}} = \{\widehat{\mathbf{z_n}}\} = \{\nabla u_n\} + \{\mathbf{R}_h \left(\mathcal{J}^0 u_n \right)\}. \tag{19.126}
$$

The appearance of the average lift $\mathbf{R}_h$ in Eq. (19.124) implies that for an arbitrary edge the numerical vector flux $\widehat{\mathbf{z_h}}$ depends not only from the jump of u_h on that edge but also from the jumps on all the edges belonging to the elements e^- and e^+ which share this edge.

Bassi et al. [20] proposed a modification of the scheme for the discretization of the elliptic operators that follows more closely the numerical flux function ideas usually employed in finite volume (FV) schemes. A typical treatment of an elliptic operator in FV schemes is often based on the definition of auxiliary staggered control volumes enclosing the boundaries of the primal control volumes that are used to construct the diffusive terms, which are the analogue to the vector flux $\widehat{\mathbf{z_h}}$ used in the DG formulation.

For each interface boundary edge $\sigma \in \Sigma_h^0$, Bassi et al. [20] consider the local lift operator $\mathbf{r}_h^\sigma$ defined as

$$
\int_{\Omega_h^\sigma} \mathbf{g}_h \cdot \mathbf{r}_h^\sigma \left(\mathcal{J}^0 u_n \right) \, d\Omega = - \int_\sigma \{\mathbf{g}_h\} \cdot \mathcal{J}^0 u_n \, dS \tag{19.127}
$$

$$
\forall \, \sigma \in \Sigma_h^0, \, \forall \, \mathbf{g}_h \in \mathbf{G}_h
$$

and since $\mathbf{g}_h$ is an arbitrary function

$$
\mathbf{R} \left(\mathcal{J}^0 u_n \right) = \sum_{\sigma \in \Sigma_n^0} \mathbf{r}_h^\sigma \left(\mathcal{J}^0 u_n \right). \tag{19.128}
$$

Using these modifications the new scheme that replaces Eq. (19.127) is

$$
-\int_{\Omega_n} \nabla v_h \cdot \nabla u_n \, d\Omega + \int_{\Sigma_n^0} \left[\{\nabla v_h\} \cdot \mathcal{J}^0 u_n + \{\nabla u_n\} \cdot \mathcal{J}^0 u_h \right] \, dS
$$

$$
+ \sum_{\sigma \in \Sigma_0^h} \int_\sigma \left\{ \mathbf{r}_\mathbf{h}^\sigma \left(\mathcal{J}^0 u_h \right) \right\} \cdot \mathcal{J}^0 v_h \, dS \qquad (19.129)
$$

$$
- \alpha \int_{\Omega_n} v_h u_h \, d\Omega = \int_{\Omega_n} v_h s \, d\Omega - \int_{\Gamma_h^n} z^b v_h \, dS
$$

where the vector numerical flux $\widehat{\mathbf{z_h}}$ is

$$
\widehat{\mathbf{z_h}}|_\sigma = \{\nabla u_h\}|_\sigma + \left\{ \mathbf{r}_\mathbf{h}^\sigma \left(\mathcal{J}^0 u_h \right) \right\} . \qquad (19.130)
$$

19.13   The DG Variational Multiscale (VMS) Method

The VMS method is a new approach for LES proposed by Hughes et al. in a series of papers [95], [96], [97]. The VMS method was subsequently further clarified by Collis [42] and applied with the DG method [43], [134] for LES of compressible turbulence. More recently the VMS method was further developed [148] for mixed-type, Fourier-spectral/finite volume formulation. The ideas behind the VMS method and its implementation with the DG space discretization method are described in this chapter.

The dynamics of turbulent shear flows are dominated by the motions of a small number of relatively large-scale structures. The separation between the large, energy containing scales and the smallest turbulent scales increases as the Reynolds number increases. The increase in range of scales prevents DNS from being a viable tool beyond very simple, low Reynolds number flows. In contrast, LES attempts to exploit the scale separation in turbulent shear flows by representing the largest scales as accurately as possible on the computational mesh and using a model to account for the influence of the unresolved smaller scales. Important progress in turbulent flow simulations with LES was made, primarily with the application of the so-called dynamic model (see [64], [123], [14], [65]). It was, however, soon recognized that the Reynolds numbers for which LES can be applied using this approach are still far too low. As a result, LES is not still an economically feasible alternative for the simulation of the vast majority of engineering flows. The deficiencies of RANS or even DES to accurately predict complex separated flows and the high computational cost of LES for flows of practical interest lead to the quest of new approaches [95], [109] that overcome the weaknesses of LES and RANS while providing consistency with DNS. The Discontinuous Galerkin/Variational Multi-Scale (DG/VMS) method [43], which merges VMS turbulence modeling [95] with the high-order accurate DG discretization, is a particularly synergistic combination that offers a number of advantages over traditional methods.

The advantages of the VMS/DG approach for LES are:

1. Variational projection with a priori scale separation [95] avoids problems associated with spatial filters.

2. The method converges to the exact DNS with h, p or $h - p$ refinement.

3. The method is high-order with the potential for exponential (spectral) convergence.

4. VMS/DG is insensitive to grid quality and suitable for complex domains.

5. The method allows for different models to be used in different regions of the flow while still retains formal convergence to the exact solution.

6. The VMS/DG approach for turbulence simulations unifies traditional DNS, LES and RANS approaches in a single computational tool.

The VMS/DG method is briefly described next.

Consider the compressible Navier-Stokes equations in strong conservation law form

$$\mathcal{N}\left(U\right) = U_t + F^i_{j,j} - F^v_{j,j} = 0 \tag{19.131}$$
$$U\left(\mathbf{x}, 0\right) = U_0\left(\mathbf{x}\right)$$

where $U = [\rho, \rho\mathbf{u}, \rho E]^T$ is the conservative variable vector $F^i_j\left(U\right)$ is the inviscid flux vector in the j coordinate direction, and $F^v_j\left(U\right)$ is viscous flux vector in the j coordinate direction.

The weak form of Eq. (19.131) is

$$\int_{\Omega_e} \mathbf{W}^T\, U_t\, d\mathbf{x} + \int_{\Omega_e} \mathbf{W}^T_j\left(\mathbf{F}^u_j - \mathbf{F}^i_j\right) d\mathbf{x} + \tag{19.132}$$
$$\oint_{\partial\Omega_e} W^T\left(\mathbf{F}^i_{n_e} - \mathbf{F}^v_{n_e}\right) d\mathbf{S} = 0$$

where $F_n = F_j\, n_j$, n_e is the outward unit normal vector on $\partial\Omega_e$, and $\mathbf{W}$ is a continuous in Ω_e weighting or test function.

Following the VMS approach of [42] the following three-level multiscale framework is used to allow direct monitoring of unresolved scales.

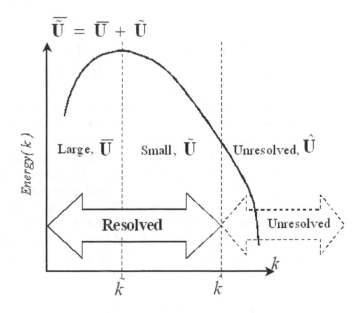

FIGURE 19.8: Wave space decomposition for the VMS method.

$$\mathbf{U} = \overline{U} + \tilde{U} + \widehat{U} \qquad (19.133)$$

The partition of the exact solution $\mathbf{U}$ of Eq. (19.133), introduces $\overline{U}$, which represents the large scales, $\tilde{U}$ for the small scales, and $\widehat{U}$ for the unresolved scales. This partition into scales is depicted in Fig. 19.8 as a range of Fourier modes in wavespace. The large scale equations have no modeling terms while the small scale equations have modeling terms that can range from Smagorinsky closure to a full Reynolds stress model. The large and small scales are referred to as resolved scales.

Assuming that the basis is orthogonal (the analysis for the general nonorthogonal basis can be found in [42]) and substituting the partition of Eq. (19.133) in the weak form of Eq. (19.132), which in condensed form is denoted as $B\ (\mathbf{W}, \mathbf{U}) = 0$, obtain the following week forms for the large- and small-scales.

The large-scale equations

$$B\ (\overline{W}, \overline{U}) - C\ (\overline{W}, \overline{U}, \tilde{U}) - R\ (\overline{W}, \tilde{U}) = \qquad (19.134)$$
$$C\ (\overline{W}, \overline{U}, \widehat{U}) + R\ (\overline{W}, \widehat{U}) + C\ (\overline{W}, \tilde{U}, \tilde{U})$$

The small-scale equations

$$B' \ (\widetilde{W}, \overline{U}, \widetilde{U}) - R \ (\widetilde{W}, \widetilde{U}) = R \ (\widetilde{W}, \overline{U}) \tag{19.135}$$
$$+C \ (\widetilde{W}, \overline{U}, \widehat{U}) + R \ (\widetilde{W}, \widehat{U}) + C \ (\widetilde{W}, \widetilde{U}, \widehat{U})$$

In Eqs. (19.134) and (19.135) $R \ (\widetilde{W}, \widetilde{U})$ is the generalized Reynolds stress, $C \ (\widetilde{W}, \widetilde{U}, \widehat{U})$ is the generalized cross stress, $B' \ (\widetilde{W}, \overline{U}, \widetilde{U})$ is the operator $B \ (\widetilde{W}, U)$ linearized about $\overline{U}$ for the linear perturbation $\widetilde{U}$.

The effect of the unresolved scales on the large scales is seen in Eq. (19.134) which contains the unresolved Reynolds stress projection onto the large scales simultaneously with the large-unresolved and small-unresolved generalized cross stresses projections onto the large scales. The small scale equation also contains the unresolved Reynolds stresses and cross stresses. For truncated or discrete approximations the combined small and large scales are identified as the resolved scales and denoted as $\widetilde{\overline{U}} = \overline{U} + \widetilde{U}$. Therefore, the resolved scale equation (Eq. (19.134)) is written more compactly as

$$B \ (\widetilde{\overline{W}}, \widetilde{\overline{U}}) = R \ (\widetilde{\overline{W}}, \widehat{U}) + C \ (\widetilde{\overline{W}}, \widetilde{U}, \widehat{U}). \tag{19.136}$$

The equations for small- (see Eq. (19.135)) and large-scales (see Eq. (19.134) or (19.136)) indicate the need to model unresolved Reynolds and cross terms appearing on the right-hand side. The modeling assumptions introduced are:

1. The unresolved scales $\widehat{U}$ have negligible direct influence on the dynamic evolution of the large scales. For sufficient a priori scale separation the right hand side term $C \ (\overline{W}, \overline{U}, \widehat{U}) + R \ (\overline{W}, \widehat{U}) + C \ (\overline{W}, \widetilde{U}, \widehat{U})$ of Eq. (19.134) is small.

2. The unresolved scales, however, are expected to significantly influence the small scales. Therefore, an appropriate model is needed for the right hand side terms $C \ (\widetilde{W}, \overline{U}, \widehat{U}) + R \ (\widetilde{W}, \widehat{U}) + C \ (\widetilde{W}, \widetilde{U}, \widehat{U})$ of Eq. (19.135).

The VMS formulation can support a wide range of models. In addition, a key feature of the VMS method is that different models can be used in different spatial locations. As a result, the model used depends on the particular flow characteristics, the desired fidelity, and the location of the particular subdomain under consideration.

Using these modeling assumptions the models for the large and small scales are:

$$B \ (\overline{W}, \overline{U}_h) - C \ (\overline{W}, \overline{U}, \widetilde{U}) - R \ (\overline{W}, \widetilde{U}) = R \ (\overline{W}, \widetilde{U}_h) \tag{19.137}$$

$$B' \ (\widetilde{W}, \overline{U}_h, \widetilde{U}_h) - R \ (\widetilde{W}, \widetilde{U}_h) = R \ (\widetilde{W}, \overline{U}) + M \ (\widetilde{W}, \overline{U}_h, \widetilde{U}_h). \tag{19.138}$$

The subscript h in Eqs. (19.137) and (19.138) denotes that the equations contain modeling errors and are the approximate numerical solution introduces discretization errors.

The large-scale equation is modeled because the effect of the unresolved scales has been ignored. The modeled large-scale equation takes the form of the exact equation only when all scales of motion are contained within the resolved scales. In this case, the VMS method is a DNS. However, by neglecting the influence of the unresolved scales on the large scales, the modeled large-scale equation has no direct modeling terms. The model in the small scales indirectly influences the large scales. This is achieved through the small-scale Reynolds and cross stresses. Thus, if the exact solution is fully represented by the large scales, then the solution to the modeled equations (19.137) is exact. This consistency is an important advantage over classical methods [95]. The model applied to the large-scale equation is nothing more than the standard approach used in a Galerkin method where the projection of the residual of the unresolved scales onto the large scales is zero. Considering the simplified case where the bases are orthogonal, this amounts to weak enforcement of zero unresolved Reynolds/cross stresses on the large scales. This indicates the particular discretizations play a role in altering the model and therefore the results.

In the small-scale equation, it is the projection of the unresolved Reynolds and cross stresses onto the small scales that is modeled using a weak implementation of a subgrid-scale model. Again, use of different methods for the small-scale discretizations is expected to alter the model. This is one of the main advantages of the VMS framework. Although the model is specified without regard to the specific discretization, the influence of discretization is obvious in the modeled equations. This fact has not been completely appreciated in the traditional turbulence modeling community until recently when it was realized that differences between the discretization with the "same model" might be as large or larger than differences in "models" using the same discretization. In the VMS framework, the influence of different discretizations is evident and the choice of discretization clearly plays an important role in the success of particular models. Therefore an important area for future research is to explore different bases for use in defining large and small scales. The modeled large and small scales in Eqs. (19.137) and (19.138) are combined as follows:

$$B\left(\widetilde{\overline{W}}, \widetilde{\overline{U}}_h\right) = M\left(\widetilde{\overline{W}}, \overline{U}_h, \widetilde{U}_h\right) + \left(\widetilde{\overline{W}}, S\right) \text{ on } \Omega_e. \tag{19.139}$$

The model of Eq. (19.139) can be implemented with VMS by incorporating the additional model term into the small scale equations. This model term can depend on both the large and small scales. Therefore it can take forms ranging from classical Smagorinsky model to models used in DES.

In the previous paragraphs, the variational multiscale method was presented for a typical subdomain Ω_e. Summing Eq. (19.139) over all the subdomains in the partition P_h of Ω and introducing typical continuous smooth finite dimensional spaces for U_h and the test function $\mathbf{W_h}$ then, subject to appropriate boundary conditions and the particular choice of model, this leads to the standard variational multiscale method for classical finite elements, spectral elements, or global spectral methods (see e.g., Refs. [95], [104] and [97]). Results from these methods are quite new, they have shown tremendous potential including the ability to accurately simulate wall-bounded and non-equilibrium turbulence using a very simple constant coefficient Smagorinsky model on the small scales [97]. However, most of the results presented to date use global spectral methods, which have a rich function space but are only feasible for very simple geometries [96], [97]. The preliminary work of Jansen [104] and co-workers offsets this by using a low-order (cubic and lower) hierarchal basis within a C° finite element method. While this method can be applied to complex geometries, the relatively low-order function spaces may not have a sufficient scale separation for effective turbulence simulation. Perhaps of even greater importance however, is that the reliance of prior approaches on C° or smoother function spaces limits ones ability to alter the large/small partition or change the form of the model as a function of space. For example, in laminar regions of a flow no model should be used, while in boundary layers a RANS type model may be appropriate (if one is only concerned with the mean flow) while in a wake region an LES type model may be used to capture the large-scale unsteadiness.

To address these limitations, the DG/VMS method introduced in [43] combines VMS turbulence modeling with a discontinuous Galerkin method in space. The combination of these two methods is particularly synergistic and the combined DG/VMS method possesses the following characteristics:

- high-order (even exponential) convergence on highly irregular unstructured meshes;

- discretization, large/small partitioning, and model equations can be changed on each subdomain, Ω_e;

- all boundary conditions are set weakly through boundary fluxes including fluxes of turbulence stress. This enables one to directly enforce zero turbulent stress at solid walls - a feature not present in prior approaches;

- the method is highly localized leading to a great degree of parallelism which is required for the large-scale turbulence simulations we are targeting.

The additional complication with the VMS/DG method is that the typical approach for diffusive problems has been to introduce auxiliary variables for the viscous fluxes and re-write the equations of motion as an extended first-order system of equations [38], [18], as was described in a previous chapter. This mixed approach has been demonstrated by Lomtev et al. [130]

for two- and three-dimensional unsteady flows. Unfortunately, the mixed approach in three-dimensions requires 6 additional unknowns and equations for three-dimensional Navier-Stokes flows. Furthermore, in applications, the model terms, $M\,(\widetilde{\mathbf{W}}, \overline{\mathbf{U}}, \tilde{\mathbf{U}})$, often take the form of diffusive terms (eddy diffusivity models) that may require the addition of even more unknowns. Over the past few years, there has been extensive research on the use of DG methods for elliptic and mixed hyperbolic/elliptic problems (see for example [38], [18], [137], [21]). Furthermore, Arnold et al. [10] show that the flux formulation, commonly used in DG methods, can be readily converted to the primal formulation.

With this background, one can merge the variational multiscale method described above with a DG method. Denoting the boundary of the domain Ω as $\partial\Omega = \Gamma_D \bigcup \Gamma_N$ where Γ_D is the portion of the boundary where Dirichlet conditions are specified and Γ_N is the portion of the boundary where Neumann conditions are set. The element boundary is denoted as $\Gamma = \{\Gamma_D, \Gamma_N, \Gamma_0\}$ where Γ_0 are the inter-element boundaries. Let Ω_1 and Ω_2 be two adjacent elements. Furthermore, let $\Gamma_{12} = \partial\Omega_1 \bigcap \partial\Omega_2$; and $\mathbf{n}^{(1)}$ and $\mathbf{n}^{(2)}$ be the corresponding outward unit normal vectors at that point. Denoting $\mathbf{U}^{(e)}$ and $\mathbf{F}_i^{(e)}$ be the state vector $\mathbf{U}$ and flux vectors $\mathbf{F}_i$, respectively, on Γ_{12}. Then, define the average $\langle . \rangle$ and jump $[.]$ operators on Γ_{12} as

$$
\begin{aligned}
(a) &\quad [\mathbf{U}n_i] = \mathbf{U}^{(1)}\,n_i^{(1)} + \mathbf{U}^{(2)}\,n_i^{(2)}, \\
(b) &\quad [\mathbf{F}_n] = \mathbf{F}_i^{(1)}\,n_i^{(1)} + \mathbf{F}_i^{(2)}\,n_i^{(2)}, \\
(c) &\quad \langle \mathbf{U} \rangle = \frac{1}{2}\,(\mathbf{U}^{(1)} + \mathbf{U}^{(2)}), \\
(d) &\quad \langle \mathbf{F}_i \rangle = \frac{1}{2}\,(\mathbf{F}_i^{(1)} + \mathbf{F}_i^{(2)}),
\end{aligned}
\tag{19.140}
$$

where $\mathbf{F}_n = \mathbf{F}_i\,n_i$.

With this notation and Eq. (19.139) the discontinuous Galerkin formulation of the $B\,(\mathbf{W}, \mathbf{U})$ term, defined in Eq. (19.134), is applied to obtain the DG formulation for the Navier-Stokes equations as

$$
\begin{aligned}
B_{DG}\,(\mathbf{W}, \mathbf{U}) = \sum_{\Omega_e} \int_{\Omega_e} & (\mathbf{W}^T\,\mathbf{U},_t + \mathbf{W}^T,_i\,(\mathbf{F}_i^v - \mathbf{F}_i))\,dx \\
& - \int_{\Gamma} ([\mathbf{W}^T\,n_i]\,\langle \hat{\mathbf{F}}_i^v - \hat{\mathbf{F}}_i \rangle - \\
& - \langle (\mathbf{D}_i\,\mathbf{W})^T \rangle\,[(\hat{\mathbf{U}} - \hat{\mathbf{U}})\,n_i])\,ds \\
& - \int_{\Gamma_0} (\langle \mathbf{W}^T \rangle\,[\hat{\mathbf{F}}_n^v - \hat{\mathbf{F}}_n] - \\
& \quad [(\mathbf{D}_n\,\mathbf{W})^T]\,\langle (\hat{\mathbf{U}} - \mathbf{U})\,n_i \rangle)\,ds
\end{aligned}
\tag{19.141}
$$

where

$$\mathbf{F}_n\,(\mathbf{U}) = \mathbf{F}_i\,(\mathbf{U})\,n_i, \tag{19.142}$$

$$\mathbf{F}_n^v\,(\mathbf{U}) = \mathbf{F}_i^v\,(\mathbf{U})\,n_i = \mathbf{D}_n\,\mathbf{U}. \tag{19.143}$$

Quantities with a hat in Eq. (19.141) are numerical fluxes that must be appropriately defined. For example, the term $\mathbf{F}_i\,(\mathbf{U}^-,\,\mathbf{U}^+)$ is an appropriate approximate Riemann flux (see previous sections of the various options). The particular choice of Riemann flux plays an important role in determining the dispersion/dissipation characteristics of the method [88]. For illustrative purposes, we use the Steger-Warming flux-vector splitting where $\mathbf{F}_n\,(\mathbf{U})$ is split into inflow and outflow components $\mathbf{F}_n^-$ and $\mathbf{F}_n^+$ as follows:

$$\mathbf{F}_n^\pm\,(\mathbf{U}) = \mathbf{R}\lambda^\pm\,LU, \quad \mathbf{\Lambda}^\pm = \frac{1}{2}\,(\Lambda \pm |\Lambda|) \tag{19.144}$$

and the approximate Riemann flux in this case is simply $\widehat{\mathbf{F}}_{\mathbf{n}}\,(\mathbf{U}^-,\mathbf{U}^+) = \mathbf{F}_n^+\,(\mathbf{U}^-) + \mathbf{F}_n^-\,(\mathbf{U}^+)$ where $\mathbf{U}^\pm = \lim_{\varepsilon\to 0} +\mathbf{U}\,(\mathbf{x}\pm\varepsilon\mathbf{n})$. Similarly, various options are available for the numerical viscous fluxes [10] and a particularly simple approach is the interior penalty method $\widehat{\mathbf{U}} = \langle\mathbf{U}\rangle$, $\widehat{\mathbf{F}}^v = \langle\mathbf{F}_i^v\rangle - \mu\,[\mathbf{U}n_i]$ where $\mu > 0$ is a stabilization parameter (see for example [196] and [105]. Thus, the discontinuous Galerkin method is:

Given $\mathbf{U}_0 = \mathbf{U}_0\,(\mathbf{x})$, for $t \in\,(0,T)$, find $\mathbf{U}\,(\mathbf{x},t) \in \mathbf{V}\,(P_h)\,\times H^1\,(0,T)$ such that $\mathbf{U}\,(\mathbf{x},0) = \mathbf{U}_0\,(\mathbf{x})$

and

$$B_{DG}\,(\mathbf{W},\mathbf{U}) = M_{DG}\,(\widetilde{\mathbf{W}},\overline{\mathbf{U}},\widetilde{\mathbf{U}}) + (\mathbf{W},\mathbf{S}), \tag{19.145}$$

$\forall\,\mathbf{W} \in \mathbf{V}\,(P_h)$ where $\mathbf{V}\,(P_h)$ is the broken space defined in [21]. If $\mathbf{V}\,(P_h)$ is restricted to a space of continuous functions, then one recovers the classical Galerkin approximation.

19.13.1 Turbulence Modeling for VMS

It remains to define the model term $M_{DG}\,(\widetilde{\mathbf{W}},\overline{\mathbf{U}},\widetilde{\mathbf{U}})$ appearing in Eq. (19.145). Considering orthogonal bases, it can be shown that the model must represent the projection of the generalized Reynolds and cross stresses onto the small scales. For incompressible flows, the projection is performed directly to the Reynolds and cross stresses. For compressible flows, there are additional terms arising from the variable density in the Reynolds stresses as well as from terms in the energy equation. For a thorough discussion of LES modeling issues in compressible flows see [66] and references therein. The general approach is illustrated here by assuming that the model takes the form of a generalized eddy diffusivity on each subdomain, Ω_e, of the form

$$M\left(\widetilde{\mathbf{W}}, \overline{\mathbf{U}}, \widetilde{\mathbf{U}}\right) = \int_{\partial\Omega_e} \widetilde{\mathbf{W}}^T \, \mathbf{F}_{n_e}^m\left(\overline{\mathbf{U}}, \widetilde{\mathbf{U}}\right) ds$$

$$- \int_{\Omega_e} \widetilde{\mathbf{W}}_{,i}^T \, \mathbf{F}_i^m\left(\overline{\mathbf{U}}, \widetilde{\mathbf{U}}\right) dx \qquad (19.146)$$

where the model flux $\mathbf{F}_i^m\left(\overline{\mathbf{U}}, \widetilde{\mathbf{U}}\right)$ is one of the form $\mathbf{F}_i^m\left(\overline{\mathbf{U}}, \widetilde{\mathbf{U}}\right) = \mathbf{D}_i^m\left(\overline{\mathbf{U}}, \widetilde{\mathbf{U}}\right) \widetilde{\mathbf{U}}$ and the matrix $\mathbf{D}_i^m\left(\overline{\mathbf{U}}, \widetilde{\mathbf{U}}\right)$ is a, possibly nonlinear, differential operator. The standard eddy diffusivity model can be put in this form

$$\mathbf{F}_i^m\left(\overline{\mathbf{U}}, \widetilde{U}\right) = 2\widetilde{v}_T \begin{bmatrix} 0 \\ (\nabla^s\widetilde{\mathbf{u}}) : i \\ \widetilde{T}_{,i} / Pr_t \end{bmatrix} \qquad (19.147)$$

where $\nabla^s\mathbf{u}$ is the symmetric part of the gradient tensor [i.e., $(\nabla^s\mathbf{u})_{ji} = (u_{i,j} + u_{j,i})/2$] and $(\nabla^s\mathbf{u})_{:i}$ is the ith column of this tensor. The Smagorinsky eddy diffusivity defined on the small-scales is $\widetilde{v}_T = (C_S\widetilde{\Delta})^2|\nabla^s\mathbf{u}|$ where C_S is the Smagorinsky coefficient, $\widetilde{\Delta}$ is a representative length scale for the small scales, and Pr_t is the turbulent Prandtl number. In a simple implementation, the Smagorinsky coefficient and the turbulent Prandtl number may be set as constants.

Extending equation (19.146) to a form compatible with discontinuous Galerkin leads to

$$M_{DG}\left(\widetilde{\mathbf{W}}, \overline{\mathbf{U}}, \widetilde{\mathbf{U}}\right) = \int_{\Gamma_0} \left([\widetilde{\mathbf{W}}^T] \left\langle \mathbf{F}_n^m\left(\overline{\mathbf{U}}, \widetilde{\mathbf{U}}\right)\right\rangle\right)$$

$$+ \int_{\partial\Omega} \left(\widetilde{\mathbf{W}}^T \, \mathbf{F}_n^m\left(\overline{\mathbf{U}}, \widetilde{\mathbf{U}}\right)\right) \qquad (19.148)$$

$$- \sum_{\Omega_e \in P_h} \int_{\Omega_e} \widetilde{\mathbf{W}}^T_{,i} \, \mathbf{F}_i^m\left(\overline{\mathbf{U}}, \widetilde{\mathbf{U}}\right) dx$$

which clearly simplifies to a classical weak Galerkin approximation for continuous functions. On inter-element boundaries, an averaged flux is used while on the domain boundary one obtains a weighted integral of the modeled turbulent flux across the boundary. This last integral marks a dramatic difference between discontinuous Galerkin and standard Galerkin approximations [95], [42] on solid surfaces.

In a standard Galerkin formulation the trial functions are assumed to satisfy the Dirichlet boundary conditions and so the weighting functions for velocity on a wall boundary would be zero, which prevents us from setting the flux of modeled turbulent stress to be zero on the wall. The fact that Hughes et al. [97] obtain reasonable solutions for solid surfaces using such a method is a matter open research. However, in the discontinuous Galerkin framework, since all boundary conditions are treated weakly through boundary flux integrals, we are able to weakly enforce zero turbulent flux by setting the second integral in Eq. (19.148) to zero on solid surfaces. Likewise, this integral can be set to particular values on inflow domains to represent the inflow of unresolved turbulent stress if desired.

It can be seen from Eq. (19.148) that the partition between large and small scales on different subdomains can be easily changed. Likewise, the particular model for the turbulent flux can be changed on each domain. Thus, the model term can be written as

$$
\begin{aligned}
M_{DG}\left(\widetilde{\mathbf{W}}, \overline{\mathbf{U}}, \widetilde{\mathbf{U}}\right) = &\int_{\Gamma_0}\left([\widetilde{\mathbf{W}}_e^T]\left\langle \mathbf{F}_n^{m_e}\left(\overline{\mathbf{U}}_{\mathbf{e}}, \widetilde{\mathbf{U}}_{\mathbf{e}}\right)\right\rangle\right) \\
&+ \int_{\partial\Omega}\left(\widetilde{\mathbf{W}}_e^T\,\mathbf{F}_n^{m_e}\left(\overline{\mathbf{U}}_{\mathbf{e}}, \widetilde{\mathbf{U}}_{\mathbf{e}}\right)\right) \qquad (19.149) \\
&- \sum_{\Omega_e \in P_h}\int_{\Omega_e}\frac{\partial\widetilde{\mathbf{W}}_e^T}{\partial x_i}\,\mathbf{F}_i^{m_e}\left(\overline{\mathbf{U}}_{\mathbf{e}}, \widetilde{\mathbf{U}}_{\mathbf{e}}\right)dx
\end{aligned}
$$

where the modeled turbulent flux and the solution space partitioning are dependent on the element index e. Across element boundaries, the first integral communicates the unresolved turbulent flux between neighboring elements thereby automatically converting from one partitioning to another and from one turbulent flux model to another. It is this novel capability of the DG/VMS that makes it particularly attractive for turbulence modeling in complex flows.

19.14   Implicit Time Marching of DG Discretizations

An implicit time marching algorithm based on the backward Euler integration scheme was presented in [150] for DG discretizations. This algorithm is unconditionally stable in two dimensions and provides time accuracy only when its time step resolves the temporal scales of the problem. The advantage of the unconditional stability is lost in the case of DNS of transitional and turbulent flows where the time scales that need to be resolved are often comparable to the time step imposed by explicit schemes. Unconditional stability is exploited to obtain steady-state solutions with the DG discretization.

Application of backward Euler time integration to the weak form of the Euler or Navier-Stokes equations with the DG discretizations yields

$$\int_{\Omega_i} \frac{\partial \mathbf{U}_i^n}{\partial t} \, \mathbf{v}_l^i \, d\Omega - \int_{\Omega_i} \mathbf{F} \, (\mathbf{U}_i^n + \Delta \mathbf{U}_i^n) \cdot \nabla \mathbf{v}_l^i \, d\Omega$$

$$+ \sum_j \oint_{\partial \Omega_{ij}} \mathbf{F}_{nun} \, (\mathbf{U}_i^n + \Delta \mathbf{U}_i^n, \, \mathbf{U}_j^n + \Delta \mathbf{U}_j^n, \, \mathbf{n}) \, \mathbf{V}_l^i \, dS = 0 \qquad l = 1, ..., N$$

$$(19.150)$$

where the superscript for the state variable indicates the time level and

$$\Delta \mathbf{U}_i^n = U_i^{n+1} - U_i^n. \tag{19.151}$$

Performing linearization of the fluxes $\mathbf{F}$ and $\mathbf{F}_{nun}$ obtain:

$$\mathbf{F} \, (U_i + \Delta U_i) = \mathbf{F} \, (U_i) + A \, (\mathbf{U}_i) \, \Delta U_i + \partial \, (\Delta t^2)$$
$$\mathbf{F}_{nun} \, (\mathbf{U}_i + \Delta U_i, U_j + \Delta U_j, n) \tag{19.152}$$
$$= \mathbf{F}_{nun} \, (\mathbf{U_i}, \mathbf{U_j}, \mathbf{n}) + A_{ij}^1 \, \Delta U_i + A_{ij}^2 \, \Delta U_j + \partial \, (\Delta t^2)$$

where

$$A \, (\overline{U}) = \frac{\partial \overline{F} \, (\overline{U})}{\partial \overline{U}}, \quad \begin{array}{l} A^1 = \frac{\partial \overline{F}_{nun} \, (\overline{u}, \overline{v}, \overline{n})}{\partial \overline{u}}, \quad A^2 = \frac{\partial \overline{F}_{nun} \, (\overline{u}, \overline{v}, \overline{n})}{\partial \overline{v}} \\[2mm] A_{ij}^1 = A^1 \, (\overline{u}_i, \overline{v}_j, \overline{n}), \, A_{ij}^2 = A^2 \, (\overline{u}_i, \overline{u}_j, \overline{n}) \end{array}$$

with these definitions dropping terms of higher than second order in Eq. (19.150) obtain

$$\int_{\Omega_i} \frac{\partial \overline{U}_i^n}{\partial t} V_l^i \, d\Omega - \int_{\Omega_i} A \, (\overline{U}_i^n) \, \Delta \overline{U}_i^n \cdot \nabla \overline{V}_l^i \, d\Omega \tag{19.153}$$

$$+ \sum_j \int_{\partial \Omega_i} (A_{ij}^1 \, \Delta \overline{U}_i^n + A_{ij}^2 \Delta \overline{U}_j) \, \overline{V}_l^i \, dS = R_{i,l} \, (\overline{U}^n)$$

$$R_{i,l} \, (\mathbf{U}^n) = \int_{\Omega_i} \overline{F} \, (u_i^n) \cdot \nabla V_l^i \, d\Omega - \sum_j \oint_{\partial \Omega_{ij}} \mathbf{F}_{nun} \, (\overline{u}_i^n, \overline{u}_j^n, \overline{n}) \, \overline{V}_l^2 \, dS. \tag{19.154}$$

Let $\widehat{\mathbf{U}} = (\widehat{\overline{U}}_1, ..., \widehat{\overline{U}}_{N_e})$ be the expansion coefficient of the approximate solution where $\widehat{U}_i = (\widehat{\overline{U}}_{i,1}, ..., \widehat{\overline{U}}_{i,N})$.

$$\overline{U}_i \, (\mathbf{x}) = \sum_{l=0}^N \widehat{\overline{U}}_{i,l} \, \overline{V}_l^i \, (\mathbf{x}) \, , \quad i = 1, ..., N_e. \tag{19.155}$$

The implicit scheme of Eq. (19.153) is written in matrix form as

$$M\left(\mathbf{U}^n\right) \Delta \widehat{U}^n = R\left(\mathbf{U}^n\right) \tag{19.156}$$

$$M\left(\overline{U}^n\right) = \frac{D}{\Delta t} - \frac{\partial R\left(\overline{U}^n\right)}{\partial \widehat{\mathbf{U}}}$$

where D denotes the block diagonal mass matrix $D = \text{diag}\left(d_1, ..., d_{N_e}\right)$.

$$[d_i]_{kl} = \int \mathbf{V}_k^i \, \mathbf{V}_l^i \, d\Omega. \tag{19.157}$$

The spatial accuracy of the solution depends on the discretization of the residual as for the implicit schemes in the finite difference or the finite volume context. Quadratic convergence and large time steps are possible only when the spatial discretization of the left-hand side of Eq. (19.156) is consistent with the discretization of the right–hand side. The linear system of Eq. (19.156) can be with the generalized minimum residual (GMRES) Krylov method [159] with single- or two-level preconditioning.

A second-order accurate in time implicit Runge-Kutta method can also be used for time integration of DG discretizations [20] as follows:

Considering the system of the DG discretization

$$D \frac{d\mathbf{u}}{dt} - R\left(\mathbf{u}\right) = 0 \tag{19.158}$$

where D is the mass matrix the second order implicit Runge-Kutta method involves two backward Euler steps and can be written as

$$\mathbf{u}_j^{n+1} = \mathbf{u}_i^n + \gamma_1 \mathbf{K}_1 + \gamma_2 \mathbf{K}_2 \tag{19.159}$$

$$\left[\frac{D}{\Delta t} + \alpha \frac{\partial R\left(\mathbf{u}^n\right)}{\partial \mathbf{u}}\right] \mathbf{K}_1 + R\left(\mathbf{u}^n\right) = 0 \tag{19.160}$$

$$\left[\frac{D}{\Delta t} + \alpha \frac{\partial R\left(\mathbf{u}^n\right)}{\partial \mathbf{u}}\right] \mathbf{K}_2 + R\left(u^n + \beta \, \mathbf{K}_1\right) = 0 \tag{19.161}$$

the constants α, β, γ_1 and γ_2 corresponding to an optimal second-order accurate scheme [98] are

$$\alpha = \frac{2 - \sqrt{2}}{2} \ , \quad \beta = 8\alpha\left(0.5 - \alpha\right) , \quad \gamma_1 = 1 - \frac{1}{8\alpha} \ , \quad \gamma_2 = 1 - \gamma_1. \tag{19.162}$$

The backward Euler scheme of Eq. (19.155) is recovered by $\gamma_1 = 1$, $\gamma_2 = 0$ and $\alpha_1 = 1$. The Crank-Niconson algorithm is obtained for $\gamma_1 = 1$, $\gamma_2 = 0$, $\alpha = 1/2$.

19.15   p-Type Multigrid for DG

A p-type multigrid acceleration method was applied [57] for convergence acceleration with the DG method. A simple backward Euler discretization in time was used [57] so that the discrete equation for time advancement is

$$\mathcal{M} \, \frac{1}{\Delta t} \, (\overline{u}^{n+1} - \overline{u}^{n}) + \mathbf{R} \, (\overline{u}^{n+1}) = 0 \qquad (19.163)$$

where $\mathcal{M}$ is the mass matrix, $\mathbf{R}$ is the residual vector, and $\overline{u}$ are the degrees of freedom to be evolved in time from time level n to $n+1$.

For steady-state solutions the nonlinear system $\mathbf{R}\,(\mathbf{U}) = 0$ is solved using a p-multigrid scheme with a linear Jacobi smoother. A generic iterative scheme for Eq. (19.163) can be written as

$$\overline{u}^{n+1} = u^{n} - \mathbf{P}^{-1} \, \mathbf{R} \, (u^{n}) \qquad (19.164)$$

where the preconditioner, $\mathbf{P}$, is an approximation to $\partial R/\partial U$. In p-multigrid the low frequency error modes can be effectively corrected by smoothing with lower order expansion of the approximate solution that serve as the worse grids [85]. p-multigrid fits naturally with the frame work of high-order DG discretizations. There is no need to store additional grid information since the same spatial grid is used by all levels. The transfer operators between the different p expansions, prolongation and restriction, are local and are stored for the reference element. It was found, however, that p-multigrid is not effective and difficult to apply. Therefore a more effective $h-p$ multigrid approach [135] was developed for the compressible Euler equations.

19.16   Results with the DG Method

The performance of the DG method was demonstrated [180] for linear problems computed with second- fourth- and sixth-order accuracy on triangular meshes. The accuracy of the numerical solutions is evaluated by comparing the computed results with exact solutions. The first test problem with an exact solution [75] is propagation and reflection from a solid wall of a Gaussian pressure pulse given by $p\,(x,y) = \exp\{-\ln2[x^2 - (y - y_0)^2]/W\}$, where W is the width of the pulse and y_0 is the distance from the wall. The second problem with an exact solution is scattering of a similar Gaussian pressure pulse from the surface of a cylinder [176].

Computed pressure field at T = 25

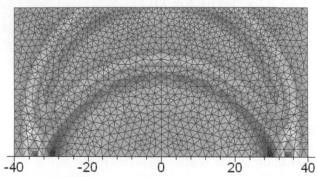

FIGURE 19.9: Solution computed with an unstructured, triangular mesh and third-order polynomial basis.

A solution of the linearized Euler equations computed with third-order polynomial bases (fourth-order accurate) on a relatively coarse, fully unstructured mesh is shown in Fig. 19.9. At the far field boundaries, the radiation boundary condition [175] was used. It can be seen that the pressure waves exit the computational domain undistorted and there are no reflections in the interior from the computational boundaries. Acoustic disturbance propagation is isotropic and does not require use of meshes with pattern [88]. Indeed, it appears that the unstructured mesh of Fig. 19.9 is more appropriate for acoustic wave propagation. Further evaluation of the numerical method is performed, on triangular meshes obtained from triangulation of structured Cartesian-type grids. The elements of the mesh for the solution of Fig. 19.10 follow a uniform triangular-mesh generating pattern. It was shown in [88] that the accuracy of the computed solution depends on the triangular-mesh generating pattern. Comparing the computed solutions of Figs. 19.9 and 19.10 it appears that the coarse, canonical grid solution ($\Delta x = 2$) obtained with the third order P^3 polynomial basis shows more distortion than the fully unstructured mesh of Fig. 19.9 that has comparable resolution. The upper part of Fig. 19.10 shows, however, that for fine grids and high order accuracy the bias introduced by the triangular-mesh generating pattern is very small. This is again consistent with the theoretical analysis of [88].

Solutions shown in the following paragraphs were computed until final time $T = 25$ on triangular canonical meshes with mesh generating pattern (see Fig. 19.10). Comparisons are carried out with the exact solution given in [75]. A comparison of the solutions computed with grid spacing $\Delta x = 2.0$ is only shown in Fig. 19.11 because within plotting accuracy for $\Delta x = 1.0$ the differences between the fourth- and sixth order-accurate solutions are not

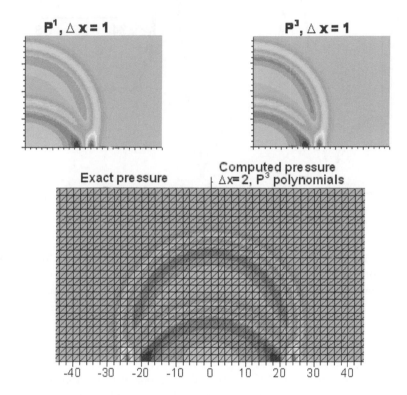

FIGURE 19.10: Comparison of the pressure field computed with $\Delta x = 2$, triangular elements, and third-order polynomial basis (right) with the exact solution (left).

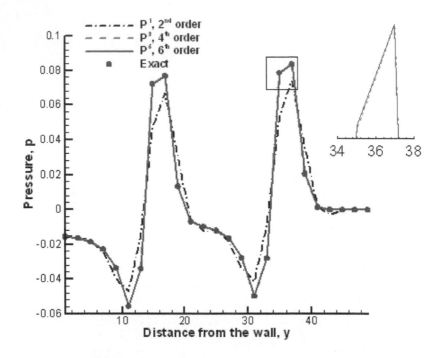

FIGURE 19.11: Comparison of the pressure computed with first, third, and fifth order polynomial bases with the exact solution in all cases the same canonical triangular mesh with $\Delta x = 2$ are used.

visible. The same time step $\Delta t = 0.01$, which is below the stability limit of the Runge-Kutta method was used for all solutions. The comparison with the exact result of [75] is shown along the symmetry line, which is normal to the wall at $x = 0$. For the same location, the error $\varepsilon_r = (p_{com} - p_{ex})$ of the computed solutions is shown in Fig. 19.12. Clearly, only for $\Delta x < 1.0$ the results computed with the first-order polynomial basis (2nd-order accurate solution) provide the accuracy level needed in aeroacoustic computations.

In Fig. 19.13, the grid convergence of the second-, fourth-, and sixth-order accurate solutions is shown. The error norm in Fig. 19.13 was computed on the symmetry line, where the error norm is expected to be the largest, and not for the entire domain. Norms of the error computed in the full domain do not differ much from the error norm obtained from the deviation of the computed solution from the exact result only on the symmetry line. In both cases, the error norm was computed from

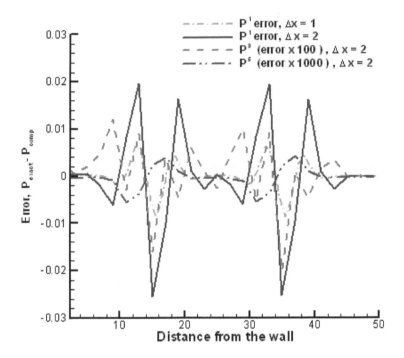

FIGURE 19.12: Error of the solutions computed with first, third, and fifth order polynomial bases and canonical triangular meshes with $\Delta x = 2$.

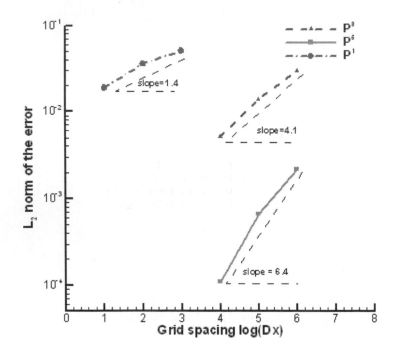

FIGURE 19.13: Grid convergence (in the L_2 norm of the computed pressure) of the second-, the fourth-, and sixth-order accurate solutions computed with triangular elements.

$$L_2 = \left\{ \frac{1}{N_{el}} \sum_{el=1}^{N_{el}} \int_{el} [p_{comp}\,(x,y,t) - p_{exact}\,(x,y,t)]^2 \right\}^{\frac{1}{2}} \qquad (19.165)$$

The grid convergence plot of Fig. 19.13 shows that all solutions achieve the expected order of accuracy. Once again it is evident that the desired order of accuracy for CAA can only be achieved with higher order methods. Furthermore, it can be seen that for this simple problem the solution computed with fourth- or sixth-order accuracy and single precision practically converges when $\Delta x \simeq 1.0$ and the remaining errors are mainly due to time integration. In Fig. 19.14, the reduction of the average error is plotted versus the required computational time for the solution. It can be seen that use of higher-order accuracy yields savings in computational time when the solution must reach certain error level.

Solutions computed with quadrilateral meshes are presented next. A com-

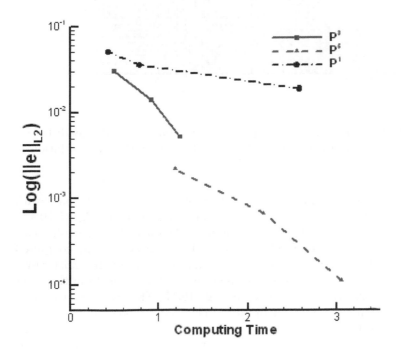

FIGURE 19.14: Computational time required in order to achieve certain error level (L_2 norm of the computed pressure) with the second-, the fourth-, and sixth-order accurate solutions.

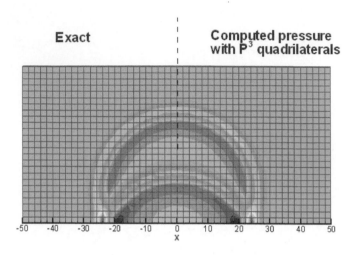

FIGURE 19.15: Comparison of the pressure field computed with quadrilateral elements $\Delta x = 2$ and third-order polynomial basis with the exact solution.

parison of the pressure field computed with third-order polynomial basis with the exact result is shown in Fig. 19.15. It can be seen (compare Fig. 19.10 and Fig. 19.15) that for equivalent grid resolution, the quadrilateral mesh yields better isotropy of the computed solution. The deviations of the computed solutions with first- and third-order polynomial bases from the exact result are shown in Fig. 19.16. The error of the computed solutions from the exact result on the symmetry line is shown in Fig. 19.17. The grid convergence of the solutions computed with quadrilateral elements is shown in Fig. 19.18. For comparison, the convergence achieved with solutions computed with triangular elements is shown in the same plot. It appears that the solutions computed with quadrilateral elements do not achieve the (k+1)th order of accuracy. This is expected because in a computation with third order basis, for example, the unit area (for $\Delta x = \delta y = 1$) is spanned by sixteen polynomials in quadrilateral elements while for triangular element discretization the same area is covered by two elements and spanned by twenty $(2 \times P^3_{j=1,...,10} = 20)$ polynomials. In terms of computational time the solution of obtained with quadrilateral elements requires slightly less time.

Solutions for more complex domains are presented next. First, scattering of sound waves from the surface of a cylinder, which is one of the benchmark problems of [176], is considered. The computed solution with third-order polynomial basis and quadrilateral elements at $T = 8$ is compared with the exact result in Fig. 19.19. Very good agreement with exact result is achieved for the solutions computed with both triangular and quadrilateral elements.

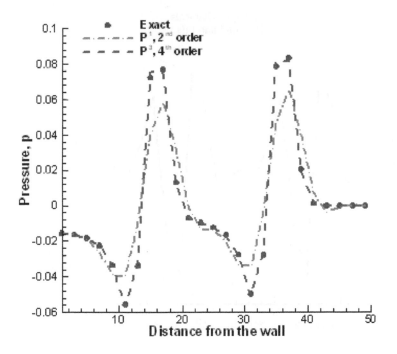

FIGURE 19.16: Comparison of pressure field computed with quadrilateral canonical meshes and first or third polynomial bases with the exact solution.

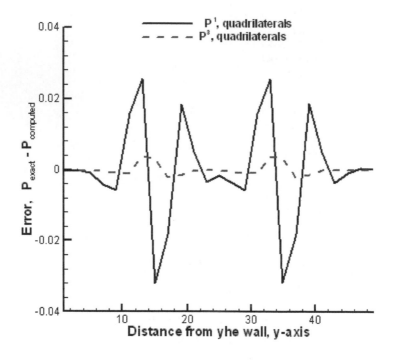

FIGURE 19.17: Error of the solutions computed with and polynomial bases on quadrilateral meshes, in both cases canonical quadrilateral meshes $\Delta x = 2$ are used.

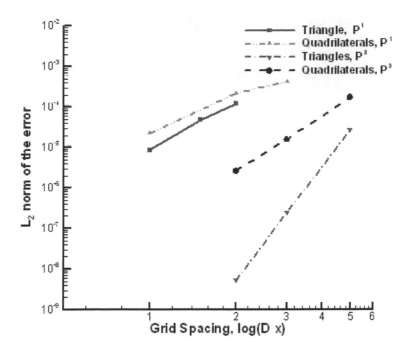

FIGURE 19.18: Comparison of the grid convergence (in the L_2 norm of the computed pressure) of solutions computed with triangular and quadrilateral elements.

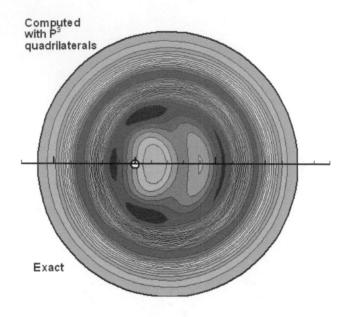

FIGURE 19.19: Comparison of the computed pressure field obtained with third order quadrilateral elements with the exact solution for scattering from a cylinder surface.

The computational domain for the solution obtained with triangular elements contains 5437 elements or 2810 vertices. Discretization of the same domain with isoparametric quadrilaterals includes 2000 elements or 2091 vertices. In both cases the cylinder was represented by 40 elements. A comparison of the computed pressure at $T = 5$ that is obtained with the triangular element solution for selected radial directions ($\phi = 0$ deg. and $\phi = 90$ deg.) with the exact result is shown in Fig. 19.20. The computed solution along the radial direction was interpolated from surrounding nodes. The interpolated computed solution is in good agreement with the exact result. The time variation of the computed solution for $0 \leq T \leq 8$ is shown in Fig. 19.21. Again good agreement with the exact result [176] is achieved.

Other numerical solutions obtained with the DG finite element method for benchmark aeroacoustic problems demonstrated that only p^3 (fourth-order) or higher-order accurate discretizations provide the required resolution for CAA. In terms of computational time, solutions with equivalent order of accuracy obtained on triangular meshes require about 10% more resources than the solutions obtained with quadrilateral meshes. In both cases, the CFL stability

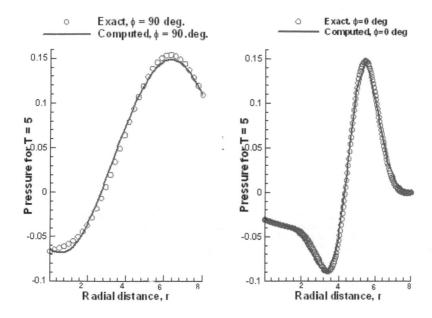

FIGURE 19.20: Comparison of the computed pressure at final time T=5 for $\phi = 0$ deg., and $\phi = 90$ deg. with the exact solution.

limitations become more stringent c.f. [12] with the increase of the order of spatial accuracy. For isotropic meshes, the resolving ability of both triangular and quadrilateral space discretizations is approximately equivalent. For two-dimensional problems, it is feasible to achieve spatial accuracy up to sixth order. Sixth- or higher-order of accuracy is possibly advantageous for long time propagation of complex waveforms. Numerical tests for pure convection $(u_t + u_x = 0)$ of complex one-dimensional waves of the form $u\,(x,t) = 0 = [2 + \cos\,(\alpha x)]\exp\,[-\ln 2\,(x/10)^2]$, $\alpha = 1.7$, using up to tenth order accurate DG discretizations showed that for long time integration $(T > 400)$ at least sixth-order accuracy is needed. Application of the DG finite element method for three-dimensional CAA applications is straightforward using the available finite element framework [91] but beyond the scope of the present paper. It appears, however, that three-dimensional CAA applications with the DG method are feasible in terms of the required computational resources but discretizations with polynomial bases of order higher than three are possibly too intensive computationally.

Computations of compressible flows over realistic aerodynamic configurations with the DG method were carried out by van der Vegt and van der Ven [182] and more recently in [183]. Examples of the recent computations of transonic steady and unsteady airfoil flows of [183] using the space time DG method and stabilization operators of Section 19.10 of Chapter 19, instead of

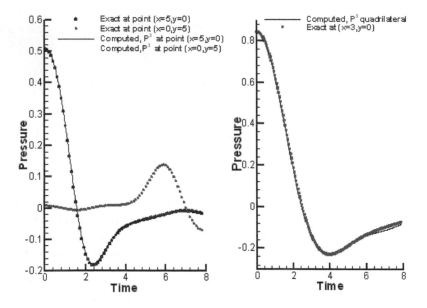

FIGURE 19.21: Comparison of the computed pressure at $r = 5$ and $0 \leq T \leq 8$ for $\phi = 0$ deg., and $\phi = 90$ deg. with the exact solution.

slope limiters are shown in Fig. 19.22 and Fig. 19.23. Very good resolution of transonic shocks was obtained with the stabilization operators. Adaptive grid refinement with "hanging nodes" that is possible for DG method yielded very good resolution of the shocks.

Computations of viscous turbulent flows with the DG method were recently presented by Bassi et al. [20]. Examples from these computations are shown in Fig. 19.26, Fig. 19.27, and Fig. 19.28. Fig. 19.26 shows the computational mesh over turbomachinery blades. An implicit Runge-Kutta method similar to the implicit scheme of Section 19.15 of Chapter 19, was used for time marching in order to avoid the stringent stability limitations imposed by the the small grid spacing required for the resolution of the near wall viscous phenomena. The unsteady flow field is shown with snapshots in Fig. 19.27, and Fig. 19.28. additional results with the DG method can be found in other recent publications [124], [149], [47], [205], [152], [19], [22], [129], [154], and [161].

References

[1] R. Abgrall. On essentially non-oscillatory schemes on unstructured meshes: analysis and implementation. *J Computational Physics*, 114(1):45–58, 1994.

[2] R. Abgrall and PL. Roe. High order fluctuation splitting schemes on triangular mesh. *Journal of Scientific Computing*, 19:3–36, 2003.

[3] N.A. Adams and K.A. Shariff. A high-resolution hybrid compact-ENO scheme for shock-turbulence interaction problems. *J Computational Physics*, 127(1):27–51, 1996.

[4] N.A. Adams. Direct numerical simulation of turbulent compression ramp flow. *Theoretical and Computational Fluid Dynamics*, 12:109–129, 1998.

[5] N.A. Adams. Direct simulation of the turbulent boundary layer along a compression ramp at $M = 3$ and $R_e = 1685$. *J Fluid Mechanics*, 420:47–83, 2000.

[6] J.U. Ahmad and P.C. Strawn. Hovering Rotor and Wake Calculations with an Overset-Grid Navier-Stokes Solver. *55th Annual Forum of the American Helicopter Society* Montreal, Canada, 25–27, 1999.

[7] M. Ainsworth. Dispersive and dissipative behavior of high order dis-

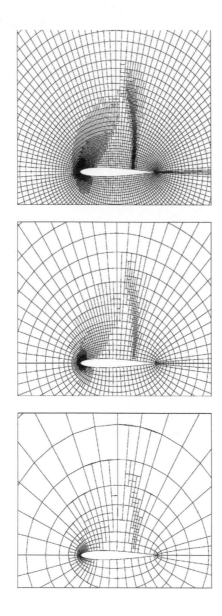

FIGURE 19.22: Grid levels in an adapted grid solution about the NACA 0012 airfoil (from Ref. [183]).

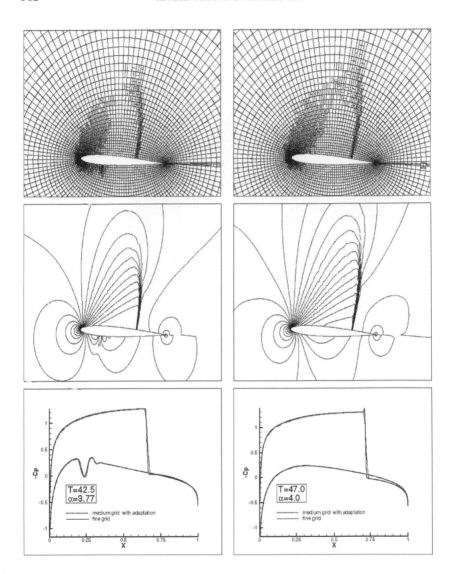

FIGURE 19.23: Adapted mesh around an oscillating NACA0012 airfoil, contours of density and surface pressure coefficient Cp for $\alpha = 3.37$ deg. (pitching upward) and $\alpha = 4.0$ deg. (pitching downward); $M = 0.8, \omega = \pi/10$ (from Ref. [183]).

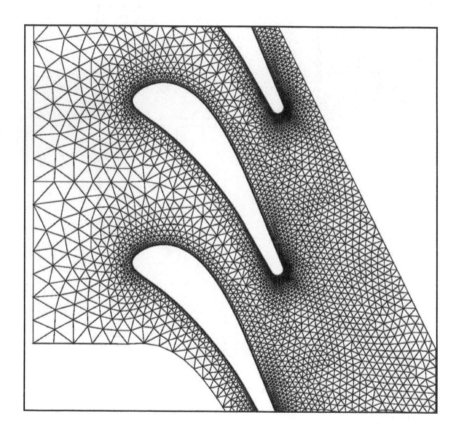

FIGURE 19.24: Global view and detail of the grid around the blade (from Ref. [20]).

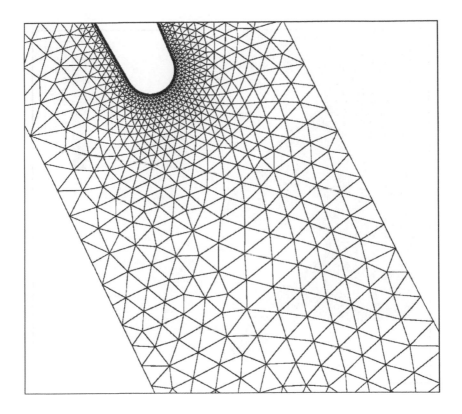

FIGURE 19.25: Snapshot of Mach number and turbulence intensity computed with P2 elements (from Ref. [20]).

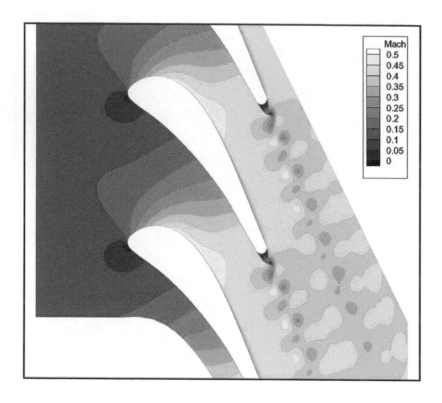

FIGURE 19.26: Shedding from the blade trailing edge shown with density contours (from Ref. [20]).

continuous Galerkin finite element methods. *J Computational Physics*, 198(1):106–130, 2004.

[8] D.A. Anderson, J.C Tannehill, and R.H. Pletcher. *Computational Fluid Mechanics and Heat Transfer*. McGraw-Hill, New York, 1984.

[9] D.N. Arnold. An interior penalty finite element method with discontinuous elements. *SIAM J Numer. Anal.*, 19:742–760, 1982.

[10] D.N. Arnold, F. Brezzi, B. Cockburn, and LD. Marin. Unified analysis of discontinuous Galerkin methods for elliptic problems. *SIAM J Numer. Anal.*, 39: 1749–1779, 2002.

[11] J. Astley, K. Gerdes, D. Givoli, and I. Harari. Special issue on finite elements for wave problems-Preface. *J Computational Acoustics*, 8: vii–ix, 2000.

[12] H.L. Atkins and C.W. Shu. Quadrature-free implementation of the discontinuous Galerkin method for hyperbolic equations. *AIAA Journal*, 36(5):775–782, 1984.

[13] J.A. Baker. *Finite Element Computational Fluid Mechanics*. Hemisphere Publication Corporation, New York, 1983.

[14] E. Balaras, C. Benocci, and U. Piomelli. Finite-difference computations of high-Reynolds number flows using dynamic subgrid-scale model. *Theoretical and Computational Fluid Dynamics*, 7:207–216, 1995.

[15] D.S. Balsara and C.W. Shu. Monotonicity preserving weighted essentially non-oscillatory schemes with increasingly high order of accuracy. *J Computational Physics*, 160 (2):405–452, 2000.

[16] T.J. Barth, and P. Frederickson. High rder solution of the Euler equations on unstructured grids using quadratic reconstruction. *AIAA Paper*, 90–0013, 1990.

[17] T.J. Barth and H. Deconinck. *High-Order Methods for Computational Physics*. Springer Verlag, 1999.

[18] F. Bassi and S. Rebay. A high-order accurate discontinuous finite element method for the numerical solution of the compressible Navier-Stokes equations. *J Computational Physics*, 131(1):267–279, 1997.

[19] F. Bassi and S. Rebay. Numerical evaluation of two discontinuous Galerkin methods for the compressible Navier-Stokes equations. *Int. J Num Meth Fluids*, 40(1):197–207, 2002.

[20] F. Bassi F, A. Crivellini, S. Rebay, and M. Savini. Discontinuous Galerkin solutions of the Reynolds-averaged Navier-Stokes and $K - \omega$ turbulence model equations. *Computers and Fluids*, 34(4-5):507–540, 2005.

[21] C.E. Baumann, and J.T. Oden. A discontinuous hp finite element method for convection-diffusion problems. *Comput. Methods Appl. Mech. Eng.*, 175:311–341, 1999.

[22] C.E. Baumann and T.J. Oden. A discontinuous hp finite element method for the Euler and Navier-Stokes equations. *Int. J Num Meth Fluids*, 31(1):79–95, 1999.

[23] J.A. Benek, P.G. Bunning, and J.L. Steger. A 3-D Chimera grid embedding technique. *AIAA Paper* 85–1523, 1985.

[24] K.S. Bey and J.T. Oden. hp-version discontinuous Galerkin methods for hyperbolic conservation laws. *Comput. Methods Appl. Mech. Engrg.*, 133:259–286, 1996.

[25] R. Biswas, K.D. Devine, and J.E. Flaherty. Parallel, adaptive finite element methods for conservation laws. *Appl. Numer. Math.*, 14:255–283, 1994.

[26] C. Canuto, M.Y. Hussaini, A. Quarteroni and T.A. Zang. *Spectral Methods in Fluid Dynamics*. Springer-Verlag, New York, 1987.

[27] M.H. Carpenter, D. Gottlieb and S. Abarbanel. The stability of numerical boundary treatments for compact high-order finite-difference schemes. *J Computational Physics*, 108(2):272–295, 1993.

[28] M.H. Carpenter and D. Gottlieb. Spectral methods on arbitrary grids. *J Computational Physics*, 129(1):74–86, 1996.

[29] J. Casper and H.L. Atkins. A finite volume high-order ENO scheme for two-dimensional hyperbolic system. *J Computational Physics*, 106(1):62–76, 1993.

[30] J. Casper and H.L. Atkins. Comparison of two formulations for high-order accurate essentially nonoscillatory schemes. *AIAA J*, 32(4):1970–1977, 1994.

[31] S.R. Chakravarthy, D.A. Anderson, and M.D. Salas. The split-coefficient matrix method for hyperbolic systems of gas dynamic equations. *AIAA Paper*, 80–0268, 1980.

[32] H. Chen and S.M. Liang. Planar blast/vortex interaction and sound generation. *AIAA J*, 40 (11):2298–2304, 2002.

[33] Y.M. Chung and P.G. Tucker. Accuracy of higher-order finite-difference schemes on nonuniform grids. *AIAA J*, 41(8):1609–1611, 2003.

[34] B. Cockburn, and C.W. Shu. TVB Runge-Kutta local projection discontinuous Galerkin finite element method for conservation laws II: General framework. *Math. Comp.*, 52:411–435, 1989.

[35] B. Cockburn, S.Y. Lin, and C.W. Shu. TVB Runge-Kutta local projection discontinuous Galerkin finite element method for conservation laws III: one-dimensional systems. *J Computational Physics*, 84(1):90–113, 1989.

[36] B. Cockburn, S. Hou, and CW. Shu. TVB Runge-Kutta local projection discontinuous Galerkin finite element method for conservation laws IV: The multidimensional case. *Math. Comp.*, 54: 545–581, 1990.

[37] B. Cockburn and P.A. Grenaud. Error estimates for finite element methods for nonlinear conservation laws. *SIAM J Numer. Anal.*, 33:522–547, 1996.

[38] B. Cockburn and C.W. Shu. The Runge-Kutta discontinuous Galerkin method for conservation laws V: multidimensional systems. *J Computational Physics*, 141(2):199–224, 1998.

[39] B. Cockburn, G.E. Karniadakis, and C.W. Shu. *Discontinuous Galerkin Methods*. Springer, 1999.

[40] B. Cockburn and C.W. Shu. Runge-Kutta discontinuous Galerkin methods for convection-dominated problems. *J Scientific Computing*, 16(3):173–261, 2001.

[41] P. Colella and P. Woodward. The piecewise parabolic method for gas-dynamical simulations. *J Computational Physics*, 54(1):174–201, 1984.

[42] S.S. Collis. Monitoring unresolved scales in multiscale turbulence modeling. *Physics of Fluids*, 13(6):1800–1806, 2001.

[43] S.S. Collis. The DG/VMS method for unified turbulence simulation. *AIAA Paper*, 3124, 2002.

[44] J. Dacles-Mariani, G.G. Zilliac, J.S. Chow, and P. Bradshaw. Numerical experimental wingtip vortex in the near field. *AIAA J*, 33(9):1561–1568, 1995.

[45] M. Delanaye and Y. Liu. Quadratic reconstruction finite volume schemes on 3D arbitrary unstructured polyhedral grids. *AIAA Paper*, 99–3259, 1999.

[46] F. Dexun and M. Yanwen. Analysis of super compact finite difference method and application to simulation of vortex-shock interaction. *Int. J Num Meth Fluids*, 36(7):773–805, 2001.

[47] V. Doleji. On the discontinuous Galerkin method for the numerical solution of the Navier-Stokes equations. *Int. J Num Meth Fluids*, 45 (10):1083–1106, 2004.

[48] M. Dubiner. Spectral methods on triangles and other domains. *J Sci. Comp.*, 6(2):345–390, 1991.

[49] M. Dumbser and C. Munz. Building blocks for arbitrary high order discontinuous Galerkin schemes. *J. Sci. Comp.*, 27(1-3): 215–230, 2006.

[50] R. Dyson. Technique for very high order nonlinear simulation and validation. *J Computational Acoustics*, 10:211–229, 2002.

[51] J.A. Ekaterinaris. Effects of spatial order of accuracy on the computation of vortical flowfields. *AIAA J*, 32(12):2471–2474, 1994.

[52] J.A. Ekaterinaris. Upwind scheme for acoustic disturbances generated by low-speed flows. *AIAA J*, 34(9):1448–1455, 1997.

[53] J.A. Ekaterinaris. Numerical simulation of incompressible two-blade rotor flowfields. *AIAA J of Propulsion and Power*, 14(3):367–374, 1998.

[54] J.A. Ekaterinaris. Performance of high-order accurate low-diffusion numerical schemes for compressible flow. *AIAA J*, 42(3):493–500, 2004.

[55] B. Engquist, P. Loetstedt, and B. Sjoergreen. Nonlinear Filters for Efficient Shock Computation. *Mathematics of Computation*, 52:232–248, 1989.

[56] E. Fatemi, J. Jerome, and S. Osher. Solution of the hydrodynamic device model using high order non-oscillatory shock capturing algorithms. *IEEE Transactions on Computer-Aided Design of Integrated Circuits and Systems*, 10:232–244, 1991.

[57] K.J. Fidkowski, and D.L. Darmofal. Development of a higher-order solver for aerodynamic applications. *AIAA Paper* 2004–0436, 2004.

[58] O. Friedrich. Weighted essentially non-oscillatory schemes for the interpolation of mean values on unstructured grids. *J Computational Physics*, 144(1):194–212, 1998.

[59] D. Gaitonde, and J.S. Shang. Optimized compact-difference-based finite-volume schemes for linear wave phenomena. *J Computational Physics*, 138(2):617–643, 1997.

[60] D.V. Gaitonde, and M.R. Visbal. Pade-Type High-Order Boundary Filters for the Navier-Stokes Equations. *AIAA J*, 38(11):2103–2112, 2000.

[61] E. Garnier, M. Mossi, P. Sagaut, P. Comte, and M. Deville. On the Use of Shock-Capturing Schemes for Large-Eddy Simulations. *J Computational Physics*, 153(2):273–311, 1999.

[62] E. Garnier, P. Sagaut, and M. Deville. A Class of Explicit ENO Filters with Application to Unsteady Flows. *J Computational Physics*, 170 (1):184–204, 2001.

[63] E. Garnier, P. Sagaut, and M. Deville. Large eddy simulation of shock/boundary-layer interaction. *AIAA J*, 40 (10):1933–1944, 2002.

[64] M. Germano, U. Piomelli, P. Moin, and W.H. Cabot. A dynamic subgrid-scale eddy viscosity model. *Phys. Fluids A* 3(7):1760–1765, 1991.

[65] S. Ghosal, T.S. Lund, P. Moin, and K. Akselvoll. A dynamic localization model for large-eddy simulation of turbulent flows. *J Fluid Mechanics*, 286:229–255, 1995.

[66] S. Ghosal and P. Moin. The basic equations for large-eddy simulation of turbulent flows in complex geometry. *J Computational Physics*, 118(1):24–37, 1995.

[67] Z.K. Godunov, A.W. Zabrodyn, and G.P. Prokopov. A Difference scheme for two-Dimensional Unsteady Problems of Gas Dynamics and Computation of Flow with Detached Shock Wave. *Mat I Mat Fyz*, 1:1020, 1961.

[68] D. Gottlieb, and C.W. Shu. Total variation diminishing Runge-Kutta schemes. *Math Comput*, 67:73–85, 1998.

[69] D. Gottlieb, and J. Hesthaven. Spectral methods for hyperbolic problems. *J Comput. Appl. Math.*, 128:83–131, 2001.

[70] F. Grasso, and S. Pirozzoli. Shock-wave vortex interactions: Shock and vortex deformations, and sound production. *Theoretical and Computational Fluid Dynamics*, 13(6):421–456, 2000.

[71] F. Grasso, and S. Pirozzoli. Simulations and analysis of the coupling process of compressible vortex pairs: free evolution and shock induced coupling. *Physics of Fluids*, 13(5):1343–1366, 2001.

[72] B. Gustafsson, HO. Kreiss, and A. Sundstrom. Stability theory of difference approximations for mixed initial boundary value problems II. *Math. Comput.*, 26:649–686, 1972.

[73] B. Gustafsson, and O. Olsson. Fourth-order difference methods for hyperbolic IBVPs. *J Computational Physics*, 117(2):300–317, 1995.

[74] A. Hadjadj, A.N. Kudryavtsev, and M.S. Ivanov. Numerical investigation of shock-reflection phenomena in overexpanded supersonic jets. *AIAA J*, 43(3):570–577, 2004.

[75] J.C. Hardin, J.R. Ristorcelli, and C.K. Tam. *ICASE/LaRC Workshop on Benchmark Problems in Computational Aeroacoustics (CAA)*, NASA CP 3300, 1995.

[76] N. Hariharan, and L.N. Sankar. High-Order Numerical Simulations of Rotor Flow Field. *Proceeding of 50th Annual Helicopter Society Forum*, American Helicopter Society, Washington, DC, 2:1275, 1992.

[77] A. Harten. The Artificial Compression Method for Computation of Shocks and Contact Discontinuities, III self–adjusting hybrid schemes. *Mathematics of Computation*, 32:363–389, 1978.

[78] A. Harten. On the symmetric form of systems of conservation laws with entropy. *J Computational Physics*, 49(1):151–164, 1983.

[79] A. Harten, and J.M. Hyman. A self-adjusting grid for the computation of weak solutions of hyperbolic conservation laws. *J Computational Physics*, 50(2):235–269, 1983.

[80] A. Harten. On a class of high resolution total-variation-stable finite-difference schemes. *SIAM J Numer. Anal.*, 21:1–23, 1984.

[81] A. Harten, and S. Osher. Uniformly high-order accurate nonoscillatory schemes. *I. SIAM J Num. Anal.*, 24(2):279–309, 1987.

[82] A. Harten, S. Osher S, B. Engquist, and S. Chakravarthy. Uniformly high order essentially non-oscillatory schemes. *III. J Computational Physics*, 71:231–303, 1987.

[83] O. Hassan, K. Morgan, and J. Peraire. *An implicit finite element method for high speed flows.* AIAA Paper 90-0402, 1990.

[84] G. Hauke and TJR. Hughes. A comparative study of different sets of variables for solving compressible and incompressible flows. *Computer Meth. in Appl. Mech. and Eng.*, 153:1–44, 1998.

[85] B. Helenbrook, DJ. Mavriplis, and H.L. Atkins. Analysis of p-multigrid for continuous and discontinuous finite element discretizations. *AIAA Paper*, 2003–3989, 2003.

[86] J.S. Hesthaven and C.H. Teng. Stable spectral methods on tetrahedral elements. *SIAM Journal of Scientific Computing*, 21:2352–2380, 2000.

[87] C. Hirsch. *Numerical Computation of Internal and External Flows.* John Wiley and Sons, 1990.

[88] F.Q. Hu, M.Y. Hussaini, and P. Rasetarinera. An analysis of the discontinuous Galerkin method for wave propagation problems. *J Computational Physics*, 151(2):921–946, 1999.

[89] F.Q. Hu and H.L. Atkins. Eigensolution analysis of the discontinuous Galerkin method with non-uniform grids. Part 1: One space dimension. *J Computational Physics*, 182(2):516–545, 2002.

[90] C. Hu and C.W. Shu. Weighted essentially non-oscillatory schemes on triangular meshes. *J Computational Physics*, 150(1):97–127, 1999.

[91] T.J.R. Hughes. *The Finite Element Method.* Prentice Hal, New Jersey 1987.

[92] T.J.R. Hughes and M. Mallet. A new finite element formulation for CFD: IV A discontinuity-capturing operator for multidimensional advective-diffusive systems. *Comput Methods Appl Mech Eng*, 58 (3):329–356, 1986.

[93] T.J.R. Hughes, L.P. Franka, and G.M. Hulbert. A new finite element formulation for computational fluid dynamics: VIII The Galerkin/least-squares method for advective-diffusive systems. *Comput Methods Appl Mech Eng*, 73(2):173–189, 1989.

[94] T.J.R. Hughes, G. Engel, L. Mazzei, and M.G. Larson. The continuous Galerkin method is locally conservative. *J Computational Physics*, 163(2):467–488, 2000.

[95] T.J.R. Hughes, L. Mazzei, and K.E. Jansen. Large eddy simulation and the variational multiscale method. *Computing and Visualization in Science*, 3:47–59, 2000.

[96] T.J.R. Hughes, L. Mazzei, A.A. Oberai, and A.A. Wray. The multiscale formulation of large eddy simulation: Decay of homogeneous isotropic turbulence. *Physics of Fluids*, 13(2):505–512, 2001.

[97] T.J.R. Hughes, A.A. Oberai, and L. Mazzei. Large eddy simulation of turbulent channel flows by the variational multiscale method. *Physics of Fluids*, 13(6):1755–1754, 2001.

[98] G.S. Iannelli, and A.J. Baker. A stiffly-stable implicit Runge-Kutta algorithm for CFD applications. *AIAA Paper* 88–0416, 1988.

[99] F. Ihlenburg. *Finite Element Analysis of Acoustic Scattering (Applied Mathematical Sciences)*, Springer-Verlag, NY, 1998.

[100] O. Inoue, and N. Hatakeyma. Sound generation by a two-dimensional circular cylinder in a uniform flow. *J Fluid Mech.*, 471: 285–314, 2002.

[101] J. Jaffre, C. Johnson, and A. Szepessy. Convergence of the Discontinuous Galerkin finite element method for hyperbolic conservation laws. *Math. Models Meth. Appl. Sci.*, 5:367–381, 1995.

[102] A. Jameson, W. Schmidt, and E. Turkel. Numerical simulation of the Euler equations by finite volume methods using Runge-Kutta time stepping schemes. *AIAA Paper* 81–1259, 1981.

[103] A. Jameson. Solution of the Euler equations for two dimensional transonic flow by a multigrid method. *Applied Mathematics and Computation*, 13:327–356, 1983.

[104] K. Jansen. The effect of element topology on variational multiscale methods for LES. *Bull. Am. Phys. Soc.*, 45(9):56, 2000.

[105] J.D. Jr, and B.L. Dupont. *Interior penalty procedures for elliptic and parabolic Galerkin methods.* Lecture notes in Physics volume 58, Springer-Verlag, 1976.

[106] G.S. Jiang and C.W. Shu. Efficient Implementation of Weighted ENO Schemes. *J Computational Physics*, 126(1):202–228, 1996.

[107] G.S. Jiang and C.C. Wu. A High-order WENO finite difference scheme for the equations of ideal magnetohydrodynamics. *J Computational Physics*, 150(2):561–594, 1999.

[108] Y. Kallinderis, A. Khawaja, and H. McMorris. Hybrid prismatic/tetrahetral grid generation for complex geometries. *AIAA J*, 34(2):291–298. 1996.

[109] G.S. Karamanos and G.E. Kraniadakis. Spectral vanishing viscosity method for large-eddy simulations. *J Computational Physics*, 163(1):22–50, 2000.

[110] G.E. Karniadakis and S.J. Sherwin. *Spectral/hp Element Method for CFD*. Oxford University Press, 1999.

[111] D.A. Kopriva. A conservative staggered-grid Chebyshev multidomain method for compressible flows, II. A semi-structured method. *J Computational Physics*, 128(2):475–488, 1996.

[112] D.A. Kopriva. Multidomain spectral solution of the Euler gas-dynamics equations. *J Computational Physics*, 96(2):428–450, 1991.

[113] D.A. Kopriva. Multidomain spectral solution of compressible viscous flows. *J Computational Physics*, 115(1):184–199, 1994.

[114] D.A. Kopriva, and J.H. Kolias. A conservative staggered-grid Chebyshev multidomain method for compressible flows. *J Computational Physics*, 125(1):244–261, 1996.

[115] H.O. Kreiss. Stability theory for difference approximations of mixed initial boundary value problems. *I. Math. Comput.*, 22:703–714, 1968.

[116] H.O. Kreiss, J. Oliger. Methods for approximate solutions of time dependent problems. World Meteorological Organization/International Council of Scientific Unions, Geneva, 1973.

[117] P. Lax. *Hyperbolic Systems of Conservation Laws and the Mathematical Theory of Shock Waves*. 2nd edition, Vol. 11, SIAM Series on Applied Mathematics, SIAM 1973.

[118] S. Lee, S.K. Lele, and P. Moin. Interaction of isotropic turbulence with shock waves: effect of shock strength. *J Fluid Mechanics*, 340:225–247, 1997.

[119] S.K. Lele. Compact finite difference scheme with spectral-like resolution. *J Computational Physics*, 103(1):16–42, 1992.

[120] R.J. LeVeque. *Finite Volume Methods for Hyperbolic Problems*. Cambridge Text in Applied Mathematics. Cambridge Univ. Press, 2002.

[121] S.M. Liang, C.T. Chen, and H. Chen. Numerical study of supersonic, underwater flows past a blunt body. *AIAA J*, 39(6):1123–1126, 2001.

[122] S.M. Liang, J.L. Hsu, and J.S. Wang. Numerical study of cylindrical blast-wave propagation and reflection. *AIAA J*, 39(6):1152–1158, 2001.

[123] D.K. Lilly. A proposed modification of the Germano subgrid-scale closure method. *Physics of Fluids A*, 4(3):633–635, 1992.

[124] S.Y. Lin and Y.S. Chin. Discontinuous Galerkin finite element method for Euler and Navier-Stokes equations. *AIAA J*, 31(11):2016–2026, 1993.

[125] S.Y. Lin and J.J. Hu. Parameter study of weighted essentially non-oscillatory schemes for computational aeroacoustics. *AIAA J*, 39(3):371, 2001.

[126] X.D. Liu, S. Osher, and T. Chan. Weighted essentially non-oscillatory schemes. *J Computational Physics*, 115(1):200–212, 1994.

[127] Y. Liu. Fourier analysis of numerical algorithms for the Maxwell equations. *J Computational Physics*, 124(2):396–416, 1996.

[128] D.P. Lockard, K.S. Brentner, and HL. Atkins. High-accuracy algorithms for computational aeroacoustics. *AIAA J*, 33(2):246–253, 1995.

[129] I. Lomtev, and G.E. Karniadakis. A discontinuous Galerkin method for the Navier-Stokes equations. *Int. J Num Meth Fluids*, 29(5):587–603, 1999.

[130] I. Lomtev, R.M. Kirby, and G.E. Karniadakis. A discontinuous Galerkin ALE method for compressible viscous flows in moving domains. *J Computational Physics*, 155(1):128–159, 1999.

[131] H. Luo, J. Sharov, D. Baum, and R. Lohner. On the computation of compressible turbulent flows on unstructured grids. *AIAA Paper* 2000–0927, 2000.

[132] K. Mahesh. A family of high-order finite difference schemes with good spectral resolution. *J Computational Physics*, 145(1):332–358, 1998.

[133] F.R. Menter, M. Kuntz, and R. Benter. A Scale-Adaptive Simulation Model for Turbulent Flow Predictions. *AIAA Paper-0767*, 2003.

[134] E.A. Munts, S.J. Hulshoff, and R. de Borst. A space-time variational multiscale discretization for LES. *AIAA Paper* 2004–2132, 2004.

[135] C.R. Nastase, D.J. Mavriplis. High-order discontinuous Galerkin methods using an hp-multigrid approach. *J Computational Physics*, 213(1):330–357, 2006.

[136] N.V. Nikitin, F. Nicoud, B. Wasistho, K.D. Squires, and P.R. Spalart. Approach to wall modeling in Large-eddy Simulation. *Physics of Fluids*, 12(7):1629–1632, 2000.

[137] J.T. Oden, I. Babuska, and C.E. Baumann. A discontinuous hp finite element method for diffusion problems. *J Computational Physics*, 146(1):491–519, 1998.

[138] S. Osher and S. Chakravarthy. Upwind schemes and boundary conditions with applications to Euler equation in general geometries. *J Computational Physics*, 50(3):447–481, 1983.

[139] A. Patera. A spectral element method for fluid dynamics: Laminar flow in a channel with an expansion.*J Computational Physics*, 54(2):468–488, 1984.

[140] P.O. Persson and J. Peraire. Sub-cell shock capturing for discontinuous Galerkin methods. *AIAA Paper–0112*, 2006.

[141] S. Pirozzoli, F. Grasso, and D. D'Andrea. Interaction of a shock wave with two counter-rotating vortices: shock dynamics and sound production. *Physics of Fluids*, 13(11):3460–3481, 2001.

[142] S. Pirozzoli. Conservative hybrid compact-WENO schemes for shock-turbulence interaction. *J Computational Physics*, 178(1):81–117, 2002.

[143] S. Pirozzoli. Dynamics of ring vortices impinging on planar shock waves. *Physics of Fluids*, 16(5):1171–1185, 2004.

[144] D. Ponziani, S. Pirozzoli, and F. Grasso. Development of optimized weighted-ENO schemes for multiscale compressible flows. *Int. J Num Meth Fluids*, 42(9):953–977, 2003.

[145] M.M. Rai. Navier-Stokes Simulations of Blade-Vortex Interaction Using High-Order Accurate Upwind Schemes. *AIAA Paper*, 1987–0543, 1987.

[146] M.M. Rai and P. Moin. Direct simulations of turbulent flow using finite-difference schemes. *J Computational Physics*, 96(1):15–53, 1991.

[147] M.M. Rai and P. Moin. Direct numerical simulation of transition and turbulence in a spatially evolving boundary layer. *J Computational Physics*, 109(2):169–192, 1993.

[148] S. Ramakrishnan and S.S. Collis. Turbulence control simulation using the variational multiscale method. *AIAA J*, 42(4):745–753, 2004.

[149] P. Rasentarinera, D.A. Kopriva, and M.Y. Hussaini. Discontinuous spectral element solution of acoustic radiation from thin airfoils. *AIAA J*, 39(11):2070–2075, 2001.

[150] P. Rasetarinera and M.Y. Hussaini. An efficient implicit discontinuous Galerkin method. *J Computational Physics*, 172(2):718–738, 2001.

[151] W.H. Reed and T.R. Hill. *Triangular Mesh Methods for the Neutron Transport Equation*. Technical Report LA-UR-73-479, Los Alamos Scientific Laboratory, 1973.

[152] M. Remaki and W.G. Habashi. A discontinuous Galerkin method/HLLC solver for the Euler equations. *Int. J Num Meth Fluids*, 43 (12):1391–1405, 2003.

[153] Y.X. Ren, M. Liu, and H. Zhang. A characteristic-wise hybrid compact-WENO scheme for solving hyperbolic conservation laws. *J Computational Physics*, 192(2):365–386, 2002.

[154] W.J. Rider and R.B. Lowrie. The use of classical Lax-Friedrichs Riemann solvers with discontinuous Galerkin methods. *Int. J Num Meth Fluids*, 40(3):479–486, 2002.

[155] D.P. Rizzetta, M.R. Visbal, and D.P. Gaitonde. Large-eddy simulation of supersonic compression ramp flow by a high-order method. *AIAA Journal*, 39(12):2283–2292, 2001.

[156] P.L. Roe. Approximate Riemann solvers, parameter vectors, and difference schemes. *J Computational Physics*, 43(2):357–371, 1981.

[157] S.E. Rogers, and D. Kwak. An upwind difference scheme for time-accurate incompressible Navier-Stokes equations. *AIAA Paper*, 88-2883, 1988.

[158] A. Rogerson, and F. Meiberg. A numerical study of the convergence properties of ENO schemes. *J Scientific Computing*, 5:151–167, 1990.

[159] Y. Saad and M.H. Schultz. GMRES: A generalized minimum residual algorithm for solving nonsymmetric linear systems. *SIAM J Sci. Stat. Comp.*, 7:865–873, 1986.

[160] P. Sagaut, and M. Germano. *Large Eddy Simulation for Incompressible Flow Flows*, Springer Verlag 2001.

[161] K. Shahbazi, P.F. Fischer, and C.R. Ethier. A high-order discontinuous Galerkin method for the unsteady incompressible Navier-Stokes equations. *J Computational Physics*, 222(1):391–407, 2007.

[162] F. Shakib, T.J.R. Hughes, and Z. Johan. A multi-element group preconditioned GMRES algorithm for nonsymmetric problems arising in finite element analysis. *Comput Methods Appl Mech Eng*, 75(1-3):415–456, 1989.

[163] J. Shi, Y.T. Zhang, and CW. Shu. Resolution of high order WENO schemes for complicated flow structures. *J Computational Physics*, 186(2):690–696, 2003.

[164] C.W. Shu and S. Osher. Efficient implementation of essentially non-oscillatory shock-capturing schemes. *J Computational Physics*, 77(2):439–471, 1988.

[165] C.W. Shu and S. Osher. Efficient implementation of essentially non-oscillatory shock capturing schemes II. *J Computational Physics*, 77(2): 439–471, 1988.

[166] C.W. Shu. Numerical experiments on the accuracy of ENO and modified ENO schemes. *J Scientific Computing*, 5:127–149, 1990.

[167] C. Shwab. *p- and hp- Finite Element Methods. Theory and Applications in Solid and Fluid Mechanics*. Oxford, 1998.

[168] B. Sjogreen, and H.C. Yee. *Multiresolution Wavelet based adaptive numerical dissipation control for shock-turbulence computation*. RIACS Report 01.01, NASA Ames Research Center, 2000.

[169] B. Sjogreen, and H.C. Yee. Grid convergence of high order methods for multiscale complex unsteady viscous compressible flows. *J Computational Physics*, 185(1): 1–26, 2003.

[170] P.R. Spalart. Strategies for turbulence modeling and simulations. *Inter J Heat Fluid Flow*, 21(3):252–263, 2000.

[171] J.L. Steger. Implicit finite-difference simulations of flow about arbitrary two-dimensional geometries. *AIAA J*, 17(7):679–686, 1978.

[172] M. Stelers. Detached eddy simulation of massively separated flows. *AIAA Paper*, 2001–0897, 2001.

[173] R. Strawn and J. Ahmad. Computational Modeling of Hovering Rotors and Wakes. *AIAA Paper*, 2000-0110, 2000.

[174] R.C. Strawn and T.J. Barth. *A finite-Volume Euler Solver for Computing Rotary-Wing Aerodynamics of Unstructured Meshes*. Proceeding of 48th Annual Helicopter Society Forum, American Helicopter Society, Washington, DC, 1: 419, 1992.

[175] C.K.W. Tam and J.C. Webb. Dispersion-relation-preserving finite difference schemes for computational acoustics. *J Computational Physics*, 107(1):262–281, 1993.

[176] C.K. Tam and J.C. Hardin. *Second Computational Aeroacoustics (CAA) Workshop on Benchmark Problems*. NASA CP 3352, 1997.

[177] C.W.K. Tam and N.N. Pastouchenka. Fine scale turbulence noise from dual stream jets. *AIAA Paper*, 2004–2871, 2004.

[178] L. Thompson and P. Pinsky. Complex wavenumber Fourier analysis of the p-version finite element method. *Comput. Mech.*, 13:255–275, 1994.

[179] V.A. Titarev, and E.F. Toro. ADER: Arbitrary high order Godunov approach. *J Scientific Computing*, 17:609–618, 2002.

[180] I. Toulopoulos and J.A. Ekaterinaris. High-order discontinuous-Galerkin discretizations for Computational aeroacoustics in complex domains. *AIAA J*, 44(3):502–511, 2006.

[181] E. Usta. *Application of a Symmetric TVD Scheme to Aerodynamics of Rotors*. Ph.D. Thesis, *GA Institute of Technology*, Atlanta, GA, 2002.

[182] J.J.W. van der Vegt and H. van der Ven. Discontinuous Galerkin finite element method with anisotropic local grid refinement for inviscid compressible flow. *J Computational Physics*, 141 (1):46–77, 1998.

[183] J.J.W. van der Vegt and H. van der Ven. Space time discontinuous Galerkin finite element method with dynamic grid motion for inviscid compressible flows: I. General formulation. *J Computational Physics*, 182(2):546–585, 2002.

[184] B. van Leer. Towards the ultimate conservative difference scheme II. Monotonicity and conservation combined in a second-order scheme. *J Computational Physics*, 14(4):361–370, 1974.

[185] B. van Leer. Towards the ultimate conservative difference scheme V. A second order sequel to Godunov's method. *J Computational Physics*, 32(1):101–136, 1979.

[186] V. Venkatakrishnan and D.J. Mavriplis. Implicit solvers for unstructured meshes. *J Computational Physics*, 105(1):83–91, 1993.

[187] V. Venkatakrishnan. Perspective on unstructured grid flow solvers. *AIAA J*, 34(3):533–547, 1996.

[188] R. Vichnevetsky. Wave propagation analysis of different schemes for hyperbolic equations: a review. *Int. J Numer. Meth. in Fluids*, 7:409–452, 1987.

[189] R. Vichnevetsky and J.B. Bowles. *Fourier analysis of numerical approximations of hyperbolic equations*. SIAM Studies in Applied Mathematics 5, 1995.

[190] B.E. Wake and D. Choi. Investigation of high-order upwinded differencing for vortex convection. *AIAA J*, 34(2):332–337, 1996.

[191] Z.J. Wang and R.F. Chen. Anisotropic Cartesian grid method for viscous turbulent flow. *AIAA Paper* 2000–0395, 2000.

[192] Z.J. Wang and R.F. Chen. Optimized weighted essentially non-oscillatory schemes for linear waves with discontinuity. *J Computational Physics*, 174(1):381–404, 2001.

[193] Z.J. Wang. Spectral (finite) volume method for conservation laws on unstructured grids. *J Computational Physics*, 178(1):210–251, 2002.

[194] T. Warburton, I. Lomtev, S. Du, S. Sherwin, and G. Karniadakis. Galerkin and discontinuous Galerkin spectral/hp methods. *Comput. Methods Appl. Mech. Engrg.*, 175:343–359, 1999.

[195] V.G. Weirs and G.V. Candler. Optimization of weighted ENO schemes for DNS of compressible turbulence. *AIAA Paper*, 97–1940, 1997.

[196] M.F. Wheeler. An elliptic collocation-finite element method with interior penalties. *SIAM J Numer. Anal.*, 15:152–161, 1978.

[197] D.C. Wilcox. *Turbulence modeling for CFD*. DCW Industries Inc., 1993.

[198] P. Woodward and P. Colella. The numerical simulation of two-dimensional fluid flow with strong shocks. *J Computational Physics*, 54(1):115–173, 1984.

[199] J.Y. Yang, Y.C. Perng, and R.H. Yen. Implicit weighted essentially nonoscillatory schemes for the compressible Navier-Stokes Equations. *AIAA J*, 39(11):2082–2090, 2001.

[200] H.C. Yee. On symmetric and upwind TVD schemes. *Proceedings of the 6th GAMM-Conference on Numerical Methods in Fluid Mechanics*, 1985.

[201] H.C. Yee and A. Harten. Implicit TVD schemes for hyperbolic conservation laws in curvilinear coordinates. *AIAA J*, 25(2):266–274, 1987.

[202] H.C. Yee. Construction of explicit and implicit symmetric TVD schemes and their applications. *J Computational Physics*, 68(1):151–179, 1987.

[203] H.C. Yee. Explicit and implicit multidimensional compact high-resolution shock-capturing methods: Formulation. *J Computational Physics*, 131(2):216–232, 1997.

[204] H.C. Yee, N.D. Sandham, and M.J. Djomehri. Low-dissipative high-order shock capturing methods using characteristic-based filters. *J Computational Physics*, 150(1):199–238, 1999.

[205] S.Y. Yuzhi and Z.J. Wang. Evaluation of discontinuous Galerkin and spectral volume methods for scalar and system conservation laws on unstructured grids Int. *J Num Meth Fluids*, 45(8):819–838, 2004.

[206] M. Zhang and C.W. Shu. An analysis of three different formulations of the discontinuous Galerkin method for diffusion equations. *Math Models and Methods in Applied Sciences*, 13(3):395–413, 2003.

[207] X. Zhong. High-order finite-difference schemes for numerical simulation of hypersonic boundary-layer transition. *J Computational Physics*, 144(2):662–709, 1998.

[208] M. Zhuang and R.F. Chen. Optimized upwind dispersion-relation-preserving finite difference schemes for computational aeroacoustics. *AIAA J*, 36(12):2146–2148, 1998.

[209] D.W. Zingg. A review of high-order and optimized finite-difference methods for simulating linear wave phenomena. *AIAA Paper* 97–2088, 1997.

Index